Sixth Edition

Maribeth Price
South Dakota School of Mines and Technology

Mastering ArcGIS

The McGraw·Hill Companies

MASTERING ARCGIS, SIXTH EDITION

Published by McGraw-Hill, a business unit of The McGraw-Hill Companies, Inc., 1221 Avenue of the Americas, New York, NY, 10020.

This book is printed on acid-free paper.

1 2 3 4 5 6 7 8 9 0 QTN/QTN 1 0 9 8 7 6 5 4 3

ISBN: 978-0-07-802142-8
MHID: 0-07-802142-1

Senior Vice President, Products & Markets: *Kurt L. Strand*
Vice President, Product & Markets: *Marty Lange*
Vice President, Content Production & Technology Services: *Kimberly Meriwether David*
Managing Director: *Thomas Timp*
Brand Manager: *Michelle Vogler*
Marketing Manager: *Matthew Garcia*
Director, Content Production: *Terri Schiesl*
Senior Project Manager: *Melissa M. Leick*
Buyer: *Susan K. Culbertson*
Media Project Manager: *Prashanthi Nadipalli*
Cover Designer: *Studio Montage, St. Louis, MO*
Cover Image: *Design Pics / Don Hammond*
Typeface: *10/12 Times Roman*
Compositor: *S4Carlisle Publishing Services*
Printer: *Quad/Graphics*

Library of Congress Cataloging-in-Publication Data

Price, Maribeth Hughett, 1963-
Mastering ArcGIS / Maribeth Price. -- Sixth edition.
pages cm
Includes index.
ISBN 978-0-07-802142-8
1. ArcGIS. 2. Geographic information systems. I. Title.
G70.212.P74 2012
910.285'53--dc23

2012050354

www.mhhe.com

Table of Contents

Preface

Welcome to *Mastering ArcGIS*, a detailed primer on learning the ArcGIS™ software by ESRI®, Inc. This book is designed to offer everything you need to master the basic elements of GIS.

Notice: ArcGIS™, ArcMap™, ArcCatalog™, ArcGIS Desktop™, ArcInfo Workstation™, and the other program names used in this text are registered trademarks of ESRI, Inc. The software names and the screen shots used in the text are reproduced by permission. For ease of reading only, the ™ symbol has been omitted from the names; however, no infringement or denial of the rights of ESRI® is thereby intended or condoned by the author.

What's new in the sixth edition?

The primary goals for this edition included updating the book for ArcGIS Version 10.1 and addressing issues caused by the enormous changes in the fifth edition. This time I have tried to keep the format stable and work on incremental improvements. The content sections are largely unchanged except for those dealing with changes to ArcGIS. The organization and order of the chapters have remained constant from the fifth edition.

I thank my readers for their patience with the fuzzy figures in the fifth edition, which were caused by changes in the Microsoft Office 2007 graphics engine and proved unfixable despite weeks looking for a workaround. Office 2010 is better, but only if screen shots are reproduced at 100%. I have to shoehorn the bitmaps into a reasonable space, and often the figure is not exactly like the actual window, although the main features are there, and they are clear! The Skills Reference section still has the old figures except where version 10.1 required changing them; getting them right takes more time now, and it proved impossible to redo them all in a single edition.

A few other improvements have been implemented:

- The size of the mgisdata folder has been reduced from 1 GB to 350 MB, although virtually everything is still there. How is my secret…
- The first chapter includes new material on cloud-based GIS, web maps, and ArcGIS Online. The citations section has been updated to include citing map services and other online data sources.
- Chapter 1 tutorial was reorganized to use only the Catalog tab—I find that we hardly use ArcCatalog separately any more. The Skills Reference section on ArcCatalog was retained, however, and ArcCatalog is still the preferred program for working with metadata in Chapter 15.
- The mention and use of GIS Services have been introduced throughout the text where appropriate.
- The concepts section of Chapter 11 on coordinate systems has been revised, and new figures added. I am always looking for better ways to explain this complex topic.

I would like to thank the many people who have used and commented on this book, and I hope that it continues to serve their needs in the rapidly evolving world of GIS.

Previous experience

This book assumes that the reader is comfortable using Windows™ to carry out basic tasks such as copying files, moving directories, opening documents, exploring folders, and editing text and word processing documents. Previous experience with maps and map data is also helpful.
No previous GIS experience or training is necessary to use this book.

Elements of the package

This learning system includes a textbook with a DVD-ROM, including

- Fifteen chapters on the most important capabilities of ArcGIS
- Comprehensive tutorials in every chapter to learn the skills, with each step demonstrated in a video clip
- A set of exercises, map documents, and data for practicing skills independently
- Reference sections on skills with video clips demonstrating each one

This book assumes that the student has access to ArcGIS Desktop Basic (formerly ArcView). A few optional topics are introduced that require an ArcGIS Desktop Standard (formerly ArcEditor) license. The Spatial Analyst extension is required for Chapter 8.

Philosophy

This text reflects the author's personal philosophies and prejudices developed from 18 years of teaching GIS at an engineering school. The main goal is not to train geographers but to provide students in any field with GIS skills and knowledge. It is assumed that most students using this book already have a background of discipline-specific knowledge and skills upon which to draw and are seeking to apply geospatial techniques within their own knowledge domains.

- **GIS is best learned by doing it, not by studying it.** The laboratory is THE critical component of the book, and theory is introduced sparingly and integrated with experience. Hence, this book is heavy on experience and lighter on theory.

- **Independent work and projects are critical to learning GIS.** This book includes a wealth of exercises in which the student must find solutions independently without a cookbook recipe of steps. A wise instructor will also require students to develop an independent project.

- **GIS analysis is best taught before data development.** Although this is the opposite of how a project unfolds, analysis is more interesting and helps students gain incentive to tackle creating their own data. Also, database development rules make more sense to someone who has worked with GIS data already.

Chapter sequence

The book contains an introduction and 15 chapters. Each chapter includes roughly one week's work for a three-credit semester course. This book intentionally contains more material than the average GIS class can cover during a single semester; instructors may choose what to emphasize.

An introductory chapter describes GIS and gives some examples of how it is used. It also provides an overview of GIS project management and how to develop a project and a proposal. The remaining chapters can be divided into three sections. Sections II and III could be done in either order, depending on instructor preference for covering data management or analysis first. Chapter 11 may be moved to any position after Chapters 1 through 4 for instructors wishing to cover coordinate systems earlier in a course.

I. GIS Data and Maps: Chapters 1–4 introduce basic skills for working with GIS data, creating maps and layouts, and working with tables.

II. GIS Analysis: Chapters 5–10 introduce different analysis techniques, including queries, spatial joins, overlay, raster analysis, geocoding, and network analysis.

III. Data Management: Chapters 11–15 cover coordinate systems, editing, geodatabases, and metadata.

IV. Skills Reference: This section contains a thematically organized compendium of step-by-step instructions for the most frequent and important tasks.

Chapter layout

Each chapter is organized into the following sections:

- **Concepts:** provides basic background material for understanding geographic concepts and how they are specifically implemented within ArcGIS. A set of review questions and important terms follows the concepts section.

- **Tutorial:** contains a step-by-step tutorial demonstrating the concepts and skills. The tutorials begin with detailed instructions, which gradually become more general as mastery is built. Every step in the tutorial is demonstrated by accompanying video clips.

- **Exercises:** presents a series of problems to build skill in identifying the appropriate techniques and applying them without step-by-step help. Through these exercises, the student builds an independent mastery of GIS processes. Answers or solution methods are included for selected exercises. A full answer and methods document is available for instructors at the McGraw-Hill Instructor web site.

The DVD-ROM contains all the data needed to follow the tutorials and complete the exercises.

Instructors should use judgment in assigning exercises, as the typical class would be stretched to complete all the exercises in every chapter. Very good students can complete the entire chapter in 3–6 hours, most students would need 6–8 hours, and a few students would require 10 or more hours. Students with extensive computer experience generally find the material easier than others.

Using this text

In working through this book, the following sequence of steps is suggested:

- READ through the Concepts sections to get familiar with the principles and techniques.
- ANSWER the Chapter Review Questions to test comprehension of the material.
- WORK the Tutorial section for a step-by-step tutorial and explanation of key techniques.
- REREAD the Concepts section to reinforce the ideas.
- PRACTICE by doing the Exercises.

Using the tutorial

The tutorial provides step-by-step practice and introduces details on how to perform specific tasks. Students should be encouraged to think about the steps as they are performed and not just race to get to the end.

It is important to follow the directions carefully. Skipping a step or doing it incorrectly may result in a later step not working properly. Saving often will make it easier to go back and correct a mistake in order to continue. Occasionally, a step will not work due to differences between computer systems or software versions. Having an experienced user nearby to identify the problem can help. If one isn't available, however, just skip the step and move on without it.

Using the videos

The DVD distributed with the book contains two types of videos. The Tutorial Videos demonstrate each step of the tutorial. They are numbered in the text for easy reference. The Skills Reference Videos show how to perform generic tasks, such as deleting a file. The videos are intended as an alternate learning strategy. It would be tedious to watch all of them. Instead, use them in the following situations:

- When the student does not understand the written instructions or cannot find the correct menu or button
- When a step cannot be made to work properly
- When a reminder is needed to do a previously learned skill in order to complete a step
- When a student finds that watching the videos enhances learning

The DVD contains an index file with hyperlinks to each clip. As you work through a chapter, keep the index on the screen, and click the appropriate link to see a video. The tutorial clips are distinguished by numbered steps and the Skills Reference Videos by their headings.

Installing the DVD-ROM components

The DVD-ROM includes HTML index documents, a folder of video clips, and a self-installing archive of the tutorial data. See the README file on the DVD for any last-minute changes to the installation instructions. A splash screen with a use agreement and hyperlinks to the data and tutorials should appear when the DVD is inserted into the drive. It can be started manually from the DVD, if necessary, by double-clicking the Start_Here.exe file.

What if I don't see the splash screen? Open the DVD in Windows Explorer. To install the data, find the mgisdata6.exe file and double-click it to start extracting. To run the videos, find the VideoIndex.html file, and double-click to open it.

To view the videos:

The tutorial video links appear in blue in a table, and the Skills Reference videos appear underneath them. Click on a blue video hyperlink to start a video. For best resolution, size the video window as large as possible. See the instructions at the start of the Chapter 1 tutorial.

To install the training data:

The mgisdata.exe archive contains a folder with the documents and data needed to do the tutorials and exercises. The student must copy this folder to his own hard drive. If more than one person on the computer is using this book, then each person should make her own copy of the data in a separate folder. The data on the DVD are a self-extracting zip file that requires approximately 350 MB of disk space. Follow these instructions to install the data:

- Place the book DVD in your computer's DVD-ROM drive and wait for the splash screen to appear. Indicate your acceptance of the license agreement.
- Click on the link to Install Data. If a dialog window appears asking whether to Run or Save the data, choose Run.
- When a dialog box appears asking whether to Open or Save the data, choose Open. Don't choose Save because it will only copy the data archive instead of extracting it.
- Click the Browse button to set the folder to which to extract the data (a). The data will be placed in a folder called mgisdata in whatever location you choose. In other words, if you select C:\gisclass as the target folder, then the data will be placed in C:\gisclass\mgisdata.
- Click Start to begin installing the data. It may take several minutes. Wait until you see the Finished window and then click OK.

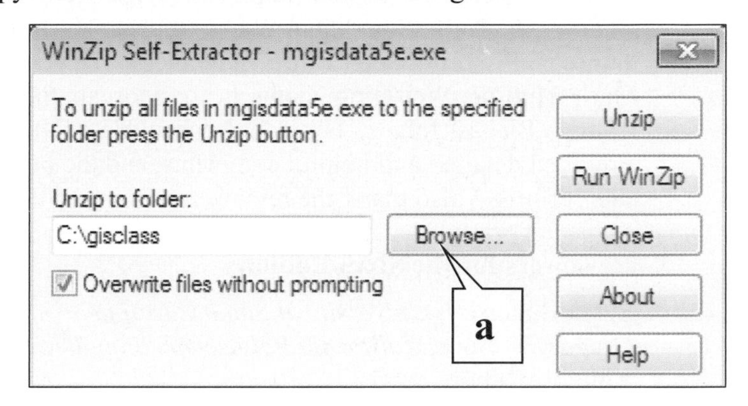

Installing the training data

System requirements

To use the tutorials and do the exercises in this book, the student must have access to a computer with the following characteristics:

Minimum hardware:

PC-Intel™-platform computer with 2.2-GHz processor or better and 2 GB RAM

Suitable sound/graphics card with 24-bit color depth and 1024 × 768 minimum screen

Software:

Windows 7™ (Ultimate, Enterprise, Professional, Home Premium), Windows Vista™ (Ultimate, Enterprise, Business, Home Premium), Windows 2000™, or Windows XP™ (Home Edition and Professional)

A web browser, such as Netscape or Internet Explorer, or Microsoft® Word

A media player that is able to display the .mp4 video format (such as Windows Media Player 12 [Windows 7 only] or QuickTime™)

ArcGIS Desktop™ 10.1 or higher (Basic Level); Standard Level and Spatial Analyst extension required for some exercises

For assistance in acquiring or installing these components, contact your system administrator, hardware/software provider, or local computer store.

Acknowledgments

I would like to thank many people who made this book possible. Governor Janklow of South Dakota funded a three-month summer project in 2000 that got the book started, as part of his Teaching with Technology program. Many students in my GIS classes between 2000 and 2012 tested the text and exercises and helped immensely in making sure the tutorials were clear and worked correctly. Reviewers of previous editions, including Richard Aspinall, Joe Grengs, Tom Carlson, Susan K. Langley, Henrietta Loustsen, Xun Shi, Richard Lisichenko, John Harmon, Michael Emch, Jim Sloan, Sharolyn Anderson, Talbot Brooks, Qihao Weng, Jeanne Halls, Mark Leipnik, Michael Harrison, Ralph Hitz, Olga Medvedkov, James W. Merchant, Raymond L. Sanders, Jr., Yifei Sun, Fahui Wang, Michael Haas, Jason Kennedy, Dafna Kohn, Jessica Moy, James C. Pivirotto, Peter Price, Judy Sneller, Dave Verbyla, Birgit Mühlenhaus, Jason Duke, Darla Munroe, Wei-Ning Xiang, L. Joe Morgan, Samantha Arundel, Christopher A. Badurek, Tamara Biegas, John E. Harmon, Michael Hass, Nicholas Kohler, David Long, and Jaehyung Yu, provided detailed and helpful comments, and the book is better than it would have been without their efforts. I also thank the reviewers who provided valuable advice for the sixth edition.

Reviewers for the Sixth Edition:

Sarah Battersby, *University of South Carolina*
Gregory S. Bohr, *California Polytechnic State University, San Luis Obispo*
Kelly R. Dubure, *Eckerd College*
Colleen Garrity, *State University of New York–Geneseo*
Raymond Greene, *Western Illinois University*
Joe Grengs, *University of Michigan*
Eileen Johnson, *Bowdoin College*
James Leonard, *Marshall University*
Tao Tang, *State University of New York–Buffalo State College*

ESRI, Inc., was prompt and generous in its granting of permission to use the screen shots, data, and other materials throughout the text. They also provided beta and prerelease versions of ArcGIS 10.1 for early development of the text. I extend heartfelt thanks to the City of Austin, Texas, for putting their fine GIS data sets in the public domain. I thank George Sielstad, Eddie Childers, Mark Rumble, Tom Junti, and Patsy Horton for their generous donations of data. I am grateful to Tom Leonard and Steve Bauer for their long-term computer lab administration, without which I could not have taught GIS courses or developed this book. I thank Linda Heindel for organizing student feedback and assisting with the initial round of edits on the first draft. I thank editors Michelle Vogler and Melissa Leick of McGraw-Hill for their unfailing encouragement and enthusiasm about the book as it took shape, as well as for their excellent feedback. I thank the McGraw-Hill team working on the sixth edition, especially Wendy Langerud. I am grateful to Daryl Pope, who first started me in GIS, and to John Suppe, who encouraged me to return to graduate school and continue doing GIS on a fascinating study of Venus. I thank my partner, David Stolarz, who provided unfailing encouragement when it seemed as though the editing on this edition would never end. Last, and certainly not least, I thank Curtis Price and my daughters, Virginia and Madeleine, for their understanding and support during the many, many hours I spent working on this book.

Introduction

What Is GIS?

Objectives

- Developing a basic understanding of what GIS is, its operations, and its uses
- Getting familiar with GIS project management
- Learning to plan a GIS project
- Finding resources to learn more about GIS

Concepts

What is GIS?

GIS stands for Geographic Information System. In practical terms, a GIS is a set of computer tools that allows people to work with data that are tied to a particular location on the earth. Although many people think of a GIS as a computer mapping system, its functions are broader and more sophisticated than that. A GIS is a database that is designed to work with map data.

Consider the accounting department of the local telephone company. They maintain a large computer database of their customers, in which they store the name, address, phone number, type of service, and billing information for each customer. This information is only incidentally tied to where customers live; they can carry out most of the important functions (billing, for example) without needing to know where each house is. Of course, they need to have addresses for mailing bills, but it is the post office that worries about where the houses actually ARE. This type of information is called **aspatial data**, meaning that it is not tied, or is only incidentally tied, to a location on the earth's surface.

Employees of the service department, however, need to work with **spatial data** to provide the telephone services. When hundreds of people call in after a power outage, the service department must analyze the distribution of the calls and isolate the location where the outage occurred. When a construction company starts work on a street, workers must be informed of the precise location of buried telephone cables. If a developer builds a new neighborhood, the company must be able to determine the best place to tie into the existing network so that the services are efficiently distributed from the main trunk lines. When technicians prepare lists of house calls for the day, they need to plan the order of visits to minimize the amount of driving time. In these tasks, location is a critical aspect of the job, and the information is spatial.

In this example, two types of software are used. The accounting department uses special software called a *database management system*, or *DBMS*, which is optimized to work with large volumes of aspatial data. The service department needs access to a database that is optimized for working with spatial data, a Geographic Information System. Because these two types of software are related, they often work together, and they may access the same information. However, they do different things with the data.

A GIS is built from a collection of hardware and software components.

- **A computer hardware platform.** Due to the intensive nature of spatial data storage and processing, a GIS was once limited primarily to large mainframe computers or expensive workstations. Today, it can run on a typical desktop personal computer.

- **GIS software.** The software varies widely in cost, ease of use, and level of functionality but should offer at least some minimal set of functions, as described in the next few paragraphs. In this book, we study one particular package that is powerful and widely used, but others are available and may be just as suitable for certain applications.

- **Data storage.** Some projects use only the hard drive of the GIS computer. Other projects may require more sophisticated solutions if large volumes of data are being stored or multiple users need access to the same data sets. Today, many data sets are stored in digital warehouses and accessed by many users over the Internet. Compact disc writers and/or USB portable drives are highly useful for backing up and sharing data.

- **Data input hardware.** Many GIS projects require sophisticated data entry tools. Digitizer tablets enable the shapes on a paper map to be entered as features in a GIS data file. Scanners create digital images of paper maps. An Internet connection provides easy access to large volumes of GIS data. High-speed connections are preferred, as GIS data sets may constitute tens or hundreds of megabytes or more.

- **Information output hardware.** A quality color printer capable of letter-size prints provides the minimum desirable output capability for a GIS system. Printers that can handle map-size output (36 in. × 48 in.) will be required for many projects.

- **GIS data.** Data come from a variety of sources and in a plethora of formats. Gathering data, assessing their accuracy, and maintaining them usually constitutes the longest and most expensive part of a GIS project.

- **GIS personnel**. A system of computers and hardware is useless without trained and knowledgeable people to run it. The contribution of professional training to successful implementation of a GIS is often overlooked.

GIS software varies widely in functionality, but any system claiming to be a GIS should provide the following functions at a minimum:

- Data entry from a variety of sources, including digitizing, scanning, text files, and the most common spatial data formats; ways to export information to other programs should also be provided

- Data management tools, including tools for building data sets, editing spatial features and their attributes, and managing coordinate systems and projections

- Thematic mapping (displaying data in map form), including symbolizing map features in different ways and combining map layers for display

- Data analysis functions for exploring spatial relationships in and between map layers

- Map layout functions for creating soft and hard copy maps with titles, scale bars, north arrows, and other map elements

Geographic Information Systems are put to many uses, but providing the means to collect, manage, and analyze data to produce information for better decision making is the common goal and the strength of each GIS. This book is a practical guide to understanding and using a particular Geographic Information System called ArcGIS. Using this book, you can learn what a GIS is, what it does, and how to apply its capabilities to solve real-world problems.

A history of GIS

Geographic Information Systems have grown from a long history of cartography begun in the lost mists of time by early tribesmen who made sketches on hides or formed crude models of clay as aids to hunting for food or making war. Ptolemy, an astronomer and geographer from the second century B.C., created one of the earliest known atlases, a collection of world, regional, and local maps and advice on how to draw them, which remained essentially unknown to Europeans until the 15th century. Translated into Latin, it became the core of Western geography, influencing cartographic giants, such as Gerhard Mercator, who published his famous world map in 1569. The 17th and 18th centuries saw many important developments in cartography, including the measurement of a degree of longitude by Jean Picard in 1669, the discovery that the earth flattens toward the poles, and the adoption of the Prime Meridian that passes through Greenwich, England. In 1859, French photographer and balloonist Gaspard Felix Tournachon founded the art of remote sensing by carrying large-format cameras into the sky. In an oft-cited early example of spatial analysis, Dr. John Snow mapped cholera deaths in central London in September 1854 and was able to locate the source of the outbreak—a contaminated well. However, until the 20th century, cartography remained an art and a science carried out by laborious calculation and hand drawing. In 1950, Jacqueline Tyrwhitt made the first explicit reference to map overlay techniques in an English textbook on town and country planning, and Ian McHarg was one of the early implementers of the technique for highway planning.

As with many other endeavors, the development of computers inspired cartographers to see what these new machines might do. The early systems developed by these groups, crude and slow by today's standards, nevertheless laid the groundwork for modern Geographic Information Systems. Dr. Roger Tomlinson, head of an Ottawa group of consulting cartographers, has been called the "Father of GIS" for his promotion of the idea to use computers for mapping and for his vision and effort in developing the Canada Geographic Information System (CGIS) in the mid-1960s. Another pioneering group, the Harvard Laboratory for Computer Graphics and Spatial Analysis, was founded in the mid-1960s by Howard Fisher. He and his colleagues developed a number of early programs between 1966 and 1975, including SYMAP, CALFORM, SYMVU, GRID, POLYVRT, and ODYSSEY. Other notable developers included professors Nystuen, Tobler, Bunge, and Berry from the Department of Geography at Washington University during 1958–1961. In 1970, the US Bureau of the Census produced its first geocoded census and developed the early DIME data format based on the CGIS and POLYVRT data representations. DIME files were widely distributed and were later refined into the TIGER format. These efforts had a pronounced effect on the development of data models for storing and distributing geographic information.

In 1969, Laura and Jack Dangermond founded the Environmental Systems Research Institute (ESRI), which pioneered the powerful idea of linking spatial representation of features with attributes in a table, a core idea that revolutionized the industry and launched the development of Arc/Info, a program whose descendants have captured about 90% of today's GIS market. Other vendors are still active in developing GIS systems, which include packages MAPINFO, MGA from Intergraph, IDRISI from Clark University, and the open-source program GRASS.

What can a GIS do?

A GIS works with many different applications: land use planning, environmental management, sociological analysis, business marketing, and more. Any endeavor that uses spatial data can benefit from a GIS. For example, researchers at the US Department of Agriculture Rocky Mountain Research Station in Rapid City conducted a study of elk habitat in the Black Hills of South Dakota and Wyoming by placing radio transmitter collars on about 70 elk bulls and cows (Fig. I.1c). Using the collars and a handheld antenna, they tracked the animals and obtained their locations. Several thousand locations were collected (Fig. I.1a) for the scientists to study the characteristics of the habitat where elk spend time.

The elk locations were entered into a GIS system for record keeping and analysis. Each location became a point with attached information, including the animal ID number and the date and time of the sighting. Information about vegetation, slope, aspect, elevation, water availability, and other site factors were derived by overlaying the points on other data layers, allowing the biologists to compare the characteristics of sites utilized by the elk. Figure I.1b shows a map of distances to major roads in the central part of the study area. The elk locations clustered in the darker roadless areas, and statistical analysis demonstrated this observation empirically.

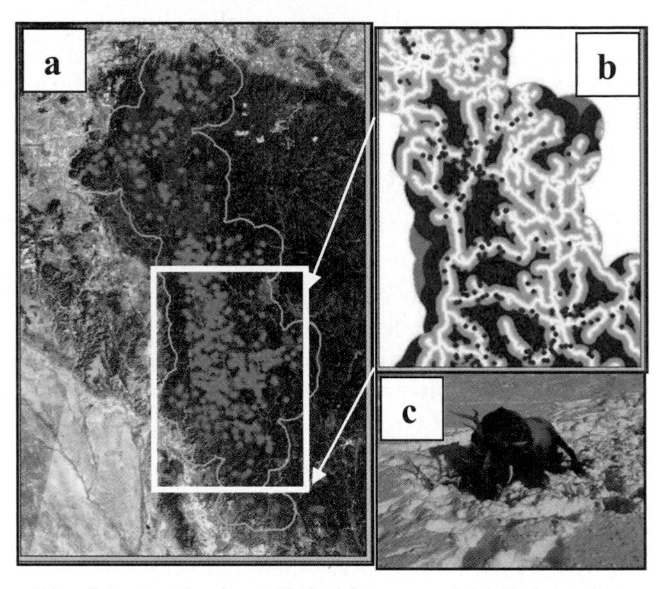

Fig. I.1. Analyzing elk habitat use: (a) elk locations and study area; (b) locations on a map of distance to nearest road; (c) collaring an elk

There are many applications of GIS in almost every human endeavor, including business, defense and intelligence, engineering and construction, government, health and human services, conservation, natural resources, public safety, education, transportation, utilities, and communication. In August 2012, the Industries section of the ESRI web site at www.esri.com listed 58 different application areas in these categories, each one with examples, maps, and case studies. Instead of reading through a list here, go to the site and see the latest applications.

New trends and directions in GIS

The GIS industry has grown exponentially since its inception. Starting with the mainframe and then the desktop computer, GIS began as a relatively private endeavor focused on small clusters of specialized workers who spent years developing expertise with the software and data. Since then, the development of the Internet and the rapidly advancing field of computer hardware have been driving some significant changes in the industry.

Proliferation of options for data sharing

Instead of storing data sets locally on individual computers or intraorganization network drives, more people are serving large volumes of data over the Internet to remote locations within an organization, across organizations, and to the general public. In the past, large collections of data were hosted by various organizations through clearinghouses sponsored by the National Spatial Data Infrastructure (NSDI) organization. The data usually existed as GIS data files in various formats and required significant expertise and the right software to download and use. Today, *GIS Servers* are designed to bring GIS data to the general public within a few clicks. Another class of data providers that give wide access to spatial data, although not specifically GIS applications, include web sites such as GoogleEarth, MapQuest, and Microsoft Virtual Earth.

Like many other computer industries, GIS is heading for the clouds. A cloud is a gigantic array of large computers that customers rent pieces of by the hour, instead of purchasing their own physical hardware. ArcGIS Online is a cloud-based platform for users to collaborate and share GIS data with each other, making it suddenly very easy to share data with a colleague, or with the world, even for those with no particular GIS expertise.

Proliferation of options for working with GIS data

In the early days, people who wanted to do GIS had to purchase a large, expensive program and learn to use it. Now, a wide variety of scaled applications permits different levels of use for different levels of need. Not every user must have the full program. There are map servers for people who just need to view and print maps, free download software for viewing interactive map publications, and scaled-down versions of the full program with fewer options.

Many organizations are turning to Server GIS as a less expensive alternative to purchasing large numbers of GIS licenses. Many workers need GIS, but they only use a small subset of GIS functions on a daily basis. A Server GIS can provide both data and a customizable set of viewing and analysis functions to users without a GIS license. All they need is a web browser. Because of more simple and low-cost ways of accessing GIS data and functions, the user base has expanded dramatically.

Expansion of GIS into wireless technology

More people are collecting and sharing data using handheld wireless devices, such as handheld computers, cell phones, and global positioning system (GPS) units. These units can now access Internet data and map servers directly so that users in the field can download background data layers, collect new data, and transmit them back to the large servers. Cell phone technology is advancing rapidly with new geolocation options, web access, and geo-applications arriving daily.

Emphasis on open-source solutions

Instead of relying on proprietary, specialized software, more GIS functions are now implemented within open-source software and hardware. GIS data are now more often stored using engines from commercial database platforms and utilize the same development environments as the rest of the computer industry. This trend makes it easier to have the GIS communicate with other programs and computers and enhances interoperability between systems and parts of systems.

Customization

With emphasis on open-source solutions, new opportunities have developed in creating customized applications based on a fundamental set of GIS tools, such as a hydrology tool or a wildlife management tool. These custom applications gather the commonly used functions of GIS into a smaller interface, introduce new knowledge, and formalize best practices into an easy-to-use interface. Customization requires a high level of expertise in object-oriented computer programming.

Enterprise GIS

Enterprise GIS integrates a server with multiple ways to access the same data, including traditional GIS software programs, web browser applications, and wireless mobile devices.

The goal is to meet the data needs of many different levels of users and to provide access to non-traditional users of GIS. The Enterprise GIS is the culmination of the other trends and capabilities already mentioned. The costs and challenges in developing and maintaining an Enterprise GIS are significant, but the rewards and cost savings can also be substantial.

What do GIS professionals do?

It's getting so easy to create a map these days that one may wonder why one should bother to learn GIS. However, the easy solutions are based on the work of experts who provide the data and the software systems to handle them. It may not be that hard to learn to use a cell phone, but you still need the engineers and software developers to build the phones and the infrastructure to make them work. GIS is much the same. These days, GIS professionals play a variety of roles. A few broad categories can be defined.

Primary Data Providers create the base data that form the backbone of many GIS installations. Surveyors and land-planning professionals contribute precise measurement of boundaries. Photogrammetrists develop elevation and other data from aerial photography or the newer laser altimetry (LIDAR) systems. Remote sensing professionals extract all kinds of human-made and natural information from a variety of satellite and airborne measurement systems. Experts in global positioning systems (GPS), which provide base data, are also important.

Applications GIS usually involves professionals trained in other fields, such as geography, hydrology, land use planning, business, and utilities, who utilize GIS as part of their work. Specialists in mathematics and statistics develop new ways to analyze and interpret spatial data. For these professionals, GIS is an added skill and a tool to make their work more efficient, productive, and valuable.

Development GIS involves skilled software and hardware engineers who build and maintain the GIS software itself, as well as the hardware components upon which it relies—not only computers and hard drives and plotters but also GPS units, wireless devices, scanners, digitizers, and other systems that GIS could not function without. This group also includes an important class of GIS developers who create customized applications from the basic building blocks of existing GIS software.

Distributed Database GIS involves computer science professionals with a background in networking, Internet protocols, and/or database management systems. These specialists set up and maintain the complex server and network systems that allow data services, Server GIS, and Enterprise GIS to operate.

GIS project management

A GIS project may be a small effort spanning a couple of days by a single person, or it can be an ongoing concern of a large organization with hundreds of people participating. Large or small, however, projects often follow the generalized model shown in Figure I.2, and most new users learn GIS through a project approach. A project usually begins with an assessment of needs. What specific issues must be studied? What kind of information is needed to support decision making? What functions must the GIS perform? How long will the project last? Who will be using the data? What funding is available for start-up and long-term support?

Without a realistic idea of what the system must accomplish, it is impossible to design it efficiently. Users may find that some critical data are absent or that resources have been wasted acquiring data that no one ever uses. In a short-term project the needs are generally clear-cut.

A long-term organizational system will find that its needs evolve over time, requiring periodic reassessment. A well-designed system will adapt easily to future modification. The creator of a haphazard system may be constantly redoing previous work when changes arise.

In studies seeking specific answers to scientific or managerial questions, a methodology or **model** must be chosen. Models convert the raw data of the project into useful information using a well-defined series of steps and assumptions. Creating a landslides hazards map provides a simple example. One might define a model such that, if an area has a steep slope and consists of a shale rock unit, then it should be rated as hazardous. The raw data layers of geology and slope can be used to create the hazard map. A more complex model might also take into account the dip (bedding angle) of the shale units. Models can be very simple, as in this example, or more complex with many different inputs and calculations.

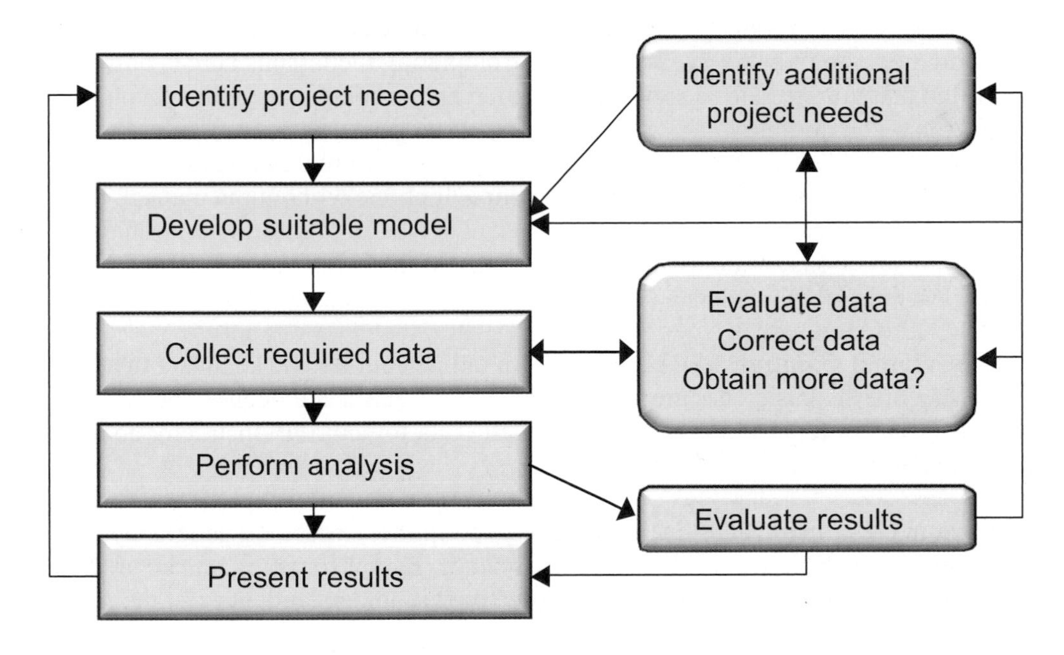

Fig. I.2. Generalized flow diagram representing steps in a GIS project

Once needs are known and the appropriate models have been designed, data collection can begin. GIS data are stored as layers, with each layer representing one type of information, such as roads or soil types. The needs dictate which data layers are required and how accurate they must be. A source for each data layer must be found. In some cases data can be obtained free from other organizations. General base layers, such as elevation, roads, streams, political boundaries, and demographics, are freely available from government sources (although the accuracy and level of detail are not always what one might wish). More specific data must often be developed in-house. For example, a utility company would need to develop its own layers showing electric lines and substations—no one else would be likely to have it.

The spatial detail and accuracy of the data must be evaluated to ensure that they are able to meet the needs of the project. For example, an engineering firm creating the site plan for a shopping mall could download elevation contours of the site for free. However, standard 10-foot contours cannot provide the detailed surface information needed by the engineers. Instead, the firm might contract with a surveying firm to measure contours at half-foot intervals.

After the data are assembled, the analysis can begin. During the analysis phase, it is not unusual to encounter problems that might require making changes to the model and/or data. Thus, the steps of model development, data collection, and analysis often become iterative, as experience gained is used to refine the process. The final result must be checked carefully against reality in order to recognize any shortcomings and to provide guidelines for improving future work.

Finally, no project is complete until the results have been communicated, whether shown informally to a supervisor, published in a scientific journal, or presented during a heated public meeting. Presentations may include maps, reports, slides, or other media.

Learning more about GIS

As you work your way through this text, you will be amazed at all the things that a GIS can do. At times, the abundance of tasks and the flexibility of options may be overwhelming. Even so, this book gives only a bare introduction to what is possible and covers a small portion of what GIS can do. Moreover, the industry evolves rapidly. Completing the lessons here is only the first step; you will find that you need to seek new information and training constantly as you develop your skills. How can you do this?

- **Read the Help.** Maybe you can figure out your cell phone without the manual, but if you ignore the Help files you are turning your back on a wealth of information. They don't just have instructions for doing things but also discussions, diagrams, references, and other ways to help you understand the concepts behind GIS as well as its implementation.

- **Use the Virtual Campus.** ESRI has dozens of online courses and seminars to help you learn. Many of the basic ones are free. If your campus has a site license, the GIS administrator can request others for free as well. Talk to your GIS instructor about getting access to these courses.

- **Build your GIS library.** GIS integrates many disciplines, including geography, surveying, cartography, statistics, computer science, spatial statistics, and so on. The more you know of these disciplines, the more you will understand, not just the *How* of doing things but also the *Why.*

- **Join a professional organization.** Many professional societies cater to GIS and remote sensing. They have newsletters and journals, conferences, and lots of professionals who can help answer your questions and share their experiences.

- **Join an online forum.** ESRI has online forums for many aspects of GIS, as well as other useful links, at http://support.esri.com/en/. You can search for answers to questions, and if nothing is there, you can ask a question and generally someone will answer within a few days. It is a great place to go when you get stuck with no one else to ask. You will also learn about bugs and workarounds.

- **Join ArcGIS Online.** Not only is it easy and fun to create maps and share them, but you can also search for courses, documents, powerpoints, videos, and other GIS-related materials.

Chapter 1. GIS Data

Objectives

- Understanding how real-world features are represented by GIS data
- Knowing the differences between the raster and vector data models
- Getting familiar with the basic elements of data quality and metadata
- Learning the different types of GIS files used by ArcGIS
- Using ArcMap to view GIS data
- Learning about layers and their properties

Mastering the Concepts

GIS Concepts

Representing real-world objects on maps

To work with maps on a computer requires developing methods to store different types of map data and the information associated with them. Objects in the real world, such as cities, roads, soils, rivers, and topography, must first be portrayed as map objects, such as those on a paper topographic map. These map objects must then be encoded for storage on a computer.

Many different data formats have been invented to encode data for use with GIS programs; however, most follow one of two basic approaches: the **vector** model or the **raster** model. In either approach, the critical task includes representing the information at a point, or over a region in space, using *x* and *y* coordinate values (and sometimes *z* for height). The *x* and *y* coordinates are the spatial data. The information being represented, such as a soil type or a chemical analysis of a well, is called the attribute data. Raster and vector data models both store spatial and attribute data, but they do it in different ways.

Both data systems are **georeferenced**, meaning that the information is tied to a specific location on the earth's surface using *x-y* coordinates defined in a standard way: a **coordinate system**. One can choose from a variety of coordinate systems, as we will see in Chapter 11. As long as the coordinate systems match, we can display any two spatial data sets together and have them appear in the correct spatial relationship to one another.

The vector model

Vector data use a series of *x-y* locations to store information (Fig. 1.1). Three basic vector objects exist: points, lines, and polygons. These objects are called **features**. **Point** features are used to represent objects that have no dimensions, such as a well or a sampling locality. **Line** features represent objects in one dimension, such as a road or a utility line. **Polygons** are used to represent two-dimensional areas, such as a parcel or a state. In all cases, the features are represented using one or more

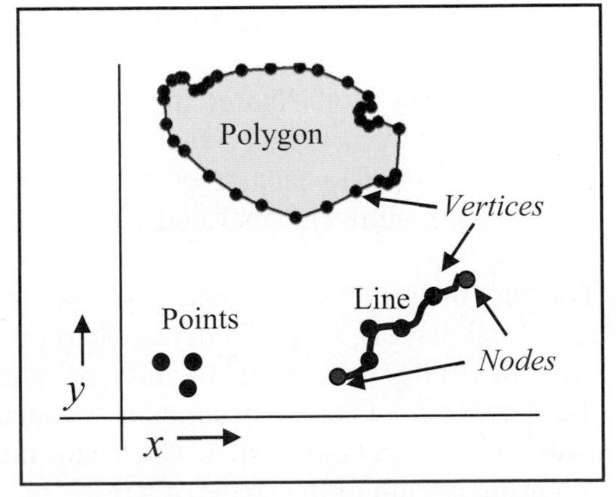

Fig. 1.1. The vector data model uses a series of *x-y* locations to represent points, lines, and polygon areas.

x-y coordinate locations (Fig. 1.1). A point consists of a single *x-y* coordinate pair. A line includes two or more pairs of coordinates—the endpoints of the line are termed **nodes**, and each of the intermediate points is called a **vertex**. A polygon is a group of **vertices** that define a closed area.

The type of object used to represent features depends on the scale of the map. A river would be represented as a line on a map of the United States because at that scale it is too small for its width to encompass any significant area on the map. If one is viewing a USGS topographic map, however, the river encompasses an area and would be represented as a polygon.

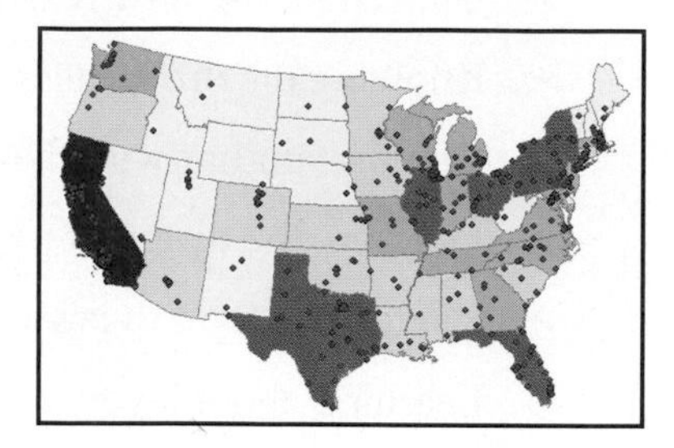

Fig. 1.2. A states feature class and a cities feature class

In GIS, like features are grouped into data sets called **feature classes** (Fig. 1.2). Roads and rivers are different types of features and would be stored in separate feature classes. A feature class can contain only one kind of geometry—it can include point features, line features, or polygon features but never a combination. In addition, objects in a feature class have information stored about them, such as their names or populations. This information is called the **attributes** and is stored in a table (Fig. 1.3). A special field, called the Feature ID (**FID**) or ObjectID (**OID**), links the spatial data with the attributes. Each feature's attributes are stored in one row of the table, and each column is a different type of information, such as population or area. A river and a highway would not be found in the same feature class because their information would be different—flow measurements for one versus pavement type for the other—and would need to be stored in different tables with different columns.

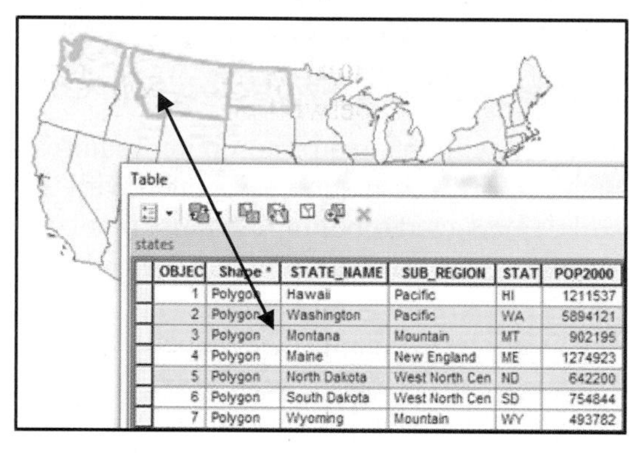

OBJEC	Shape *	STATE_NAME	SUB_REGION	STAT	POP2000
1	Polygon	Hawaii	Pacific	HI	1211537
2	Polygon	Washington	Pacific	WA	5894121
3	Polygon	Montana	Mountain	MT	902195
4	Polygon	Maine	New England	ME	1274923
5	Polygon	North Dakota	West North Cen	ND	642200
6	Polygon	South Dakota	West North Cen	SD	754844
7	Polygon	Wyoming	Mountain	WY	493782

Fig. 1.3. Each state is represented by a spatial feature (polygon), which is linked to the attributes.

When a state is highlighted on the map, its matching attributes are highlighted in the table, and vice versa. It is this live link between the spatial and attribute information that gives the GIS system its power. It enables us, for example, to create a map in which the states are colored based on an attribute field, such as population (Fig. 1.2). This technique is called **thematic mapping** and is one example of how linked attributes can be used to analyze geographic information.

Feature classes can be stored in several different formats. Some data formats only contain one feature class. Others, called **feature datasets**, can contain multiple feature classes that are in some way related to each other. For example, a feature dataset called Transportation might contain the feature classes Roads, Traffic Lights, Railroads, and Canals.

The benefits of the vector data model are many. First, it can store individual features, such as roads and parcels, with a high degree of precision. Second, the linked attribute table provides great flexibility in the number and type of attributes that can be stored about each feature. Third, the vector model is ideally suited to mapmaking because of the high precision and detail of features that can be obtained. The vector model is also a compact way of storing data, typically requiring a tenth of the space of a raster with similar information. Finally, the vector model is ideally suited to certain types of analysis problems, such as determining perimeters and areas, detecting whether features overlap, and modeling flow through networks.

However, the vector model has some drawbacks. First, it is poorly adapted to storing continuously varying surfaces, such as elevation or precipitation. Contour lines (as on topographic maps) have been used for many years to display surfaces, but calculating derived information from contours, such as slope, flow direction, and aspect, is difficult. Finally, some types of analysis are more time-consuming to perform with vectors.

Modeling feature behavior with topology

Two basic vector models exist: **spaghetti models** and **topological models**. A spaghetti model stores features of the file as independent objects, unrelated to each other. Simple and straightforward, this type of model is found in many types of applications that store spatial data. It is also commonly used to transfer vector features from one GIS system to another.

A topological data model stores features, but it also contains information about how the features are spatially related to each other. Many types of spatial relationships might be of interest, for example, whether two parcels share a common boundary (**adjacency**), whether two water lines are attached to each other (**connectivity**), whether a company sprayed pesticide over the same area on two different occasions (**overlap**), or whether a highway connects to a crossroad or has an overpass (**intersection**). Although computer algorithms can determine whether these spatial relationships exist between features in a spaghetti model, storing explicit information about the relationships can save time if the relationships must be used repeatedly.

Another application of topology involves analyzing the **logical consistency** of features. Logical consistency evaluates whether a data model or data set accurately represents the real-world relationships between features. For example, two adjacent states must share a common boundary that is exactly the same (the real-world situation), even though the states are stored in the data model as two separate features with two boundaries that coincide (Fig. 1.4). Lines representing streets should connect if the roads they represent meet. A line or a polygon boundary should not cross over itself.

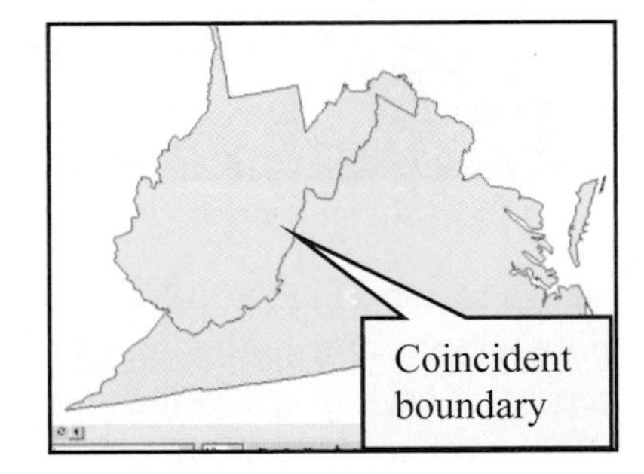

Fig. 1.4. A coincident boundary gets stored twice but is the same for both features.

Finally, topology can be used to better model the real-world behavior of features. In a network topology, for example, the connections between features are tracked so that flow through the network can be analyzed. Applications of networks include water in streams, traffic along roads, flights in and out of airline hubs, and utilities through pipes or electrical systems.

The raster model

The raster model has the benefit of simplicity. A set of spatial data is represented as a series of small squares, called **cells** or **pixels** (Fig. 1.5). Each pixel contains a numeric code indicating a single attribute, and the raster is stored as an array of numbers.

Vector features, such as roads or land use polygons, can be converted to raster format by selecting a single attribute to be stored in the cells. In Figure 1.5a, the cells store numeric values representing a land cover type, such as 46 for conifer forest or 23 for hardwood forest. Each value is given a different color for display. The roads shown in Figure 1.5b were originally vector line features with a text attribute indicating a primary, secondary, or primitive road type. When converted to a raster, the number (1, 2, or 3) is used to represent each road that passes through a cell. The cells that don't contain roads are given a null value. Rasters that store vector features in a raster format are sometimes called **discrete** rasters, because they represent discrete objects (roads, pipelines, land use polygons).

However, rasters may also be used to store a map quantity rather than features. Map quantities are values or variables that change over the earth's surface. A **digital elevation model** (DEM), for example, stores elevation values (Fig. 1.5c). Cells are unlikely to have the same elevation as their neighbors, and the values range smoothly into one another, forming a continuous surface (or continuous field). Therefore, they are commonly called **continuous** rasters.

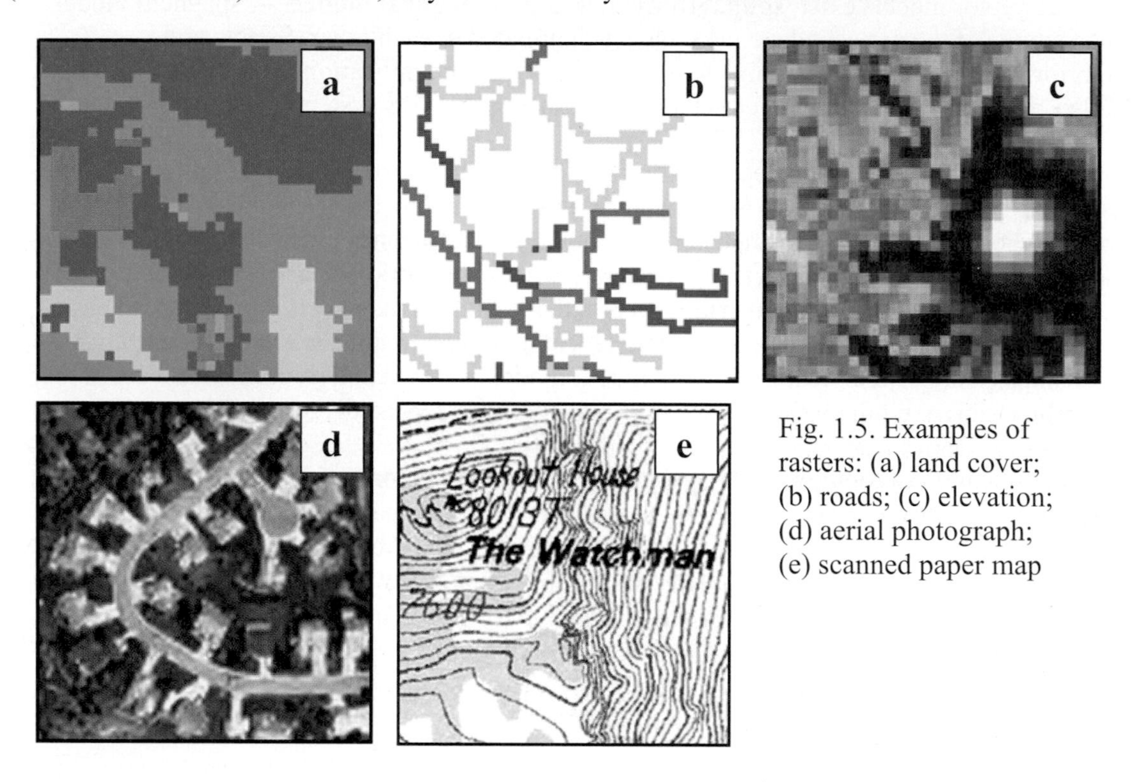

Fig. 1.5. Examples of rasters: (a) land cover; (b) roads; (c) elevation; (d) aerial photograph; (e) scanned paper map

Image rasters (Fig. 1.5d) store brightness values and are commonly used to store aerial photographs or satellite images. Images may contain multiple arrays of values, called **bands**, to represent each pixel. Color images often contain a red band, a green band, and a blue band for each pixel, and the mixture of values in each band defines the color of the pixel. One can also do a digital scan of a paper map and store it as a raster, such as the US Geological Survey topographic map shown in Figure 1.5e, known as a **digital raster graphic** (**DRG**). Each cell stores an index code representing a different color, such as 5 for the brown contours and 1 for the white background.

A raster data set is laid out as a series of rows and columns. Each pixel has an "address" indicated by its position in the array, such as row = 3 and column = 6. Georeferencing a map in an *x-y* coordinate system requires four numbers: an *x-y* location for one pixel in the raster data set and the size of the pixel in the *x* and *y* directions. Usually the upper-left corner is chosen as the known location, and the *x* and *y* pixel dimensions are the same so that the pixels are square. From these four numbers, it is possible to calculate the coordinates of every other pixel based on its row and column position. In this sense, the georeferencing of the pixels in a raster data set is implicit—one need not store the *x-y* location of every pixel.

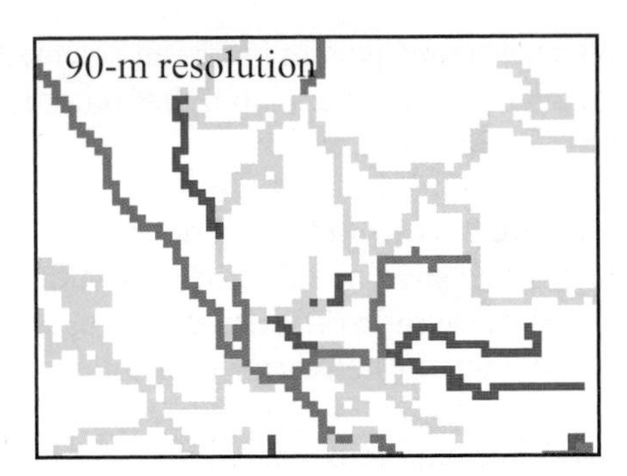

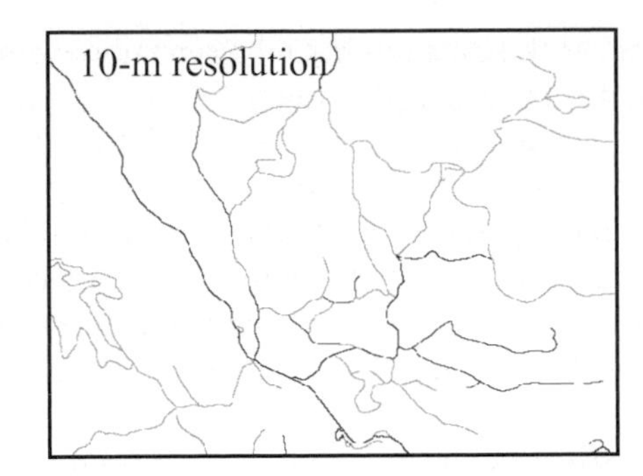

Fig. 1.6. Impact of raster resolution. The 10-meter resolution raster stores the roads more precisely, but it takes 81 times as much space.

The *x* and *y* dimensions of each pixel define the **resolution** of the raster data. The higher the resolution, the more precisely the data can be represented. Consider the 90-meter resolution roads raster in Figure 1.6. Since the raster cell dimensions are 90 meters, the roads are represented as much wider than they actually are, and they appear blocky. A 10-meter resolution raster could represent the roads more accurately; however, the file size would increase by 9 × 9, or 81 times.

The raster model mitigates some of the drawbacks of vectors. It is ideally suited to storing continuous and rapidly changing discontinuous information because each cell can have a value completely different from its neighbors. Many analyses are simple and rapid to perform, and an extensive set of analysis tools for rasters far outstrips those available for vectors.

The drawbacks of rasters lie chiefly in two areas. First, they suffer from trade-offs between precision and storage space to a greater extent than vectors do. The second major drawback of rasters is that they can store only one numeric attribute per raster. Vector files can store hundreds of attribute values for each spatial feature and can handle text data more efficiently.

Coordinate systems

Both raster and vector data rely on *x-y* values to locate data to a particular spot on the earth's surface. The *x-y* values of the coordinate pairs can vary, however. The choice of values and units to store a data set is called its **coordinate system**. Consider a standard topographic map, which actually has three different coordinate systems marked on it. The corners are marked with degrees of latitude and longitude, which is termed a **geographic coordinate system (GCS)**. Another set of markings indicates a scale in meters representing the UTM, or Universal Transverse Mercator, coordinate system. A third set of markings shows a scale in feet, corresponding to a State Plane coordinate system. Any location on the map can be represented by three different *x-y* pairs corresponding to one of the three coordinate systems (Fig. 1.7). A global positioning system (GPS) unit also has this flexibility. It can be set to record a location in GCS degrees, UTM meters, State Plane feet, or other coordinate systems.

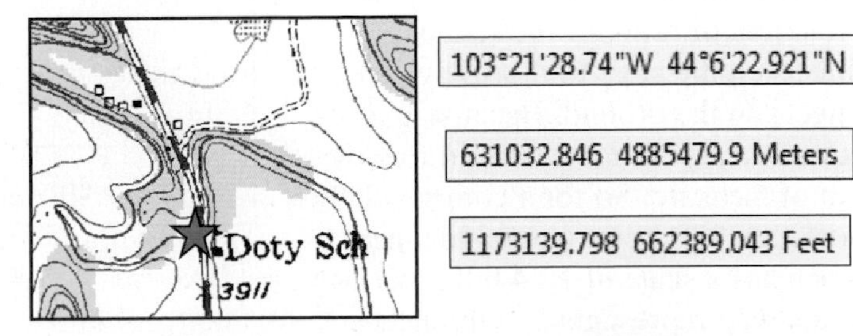

Fig. 1.7. A location can be stored using different coordinate systems and units. The *x-y* location of this school is shown in degrees, UTM meters, and State Plane feet.

When creating a vector or raster data set, one must choose a coordinate system and units for storing the *x-y* values. It is also important to label the data so that the user knows which coordinate system has been selected and what the units for the *x-y* values are. Thus, every GIS data set must have a label that records the type of coordinate system and units used to store the *x-y* data inside it.

In older GIS systems, the coordinate systems of different data sets had to match in order for them to be drawn together in the same map. UTM data could only be shown with other UTM data, State Plane data with other State Plane data, and so on. If data were in different coordinate systems, they would need to be converted to the same coordinate system prior to display.

Although it is still true that coordinate systems must match for data to be displayed together, many GIS systems can now perform the conversion on-the-fly. This feature allows data to be stored in different coordinate systems yet to be drawn together. In ArcMap, the user defines a coordinate system for the map, and all of the data are converted to match (Fig. 1.8). The units defined for the map coordinate system, whether they are meters, feet, or degrees, become the **map units** and may differ from the stored units in the files. In Figure 1.8, the UTM data use meters to store the *x-y* coordinates, the GCS data use degrees, and the State Plane data use feet. The Oregon Statewide Lambert coordinate system uses meters, so meters become the map units.

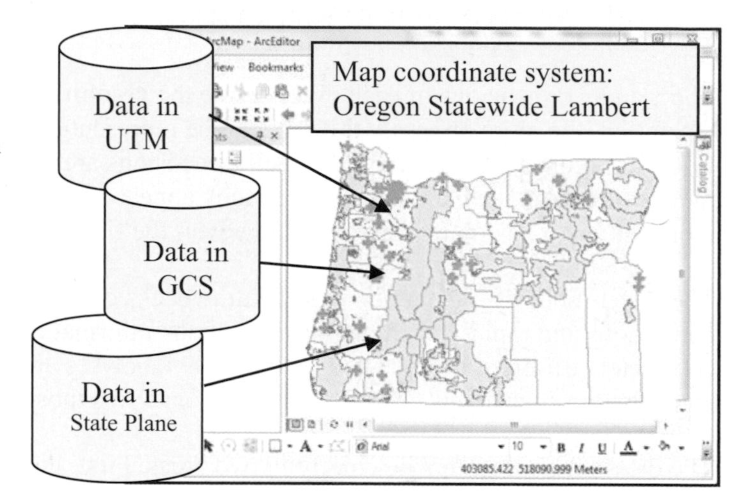

Fig. 1.8. Data in any coordinate system can be displayed together by setting the data frame coordinate system.

Map scale

The act of taking a set of GIS features with *x-y* coordinate values and drawing them on a screen or printing them on a piece of paper establishes a map scale. On a paper map, the scale is fixed at the time of printing. Within a computer system that allows interactive display, the scale changes every time the user zooms in or out of the map.

What is map scale?

Map scale is a measure of the size at which features in a map are represented. The scale is expressed as a fraction, or ratio, of the size of objects on the page to the size of the objects on the ground. Because it is expressed as a ratio, it is valid for any unit of measure. So for a common US Geological Survey topographic map, which has a scale of 1:24,000, one inch on the map represents 24,000 inches on the ground. Imagine a map made to the scale of 1:100,000. You can use the map scale and a ruler to determine the true distance of any feature on the map, such as the width of a lake (Fig. 1.9). Measure the lake with a ruler and then set up a proportion such that the map scale

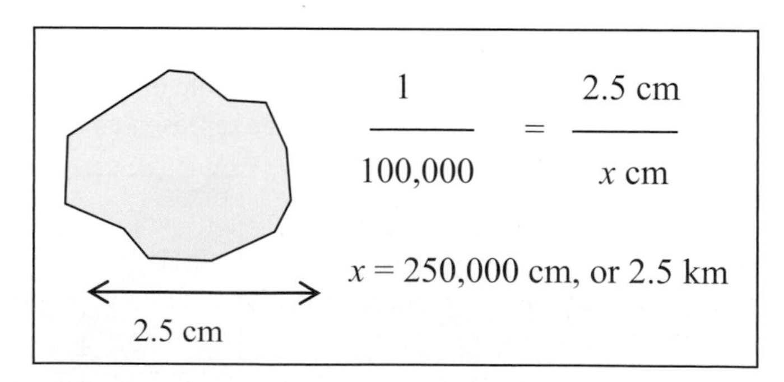

Fig. 1.9. Solving for the size of a lake.

equals the measured width over the actual width (x). Then solve for x. Keep in mind that the actual width and the measured width will have the same units. You can convert these units, if necessary.

Often people or publications refer to large-scale maps and small-scale maps. A large-scale map is one in which the *ratio* is large (i.e., the denominator is small). Thus, a 1:24,000 scale map is larger scale than a 1:100,000 scale map. Large-scale maps show a relatively small area, such as a quadrangle, whereas small-scale maps show bigger areas, such as states or countries.

Scales for GIS data

When data are stored in a GIS, they technically do not have a scale because only the coordinates are stored. They acquire a scale once they are drawn on the screen or on a piece of paper. However, most data layers have an intrinsic scale at which they were created. A 1:1 million scale paper map that is scanned or digitized cannot effectively be used at larger scales. The map in Figure 1.10 shows congressional districts in pink and the state outlines in thick black lines. The state boundaries are more angular and less detailed than the districts because they were digitized at a smaller scale. Thus, although it is possible to take small-scale data and zoom in to large scales, the accuracy and detail of the data will suffer. The original scale of a data set is an important attribute and is included when documenting it.

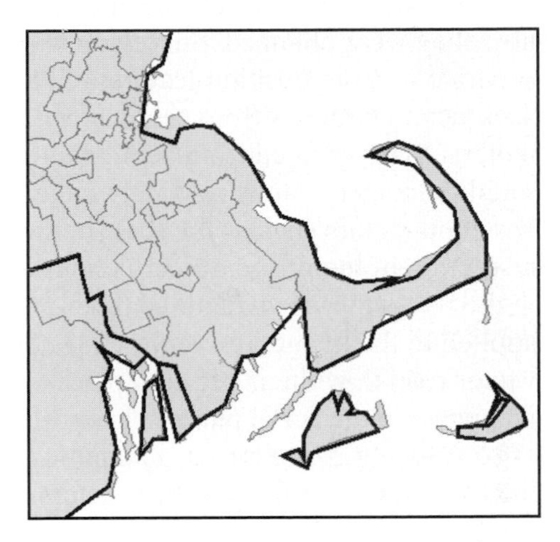

Fig. 1.10. These two layers showing Massachusetts originated from maps with different scales.

One should exercise caution in using data at scales very different from the original source. Zooming in to a data set may give a false impression that the data are more precise than they actually are. A pipeline digitized from a 1:100,000 scale map has an uncertainty of about 170 feet in its location due to the thickness of the line on the paper. Displaying the pipeline on a city map at 1:60,000 might look fine, but zooming to 1:2000 would not help, should you desire to locate the pipeline by digging.

From looking at Figure 1.10, one might conclude that it is desirable to always obtain and use data at the largest possible scale. However, large-scale data require more data points per unit area, increasing data storage space and slowing the drawing of layers. Every application has an optimal scale, and little is gained by using information at a higher scale than needed.

Data quality

Representing real-world objects as points, lines, polygons, or rasters always involves some degree of generalization. No data file can exactly capture all the spatial or attribute qualities of any object. The degree of generalization often varies with the scale. On a standard topographic map, a river has a width and can be modeled as a polygon with two separate banks. A city would be shown as a polygon area. For a national map, however, the river would simply be shown as a line, and a city would be shown as a point.

Even a detailed representation of an object is not always "true." Rivers and lakes can enlarge in size during a spring flood event or shrink during a drought. The boundary of a city changes over time as the city grows. Users of GIS data must never forget that the data they collect and use will contain flaws, and that the user has an ethical and legal responsibility to ensure that the data used for a particular purpose are appropriate for the task. When evaluating the quality of a data set, geospatial professionals consider the following aspects.

Geometric accuracy refers to the *x-y* values of a feature class or raster. How closely do the locations correspond to the actual location on the earth's surface? Geometric accuracy is usually a function of the original scale at which data are collected and of how they were obtained. Surveying is one of the most accurate ways to position features. GPS units have an accuracy that ranges from centimeters to tens of meters. Maps derived from aerial photography or satellite imagery can vary widely in geometric quality based on factors such as the scale of the image, the resolution of the image, imperfections and distortions in the imaging system, and the types of corrections applied to the image. In Figure 1.11, notice that the vector road in white is offset in places from the road as it appears in the aerial photo. These differences can arise from digitizing errors, geometric distortions from the camera or satellite, or other factors.

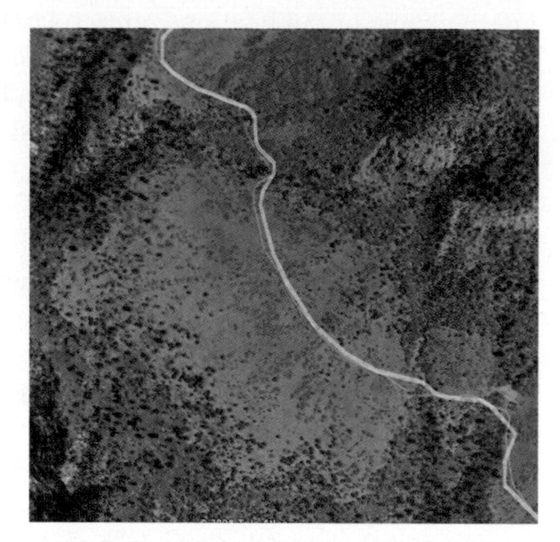

Fig. 1.11. Aerial photo near Woodenshoe Canyon, Utah. Source: Google Earth and TeleAtlas.

Moreover, not every boundary can be as precisely located as a road. Imagine that you wish to delineate the land-cover types: *forest*, *shrubland*, *grassland*, and *bare rock* in this photo. Where would you draw the line between *shrub* and *grassland*? At what point does the *shrubland* become *forest*? Six people given this photo would come up with six different maps. Some boundaries would match closely; others would vary as each person made a subjective decision about where to place each boundary.

Thematic accuracy refers to the attributes. Some types of data are relatively straightforward to record, such as the name of a city or the number of lanes in a road. Even in this situation, the value of a feature might be incorrectly recorded. Other types of information can never be known exactly. Population data, for example, are collected through a process of surveying and self-reporting that takes many months. It is impossible to include every person. Moreover, people are born and die during the survey process, or are moving in and out of towns. Population data can never be more than an estimate. These difficulties don't mean that it is pointless to collect the data. However, it is important to understand the limitations and potential biases associated with thematic data.

Resolution refers to the sampling interval at which data are acquired. Resolution may be spatial, thematic, or temporal. Spatial resolution indicates at what distance interval measurements are taken or recorded. What is the size of a single pixel of satellite data? If one is collecting GPS points by driving along a road, at what interval is each point collected? Thematic resolution can be impacted by using categories rather than measured quantities: if one is collecting information on the percent crown cover in a forest, is each measurement reported as a continuous value (32%) or as a classified range (low, medium, high)? Temporal resolution indicates how frequently measurements are taken. Census data are collected every 10 years. Temperature data taken at a climate station might be recorded every 15 minutes, but it might also be reported as a monthly or yearly average.

Precision refers to either the number of significant digits used to record a measurement or the statistical variation of a repeated single measurement. Many people confuse precision with accuracy, but it is important to understand the distinction. Imagine recording your body temperature with an oral digital thermometer that records to a thousandth of a degree and getting the value of 99.894 degrees Fahrenheit. This measurement would be considered precise. However, imagine that you take the reading immediately after drinking a cup of hot coffee. This action throws off the thermometer reading so that it does not record your true body temperature. Thus, the measurement is precise, but it is not accurate.

Evaluating the quality of a data set can be difficult, especially if the data were created by someone else. Professionals who create data incur an obligation to evaluate the quality of the data and to provide a report that summarizes the spatial and thematic accuracy so that users can determine whether a data set is suited to a particular purpose. Producers should also provide information on other aspects of a data set, such as what geographic area it covers, what coordinate system it uses, what the information in the attribute tables means, how a potential user can access the data, and more. If the original data were created or compiled by others, the producer must also give proper credit to the originators. Such information about a data set is called **metadata** (Fig. 1.12). The content and format of metadata are established by international standards.

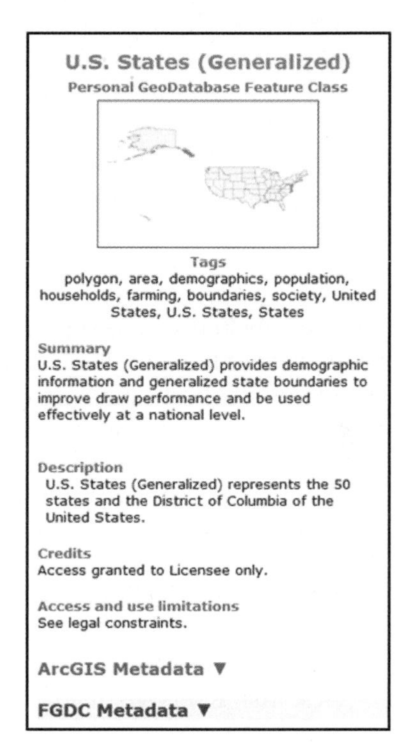

Fig. 1.12. Metadata

Organizations assemble collections of metadata to allow potential users to search for and evaluate data sets before they are downloaded or purchased, like an electronic library catalog. Once a candidate data set is identified, the user can explore the full metadata record to determine if the data set is appropriate for the particular application. If so, the metadata tell the user where the data are located, how they can be obtained, and what the cost might be.

Metadata also record the access and use constraints on GIS data. GIS data can be copyrighted and their uses restricted to certain people or certain actions. Some GIS data, including most data sets derived from federal agencies, can be freely copied and redistributed with credit given to the originating agency. Other data are developed by companies, and the rights are licensed to specific users. Often the license includes the right to distribute maps or other static copies derived from the data, but not the data themselves. Every user is responsible for understanding the applicable use constraints placed on any data set and for abiding by them. Failure to do so can result in civil and criminal penalties against the individual or the organization he or she works for.

Citing GIS data sources

Ethical and professional considerations require that any map, publication, or report should cite the data source(s) used and give proper credit to the originators of the data. The metadata, or the site from which the data were obtained, are good sources of information for citations. The best practice is to record the citation when the data are obtained so the information is available when needed. Generally one cites only data that are publically available (free or purchased). Data created internally within the workplace need not be cited, although often the company name or logo will appear on the map. A data set provided once in response to a personal request should be cited as a personal communication.

Keep in mind, when citing, that the place you found the data may not always be its source. Your GIS administrator may have placed often-used data sets, such as ESRI Data and Maps, on a workplace fileserver for easy access, but the fact that they were obtained internally does not free you from the need to cite them, and you must cite the original source, not the local server.

General format

The purpose of a citation is to allow people to obtain the data if they wish. The following general format for citations may be used:

Data set name (Year published) [source type]. Producer name, producer contact information. *Resource URL: [Date accessed].*

Data set name. The name is assigned by the creator or provider of the data.

Year published. Some data sets are assembled and provided once or at long intervals, and these are considered to have a publication date. For example, the ESRI Data and Maps product is released in revised form with each version of the software and carries a publication year. Aerial photography is flown on a particular date (although mosaics such as Google Earth use multiple sources spread over several years). It may take a little hunting or a few questions to find the publication date. Some data sets are updated at shorter intervals, or are even live. These should be assigned the current year.

Source type. Indicate the format in which the data are available. Types might include physical media (DVD, CD-ROM, USB stick), a file downloaded from the Internet, or a service that provides live data on demand. Different types of services exist, such as database service, map service, image service, map package, and layer package. New types are being added all the time.

Producer name. Give the name of the person or agency that makes the data available. In some cases, this may be different from the originator of the data. For example, ESRI publishes Data and Maps using data from many different sources. ArcGIS Online serves public data.

Producer contact information. For a company or small agency, the city and state should be included. For large agencies, particularly those with many offices but a unified web presence, the name itself is sufficient. Indicate a clearing house name, such as ArcGIS Online, if appropriate.

Resource URL. This entry is optional; include when appropriate. Use only static URLs even if it means the user has to hunt for the data. (A static URL always has the same form, whereas a dynamic URL is generated automatically based on search strings or other information. Dynamic URLs generally contain gibberish and/or characters like %.)

Date Accessed. This entry is optional because it mainly applies to online data sets. Include the year and month when you accessed or downloaded the data.

Examples of citations

Black Hills National Forest Database (2008) [downloaded file]. Black Hills National Forest, Custer, SD. URL: http://www.fs.usda.gov/main/blackhills/landmanagement/gis [August, 2010].

ESRI Data and Maps (2010) [DVD]. ESRI, Inc., Redlands, CA.

National Hydrology Dataset (2012) [downloaded file] United States Geological Survey on the National Map Viewer. URL: http://viewer.nationalmap.gov/viewer/ [July 23, 2012].

Geographic Names Information System (2008) [downloaded file]. United States Geological Survey. URL: http://geonames.usgs.gov/domestic/download_data.htm [May 21, 2009].

USA Topo Maps (2009) [map service]. ESRI on ArcGIS Online. URL: http://server.arcgisonline.com/ arcgis/services/USA_Topo_Maps/MapServer [January 1, 2012].

EIA Coalbed Methane Field Boundaries (2011) [map service]. US Department of Energy on ArcGIS Online. URL: http://arcgis.com [August, 2013].

Mineral Operations of Africa and the Middle East (2010) [layer package]. J.M. Eros and Luissette Candelario-Quintana on ArcGIS Online. URL: http://ArcGIS.com [July, 2012].

Badlands National Park GIS Database (2012) [CD-ROM]. Interior, South Dakota: National Park Service—Badlands National Park, personal communication.

About ArcGIS

ArcGIS overview

ArcGIS is developed and sold by Environmental Systems Research Institute, Inc. (ESRI). It has a long history and has been through many versions and changes. Originally developed for large mainframe computers, in the last 15 years it has metamorphosed from a system based on typed commands to a user-friendly graphical user interface (GUI). Data models, too, have changed over time, so that one is likely to encounter data sets in different formats. Knowing this background helps a student of GIS understand the nature of the ArcGIS system and its data.

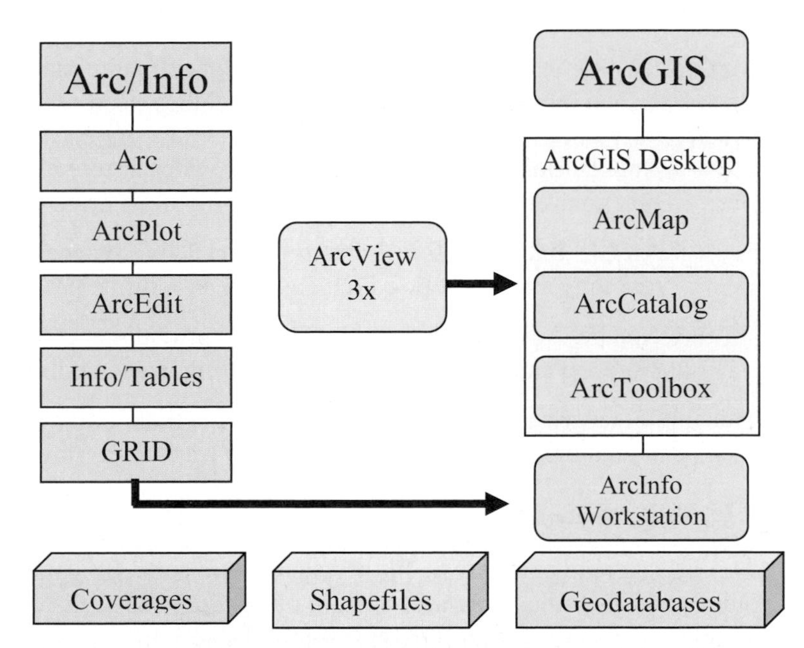

Fig. 1.13. Relationship between ESRI products and data formats

The old core of the ArcGIS system was called Arc/Info and included three programs: Arc, ArcEdit, and ArcPlot (Fig. 1.13), which utilized a data model known as a **coverage** and incorporated a database program called INFO, which appears primitive today. All of the programs were command based, meaning that the user typed commands into a window to make the program work.

The difficulty of learning Arc/Info prompted ESRI to create ArcView in 1992, which was easier to use but not as powerful as Arc/Info. ArcView was designed primarily to view and analyze spatial data rather than create them. ArcView used a simpler data model, called the shapefile, although it could read coverages and convert them to shapefiles.

ArcGIS Desktop, released in 2001, combines power and ease of use. It contains two main programs.

- ArcMap provides the means to display, analyze, and edit spatial data and data tables.
- ArcCatalog facilitates viewing and managing of spatial data files. It should *always* be used to delete, copy, rename, or move spatial data files.

In addition, ArcGIS Desktop contains ArcToolbox, a collection of tools and functions for operations in ArcCatalog and ArcMap, such as converting between data formats, managing map projections, and performing analysis (Fig. 1.14). Users may create and add their own tools or scripts for special or often-used tasks. A program called Model Builder lets users graphically arrange and run sequences of steps, and save them to be used over and over.

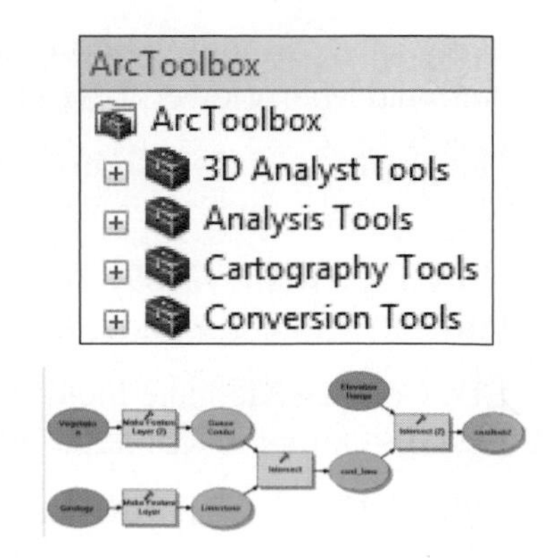

Fig. 1.14. ArcToolbox and a model

ArcGIS Desktop provides different levels of functionality that all use the same basic interface. Users can save money by buying only the level they need.

- ArcGIS Basic (formerly ArcView) provides a suite of mapping, editing, and analysis functions and is the level of functionality most users will require on a regular basis.
- ArcGIS Standard (formerly ArcEditor) adds advanced editing capabilities, such as topology and network editing, as well as additional data conversion tools.
- ArcGIS Advanced (formerly ArcInfo) provides access to the full functionality and adds more advanced tools for analysis, editing, data handling, and cartography.

This book focuses almost exclusively on the functions available with an ArcGIS Basic license. Users can read the software documentation to learn more about the advanced capabilities.

Data files in ArcGIS

ArcGIS can read a variety of file formats. Some come from older versions of the software, or from other programs such as computer-aided design (CAD) systems. Table 1.1 lists many of the data sets that can be used in ArcGIS.

Shapefiles

Shapefiles are spaghetti data models developed for the early version of ArcView. A shapefile contains one feature class composed of points or lines or polygons but never a mixture. The attributes are stored in a **dBase** file. Shapefiles can, however, store **multipart features**, which are single features made of multiple objects. For example, Hawaii requires multiple polygons to represent the different islands, but the state can be stored as a multipart feature so that it has only one record in the attribute table.

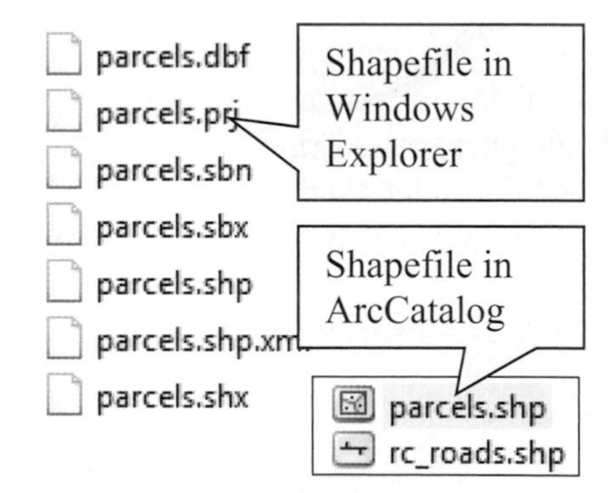

Fig. 1.15. Shapefiles are groups of files but appear as single entries in ArcCatalog.

Although a shapefile appears as one icon in ArcGIS, it is actually composed of multiple data files, as can be seen in Windows Explorer (Fig. 1.15). The rivers shapefile consists of seven different files. The .shp file stores coordinate data, the .dbf file stores attribute data, and the .shx file stores a spatial index that speeds drawing and analysis. These first three files are required for every shapefile to function. Additional files may also be present: the .prj file stores projection information, the .avl file is a stored legend, and the .xml file contains metadata. To copy a shapefile, all of these files must be included. ArcCatalog takes care of this automatically, but Windows Explorer does not—one reason it is wise to manage GIS data using ArcCatalog.

.xls

In a shapefile attribute table, the first two columns of data are reserved for storing the feature identification code (FID) and the coordinate geometry (Shape) field. These fields are created and maintained by ArcGIS and cannot be modified by the user. All other fields are added by the user.

Table 1.1. Types of files and data sources used by ArcGIS

File type	Description
Shapefiles	Shapefiles are vector feature classes developed for the early version of ArcView and have been carried over into ArcGIS.
Coverages	A coverage is the vector data format developed for Arc/Info and is the oldest of the data formats.
Geodatabases	Geodatabases represent an entirely new model for storing spatial information with additional capabilities.
Database connections	Database connections permit users to log in to and utilize data from an RDBMS geodatabase.
Layer files	A layer file references a feature class and stores information about its properties, such as how it should be displayed.
Rasters	Rasters represent map data or imagery using arrays of regular cells, or pixels, containing numeric values.
Tables	Tables can exist as separate data objects that are unassociated with a spatial data set.
GIS servers	GIS Servers provide GIS data over an Internet connection as maps, features, or images
TINs	TINs are Triangulated Irregular Networks that store 3D surface information, such as elevation, using a set of nodes and triangles.
CAD drawings	Data sets created by CAD programs can be read by ArcGIS, although they cannot be edited or analyzed unless they are converted to shapefiles or geodatabases.

Geodatabases

A geodatabase can contain many different objects, including multiple feature classes, geometric networks, tables, rasters, and other objects. Figure 1.16 shows a geodatabase named rapidnets. Feature classes may exist as individual objects in a geodatabase (as do the restaurants or schools), or they may be grouped into feature datasets. A feature dataset contains a collection of related feature classes with the same coordinate system, such as the Utilities feature dataset in Figure 1.16.

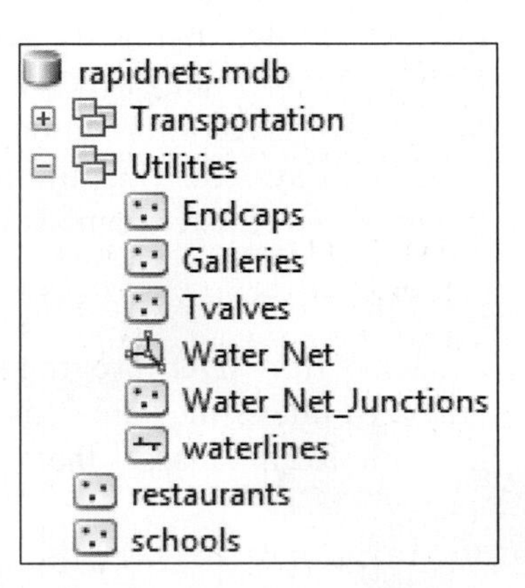

Fig. 1.16. A geodatabase

A feature dataset can also store topological associations between feature classes. The Utilities feature dataset in Figure 1.16

contains a **network topology** stored in the Water_Net and Water_Net_Junctions feature classes. Chapter 9 covers some special analysis functions that can be used with networks. Feature datasets may also contain **planar topology**, which contains rules about spatial relationships within or between layers. Special tools help find and fix errors that violate the topology rules.

Finally, geodatabases may contain rules that assist in entering and validating attribute data. Called **domains**, these rules specify which values or range of values may be entered in a particular field; a percent field, for example, should contain only numbers between 0 and 100.

Three types of geodatabases are used by ArcGIS: personal geodatabases, file geodatabases, and SDE geodatabases. The behavior of the three types is similar, but the data storage formats and capabilities differ. They are described in Chapter 14.

Coverages

A **coverage** is the oldest vector format, developed for Arc/Info. ArcGIS Desktop has limited functions for managing coverages, so most users will encounter them simply as an old data format that must be converted to a shapefile or exported to a geodatabase. Several things are helpful to know in this process.

Coverages contain multiple feature classes, which may store points, arcs, polygons, and polygon labels. Coverages also store topology, and the tables have several attribute fields reserved for this purpose. Figure 1.17 shows these fields for a coverage called LANDUSE (some fields use the coverage name as part of the field name). It makes sense to delete these fields during or after the conversion, for they serve no useful purpose afterwards.

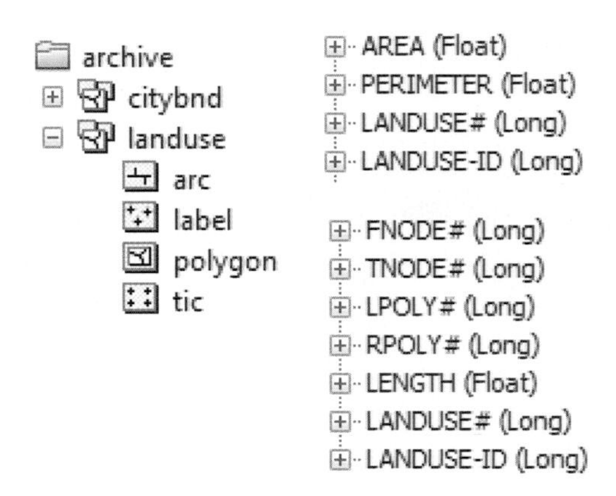

Fig. 1.17. Coverages contain fields of little value once they are converted.

VERY IMPORTANT TIP: Do not use Windows to copy or delete coverages, shapefiles, and geodatabases. Always use ArcCatalog to delete or copy spatial data sets.

Rasters

Rasters in ArcGIS can take a variety of formats, such as JPEG, TIF, GEOTIF, BMP, MrSID, and raw binary files (BIP, BIL, BSQ). A list of supported raster formats can be found in the ArcMap Help under the index heading "rasters, formats." Most rasters consist of the spatial data plus a header that gives information about the file, such as its number of rows and columns and its coordinate system. This information may be stored in a separate file or as the first part of the binary raster. Rasters can also be stored inside geodatabases.

CAD files

Data sets from CAD programs can be read by ArcGIS, although they cannot be edited or analyzed unless they are converted to shapefiles or geodatabases. A CAD file may contain multiple feature classes that correspond to layers of the drawing and can be opened separately.

Database connections

A user can connect to a database management system (DBMS) on a network through a database connection. This connection can be open, or it can require a login and password. Once inside, the user can access the tables within the database. Database connections may be available on a local network through a mounted drive, but some can be accessed online as database services.

Internet data services

Many organizations make data available over the Internet. Free data services, such as GoogleEarth and MapQuest, provide access to huge volumes of image and map data. These services are used online in a web interface and cannot be viewed in ArcMap. Although the data quality and documentation procedures are not designed for professional-level work, the volume and popularity of these sites introduce many people to GIS techniques and data. Other sites, such as the United States Geological Survey's National Map, do allow users to download data.

GIS servers

GIS servers provide geospatial data over Internet connections. Hosting GIS data on a server requires software that responds to requests from users for specific maps or data in the user's current window. Some GIS services are open and free, designed to provide public access to data and maps. Others may be locked down to a particular organization or group of users. Many organizations use servers to make in-house data available to personnel out in the field, or to allow employees in different locations to access the same company data sets. Servers provide data in a client-neutral format; as long as the client program knows how to use the service, the data can be used in ArcGIS Desktop, in a web application, on a tablet, or on a smartphone.

Several types of services may be offered. A **map service** renders map layers as tiles and sends them to the user as static images. This type of service is efficient and fast, but the user cannot modify how the map is symbolized. A **feature service** shares the requested data features; it tends to be slower, but the user can change how the features look and potentially edit or analyze them. A **layer package** or **map package** sends the features to the client, where they are stored locally during use. An **image service** provides access to large mosaics of satellite imagery or aerial photography. GIS servers can also provide analysis tools in addition to data. A **geoprocessing service** makes available certain computations and functions, so that users can perform the designated analysis through a web site, even if they do not have GIS software installed.

Cloud-based services

Setting up and managing a GIS server is a complex and expensive task requiring suitable space, equipment, software, and expertise. New cloud-based services are lowering this barrier. A **cloud** consists of warehouses of giant computers and hard drives managed by a company that rents processing power and disk space to clients. Some companies, for example, offer space to individuals to back up their computers or store their movies and music files in one place that is accessible to all of their devices.

Another type of cloud service is the virtual machine, or VM, created by setting aside part of a large computer with its own operating system and software, allowing it to be managed and used as an ordinary physical computer. The VM can be scaled from a standard desktop to a powerful machine capable of running intensive computations or serving thousands of web requests. The client rents the VM for an hourly fee. Furthermore, cloud services can be configured to automatically add more VM units when demand is high and scale back when usage is light.

Many organizations have moved to the cloud to host their GIS data, either as a complement to or even instead of housing their own GIS servers. Advantages of cloud services include the speed and ease of deployment, the ability to scale up and down to meet changing demand, and the security benefits of hosting public information outside the organizational firewall.

ArcGIS Online and web maps

ArcGIS Online is a cloud-based platform that provides an environment to create and share maps. It makes data available to people with little to no GIS training, yet it also addresses the needs of professional users. Anyone may use the data, and those willing to set up a free account may also share their data. The user controls whether a published data set is visible to a few selected people or to the general public. The platform also contains easy tools for creating web pages that include interactive maps. Organizations may purchase a subscription account, which allows them to manage security, designate users allowed to access or publish data, and host data in the cloud without needing to configure their own GIS servers.

ArcGIS Online is designed around the **web map**, an interactive map based solely on GIS services. Web maps generally perform a restricted set of basic functions, such as zoom and query; however, they are device independent and can be used in ArcGIS Desktop, in web browsers, on mobile devices (such as smartphones or tablets), and even within social media sites. They are relatively simple to create and to share.

Summary

- A GIS is a database system that uses both spatial and attribute data to answer questions about where things are and how they are related. It has many functions, including creating data, making maps, and analyzing relationships.

- Raster data employ arrays of values representing conditions on the ground within a square called a pixel. The array is georeferenced to a ground location using a single *x-y* point.

- Vector data use sequences of *x-y* coordinates to store point, line, or polygon features. Every feature is linked to an attribute table containing information about the feature.

- Every GIS data set has a coordinate system defined for stored *x-y* coordinate values. Many different coordinate systems are used, so each data set must be labeled with information about the coordinate system.

- Data are stored as simple spaghetti models or as topological models. Topological models can better model feature behavior and aid in locating and correcting geometric errors.

- Every GIS user has a responsibility to ensure that data are suitable for the proposed application. Data quality is measured in terms of geometric accuracy, thematic accuracy, resolution, and precision.

- Metadata store information about GIS data layers to help people use them properly.

- ArcGIS Desktop employs a menu-based interface for two programs called ArcMap and ArcCatalog. It uses many data formats, including shapefiles, coverages, geodatabases, rasters, images, TINs, CAD drawings, and Internet-based data services.

- ArcToolbox contains functions for processing, managing, and analyzing GIS data. Users may customize it by building models or writing scripts to repeat often-used sequences.

IMPORTANT TIP: Do not use Windows to copy or delete coverages, shapefiles, and geodatabases. Always use ArcCatalog to delete or copy spatial data sets to prevent problems. It is also prudent to avoid using spaces in folder names.

Important Terms

adjacency	discrete	layer	point
attributes	domain	layer file	polygon
band	feature	layer package	precision
cell/pixel	feature class	line	raster
cloud	feature dataset	logical consistency	resolution
connectivity	feature service	map service	scale range
continuous	FID/OID	map units	spaghetti model
coordinate system	folder connection	map package	Table of Contents
coverage	geographic coordinate system (GCS)	metadata	thematic accuracy
data frame	geometric accuracy	multipart feature	thematic mapping
datum	geoprocessing service	network topology	topological model
dBase	georeferenced	node	vector
digital elevation model	intersection	overlap	vertex/vertices
digital raster graphic		planar topology	web map

Chapter Review Questions

1. Explain the difference among the terms *feature*, *feature class*, and *feature dataset*.
2. Imagine you are looking at a geodatabase that contains 50 states, 500 cities, and 100 rivers. How many feature classes are there? How many features? How many attribute tables? How many total records in all the attribute tables?
3. If the following data were stored as rasters, which ones would be discrete and which would be continuous: rainfall, soil type, voting districts, temperature, slope, and vegetation type?
4. John and Mary are collecting GPS data together. John's GPS says their location is at (631058, 4885805). Mary's GPS says their location is at (1204817, 663391). Explain what is going on. What must be done to make the GPS units agree?
5. Would raster or vector be a better format for storing land ownership parcels? Give at least three reasons for your choice.
6. You measure a football field (100 yards long) on a large-scale map and find that it is 0.5 inch long. What is the scale of the map?
7. Scott is walking the boundary of a wetland area to map it. His expensive GPS records locations to the nearest 0.10 meter. Is the boundary he creates accurate? Is it precise? Explain your reasoning.
8. Imagine a feature class of agricultural fields with attributes for the crop and the organic matter content of the soil. What issues might impact the thematic accuracy of each attribute?
9. Explain some ways that GIS services are different from data that reside on your hard drive.
10. Construct an appropriate citation for the data that come with this book.

Mastering the Skills

Preparing to begin

The tutorials in this book use a data set that comes with the text on a DVD. These data must be installed before you start using the tutorials. Each step of the tutorials and all skills learned in this book are illustrated by video clips on the DVD. For instructions on installing the data and using the video clips, consult the Preface and/or the readme.txt documentation on the DVD.

As you are working through the tutorials, you will learn better if you think carefully about each step. Be sure to follow the directions exactly, or you may find that subsequent steps look or behave differently than the book's instructions.

A **TIP** gives useful information about the program or settings that will save time or that are important to know. Pay attention to them.

A **SKILL TIP** tells about a technique that is not covered in a tutorial but that you may wish to learn about later using the Skills Reference at the back of the book. The section of the Skills Reference to consult is indicated in parentheses (ArcMap Basics).

TIP: In these tutorials, values you must enter are shown in this font: type this.

Teaching Tutorial

The following examples provide step-by-step instructions for doing basic tasks and solving basic problems in ArcGIS. The steps you need to do are highlighted with an arrow ➔; follow them carefully. Click on the video number in the VideoIndex to view a demonstration of the steps.

TIP: This book comes with a folder of data called mgisdata, and you need to know where you installed it. In the examples we use the pathname C:\gisclass\mgisdata, but your folder may be elsewhere. Write down the pathname to its location. ____________

Adding data to ArcMap

1➔ Start ArcMap.

1➔ In the Getting Started window, click the New Maps text on the left side of the window, indicating that you want to start with a new, empty map document.

1➔ Click on the Blank Map template icon, and click OK.

TIP: If using ArcGIS on a network, you may see a different configuration of toolbars. Choose Customize > Toolbars from the main menu, and you will see a list. Only these toolbars should be checked: Draw, Standard, and Tools. If you see others checked, click the names to turn them off.

2➔ Click the Add Data button.

2➔ Click on the *Look in* drop-down button to show available data folders (Fig. 1.18).

ArcGIS is able to work with many kinds of data, including files on your hard drive as well as online data. Data on your local computer, or on a network drive mounted on your computer, will be shown in **Folder Connections**, which stores shortcuts to locations containing GIS data.

2➔ Click the *Look in* drop-down button again, if necessary, and select Folder Connections.

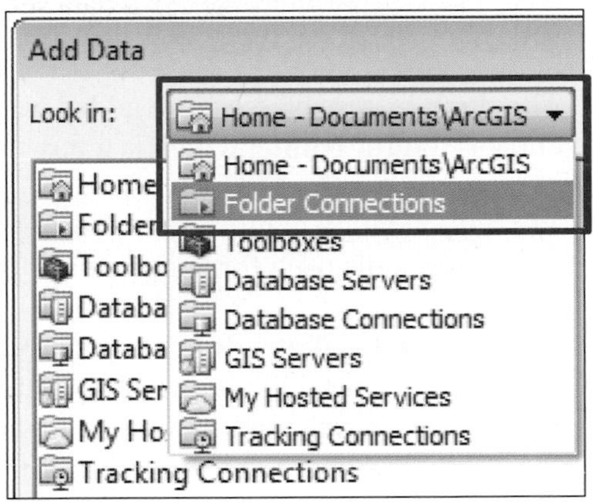

Fig. 1.18. Sources of data

If this is the first time ArcGIS has been used on the computer, this folder will be empty. Otherwise, it may contain connections, which will appear as pathname entries (Fig. 1.19a). You must add a connection to your mgisdata folder to start using your data.

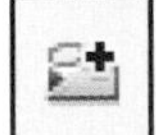

2➔ To add a connection, click the Connect to Folder button in the Add Data window.

2➔ Navigate to the directory containing your mgisdata folder and click once on the mgisdata folder icon to select it (Fig. 1.19b; yours may be in a different location). Do not select any of the subfolders in the mgisdata folder. You are adding a connection right now; you will access the data later.

2➔ Click OK to finish adding the connection.

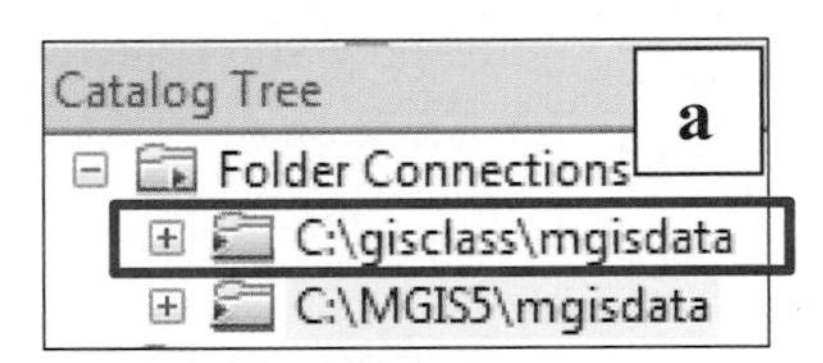

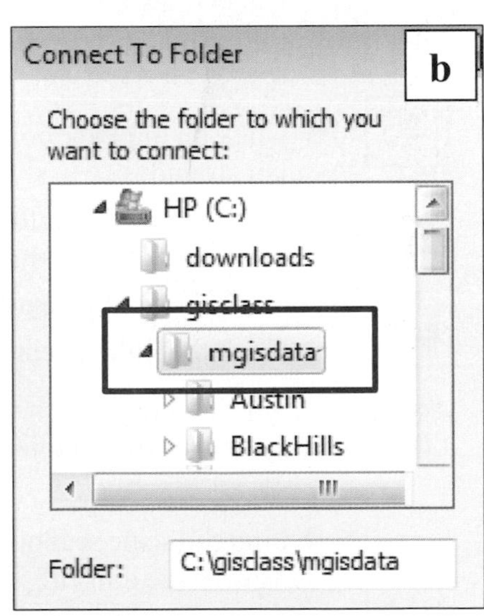

Fig. 1.19. (a) Connection to mgisdata; (b) connecting to a folder

TIP: Once you add a connection it will remain until you disconnect it. Connections tend to build up over time, so disconnect older ones no longer in use. Avoid having multiple pathways to the same folder. Keep connections as uncluttered and simple as possible.

Now that the connection is added, you are ready to start adding data to your map.

3➔ Click the Look in drop-down one more time to find the connection you added. Click on it to open it.

3➔ Double-click the Oregon folder to open it.

3➔ Double-click the gray oregondata icon to open the geodatabase.

3➔ Click the counties feature class to highlight it; then click Add to add it to the map.

4➔ Click the Add Data button again. This time, double-click the Transportation feature dataset to open it.

4➔ Click the airports feature class; then hold down the Ctrl-key and click on the highways feature class. Click Add to add them both.

Viewing the map

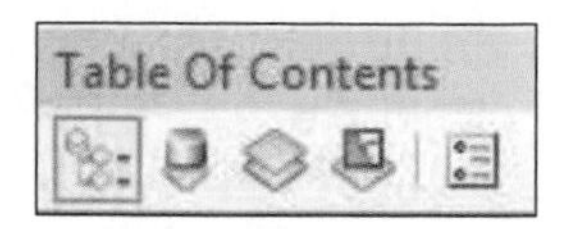

Examine the ArcMap window. The area on the left is called the **Table of Contents**. Examine the first four icons at the top, which present the layers in four different ways. For now, make sure the leftmost icon is clicked to display the layers in the order in which they are drawn, from bottom to top.

The three feature classes you added were placed underneath a heading called Layers. The Layers entry is a **data frame**, which is a box to contain map data to be displayed together.

We will begin by exploring the Tools toolbar. Figure 1.20 shows the tools used for zooming and panning the map.

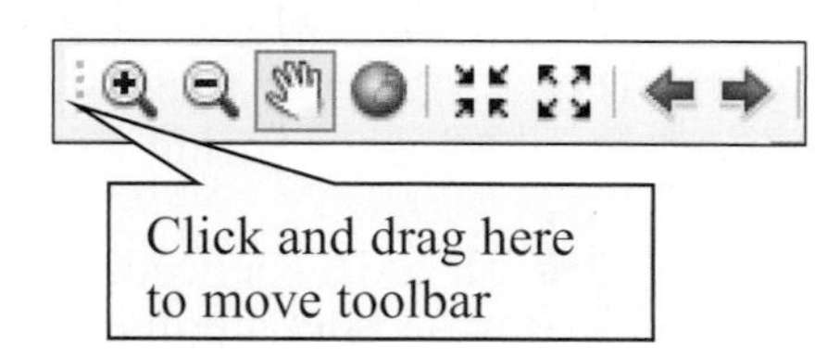

Fig. 1.20. Part of the Tools toolbar

5➔ Click the Zoom In tool. Place the cursor at the upper-left corner of the state, then click and hold the mouse button down to drag a box around a few counties. When finished, let go of the mouse button.

5➔ Click once in the lower-right corner of the map to zoom in again.

5➔ To return to the full extent of the map (useful if you made a mistake or do not like the area you selected), click on the Full Extent button.

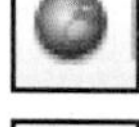

5➔ Click on the Zoom Out tool and click in the upper-left corner of the state again. The view zooms out with the point clicked at the center.

TIP: You can also draw a box using the Zoom Out tool. If a large box is drawn, the view zooms out a little bit. If a small box is drawn, the view zooms out a large amount.

6➔ Click on the Pan tool and then click and drag inside the display window to move the map around.

6➔ Click the Go Back to Previous Extent button. It allows you to undo a zoom that you did not intend, or return to an earlier zoom level.

6➔ Click the Go to Next Extent button. It reverses the effect of the Go Back button.

6➔ Click the Full Extent button to return to the view of all the counties.

TIP: Hover the cursor over a tool for a moment to find out the tool name and what it does.

7➔ Find the scale readout on the main toolbar and click the drop-down arrow to see a list of preset scales. Choose 1:3,000,000.

7➔ Click on the Fixed Zoom In button two or three times. It zooms in to the next preset scale in the list.

7➔ Click on the Fixed Zoom Out button two or three times. It zooms to the next scale.

7➔ When finished experimenting, click the Full Extent button to return to the full extent of the Oregon data.

SKILL TIP: Learn to use the Measure tool to determine areas, lengths, and perimeters in an assortment of different units (ArcMap Basics).

The Identify tool can be used to examine the attributes of a feature.

8 ➔ Locate the Identify tool on the Tools toolbar and click on it.

8➔ Place the tool on top of one of the counties, staying away from an airport or a highway, and click on it.

8➔ The county will flash on the screen, and the Identify Results box will appear. The attributes of the county are displayed in the Identify window.

8➔ Use the scroll bar to examine the attributes in the Identify window.

8➔ Click another county and examine its attributes.

1. What is the name of the county in the northeast corner of Oregon? ____________________

9➔ In the Identify window, change the *Identify from* drop-down to highways.

9➔ Click on a highway and examine its attributes.

9➔ Close the Identify window by clicking the X in the upper-right corner.

Viewing attribute tables

To view the attributes for all of the features in a feature class, you can open the table.

10➔ Right-click the counties layer in the Table of Contents and choose Open Attribute Table from the context menu that appears.

10➔ Scroll to the right, to the end of the table and back, noting all of the fields.

2. How many records (rows) are in this table? ____________________

10➔ Hold the cursor over the right edge of the NAME field until it turns into a double arrow bar. Click and drag the edge to increase or reduce the width of the column.

Tables have an options button with various tasks, such as finding text or exporting the table.

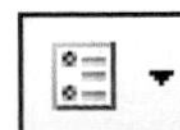

11➔ Click the Table Options button and choose Find and Replace. Click the Find tab.

11➔ Type Hood in the *Find what* box and click Find Next.

11➔ The cursor jumps to the record for Hood River and outlines the data cell.

11➔ Examine the other Find options and then close the Find and Replace window.

A context menu gives access to several commands relating to the individual fields.

12➔ Right-click the field name POP2000 to display a context menu. Choose Sort Descending. (Make sure you scroll back to the top of the table to see the most populous county, Multnomah.)

3. Which county has the smallest 2000 population? ____________________

12➔ Right-click the POP2000 field and choose Statistics to see basic statistics and a frequency diagram of the values. Close the Statistics window when done looking.

12➔ Right-click the NAME field and choose Freeze/Unfreeze Column. Scroll to the right, noting that the NAME field stays put now.

12➔ Close the counties table by clicking the X in the upper- right corner.

TIP: More than one field can be frozen at a time. Unfreezing allows the field to scroll again, although it remains on the left side until the field settings are reset.

TIP: You can sort, freeze, and do statistics without changing the source data.

Choosing map symbols

Feature classes contain only locations and attributes and do not store symbols or labels. When a feature class is loaded into ArcMap, it becomes a **layer**. A layer is a specification for how the feature class will look and behave in the map document, and you can set properties to control it.

13➔ Click the Add Data button. You last added data from the Transportation feature dataset, so click the Up arrow once to return to the main geodatabase.

13➔ Click the volcanoes feature class to select it and click Add.

13➔ Right-click the volcanoes layer to make the context menu appear, containing actions or commands that can be performed on the layer.

13➔ Move the cursor to the Properties entry at the bottom of the menu, and click it to open the Layer Properties window.

The tabs at the top of the window provide access to menus for setting various properties.

14➔ Click on the Symbology tab.

14➔ Click on the Symbol button that shows the current symbol. It opens the Symbol Selector window.

14➔ Click on the Triangle 1 symbol to use it. Change the color to a reddish-brown and the size to 16. Click OK and then OK again in the Layer Properties window.

TIP: Clicking Apply in a window applies any changes you have made but keeps the window open. Clicking OK applies the change and closes the window.

15➔ Right-click the airports layer and choose Properties.

15➔ Click on the current symbol button to open the Symbol Selector.

15➔ Scroll down and find the Airplane symbol. Select it.

15➔ Leave the color black, but increase the size to 25-pt, and set the angle to -45.

15➔ Click OK and OK.

SKILL TIP: Learn how to search for and add more symbols to a map than are listed in the Symbol Selector by default (Maps and Symbols).

You don't have to open the Properties to change the symbols for a layer, however.

16➔ Right-click the highways layer symbol (not the text) to bring up a color context menu. Choose the black color.

16➔ Click (left-click) on the counties symbol (not the text) to open the Symbol Selector, and choose the Beige symbol.

16➔ Click the Outline Color drop-down button and hold the cursor on top of a color box until the color name appears. Set the color to the Gray 20% symbol. Click OK.

Labels are another property of a layer. You can create a set of labels and turn them on or off as you wish.

17➔ Right-click the airports layer and open its properties. Click on the Labels tab.

17➔ Check the box to *Label features in this layer*.

17➔ The Label Field is preset to NAME, but it could be changed to something else if desired.

17➔ Change the symbol to Arial 9-pt Bold font.

17➔ Ignore the other options and click OK. The map should now look similar to Figure 1.21.

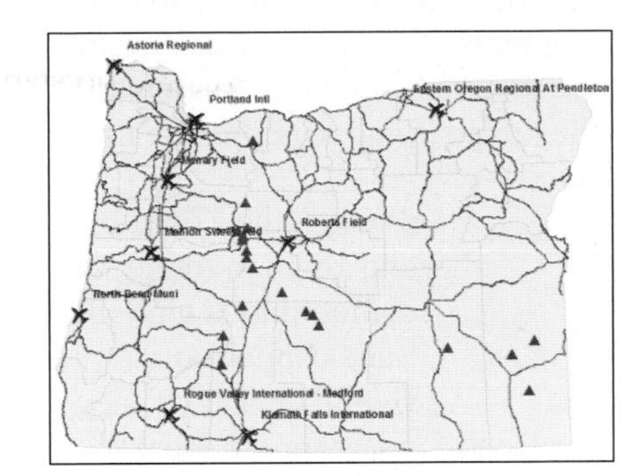

Fig. 1.21. The map after Step 17

TIP: To remove a layer from the map, right-click it and choose Remove. Removing a layer only takes it out of the map. It does not delete it from the hard drive.

This is a good time to save the map so far.

18➔ Choose File > Save As from the main menu.

18➔ Navigate to the mgisdata\MapDocuments folder, if necessary.

18➔ Enter the name **Oregonmap** and save the map document.

Layer properties and layer files

Imagine that you will create series of maps and want airports on all of them. The symbols and labels you just set will appear only in this map. Instead of setting them again each time, you can save these settings as a **layer file**. A layer file does not contain spatial data, but it points to the original feature class and stores instructions on how to display it.

19➔ Right-click the airports layer name and choose Save As Layer File.

19➔ Navigate to the mgisdata\Oregon folder.

19➔ Enter **myairports.lyr** as the name of the layer file and click Save.

Nothing obvious happens, but the layer file has been stored on disk for you, ready for the next time you want the airports symbolized this way on a map. We will use it later.

Now let's examine some of other properties, besides symbols and labels, that can be set for layers. Instead of starting from scratch, we will examine properties already set and stored in a group layer file, which is similar to a layer file but contains properties for multiple layers.

SKILL TIP: Learn to create group layers and store them as group layer files (BASICS: General).

20➔ Click the New Map File button on the main toolbar to open a new map document.

20➔ Choose the Blank Map icon and click OK.

20➔ Click the Add Data button. Navigate to the mgisdata\Usa folder.

20→ Click on the US Example layer file and choose Add.

Although only the states show in the map window, many layers are listed in the Table of Contents. The check boxes for these layers are gray and cannot be turned on or off. The creator of the layer file applied a **scale range** to these layers, defining a specific range of scales at which they appear and making sure that they appear only when appropriate.

21→ Click the Zoom In tool and zoom in to the western conterminous United States. State capitals and rivers should appear (if they do not, zoom in more).

21→ Right-click the Capitals layer and choose Properties. Click the General tab.

21→ Examine the Scale Range property (Fig. 1.22). Capitals will be shown only when the user has zoomed in below a scale of 1:40 million. Click Cancel.

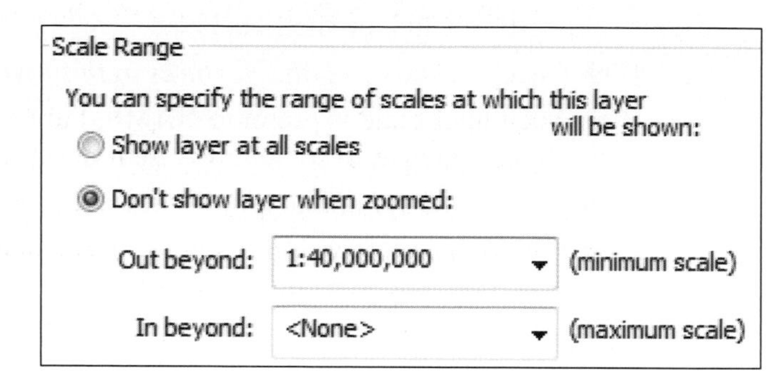

Fig. 1.22. The scale range property sets the scales at which a layer will be displayed.

1:40,000,000

22→ Locate the Map Scale box on the main toolbar, indicating the current map scale.

22→ Click in the Map Scale box and type 40,000,000 (with or without the commas). Click Enter. The map zooms to that scale.

22→ Click the Fixed Zoom Out button. The capitals and rivers disappear.

22→ Click the Fixed Zoom In button once. They appear again.

22→ Click the Fixed Zoom In button a few more times until the labels for the capitals appear. Labels can have a different scale range than the features they represent.

23→ Click the Fixed Zoom In button until the interstates and the river labels appear.

23→ Continue to zoom in until the counties appear, symbolized by native population.

23→ Place the cursor on top of a county and wait. A small label pops up to show the county name. These labels are called MapTips.

MapTips are another property of a layer. They can be set to show the contents of a specified field when the cursor is on top of the feature.

TIP: Double-clicking a layer's name also opens its Properties.

24→ Double-click the Counties layer to open its properties and click the Display tab.

24→ Examine the settings on this tab. Find the *Show MapTips* check box.

4. What is the name of the field that is being displayed in the MapTips? ____________

24→ Click the Symbology tab for the Counties layer properties.

24➔ Instead of a single symbol, this layer is set to display the percentage of Native Americans in the counties. Examine the settings but don't change anything.

24➔ Click Cancel to close the Properties window.

Using the Catalog tab

The Table of Contents lets you work with layers of your map. The Catalog tab allows you to examine and manage your GIS files. The Catalog tab is an instance of ArcCatalog that runs inside ArcMap rather than as a separate program.

25➔ Click the Catalog window button on the main toolbar to open it.

25➔ If it opens docked on the right side of the ArcMap window, leave it there.

25➔ If it opens elsewhere, click the bar at the Catalog window's top and drag it to the blue arrow that appears on the right side of the ArcMap window. Release the mouse to dock it.

SKILL TIP: Learn how to move and arrange the toolbars, as well as how to use the docking icons and pins to manage windows in ArcGIS (General).

26➔ Examine the Catalog. A list of folders and data is shown (Fig. 1.23), called the Catalog Tree.

26➔ Clicking a plus sign expands a folder to show its contents; clicking a minus sign collapses it.

26➔ If any folders are expanded right now, click the minus signs to close them all.

26➔ Examine Catalog Tree and find the entry titled Folder Connections. Expand it.

26➔ The connection to your mgisdata folder, added in Step 2, is already shown. Other connections may be present if ArcGIS has been used on the computer before.

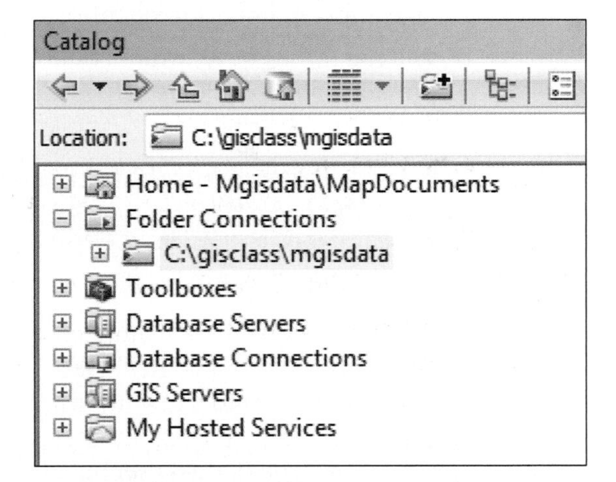

Fig. 1.23. The Catalog tab

TIP: The Folder Connections entry shows data from your computer's hard drives or mapped network drives. The Toolboxes folder contains additional commands and tools. The remaining four entries are used to access data from databases or online data sources.

27➔ Expand the mgisdata folder connection to see its contents.

27➔ Expand the Rapidcity folder. Inside it, expand the archive folder to examine its contents (Fig. 1.24).

5. How many coverages are there in the archive folder? _____ How many tables? _____ How many rasters? _____ How many layer files? _____ How many shapefiles? _____

27➔ Collapse the archive folder contents.

27➔ Expand the Oregon folder.

27➔ Expand the oregondata geodatabase (Fig. 1.25). Expand the Transportation feature dataset to see its feature classes.

Name	Type
citybnd	Coverage
landuse	Coverage
buildings.shp	Shapefile
connects.shp	Shapefile
lucodes.dbf	dBASE Table
parcels.shp	Shapefile
rc_roads.shp	Shapefile
rceast_nw.sid	Raster Dataset
sdschools.shp	Shapefile
stategeol.lyr	Layer
stategeol.shp	Shapefile
watersheds.shp	Shapefile

Fig. 1.24. Data sets in ArcCatalog

SKILL TIP: Learn to create, copy, rename, and delete GIS files and folders (Files and Geodatabases).

In addition to finding data in the Catalog window, you can get information about individual data sets.

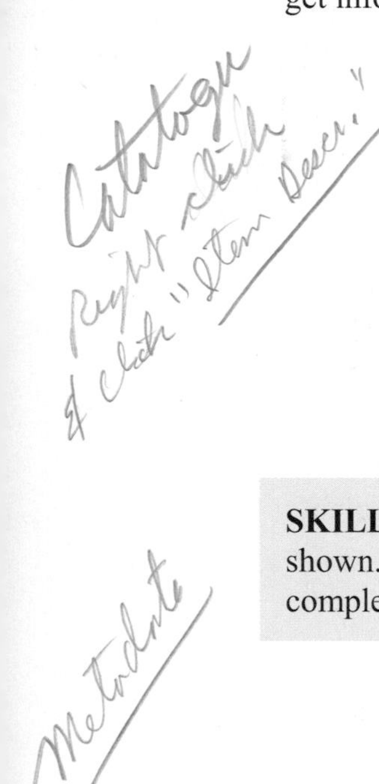

28➔ In the Catalog Tree, collapse the Oregon and Rapidcity folders.

28➔ Expand the Usa folder and the usdata geodatabase.

28➔ Right-click the rivers feature class and choose Item Description. It will take a few moments to open.

28➔ Read through the information provided about the data set.

oregondata.gdb — Geodatabase
Transportation — Feature data set
airports — Point feature class
highways — Line feature class
rail100k
Water
cities
counties — Polygon feature class
gtopo1km

Fig. 1.25. A geodatabase in ArcCatalog

SKILL TIP: By default only a brief description is shown. Learn how to make ArcCatalog display the complete metadata information (Metadata).

29➔ Click the quakehis feature class in the Catalog Tree. The Item Description window updates to show information on this feature class.

You can not only read about a data set but also preview it.

29➔ Click on the Preview tab in the Item Description window.

29➔ Use the Zoom or Pan tool in the Item Description window to get a closer look.

30➔ Change the Geography drop-down to Table to preview the attributes. This Preview window does many of the same things as the Table window in ArcMap.

30➔ Right-click the MAG field and choose Sort Descending.

30➔ Use the Table Options menu to open Find and Replace. Then search for San Francisco. The cursor in the table highlights the cell when it finds it.

30➔ Close the Find/Replace and the Item Description window.

Now that you know it's the one you want, it is easy to add it to the map.

31➔ Click on the quakehis feature class in the Catalog Tree and drag it into the map window. Release the mouse button to drop it in the map.

31→ Right-click the quakehis layer symbol in the Table of Contents (not the Catalog Tree) and choose a bright pink color for the earthquakes so that they show up well.

TIP: Be careful to select the correct window when right-clicking an item. The Table of Contents window lists map layers, and making changes only affects how the map appears. The Catalog Tree in the Catalog tab lists folders and data sets, and it can permanently change them (usually with no Undo). Be especially careful when working in the Catalog Tree.

Since the Catalog Tree has the potential to change or damage our data if we are not careful, we are going to make it less easy to use it by mistake, while keeping it handy for when we need it. It will also let us see more of our map.

32→ Click the Auto Hide button on the Catalog window to hide it, creating a Catalog tab where the window was docked.

32→ Hold the cursor over the Catalog tab (no click needed) to open the Catalog window. Move off the window to hide it again.

TIP: The Auto Hide button is a toggle switch. If you wanted the Catalog tab to remain open, you would click the Auto Hide button again. Try it now, if you like, but leave it hidden when done.

Coordinate systems in ArcCatalog and ArcMap

Both vector and raster data are referenced to the earth's surface by an *x-y* coordinate system. Every data set should have a label that indicates the coordinate system used to store its *x-y* values.

33→ Right-click the Capitals layer in the Table of Contents and choose Properties.

33→ Click on the Source tab and read the information in the Data Source window.

Two items of interest are shown. First appears the name and pathname of the feature class upon which this layer is based. After it comes the name of the coordinate system, North_America_ Equidistant_Conic. Most of the coordinate system information will not make sense until you study Chapter 11, but do note the Linear Unit entry, which is Meter. The *x-y* values in this data set are stored in meters.

33→ Click Cancel to close the Layer Properties window.

33→ Open the properties for the Oregon Highways layer.

6. What is the name of this coordinate system? ______________________________
   What is the linear unit? ______________

These two layers are stored in different coordinate systems, so which coordinate system is being used for the map? Notice the data frame name, Layers, at the top of the Table of Contents. The data frame has properties also, including the coordinate system used to display its layers.

34→ Close the Oregon Highways layer properties.

34→ Right-click the Layers data frame name and choose Properties.

34→ Click the Coordinate System tab and read the information in the lower panel.

The data frame coordinate system is set to North_America_Equidistant_Conic. The Oregon Highways layer is being converted on-the-fly to match it. You can change the data frame coordinate system to view layers in any coordinate system that you choose.

34➔ On the Coordinate System tab, find the Geographic Coordinate Systems folder.

34➔ Expand Geographic Coordinate Systems > North America and select the NAD 1983 entry (the simple one in a group is usually the right choice).

34➔ The Current coordinate system name in the lower box changes to GCS_North _American_1983. Click OK.

The shapes of the map features change, stretching in the east-west direction. North-south lines that were previously angled become vertical. Changing coordinate systems can dramatically affect the appearance of a map.

35➔ Open the Layers data frame properties again to choose another coordinate system.

35➔ Expand the Projected Coordinate Systems folder and navigate to the Continental > North America > USA Contiguous Albers Equal Area Conic coordinate system. Select it and click OK.

The map changes again. Every coordinate system causes distortion of areas, distances, shapes, or angles. Each one is designed to minimize or eliminate different kinds of distortion, and the name often gives a hint. An equidistant coordinate system displays accurate distances. An equal area coordinate system displays accurate areas. The best one to use depends on the purpose of the map, which is why data are stored and used in many coordinate systems instead of just one.

TIP: Changing the coordinate system of the data frame only changes the map display. It does not affect the coordinate systems used to store the feature classes.

Using Internet map services

Your hard drive or office network drive is not the only source of GIS data. Many organizations host services that provide maps, imagery and data, and individuals can share content on the ArcGIS Online platform. Before we explore these riches, however, let's see how ArcGIS Help can be a valuable asset in learning more about GIS.

36➔ Choose Help > ArcGIS Desktop Help from the main menu bar.

36➔ Click on the Contents tab, if necessary.

36➔ Expand the entry for *Mapping* and for *Using web maps and GIS services.*

36➔ Click on the entry *What are web maps?* and read through the information.

You can reinforce concepts and learn new things by reading Help. An amazing amount of information is available, equivalent to many books. The Search function helps find the information that you need. Let's see what it has to say about ArcGIS Online.

37➔ Click the Search tab in the Help window.

37➔ Type ArcGIS Online into the box at the top and click Ask.

37➔ Click on *Using ArcGIS Online* and read through the entry.

37➔ At the bottom of the entry, click on the link *A quick tour of web maps and GIS services*. Read through it.

37➔ At the bottom of the entry, click on *What are GIS services?* Read about them.

37➔ Close ArcGIS Help.

TIP: Use the Search function in Help during a tutorial whenever you'd like to learn more about a tool or concept. Your GIS knowledge and skills will improve dramatically.

Now let's explore ArcGIS Online. You need a high-speed Internet connection for the next steps. If you are not connected, you may end the tutorial here.

38➔ Click the New Map File button on the main toolbar and start with a Blank Map. You do not need to save your changes to the previous map.

Add Data...
Add Basemap...
Add Data From ArcGIS Online...

38➔ Click the Add Data drop-down triangle and choose Add Basemap.

38➔ Select the Streets basemap and click Add.

TIP: If ArcMap crashes when you try to load a basemap, start ArcMap again. Go to Customize > ArcMap Options and click the Data View tab. Uncheck the box to *Enable hardware acceleration.*

39➔ Zoom to your home state and then to your home town.

39➔ Zoom to your home neighborhood and find your street.

TIP: Be patient when using Internet services because they can take time to draw. Sometimes they may not be available. If this one does not work, try another, or try again later.

Notice that the map becomes more detailed as you zoom in; map services generally employ scale ranges to show more detailed layers at larger scales. The basemap is only the beginning. Many agencies and individuals provide map data for others to use.

39➔ Zoom back out so that the conterminous United States is being shown.

40➔ Click the Add Data drop-down again and choose Add Data from ArcGIS Online.

40➔ Type `weather` in the search box and click the Search button. Examine the results.

40➔ Examine the US Weather Warnings service, which should be near the top.

40➔ Click on the Details link and review the information.

7. Which agency provides the data for this service? ______________________________
   Which agency hosts the GIS server that provides this service? ____________________

40➔ Click the Add link to add the service to your map.

40➔ Examine the map to find where weather warnings have been issued today.

This map service is continually updated as new information is posted by the National Weather Service. You do not have as much control of a map service as your own layers, but a few display options can be set.

41➔ Right-click the US Weather Warnings group layer name and choose Properties.

41➔ Examine the Source tab. Note the service URL and the coordinate system.

41➔ Click the Advanced tab.

41➔ Move the Layer Transparency slider to about 50% and click OK.

Now you can see the underlying base map showing through the warnings.

42➔ Click the Identify tool and click on one of the warning polygons.

42➔ If the fields in the Identify window look blank, they need to be resized. Place the cursor on a field boundary to resize it and see the information within.

42➔ Examine the attributes of several polygons and then close the Identify window.

42➔ Expand the US Weather Warnings group subheadings to see legend information.

Remember the Oregon airports layer file that you created earlier? Let's add it to this map to see if any airports are in a hazard area.

43➔ Click the Add Data button and navigate to the mgisdata/Oregon folder.

43➔ Select the myairports.lyr layer file and click Add.

43➔ A warning message about the geographic coordinate system may appear. If it does, check the box *Don't warn me again in this session* and click Close. We will learn more about this warning later.

43➔ Right-click the airports layer and choose Zoom to Layer.

Notice that the airplane symbols and labels that you created for this layer were saved in the layer file and recreated when you added it to the map.

44➔ Right-click the airports layer and choose Remove.

44➔ Zoom in to your home state again.

44➔ Right-click the Layers data frame name and open its Properties.

44➔ Click the Coordinate System tab, if necessary, and examine the coordinate system of the data frame, which defaulted to the basemap's coordinate system.

You may notice that your state appears elongated in the east-west direction. The data frame defaults to the coordinate system of the first layer added, in this case, the base map. The distortion is a property of the basemap coordinate system, WGS 1984 Web Mercator Auxiliary Sphere. Let's use a different coordinate system less likely to distort your home state.

45➔ Scroll and collapse folders, if necessary, to find and expand the Projected Coordinate Systems folder and the State Plane folder.

45➔ Expand the NAD 1983 (Meters) folder.

45➔ Choose a coordinate system for your state. In many cases there will be more than one (for example Maine East and Maine West). Pick the one that best represents the part of the state containing your home town. Click OK.

45➔ The map will redraw to a less distorted view of your state.

TIP: NAD 1983 stands for North American Datum of 1983. A **datum** is a correction to the coordinate system to make it more accurate. Chapter 11 has more information on datums.

46➔ Click the Add Data drop-down button and choose Add Data from ArcGIS Online.

46➔ Enter your home state's name in the Search box and search for data.

46➔ Select an interesting map service or layer package and add it to the map.

TIP: Many ArcGIS Online data sets are contributed by ordinary folks, and not all of them work well. If you have trouble getting one to display, remove it and try another.

Let's view some imagery now. We will open a new map so that the data frame defaults to the coordinate system of the service and does not spend time trying to project the images on-the-fly.

47➔ Open a new, blank map. You may save the previous map if you wish.

47➔ Choose Add Data > Add Data from ArcGIS Online.

47➔ Type the term imagery in the search box.

47➔ Examine the details of the World Imagery map service; then add it to the map.

48➔ Click the Find button on the Tools toolbar.

48➔ Click the Locations tab.

48➔ Choose the World Geocode Service (ArcGIS Online) locator.

A locator is an example of a **geoprocessing service**. When you click Find, the string is sent to the server, which parses it, finds the location by comparing the string to GIS data located on the server, and sends the location back to you. The work is done by the server, not by your computer.

49➔ Type New York City in the *Single Line Input* box and click Find.

49➔ Right-click the New York City entry in the bottom of the window and choose Zoom To. Move the Find window out of the way but leave it open.

49➔ Set the scale in the Map Scale box to 1:100,000.

49➔ Zoom in more to see greater detail.

This map service uses a variety of imagery depending on what is available. The resolution in New York City is sub-meter, but it will not be that good everywhere. Let's try locating an address instead of a place name.

50➔ Type your full home address in the Single Line Input box, including city and state (or equivalent).

50➔ If it finds it, right-click it and choose Zoom To. Right-click again and choose Add Point. Zoom in more to see how well the service did at finding your house.

The beauty of GIS servers is that you have access to terabytes of data without having to store them, update them, or download more than the portion you need. The disadvantages are that you must be online, and you may not have access to features or be able to modify the symbols.

This is the end of the tutorial.

➔ Close ArcMap. You don't need to save your changes.

Exercises

TIP: If you need to capture a map or screen shot to submit to your instructor, you can do so by pressing Alt-Prnt Screen keys on your computer, or using a program such as Jing or Snagit. Then paste the capture into a Word document. For help, ask your instructor.

Use ArcMap, the Catalog tab, and the mgisdata folder to answer the following questions. Be careful not to make any changes to properties when in the Catalog tab.

1. How many feature datasets are there in the oregondata geodatabase in the mgisdata\Oregon folder? List their names. How many total feature classes does the geodatabase have? How many rasters?

2. What is the coordinate system of the country shapefile in the mgisdata\World folder? Of the parks feature class in the oregondata geodatabase?

3. What type of information does the feature class cd111 in the usdata geodatabase contain? On what date was the information current? Is it current now?

4. Which is the largest lake in the United States? What is its area?

5. Which state has a county named Itawamba? (**Hint:** Use Find in the Table Options menu.)

6. What is the minimum, maximum, and average 2010 population density of census tracts in the city of Austin, TX? (**Hint:** Use the Austin geodatabase, not the usdata geodatabase.)

7. Use ArcGIS Help to find a discussion of feature class basics and read it to expand your knowledge. Although this text presents three basic types of feature classes (point, line, polygon), several others can be stored in a geodatabase. List the other four types.

8. How many rows and columns does the Landsat image L720021127 have? (**Hint:** It shows Crater Lake in Oregon.) What is the cell size (including the units)? How many bands does it have? What is its coordinate system?

9. Add the country and latlong shapefiles from the World folder to a new, blank map. What is the coordinate system in which these shapefile *x-y* coordinates are stored? In what units are they stored?

10. Change the data frame coordinate system to view the layers in the World Robinson projection. **Capture** the map.

Challenge Problem

Search ArcGIS Online and find a map service that you think is interesting. Combine it with at least one data set from the mgisdata folder to create a map for a place that you know. Choose an appropriate coordinate system.

Capture the map in a Word document. Underneath it, construct appropriate citations for the sources of the data. Also include a statement of the coordinate system used for the map.

Chapter 2. Mapping GIS Data

Objectives

- Distinguishing among nominal, categorical, ordinal, and ratio/interval data
- Creating maps from attributes using different map types
- Selecting appropriate classification methods when displaying numeric attributes
- Displaying thematic and image rasters
- Understanding how map documents save and represent data
- Distinguishing between a feature class and a layer

Mastering the Concepts

GIS Concepts

Geographic Information Systems are used for many purposes, but creating maps is one of the most common. The practice of cartography, or mapmaking, has a long history. Although a computer makes the process easier, and offers the ability to explore and edit designs, the cartographer must still understand the basic principles behind portraying spatial data and the aesthetic challenges in designing an effective and attractive way to communicate ideas.

Mapping features

Cartographers may choose a variety of strategies for symbolizing features in a map. A data layer may simply be portrayed using one symbol (Fig. 2.1), or different features can be assigned different symbols depending on the value of an attribute field. Point data are shown using marker symbols, line features with linear symbols, and polygon features with shaded area symbols (Fig. 2.2). When creating maps, it is important to distinguish among **nominal, categorical**, and **numeric data** because the type of data will influence the kind of map that can be made.

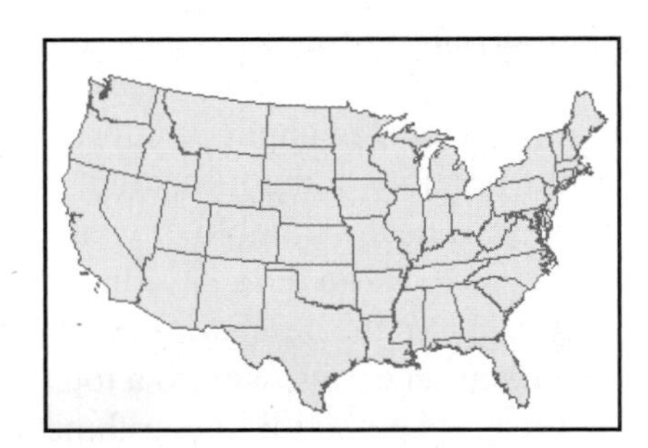

Fig. 2.1. Single-symbol map

Maps for nominal and categorical data

Nominal data name or identify objects, such as the name of a state. Nearly every feature will have a different name. Nominal data aren't always text. A parcel number or a tax identification number also serves the purpose of uniquely identifying an object. The US Census uses such data in its Federal Information Processing Standards Codes, or FIPS codes. Each state, county, city, block, and so on is assigned a unique identifying number that is consistent from product to product. Nominal data are usually portrayed on a map by labels.

Categorical data separate features into distinct groups or classes. Figure 2.2 shows examples of categorical data for point features (volcano type), linear features (road class), and polygon features (landcover). Other examples include soil types, ethnic groups, and geological rock formations. Categorical data are often stored as text, but it is also possible to represent the categories with numeric codes, such as Commercial Services = 20 and Industrial = 40. Even though the values are numeric, they still represent categories. Categorical data are represented by a **unique values map**, which gives each category a different symbol (Fig. 2.3).

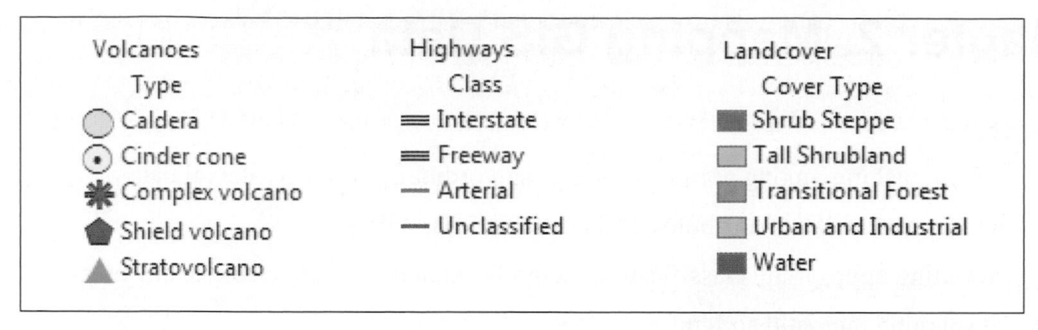

Fig. 2.2. Examples of categorical data representation for points, lines, and polygons

Ordinal data have categories that are ranked based on some quantitative measure, although the measure may not be linear. For example, cities might be classified as small, medium, or large. Students are assigned grades of A, B, C, D, or F. Soils are designated as A, B, C, or D depending on their infiltration. Ordinal data must be represented by unique values maps if the values are text, but the colors should be the same shade and give a sense of increase. Numeric ordinal data may be represented either with unique values maps or graduated color maps.

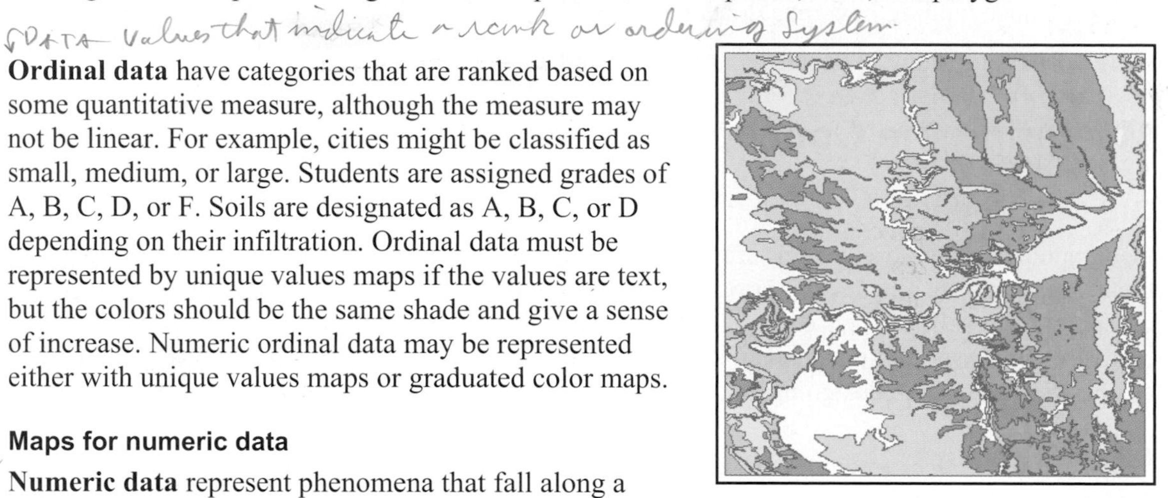

Fig. 2.3. A unique values map based on geological units

Maps for numeric data

Numeric data represent phenomena that fall along a regularly spaced measurement scale, such as distance or rainfall. Equal changes in interval involve equal changes in the quantity being measured. For example, the energy needed to heat a thimble of water from 6 to 7 degrees is the same as heating it from 96 to 97 degrees. **Ratio data** have the added property that the measurements have a meaningful zero point that indicates the absence of the thing being measured. Precipitation is an example of ratio data; zero precipitation corresponds to a total lack of rain, and two inches of rain is twice as much as one inch. Ratio data support all four arithmetic operations of addition, subtraction, multiplication, and division. **Interval data** have a regular scale but are not related to a meaningful zero point. Temperature data measured in the familiar Celsius or Fahrenheit scale are interval data because a temperature of zero does not correspond to a complete lack of temperature (heat energy). Any scale that can have negative values, such as elevation, is an interval scale. Interval scales support only addition and subtraction. Today may be 40 degrees hotter than yesterday's high of 40 degrees, but it is only a figure of speech if I tell you that it is twice as hot today.

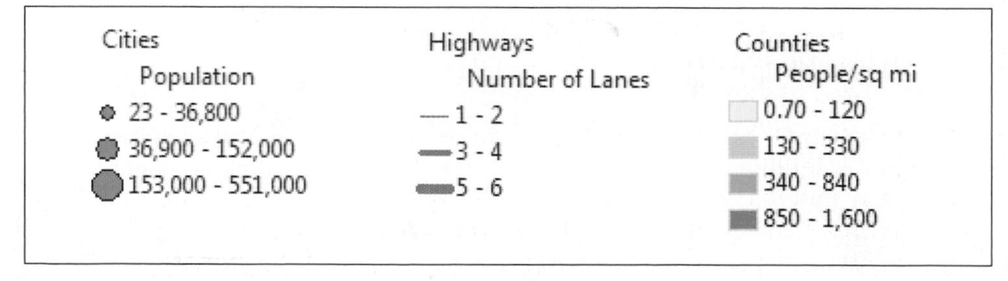

Fig. 2.4. Examples of numeric data representations for points, lines, and polygons

Numeric data take on values along a continuous scale of possibilities. Think about state populations, for example. Each state has a number representing its population, and it would be unlikely for any two states to have the same population. In order to symbolize numeric data, the values must be partitioned into groups with specific ranges, such as those shown in Figure 2.4. These ranges are called classes, and the maps are called classified maps. In Figure 2.4, the cities and roads have three classes of values, and population density has four classes. Each class is symbolized using variations in symbol size, thickness, or color.

Point or line data are usually displayed by varying symbol size or thickness and are portrayed using a **graduated symbol map**. However, point or line data can also be portrayed using a **proportional symbol map** in which the numeric value is used to proportionally determine the size of the symbol. Instead of a few size classes, the map has a continuous range of symbol sizes. This style is often referred to as an unclassed map. Figure 2.5 compares graduated symbol and proportional symbol maps for the populations of state capitals.

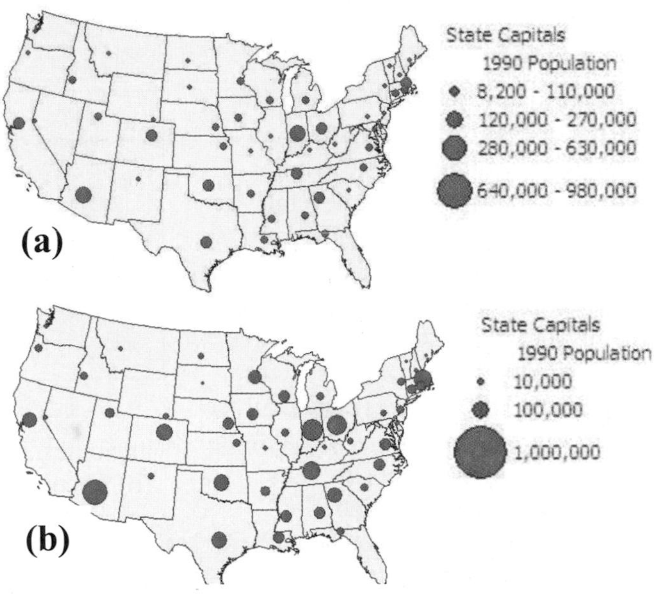

Fig. 2.5. Comparison of (a) a graduated symbol map and (b) a proportional symbol map showing the populations of state capitals

Numeric data classes for polygons are represented by changes in hue, saturation, or intensity of color using shaded symbols (Fig. 2.6), called a **graduated color map**, or a **choropleth map**. The maps are usually symbolized using a single color that changes in intensity (monochromatic color ramp). Color ramps with two or more colors may become difficult to interpret unless the symbols are chosen carefully (such as the precipitation map in Figure 2.6, which implies a dry to wet regime).

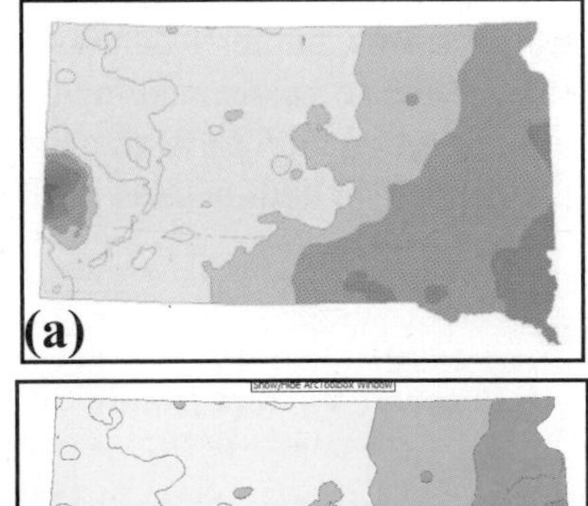

Fig. 2.6. A graduated color map showing annual precipitation with (a) a monochromatic color ramp and (b) a two-toned color ramp

For some attributes, the size of the feature influences the value of the attribute reported for that feature. For example, counties with larger areas tend to have more acres of farmland or larger populations. Such data should be **normalized** to the quantity per unit area when they are mapped, in order to prevent the size of the measurement unit from influencing the map appearance. Thus, population data are usually normalized to show the number of people per unit area (population density). Some attributes don't need normalization. In a precipitation map, each value already represents inches of rain falling at any location. The median rent for a county does not depend directly on the area of the county. These attributes would not be normalized.

Another way to normalize values to the area of a polygon is to create a **dot density map**, which uses randomly placed dots to show the magnitude of a value in the attribute table. Each dot represents a certain number, for example, one million people. In that case, California would have about 33 dots in it, and Wyoming would barely have one. In Figure 2.7, one dot equals 100,000 people. Note that the locations of the dots, which are randomly placed, do not necessarily represent the actual distribution of people in the state.

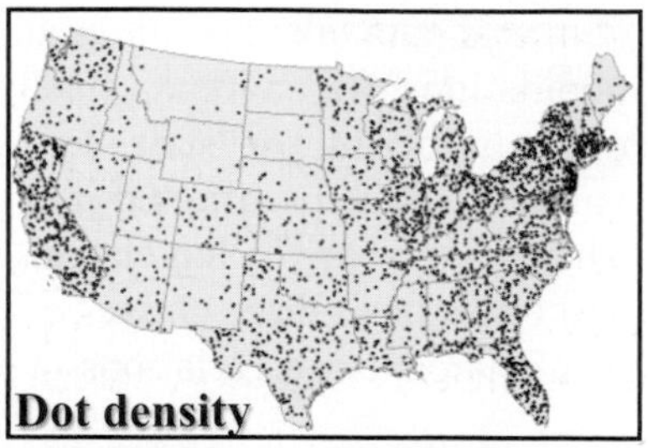

Fig. 2.7. A dot density map showing state populations

Chart maps

A chart map expands the number of attributes that can be displayed on a map by replacing a single symbol with a chart representing several attributes. The chart could be a pie chart, bar chart, or stacked bar chart. Figure 2.8 shows a pie chart map with the proportions of Caucasians, African Americans, and Hispanics in each state. The pie sizes can be all the same (Fig. 2.8a) or proportional to the sum of the three categories, thus showing the relative number of people in the state as well (Fig. 2.8b). Notice how the pie colors are chosen to facilitate recognition of the classes: cream for Caucasians, dark brown for African Americans, and reddish for Hispanics. Such details help make maps easier for the reader to interpret, and they lessen the need for the reader to look back and forth between the legend and the map.

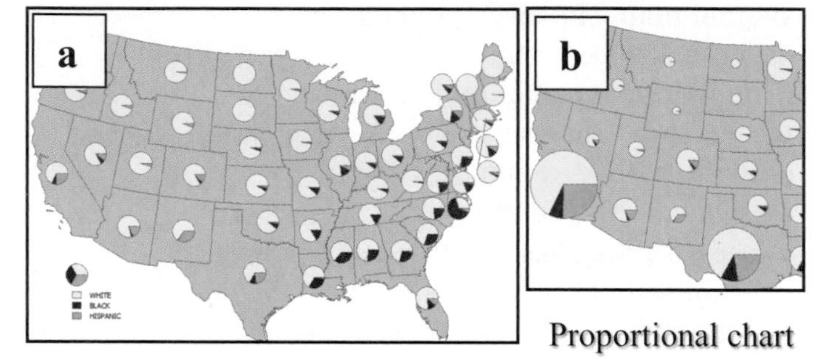

Fig. 2.8. (a) A chart map can represent several attributes, such as the proportion of ethnic groups. (b) The pie size can represent the total population.

Displaying rasters

Recall from Chapter 1 that rasters are a cell-based data model in which an array of cells or pixels store numeric values relating to some feature or quantity on the earth's surface. Rasters may be broadly grouped into three main types, each of which is displayed differently: thematic rasters, image rasters, and indexed rasters.

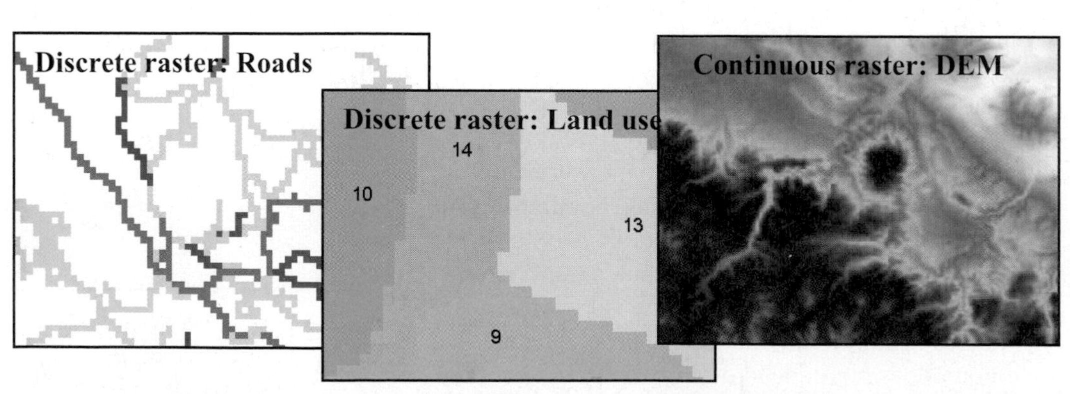

Fig. 2.9. Thematic rasters. Discrete rasters store feature data, such as land use or road types. Continuous rasters store data that vary smoothly, such as elevation or rainfall.

Thematic rasters

A **thematic raster** represents map features or quantities, such as roads, geology, elevation, or vegetation density (Fig. 2.9). When rasters are used to store objects, such as roads or land use polygons, they are called **discrete** rasters. Typically they are created by converting features from a vector data set to a raster format, using a single attribute field as the value to be placed in the cells. Discrete rasters contain groups of adjacent cells with the same value. In contrast, **continuous** rasters store values that represent a continuously varying quantity, such as elevation.

Values in a raster are always stored as numbers and may be characterized as categorical, ordinal, or interval/ratio. (Nominal data are usually not stored in rasters—it would be inefficient.) The methods used to display thematic rasters are similar to those used for displaying vector data and, likewise, depend on the data type.

A raster representing categorical or ordinal data is displayed using a unique values map. Just like the unique values map used for feature classes, each value of the raster receives its own color, as in the geology map shown in Figure 2.10a. Color schemes for unique values maps have 32 possible values and work best with data that have relatively few categories. When representing ordinal data, a monochromatic color scheme is used to communicate a sense of increase.

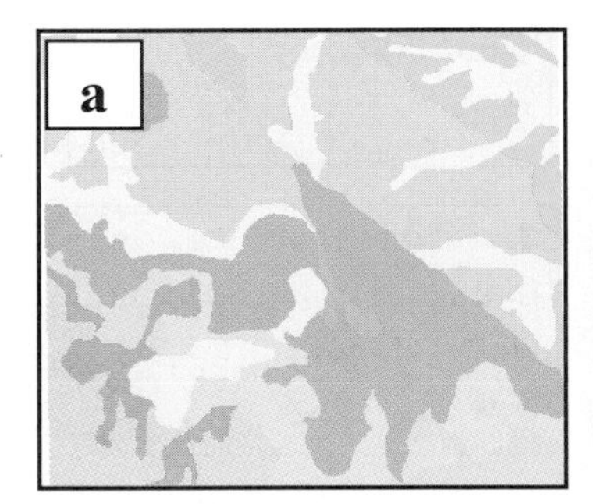

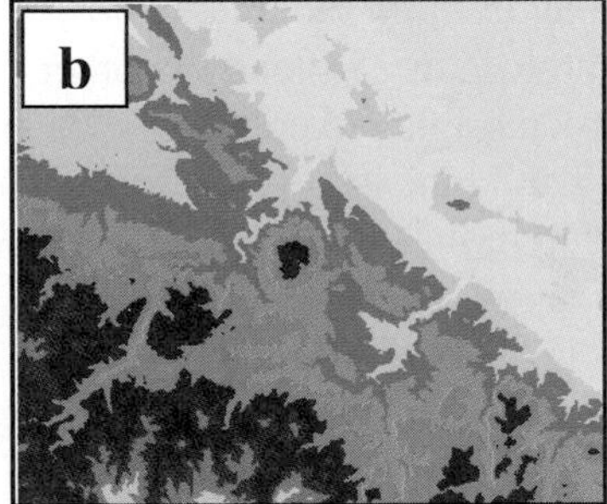

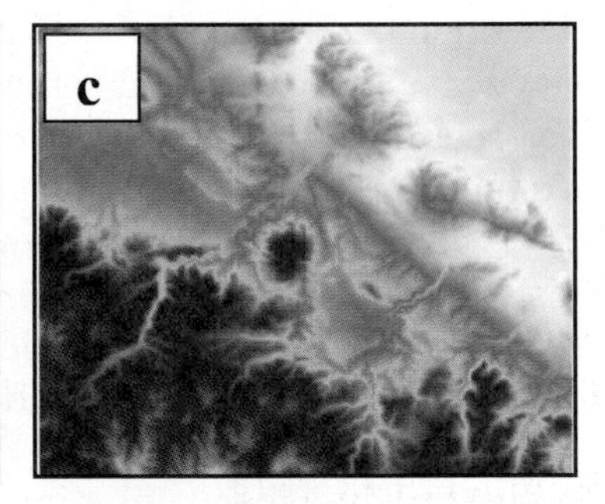

Fig. 2.10. Display methods for thematic rasters: (a) unique values geology; (b) classified elevation; (c) stretched elevation

A thematic raster containing interval or ratio data must be classified into ranges in one of two ways. The **classified** display method divides the values into a small number of bins, similar to a graduated color map. A color ramp is chosen to assign the colors to each bin. The elevation map in Figure 2.10b has 12 classes represented by 12 colors from a color ramp. The **stretched** display method scales the image values to a color ramp with 256 shades (Fig. 2.10c). The raster is first subjected to a **slice**, which rescales the elevation values (ranging from 800 to 1600 meters) into 256 bins, then matches the bins to the 256 shades in the color ramp. You can think of a slice as a classification that creates 256 classes.

Image rasters

Image rasters include aerial photography and satellite data. The pixels represent degrees of brightness caused by light reflecting from materials on the surface. The brightness values are often placed on a scale of 0–255 DN (digital numbers). A dark shadow would have a low brightness and a low DN, and a white cement road would have a high brightness and a high DN.

An image raster may contain one or more bands of information. Imagine a black-and-white digital camera that measures brightness and stores a value between 0 and 255 for each pixel on its screen. This camera would produce a single-band raster that could be displayed with a grayscale color ramp from black to white, as shown in Figure 2.11a. Such images are displayed using the same **stretched** method that is applied to continuous thematic rasters. Images based on the 0–255 scale do not need to be sliced, but some images contain larger values and do require slicing.

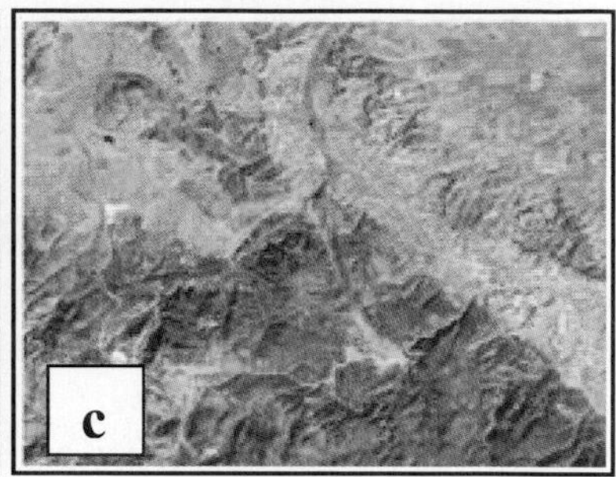

Fig. 2.11. Raster display methods for images: (a) single-band stretched image; (b) RGB composite image (true color); (c) RGB composite with Landsat bands 7, 4, and 1

To store color images, at least three bands are needed. A color digital camera measures brightness separately for three wavelengths of light (red, green, and blue) and stores three brightness values for each pixel in three separate bands. If the value in the red band is high (close to 255) and the other two bands have low values near zero, a bright red color will result. If red and green are both high and blue is low, a mixture of red and green will create yellow. All colors may be represented by varying proportions of red, green, and blue light, as dictated by the brightness values in the bands. Figure 2.12 shows two possible mixtures and the resulting colors.

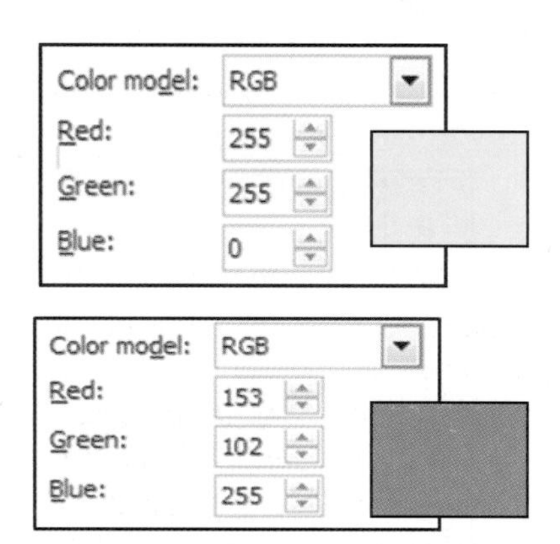

Fig. 2.12. Computers store and identify colors as mixtures of red, green, and blue light on a scale from 0 to 255.

To display a multiband color image, an RGB Composite method is used. One band is assigned to each of the three color guns on the computer monitor (red, green, and blue) to produce a color image (Fig. 2.11b) from the brightness levels stored in the three bands.

Many satellites and airborne cameras measure brightness in more than three wavelengths and produce more than three bands. Landsat images, for example, have seven bands. Multiband images also use the RGB Composite display method, but any of the bands may be applied to the RGB guns on the computer. In Figure 2.11b, the Landsat bands 3, 2, and 1 representing red, green, and blue light are used, producing a true-color image. In Figure 2.11c, bands 7, 4, and 1 are used to contrast green vegetation with dry or urban areas.

Stretching image values

Image DN values are often normally distributed. Figure 2.13a shows a typical **histogram** of cell values, with the horizontal axis showing the range of values with the corresponding grayscale color ramp, and the vertical axis showing the number of cells for each value (color). The image has few values near 0 or 255, with most of the pixels occurring in the middle gray range. The image is shown in Figure 2.13b, and it appears dim and featureless with little contrast.

The **stretched** display method can improve the display of normally distributed values by ignoring the tails of the distribution. One method assigns the lowest and highest image values to 0 and

255, respectively, and stretches the remaining colors along the ramp. This is called a minimum-maximum stretch. Even greater contrast can be developed using a standard-deviation stretch, which uses only the values within two standard deviations of the mean (Fig. 2.13c).

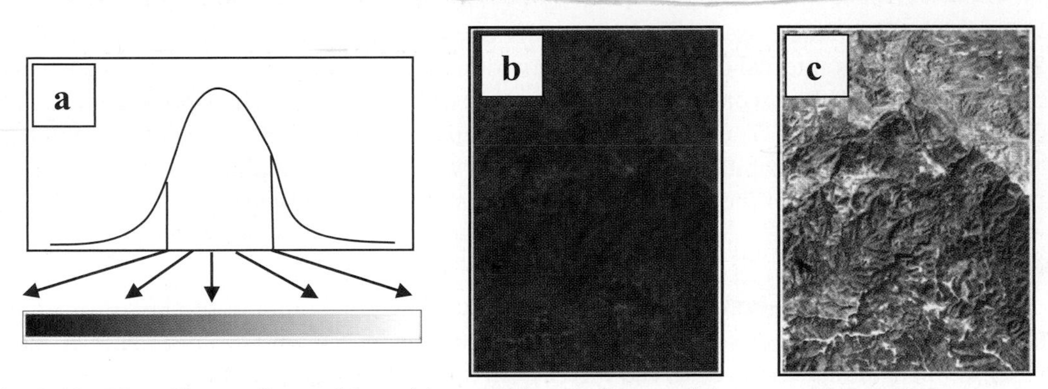

Fig. 2.13. The effects of stretching: (a) stretching an image; (b) no stretch applied; (c) a standard-deviation stretch

Stretches can be applied to thematic continuous rasters as well as images and can be helpful when the image values are not evenly distributed. The RGB composite display method permits the user to stretch each individual band.

Indexed color rasters and colormaps

The RGB color model is capable of storing millions of different colors and is best for fine color rending of images. However, it uses large amounts of memory. An alternate method for storing color identifies a restricted set of colors actually appearing in the image. It assigns an integer from 0 to 255 to each color and stores the RGB proportions needed to make it. This list of colors is called the **colormap** and is stored with the image. Each value in the raster is portrayed using its assigned color.

Fig. 2.14. An indexed color raster stores a colormap that assigns a color to each value in the raster.

Figure 2.14 shows part of a digital raster graphic (DRG). A DRG is a scanned US Geological Survey topographic map. The contours are represented by pixels with the value 4 and displayed using the assigned brown color. Every pixel value and defined color used in the map is shown in the accompanying colormap: green for vegetation, black for text, etc.

Indexed color rasters are commonly produced when scanning large color maps. Storing them in RGB could use hundreds of megabytes or more. An indexed color map typically uses less than a tenth of the space, and the human eye cannot easily distinguish between the two.

Classifying numeric data

Mapping numeric data, raster or vector, requires classifying a range of values into a small number of groups, each of which can be represented by a different color or symbol size. This process is called **classification**. There are many ways to classify data, and the choice of method affects the appearance of the map and the message that it portrays. Some common methods are compared in Figure 2.15 using the same data set of average farm size by state.

Natural breaks (Jenks method)

The **Jenks method** sets the class breaks at naturally occurring gaps between groups of data (Fig. 2.15a). Each class interval can have its own width, and the number of features in each class will vary. The Jenks method works well on unevenly distributed data, such as populations of the capitals shown in Figure 2.5. There are many low-population and medium-population capitals and a few very high-population capitals. The Jenks method works well with almost any data set, making it a natural choice for the default classification scheme in ArcMap.

Equal interval

The **equal interval** classification divides the values into a specified number of classes of equal size (Fig. 2.15b). This method is useful for ratio data, such as income or precipitation, because it gives a sense of regularity to the observed increases. However, it is hard to predict how many features will end up in each class. Notice in the farm size example that nearly all of the states fall into the first class. Compare this map to Figure 2.15a.

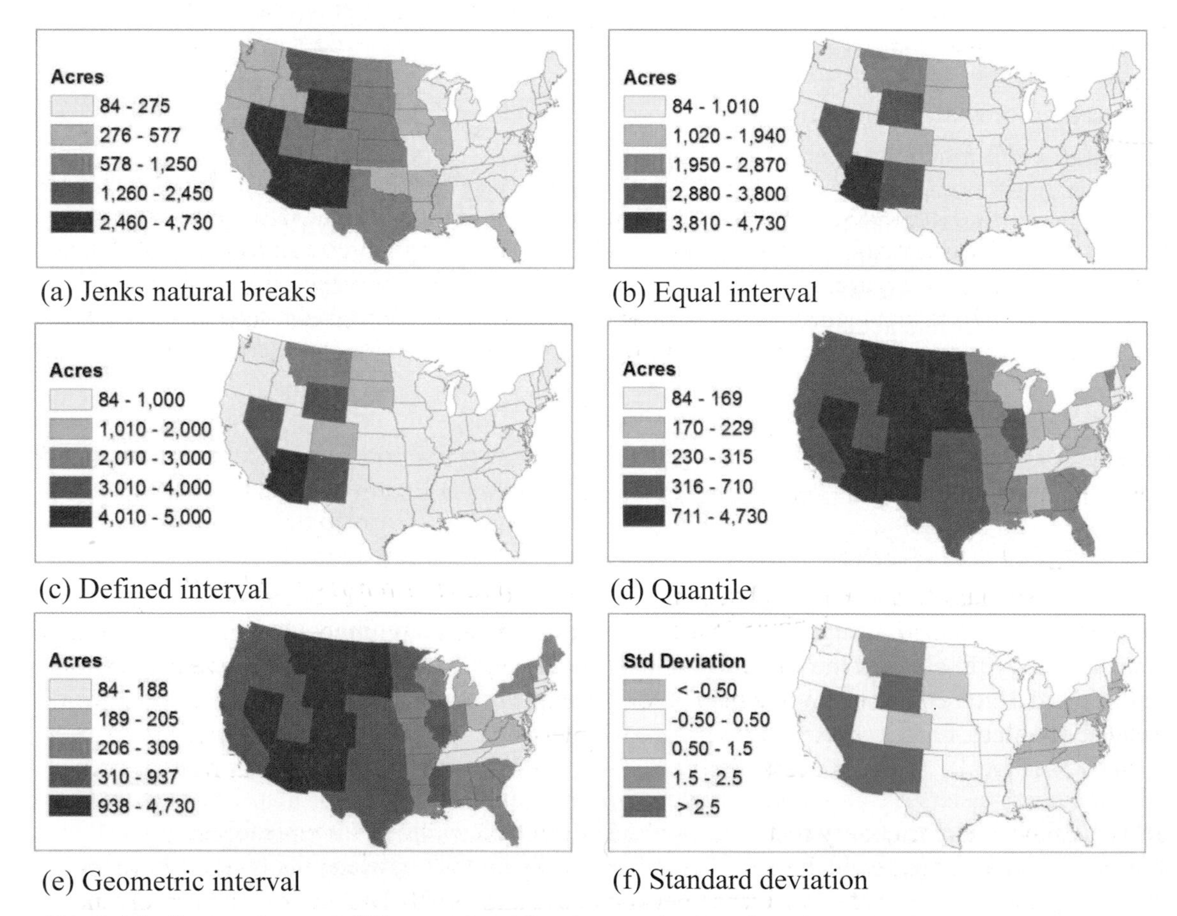

(a) Jenks natural breaks (b) Equal interval
(c) Defined interval (d) Quantile
(e) Geometric interval (f) Standard deviation

Fig. 2.15. Comparison of different classification methods using the same data set: average farm size in acres by state

Defined interval

A **defined interval** classification is similar to an equal interval one, except that the user specifies the size of the class interval, and the number of classes then depends on the range of values (Fig. 2.15c). This method will create rounded values in the classes that are easy to interpret. Defined interval maps are ideal when comparing classes composed of percentages, dollars, temperatures, and other values when specific break values are desired (100, 200, etc.). It does, however, suffer the same disadvantages as the equal interval classification.

Quantile

A **quantile** classification puts about the same number of features in each class (Fig. 2.15d). This method will create a very balanced map with all classes equally well represented, but some of the features in the same class could have very different values, and features in different classes could have similar values. Quantile classifications are best applied to linearly distributed data. Notice that a quantile classification highlights differences between the eastern states that are hidden in the Jenks, equal interval, and defined interval classifications.

Geometric interval

A **geometric interval** classification (Fig. 2.15e) bases the class intervals on a geometric series in which each class is multiplied by a constant coefficient to produce the next higher class. A geometric interval classification is designed to work well with continuous data like precipitation and to provide about the same number of values in each class range. It works especially well with positively skewed data distributions.

Standard deviation

A **standard deviation** classification apportions the values based on the statistics of the field. The user selects the class breaks as the number of standard deviations, and the data range determines the number of classes needed. This method excels at highlighting which values are typical and which are outliers. In Figure 2.15f, the yellow states are close to the mean farm size, several eastern states appear to have smaller than average farms, and some western states have much larger farms than average. This map is best applied to normally distributed data.

Manual class breaks

Finally, if none of the above options gives the desired map, the user can manually set class break points to any chosen values. This option works for assigning a logarithmic scale, as well.

Choosing the classification method

The choice of classification method depends on the mapmaker's purposes and on the type of data. Jenks shows the "nearest neighbors" in a distribution, whereas a defined interval or equal interval map does a better job of portraying relative magnitude. Notice the difference between the maps in Figures 2.15a and 2.15b. The Jenks map gives the impression that many states have large farms because the class sizes increase slowly; an equal interval map shows that farms in most states actually average 1000 acres or less. Figure 2.15b makes this observation clear at first glance.

Some data possess the quality that the magnitude of the values has an intrinsic meaning. Percentage data, for example, have a physical and a psychological meaning—people have an intuitive understanding of the difference between 50% and 100%. Differences in median rent occur in dollar values to which people can attach meaning. When dealing with such data, it is wise to use a defined interval map to choose classes with logical breakpoints, such as 20%, 5 inches of rain, or $200, rather than 12.6%, 1.47 inches of rain, or $187. The reader can more effectively interpret the classes.

The distribution of values has impact also. Jenks and geometric interval classifications are designed to work well with unevenly distributed data. Equal interval, defined interval, and quantile maps can be used with any data but show better results with evenly distributed data. The statistics behind standard deviation maps assume that the data are normally distributed.

About ArcGIS

Map documents

The ArcMap program has two main functions. First, it provides a place in which data sets may be viewed together and analyzed. Second, it is used to develop a map layout, or a map page designed for printing. ArcMap creates map documents, a collection of spatial data layers and tables and their properties. It also stores a map layout intended for printing.

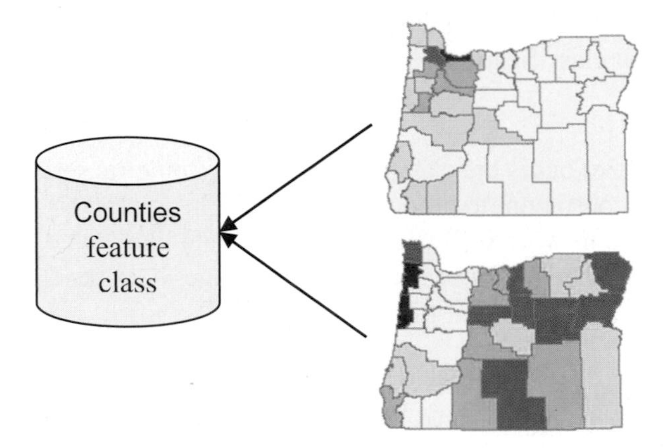

Fig. 2.16. Layers and feature classes. The counties feature class is stored on a drive and contains the spatial and attribute information. The layers point to the feature class but present the data two different ways: as a graduated color map of population density (top) or the vacancy rate of housing units (bottom).

When a feature class or a raster is added to ArcMap, it becomes a **layer**. This distinction is important to grasp. A feature class is a stored set of spatial data. *A layer points to a feature class and stores information about how to display it and use it* (Fig. 2.16). It is like a recipe that explains how to turn a feature class into different kinds of maps, just as a cooking recipe explains how to turn chicken into different dishes. Unlike a food recipe, though, a layer does not "use up" the ingredients. The same feature class can be accessed by multiple layers to present different views of the same data, even within a single map document (Fig. 2.16). Layers are held in memory and stored when the map document is saved. The information in the layer, such as which symbols should be used and whether the features have labels, are called the layer properties. Creating a map involves manipulating these properties in menus to produce the desired result.

Map documents organize layers using **data frames**, windows that contain groups of layers that are drawn together. Maps can contain multiple data frames (Fig. 2.17). One frame at a time is designated the **active frame**. All commands and actions, such as adding layers or creating a scale bar, occur in the active frame.

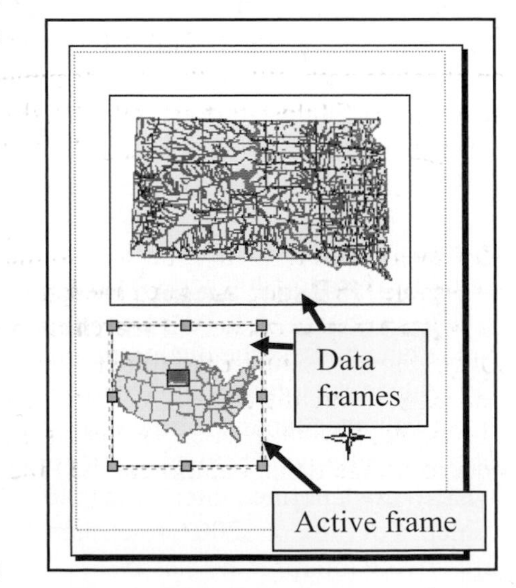

Fig. 2.17. Data frames

A map document does not store the GIS data files within it. Instead, the map stores the name and disk location of the spatial data—its **source**. Most changes to the map document do not affect the source data, only its properties within the document. Some functions do change the source data, such as adding attributes or editing boundaries. Users need to learn which are which.

Because the source data are stored outside the map document, *changes to source data affect all map documents using the same data set.* Because data are often shared in a GIS system, precautions are needed to ensure that an editing mistake does not affect dozens of people. Often the source files are write-protected so that only one or two users are authorized to make changes.

Storing the data separately from the map document has several benefits. First, when a shared database is updated, the changes appear in every map document based on that database. Second, it saves space. GIS data sets can be very large, but the map documents that reference them don't have to be. However, the practice has implications for sharing map documents. Giving a colleague a copy of a map document by e-mail works only if the recipient has access to exactly the same data in the same disk location as the original user.

Map documents and pathnames

Map documents keep track of the source files by storing the location of each file as a **pathname**, which lists the successive folders traversed to a data set, each separated by backslashes. The pathname C:\mgisdata\usa\states.shp refers to the feature class states.shp, which resides in the Usa folder, which in turn resides in the mgisdata folder, which is found on drive C:\ (Fig. 2.18).

The pathname C:\mgisdata\Usa\States.shp is termed an **absolute pathname** because it starts at the drive letter and proceeds downward to the file. When searching for a file using an absolute pathname, the search begins at the drive level. A **relative pathname** is used to indicate that the search should begin in the folder containing the map document. A double dot indicates movement up one folder. The pathname ..\Usa\States.shp indicates a search up one folder and then down into the Usa folder and then to the States feature class (Fig. 2.18).

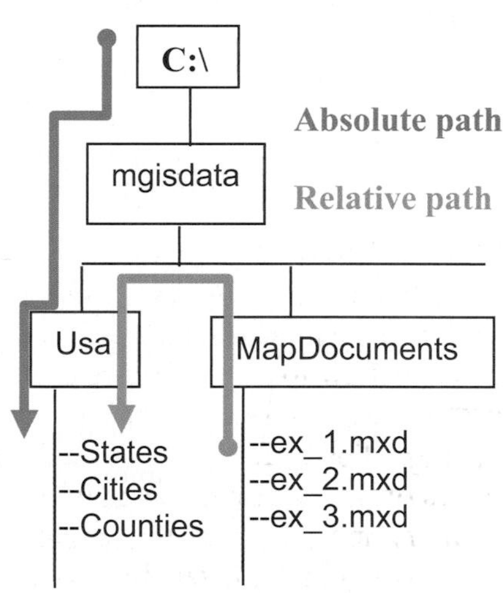

Fig. 2.18. Pathnames

Map documents can store either absolute or relative pathnames as chosen by the user, with absolute pathnames as the default. Imagine that the map document ex_1.mxd in Figure 2.18 refers to the States file in the Usa folder using an absolute path. When the map document is opened, it searches for States starting at the C:\ drive down through C:\mgisdata\Usa\States.shp in order to locate and draw the states in the map.

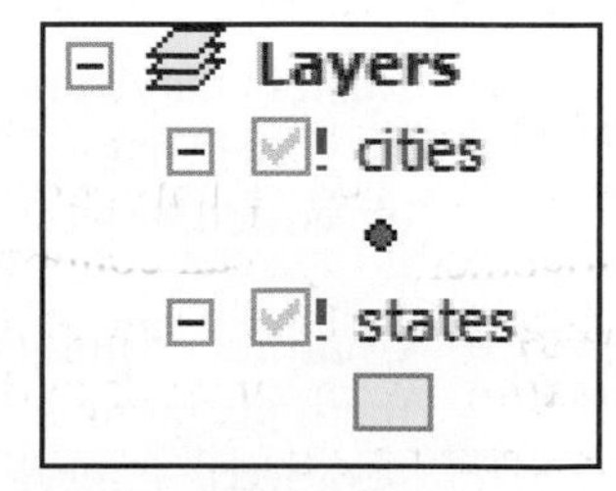

Fig. 2.19. The exclamation point indicates that the pathname link is broken.

However, what if the user decides to move the mgisdata folder to a portable USB stick assigned the letter K:\? Now when the ex_1.mxd map is opened, it searches down the C:\ path but can no longer locate the mgisdata folder. The links to the data are broken, and in ArcMap, the States layer will appear with a red exclamation point (Fig. 2.19). The user can manually locate the data to fix the links, but this can be time consuming. (ArcCatalog can fix multiple links efficiently.)

A better solution is to direct the map document to store relative pathnames. In this case, the States pathname is stored as ..\Usa\States.shp, and it no longer matters whether the mgisdata folder exists on the C:\ drive or the K:\ drive. As long as the relative positions of the MapDocuments and Usa folders remain the same, the map document can always locate the data.

The choice of absolute or relative pathname depends on the situation. If the map documents will always refer to a central data server that all users on a network have access to, then absolute pathnames will more reliably locate the data layers, even when e-mailing documents. On the other hand, if a user plans to move the map documents and data together as a unit, then relative pathnames work better. The map documents on the DVD that comes with this book are stored with relative pathnames so that the links to the data layers continue to work no matter where you install the mgisdata folder.

VERY IMPORTANT TIP: Pathnames, including the names of map documents and spatial data sets, should not contain spaces or special characters, such as #, @, &,*, and so on. Use the underscore character _ to create spaces in names, if necessary. Although ArcCatalog and ArcMap allow spaces in names, spaces create problems for certain functions. It is wise to avoid them.

Summary

- Attribute data have a data type designation: nominal, categorical, ordinal, interval, or ratio. The data type determines the kind of map and even the types of analysis that may be performed on that attribute.

- Nominal data name things or uniquely identify them and may be text or numbers. Each feature usually has its own value, and repeats are the exception.

- Categorical data group objects into smaller sets identified by a unique value. Numbers may be categorical when they are used as codes. Ordinal data consist of categories that are ranked in some way.

- Interval data are measured on a regular scale, and ratio data are measured on a scale with a meaningful zero point.

- Single symbol maps and labels are used to map nominal data. Unique values maps are used for categorical or ordinal data. Interval or ratio data are displayed using graduated color, graduated symbol, proportional symbol, dot density, or chart maps.

- Rasters may contain thematic or image data. Discrete thematic rasters can be displayed using a unique values method. Continuous thematic rasters may be classified or stretched. Image rasters use a stretched or RGB composite display.

- Continuous numeric data are classified before being mapped. Classification methods include Jenks natural breaks, equal interval, defined interval, quantile, geometric interval, standard deviation, and manual. The best classification method depends on the type of data and the data distribution.

- Map documents contain layers, which point to feature classes or rasters stored on the computer and contain information on how to display them. The stored feature classes or rasters may be used multiple times to create different maps.

- GIS data are stored outside the map document, and many map documents can refer to the same data. Caution is needed to ensure that a document does not lose track of where to find its data.

- Map documents may store either absolute or relative pathnames for data files. Absolute pathnames work better when the data will always be in the same place for every user. Relative pathnames are best if the data will be moved about.

Important Terms

absolute pathname	defined interval	Jenks method	ratio data
active frame	discrete	layer	relative pathname
categorical data	dot density map	Layout view	RGB composite
chart map	dynamic labels	nominal data	slice
choropleth map	equal interval	normalized data	source
classification	geometric interval	numeric data	standard deviation
classified	graduated color map	ordinal data	stretched
colormap	graduated symbol map	pathname	thematic raster
continuous	histogram	proportional symbol map	unique values map
data frame	image	pyramids	
Data view	interval data	quantile	

Chapter Review Questions

1. A 1:20,000,000 scale map of the United States displays the interstates with a line symbol that is 3.4 points wide. There are 72 points to an inch. What is the uncertainty in the location of the road due to the width of the line used to represent it? Give the answer in feet and miles.

2. For each of the following types of data, state whether it is nominal, categorical, ordinal, interval, or ratio. Explain your reasoning.

   bushels of wheat per county
   vegetation type
   average maximum daily temperature
   parcel street address (e.g., 51 Main St.)
   parcel ID number (e.g., 1005690)
   pH measurement of a stream
   state rank for average wage
   number of voters in a district
   student grade in a class (e.g., A, B+)
   soil type

3. For each of the following attributes, state whether a single symbol, graduated color, or unique values map would be most appropriate. Explain your reasoning.

   precipitation
   geological unit
   acres of corn planted per county
   rivers
   land use
   household income

4. If mapping the following attributes for counties, indicate which ones would generally be normalized and which would not.

average daily temperature	median rent
number of Hispanics	total river miles
square miles of parkland	sales tax rate

5. State whether you would use a unique values, classified, stretched, or RGB composite display method for each of the following rasters.

geological units	black-and-white aerial photo
4-band satellite image	precipitation
landslide hazard (high, medium, low)	slope in degrees

6. Which would take more storage space, a layer file showing all the US counties or a layer file showing all the US states?

7. (a) Jill creates a map document and e-mails it to a colleague in another state. He calls and tells her the document opens, but no map appears. Explain what is wrong. (b) Jill creates a map document using data from a central fileserver and e-mails it to a colleague in her office. Again, no map appears. What is wrong this time?

8. What characters should be avoided when naming GIS files, folders, and map documents?

9. Explain the difference between thematic rasters and image rasters.

10. What does it mean if you find a red exclamation point next to a map layer? How would you fix it?

Mastering the Skills

Teaching Tutorial

The following examples provide step-by-step instructions for doing basic tasks and solving basic problems in ArcGIS. The steps you need to do are highlighted with an arrow ➔; follow them carefully. Click on the video number in the VideoIndex to view a demonstration of the steps.

The first time we opened ArcMap we started with a default blank map. We can also use templates to specify the size and layout of the map layout.

1➔ Start ArcMap. In the Getting Started window, click on the Templates > North American entry.

1➔Scroll down to the Letter (ANSI A) Portrait entry and select it. Do not click OK yet.

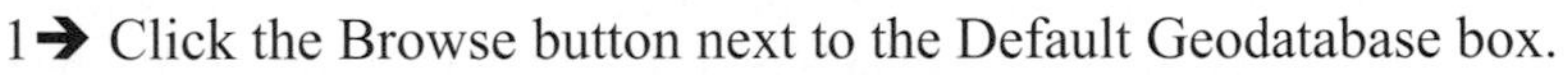

1➔ Click the Browse button next to the Default Geodatabase box.

1➔ Navigate to the mgisdata\Oregon folder and click the oregondata geodatabase to highlight it. Click Add.

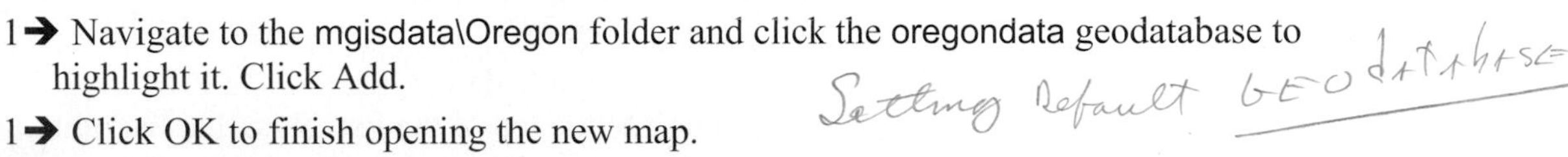

1➔ Click OK to finish opening the new map.

TIP: The default geodatabase can be set for each map document. When you add data, it will take you there first. If new feature classes or rasters are created, they will be stored there.

TIP: The names of map documents, spatial data sets, and folders should not contain spaces or special characters, such as #, @, &, *, and so on. Use the underscore character _ to create space.

To ensure that moving the mgisdata folder to a different place will not break any data links, we will set the map document to store relative pathnames. We will also make relative pathnames the default for any new map documents that we create.

2➔ Choose File > Map Document Properties from the main menu bar.

2➔ Check the box to *Store relative pathnames to data sources.*

2➔ Click OK. Now as long as the map document stays in the same folder relative to its source data, the links to the data will stay valid.

2➔ Choose File > Save As from the menu. Navigate to the mgisdata\MapDocuments folder and save the map document using the name ex2_mine.mxd. Then save periodically as you work, in case you encounter an error and have to backtrack.

Adding Layers to ArcMap

We begin by adding some feature classes from the Oregon geodatabase.

3➔ Click the Add Data button.

3➔ Find the Default Geodatabase button and click it to go directly to the default geodatabase, oregondata.

3➔ Add the counties and volcanoes feature classes from the geodatabase. (Remember to hold down the Ctrl-key to select more than one feature class.)

4➔ Click Add Data again and expand the Transportation feature dataset. Add the airports and highways also.

4➔ Add the rivers feature class from the Water feature dataset.

Notice the icons that appear directly below the Table of Contents title. These icons control the appearance of the Table of Contents and how it arranges the layers. The first icon is currently highlighted.

5➔ Hover the cursor above the highlighted icon to see the icon name.

5➔ Examine the names of the other icons as well.

5➔ You may click on the other options to see what they look like, if you wish, but make sure afterward that the first icon is again highlighted.

The List by Drawing Order icon shows the layers in the order in which they are drawn, from bottom to top. Notice that the point and line layers that you just added were on top of the polygon layer. The draw order can be changed if needed.

6➔ Click the counties layer and drag it to the top of the list. Note that the other layers disappear from the map, as the polygons are drawn on top of them now.

6➔ Move the counties layer back down to the bottom of the list.

6➔ Place the airports above the volcanoes, if they are not already there.

When you load a feature class into ArcMap, it becomes a layer. A layer points to a feature class stored outside the map document, but it allows you to set properties that control how the layer is displayed and used. Let's take a look at some of the layer properties.

7➔ Right-click the volcanoes layer and open its properties, or double-click it.

7➔ Find the General tab and click on it. Examine the items in this window.

The layer name and description information are copied from the feature class. You can make changes that will be stored with this layer in ArcMap, but they will not affect the original file.

7➔ Change the layer name to Volcanoes so that it looks better.

7➔ Click Apply to enact the change and keep the window open.

7➔ Click on the Source tab next. Examine the information in the Data Source box.

This box contains information about the feature class used by the layer. The layer points to the feature class by storing the pathname, such as C:\gisclass\mgisdata\Oregon\oregondata.mdb. (Your pathname depends on where you installed the mgisdata folder.) You can also view the coordinate system in this tab (the one in which the data are stored).

1. What is the name of the coordinate system for volcanoes? ______________________

7➔ Click Cancel to close the Properties window without making any changes.

Working with data frames

Now we will learn to manage data frames. Our goal is to create a map layout with two different views of Oregon showing its geography and population, so we will need two data frames.

The map window has two view modes. In **Data view**, the user works with a single data frame. The **Layout view** shows how the map document will look when printed, including titles, scale bars, and so on.

8➔ Click the Data View icon in the lower left of the display area to switch to Data view (Fig. 2.20). Then click the Layout view button.

8➔ Click the Select Elements tool on the Tools toolbar or the Draw toolbar. This tool is used to select and work with graphic elements, such as text or data frames.

8➔ Click inside the map, on top of Oregon. Blue square handles appear on the border of the data frame. The handles indicate that the frame is currently selected by the Select Elements tool.

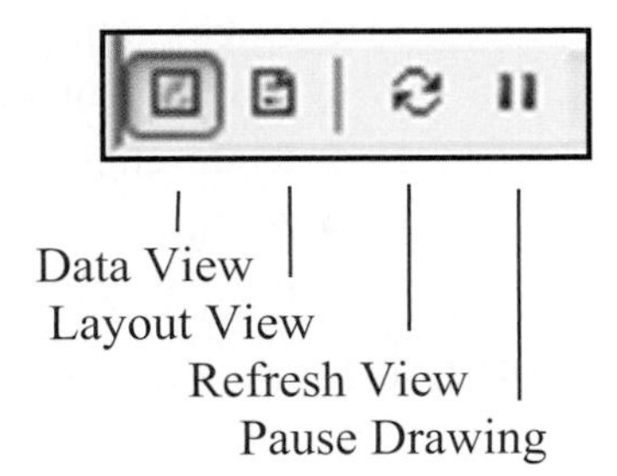

Fig. 2.20. View buttons

8➔ Place the cursor on top of the middle lower blue handle, and click and drag it to size the data frame to occupy only the top half of the page.

9➔ In the Table of Contents, click once on the Layers data frame name, wait a moment, and click again. The text is highlighted, allowing you to change it.

9➔ Type Geography for the name of the data frame and press Enter.

10➔ Choose Insert > Data Frame from the main menu bar.

10➔ When the new data frame appears, click inside it and drag it below the original data frame. Use the handles to adjust its size and shape until the layout page looks similar to Figure 2.21.

10➔ Click twice slowly on the New Data Frame name in the Table of Contents, and change it to Population.

A map may have many data frames, but one is always the active frame, indicated by boldface type in the Table of Contents. Data View always shows the active frame. Data added to the map are placed in the active frame.

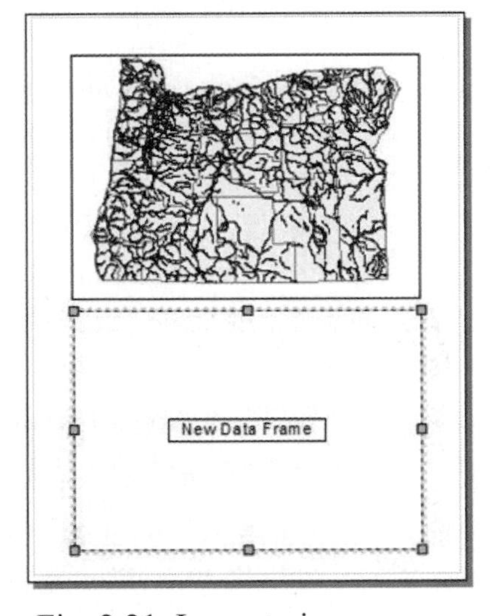

Fig. 2.21. Layout view

11➔ Make sure that the Population data frame is the active frame (its name is in boldface type).

11➔ Click the Add Data button. Add the counties feature class from oregondata. It appears in the Population data frame.

Notice that we now have two layers based on the counties feature class. Both point to the same file, but each one has properties that can be set separately, allowing two views of the data.

12➔ Right-click on the Geography data frame name in the Table of Contents and choose Activate from the context menu. (The blue handles don't move from the Population data frame, but the Geography frame is active, as shown by the boldface text.)

12➔ Click the Data View button in the lower-left corner of the map window to return to data view. You should now see the Geography map with airports and volcanoes.

13➔ Click on the airports symbol to open the Symbol Selector. Choose the Airplane symbol and make it 25 pt. and rotated −45 degrees.

13➔ Right-click on the symbol for counties and change the color to light yellow.

13➔ Right-click on the line symbol for rivers and change the color to deep blue.

SKILL TIP: Use the Symbol Editor to create and manage new symbols (Maps and Symbols).

The rivers seem to be cluttering the map at this scale, so they would benefit from a scale range.

14➔ Right-click the rivers layer and open its properties.

14➔ Click the General tab and fill the button for *Don't show layer when zoomed out beyond* and choose the scale 1:3,000,000. Click OK.

Labeling features

Nominal data are usually portrayed using labels. **Dynamic labels** can be created for a layer using an attribute field and can be turned on or off for the entire layer. They use an auto-placement function to avoid overlaps, and not all labels may be shown. Dynamic labels are redrawn each time the map is redrawn and may change as the scale changes or when the map is printed.

15➔ Click twice slowly on the airports layer and rename it Airports.

15➔Double-click on the Airports layer name to open its properties.

15➔ Click on the Labels tab.

15➔ Check the box to *Label features in this layer* and leave the method set to *Label all features the same way*.

15➔ Make sure that the Label field is set to NAME.

15➔ Choose Arial 9 pt. bold as the font.

15➔ Click Apply to make the changes, but keep the window open. Move it aside, if necessary, to see the labels.

Basic labels are easy to set, but they often are not quite what is wanted. There are several additional options to control dynamic labels, such as adding a mask to help them stand out better.

16➔ Click the Symbol button in the Label Properties window. Notice the different pre-set symbols available, but don't choose any.

16➔ Click the Edit Symbol button. You can do very detailed editing here, but we just want the mask.

16➔ Click the Mask tab. Choose Halo, leave the size at 2 pt., and click OK and OK to close the symbol windows. Click Apply to enact the change.

Some control is given over where the symbols are placed relative to the feature.

17➔ Click the Placement Properties button in the Label Properties window. Make sure that the Placement tab is clicked.

2	2	1
3		2
3	3	2

17➔ Examine the diagram that shows the placement priority around the point. Click Change Location to examine the other options, but don't choose one. Click Cancel.

17➔ Examine some of the other options, such as placing the label on top of a point, or placing it at an angle. Click Cancel.

17➔ Click OK to close the layer properties. We are finished with the airports.

Line features have different placement options.

18➔ Rename the rivers layer **Rivers**, and open its properties.

18➔ Click the Labels tab, if necessary. Check the box to label the features, and make sure that the Label Field is set to the NAME field.

18➔ Change the symbol to Arial 8 pt. dark blue italic, and click OK.

18➔ The Rivers layer has a scale range, so set the map scale to 1:3,000,000 using the drop-down button on the main toolbar.

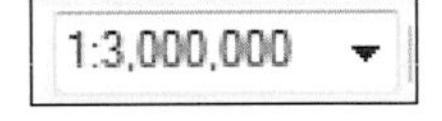

Even at this scale there are too many rivers to label effectively, and the map is cluttered. You can set a scale range for labels separately from that for the layer.

19➔ Open the Rivers properties again. On the Labels tab, click the Scale Range button.

19➔ Change the settings to *Don't show labels when zoomed out beyond* 1:1,000,000. Click OK to return to the Label properties, and click OK.

19➔ Set the map scale to 1:1,000,000 to examine the river labels now.

It would also look nice to curve the labels along the rivers instead of using straight text.

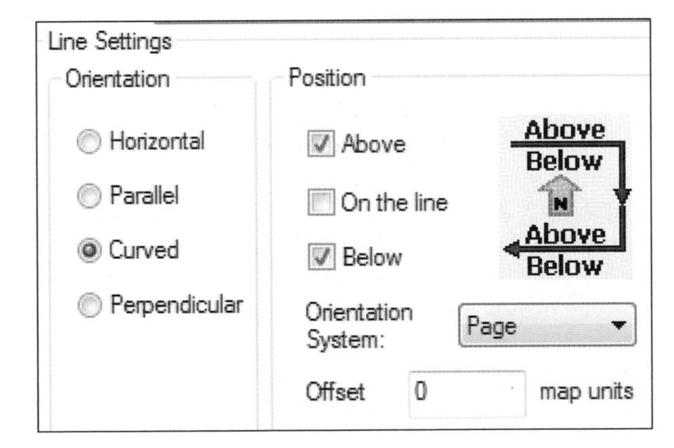

Fig. 2.22. Setting curved labels along rivers

20➔ Open the Rivers label properties again. Click the Placement Properties button.

20➔ Choose the Curved orientation button (Fig. 2.22). Check the boxes to place labels *Above* or *Below* the line.

20➔ Examine the other settings, but leave them alone. Click OK to finish setting the placement options, and then OK to finish.

20➔ Examine the labels, and then use the Full Extent button to zoom to all of Oregon.

SKILL TIP: The Label Manager makes it easier to create and edit labels for several layers at once (Labels and Annotation).

Symbols and labels may be combined when labeling.

21➔ Rename the highways layer **Highways**, and give it a thick dark brown symbol.

21➔ Open the Highways properties. On the Labels tab, turn on the labels and make sure the Label Field is set to HWY_SYMBOL.

21➔ Click the Symbol button. In the Symbol Selector, scroll down and choose the US Route HWY symbol. Click OK.

22➔ Set the Scale Range to not show the symbols when zoomed out beyond 1:3,000,000. Click OK.

22➔ Click the Placement Properties tab. Choose the Horizontal orientation. Click OK and OK.

22➔ Zoom in to 1:3,000,000 to see the labels. Then return to the Full Extent of the map.

SKILL TIP: You can create classes for labels and label each class differently, such as giving the interstates one symbol and the highways another (Labels and Annotation).

23➔ Rename the counties layer Counties.

23➔ Use what you have learned so far to give the counties Arial 8 pt. bold labels with a 2-pt. halo. Don't assign a scale range.

Creating maps from attributes for points

Currently, each feature class is displayed with a single symbol. Next, we will learn how to create maps based on attributes containing categorical, ordinal, interval, or ratio data. (If you need to review the data types, go back to the Concepts section.) Let's examine the fields in the table.

24➔ Right-click the Volcanoes layer and choose Open Attribute Table.

2. What is the data type of these two attribute fields in the volcanoes feature class?

ELEVATION ____________ TYPE ____________

The ELEVATION field contains numeric data, so it must be portrayed using a quantities map, and the values must be classified.

24➔ Close the Table window, and open the Volcanoes layer properties. Click on the Symbology tab.

24➔ Click the Quantities text on the left to open the map types. Choose Graduated colors (Fig. 2.23).

24➔ Set the Value field to ELEVATION. A set of classes appears below with the default to five classes and the Jenks Natural Breaks classification.

24➔ Set the color ramp to the one shown in Figure 2.23. Click OK.

The symbols now show which volcanoes have lower and higher elevations. However, the small symbol size makes the colors difficult to see—a common problem with graduated color maps for point data. We can edit the base symbol to make the map easier to interpret.

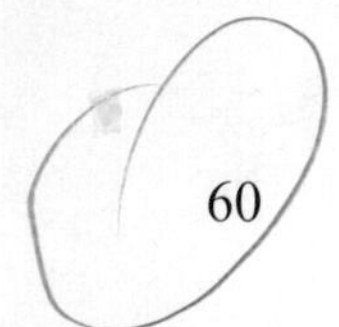

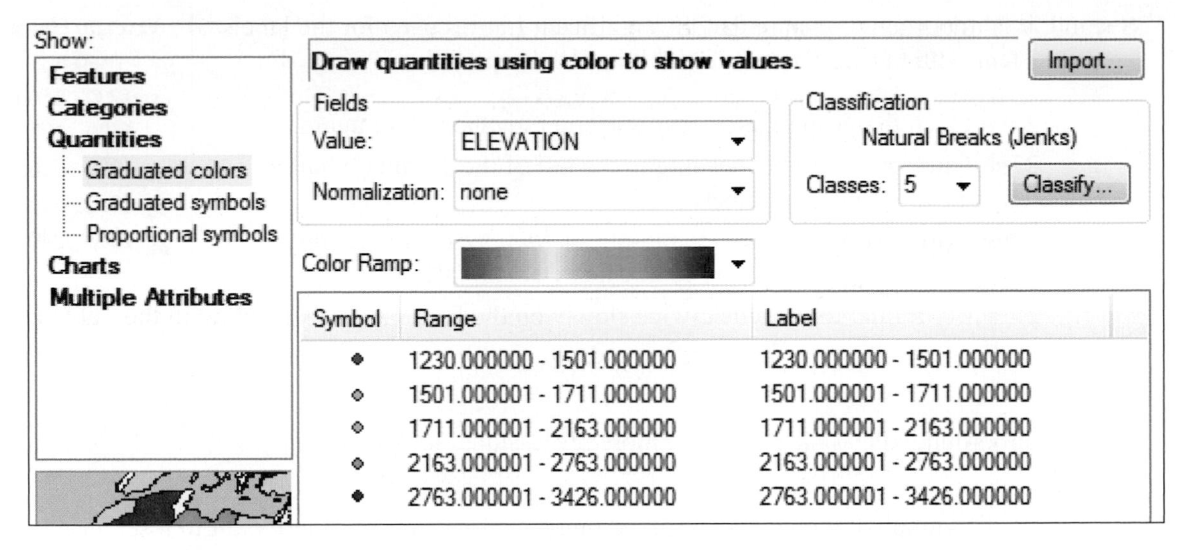

Fig. 2.23. The Symbology window for creating a graduated color map

25➔ Open the Volcanoes layer properties.

25➔ Under the color ramp setting, click on the Symbol heading to reveal a menu. Choose Properties for All Symbols (Fig. 2.24).

25➔ Choose the Triangle 2 symbol. Click OK.

25➔ To update the symbol colors, set the color ramp again. Click OK.

25➔ Uncheck the airports layer box in the Table of Contents to turn it off.

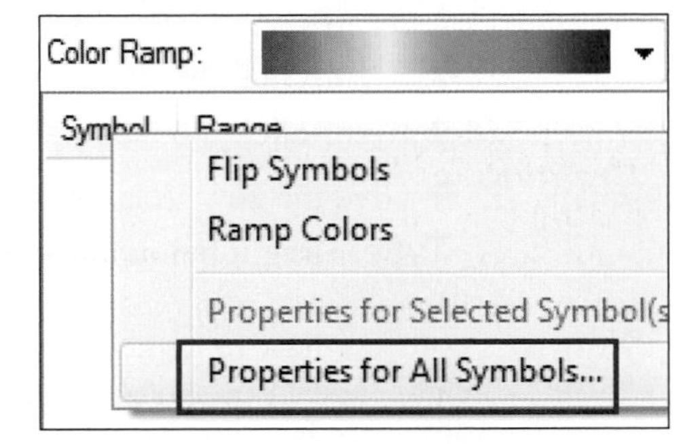

Fig. 2.24. Changing properties for graduated color symbols

You can now see that the eastern volcanoes are lower in elevation, and the highest volcanoes lie in the north. But let's try a graduated symbol map instead.

26➔ Open the Volcanoes symbol properties again.

26➔ Change the Quantities map type to Graduated symbols. Keep the Value field set to ELEVATION. Circles are the default symbols, but triangles would make more sense.

26➔ Look right and click the Template button, which is used to set the base symbol.

26➔ Choose the Triangle 2 symbol again and set its color to a reddish brown. Click OK.

26➔ The smallest class symbol is barely visible. Change the minimum size value to 6 pt., but leave the maximum size at 18 pt. Click OK.

It is often possible to portray the same data several different ways, and the mapmaker must choose the method that best communicates the idea being presented. In your opinion, did the graduated color or graduated symbol map do a better job of showing the differences in elevation?

27➔ Click the Volcanoes layer name twice, slowly, and type in the new name **Volcano Elevation**. Press Enter.

Examine the elevation values in the Table of Contents, and notice that they have many decimal values, which is undesirable for two reasons. First, they are pointless because they are all zeros.

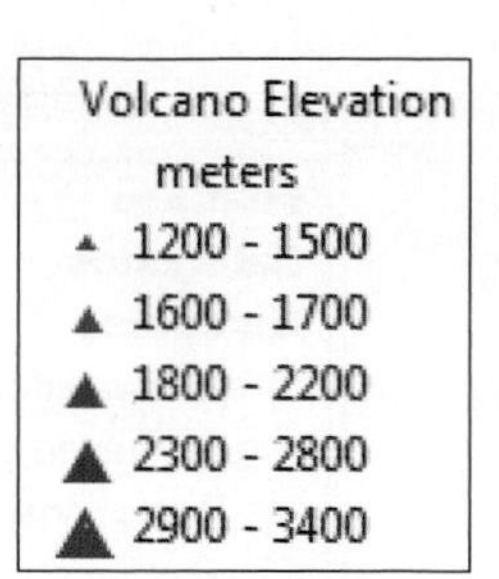

Second, it is important to ensure that the significant figures used for the labels are consistent with the precision of the values being reported.

27➔ Open the properties for the Volcano Elevation layer again.

27➔ Click on the Label heading to the left of the Template button, and choose Format Labels.

27➔ Under Rounding, fill the button for *Number of significant digits*, and set the number to 2. Click OK and OK.

27➔ For a final touch, click twice slowly on the ELEVATION heading in the Table of Contents and type (meters) to indicate the units.

TIP: Evenly rounded class labels give a more professional look to a map.

Next, we will symbolize the volcanoes based on type. However, we don't want to lose the elevation map, so we will create a copy of the layer. One of the advantages of layers is the ability to create different views from the same feature class.

28➔ Right-click the Volcano Elevation layer and choose Copy.

28➔ Right-click the Geography data frame name and choose Paste Layer(s).

28➔ Give the new Volcano Elevation layer (above the old one) a new name, Volcano Type. Drag it below the Airports layer, if necessary.

Now let's consider how to symbolize the volcano TYPE field. It contains repeating entries like "Stratovolcano" and "Volcanic field," which represent categorical data. The proper type of map for categorical data is a unique values map.

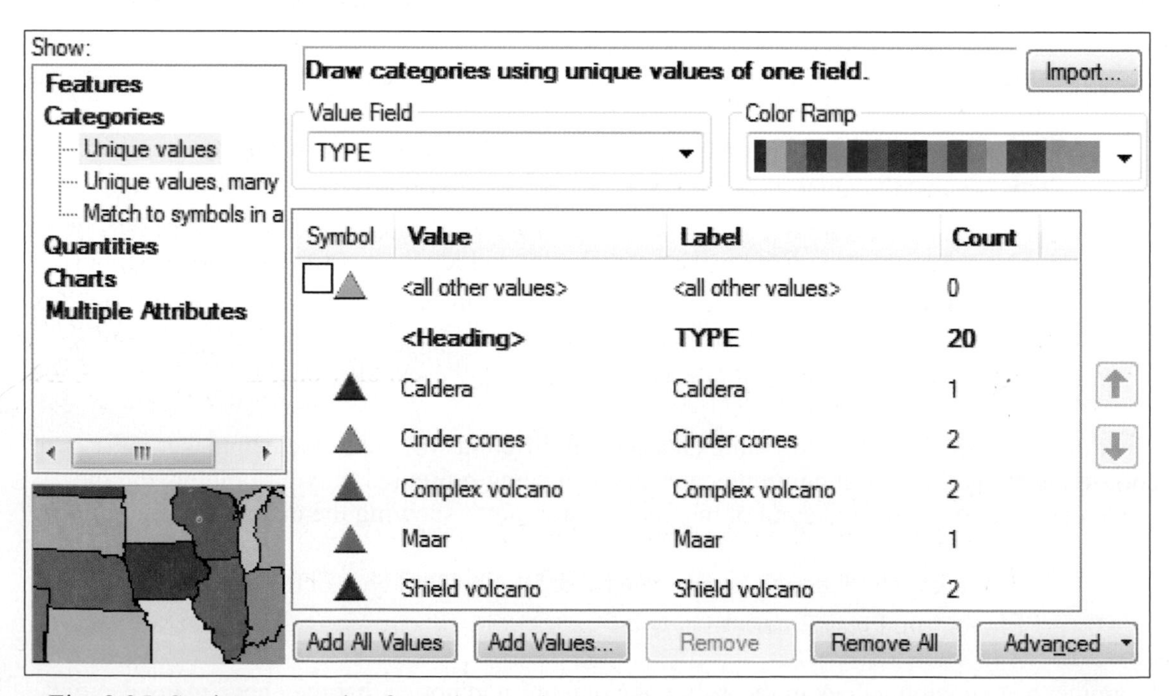

Fig. 2.25. Setting properties for unique values maps

29➔ Open the Volcano Type symbol properties. Click Categories to change the map type and select the Unique values map (Fig. 2.25).

29➔ Change the Value field to TYPE. Click the Add All Values button. It will not show the categories until you do.

29➔ Notice the *<all other values>* entry at the top of the categories list. This offers the option of grouping small categories together. In this case there are no features in this class, so uncheck the box to turn off display of this category.

29➔ Click the Symbol heading to access the Properties for All Symbols menu (as shown in Fig. 2.24). Set the symbol to Triangle 2.

29➔ Change the color ramp to one with dark, bold colors that will show up well.

29➔ Examine the window to make sure it looks like Figure 2.25. Then click OK.

29➔ Turn off the Volcano Elevation layer by unchecking it in the Table of Contents.

TIP: If you don't like the color assigned to each volcano, right-click a symbol in the Table of Contents to change the color.

Creating maps for attributes of lines

Next we turn our attention to displaying the highways with different symbols for each type. First, we need to check the attribute table to see what fields are available.

30➔ Right-click the Highways layer and choose Open Attribute Table.

30➔ Examine in particular the HWY_TYPE field.

3. What three values are found in this field? S, U, I________ Can you decipher what they mean by examining the other fields for clues?

Again we have categorical data, and the unique values map type is the appropriate choice.

30➔ Close the Highways attribute table.

30➔ Open the symbol properties for Highways.

30➔ Choose the Categories heading and select the Unique values map type.

30➔ Set the Value field to HWY_TYPE and click Add All Values.

30➔ Click the Count heading above the categories to force it to count the number of features in each category. Notice the lone H value and the 58 unlabeled roads.

It is reasonable to assume that "I" stands for interstates, "S" stands for state highways, and "U" stands for US highways, but what could the "H" represent? With only one, there is a chance it is simply a data entry error. We also have no idea what to label the blank highways. However, we can edit the legend categories to create a clearer legend.

31➔ Click on the row containing the "H" value and click the Remove button. It removes the category and places the feature in the <all other values> entry.

31➔ Also click on the row with the unlabeled features and Remove it as well.

31➔ Click on the "U" under the Label heading to highlight it. Type in US Highways.

31➔ Click on the "S" and type in State Highways.

31➔ Click on the "I" and type Interstates.

31➔ Change the HWY_TYPE label to Road Class.

31➔ Change the <all other values> label to Unclassified.

31➔ Click on the US Highways symbol to highlight the row. Then click the Up arrow to move it below the Interstates.

Now we have a map of the highway type, but the thin colored lines do not distinguish the types very well. When there are only a few classes, it can be convenient to set each symbol manually.

32➔ Double-click the symbol for the Interstates to open the Symbol Selector.

32➔ Choose the Expressway symbol and click OK.

32➔ Double-click the symbol for the US Highways. Select the Major Road symbol and change the color to dark green. Click OK.

32➔ Double-click the symbol for the State Highways. Choose the Major Road symbol and set its color to a light brown.

32➔ When finished, the legend should look as shown in Figure 2.26. Click OK.

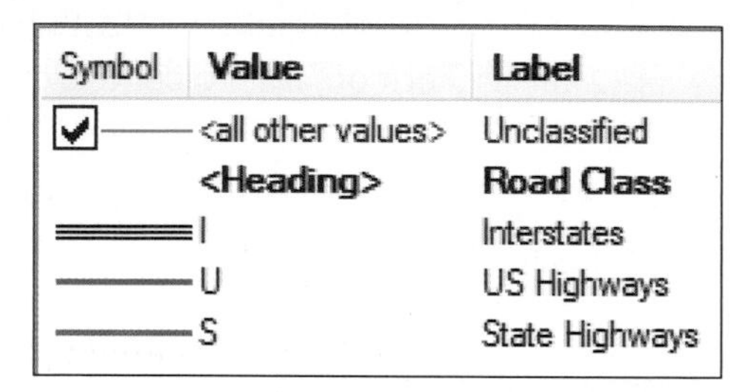

Fig. 2.26. Final legend for highways

Creating maps for attributes for polygons

Let's move on to creating maps based on polygon attributes using the counties layer in the other data frame. We will start with population.

4. What kind of data does population represent? ____________ What kind of map should be used to display it? ____________

33➔ Click the minus sign by the Geography data frame name to collapse the layers.

33➔ Right-click the Population data frame name and choose Activate.

34➔Open the symbol properties for the counties layer.

34➔ Choose Quantities: graduated colors for map type and POP2010 for the Value field.

Remember, population data are usually normalized to the area of the measurement unit.

34➔ Choose SQMI as the Normalization field (near the bottom of the list).

34➔Change the color ramp to a monochromatic one, such as light to dark orange.

34➔ Click on the Label heading above the classes and choose Format labels. Give them two significant digits and check the box to *Show thousands separators*. Click OK and OK to view the map.

We will create new layers for each new map so that we can keep the previous ones.

35➔ Click twice slowly on the counties layer name and rename it Population Density.

35➔ Right-click on the Population Density layer name and choose Copy.

35➔ Right-click on the Population data frame name and choose Paste Layer(s).

Next we will map the median age of the counties. Think about this attribute for a moment. Whereas population values may vary in part based on the size of the polygon, the median age should be independent of the area being measured. This attribute should not be normalized.

36➔ Rename the new layer (on top) **Median Age**.

36➔ Create a graduated color map based on the MED_AGE field. Change the normalization field to none. Format the labels to have two significant digits.

Save 12-25-13

Classifying data

We have been using the default Natural Breaks classification so far, but it is not always the best method. Examine the uneven class ranges for the Median Age layer. The legend might make more sense if the age ranges are evenly separated by five years.

37➔ Open the symbol properties of the Median Age layer.

37➔ Click the Classify button. Take a moment to examine the histogram, and note that these data are fairly uniformly distributed.

37➔ Change the Classification method to Defined Interval. If you get an error, just close the warning and continue. It is a bug. Set the Interval Size to 5.

37➔ Click OK and OK. Examine the new ranges in the Table of Contents.

Most readers would find this classification easier to understand than the uneven classes produced by the Jenks method. Next, symbolize the population of the cities.

38➔ Add the cities feature class from the mgisdata\oregondata geodatabase.

38➔ Open the symbol properties for the cities layer and choose the Quantities: graduated symbols map type.

38➔ Set the Value field to POP2007 and click the Classify button.

This data set is skewed, with many small cities and one very large one. The default Jenks Natural Breaks is probably the best classification choice. Five classes aren't needed, though.

39➔ Change the number of classes to three, and click OK.

39➔ Click the Template button and set the color to a light one. They will show up better when the symbols overlap.

39➔ Format the labels to three significant figures with thousands separators. Click OK in all of the windows to view the map.

SKILL TIP: Unclassed maps, such as proportional symbol or dot density maps, don't require classification and can present an unbiased view of the data (Maps and Symbols).

Displaying thematic rasters

Thematic rasters portray map data, such as soil types, roads, elevation, or precipitation. We'll use the Geography data frame to experiment. We will begin with a digital elevation model (DEM), a raster that portrays an elevation surface with a pixel size of one kilometer.

40➔ Activate the Geography data frame and expand its contents.

40➔Collapse the individual layer legends by clicking the minus sign next to them, so that only the layer name appears.

40➔ Turn off any layers that are currently being displayed.

TIP: The first time that a raster is displayed, the user is asked if she wishes to build **pyramids**, which help speed the display of the raster on subsequent occasions. You can click Yes.

41➔ Add the gtopo1km raster from the oregondata geodatabase.

41➔ Open the gtopo1km layer properties and click on the General tab. Note that the elevations are in meters.

41➔ Click on the Source tab and examine the properties of the raster. Examine the cell size and, scrolling down, the spatial reference (coordinate system).

41➔ Click on the Symbology tab and examine its settings.

5. What kind of data type is elevation?________________ What kind of raster display method(s) could be used to display it?____

When a raster is added to a map, ArcMap chooses an initial display method based on the raster type. In this case, it chose the Stretched method with a grayscale color ramp. Notice the minimum and maximum elevation values, from 1 to 3124 meters. The Stretched method slices these values into 256 bins for use with the 256-grayscale ramp.

42➔ Close the Properties window.

42➔ Click the Identify tool and click on the raster to view values for a pixel.

Field	Value
Stretched value	241
Pixel value	2272
OBJECTID	2272
COUNT	14

Fig. 2.27. Identifying pixels in the DEM

Figure 2.27 shows the results when a white pixel is clicked. The Pixel value is the elevation, 2272 meters. The Stretched value indicates the bin number, 241. This bin is displayed as nearly pure white. The COUNT field tells how many pixels have this value.

42➔ Click on more pixels in the raster, choosing black, white, and gray ones. Compare the pixel values and the stretched values for each one.

42➔ Close the Identify window.

43➔ Open the properties for gtopo1km and examine the Symbology tab. A Standard Deviation stretch is being used with 2.5 standard deviations.

43➔ Move the Properties window aside so that the map can be seen. Change the Stretch type to None and click Apply. Examine the map.

43➔ Change the Stretch type back to Standard Deviations. Click Apply.

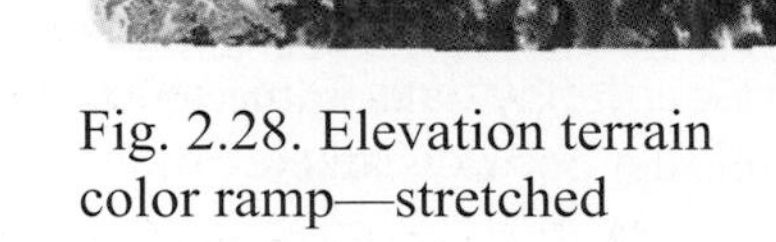

Fig. 2.28. Elevation terrain color ramp—stretched

44➔ Try experimenting with different color ramps, clicking Apply after each.

44➔ End by choosing the color ramp shown in Figure 2.28. Click OK.

The Stretched method utilizes 256 equal interval classes. The classified display method gives control of the number of classes and the classification method.

45➔ Rename the gtopo1km layer Elev-Stretched.

45➔ Right-click the Elev-Stretched layer and choose Copy.

45➔ Right-click the Geography data frame name and choose Paste Layer(s).

45➔ Rename the new layer Elev-Classified and open its symbol properties.

46➔ Change the raster display method to Classified. Find the same color ramp used for Figure 2.28 and select it.

46➔ Set the number of classes to 10. Continuous data, because they often have underlying spatial patterns, can use more classes than vector data. Click OK.

46➔ Use the Identify tool to examine the pixel values and the class values (which go from 0 to 9 rather than 1 to 10). Close the Identify window when finished.

One advantage of using the classified display method is controlling the classification. It defaulted to Jenks Natural Breaks. However, elevations are best displayed in even classes with rounded numbers to appear like a contour map.

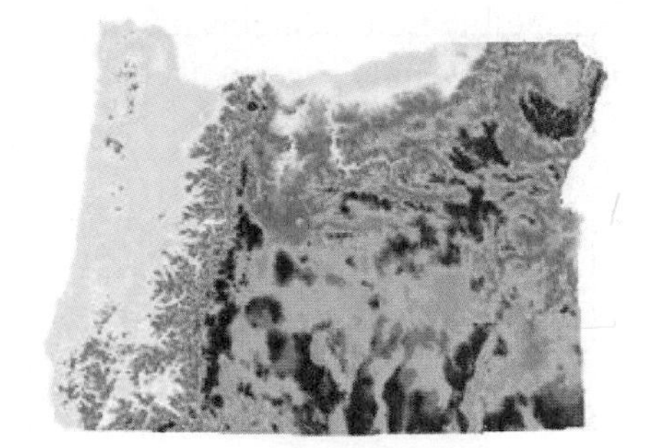

Fig. 2.29. Elevation terrain color ramp—classified

47➔ Open the Symbology properties for the Elev-Classified layer. Click the Classify button.

47➔ Change the classification method to Defined Interval. Set the interval to 200 meters.

47➔ Notice the histogram. Most of the elevations fall below 2000, with very few pixels in the higher elevations. Click OK and OK.

Notice how few white pixels show now (Fig. 2.29). The Stretched method and the Jenks classification overemphasized high elevations. The defined interval classification presents a more realistic view, with highest elevations restricted to a few peaks.

48➔ Add the gtoposhd raster from the oregondata geodatabase.

This type of raster is called a hillshade raster. It is derived from a DEM by modeling how the illuminated surface might appear and contains brightness values from 0 to 255. Hillshade rasters provide an intuitive and detailed view of the surface and are useful for providing a base for transparent overlays.

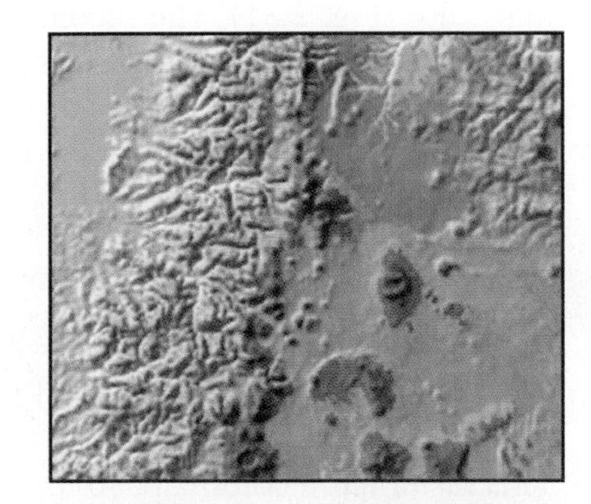

Fig. 2.30. Transparent elevation displayed over a hillshade raster

48➔ Click on the Elev-Classified layer and drag it above the gtoposhd layer in the Table of Contents. Remember, layers are drawn from bottom to top.

48➔ Open the Elev-Classified layer properties.

48➔ Click on the Display tab. Set the Transparency to 50% and click OK (Fig. 2.30).

49➔ Use the minus buttons to collapse the legends for the three elevation layers.

49➔ Add the slopeclass raster from the oregondata geodatabase. It shows areas of low, medium, and high slope indicated by the values 1, 2, and 3.

6. What type of data does this raster contain? _______________ What type of map should be used to display it?_________________________________

50➔ Open the properties for the slopeclass layer and click the Symbology tab.

50➔ Notice that the unique values map was initially chosen. This choice is fine, but for ordinal data, a monochromatic color ramp will show the data better.

50➔ Click on the Color Scheme choice and select a monochromatic ramp. Click Apply and experiment until you find one that you like. Then click OK.

51➔ Add the landcover raster from the oregondata geodatabase.

51➔ Right-click the landcover layer name and choose Open Attribute Table.

The Value field contains the numeric values stored in the raster, and the Count field indicates how many cells contain each value. All raster attribute tables contain those two fields. The other fields were added to provide information about what the values mean.

TIP: Rasters with fewer than 2000 unique values will automatically be given an attribute table.

51➔ Close the Table window and open the symbol properties for the landcover layer.

51➔ Scroll down and examine the colors and values.

A new Colormap method appears in the list, which assigns preselected colors to each raster value. If a raster has a colormap, then this display method will be initially used.

52➔ Close the Properties window and open the Identify tool.

52➔ Click on different colors of the landcover layer.

Field	Value
Color Index	77
Color(a,r,g,b)	255,150,210,150
OBJECTID	76
Count	581164
GAPCODE	5257
ORCODE	5257
DISPLAY	5257 Big Sagebrush Sh
HABNAME	Sagebrush Shrublands

Fig. 2.31. Identifying land cover pixels

The first row shows the color index, the number assigned to that color (Fig. 2.31). The second row is the color specification, indicated as mixtures of red, green, and blue. This raster has 150 categories of land cover. The raster contains another field, HABNAME, which groups the cover classes into fewer categories. Let's display the land cover using this field.

52➔ Close the Identify window and open the symbol properties for landcover.

52➔ Change the display method to Unique Values.

52➔ Change the Value Field to HABNAME and click OK.

This raster has a colormap, so its default colors are the best choice. Transparency can be used with any raster or vector layer to enhance the display.

53➔ Collapse the legend for landcover layer.

53➔ Click and drag the gtoposhd layer just below the landcover layer in the Table of Contents and turn it on.

53➔ Open the properties for the landcover layer. Click the Display tab and set the Transparency to 50%. Click OK.

Displaying image rasters

Images often represent brightness of land features as measured by a camera or spectrometer. They can have one band or multiple bands. This Landsat image over Crater Lake has eight bands.

54➔ Turn off all layers in the Geography data frame.

54➔ Add the L720021127av2 image from the mgisdata\Oregon folder. Click Close to ignore the coordinate system warning.

54➔ Rename the image layer Landsat.

54➔ Right-click the Landsat layer in the Table of Contents and choose Zoom to Layer.

54➔ Open the Landsat layer properties and click the Symbology tab.

Notice the two ways to display the image. Stretched is appropriate for single-band images or to display one band of a multiband image at a time.

55➔ Click Stretched and select Band 1 (blue light) as the band to display.

55➔ Keep the default Standard Deviation Stretch and leave the other settings.

55➔ Click Apply and move the Properties window aside to examine the image.

RGB composites display three bands at a time, one in each RGB color gun of the monitor. You can experiment with assigning different bands to each color.

56➔ Open the Symbol properties, if necessary, and choose RGB Composite.

56➔ Choose Band 3 for the red channel, Band 2 for the green, and Band 1 for the blue. The Landsat bands will match the color guns to create a true-color image.

56➔ Click Apply and examine.

56➔ Change the “n” for the Standard Deviation stretch from 2.5 to 1. Click Apply.

57➔ Choose Band 4 for the red channel, Band 3 for the green, and Band 1 for the blue. Click OK.

Band 4 shows near-infrared light, which is strongly reflected by vegetation. This combination of bands is known as a false-color image, and it is helpful in assessing vegetation density and health. You can experiment with other band combinations if you have time, such as the popular 7-4-1 combination. Try out your own.

This is the end of the tutorial.

➔ Close ArcMap. You can save the map document if you wish.

Exercises

Use ArcMap and the data in mgisdata\Oregon\oregondata to create a map document showing various aspects of Oregon, as described below. Use what you have learned to make each map as aesthetic and legible as you can.

1. Start with a new, blank map using the Letter (ANSI A) Landscape template. Create four data frames, each of which takes up a quarter of the page. Name them Volcanic Hazards, Farms, Housing, and Physiography.

In the Volcanic Hazards data frame:

2. Create a map showing the population density of the counties. Also show the hospitals, marked with blue crosses.
3. Create a proportional symbol map of the volcanoes based on the KNOWN_ERUP field.

In the Farms data frame:

4. Create a map of counties showing the density of farms. Label the county names.
5. Create a map showing the transportation routes, symbolized by type (road type, rail, and air).

In the Housing data frame:

6. Create a map showing the vacancy rate of the counties. What would you normalize by in this case? Label the county names.
7. Show the major cities feature class (majcities) symbolized by population.

In the Physiography data frame:

8. Add the state boundary and the county boundaries, and then add the World Imagery service from ArcGIS Online. Use a hollow symbol and a contrasting outline color for the state and county boundaries so that you can see them well against the imagery.

For all the frames:

9. Examine the legends in the Table of Contents. Make sure you used good classification schemes, rounded label values, and used appropriate significant digits or decimal places. Be sure that each layer has an informative capitalized name.
10. Capture a screen shot of each data frame in Data View, including the Table of Contents showing the corresponding legends to each frame. Also capture a screen shot of the entire layout with all four frames.

Challenge Problem

Start with the map created in the Volcanic Hazards data frame and turn off the hospitals. Consult the Skills Reference section to learn how to create label classes to label features differently based on an attribute. Label the active volcanoes (KNOWN_ERUP > 0) with large red labels and the inactive volcanoes (KNOWN_ERUP = 0) with smaller green labels.

Chapter 3. Presenting GIS Data

Objectives

- Learning basic elements of creating good maps
- Choosing appropriate coordinate systems for mapping
- Creating text and labels on maps
- Creating map layouts and printed maps

Mastering the Concepts

GIS Concepts

GIS analysis often results in sharing information with others in the form of maps. Whether you are creating a large poster-style or a page-sized map, a few guidelines help in devising a map design that expresses the essence of the data and gets the message across. This chapter introduces some basic ways to communicate ideas to others, but it cannot cover the full range of concepts included in the art and science of cartography. Taking a cartography class or reading some map design texts are good ways to improve your skills.

Basic elements of map design

A map can be a work of art as well as a vehicle of information, and these two aspects support each other. By designing the map carefully and using some basic guides to style, you can maximize the transfer of information and present yourself as a competent professional. A poorly designed map gives the opposite impression.

When pen and ink were involved, people spent more time on design to make a good map with minimum effort. Now that computers can quickly create and modify maps, the design step is often ignored. People may not even take advantage of the computer's wonderful ability to try out different designs, and many bad maps result. High-quality maps require planning, attention to detail, and aesthetic sense. The basic process follows these steps:

- **Determine the objectives of the map.** What is being communicated? Who is likely to use the map and under what circumstances? The use of the map will guide some important design decisions, such as how big the text must be. A map being projected on a screen to a large audience needs simplicity, large text, and bold colors. A highway travel map needs more details.

- **Decide on the data layers to be included.** Computers provide a big advantage here because one can easily add or remove layers as the map takes shape and can change a design decision if it does not appear to work in practice.

- **Plan the layout.** Be sure to include all the frames and other **map elements** (legends, etc.) needed. Following some simple guidelines about placing the elements yields more elegant, pleasing maps.

- **Choose colors and symbols.** Select ones that create the right effect and maximize readability. Symbols with psychological relevance can be used to make maps easier to interpret, such as making water features blue. The choice of symbols can enhance a message or hide it.

- **Create the map.** Pay attention to details, colors, layout, and aesthetic qualities.

Determine the map objectives

Maps serve a variety of purposes. A one-page map can present information in a report or a paper. A poster-sized map can display important information at a public meeting or an information booth. Some maps are presented to the public and must be meticulously planned and executed. Others are rough copies to take into the field for collecting more data. Some questions to help guide in map planning are suggested in the following paragraphs.

Who will be using the map? Maps for the general public should be simple and carefully explained using legends and supporting text. These maps can benefit from orthophotos or shaded relief backgrounds that show land surfaces and help the reader easily identify landmarks and other features of interest (Fig. 3.1). Consideration should be given to possible human disabilities in mapmaking, such as poor eyesight or color blindness. Maps for professional interest may be more complex and make more assumptions about the reader's level of knowledge.

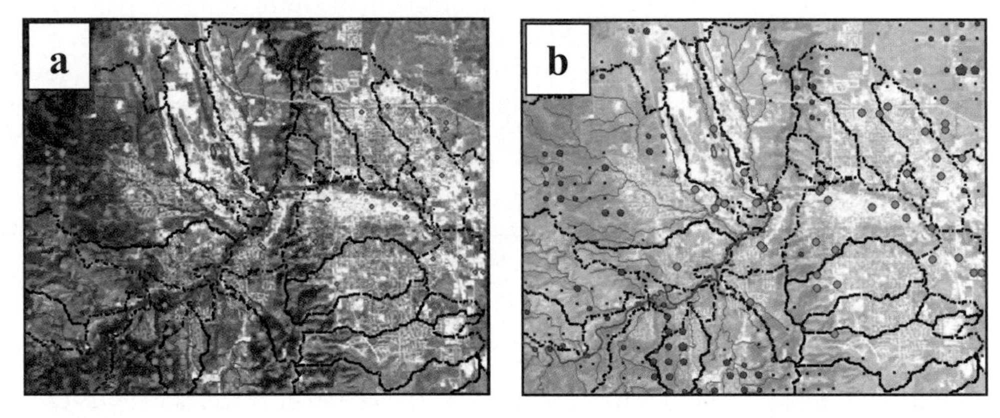

Fig. 3.1. Watersheds, septic systems (orange), and gas stations (green). (a) The satellite image dominates. (b) A transparent image lets the other symbols show up better.

Under what circumstances will the map be used? A page-sized map in a report should be legible to most readers with normal eyesight in good light. A travel map might be used at night or while driving a car, so enhanced readability helps its purpose. A poster map should be tested to ensure that all its text is legible to a person standing several feet away—an 18-pt. font is considered a minimum size for this type of map. A map used for field work might have many details to help the user find the right locations easily. One issue that is seldom considered is the distribution of the map. Maps in reports or articles are often read as a facsimile—consider designing the map so that its main features are apparent even as a black-and-white copy. Designing a good black-and-white map is even more challenging than using all of those beautiful colors.

What objectives should the map achieve? Maps do not always simply convey information. The design can contribute to the messages intended by the map creator or can detract from it. For example, a city map showing point-source pollution sites in big red symbols suggests danger and may unnecessarily alarm nearby residents. Such a map might cause unwanted publicity and action by interest groups out of proportion to the actual risk. On the other hand, the map could alert the city council to a real danger in order to encourage appropriate action.

How sensitive is the map information? Are privacy issues involved? Many GIS mapmakers, carried away with the ability to create dozens of maps, do not stop to consider the possible effect of certain types of information, especially when layers are put together. For example, laying a fault map on top of the city parcels layer might have serious consequences for the property values of parcels near faults. The public may not be aware that most faults are inactive and do not pose a threat. Maps of fossil sites or archeological digs must ensure that the exact locations of sensitive material remain out of the public domain. One must also be cautious in putting private information on public maps, such as the names of parcel owners or property values.

Choosing the map layers

The objectives of the map usually dictate the choice of the main layers, but the addition of other layers can often improve a map (or diminish it). Generally, the more ancillary information on a map, such as roads, towns, landmarks, topography, and vegetation features, the easier it is to find locations on the map and relate it to what we know. A geological map without topography or roads does little to help a geologist evaluate hazards in a development area. However, too much information can drown the message or make the map difficult to interpret.

A good design rule of thumb is to present the most important layers with the clearest and largest symbols so that the critical aspects of the layer are emphasized. Minor layers can be presented with less obtrusive symbols and smaller text. One should not have to hunt to be able to see the major patterns of import in the map. Figure 3.1a attempts to show the distribution of septic systems and gas stations with respect to watersheds in Rapid City, South Dakota. The satellite image in the background helps people relate the watersheds to known landmarks and neighborhoods, but it interferes with the interpretation of the map. A redesigned map in Figure 3.1b uses the transparency function to tone down the satellite image and changes the size and color of the marker symbols for enhanced visibility. Now one can easily identify which watersheds are most at risk of contamination.

Planning the layout

The basic additions to a map include a title, a legend, an indication of north, and scale. Other elements include graphics, pictures, neatlines, and graticules (latitude-longitude markers).

Balance and alignment are very important in designing the map. Balance means that the elements are evenly arranged on the page and are a good size relative to each other. Filled space should be evenly distributed with empty space. Take a look at the map shown in Figure 3.2. All of the basic elements are there, but notice the poor layout. The elements are crowded into the lower-left corner. Their edges are uneven. Alaska is cut off in the corner. The legend is large compared to the map, and the legend boxes are large compared to the legend text. The scale is almost as long as the map. The elements are not aligned but form arbitrary stair-step lines along the map edges. This is a poorly designed map. It looks like someone made it at 2:00 a.m. the morning before the report was due. However, a few simple changes can dramatically improve its appearance (Fig. 3.3).

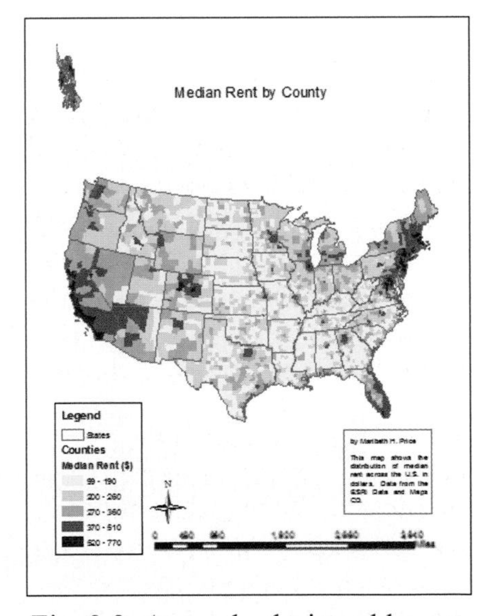

Fig. 3.2. A poorly designed layout

- Choose a landscape layout to complement the east-west length of the United States and reduce blank space. This choice also increases the size of the map for better visibility.

- Distribute the elements evenly on the page.

- Reduce the size of the legend and scale to emphasize the importance of the map.

- Use **neatlines** (boxes around map elements) to enclose and balance the space. Make sure that the neatline edges align with each other.

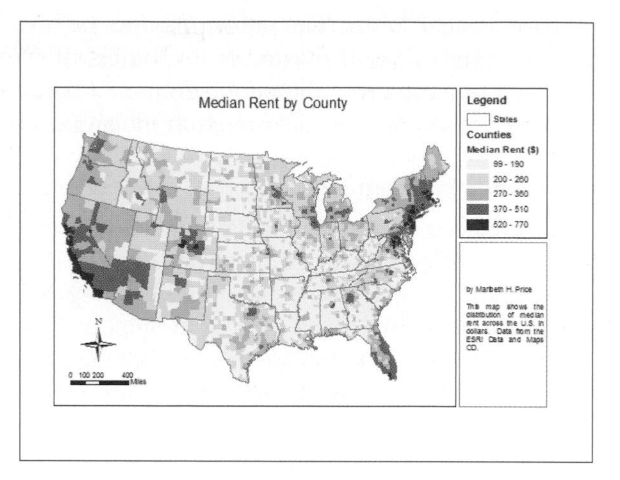

Fig. 3.3. A revision of the map in Figure 3.2 enhances its professionalism.

Choosing symbols

Symbols can highlight or obscure the meaning of a map. Here are some important guidelines when choosing symbols:

- Earth tones, such as browns, greens, blues, and tans, are more pleasing and easy on the eye than strident reds, hot pinks, and wild purples. Such colors should be saved for special emphasis or contrast and should occupy a small area of the map.

- Light pastels are best for large areas and should take up most of the map. Dark or bold colors should be used sparingly to highlight important features.

- Take advantage of the psychological aspects of symbols. Blues and greens are cool and calming and are easily understood when applied to water and vegetation, respectively. Reds and oranges tend to denote danger, heat, and agitation. Browns are steady and solid or can suggest dryness when combined with greens for vegetation.

- Use symbols that mimic the phenomenon being mapped. Water should usually be blue. A reader viewing the map in Figure 3.4 would probably at first glance interpret the Gulf of Mexico as land. NDVI maps showing the amount of vegetation are often displayed with browns and tans in low-vegetation areas, light greens in moderate areas, and dark greens in heavy areas, mimicking the actual appearance of the ground conditions.

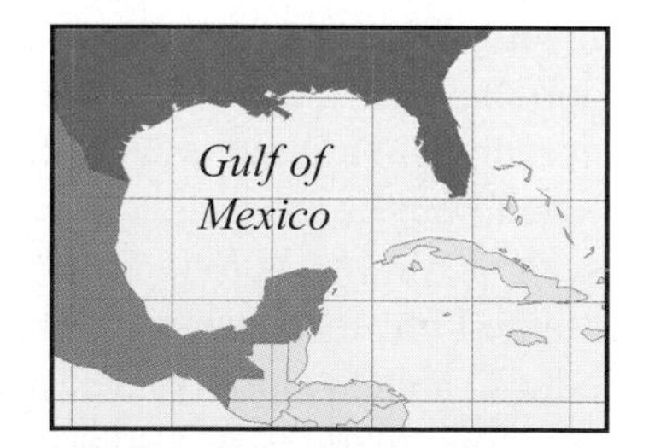

Fig. 3.4. Blue makes features look like water.

- Maps showing numeric values, such as county population, city size, or road traffic loads, should have symbols that follow an easily recognized pattern. Increases for polygons are best shown using a single or dichromatic color ramp. For lines, increasing thickness can illustrate larger values, and increasing marker size is typically used for points.

- Avoid "vibrating" color patterns, such as thick, alternating white and black lines (Fig. 3.5), because they are tiring and unpleasant to look at. They are called moiré patterns. However, thinning the lines, increasing the separation, and making the white background transparent creates an effective symbol to show the extent of urban area (Fig. 3.6).

- Emphasize important information by using dark and bold colors, large sizes, or high contrast with surrounding map elements. Deemphasize less important information using pale colors, turning black to gray, or using transparency.

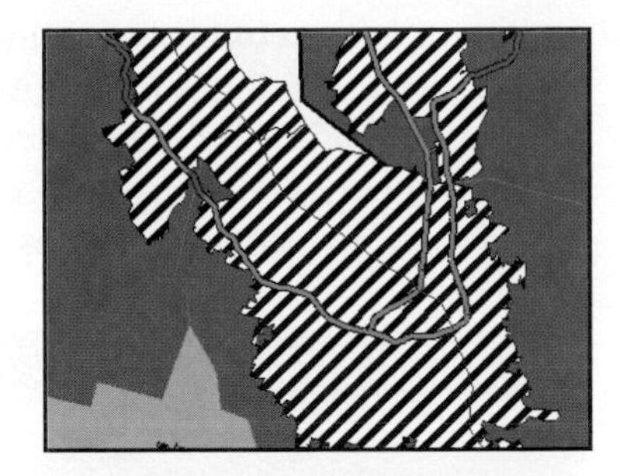

Fig. 3.5. A moiré pattern

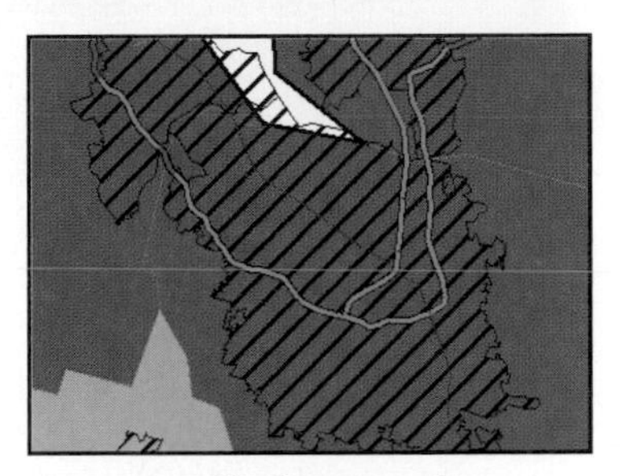

Fig. 3.6. Eliminating the white background provides a useful see-through pattern.

Examples

In Figure 3.7, we see the population density and highways of the San Francisco area. In Figure 3.7a, a rainbow color ramp is used, but the sense of increase is not obvious. The lower-density counties are dramatized in dark red, drawing inappropriate attention to them. The preponderance of mixed bold colors is also tiring to look at. This map is aesthetically poor and difficult to interpret.

In Figure 3.7b, the rainbow color ramp is replaced with graduated tan shades, making it obvious which counties have high populations. The color balance seems far more restful and pleasant. The highways are shown in a contrasting green for good legibility, and the urban highways, which are shorter and more difficult to spot, have been rendered in bright green to enhance their visibility. The major interstates are shown as thick lines and the highways as thin lines, delineating the difference at a glance. In Figure 3.7c, the county outlines are light gray to deemphasize them relative to the roads. The final map is clearer, more attractive, and less cluttered than the original.

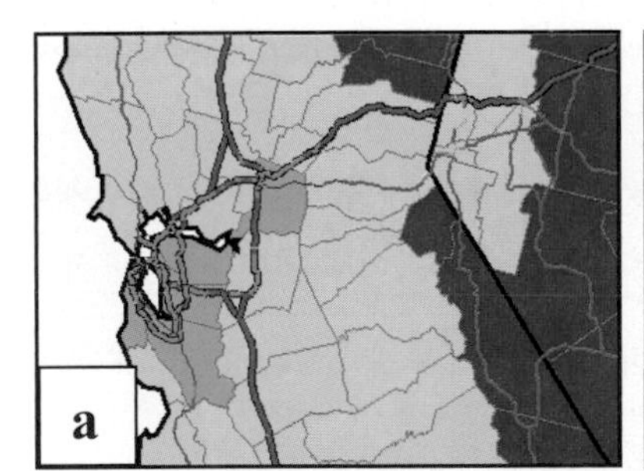

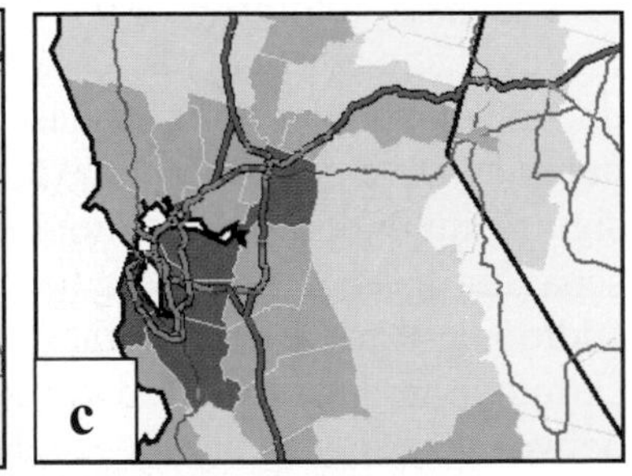

Fig. 3.7. Map of county populations near San Francisco. (a) A map with poorly chosen symbols. (b) A monochromatic ramp highlights the patterns instead of obscuring them. (c) Gray county boundaries emphasize the more important roads.

The world population map in Figure 3.8a illustrates other guidelines. The ocean is blue, but the mapmaker has disregarded the guideline to reserve dark, bold colors for highlighting important areas. The red-orange graduated color highlights the population but contrasts noisily with the dark blue. The latitude-longitude lines, which are not the focus of the map, crisscross the population data and create a busy feeling. Let's see how a redesign can improve this map.

In Figure 3.8b, a light blue color for the oceans instantly creates a more relaxed, agreeable atmosphere. The latitude-longitude lines have been deemphasized by making them gray and drawing them behind the continents instead of on top of them. The graduated color shades contrast nicely with the ocean, are brown/orange to suggest earth, and are predominantly light. The darker, populous countries are naturally emphasized.

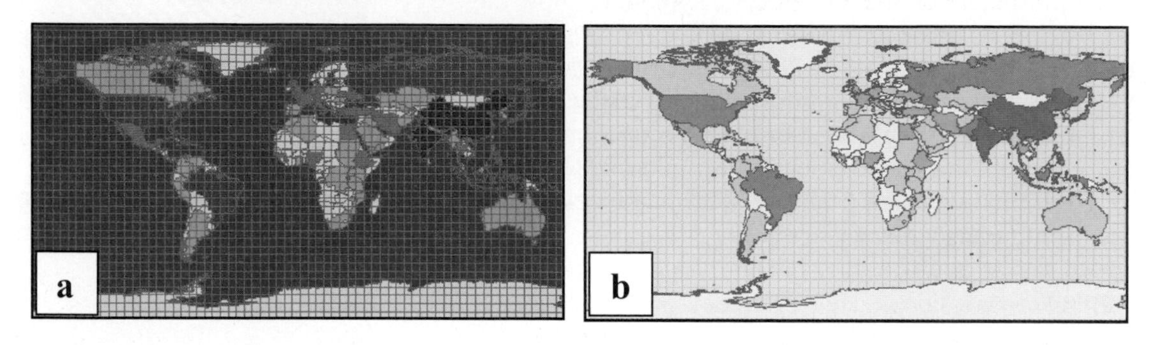

Fig. 3.8. A map of the world. (a) Dark colors and busy lines give this map a heavy and chaotic feel. (b) The pastel colors and deemphasized graticules are far more pleasant.

The layout of the legend often reveals the difference between an amateur and a professional map. The legend in Figure 3.9a illustrates the kind of effort to avoid. No self-respecting cartographer would use so many significant figures in a legend: not only are they hard to read, but population data are estimates at best, and such precision is unjustified. The file name cities.shp makes a poor heading—the reader does not care what the name of the file is. The POP1990 label might not be understood by some people as population data. The outline crowds the text. This legend demonstrates some classic mistakes made by beginners or an amateur in a hurry.

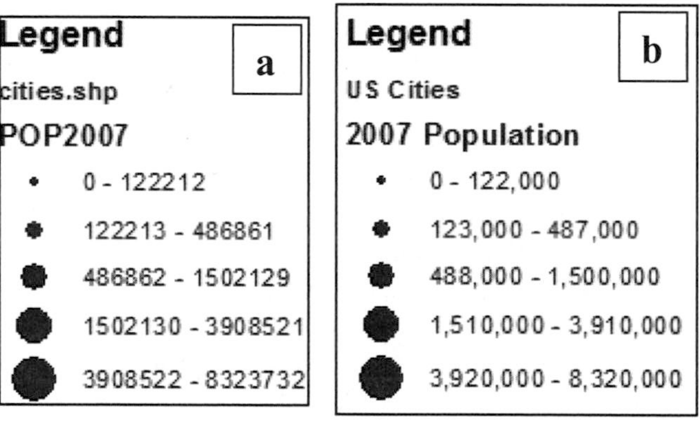

Fig. 3.9. (a) A poor legend; (b) a better legend

The redesign in Figure 3.9b has replaced the layer and attribute names with understandable labels. The class numbers have been formatted to three significant digits, and the thousands separator has been included for additional clarity. Finally, the cartographer has increased the gap between the legend neatline and its contents. This legend is clear, balanced, and professional.

Choosing coordinate systems

Another critical choice for the cartographer involves the coordinate system used to present the map. In Chapter 1, we introduced the idea of a coordinate system and learned that data may be stored in different coordinate systems, such as UTM or State Plane. We also discovered that all layers within a data frame are projected on-the-fly to the coordinate system specified by the user. But what coordinate system should be used? The choice of the coordinate system affects the appearance and use of a map. Figure 3.10 shows the United States using two alternative coordinate systems. Why do they look so different?

The map in Figure 3.10a is based on the familiar latitude-longitude coordinate system, often called a geographic coordinate system, or GCS, which measures angles and thus has units of degrees. When degrees of latitude and longitude are displayed by a GIS, they are treated as if they were planar coordinates instead of spherical coordinates. This practice introduces a distortion to the features and elongates them in the east-west direction (Fig. 3.10a). Thus, although most people are familiar with the latitude-longitude values, a GCS is a poor choice for making maps.

Fig. 3.10. An (a) unprojected and a (b) projected coordinate system

The map in Figure 3.10b utilizes a **projection**, which takes the 3D angular coordinates of a GCS and uses mathematical equations to displace them to a cylindrical, conic, or planar surface, which is then unwrapped to a plane (Fig. 3.11). During this process, the angular units in degrees are converted to planar units of feet or meters. Four properties of map features may be distorted by projections: *area, distance, shape,* and *direction.* Usually map projections reduce or eliminate certain distortions at the expense of the others. Maps based on cylindrical projections (Fig. 3.11a) typically preserve direction and shape at the expense of distance and area. Notice that the longitude lines point north-south, indicating the direction correctly, but Alaska and Greenland are enlarged. Maps based on conic projections typically preserve area and distance at the expense of direction and shape (Fig. 3.11b). The longitude lines no longer point north-south, but the areas are better preserved. The map in Figure 3.10b is in a conic projection. Maps based on azimuthal projections typically preserve areas as well as distances (Fig. 3.11c).

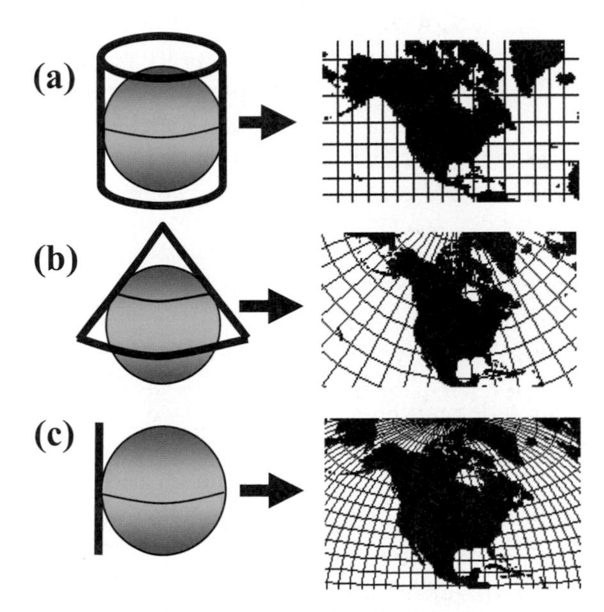

Fig. 3.11. Projections: (a) cylindrical; (b) conic; (c) azimuthal

Choosing a suitable coordinate system is part of developing a good map. National-scale and world-scale maps are impossible to portray without distortion, so the map's purpose drives the choice. If the project includes measuring distances and areas, it is important to choose equidistant or equal-area map projections. If the maps are used for navigating by compass, maps that correctly display direction are important. The cylindrical Mercator projection has been popular for world maps (Fig. 3.12a), but recently critics have charged that this projection enlarges the size of the prosperous countries in Eurasia and North America with respect to developing countries nearer the equator. Attempts have been made to popularize equal-area projections that don't exaggerate the importance of the northern hemisphere (Fig 3.12b).

In Chapter 11 you will learn more about coordinate systems, but for now a few simple guidelines can help you decide on the best projection for a map. Often the name of the coordinate system is a clue, containing "equidistant" or "equal-area," indicating which property is preserved. "Conformal" indicates that shape is preserved. Generally, conic maps tend to preserve distance and area. Mercator and other cylindrical projections tend to preserve shape and direction. Regional data at the state, county, or local scale generally use UTM or State Plane coordinate systems that preserve all four properties over small areas. The inside front cover of the text summarizes some of the more common map projections along with the properties and uses of each.

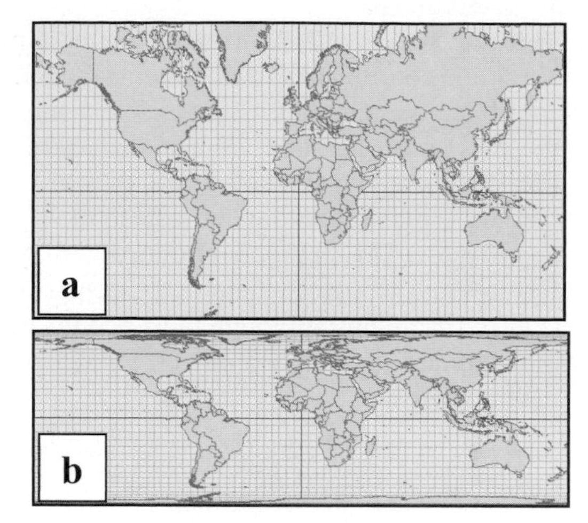

Fig. 3.12. (a) A Mercator and (b) an equal-area cylindrical projection

TIP: A GCS employs an approximation to the shape of the earth, called a datum. A projection inherits this approximation, and therefore you will see the datum included in many coordinate system names. The **North American Datum** of 1983 (NAD 1983) and World Geodetic System (WGS 1984) are two of the most common datums.

The map projection may influence some of the map elements placed in a layout. A conic projection showing a large area does not have a single north direction, and a single north arrow cannot represent it. A map that distorts area or distance, such as Mercator, cannot use a single scale bar to represent distances. In these maps, a **graticule grid** is employed, which shows latitude and longitude markings (Fig. 3.13a). Two other types of grids are often used on maps. A **measurement grid** shows the map units present in the coordinate system (Fig. 3.13b). A **reference grid** shows letters and numbers delineating each grid square (Fig. 3.13c). A reference grid is commonly found in a telephone directory in conjunction with a street index, allowing streets to be located within squares of the map.

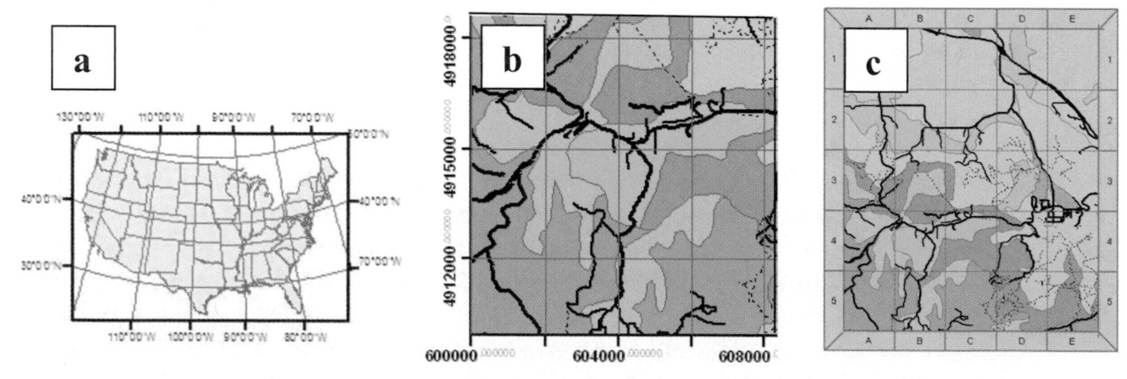

Fig. 3.13. Data frame grids: (a) graticule grid; (b) UTM measurement grid; (c) reference grid

About ArcGIS

Concepts

Maps in ArcGIS

ArcGIS provides an environment for cartography that spans the range from creating a simple map for a report to preparing sets of maps for commercial printing. This chapter covers the basic steps to create a printer-ready map, but the reader should be aware that many advanced cartographic tools are available, as well as methods for developing graphs and database report printouts; although they are not covered here, the Help files have more information about them.

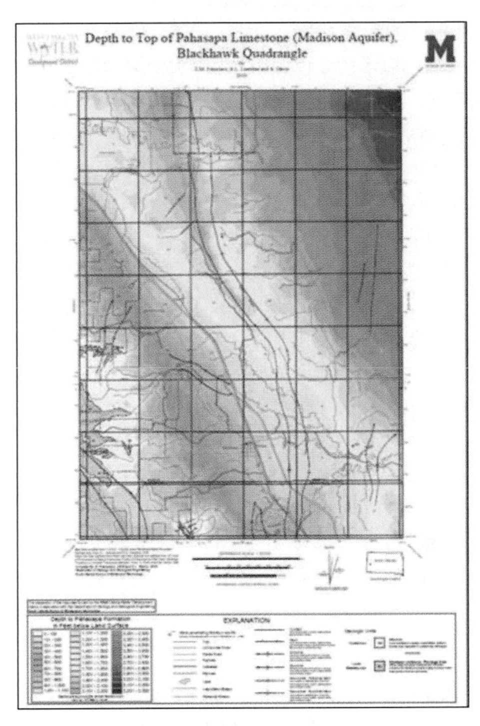

Fig. 3.14. A map layout

In ArcGIS, the draft of a map intended for printing or distribution is called a **layout**. The layout mode in ArcMap facilitates map design by incorporating data frames and map elements, such as legends, north arrows, scale bars, text, and more. Figure 3.14 shows a map from a series of 1:24,000-scale maps produced to the show properties of aquifers for 7.5-minute quadrangles in the Black Hills of South Dakota. The maps were intended to mimic the design of a standard USGS topographic quadrangle. They can be printed on a large-format plotter if one is available but are being distributed primarily as PDF-format files. The development of the data and the map layout took place in ArcGIS Desktop. (The layout has been drastically reduced in the figure, but originals of this and similar maps can be downloaded from http://aquifers.sdsmt.edu.)

Each map document can contain one map layout. Map documents may also be saved as templates, which can be used to re-create a design again and again or to give an organization's maps a common format and look. It is also possible to create map books, in which the sample map design is applied to multiple tiles covering a larger area, with the map extents, legends, titles, and other aspects generated automatically for each map using a feature called Data Driven Pages.

Assigning map scales

GIS spatial data acquire a scale as soon as they are drawn on a computer screen, printed on paper, or saved in a PDF. The first step of creating any map is generally to define the size of the layout page, which will be influenced by the page-size capabilities of the available printers. Once the page size is set, the data frames are arranged on the layout page, and the user must decide how much of the map data will be shown within the data frame. The **map extent** is defined as the range of *x*-*y* values currently displayed in the data frame. Zooming in reduces the map extent; zooming out enlarges it. The printed scale is determined by the map extent and the size of the data frame on the paper. Assigning the map scale of a data frame can follow one of three strategies.

Automatic scaling is used most frequently (Fig. 3.15a). The user sets the size and position of the data frame on the layout page and zooms in or out of the data until the desired map extent is achieved. The user does not try to impose a specific scale but uses his or her eye and aesthetic sense to balance the map frame with the other map elements.

Fixed scale is employed when the printed map must have a specific scale, such as 1:24,000. In this mode, the features always appear at the same size, and changing the data frame size changes the map extent (and may crop the map if the user is not careful) (Fig. 3.15b). Zooming in or out of the map data would change the scale and is not permitted, but panning is.

Fixed extent adopts the current map extent and prevents it from changing. The data frame can be resized, which changes the scale of the map but not its extent. (Fig. 3.15c). The aspect ratio of the data frame is also fixed. The Zoom and Pan tools are unavailable in Fixed Extent mode.

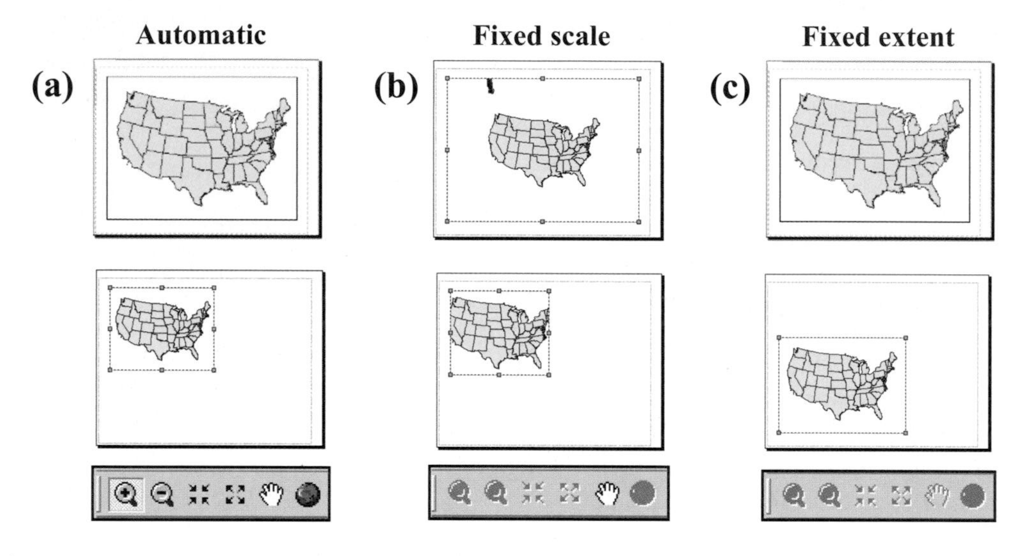

Fig. 3.15. Scaling options for data frames, shown in Layout mode. The effects of resizing the data frame differ in each case. The Zoom/Pan tools are only active in Automatic mode.

Setting up scale bars

Most maps should contain either a scale bar or a statement of the scale, such as "This map is shown at a scale of 1:62,500." ArcMap can automatically construct a suitable scale and resize it when the map scale changes. The user has to set several options, which requires some knowledge of terminology.

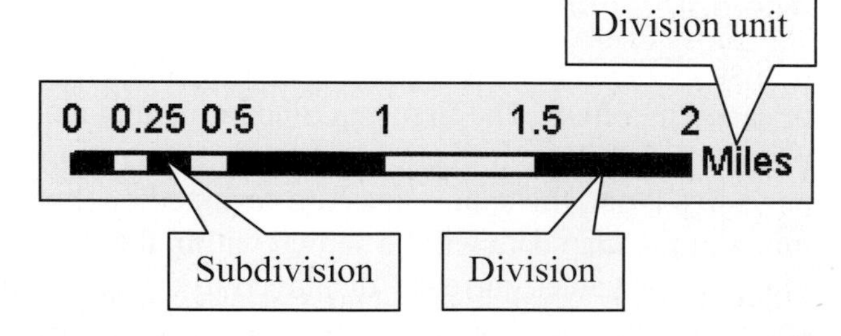

Fig. 3.16. Terminology of a scale bar

The **divisions** are the sections into which the scale is divided. The scale bar in Figure 3.16 has four divisions. The first division may be subdivided, yielding what are called **subdivisions**. This scale bar also has four subdivisions. The **division units** are the units of measurement used to divide the scale bar (miles in this case). The **division value** is the length of each division, expressed in division units. This scale bar has a division value of 0.5 mile. The division units, the map scale, the division size, and the length of the scale bar are interrelated. Zooming changes the map scale and requires the scale bar to change also. Three options control how the scale bar responds to changes in map scale.

Adjust Width keeps the division size and number of divisions constant. If the map scale changes, the entire scale bar gets larger or smaller. This method keeps nicely rounded numbers for the divisions but may result in an unacceptably small or large scale bar if the scale changes dramatically.

Adjust Division Value keeps the scale bar roughly the same length on the page and with the same number of divisions but changes the size of the divisions. If the map scale changes, the division size of the scale in Figure 3.16 might go from 0.5 mile to 0.6 or 0.4. This option preserves the width of the scale bar but may result in less evenly rounded division sizes and labels.

Adjust Number of Divisions leaves the division value constant and adds or subtracts full divisions if the scale changes. For minor changes in scale, the bar will increase or decrease slightly in size. For dramatic changes, the scale bar will remain about the same size but may have more or fewer divisions. The bar in Figure 3.16, for example, might become 1 mile long or 3 miles long, but the division size would stay 0.5.

A fourth option, to *Adjust Divisions and Division Values*, has also been added.

Labeling, text, and annotation

ArcGIS offers several ways to place text on maps. You have already learned how to create **dynamic labels**, which are derived from a feature attribute and applied to an entire layer at once. **Graphic text** is used to place a few labels, one at a time. **Annotation** is used to gain exact control of labels. We briefly summarize the key features of each type.

Graphic text

Graphic text items are constructed by the user, one at a time, to identify map features or provide titles and supporting text for a layout. Different tools are provided to create standard, splined, callout, or multiline text or to label a feature using the value from its attribute table (Fig. 3.17). All graphic text exists on the map as simple graphic elements. Items remain where they are placed but are inefficient when many labels must be made. You must be in Layout view to create graphic text. However, the Label tool is an exception; it derives the text for the label from the attribute table and must be placed in Data view.

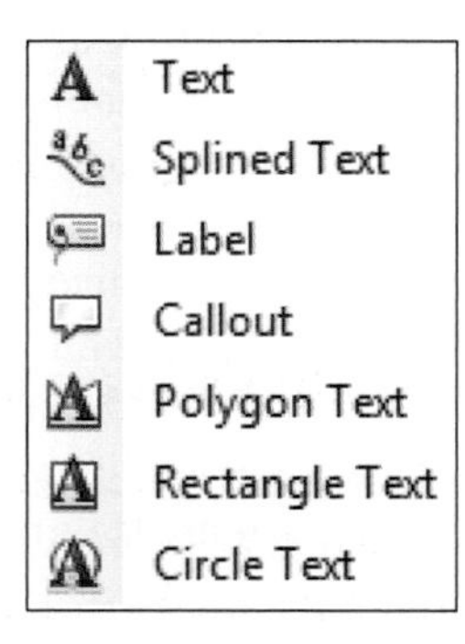

Fig. 3.17. Text tools

Dynamic labels

Dynamic labels are created from an attribute and are redrawn every time the user changes the map view. They can be turned on or off as needed. Dynamic labels utilize a placement algorithm that omits overlapping labels. The number of labels displayed can vary, depending on the zoom scale and label size.

Although largely automatic, dynamic labels provide some control options. The *Placement Properties* help manage the overlap and optimization of the labels to give the best possible result. The *Conflict Detection* options can handle priorities between different layers. *Label weights* control which layers take priority; for example, states might have a higher priority than cities and be preferentially labeled in case of a conflict between the two. *Feature weights* specify the layers whose features cannot be overlapped by labels. Finally, a label *halo* prevents new labels from encroaching on the edges of existing labels.

Because the labels are redrawn each time, printing maps with dynamic labels can bring surprises. What shows on the screen does not necessarily appear the same way on the map. For many purposes a few extra or missing labels do not matter. If they do, however, converting labels to annotation provides greater control over label placement.

Annotation

Annotation provides precise control of each label. Ordinary dynamic labels can be converted to annotation. The labels that fit on the map are automatically placed, and the overlapping labels are directed to an overflow window for interactive placement. Once placed, the annotation label always appears in the same place. Annotation may be stored in three ways.

- *In the map document.* The annotation is stored as graphic text and is edited using the Draw toolbar. This annotation can be used only in a single map document.
- *As a feature class in a geodatabase.* The annotation can be used in multiple map documents. Editing must be used to modify it, so we cover this method in Chapter 14.
- *As a feature-linked annotation in a geodatabase.* The annotation is linked to the corresponding point, line, or polygon features. If the feature is deleted, so is the annotation, and vice versa. This type is available with ArcGIS Standard or Advanced licenses.

Reference scale

When labels and annotation are created, they are assigned a size in points. For a given map, the size of the text relative to the features, such as the counties shown in Figure 3.18a, is determined by the text point size and the current map scale. But what happens if you zoom in? By default, the labels (and symbols) stay the same size when the user zooms in or out (Fig. 3.18b); 12-pt text remains 12-pt. text regardless of the scale of the map.

This behavior changes if the user sets the **reference scale**, which is a property of the data frame. The reference scale is the scale at which the labels and symbols appear at their assigned size. If a user places 12-pt. labels on a map with a scale of 1:100,000, then the labels will appear 12 points in size only as long as the map scale remains 1:100,000. If the user zooms in, the labels and symbols increase in size (Fig. 3.18c). If the user zooms out, they decrease in size.

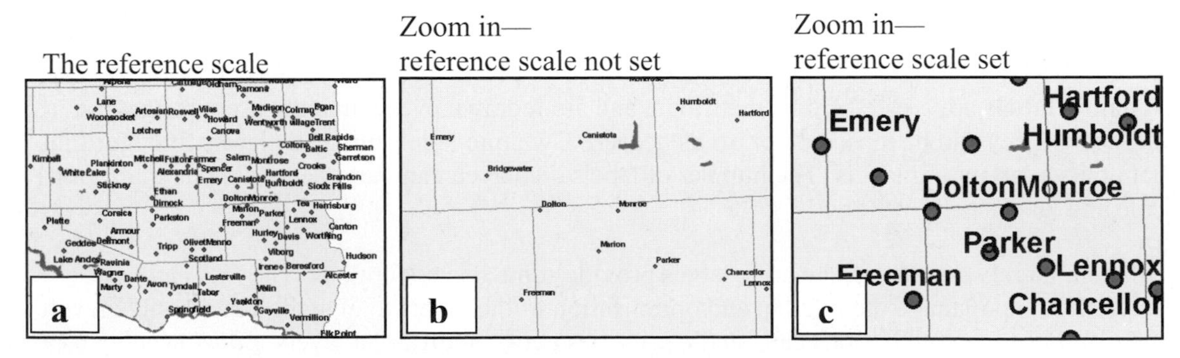

Fig. 3.18. (a) Text and symbols appear at their assigned size at the reference scale. (b) If the reference scale is not set, the symbols will always appear the same size. (c) If the reference scale is set, then zooming in or out changes the size of the text and symbols.

TIP: Be sure to understand the difference among the terms *map scale*, *reference scale*, and *visible scale range*. Map scale determines the ratio of features on the map to features on the ground. The reference scale is the scale at which symbols and text appear at their assigned size. The visible scale range determines the range of scales at which a layer will appear.

Summary

- For best results, making maps requires planning, knowledge of some basic guidelines, and attention to detail.
- The purpose and audience of a map influence the design process. However, most maps should contain at least a title and contact information, scale indication, a north arrow, and an appropriate legend.
- Maps should contain sufficient detail for interpretation, without excessive detail that detracts from the purpose.
- Layouts should be designed with attention to symmetry, size, and balance of the elements. Neatlines and shades can be used to help distribute space well. Alignment of objects along page margins is an important detail.
- Careful choice of symbols enhances the map. Guidelines for symbols include using primarily pastels and earth tones except when emphasis is desired, using larger and bolder symbols for the more important features of a map, using colors and symbols that naturally reflect the nature of the features, and avoiding clutter and clash.
- Maps should use projected coordinate systems. All projections introduce distortions in area, distance, direction, or shape. Different projections preserve some properties at the expense of the others.
- The choice of projection depends on the application and should minimize distortion in the properties important to the map function, such as using an equidistant map to measure distances.
- Scaling of the map is controlled by the data frame properties. Three scaling methods may be used: automatic scaling, fixed scale, and fixed extent.
- Three types of labeling are available. Interactive labels are used to place any kind of text. Dynamic labels are automatically generated for entire layers at a time. Annotation is created from dynamic labels but can be precisely controlled.

Additional reading on cartography

MacEachren, Alan M. 1995. *How Maps Work: Representation, Visualization, and Design.* New York: The Guilford Press.

Monmonier, Mark. 1996. *How to Lie with Maps.* Chicago: The University of Chicago Press. Second edition. First published 1991.

Robinson, Arthur H., Joel L. Morrison, Phillip C. Muehrcke, A. Jon Kimerling, and Stephen C. Guptill. 1995. *Elements of Cartography.* New York: John Wiley & Sons. Sixth edition. First published in 1953.

Important Terms

annotation	divisions	layout	projection
automatic scaling	dynamic labels	map elements	query
chart	fixed extent	map extent	reference grid
datum	fixed scale	measurement grid	reference scale
division units	graphic text	neatline	subdivisions
division value	graticule grid	North American Datum	

Chapter Review Questions

1. List four questions about map objectives that would influence the design of a map.
2. What factors should be considered in evaluating the balance of a map?
3. What types of colors generally work best for maps? How can the psychology of colors be used to enhance a map's meaning?
4. List three common pitfalls that amateurs make when creating legends.
5. What is a geographic coordinate system, and why is it a poor choice for creating maps?
6. What four properties are distorted by map projections? Which tend to be preserved by conic projections? What distortions are present in UTM and State Plane projections?
7. Examine the map projections on the inside front cover of this text. List which projection(s) might be suitable for a (a) map of a county, (b) map of the United States, (c) United States map used to calculate travel distances, and (d) United States map used to calculate areas.
8. When does a north arrow not point up? When should a north arrow not be used?
9. If you have an ArcGIS Basic license and wish to create and use annotation in different map documents, how would you need to store it?
10. What are the differences among the map scale, the scale range, and the reference scale?

Mastering the Skills

Teaching Tutorial

The following examples provide step-by-step instructions for doing basic tasks and solving basic problems in ArcGIS. The steps you need to do are highlighted with an arrow ➔; follow them carefully. Click on the video number in the VideoIndex to view a demonstration of the steps.

➔ Start ArcMap. Choose Existing Maps > Browse for more… on the splash screen, and open the ex_3a.mxd map document from the mgisdata\MapDocuments folder.

➔ Use Save As to rename the document and remember to save frequently as you work.

1➔ Switch to Layout view by clicking the Layout button in the lower-left corner of the map window.

1➔ Click the Zoom Whole Page button on the Layout toolbar, if necessary, to see the entire page.

Setting up the map page

This map document has been set up to show volcanic hazards in Washington and Oregon, and we will use it as a starting place to learn about creating layouts. First, we are going to change the page layout and turn on the grid so we can align the objects more easily.

1➔ Choose File > Page and Print Setup from the main menu bar.

1➔ Examine the options to see what these are.

Look at the two sections labeled Paper and Map Page Size. The lower Map Page Size section controls the size and layout of the map itself. Always set this section first.

TIP: Always design and edit a map at the size at which you intend to print it.

The Paper section is used to modify the printer settings at the time of printing. It should not be set until you are ready to print. In many cases, the settings will be the same as the Map Page Size. However, after you have created an 11-inch by 17-inch map, you might decide to print it on a letter-size page for a report. Setting the Paper section makes it possible.

1➔ Uncheck the box to *Use the Printer Paper settings* if it is checked.

1➔ Check the button to *Scale the map elements proportionally to changes in page size* to help in resizing the frames later.

1➔ Fill the buttons to change the Map Page Size to Landscape orientation.

1➔ Also set the Paper orientation to Landscape. Click OK.

TIP: Checking the box to use the Printer Paper settings is not recommended unless you always use the same printer. It can cause undesired changes in the layout if you switch printers.

2➔ To set the grid, choose Customize > ArcMap Options from the main menu and click the Layout View tab.

2➔ Notice the options for setting rulers: the page units (currently inches), the grid, and snapping.

A grid is a set of points a specified distance apart (now set to 0.25 inch). When snapping is on, objects within the snap tolerance (currently 0.2 inch) will be automatically snapped to the closest grid location. This feature simplifies aligning map elements.

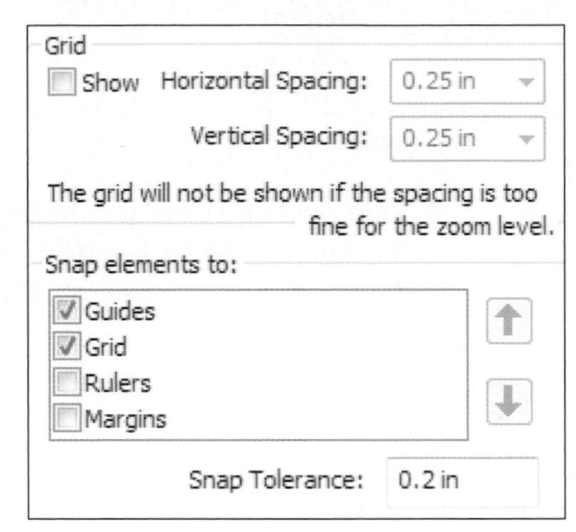

Fig. 3.19. Setting the grid options

2➔ Check the Show box under the *Grid* section to show the grid (Fig. 3.19).

2➔ Check the Grid box under the *Snap elements to* section and make sure the Snap Tolerance is 0.2 inch.

2➔ Click OK to close the Options box.

The Layout toolbar (Fig. 3.20) provides zoom tools that look similar to those on the regular Tools bar, but they affect the view of the page only, not the map inside the data frame. A demonstration will help show the difference.

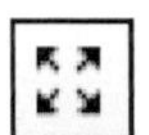

3➔ Find the Fixed Zoom Out button on the Tools toolbar and click it three times. The map gets smaller within the data frame, but the page remains fixed in the map window.

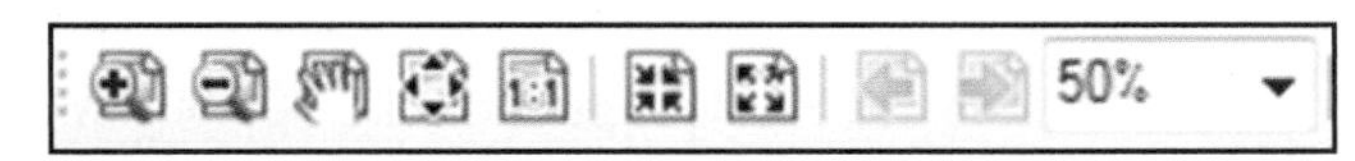

Fig. 3.20. The Layout toolbar zoom tools

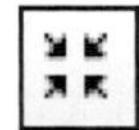

3➔ Find the Fixed Zoom In button on the Tools toolbar and click it three times to return to the initial extent.

3➔ Now find the Fixed Zoom Out tool on the Layout toolbar and click it three times. Notice that the entire page now gets smaller, but the map continues to occupy a constant percentage of the data frame.

3➔ Use the Zoom Whole Page button on the Layout toolbar to fill the map window with the layout again.

TIP: The Zoom/Pan tools on the Tools toolbar control the scale and placement of the map in the active data frame. The Zoom/Pan tools on the Layout toolbar control the view of the layout page.

Now let's see how grid snapping works.

4➔ Use the Layout Zoom In tool on the Layout toolbar to zoom in to the lower-left corner of the Hazards data frame.

4➔ Click the Select Elements tool on the Draw or Tools toolbar. Click the Hazards frame to select it, and drag it to a new location. Notice that the frame corner snaps to a grid point when it gets close.

4➔ Click the Zoom Whole Page button on the Layout toolbar to view the entire page.

4➔ Use the blue handles and snapping to give the data frame 0.75-inch margins on the top, bottom, and left sides (remember, each dot of the grid is 0.25 inch apart).

Next, we will set up the basic layout design. (Be sure to always provide ample margins to avoid cropping when the map is printed.) Although it is simple to arrange data frames by clicking, dragging, and resizing, sometimes precise sizes and locations are needed. Set these options using the data frame properties.

5➔ In the Table of Contents, right-click the Hazards data frame and choose Properties.

5➔ Click the Size and Position tab.

5➔ Check that the *X* and *Y* positions are set to 0.75 (inch) where you placed the corner earlier. The page has coordinates of inches, with 0.0 at the lower left.

5➔ In the Size window, uncheck the *As Percentage* box if needed. Set the frame size to 6 inches wide and 7 inches high. Click OK.

TIP: If the map does not redraw completely, use the Refresh button, located next to the Layout and Data view buttons, to redraw it.

5➔ Use the Zoom/Pan tools on the Tools toolbar to center the Hazards map and make it fill the frame.

SKILL TIP: Learn how to set the data frame to use the Fixed Scale or Fixed Extent scaling method (Layouts and Data Frames).

Creating a location map

Many maps, especially if they show a place that is not commonly known, include a location map to inform the viewer where the map is. To add a location map, we need another data frame.

6➔ Choose Insert > Data Frame from the main menu bar.

6➔ Click on the Select Elements tool.

6➔ Place it on the new data frame and click and drag to move the frame to the lower right of the page.

6➔ Use the blue handles and the snapping function to place the new frame 0.25 inch from the Hazards frame and 0.75 inch from the page margins (Fig. 3.21).

6➔ Find the New Data Frame name in the Table of Contents and click it twice slowly to rename it USA.

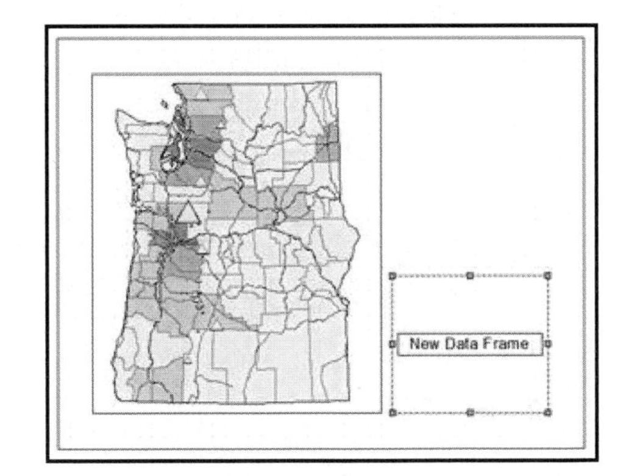

Fig. 3.21. Arrange the frames like this.

6➔ If the name appears in boldface type, USA is already the active frame. If not, right-click the name and choose Activate.

7➔ Add the states feature class from the mgisdata\Usa\usdata geodatabase.

7➔ Use the Zoom In tool from the Tools toolbar to zoom in to the conterminous US states.

7➔ Right-click the states layer symbol and choose a light yellow color.

A location map usually includes a box showing the extent of the main map. We can add this as a property of the USA data frame.

8➔ Right-click the USA data frame name and choose Properties. Click the Extent Indicators tab.

8➔ Click on the Hazards name in the window to highlight it, and click the > button to move it to the right side.

8➔ Check the box to *Use simple extent* (the outline of the data frame).

8➔ Click the Frame button and use the drop-down to choose a 3-pt. border for the extent rectangle. Change its color to a dark orange. Click OK and OK.

We have not yet considered the issue of the map projection for this frame. The Hazards frame uses an equidistant conic projection centered on the two states, minimizing distortion.

9➔ Right-click the USA frame and open its properties.

9➔ Click on the Coordinate System tab and read the information. Write the coordinate system here. North American Equidistant Conic

Although the USA frame is also using an equidistant conic projection, it is centered in the middle of the country and distorts direction at the edges of the map. Notice that north does not point straight up, and the extent rectangle is skewed. This makes the location map confusing. We should choose a projection that preserves direction for the location map.

1. Which projection(s) from the front cover of the book preserves direction? ______________

9➔ On the data frame Coordinate System tab, expand the Projected Coordinate Systems > World folder.

9➔ Choose the Mercator (world) projection and click OK.

9➔ Check the box not to warn you again in this session, and then click Yes to use the coordinate system. A datum difference at this scale is not significant.

10➔ Right-click the USA data frame name and open its properties. Click on the Size and Position tab.

10➔ Set the data frame height to exactly 2 inches and click OK.

10➔ Use the Zoom/Pan tools on the Tools toolbar to center the US map in the data frame.

Using graphic text on layouts

Dynamic labels are useful for quickly adding labels to a layer. You can also create **graphic text** on the layout using the Draw toolbar. Several styles of graphic text are available. You must be in Layout view to place graphic text.

11➔ Click the New Text tool on the Draw toolbar. The black triangle indicates that more text tools are accessible from this icon (Fig. 3.22).

11➔ In the larger Hazards frame, click offshore of the states and type Pacific Ocean. Press Enter.

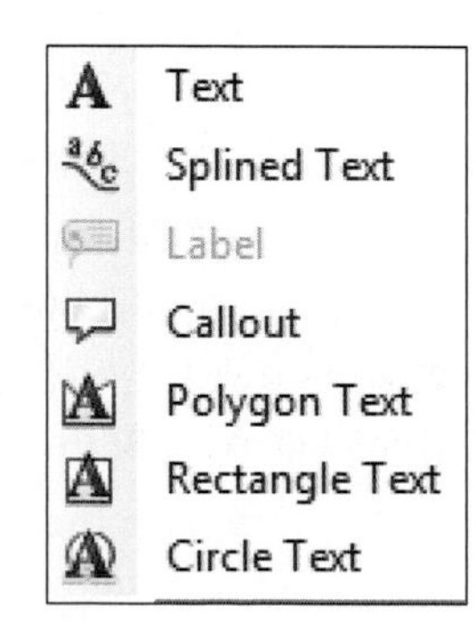

Fig. 3.22. Text tools

11➔ Click once on the text to select it. The blue dashed line around the text indicates that it is selected.

11➔ This horizontal label does not fit well, so click the Delete key to remove it.

TIP: To work with graphics, use the Select Elements tool from the Draw toolbar. To delete text, click the text to select it and press the Delete key. Use the Select Elements tool to draw a box around many text items, or hold down the Ctrl-key to select multiple text boxes.

12➔ Let's try a different text tool. Click on the black arrow next to the New Text tool and select the Splined Text tool.

12➔ Click to enter vertices of a line that curves along the coast of Oregon and Washington from south to north. Double-click to end the line.

12➔ A small text box appears. Type in Pacific Ocean and press Enter. The text appears splined along the line you entered.

12➔ Change the text size to 24 pt., the style to Italic, and the color to blue, using the buttons on the Draw toolbar.

12➔ Select and move the curved label if its position needs adjusting.

The graphic text tools require you to type the text to be displayed. The Label tool, however, can read the label from an attribute field in the feature class table. It uses the primary display field, which is one of the layer properties.

13➔ Right-click the Capitals layer and choose Properties. Click the Display tab and check the Display Expression Field to verify that it is set to NAME. Click OK.

13➔ Try to choose the Label tool from drop-down button for the interactive labels. It is dimmed because the Label tool must be used in Data view.

13➔ Click the Data view button. Activate the Hazards frame, if necessary.

14➔ Choose the Label tool. The Label Tool Options window appears.

14➔ Keep the default option to *Automatically find best placement*.

14➔ Fill the button to *Choose a style*. Scroll down and click on the Capital style from the list.

15➔ Click on the capital of Oregon and watch the "Salem" label appear. If you accidently click the wrong feature, delete the label and try again.

15➔ Click on the capital of Washington (Olympia).

15➔ Use the Select Elements tool to position the labels to be clearly seen.

15➔ Return to Layout view.

TIP: Graphic text and other types of graphics are termed Elements. To Select All Elements or Unselect All Elements, use the functions in the Edit pull-down menu on the main menu bar.

Working with dynamic labels and annotation

Dynamic labels are quick to set up, but annotation offers precise control. We will create two sets of annotation for this map: one inside the triangles to indicate the number of eruptions and one to label the volcano names. Annotation is created from dynamic labels.

To avoid clutter, we want labels only for the volcanoes with more than five eruptions. We must begin by creating a layer containing only these active volcanoes. We do this using a **query**, or an expression that selects certain features based on an attribute.

16➔ Right-click the Volcanoes layer and choose Open Attribute Table.

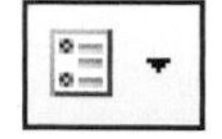

16➔ Click the Table Options button in the table window and choose Select by Attributes.

16➔ Double-click the [KNOWN_ERUP] field in the upper window to enter it in the lower window (Fig. 3.23).

16➔ Click on the ">=" button.

16➔ Enter 5 from the keyboard and click Apply.

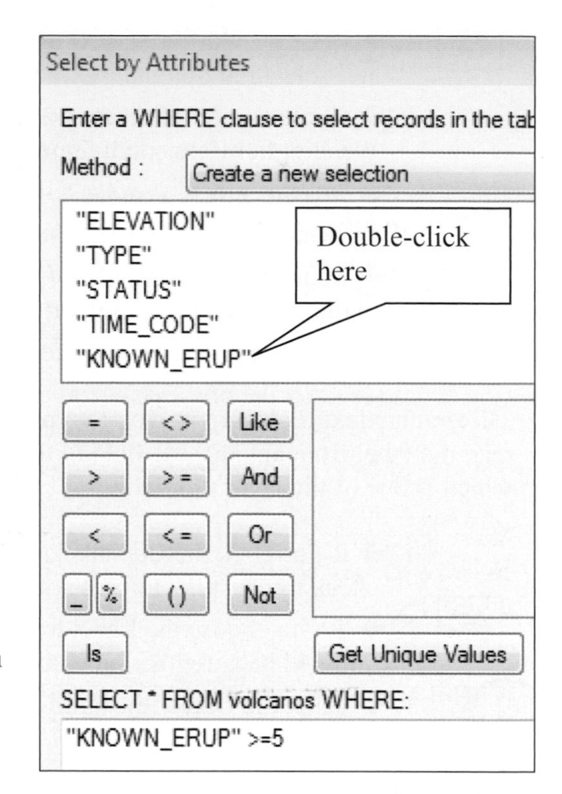

Fig. 3.23. Entering a query

The selected volcanoes are highlighted in the table and on the map. Next, we will create a layer containing only the selected volcanoes.

17➔ Close the Table and query windows.

17➔ Right-click the Volcanoes layer name in the Table of Contents and choose Selection > Create Layer from Selected Features.

17➔ A new layer appears at the top of the Table of Contents, called Volcanoes selection. Rename it Active Volcanoes.

17➔ Right-click the Volcanoes layer and choose Selection > Clear Selected Features.

Next, we'll create dynamic labels for the Active Volcanoes layer.

18➔ Open the Active Volcanoes properties and click the Labels tab.

18➔ Turn on the labels, set the label field to NAME, and make the labels Arial 10 pt. bold font. Give the labels halos, if you wish. Click OK.

SKILL TIP: Use the Label Manager to more easily set and control labels for multiple layers at once (Labels and Annotation).

19➔ Open the Volcanoes properties and click the Labels tab.

19➔ Turn on the labels, set the label field to KNOWN_ERUP, and set the symbol to Arial 8 pt.

19➔ Click the Placement Properties button and fill the button to Place label on top of the point. Click OK and OK.

TIP: Annotation is created for the data frame and includes all the labels currently set for the data frame. You must turn off any layer labels that you don't want converted.

19➔ Examine the map to make sure you set everything correctly. You should have eruption numbers in the triangles and name labels for the active volcanoes. (There are labels for the inactive volcanoes with 0 eruptions, but we will delete these later.)

20➔ Right-click the Hazards data frame name in the Table of Contents and choose Convert Labels to Annotation.

20➔ Fill the button to *Store annotation in the map*.

20➔ Notice the reference scale value listed in the upper-right corner. This is the scale at which the labels will appear at their assigned point size.

20➔ Choose to *Create annotation for features in the current extent*.

20➔ Make sure that only the layers Active Volcanoes and Volcanoes are listed.

20➔ Keep the box checked to *Convert unplaced labels to unplaced annotation*. These labels will be put in an overflow window for individual placement later.

20➔ Click Convert to create the annotation.

View the new annotation. It looks like dynamic labels, but the labels are now graphic text and are individually adjustable.

TIP: If you made a mistake and want to remove the annotation from the map, open the properties for the data frame and click the Annotation Groups tab. Highlight the Volcanoes Anno group and click Remove.

If you had unplaced labels, they were put in an overflow window (Fig. 3.24). Your list may be different than the one in the figure. We can place these labels manually. If you had no unplaced labels, go on to Step 22.

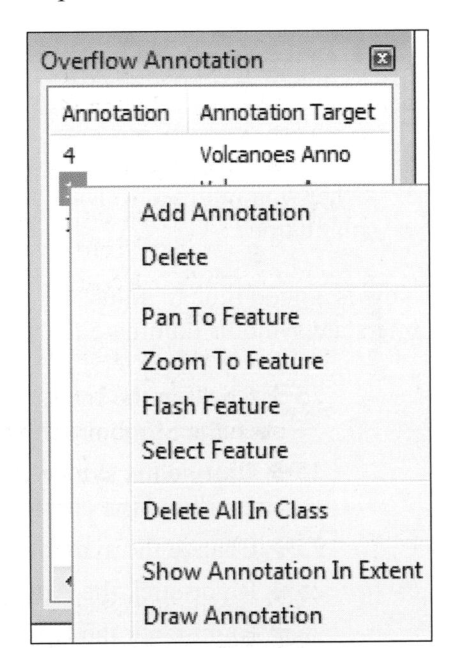

Fig. 3.24. Manually placing overflow annotation

21➔ In the Overflow window, right-click one entry and choose Flash Feature. Repeat, if necessary, until you see it flash. (You may need to move the Overflow window aside.)

21➔ We don't want the eruption labels for the smaller triangles. For each piece of annotation that is a number less than 5, right-click it in the Overflow Annotation window and choose Delete.

21➔ Right-click each remaining entry and choose Add Annotation. The text is added to the map. The placement may not be perfect, but we will adjust it in a moment.

21➔ Close the Overflow Annotation window when it is empty.

The Active Volcanoes layer symbols aren't needed as symbols, but we still need the layer turned on to see the annotation. We will set the symbols so that they don't show.

22➔ Click on the symbol for the Active Volcanoes layer to open the Symbol Selector.

22➔ Set the symbol to Circle, the size to 1, and the color to No Color.

22➔ Change the Active Volcanoes layer name to Volcano Labels.

To adjust the annotation, we need to be in Data View. For these fine adjustments, it will be helpful to turn the snapping grid off.

23➔ Open Customize > ArcMap Options on the main menu bar, click the Layout View tab, and uncheck the Grid box in the *Snap elements to* section. Click OK.

23➔ Click the Data View button. If you get the USA frame, right-click the Hazards frame name and choose Activate.

23➔ Click the Select Elements tool, if necessary. Click on the label for Baker to select it. Adjust the position of the label so that it is clearly visible.

23➔ Click on the 13 inside the Baker symbol to select it. Use the mouse or the arrow keys to move it to the center of the triangle.

24➔ Carefully select and delete any annotation in the smaller triangles with fewer than 5 eruptions.

24➔ Delete the 0 labels for the inactive volcanoes. You can draw a box around them to select them in groups—just be careful not to include any other annotation.

24➔ Adjust the remaining labels for best effect.

2. When you returned to Data view, the labels got larger. What does this tell you about annotation? ____

Now is a good time to adjust the symbols to reduce clutter and improve legibility. Notice how the interstate symbols dominate the map and should be toned down.

25➔ Click on the Interstate symbol of the Highways layer in the Table of Contents to open the Symbol Selector.

25➔ Change the symbol to Double, Plain (near the bottom). Set the color to Spruce Green (hold the cursor over a color to see its name).

25➔ Change the symbol for the Freeways to Double, Plain with Dark Olivinite color.

25➔ Right-click the Arterial symbol and change the color to Spruce Green.

25➔ Right-click the Unclassified symbol and change the color to Burnt Umber.

26➔ Open the Symbology properties for the Counties layer.

26➔ Click on the Symbol heading above the shades and choose Properties for All Symbols. Change the Outline color to Gray 20%. Click OK and OK.

Finally, let's investigate the reference scale.

27➔ Use the Zoom In tool on the Tools toolbar to zoom in to Washington.

Notice that the annotation appears larger because it had a reference scale assigned to it when it was created. However, the symbols do not scale, and the eruption numbers don't fit correctly in the triangles now. We need to set the reference scale for the symbols separately.

27➔ Return to the previous extent showing both states.

27➔ Right-click the Hazards data frame name and choose Reference Scale > Set Reference Scale.

27➔ Zoom in again and watch the symbols scale this time.

27➔ Return to the previous extent.

TIP: To set an exact reference scale, such as 1:24,000, right-click the data frame name, choose Properties, click the General tab, and type a specific reference scale in the appropriate box.

Adding a legend to the map

Now that the map is in good shape, it is time to continue creating the layout.

28➔ Click the button to return to Layout view. Use the Zoom/Pan tools on the Tools toolbar to adjust the Hazards map inside the data frame, if necessary.

28➔ Choose the Select Elements tool. Right-click the Hazards frame name and choose Activate. The legend is always based on layers in the active frame.

28➔ Choose Insert > Legend from the main menu bar. A Legend Wizard will appear.

29➔ Examine the list of layers and make sure you want them all. Volcano Labels has only labels, so select it on the right and click the < button to remove it from the list.

29➔ Now establish the order of layers. To move a layer, click on it to select it and click the Up or Down arrow until it is in the right location. Put Volcanoes first, Counties next, then Highways, then Capitals, then Rivers, then States. Click Next.

30➔ Leave the Legend Title as is, except click the button to Center it. Click Next.

30➔ Choose the 1.0-pt. border for the Legend. Make sure that the Gap is set to 10 pt. Click Next.

30➔ To choose a different style patch, click the Rivers layer and choose the Flowing Water line. Click Next.

30➔ This section gives very detailed control of the spacing between different elements of the legend. It is fine to leave the defaults on this step. Click Finish.

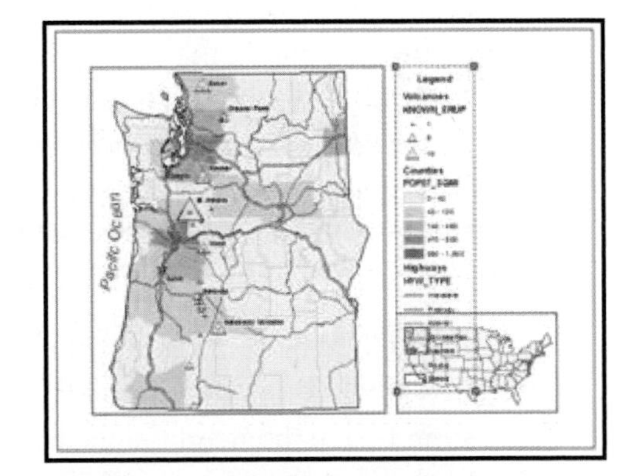

Fig. 3.25. Adding the legend

30➔ The legend appears in the middle of the map. Click and drag it next to the map frame and even with its top (Fig. 3.25).

TIP: Turn the snap grid back on to ensure that the box outlines align.

If you made a mistake creating the legend, don't worry. You can change its properties. Note that the legend is longer than the space available. We can make changes to help it fit better. The Volcanoes layer has two headings, Volcanoes and KNOWN_ERUP. We only need one.

31➔ Right-click the Legend and choose Properties. Click the Items tab.

31➔ Click on Volcanoes and click the Style button underneath.

31➔ Choose the item style shown in Figure 3.26, called *Horizontal Single Symbol Layer Name and Label.*

31➔ Click OK and OK and examine the change in the legend.

31➔ Rename the Volcanoes layer **Known Eruptions** in the Table of Contents. Notice that the name changes in the legend as well.

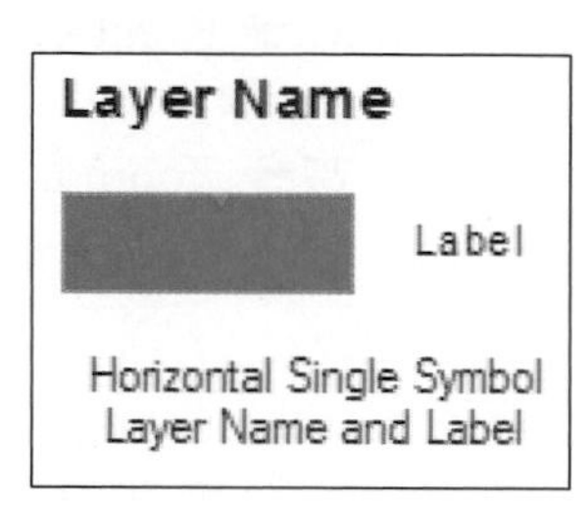

Fig. 3.26. Choose this item style.

32➔ Open the Legend properties again and click the Items tab.

32➔Click on the Highways entry and change its item style to the same one used for the Volcanoes. Click OK and OK.

SKILL TIP: Learn more about the different legend styles and how to manage them (Layouts and Data Frames).

33➔ Click twice slowly on the POP10_SQMI heading under the Counties layer in the Table of Contents and change it to **People per sq mile.**

33➔ Click twice on the Capitals layer name in the Table of Contents and change the name to **State Capitals**.

We have shortened the legend, but it still does not fit. Let's use two columns instead of one.

34➔ Double-click on the Legend to open its Properties.

34➔ Click the Items tab, if necessary. Move the Properties box to see the Legend.

34➔ Select the Highways and check the box to *Place item(s) in new column*. Click Apply.

That looks like it will work, but we want to align the edges of the legend with the location map.

35➔ Click the Size and Position tab in the Legend Properties window.

35➔ Leave the height alone but set the width to 3.275 inches. Click OK.

35➔ Move the legend into its final place, 0.25 inch above the location map and right of the Hazards frame (Fig. 3.27). Use the snap grid to align the edges.

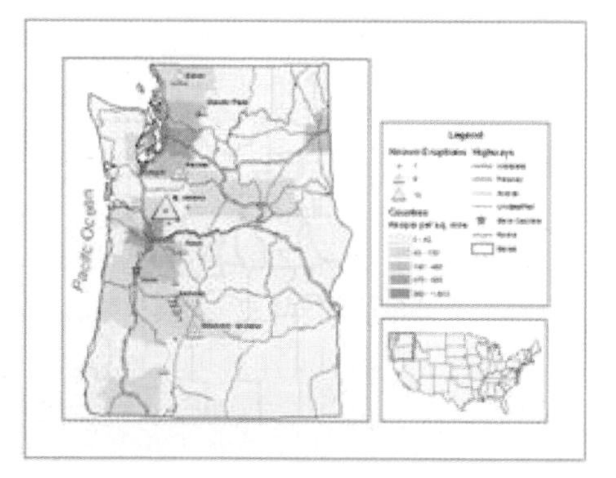

Fig. 3.27. Final position of the legend

Placing a scale bar on the map

The scale bar is placed in the active frame, so make sure that the correct frame is active before creating the scale bar. Our scale bar will refer to the Hazards data frame.

3. To review the scale bar terms, write the correct terms in the boxes of Figure 3.28.

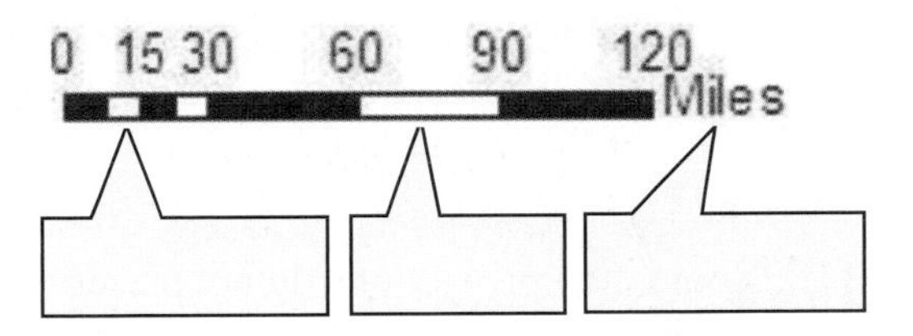

Fig. 3.28. Terminology of a scale bar

36➔ Make sure the Hazards frame is active.

36➔ Choose Insert > Scale Bar from the main menu.

36➔ Choose the Alternating Scale Bar 1 style bar and click OK.

The default scale bar probably needs editing. It looks much better, for example, to use miles or kilometers for a map of this scale, rather than the default meters.

37➔ Move the scale bar down to the lower right corner of the Hazards frame, on top of Oregon.

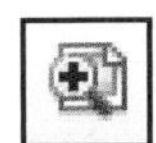

37➔ Use the Zoom In tool on the Layout toolbar to zoom in to the scale bar to examine it more closely.

The bar may be too long or too short, and it may have uneven divisions, such as 110 miles. If it appears unsuitable, try resizing the scale bar to improve it.

38➔ Click the Select Elements tool.

38➔ Click the right boundary of the scale bar (a double horizontal arrow will appear), and drag it to the left to decrease the length of the scale bar.

38➔ Repeat until the scale bar is about 120 miles long.

Unless you were lucky, you probably ended up with some fairly uneven labels instead of nice, round numbers. We can modify the scale bar to get exactly the properties we want.

39➔ Double-click the scale bar to open its properties, and click the Scale and Units tab.

Notice that the division value is dimmed out. In order to set it explicitly, we must change the way ArcMap adjusts the bar when resizing.

39➔ Under When Resizing, set the drop-down box to *Adjust number of divisions*.

39➔ Set the *Division value* to 50 miles.

39➔ Set the *Number of subdivisions* to 2. Click OK.

39➔ Examine the new scale bar. Resize it several times. Make it 150 miles long.

The size changes in even increments of division units, 50 miles. This method is a better way to handle your scale bars to prevent uneven division units.

Adding other map elements

Many other elements can be added to the map using the Insert menu. Keep in mind that the elements will refer to the data frame that is currently active.

Adding a north arrow

40➔ Click the Zoom Whole Page button on the Layout toolbar.

40➔ Choose Insert > North Arrow from the main menu bar.

40➔ Click the symbol you like best.

40➔ Click the Properties button and set the color to a dark blue. Click OK and OK.

40➔ Click the Select Elements tool and move the north arrow to the lower-right corner of the Hazards frame.

TIP: Some map projections do not preserve direction, and the north arrow should point at the appropriate angle instead of vertically. ArcMap does not tilt the arrow automatically. Use the Properties tab to set the north arrow to the correct angle.

The United States location map might look good with a graticules grid on it.

41➔ Right-click the USA data frame name in the Table of Contents and choose Properties. Click on the Grids tab.

41➔ Click New Grid. Choose the Graticule grid. Click Next.

41➔ Choose Graticule and labels. Accept the default intervals. Click Next.

41➔ Accept the defaults on the axes and labels. Click Next.

41➔ Accept the defaults on the graticules. Click Finish and OK.

The grid appears, but notice that the labels extend into the other map frame. At this point, a design decision is in order. You can resize the USA data frame so that the labels fit into the space available, or you can remove the grid. Since this frame merely contains a location map, having a scale or north direction or grid is not important. Make your own decision to keep the grid or not.

42➔ If keeping the grid, resize the USA data frame so that the grid labels fit comfortably in the space provided.

42➔ If removing the grid, open the Properties for the USA data frame, select the Graticule grid, and uncheck the Graticule box. Click OK.

Adding titles and text

43➔ Choose Insert > Title from the main menu bar. A centered text field appears with a default title.

43➔ Double-click the text to open its properties.

43➔ In the text box, type Pacific Northwest, press Enter, and type Volcanic Hazards on the next line. Click OK.

43➔ Use the Drawing toolbar to set the font size to 18 pt. and make it bold. Center it above the legend, aligning it with the top of the Hazards data frame.

44➔ Choose the Insert > Text choice in the main menu bar. Find the tiny new text box on the map, waiting for you to type the text.

44➔ Type by Your Name and press Enter.

44➔ Move the text to a centered position below the Title.

TIP: Did you notice the Dynamic text option under the Insert > Text menu choice? You can add times, dates, and other types of updatable text to the map.

45➔ To add wrapped text, click the label drop-down on the Draw toolbar to find and select the Rectangle Text tool.

45➔ Click and drag to create a box inside the blank area in the legend.

45➔ Double-click inside the new text box to open its properties.

45➔ Click the Text tab and type the following citation into the box without using any Enter keys:
ESRI Data and Maps [Download]. (2010) Redlands, CA: ESRI [October, 2012].

45➔ Make the text left-justified.

45➔ Click the Frame tab and set the border to <None>. Click OK.

45➔ Set the font size to 8 pt. Adjust the text box to fit neatly in the legend space. The legend should look like Figure 3.29.

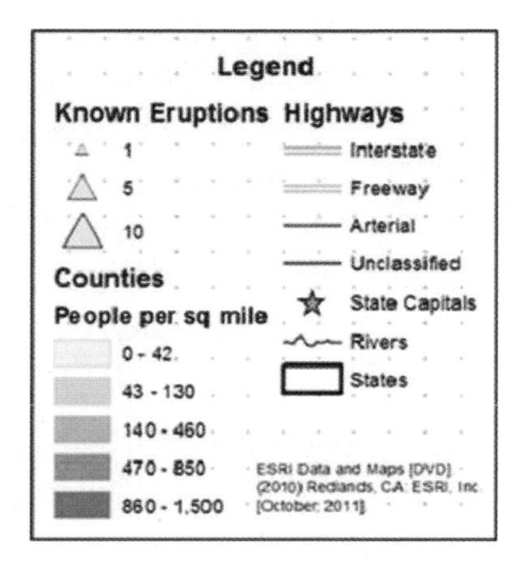

Fig. 3.29. The final legend

Adding neatlines and shades

46➔ Choose Insert > Neatline from the main menu bar.

46➔ Fill the button to *Place around all elements.*

46➔ Set the gap to 10 pt. (the gap controls the distance between the elements and the neatline).

46➔ Choose the Triple Graded border from the drop-down box.

46➔ Choose the Linear Gradient fill from the Background drop-down box and keep the default yellow color. Click OK.

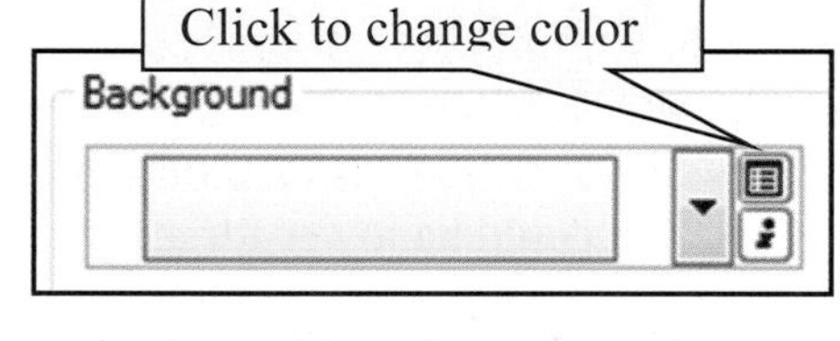

Fig. 3.30. Changing the gradient color

TIP: To change the gradient color, click the small icon to the upper right of the background symbol box (Fig. 3.30). Click Properties and then click Change Symbol. Click Edit Symbol and choose the color ramp. Click OK in all boxes.

SKILL TIP: Learn how to add pictures or graphics to a layout (Layouts and Data Frames).

Reviewing and printing the map

Now for the final review and printing.

47➔ Examine the map one last time for balance and aesthetics, and fix any problems you might see (Fig. 3.31).

47➔ It is always wise to SAVE the map document before you print.

47➔ Choose File > Print from the main menu bar.

47➔ Click Setup to change the printer or adjust the print settings.

47➔ When ready, click OK.

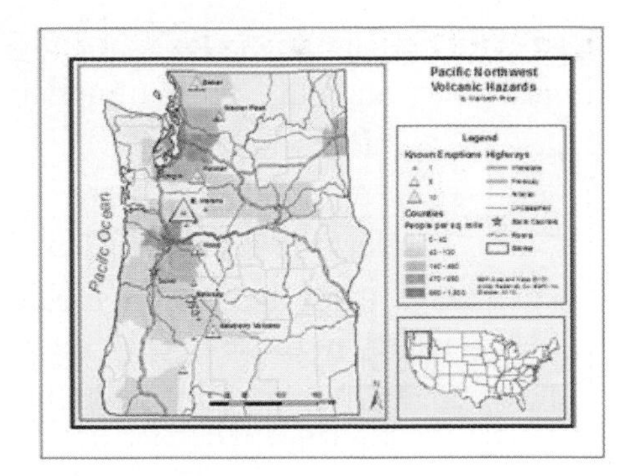

Fig. 3.31. The final layout

TIP: Colors shown on the computer screen often look different than colors printed on paper. Plan to print a draft of the map to test the color balance. Colors can also vary from printer to printer.

Maps can be saved as image files. The PDF format is a popular way to distribute maps to others.

48➔ Choose File > Export Map.

48➔ Choose the location and name for the image file.

48➔ Set the Save As type to PDF.

48➔ Click on the Options drop-down flag, if necessary, to show the Options settings.

48➔ Check the resolution. The minimum resolution for printing should be 300 dpi. It could be lower for a web or screen image.

48➔ If planning to place the map in another document, check the box to Clip Output to Graphics Extent and save as a JPEG or GIF. Click Save.

SKILL TIP: Learn to use a map template to produce a quick layout of a map, or learn to create a simple graph (Layouts and Data Frames).

This is the end of the tutorial.

➔ Exit ArcMap. You do not need to save your changes.

Exercises

Create an attractive map layout showing median age in the counties of the conterminous United States. Pay attention to details of layout, colors, balance, and so on. Export the map as a PDF file.

Challenge Problem

Modify the median age map to include additional data frames showing Alaska and Hawaii individually at larger scales. Each frame should be given its own appropriate coordinate system and scale. Label the coordinate systems used in each data frame on the map.

Consider: Ideally, each data frame should use the same classification breaks in the legend so that you need only put the legend on the map once. How do you make sure this happens?

Chapter 4. Attribute Data

Objectives

- Understanding how tabular data are stored and used
- Using queries to select records of interest
- Understanding joins and cardinality concepts
- Summarizing tables to get statistics on groups
- Defining appropriate field properties
- Editing and calculating fields in tables
- Generating point layers from *x-y* coordinates in tables

Mastering the Concepts

GIS Concepts

Overview of tables

A **table** is a data structure for storing multiple attributes about a location or an object. It is composed of rows, called **records**, and columns, called **fields** or **attribute fields**. Figure 4.1 shows an example of a table containing attributes of counties.

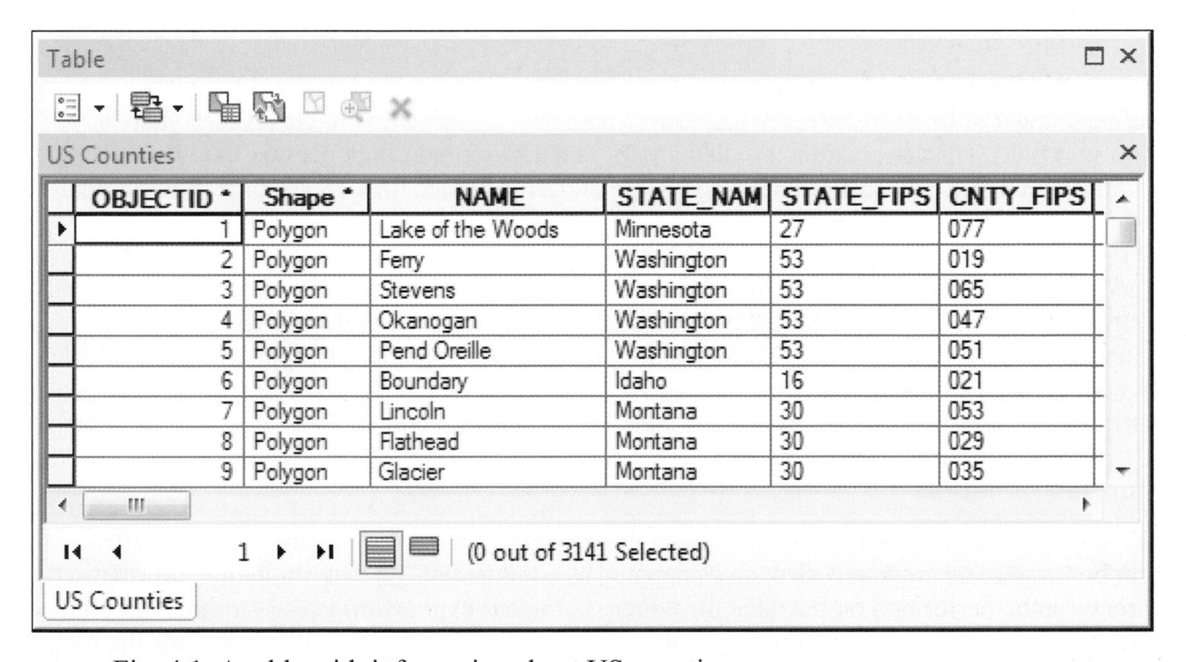

OBJECTID *	Shape *	NAME	STATE_NAM	STATE_FIPS	CNTY_FIPS
1	Polygon	Lake of the Woods	Minnesota	27	077
2	Polygon	Ferry	Washington	53	019
3	Polygon	Stevens	Washington	53	065
4	Polygon	Okanogan	Washington	53	047
5	Polygon	Pend Oreille	Washington	53	051
6	Polygon	Boundary	Idaho	16	021
7	Polygon	Lincoln	Montana	30	053
8	Polygon	Flathead	Montana	30	029
9	Polygon	Glacier	Montana	30	035

Fig. 4.1. A table with information about US counties

In ArcGIS, tabular data files fall into two main categories: **attribute tables** and **standalone tables**. An attribute table, such as the one shown in Figure 4.1, contains information about features in a geographic data set. In an attribute table, there is always one and only one row of information for each feature. In a shapefile, the row is linked to the spatial feature in a separate file using a unique ID number called a Feature ID, or FID. In a geodatabase, the file stores both the attributes and the *x-y* coordinates in the same data file, although the coordinates are not visible in the table, and it uses an Object ID, or OID, instead of an FID (Fig. 4.1). In contrast, a standalone table simply contains information about one or more objects in tabular format instead of having information about map features. A standalone table might come from a text file, an Excel® spreadsheet, a global positioning system data file, or a database. Standalone tables exist independently of a geographic data set and may be only incidentally related to map features. They also have an OID rather than an FID.

Database management systems

GIS tables share a history and many current properties of the large database programs that are routinely used in many commercial, governmental, and academic settings. Nearly every GIS system uses an underlying database management system (DBMS) to store its data. These programs are designed to store, manipulate, analyze, and protect tabular data of all kinds. Governments use them to store information about citizens, parcels, taxes, and more. Companies use them to store information about customers. Universities use them to manage information about students, classes, and faculty. Three types of databases have traditionally been used.

A **flat file database** stores rows of information in a text or binary file. Finding information requires parsing the table and selecting the records of interest. They are simple but not efficient.

A **hierarchical database** has multiple files, each of which contains different records and fields. Parent tables can be linked to child tables through a specified field called a key. For example, a table of college classes might be linked to a table of students in each of the classes through a course ID number. Relationships between tables are fixed, which makes looking up information quick. However, the relationships are inflexible, designed to permit only a small set of operations.

A **relational database** also has multiple tables stored as files. However, the relationships are not defined ahead of time. Instead, the user can temporarily associate two tables if they share a common field. This association is called a **join**, and the common field becomes the key. This database model is extremely flexible and is the preferred choice for GIS systems. It is called a relational DBMS, or RDBMS.

Because GIS data are intimately linked to an underlying database structure, it is not a far stretch to incorporate tables from database programs into GIS analysis. ArcGIS has capabilities for connecting to and working with DBMS data files directly, expanding the type of data available. If a county keeps its parcel tax records in a DBMS such as Oracle or SQL Server, for example, a GIS analyst can bring the database tables into ArcMap as standalone tables.

Queries on tables

Often one wants information about a subset of records in a table, for example, knowing the number of parcels in the city that are designated as commercial. To determine this information, a **query** can be performed on the table. In a query, a **logical expression** is used to specify certain criteria (e.g., zoning = commercial), and then the software searches the table and finds the records (parcels) that match the criteria. Those records are returned as a **selected set**. Selected records can become the input to another action, such as printing them, exporting them to a new file, or executing a GIS function on them.

Most databases use a special language called Structured Query Language (SQL) to write and execute queries. A land use table might contain a field with the zoning code. If the field containing the zoning is named ZONE, and the code for commercial property is 492, the logical expression in SQL might look like this:

```
SELECT * FROM landuse WHERE [ZONE] = 492
```

In this example, landuse is the name of the table, [ZONE] is the field name, and [ZONE] = 492 is the logical expression stating the criteria to be met. SQL queries can have multiple lines and include many criteria.

```
SELECT * FROM landuse WHERE
[ZONE] = 492 AND [VALUE] > 300000
```

This query would find all commercial parcels with a value greater than $300,000. This chapter presents methods for performing queries on tables. Chapter 5 examines queries in more detail.

Joining and relating tables

In an RDBMS and in a GIS, tables are commonly combined using a **join**, in order to bring different sets of information together. The tables are combined using a common field called a **key**. The key field must be of the same data type in both tables. When a join is performed, the two separate tables become one and contain the information from both tables (Fig. 4.2). The join is a temporary relationship and may be removed when it is no longer needed.

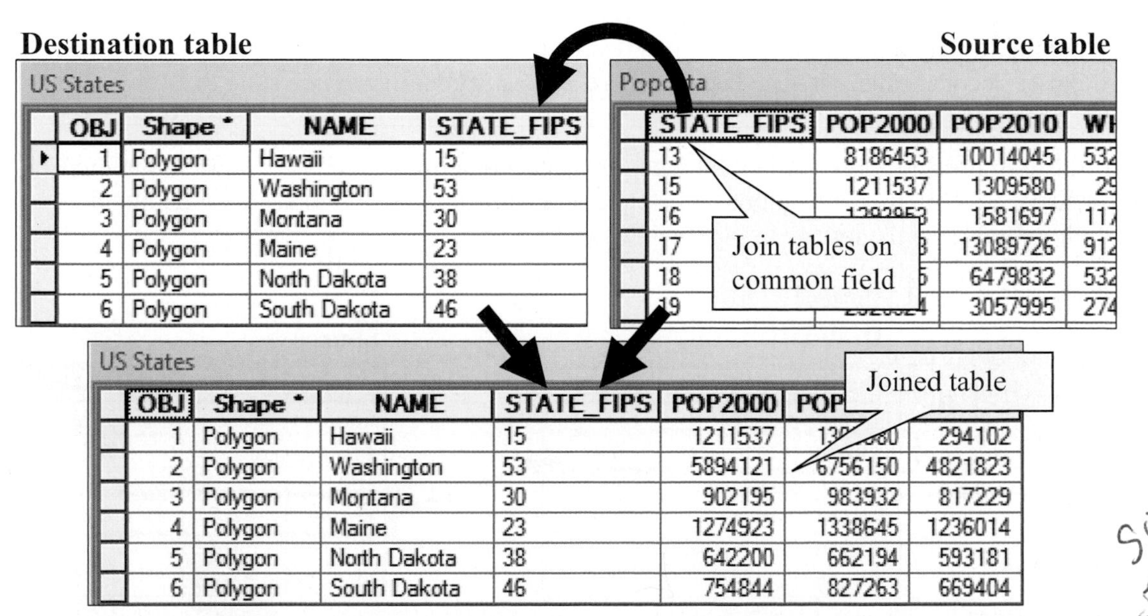

Fig. 4.2. Joining on a common field. These two tables are joined into one using the common field STATE_FIPS. The data from the source table are placed into the destination table.

Joins have a direction. The table containing the information to be appended is called the **source table**. The table that receives the appended information is called the **destination table**. In Figure 4.2, the destination table, US States, is a feature attribute table. The source table containing demographic data is a standalone table. When the two tables are joined, the demographic data are appended to the US States attribute table. Often joins are used to bring data from standalone tables into an attribute table for mapping or analysis.

Join direction matters. Once the demographics are joined to the table of the states shapefile, then the information could be used to make a map of the population data. If, however, the join had been performed in the opposite direction, with the demographics as the destination and the states as the source, then the resulting table would be a standalone table. In this case, a map showing the demographics could not be made because the demographics table is not a feature class.

Cardinality

When joining, the **cardinality** of the relationship between the tables must be considered. Cardinality is the numeric relationship between the objects in one table and their matches in the other. The simplest kind of relationship is *one-to-one*, in which each record in the destination table matches one record in the source table. In Figure 4.2, each state has one corresponding record of demographic data. In a *one-to-many* relationship, each record in the destination table could match more than one record in the source table. For example, a store location could have many employees. In a *many-to-one* relationship, many records in the destination table would match a single record in the source table, such as many cities falling within one state. Finally, a *many-to-many* relationship indicates that multiple records can appear in both tables. For example, a student may take more than one class, and most classes have more than one student.

The direction of the join must be taken into account when ascertaining the cardinality of a relationship. The destination table is the point of reference and comes first; that is, the relationship cardinality is reported as {destination} to {source}. Imagine two tables containing *states* and *counties.* If one performs a join with *states* as the destination table and *counties* as the source, the cardinality is one-to-many because each state contains many counties (and a county may belong to only one state). If the join is reversed and *counties* is the destination table, then the cardinality becomes many-to-one because there are many counties in one state.

TIP: Putting the destination table first when stating cardinality is only a convention but one adopted by ESRI in its publications and Help documents. It may help to always imagine the destination table on the left, as in Figure 4.3.

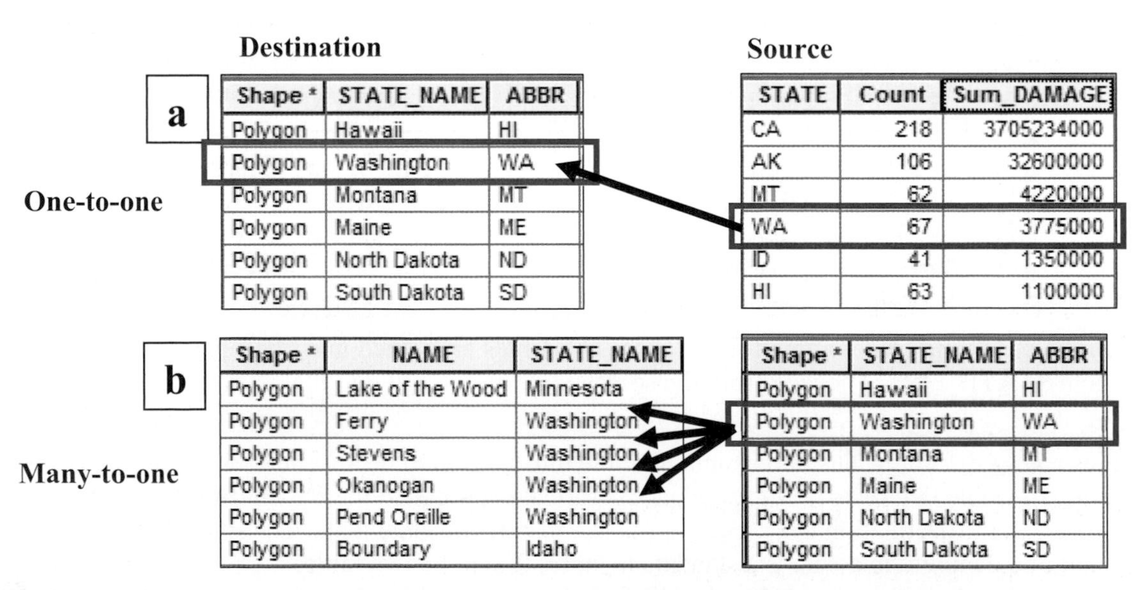

Shape *	STATE_NAME	ABBR
Polygon	Hawaii	HI
Polygon	Washington	WA
Polygon	Montana	MT
Polygon	Maine	ME
Polygon	North Dakota	ND
Polygon	South Dakota	SD

STATE	Count	Sum_DAMAGE
CA	218	3705234000
AK	106	32600000
MT	62	4220000
WA	67	3775000
ID	41	1350000
HI	63	1100000

Shape *	NAME	STATE_NAME
Polygon	Lake of the Wood	Minnesota
Polygon	Ferry	Washington
Polygon	Stevens	Washington
Polygon	Okanogan	Washington
Polygon	Pend Oreille	Washington
Polygon	Boundary	Idaho

Shape *	STATE_NAME	ABBR
Polygon	Hawaii	HI
Polygon	Washington	WA
Polygon	Montana	MT
Polygon	Maine	ME
Polygon	North Dakota	ND
Polygon	South Dakota	SD

Fig. 4.3. A cardinality of one-to-one or many-to-one permits tables to be combined without violating the Rule of Joining. (a) One-to-one cardinality. (b) Many-to-one cardinality.

The cardinality of a relationship dictates whether the tables can be joined. **The Rule of Joining stipulates that there must be one and only one record in the source table for each record in the destination table.** Consider the four tables shown in Figure 4.3.

One-to-one cardinality. In Figure 4.3a, a table containing the number of earthquakes and total damage in each state (source) is being joined to a states attribute table (destination), using the common field provided by the state abbreviation. Each state occurs once in each table so that there is no ambiguity about matching the source records to the destination records.

Many-to-one cardinality. In Figure 4.3b, a table containing state information (source) is being joined to a table of counties (destination) using the common field provided by the state name. Although there are many counties in each state, there is only one record for each state in the source table and no ambiguity in linking the source records to the destination records. The state record does get used more than once, but it still does not violate the Rule of Joining.

One-to-many cardinality. Figure 4.4 shows the reverse of the join in Figure 4.3b. Now states are the destination table and counties are the source table. Many county records in the source table match each state record in the destination table. The Rule of Joining is violated, and it becomes ambiguous which county record should be matched to the state. One cannot perform a join if a one-to-many relationship is present. Instead, we perform a different operation called a **relate**.

Destination

states

Shape *	STATE_NAME	STATE_FIPS
Polygon	Hawaii	15
Polygon	Washington	53
Polygon	Montana	30
Polygon	Maine	23
Polygon	North Dakota	38
Polygon	South Dakota	46
Polygon	Wyoming	56

?

Source

counties

Shape *	NAME	STATE_NAME	FIPS
Polygon	Lake of the Wood	Minnesota	27077
Polygon	Ferry	Washington	53019
Polygon	Stevens	Washington	53065
Polygon	Okanogan	Washington	53047
Polygon	Pend Oreille	Washington	53051
Polygon	Boundary	Idaho	16021
Polygon	Lincoln	Montana	30053

Fig. 4.4. A one-to-many relationship violates the Rule of Joining because more than one record in the source table matches a record in the destination table.

In a relate, the two tables are still associated by a common field, but the records are not joined together. The two tables remain separate. However, if one or more records are selected in one table, then the associated records can be selected in the other table. For example, selecting the state of Washington in the states table allows the selection of all the counties in Washington in the related table, as shown by the red boxes in Figure 4.4. Many relationships in the world have cardinalities of one-to-many, such as a school to its classes or a well to its yearly water quality tests, and relates provide valuable support in dealing with those features and their attributes.

Many-to-many cardinality. Many-to-many relationships exist but are difficult to deal with in a relational database. Consider a university. Each class has many students in it. However, most students are taking more than one class (Fig. 4.5). In practice, one would have to use multiple relates to represent the relationships

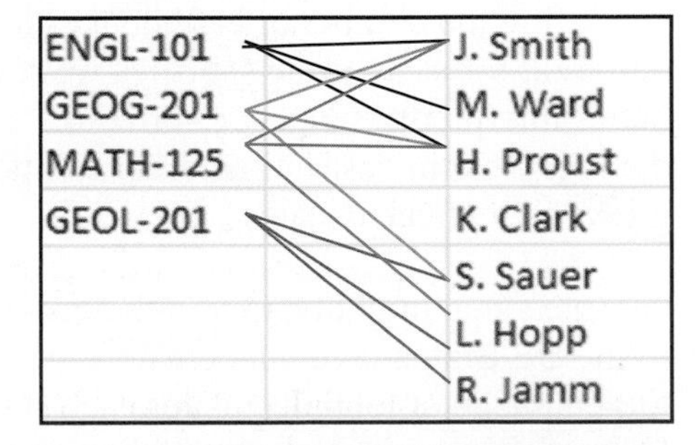

Fig. 4.5. A many-to-many relationship

and work with them one at a time. A one-to-many relate could be set up from the classes table to the students to allow all the students in each class to be found. Another one-to-many relate would allow the list of classes for each student to be generated.

To recap, a join may be used to combine tables whenever a one-to-one or many-to-one relationship exists between the destination table and the source table. If the relationship is one-to-many or many-to-many, then a relate must be used.

Summarizing tables

The Summarize function is a powerful form of statistics applied to tables. Summarize first combines the records into groups based on a categorical attribute field and then calculates statistics separately for each group. For example, one might group a table of earthquakes according to which state the earthquake is in, and generate earthquake statistics for each state separately.

quakehis

STATE	DEATHS	DAMAG	MAG	LOCATION
MO	7	0	7.88	New Madrid, Missouri
SN	51	0	7.36	Northern Sonora, Mexico
AK	0	0	8.15	Yakutat Bay, Alaska
AK	0	0	8.26	Southeast Alaska
AR	7	0	7.68	Northeast Arkansas
CA	3000	52400000	7.8	Near San Francisco, Calif
CA	12	6000000	7.48	South of Bakersfield, Calif

Fig. 4.6. Earthquake table

Figure 4.6 contains a table showing major historical earthquakes in the United States. Attributes for each quake include the state in which it occurred, the number of deaths it caused, the total damage caused, and the Richter scale intensity (MAG). We would like to know which states have suffered the most from historical earthquakes. In particular, we want to know the total deaths, total damage, and average Richter magnitude for each state. We summarize based on the state field, dividing the earthquakes into groups by state, and specify which statistics to calculate. Separate statistics are calculated for each group (state).

quakesum

STATE	Count_STATE	Sum_DEATH	Sum_DAMAG
CA	218	3777	3705234000
AK	106	125	32600000
NV	98	0	0
WA	67	15	3775000
HI	63	79	1100000
MT	62	32	4220000
UT	59	0	100000

Fig. 4.7. Summarize produces a file that contains the summary field (STATE), the number of earthquakes in each state, and the requested statistics.

Because many statistics can be returned from a Summarize, this command produces a new table. The quakesum.dbf output table, shown in Figure 4.7, has one record for each state. It also has a Count field indicating the number of earthquakes in each state and one field for each of the statistics we requested. To better interpret the data, we sorted the table by the damage field in descending order to find the most grievously affected states.

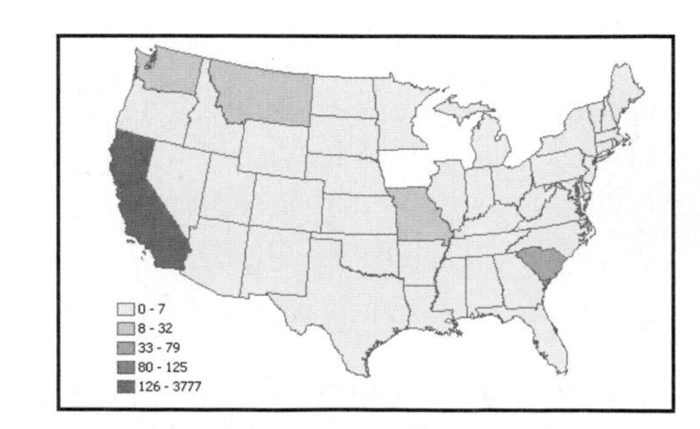

Fig. 4.8. Earthquake deaths by state

Our next step might be to join the quakesum.dbf table to the US States layer and create a map showing the total damage or total deaths for each state (Fig. 4.8). Using Summarize, we collapsed a one-to-many relationship between states and earthquakes into a one-to-one relationship between states and earthquake statistics, allowing the data to be mapped.

Field types

Creating database tables always begins with an analysis of the fields that the table will contain. Unlike a spreadsheet, in which any cell can contain any type of data, a database field (column) must contain only one type of data, perhaps text, perhaps integers, but never both. Each field must be defined, and the type of its contents must be specified, before any data are entered. Furthermore, once a **field definition** is set, it cannot be changed. The most common field types are numbers, strings, dates, and Boolean fields. Each field may also have parameters that further specify what it may contain, such as the maximum number of characters allowed in a text field.

A field [STREETNAME] would be a text field because street names contain letters. The **field length** defines how many characters can be stored in the name. If the [STREETNAME] field is given a length of 10, then any name longer than 10 characters will be truncated to the first 10 letters. Thus, "Elm Street" fits perfectly (including the space), but "Maple Street" would be truncated to "Maple Stre" to fit in the field. It is important to ensure enough space in fields to hold the longest likely value.

Numeric fields have different parameters. When defining a numeric field, the user may designate a storage width (**precision**) and the number of decimal places (**scale**). For example, a field with a precision of 5 and a scale of 0 could store a number between –9999 and 99999 (the minus sign takes one space). A field with a precision of 5 and a scale of 2 could store a number between –9.99 and 99.99 (the decimal place also takes one space). When designing a database, try to use the smallest field widths that will store every possible value because extra spaces increase the size of the database.

Numbers can be stored in a variety of ways. Most databases offer several formats for storing numeric values, and these formats can differ from database to database. The next section describes the common options in general terms.

ASCII versus binary

A **byte** is the basic unit of storage space for a computer—it is composed of a string of eight digits (bits), which may be zeros or ones. These zeros and ones represent a number in base 2; such numbers are called binary numbers. A single byte can store a binary value from 0 (00000000) to 255 (11111111). Two bytes can store values up to $2^{16} - 1$, or 65,535. The more bytes allotted to a value, the larger the number that can be stored.

All text is stored as sequences of characters using a special code called **ASCII** in which every number, every letter, and every symbol (such as $) is assigned a single-byte code between 0 and 255. To store the word *cat*, the computer stores the code for *c*, the code for *a*, and the code for *t*. Thus, it takes three bytes to store *cat*. The word *horse* requires five bytes. Numbers may also be stored using ASCII by storing the ASCII one-byte code for each numeral. Thus, it requires three bytes of data to store *147* and five bytes to store *147.6* . This scheme is a simple and standard way to store information. Text files and HTML files, among others, are stored in ASCII.

Unlike text, numbers have another storage option called **binary**. In this case, a number is stored in base 2 directly rather than being assigned one byte per character. The number *16* would be stored in binary as *00010000* in a single byte of information. Binary is a more efficient way to store numbers than ASCII. The number *14456* would require five bytes in ASCII but only two bytes in binary. It is also faster to compute with binary values because the computer by design does all its calculations in base 2. If the number is already stored in base 2, the computer does not need to convert it before calculating. Thus, it is often advantageous to use binary storage. Many types of files use a binary encoding scheme, including spreadsheets, word processing documents, and shapefiles. Raster and image data are also stored as binary data.

Precision

Very large numbers, such as 1,000,000,000,000, require many bytes to store, as do very small numbers, such as 0.0000000000001. People often use scientific notation when dealing with very large or very small numbers; these two would be written as 1.0×10^{12} and 1.0×10^{-13}. Computers can also use scientific notation, usually called exponential or floating-point data. This notation stores values composed of a **mantissa** (the decimal part of the number) and an exponent. For example, the number 123456789 stored as an exponential number would consist of a mantissa of 1.23456789 and an exponent of 8, yielding 1.23456789×10^8 (often written as 1.23456780e08). The computer truncates the mantissa at a certain level of precision—when storing values in the trillions, differences of tenths or hundredths are often of little interest. The number might become 1.2345e08. In ArcGIS, a **single-precision** floating-point field stores eight significant digits of information in the mantissa. A **double-precision** field stores 16 significant digits. Floating-point data types are much more flexible than numeric or binary types because they can store either very large or very small values using the same field.

Different types for different data formats

Every attribute must be defined before use; that is, the field type must be specified (text or numeric) and the field properties set. Once a field is defined, the definition cannot be changed. If a mistake is made defining the field, one must delete the field and redefine it. Different GIS systems may have different allowable data types, although most use the basic categories of text, integers, and floating-point values. Table 4.1 shows the specific data types used in ArcGIS. In Version 10, geodatabases also have a Raster and GUID type for advanced users.

Table 4.1. Field data types available for feature classes (*geodatabases only)

Field type	Explanation	Examples
Short	Integers stored as 2-byte binary numbers *Range of values −32,000 to +32,000*	255 12001
Long	Integers stored as 10-byte binary numbers *Range of values −2.14 billion to +2.14 billion*	156000 457890
Float	Floating-point values with eight significant digits in the mantissa	1.289385e12 1.5647894e–02
Double	Double-precision floating-point values with 16 significant digits in the mantissa	1.12114118119141e13
Text	Alphanumeric strings	'Maple St' 'John H. Smith'
Date	Date/time format for calendar dates and times	07/12/92 10/17/63 13:24:06
BLOB*	Binary large object; any complex binary data, including images, documents, etc.	

About ArcGIS

Tables in ArcGIS

Tables in ArcGIS may come from any of the underlying RDBMS programs supported by ArcGIS. The tables may be in different formats for storage, but the table itself always looks the same and has the same functions so that users don't need to learn different commands for working with different file types.

Tables are viewed and manipulated in the Table window in ArcMap (Fig. 4.9). This window contains a Table Options menu and several tools at the top. More commands are accessible from a context menu that appears if a field is right-clicked. If more than one table is opened, they all appear in the Table window. Tabs at the bottom of the window are used to switch between tables. Tables may also be viewed side by side or vertically using a menu under the Table Options.

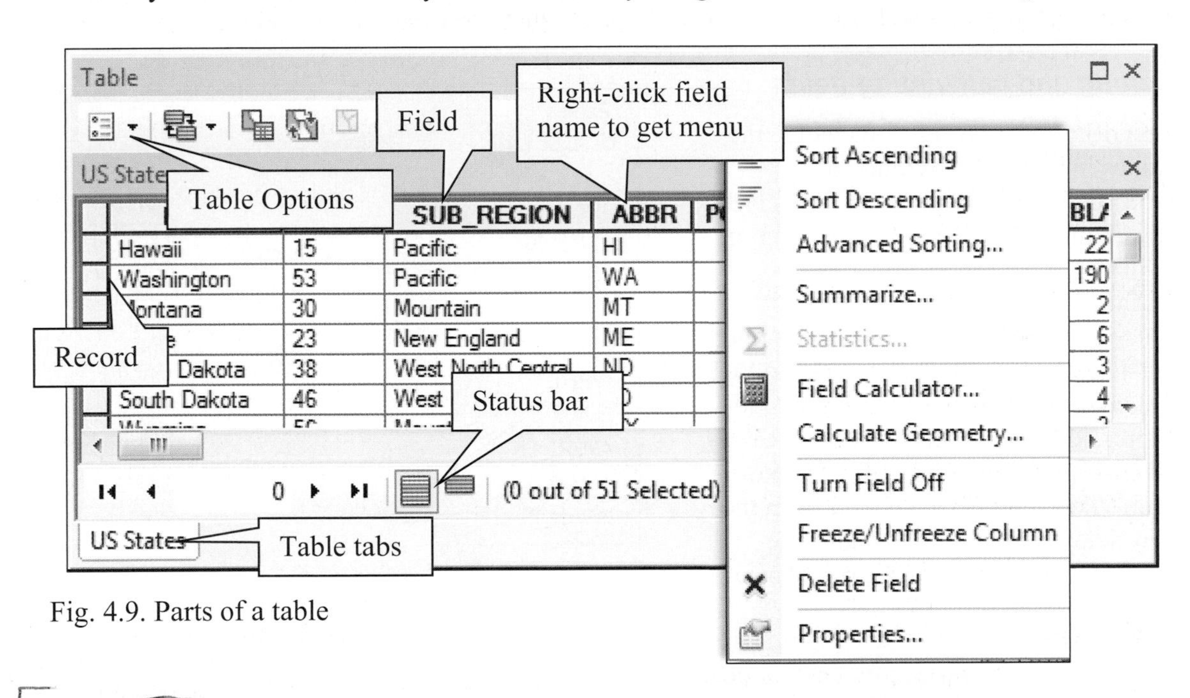

Fig. 4.9. Parts of a table

In a shapefile, the table is stored in a dBase format file and includes a unique feature identification number (FID) that links the spatial and the attribute data. In a geodatabase and in any standalone table, each record contains an Object ID (OID) analogous to the row number in a spreadsheet. Both shapefiles and geodatabase feature classes contain a Shape field that represents the *x-y* coordinate values of the feature. Geodatabase tables for line and polygon feature classes also have Shape_Area and/or Shape_Length fields, which keep track of the areas and lengths of features.

Figure 4.9 shows some terminology associated with tables. A table consists of rows and columns. A row is called a record, and it contains information about a single object or feature. A column is called a field, and it stores one type of information. Each field has a name shown in the top row. Field names must contain 13 or fewer characters and should contain only letters, numbers, and underscores. An alternative name, called an **alias**, can be set in the layer properties to give a field a more descriptive name that does not have to follow the naming rules. For example, the somewhat perplexing field MEDREN could be given a more understandable alias, such as MedianRent. Aliases form part of a layer definition. If defined in ArcMap, they persist only inside that map document. If defined as part of a layer file, they can be used in different map documents by loading the layer file instead of the feature class.

In an attribute table, some fields are required and are created and updated by the program that creates the geographic data set. For example, a shapefile has a feature ID (FID) field and a Shape field. The geodatabases feature class has an Object ID and a Shape field. **These fields should never be altered by the user.** Likewise, users of shapefiles must make sure that they never delete records in an attribute table unless the accompanying features in the spatial data set are also deleted. Editing features only within an editing session in ArcMap will meet this criterion, as does deleting rows in a geodatabase feature class.

As we learned in Chapter 1, the different GIS data formats use different underlying databases. The coverage model uses an RDBMS called INFO. Shapefiles use a dBase format file. Personal geodatabases also use the database underlying Microsoft Access, called Jet. SDE geodatabases use a large-scale commercial RDBMS, such as Oracle or SQL Server. ArcGIS can read comma-delimited text files (fields separated by commas) and tab-delimited text files (fields separated by tabs), as well as Excel™ spreadsheets if they are correctly formatted.

Editing and calculating fields

ArcGIS offers two ways to change the values in a table, by typing the information directly into the fields or by calculating the value of a field. Typing information into fields must be done during an edit session in ArcMap. Calculating fields can be done inside or outside an edit session and uses the Field Calculator to generate arithmetic expressions using fields as variables. For example, the percentage of Hispanics in each state could be calculated from the two fields containing the total state population and the number of Hispanics.

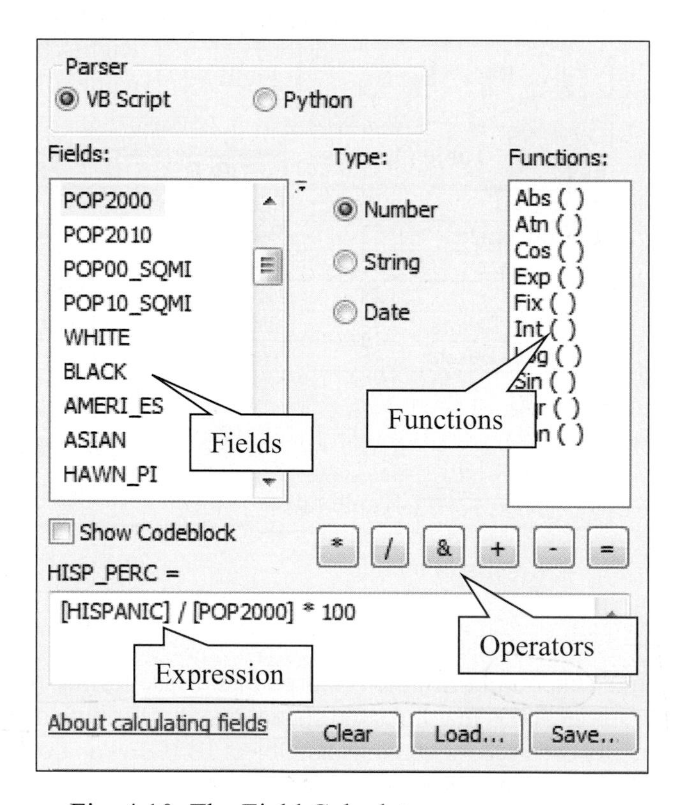

Fig. 4.10. The Field Calculator

Calculating within an edit session is safest, because a mistake can be undone. However, it can be inconvenient to start and stop an edit session each time you want to run a calculation. If you choose to calculate outside an edit session, you must be VERY careful to ensure that the field you are calculating either is empty or contains data that are easy to replace (by a different calculation, for example).

The first window in the Field Calculator (Fig. 4.10) contains the fields in the table that can be used to create the expression. The functions box contains different functions that can be used in expressions. The functions displayed will depend on whether Number, String, or Date is selected. The expression used to calculate is entered in the large box at the bottom. The operators (*, /, &, +, –, =) appear to the lower right. The Advanced button allows more complex VB Script or Python expressions to be entered. A complicated expression can be saved and loaded for use another time.

The Calculate Geometry tool (Fig. 4.11) can add information on feature areas, perimeters, lengths, or *x-y* centroids to a table. The field to contain the information must already exist. The user may choose which coordinate system and output units to use in the calculation.

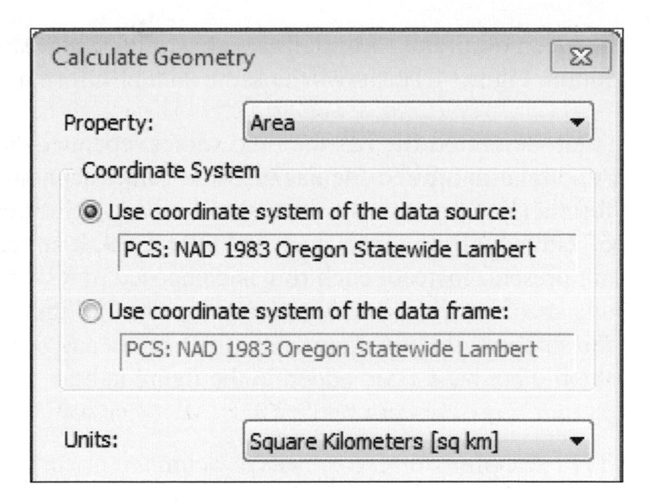

Fig. 4.11. The Calculate Geometry tool

Importing tables

Many GIS projects encounter situations when external tabular data must be imported for use in ArcGIS. Several different formats may be used, but one must always take care that the layout of the records and fields follows the requirements of ArcGIS tables. Field names, for example, must be present in the first row and must follow the ArcGIS naming conventions. Columns that are supposed to hold numbers must hold only numbers and avoid NoData or xx or Null or N/A to indicate missing values. Blank rows, merged cells, and formulas are not allowed. Most import formats are read-only within ArcGIS. They can be viewed, but not edited or changed, and some tools and functions may not work on them. Usually, the import file is exported to a dbf file or geodatabase table if changes to it are needed.

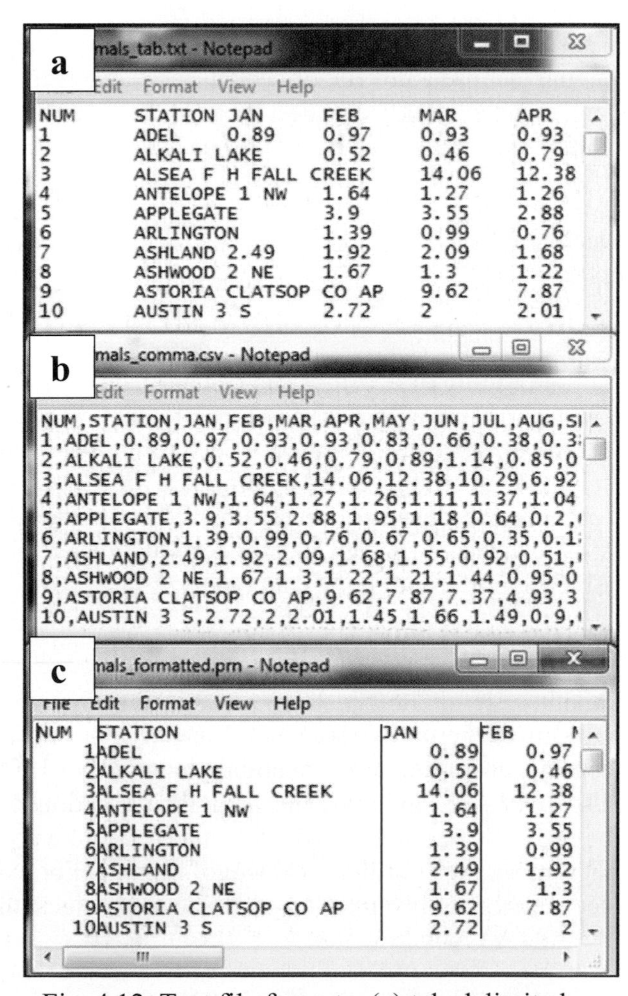

Fig. 4.12. Text file formats: (a) tab-delimited; (b) comma-delimited; (c) fixed-column

SKILL TIP: Learn about the formatting requirements for importing Excel, text, and other data types to ArcGIS (Tables).

Dbf files, or dBASE files, are commonly produced by database programs. These files can be directly imported to ArcGIS and even modified and returned to the database program. This is the only import format that can be edited within ArcGIS.

Text files, such as those shown in Figure 4.12, come from a wide variety of sources. They might come from text copied from a web site, data copied from a word processing document, or values produced from a statistical program. Nearly any program that handles tabular data has the option to save them as text, and it is one of the most basic and standard formats for transferring data between programs. All text files are stored using ASCII characters. Three primary formats are found. Figure 4.12 shows the same tabular information stored in all three formats.

A tab-delimited file has the field values separated by the ASCII tab character (Fig. 4.12a). A comma-delimited file has the field values separated by commas (Fig. 4.12b). The fixed-column format (also known as a formatted or space-delimited file) is most often encountered as the output of a computer program, such as FORTRAN. It appears similar to a tab-delimited file, but no tabs are present. Instead, each row is composed of a string of characters, and each field starts and ends at a specified character location (column). In Figure 4.12c, the first field includes columns 1–5; the station field goes from columns 6 to 30, and so on. (The red lines are not actually present in the file but have been added to the figure to help you visualize where each field starts and stops.)

TIP: ArcGIS can read either tab-delimited or comma-delimited text files. It cannot read fixed-column text files. (However, Excel™ can read fixed-column files and save them as a comma-delimited CSV file.)

Text files do not require that fields be defined. One problem with importing tabular data and saving them as a dbf file or geodatabase table is that ArcGIS must interpret the data in the field and assign a data type as best it can, and the results may be less than satisfactory. For example, a column containing zip codes will be interpreted as numbers, and any leading zeros will be eliminated (05575 becomes 5575). Putting quotes around them will enforce that they are interpreted as text. Text fields with only a few characters may be given a 255-character field in the new table. Often, too, a text file will contain formatting problems, such as field names with spaces, or an xx in a numeric column to indicate missing data. Such issues will cause the import to fail and may need to be fixed before the import can be successful.

Simple editing of text files can be done using system tools, such as the Notepad program. However, more advanced and powerful editing is available using the Microsoft Excel™ program. It can open and save a variety of text formats and can be used to translate one type to another, such as converting a fixed-column text file to a tab-delimited file. (The files shown in Figure 4.12 were all saved from an Excel™ worksheet.) Since ArcGIS can open Excel™ files directly, it can also be helpful to assign specific text or numeric formatting to the worksheet columns to aid ArcGIS in translating the data types.

Mapping x-y *coordinates from a table*

Another useful GIS function, Add XY, involves taking *x-y* coordinates from a table and converting them to locations on a map (Fig. 4.13). The *x-y* values must be given in a real-world coordinate system, such as latitude-longitude or UTM meters. The coordinate system details must also be known, including the geographic coordinate system/datum and the projection, and must be explicitly set in the Add XY window. Moreover, it should match the *x-y* values in the file, rather than the coordinate system of the map. For example, if the map is in UTM and the table *x-y* coordinates are given as longitude-latitude values, the coordinate system must be set to GCS in the Add XY window.

Global positioning systems (GPS) frequently provide tables of *x-y* data. A GPS receiver can calculate its location by triangulating distances and directions from a constellation of orbiting GPS satellites and may produce tables of *x-y* locations as output. Other examples of *x-y* data sources include benchmarks, surveyed points, and locations measured on a map. Regardless of the source, the procedure for mapping *x-y* locations from a table is identical.

The layer added from *x-y* points is called an **event layer**. This terminology comes from considering each entry in the table to be an event, such as an earthquake or a traffic accident. Not all tables contain events, of course, but the terminology remains. Event layers appear similar to point feature classes, but they do not constitute an actual feature class, for they remain a table of objects represented by spatial points. Some geoprocessing functions will not accept event layers as input. In many cases, users will export the event layer to a shapefile or a geodatabase feature class for use in a tool or for permanent storage. When exported, all the attribute fields in the original table become attributes of the feature class.

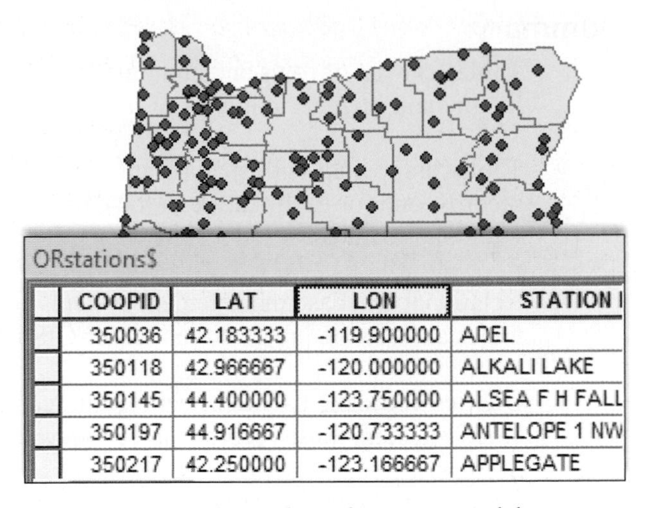

COOPID	LAT	LON	STATION I
350036	42.183333	-119.900000	ADEL
350118	42.966667	-120.000000	ALKALI LAKE
350145	44.400000	-123.750000	ALSEA F H FALL
350197	44.916667	-120.733333	ANTELOPE 1 NW
350217	42.250000	-123.166667	APPLEGATE

Fig. 4.13. Climate stations from a spreadsheet displayed as points on a map

Latitudes and longitude outside GIS systems are commonly reported in degrees-minutes-seconds and must be converted to decimal degrees before they can be used to create an event layer. This conversion is most easily done in Excel™ using the formula

$$\text{X in decimal degrees} = \text{degrees} + \text{minutes}/60 + \text{seconds}/3600$$

For example, 103° 30′ 15″ would be 103 + 30/60 + 15/3600 = 103.504167. Keep in mind that degrees are large units, and six or seven decimal places should be kept to ensure reasonable accuracy. At the equator, a meter is equal to approximately 10^{-5} degrees. Also, longitudes west of the Prime Meridian in England are negative, but the minus sign is often omitted for convenience. If this longitude is reported by a GPS unit in Nebraska, for example, the true value needed by the GIS is −103.504167.

Summary

- Tables consist of rows and columns of information. A row is associated with one feature and includes columns of information called fields.

- Tables associated with spatial data sets are called attribute tables and contain records, one for each feature in the data set. Standalone tables are not associated with map features.

- Relational database management systems construct temporary links between data tables and are the preferred model for GIS software.

- Queries allow the user to select certain records based on one or more criteria. Once selected, these records may be viewed, exported, or analyzed.

- Tables may be joined or related on a common field in order to access information in one table from another. Joins may be performed on tables with one-to-one or many-to-one cardinality. Relates must be used on tables with one-to-many or many-to-many cardinality.

- The Statistics function calculates basic statistical values for all selected records in a table. The Summarize command generates statistics about groups of features defined by a categorical field.

- Fields must be defined to contain a specific type of data, such as text, integers, dates, or floating point values. Once defined, the field type cannot change.

- ArcGIS tables can display and manipulate tabular data from a variety of sources, including dBase files, INFO files, geodatabases, SQL queries, or comma-delimited text files.

- New fields may be added to tables by defining names and field types. New tables may have data entered by typing in values or by using the Calculate function. The Calculate Geometry function can put areas and lengths in attribute tables for the corresponding features.

TIP: Field names must have 13 or fewer characters; may include letters, numbers, and the underscore character (_); and should not contain spaces or special characters, such as @, #, !, $, or %. Field names must also start with a letter, not a number.

Important Terms

alias	double-precision	key	Rule of Joining
ASCII	event layer	logical expression	scale
attribute field	field	mantissa	selected set
attribute table	field definition	precision	single-precision
binary	field length	query	source table
byte	flat file database	record	standalone table
cardinality	hierarchical database	relate	Summarize
destination table	join	relational database	table

Chapter Review Questions

1. Describe the difference between an attribute table and a standalone table.

2. Which type of database management system are GIS systems based on? How does this type of system differ from other DBMS types?

3. List the types of data sources from which tables may display data.

4. Describe how storing the number 255 in ASCII differs from storing it as a binary representation.

5. Choose the best field type for each of the following types of data in a geodatabase:

   populations of countries in the world
   precipitation in inches
   number of counties in a state
   highway name
   distances between US cities, in meters
   birthdays

6. What is the cardinality of each of the following relationships?

   students to college classes
   states to governors
   students to grades
   counties to states

7. Describe the differences between a join and a relate.

8. You have a table of states and a table of airports, both with a state abbreviation field. Can you join them if states is the destination table? If airports is the destination table? Explain your answer.

9. Describe the difference between using Statistics and using Summarize functions on a field.

10. For each of the following problems, using data sets for the United States, state whether using a query, the Statistics function, or the Summarize function would be the best approach to solving it.

    __________ Find all towns with more than 20,000 people.
    __________ Find the total number of volcanoes in each state.
    __________ Determine the total damage caused by earthquakes in the United States.
    __________ Find the states in which Hispanics exceed the number of African Americans.
    __________ Find out which subregion of the country has the most Hispanics.

Mastering the Skills

Teaching Tutorial

The following examples provide step-by-step instructions for doing basic tasks and solving basic problems in ArcGIS. The steps you need to do are highlighted with an arrow ➔; follow them carefully. Click on the video number in the VideoIndex to view a demonstration of the steps.

➔ Start ArcMap, if necessary. Navigate to the MapDocuments folder in the mgisdata directory and open the map document ex_4.mxd.

➔ Use Save As to give the document a new name and save frequently as you work.

Viewing tables

Let's begin by exploring some tables and learning some basic skills for working with them. Many table settings affect only how the tables appear and do not change the underlying data.

1➔ Right-click on the US States layer and choose Open Attribute Table from the context menu.

1➔ Narrow the STATE_NAME field to a more suitable width by clicking and dragging the right border of the field.

The OBJECTID field simply counts the rows, so hide it from sight.

1➔ Right-click the OBJECTID field name and choose Turn Field Off.

Notice that the STATE_ABBR field is also too wide for its data but that narrowing it will cut off the field name. Each field has display properties that can be set.

2➔ Right-click on STATE_ABBR and choose Properties.

2➔ Change the Alias to ABBR, and click OK.

2➔ Right-click the STATE_NAME field, open Properties, and give it the alias NAME.

Scroll to the right and look at the fields of population data from the US Census. It is difficult to match the values to the right state, however, because the state names quickly scroll out of sight.

3➔ Right-click on the NAME field and choose Freeze/Unfreeze Column.

3➔ Now scroll again and see how much easier it is to read the data.

3➔ Right-click on the NAME field and choose Freeze/Unfreeze Column again to unfreeze the field.

TIP: An unfrozen column will not automatically go back to its original place.

Now let's use the Sort function to obtain information about the largest and smallest states.

3➔ Right-click on the NAME heading and choose Sort Ascending from the menu.

TIP: Right-click a field name and choose Advanced Sorting to sort on more than one field.

Notice that the POP00_SQMI field has decimal values in it, which makes it hard to compare the population densities of different states. We can format the field so that no decimals are displayed.

4➔ Right-click on the POP00_SQMI field and choose Properties.

4➔ Click on the button next to Numeric to open the Number Format window.

4➔ Fill the button that says *Number of decimal places* and enter 0. Click OK and OK.

We can even modify which fields are shown and the order in which they appear. Like the field properties, these table properties affect only the layer, not the stored data

5➔ Right-click the US States layer and choose Properties. Click the Fields tab.

5➔ Click the Options menu and make sure that Show Field Aliases is checked.

5➔ Click on the NAME field in the list on the left. Notice the details about its appearance and definition on the right (Text data type with a length of 25 characters).

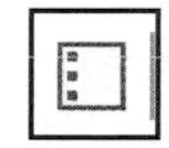

5➔ Click the Turn All Fields Off button to turn all the fields off. Click Apply and move the Properties window aside so that the table is visible.

5➔ Turn the check boxes on for the NAME and POP2010 fields, and click Apply. These two fields are the only ones shown.

5➔ Click the Turn All Fields On button to turn all the fields back on. Click Apply.

6➔ Click on the NAME field in the Properties window to highlight it.

6➔ Click on the Move Down button to move the NAME field back to its original position below the Shape field. Click Apply.

6➔ Click on the Options button and examine the menu items. Then choose Sort Ascending, and click Apply.

6➔ Click Options again and choose Reset Field Order and click Apply.

6➔ Close the US States Layer Properties window.

TIP: Use this technique to move an unfrozen field back to its original place.

7➔ Turn your attention again to the US States table in the Table window.

7➔ Examine the text at the bottom that says (0 out of 51 Selected). It tells you that there are 51 records in this table (50 states plus the District of Columbia).

7➔ Right-click the US Counties layer in the Table of Contents and choose Open Attribute Table. It opens as a new tab in the same Table window.

8➔ Click on the US States tab in the lower-left corner of the Table window to switch back to the US States table.

8➔ To see both tables at once, choose Table Options > Arrange Tables > New Vertical Tab Group. The tables now appear side by side.

8➔ Choose Table Options > Arrange Tables > Move to Previous Tab Group to return to the original layout.

8➔ Right-click the US Counties tab at the bottom of the Table window and choose Close, leaving only the US States table.

Using queries and statistics on tables

Often it is necessary to isolate a subset of records from a table and work with them alone. GIS handles this task using a query. Imagine that you are exploring the politics of the larger states. Let's select the states that had more than 5 million people in the year 2010.

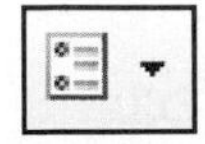

9➔ Click the Table Options menu button in the Table window and choose Select By Attributes.

9➔ Double-click the POP2010 field name in the upper box to enter it in the expression box below. Make sure that the name appears in the lower box before continuing.

9➔ Click the > button to add it to the expression.

9➔ Type the value 5000000, without commas, to complete the expression.

9➔ You should now see the expression "POP2010" > 5000000 in the lower box.

9➔ Click the Verify button to make sure that you entered the expression correctly. If it is correct, click OK. (If not, click OK, then Clear, and try again.)

9➔ Click Apply to execute the query once it is entered correctly, and close the Select By Attributes window.

TIP: Field names in queries may appear with brackets or quotes around them (like [POP2010] or "POP2010") depending on the data format used to store the table.

The selected states are highlighted in the map and in the Table window. Also, the text at the bottom of the Table window changes to read *22 out of 51 Selected*, indicating that 22 states matched the query.

10➔ Click on the Show Selected Records button at the bottom of the Table window to view only the selected records.

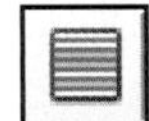

10➔ Click the Show All Records button to show them all again.

10➔ Clear the selected set by clicking the Table Options button on the table and choosing Clear Selection from the menu.

Now it is time to view some population statistics for the states.

11➔ Right-click on the POP2010 field name and choose Statistics from the menu.

11➔ Review the statistics values and examine the frequency distribution.

1. What is the population of the largest state? ____________________

11➔ Use the drop-down list in the Statistics box to select the POP10_SQMI field.

11➔ Close the Statistics window.

The Statistics command uses all records, unless a subset of records has been selected, in which case it uses only the selected subset to calculate the statistics.

12➔ Click on the Table Options button and choose Select By Attributes again.

12➔ Click Clear to delete the previous expression, if necessary.

12➔ Double-click the SUB_REGION field to enter it in the expression box.

12➔ Click the = button, and then click Get Unique Values to list the available values. When selecting text or categorical values, using this option is often faster than typing.

12➔ Double-click the 'New England' entry in the list to enter it in the expression. Notice that it is enclosed in single quotes, because it is text.

12➔ The final expression should read "SUB_REGION" = 'New England'. Click Verify to check it, and then click Apply and close the Select By Attributes window.

TIP: Beginners often find it easier to enter expressions using double-clicks and buttons, rather than trying to type the expressions manually. The program helps you get the syntax right.

13➔ Right-click the POP2010 field in the Table window and choose Statistics. Examine the Count statistic and observe that only six states are included—the New England states that you just selected.

13➔ Close the Statistics window.

13➔ Clear the selected records by clicking on the Table Options button and choosing Clear Selection.

13➔ Close the Table window.

14➔ Turn off US States in the Table of Contents and turn on 111th Congress.

14➔ Open the table for the 111th Congress layer by right-clicking the layer name and choosing Open Attribute Table.

14➔ Examine the fields, noting the field PARTY.

15➔ Open the layer properties for 111th Congress layer and click the Symbology tab.

15➔ Create a Categories: unique values map based on the PARTY field. (Remember to click the Add All Values button to show the categories). Click OK.

15➔ Right-click the symbols in the Table of Contents to change the symbol for the Democratic counties to blue, the Republican counties to red, and the Vacant districts to light gray (Fig. 4.14).

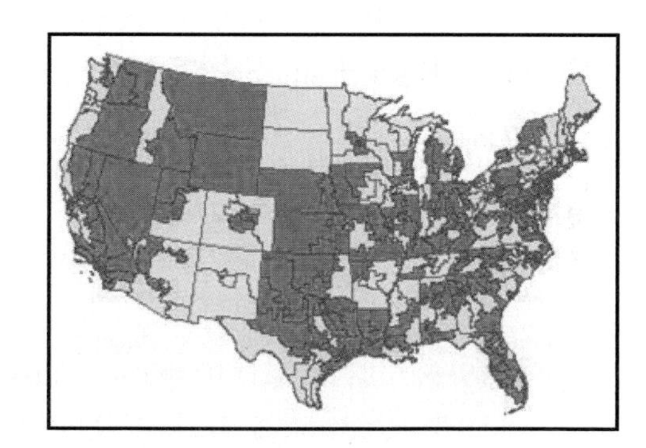

Fig. 4.14. 111th Congress districts

How many Democratic districts are there? We can find out using a query.

16➔ Open Select By Attributes from the Table Options menu.

16➔ Enter the expression "PARTY" = 'Democrat'. Use the Get Unique Values button so that you can select the 'Democrat' value instead of typing it.

16➔ Click Apply and close the Select By Attributes window.

2. How many Democratic districts are there? _______________

Finding the number of Democratic districts using a single query is fine. But suppose you wanted to know the number of districts for all three categories: Republican, Democrat, and Vacant? The Summarize command can generate statistics about multiple groups defined by a field.

16➔ Use the Clear Selection button in the table window to clear the selected records.

17➔ Right-click the Party field and choose Summarize. Note that the Party field is already entered in box #1 because you right-clicked it.

17➔ You are only interested in the number of districts, which is always counted by default, so you need not enter any statistics. Do not change anything in box #2.

17➔ Click the Browse button. Make sure that the *Save as type* is set to dBASE table and that you are currently in the Usa folder. Give the new file the name cd111numdist.dbf.

17➔ Click Save and OK, and answer Yes to add the resulting table to the map.

TIP: Because this table was created to answer a transitory question, we saved it in a folder instead of making it a permanent part of the geodatabase. Be sure to give your saved tables descriptive names so that you can remember what they contain later.

Examine the Table of Contents window. The icons at the top change how layers are listed. Originally, it was set to the first icon on the left, List By Drawing Order. Because a standalone table was added, it switched to the second view style, List By Source. Each layer is now listed under its folder or geodatabase name. This is the only view in which standalone tables are visible.

18➔ Right-click the new cd111numdist.dbf table and choose Open.

18➔ See that all three categories are listed, with the Count_PARTY field showing the number of districts in each category.

18➔ Right-click the cd111numdist tab in the Table window and choose Close.

Joining tables

Let's find out which districts may have changed their party affiliation since the 110th Congress. We need to compare the party in two different tables using a query, so we must combine the two tables using a join. To join, we need to find a key field.

19➔ Add the cd110 feature class from the mgisdata\Usa\usdata geodatabase.

19➔ Right-click the cd110 layer and open the table. Examine the fields and try to identify the appropriate key field that might work to combine the tables. The key must be unique to each record and have the same value in both tables.

3. What is the best potential key field in this table? DISTRICTID

19➔ Click the 111th Congress tab in the Table window and confirm that this field is present.

We want 111th Congress to be the destination table. It is currently the active table in the Table window, so it will become the destination table as we intend.

20➔ Open the Table Options menu and choose Joins and Relates > Join.

20➔ Make sure that the top drop-down reads *Join attributes from a table* (Fig. 4.15).

20➔ Choose DISTRICTID as the field that the join will be based on.

20➔ Choose cd110 as the table to join to the layer.

20➔ Choose DISTRICTID as the field in the table to base the join on.

20➔ Make sure that your window reads as shown in Figure 4.15, and then click OK. Choose Yes if asked to create an index.

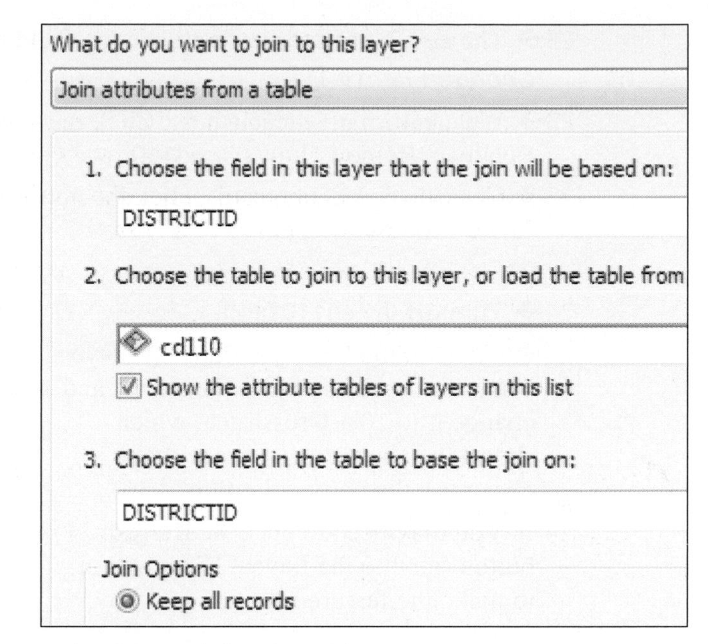

Fig. 4.15. Joining the tables

20➔ Scroll to the right, noticing the additional fields now in the table.

The fields from cd110 have been appended to the right of the fields for 111th Congress. However, it can get confusing to remember which fields came from which table.

21➔ Choose Table Options and select Show Field Aliases to uncheck it.

Now each field name is prefixed with the name of its table (the original name, not the title in the Table of Contents). Hence, the 111th Congress table fields are prefixed with the name cd111, and the cd110 fields are prefixed with cd110.

Now we can use a query to find the districts that have changed affiliation so that the cd111.PARTY field is different from the cd110.PARTY field.

22➔ Open the Select By Attributes window from the Table Options menu.

22➔ Clear the previous expression, if necessary.

22➔ Double-click the cd111.PARTY field to enter it in the expression box; then click the "< >" (not equal to) button, and double-click the cd110.PARTY field.

22➔The final expression should look like cd111.PARTY <> cd110.PARTY. Verify it and then click Apply. Close the Select By Attributes window.

22➔ Click the Table button to Show selected records.

4. Which party received most of the changed seats in the 111th Congress? ________________

Let's save the selection as a new layer for future reference and to symbolize it differently.

23➔ Right-click the 111th Congress layer in the Table of Contents, and choose Selection > Create Layer From Selected Features.

23➔ The new layer appeared near the top as 111th Congress selection. Rename it Changed Party 111.

23➔ We can turn off the join now. Click the Table Options menu, and choose Joins and Relates > Remove Join(s) > cd110.

23➔ Clear the selected records, click the button to Show All Records, and close the Table window.

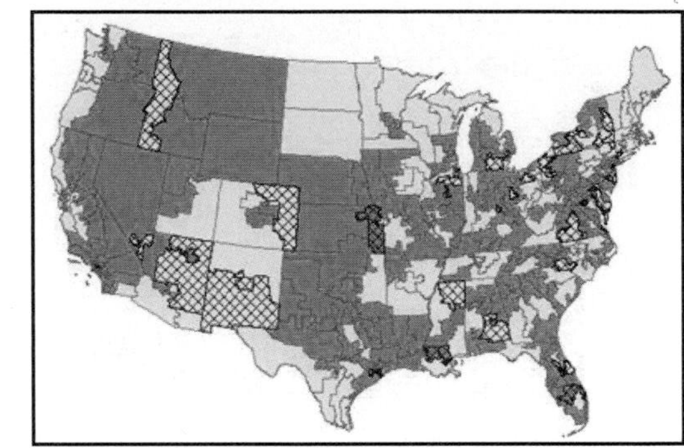

Fig. 4.16. Map of changed districts

24➔ Turn off the cd110 layer.

24➔ Click on the Changed Party 111 layer symbol to open the Symbol Selector and change it to 10% Crosshatch, which shows the changed districts without obscuring the party color (Fig. 4.16).

24➔ You may need to click the Refresh button (next to the Layout View button) to make the feature selection go away.

Summarizing tables

Some states have only one district; others have more. Suppose that we want to know many districts each state has. The Summarize function will yield this information. It will group the districts by state and calculate statistics for each group.

25➔ Open the attribute table of 111th Congress.

25➔ Right-click on the STATE_ABBR field and choose Summarize.

The Summarize command always counts the number of features in each group and will report it in the output table, so you don't need to request any other statistics.

25➔ Enter the name of the new table to be created, dists_per_state, making sure it will be saved as a dbf file in the Usa folder. Click OK.

25➔ Click Yes to add the table to the map document.

5. Is this new table an attribute table or a standalone table? ____________________

26➔ Find the new dists_per_state.dbf table in the Table of Contents and open it.

26➔ Right-click the Count_STATE_ABBR field and choose Sort Descending.

6. Use a query to determine how many states have only one district: ___9___. Clear the selection before going on.

Sometimes the goal is to generate statistics about each group in addition to counting the records. Perhaps you would like to know the total area represented by each party.

27➔ Click the 111th Congress tab in the Table window.

27➔ Right-click the PARTY field and choose Summarize.

27➔ Verify that the field to summarize is set to PARTY (Fig. 4.17).

27➔ Locate the SQMI field in the list, expand it, and choose Sum as the statistic to calculate.

27➔ Enter a location and name for the output table, calling it partyarea.dbf and placing it in the Usa folder.

27➔ Click OK and Yes.

28➔ Find the new partyarea table in the Table of Contents and open it.

28➔ Right-click the Sum_SQMI field and choose Sort Descending.

7. Which party represents the greater area, and by how much? ____________

28➔ Close the Table window before going on.

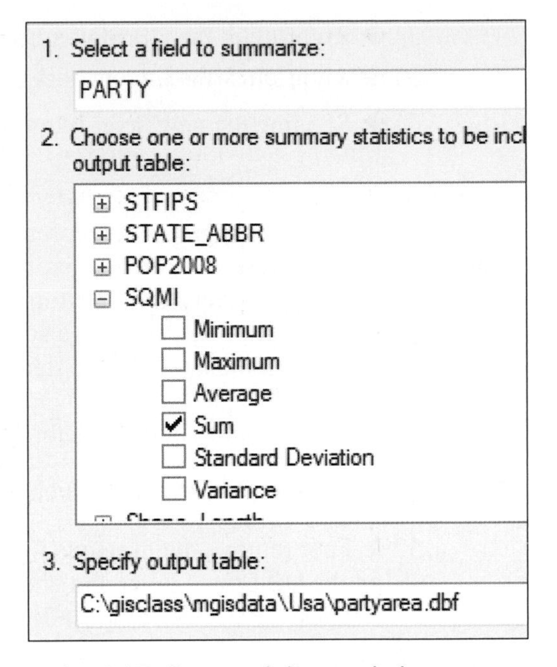

Fig. 4.17. Summarizing statistics

A map showing the number of districts in each state would be nice, but the dist_per_state table is a standalone table and can't be mapped. However, it can be joined to the US States feature class. In this join, the US States table must be the destination.

8. What is the cardinality of this join? ____________

29➔ Right-click on the US States layer and choose Joins and Relates > Join.

29➔ Choose ABBR as the field in US States to base the join on.

29➔ Choose dists_per_state as the table to join.

29➔ Choose STATE_ABBR as the field in the source table to join on.

29➔ Click OK to finish the join.

29➔ Open the US States table, and scroll to the right to find the new fields added as a result of the join.

The number of districts in each state is contained in the Count_STATE_ABBR field.

30➔ Close the US States table.

30➔ Turn off all of the layers in the map, and turn on the US States layer.

30➔ Open the US States layer properties, click the Symbology tab, and create a Quantities: graduated color map based on the Count_STATE_ABBR field (Fig. 4.18).

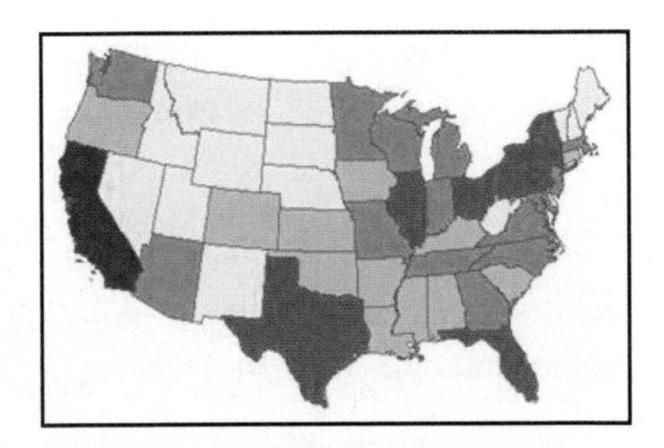

Fig. 4.18. The number of districts

31➔ Right-click the US States layer and choose Copy.

31➔ Right-click the Layers data frame and choose Paste Layer(s).

31➔ Rename the new layer Number of Districts.

Relating tables

It would be interesting to group the representatives according to subregion, as they would have similar issues to address with their constituencies. However, the SUB_REGION field is in the US States table, not the districts table. Setting up a relate between the US States table and the 111th Congress table will allow us to select a subregion and obtain a list of representatives from it. The common key between the tables is the state abbreviation.

9. What is the cardinality between subregions and districts? ____________________

10. Which is the source table, and which is the destination table? ____________________

32➔ First remove the previous join by right-clicking on the US States layer and choosing Joins and Relates > Remove Join(s) > Remove All Joins.

32➔ Right-click on the US States layer and choose Joins and Relates > Relate.

32➔ Fill out the fields as shown in Figure 4.19. Click OK.

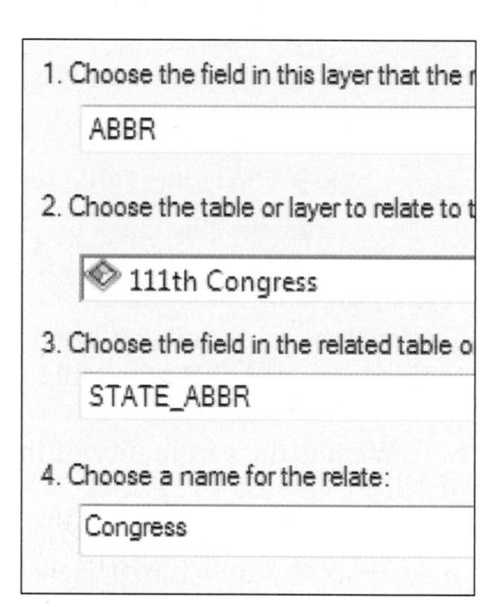

Fig. 4.19. Setting up a relate

Now let's use the relate to find out which representatives are from New England states.

33➔ Open the US States table and the 111th Congress table.

33➔ Choose Table Options > Arrange Tables > New Vertical Tab Group so that both tables are side by side. Enlarge the Table window if needed to see both tables well.

TIP: When more than one table is visible, the one with the darker bar across the top is the active table, and choices made from the Table Options menu will be applied to the active table.

33➔ Click the title bar of the US States table to make sure it is the active table.

33➔ Choose Table Options > Select By Attributes and select the states in the New England subregion.

33➔ Close the Select By Attributes window.

Now that the New England states are selected and highlighted, the relate can access the representatives from these states.

34➔ Click on the Table Options button and choose Related Tables > Congress: cd111.

34➔ The other table view highlights the related records, now also selected.

The Changed Party 111 layer is based on a selection from the 111th Congress layer, and so both of them show up in the relate. We're not interested in the changed representatives right now, and there is only one match, so let's close that table.

34➔ Right-click the Changed Party 111 tab in the Table window and choose Close.

11. How many representatives come from New England states? ____________

Exporting the selected records from the representatives table will provide a new file that can be saved permanently and given to interested constituents.

35➔ Click the bar of the 111th Congress table to make sure it is active.

35➔ Choose Table Options > Export. Make sure it says to export the *Selected records*.

35➔ Click the Browse button.

35➔ Change the *Save as type* to dBASE Table. Navigate inside the Usa folder.

35➔ Type in NE_reps.dbf as the name of the table and choose Save and OK.

35➔ Say Yes to add the table to the map.

35➔ Open the new NE_reps table and verify that it contains the correct representatives.

36➔ Choose Selection > Clear Selected Features from the main menu to clear the selection in all tables.

36➔ Click the US States table to make it active and choose Table Options > Joins and Relates > Remove Relate(s) > Remove All Relates.

36➔ Close the Table window.

Editing values in tables

Sometimes fields in a table must be created or updated manually. Imagine that you wish to add more information to the partyarea table you created earlier, indicating status as the majority or minority party.

37➔ Open the partyarea table.

37➔ Click Table Options and choose Add Field. Name it STATUS.

37➔ Set the Type to Text and set the Length to 10. Click OK.

To type values into a table, you must be in an edit session. Editing can be performed in only one folder or geodatabase at a time, so you must specify which source you want to edit.

38➔ Click the Editor Toolbar button on the Standard toolbar, to open the Editor toolbar.

38➔ Choose Editor > Start Editing from the Editor toolbar.

38➔ Click the partyarea table to highlight it and click OK.

Notice that the field headings turn white, indicating that this table is now editable.

39➔ Click in the STATUS field in the Republican row and type Minority. Click Enter.

39➔ Type Majority in the Democrat row and click Enter.

39➔ Type Vacant in the Vacant row and click Enter.

39➔ Choose Editor > Save Edits.

39➔ Choose Editor > Stop Editing to close the edit session.

39➔ Close the Table window.

TIP: You cannot add a new field to a table during an edit session.

Calculating fields in tables

Imagine that we are writing a report and need a table showing the percentage of Hispanics in each state. The US States table has only the number of Hispanics. Creating a new field and calculating the percentage of Hispanics solves the problem.

40➔ Open the US States attribute table.

40➔ Click on the Table Options button and choose Add Field.

40➔ Name the field HISP_PERC and select Float as its type. Click OK.

Now calculate the percentage of Hispanics. The minority population fields in this table represent 2000 census data, not 2010, so we will use the POP2000 field for calculating percentages.

TIP: Be VERY careful when calculating because you can accidently overwrite existing data, and there is no way to undo it unless you are in an edit session. Be sure a field is empty before you try to calculate it.

41➔ Right-click on the heading of the new, empty HISP_PERC field (located to the far right of the table), and choose Field Calculator from the context menu.

41➔ Double-click the field names and the operator buttons to the expression [HISPANIC] / [POP2000] * 100 and Click OK.

TIP: Do all of the records equal zero, or null? You might have accidentally selected a single record by clicking on the table. Check the bottom status bar to see if any records are selected. If so, choose Table Options > Clear Selection from the table and try the calculation again.

Finally, for a report you need to determine the density of Hispanics per square kilometer in the counties. You'll need to create and calculate two new fields, the area in kilometers and the Hispanic density. Because you're calculating areas, make sure the data frame coordinate system is set to an equal-area projection.

42➔ Close the Table window.

42➔ Open the Layers data frame properties and click the Coordinate System tab.

42➔ Set the coordinate system to the Projected Coordinate Systems > Continental > North America > USA Contiguous Albers Equal Area Conic projection.

43➔ Open the US Counties attribute table.

43➔ Click the Table Options button and choose Add Field.

43➔ Name the field AREA_KM and set its Type to Float. Click OK.

43➔ Add another new field named HISP_SQKM, also making it a Float type.

TIP: It is safer, but less convenient, to open an edit session before doing field calculations. Then you can undo a mistake, if you make one.

Now we will calculate the new fields. Areas are calculated with a special function.

44➔ Right-click the empty AREA_KM field heading and choose Calculate Geometry. Click Yes to calculate outside an edit session.

44➔ Set the Property to Area, fill the button to *Use the coordinate system of the data frame*, and set the units to square kilometers. Click OK.

45➔ Right-click the empty HISP_SQKM field and choose Field Calculator. Enter the expression [HISPANIC] / [AREA_KM] and click OK.

45➔ Close the Table window. Turn off all layers and then turn on the US Counties.

46➔ Open the symbology properties for US Counties and create a Quantities: graduated color map using the HISP_SQKM field.

The population density of Hispanics has an extremely skewed distribution due to the presence of densely populated areas like Los Angeles. The Jenks classification method does not show the distribution of Hispanics well.

46➔ Open the US Counties symbology properties again.

46➔ Use the Classify button and change the classification scheme to Geometrical Interval with five classes. Click OK and OK and view the new map (Fig. 4.20).

46➔ Save the map document.

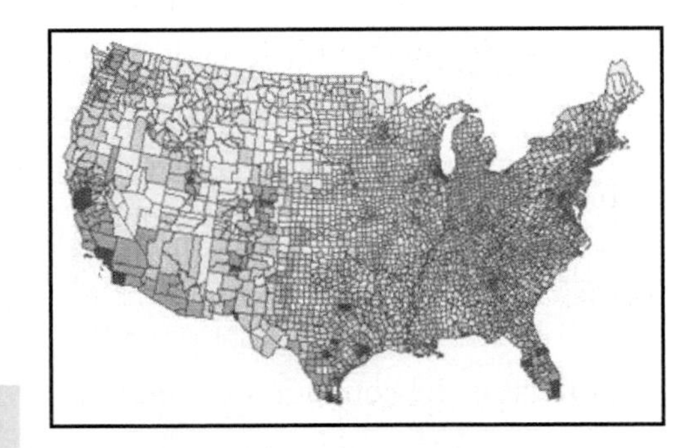

Fig. 4.20. Density of Hispanics

SKILL TIP: Learn how to create new, empty tables and how to edit values in tables (Tables).

Working with Excel and XY data

We will do a brief exercise on using Excel data in ArcGIS and displaying locations in a table as points on a map.

47➔ Click on the New Map File button.

47➔ Change the default geodatabase to mgisdata\Oregon\oregondata. Click OK to open the new blank map.

48➔ Add the gtoposhd raster from the mgisdata\Oregon\oregon geodatabase.

48➔ Click the Add Data button, navigate up to the mgisdata\Oregon folder, and double-click the ORstations.xls workbook. Select the ORstations$ worksheet and click Add.

TIP: An Excel .xls or .xlsx file is a workbook that may contain multiple worksheets. Only single worksheets may be opened in ArcMap.

48➔ Open the ORstations$ table and examine the fields. Note the LAT and LON fields containing the station location in decimal degrees.

Any table with *x-y* coordinate locations can be displayed as a point layer. You need to know the coordinate system of the *x-y* locations in the table. In this case the units are degrees of latitude and longitude, so we know that it is a geographic coordinate system (GCS). A look at the web site where these data were downloaded from the National Climatic Data Center (NCDC) would tell us that the GCS is based on the NAD 1983 datum.

49➔ Close the Table window.

49➔ Right-click the ORstations$ table name and choose Display XY Data.

49➔ Set the X Field to LON and the Y Field to LAT.

49➔ Click the Edit button to set the coordinate system.

49➔ Choose Geographic Coordinate Systems > North America > NAD 1983 and click OK and OK. Click OK in the notice.

Observe the station points, called an event layer. The precipitation values are stored in a different worksheet. We will add it and join it to the event layer so that we can map the precipitation.

50➔ Click Add Data. Navigate up to the Oregon folder and double-click the ORprecipnormals.xls file. Add the ORprecipnormals$ worksheet.

50➔ Open the ORprecipnormals$ table and examine the monthly and annual precipitation values for each station.

50➔ Open the ORstations$ table and compare it with the ORprecipnormals$ table in the Table window.

12. What is the common field in these two tables? ____________________

51➔ Close the Table window.

51➔ Right-click the new ORstations$ Events layer and choose Joins and Relates > Join. Click Yes to continue.

51➔ Choose STATION NAME as the field to base the join on.

51➔ Choose the ORprecipnormals$ as the table to be joined, and STATION as the second field. Click OK.

51➔ Open the attribute table for the ORstation$ Events layer and examine it. Close it when finished.

An event layer is a temporary layer in a map document. You can export it to save it in permanent form in the oregondata geodatabase.

52➔ Right-click the ORstations$ Events layer and choose Data > Export Data.

52➔ Choose to use the same coordinate system as the data frame so that the stations will be stored in a projection (Oregon Statewide Lambert).

52➔ Click the Browse button. Change the *Save as type* to File and Personal Geodatabase feature classes.

52➔ Navigate inside the mgisdata\Oregon\oregondata geodatabase and enter **precip** for the name of the output feature class. Click Save, OK, and Yes to add the feature class.

TIP: Exporting a joined table or layer will place the attribute fields of the source table in the output along with those from the destination table.

53➔ Right-click the ORstations$ Events layer, which you no longer need, and choose Remove.

53➔ Open the symbology properties of the new precip layer and create a Quantities: graduated symbol map based on the ANN field (Fig. 4.21).

This is the end of the tutorial.

➔ Close ArcMap. Save the map document if you wish.

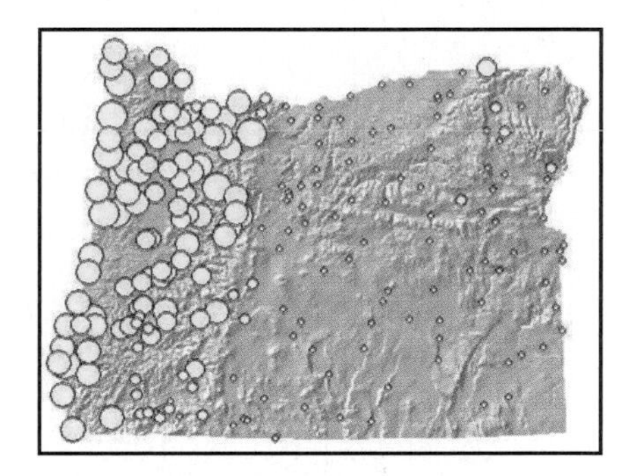

Fig. 4.21. Oregon precipitation

SKILL TIP: Not all Excel files can be successfully opened in ArcMap. Learn more about which files are suitable and how to set up a file that can be read (Tables).

Exercises

Open the ex_4.mxd map document in the MapDocuments folder of the mgisdata directory and answer the following questions.

1. How many counties in the United States have the name Washington? What is their total 2010 population? Which one has the largest area?

2. Calculate the percentage of the population in each state that is African American. (Use the BLACK field for the African Americans and the POP2000 field for the total population in each state.) Which state has the largest *number* of African Americans? Which state has the largest *percentage* of African Americans?

3. Which subregion of the country has the greatest number of African Americans? How many does it have?

4. Add the table dating from the mgisdata\Usa\usdata geodatabase to the map document. This table contains information about marital status by county. Examine the table. Which fields could you use to join this table to the US Counties table? Why would the county name field NOT work?

   In the US Counties table, examine the three fields STATE_FIPS, CNTY_FIPS, and FIPS. How are these fields related? Join the tables.

5. Join the dating table to the US Counties table. Then select the counties that *lost* population in the 20-year period from 1990 to 2010. How many are there?

   In what area of the country are these counties predominantly located? Make a map showing the "losing" counties. **Capture** the map. (**Hint:** Create a new layer from the selected features.)

6. Use Summarize to determine the number of counties in each state that lost population. Which three states had the most losing counties, and how many losing counties did each have?

7. Imagine that you are the PR director of a group seeking aid for Native American artists, and you wish to contact the *Democratic* representatives in the 111th Congress who are from states in which the number of Native Americans and Eskimos (AMERI_ES field in US States) exceeds 100,000. Which states are included in the list? How many representatives will you need to contact? (**Hint:** Create a layer from the districts in the qualified states.)

8. Using the Major Cities layer, determine how many people in the United States lived in state capitals in 2007. What is the smallest, largest, and average population of the capitals?

9. Which state has the highest percentage of owner-occupied housing units? Which one has the lowest? What are the percentages? Create a map showing the percentages using a standard deviation classification method.

10. Which subregion of the United States has the largest number of counties? How many counties does it have? Create a map of the counties showing to which subregion each belongs. **Capture** the map.

Challenge Problem

Imagine that you are planning to open a dating service in Texas. Your preliminary research indicates that such a service would expect to gross $3 for every divorced person, $2 for every single (never married) person, and $1 for every widowed person in a typical county. A table with the county populations in these categories is available in mgisdata\Usa\usdata\dating. Calculate the projected gross income for each county in Texas and create a map layout showing the projected income for each county. (Make sure your map shows only Texas counties.) What is the total projected gross income for the entire state? Is there a large cluster of high-income counties that might be a good place to establish your office?

Chapter 5. Queries

Objectives

- Understanding queries and how they are used
- Selecting features based on attributes using SQL and Boolean operators
- Selecting features based on their spatial location with respect to other features
- Applying selection options, including the selectable layers and the selection method

Mastering the Concepts

GIS Concepts

About queries

Queries are a fundamental tool in GIS, used in nearly every analysis. They serve a critical function in data exploration or in picking one's way through a data set looking for patterns. They can also serve as the preparatory step of extracting certain features prior to performing another function. Some common applications involving queries include the following:

- **Selecting features of interest.** Queries can be used to search a table and to find which features meet certain criteria. How many houses currently for sale fall into my price range? Combined with the statistics function, this tool becomes powerful. What is the average cost of three-bedroom homes in this town?

- **Exploring spatial patterns.** Creating a map from selected features and examining their spatial distribution can be illuminating. Where do the wells with high values of contaminants occur? Do they show a pattern relative to known point source pollution emissions? Are they widely spaced or clustered together?

- **Isolating features for more analysis.** Select the largest earthquakes and find out how many of them lie close to volcanoes. Choose the states that have lost population and subject them to a multivariate statistical analysis of demographic data to find out why.

- **Exploring spatial relationships.** Which parcels lie in the floodplain? Which cities are close to a volcano? Where do roads cross unstable shale units?

- **Creating raster queries.** Although we have only discussed vector queries, cells of a raster can be queried also. Which cells have a slope greater than 10 degrees? Which cells show an increase in vegetative greenness as measured by satellite images taken in 1998 and 1999? Where did land use change between 1970 and 2000?

Formally speaking, a query extracts features or records from a feature class or from a table and isolates them for further use, such as printing them, calculating statistics about them, editing them, graphing them, creating new files from them, or doing more queries on them. In the simplest kind of query, you look at a map or a table and use the mouse to select the desired record(s). After selecting the features, they are highlighted on the screen, and the corresponding records in the table are highlighted as well.

Queries fall into two main categories. An **attribute query** uses records in the attribute table to test a condition, such as finding all cities with population greater than 1 million people. Attribute queries are sometimes called aspatial queries because they do not require any information about location. A **spatial query** uses information about how features from two different layers are located with respect to one another. A spatial query might choose cities within a certain county, rivers that are within a state, or hospitals within 20 miles of an airport. Attribute queries can be performed on either attribute tables or standalone tables. A spatial query requires a spatial data layer. Spatial and attribute queries may be combined, such as selecting cities with more than 50,000 people who are within 50 miles of an airport. The spatial and attribute queries must be performed separately but can be done in any order.

Attribute queries

In this method, the user specifies a certain condition based on fields in the attribute table and selects the records that meet the criteria, such as choosing all states with a population greater than 5 million. We already introduced selecting by attributes from tables in Chapter 4. In this section we extend selection to features and discuss more advanced selection options.

Attribute queries test conditions between two or more attributes using **logical operators**, which include the familiar <, >, and = operators. The condition is written as an **expression** in a format called **Structured Query Language** (SQL), which is used by most major databases to perform selections. SQL expressions can be complex and sophisticated, or fairly simple.

SELECT *FROM cities WHERE “POP1990” >= 500000

SELECT *FROM counties WHERE [BEEFCOW_92] < [BEEFCOW_87]

The SELECT *FROM and WHERE key words must be present as dictated by the syntax rules of SQL. The lowercase words (e.g., *cities*) give the name of the table from which to select, and the terms in double quotes or brackets are the field names.

This syntax would be used in any SQL-compliant database, so SQL is a transportable language. However, minor differences between database SQL varieties can be found, such as whether field names are enclosed in brackets or quotes.

Boolean operators

When a query includes more than one condition, the additional operators AND, OR, XOR, and NOT are needed. Consider the following queries with two conditions:

SELECT *FROM accounts WHERE “Cust” = ‘COM’ AND “Balance” > 500

SELECT *FROM accounts WHERE “Cust” = ‘COM’ OR “Cust” = ‘GOV’

SELECT *FROM accounts WHERE [Balance] > 500 AND [Balance] < 1500

The first expression selects records from a customer accounts table that are held by commercial customers and have a balance over $500. The second expression finds the accounts held by commercial and governmental customers. The third expression finds accounts with balances between $500 and $1500.

TIP: SQL requires that a field name appear in every condition of the expression, even if it is the same field. SQL will not correctly evaluate the expression "Cust" = 'COM' AND 'GOV'.

AND, OR, XOR, and NOT are called **Boolean operators**, and they evaluate two input conditions, which may be either true or false, and return a set of records meeting those conditions. Figure 5.1 represents these operators graphically using **Venn diagrams**. The large ellipse labeled T represents all the records in the table. Circle A represents the subset of records meeting condition A, circle B represents the subset meeting condition B, and the blue region indicates the records selected by the different Boolean operators.

Examine the first expression in the preceding examples of Boolean operators. The first condition, "Cust" = 'COM', constitutes condition A; condition B would be "Balance" > 500. A AND B finds the accounts for which both input conditions are true (commercial customers and balance over $500). If either condition is false, the record is not selected.

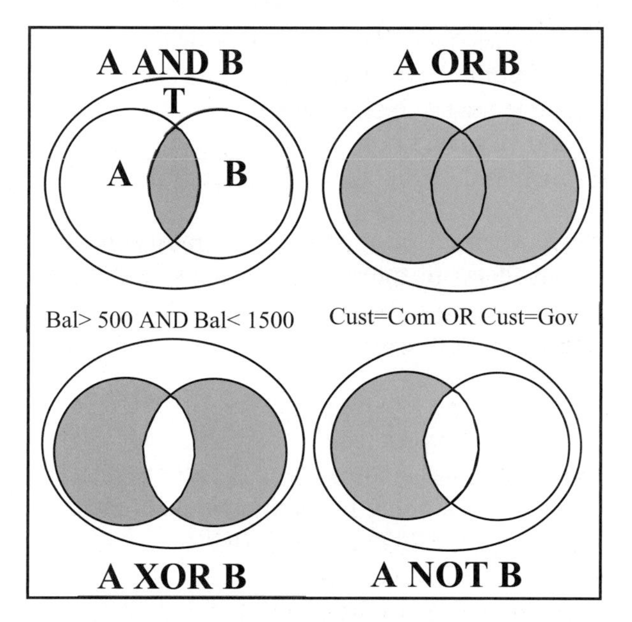

Fig. 5.1. Boolean operators

Let's examine the other operators using the same conditions. The expression A OR B selects the records for which either condition is true, finding all the commercial customers and all the accounts with balances over $500 (commercial or not). A NOT B finds all the commercial accounts but excludes the ones with balances over $500, instead finding the commercial customers with balances under $500. The "exclusive or" operator XOR is not included in the SQL calculator in ArcGIS but is another common Boolean operator used to find records for which one of the conditions holds but not both at the same time. XOR would find all the commercial customers except those with high balances, plus all the high balances except those held by commercial customers.

Novices often confuse the AND and OR operators when writing expressions, for example, in selecting both the commercial and the governmental accounts. In English we would say, "select the commercial and the governmental accounts," and it seems natural to use the expression "Cust" = 'COM' AND "Cust" = 'GOV'. However, this expression yields no records because a customer account cannot be both commercial and governmental at the same time. The OR operator should be used.

The AND operator is used to find records within a numeric range. Consider the third expression, "Balance" > 500 AND "Balance" < 1500. If a record with "Balance" = 1000 is being considered, it would test true for both conditions and be selected. Using OR is incorrect; it would return ALL of the values instead of those inside the range. A record with "Balance" = 300 would test true for the second condition and so be included. A record with "Balance" = 3000 would test true for the first condition and be included.

SQL does accept three or more Boolean conditions. Typically, parentheses must be used to enforce the correct order of evaluation. Compare these two expressions for selecting parcels:

("LUCODE" = 42 AND "VALUE" > 50000) OR "SIZE" > 50

"LUCODE" = 42 AND ("VALUE" > 50000 OR "SIZE" > 50)

In the first expression, all parcels > 50 hectares are chosen regardless of their zoning or value, and additional parcels are added if they are code = 42 and high in value. In the second expression, all of the parcels must have a zoning of 42 and can be either high in value or large in size.

Searching for partial matches in text

Sometimes queries based on text fields need to look for partial matches to a condition rather than a perfect match. Consider a table containing customer names. A user might wish to find all customers with the last name *Smith.* If the first and last names were stored in separate fields it would be easy, but if they were stored together in a single field it would be necessary to search for names that contain *Smith* as part of the full name.

Also, when selecting strings, one must sometimes allow for extraneous spaces or inconsistencies in how data are entered. For example, the string *Maple St* with a space following *St* would not match the string *Maple St* or *Maple St.* or *Maple Rd.* Often databases have little inconsistencies like this one, because it is difficult to standardize the formats perfectly or because people make typing mistakes.

To solve such problems, most GIS systems (and databases) provide an operator, such as LIKE or CONTAINS, that allows searches for a particular substring within another string. This operator is helpful in searching for *Smith* within a field containing both first and last names or *Maple* within a field of street addresses.

LIKE searches for the specified set of characters within the field and returns any record that contains those characters. When using LIKE, a wild card character (% for shapefiles; * for geodatabases) is used to stand for other letters that may appear. Consider selecting cities based on their NAME field. The two queries

"NAME" LIKE 'New %' *(shapefile)*
"NAME" LIKE 'New *' *(geodatabase)*

would select all cities beginning with the word *New*, such as New London and New Haven, but would not include Newcastle because there is no space in the latter. The expression

"NAME" LIKE '%Smith%'

would return the customers with the name *Smith.* Notice the second % sign used in the expression. One might expect it to be superfluous, and often it would be, but it does ensure that an extraneous space character at the end of the name would not prevent it from being selected.

Spatial queries

Spatial queries are a powerful tool unique to GIS because they select based on spatial relationships, such as finding wells within five miles of a river or finding parcels inside a floodplain. A spatial selection uses two layers and one spatial condition. The features of the layer being selected are compared spatially with the features of the second layer to see which ones meet the criteria, and features meeting the criteria are selected.

Panning and zooming, techniques already familiar to the reader, are actually a type of spatial query. Pans and zooms define a rectangle of interest based on *x-y* coordinates and asks that the software return all features that fall inside it.

Because feature classes vary in precision and geometric accuracy, it can often happen that two objects which coincide in the real world will not match when their GIS coordinates are compared. It is helpful to be able to specify a search radius when evaluating a spatial condition so that the features need not exactly match.

Fig. 5.2. Features for Minneapolis–St. Paul and the Mississippi River

Consider the problem of selecting cities intersecting rivers. A city at a national scale is represented by a point, and a river by a generalized line. Even if the real-life city intersects a river, it would be only luck if the point occurred on the line feature representing the river (Fig. 5.2). Using the search radius provides some slack. If one assumes that a typical large city might be represented by a square 10 kilometers on a side, then setting the search radius to 5 kilometers might produce an acceptable result. Even so, one would probably encounter some false positive or false negative results.

intersect
intersect (3d)
are within a distance of
are within a distance of (3d)
contain
completely contain
contain (Clementini)
are within
are completely within
are within (Clementini)
are identical to
touch the boundary of
share a line segment with
are crossed by the outline of
have their centroid in

Fig. 5.3. Spatial criteria operators

Users must always remember the difference between real-world entities and how they are represented by a GIS, and be prepared to mitigate problems that arise because of scale or accuracy issues. In this case, for example, the user might go ahead with the query, but then check each selected city using the online imagery to ensure that the correct relationship exists.

Attribute queries test relationships between attributes using logical operators, such as <, >, or =. Spatial queries test for relationships using **spatial operators**, including **containment**, **intersection**, and **proximity**. Consider two feature classes, A and B. The spatial operators shown in Figure 5.3 test for the relationship of each feature in A to each feature in B and return the features from A that meet the criteria with respect to B. A complete description of each operator can be found in the ArcGIS Help.

Containment tests whether one feature includes another. The test A *completely contains* B returns all features in A that fully surround the features in B. A *contains* B if B is inside A but they share a boundary. For example, both the yellow and blue counties in Figure 5.4 are contained by the state of Oregon, but only the blue counties are completely contained by the state. The *are within* and *are completely within* operators are the inverse of *contain* and *completely contain*. The *have their centroid in* operator tests whether the center of a feature in A lies inside a feature in B.

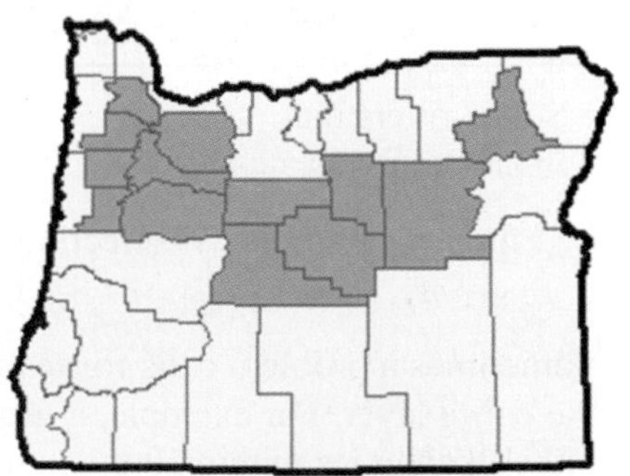

Fig. 5.4. The blue counties are completely contained by Oregon.

Intersection is the most generic operator and returns any feature in A that touches, crosses, or overlaps any part of a feature in B. The *are crossed by the boundary of* operator is a special case of intersection that returns features in A that are crossed only by the boundary of features in B, such as the parcels that are crossed by the city limits. The most stringent condition is that the feature from A equals the feature from B exactly (i.e., they have precisely the same geometry). This corresponds to the operator *are identical to*.

Proximity tests how close features in A are to features in B. The most generic test, *are within a distance of*, selects features in A that are within a certain distance of B, such as returning all parcels within two miles of a school. Adjacency is a special case of proximity where the distance goes to zero and the boundaries of the features actually touch each other. These conditions are covered by the *share a line segment with* and *touch the boundary of* operators. Figure 5.5 shows several examples of applying various operators to two different feature classes.

TIP: A complete description of the different spatial operators with graphic examples can be found in the ArcGIS Help by searching in the Index for *Select By Location, described.*

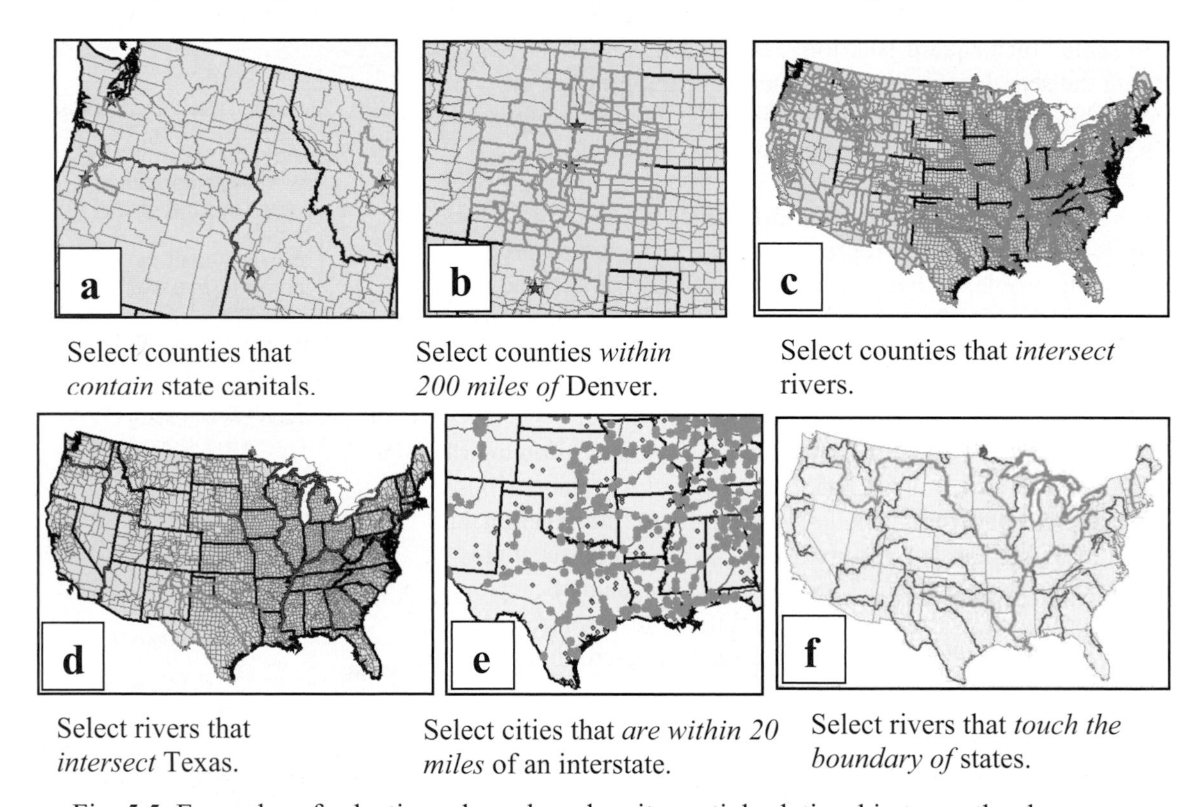

a
Select counties that *contain* state capitals.
b
Select counties *within 200 miles of* Denver.
c
Select counties that *intersect* rivers.
d
Select rivers that *intersect* Texas.
e
Select cities that *are within 20 miles* of an interstate.
f
Select rivers that *touch the boundary of* states.

Fig. 5.5. Examples of selecting a layer based on its spatial relationship to another layer

Sometimes a problem calls for testing spatial relationships using a *subset* of features rather than the entire layer. For example, a user might want to select cities within 50 miles of Interstate 80, rather than cities within 50 miles of any interstate. This procedure begins with an attribute query to isolate the desired interstate, I-80, and then a spatial query to select the cities.

It must be kept in mind that spatial queries can only select or not select an entire feature. Consider Figure 5.5d, which shows the rivers that intersect Texas. Most of the rivers extend beyond the Texas border. If the within operator is used, then only rivers inside the state are selected (in this case, none of them). A different type of function is needed to extract only the portions of rivers that fall inside Texas. They are called overlay functions and are discussed in Chapter 7.

About ArcGIS

ArcMap offers three ways to select features. **Interactive selection** uses a pointer to select features on the screen. Attribute queries are performed using **Select By Attributes**. Spatial queries are executed using **Select By Location**. After a query, the selected features are highlighted in the map and in the table.

Processing layers with selections

ArcGIS follows an important rule when processing features or records in any of its functions or tools: **after a query is performed and a subset of features or records is selected, any subsequent operation on that layer or table honors the selected set.** You have already seen this rule in action. If you were to select the New England states and calculate statistics, only the New England states would be included. If you selected parcels in a floodplain and then exported the features to a new feature class, only the parcels in the floodplain would be exported. Once a query has been executed on a layer, the layer acts as if the unselected features do not exist.

By default, a layer has no query applied. In this state, all of the features are available for processing by a command. If a Statistics command were requested in this state, all of the features would be included in the calculations. No features are highlighted in the map or the table. Executing a query on a layer pushes it into a new state. The layer has a selection, and any function applied in ArcMap or in ArcToolbox will restrict the operation of the function to the selected features or records.

Sometimes a user may execute a query that has proper syntax but yields no matching records. For example, it is possible to query for cities above 10,000 feet in elevation, but no cities actually meet the condition, and no features would be returned. When this happens, ArcGIS removes the query and sets the layer back to its default state with all features available. Essentially, it does not permit the user to execute a query that returns no records or features.

Interactive selection

In an interactive selection, you use the mouse to click on one or more objects in the map or table until you have all the desired records. For example, to get all the states beginning with A, you could sort the table by state name and then find and click on each record containing states starting with A. Or, if you wanted the states that have a Pacific Ocean coastline, you could look on the map, hold down the Shift key, and choose California, Oregon, Washington, and Alaska.

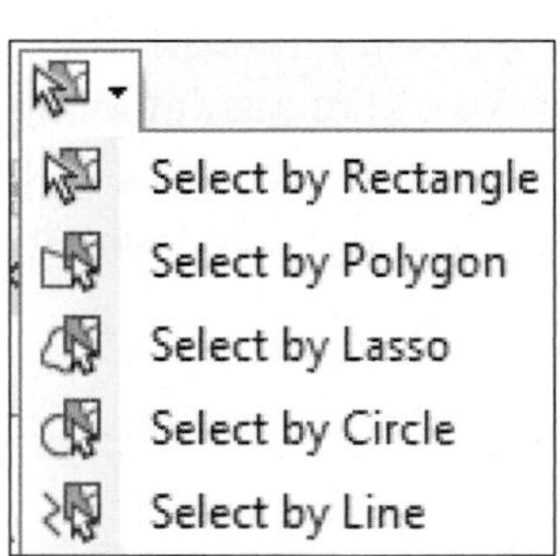

Fig. 5.6. Selecting using shapes

You can use graphics to select features, such as a box, a polygon, a circle, or a line. Two approaches can be used. The Standard toolbar has several tools that allow the user to select features by constructing a shape on the fly (Fig. 5.6). You can also select using a graphic that was already created using the tools on the Drawing toolbar, as shown in Figure 5.7.

When selecting using a shape or graphic object, features only need to touch the object, not be completely enclosed by it. Notice that three states are selected in Figure 5.7b, even though the rectangle does not completely enclose any of them. This default setting can be changed in the Selection options menu, however.

The user can also control which layers may be selected using an interactive query. By default, all layers are selectable, so that the rectangle in Figure 5.7a has selected states, counties, rivers, roads, and more. The Selectable Layers option can turn off selection for one or more layers. In Figure 5.7b, the same rectangle has been used as before, but the user has set the selectable layers to states only.

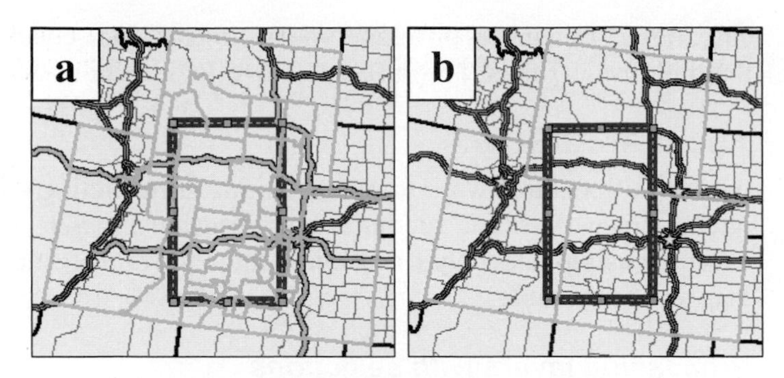

Fig. 5.7. (a) Selecting with all layers selectable; (b) selecting with only states selectable

TIP: The Selectable Layers option applies only to interactive selection with the mouse. It does not affect selecting by attributes or selecting by location using the menus.

Selecting by attributes

SQL expressions in ArcMap are constructed using the fields and the buttons in the Select By Attributes window (Fig. 5.8). The window automatically enters the first part of the expression (SELECT *FROM *layer* WHERE), so in subsequent examples we will drop this part and focus on the text to be entered. Notice a few additional options that can be used in this window.

The check box allows the user to restrict the available layers to include only the ones set as selectable. By default this box is unchecked.

The *Method* option controls what happens to previous selections when this one is executed. By default a previous selection is discarded before the new one is applied. For information about the other options, see the later section "Choosing the selection method."

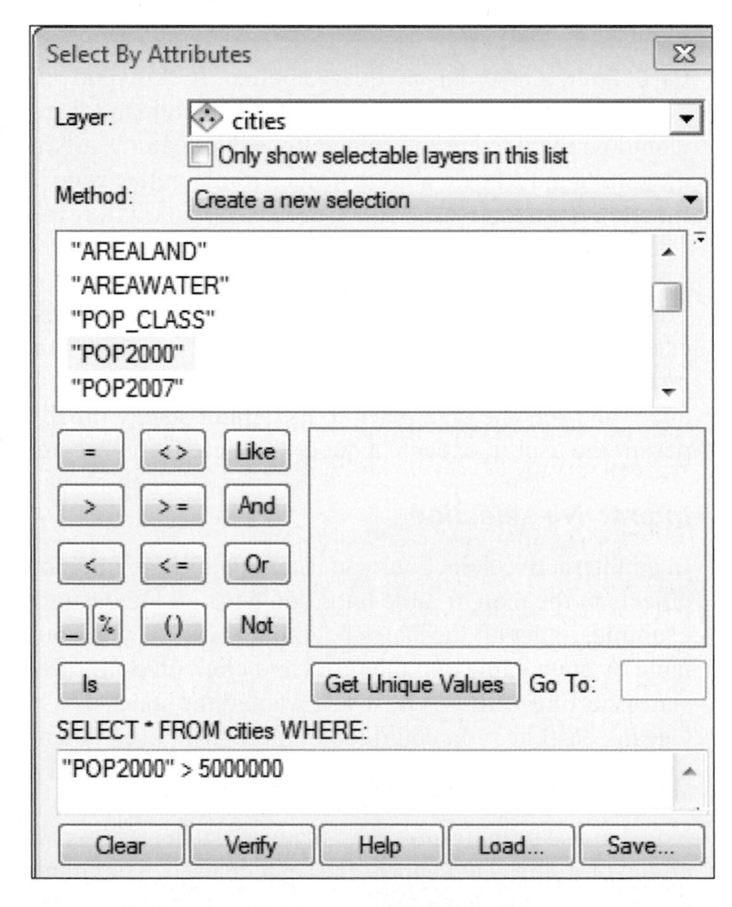

Fig. 5.8. The Select By Attributes window

The *Get Unique Values* button adds values from the table to the window where the user may select them instead of typing them. This button is useful for text and categorical data when you don't know the values to use.

The *Verify* button checks an SQL expression before executing it.

The *Save* and *Load* buttons store queries for use again later or to record what criteria were used, especially for long, complex queries with many conditions. Queries are saved in text files with the extension .exp and can be edited with a simple text editor.

The *Clear* button erases old or invalid queries and starts again.

TIP: Expressions for shapefiles and coverages enclose the field names in quotes. Expressions for feature classes in a geodatabase enclose the field names in brackets.

Selecting by location

The Select By Location window provides the means to set up a spatial query (Fig. 5.9). The target layer refers to the layer being selected, in this case Interstates. They will be the features highlighted after the query. The source layer is the one against which the target layer is being compared. The spatial selection method identifies the spatial operator used to compare the layers. This window is set up to select the interstates that intersect rivers.

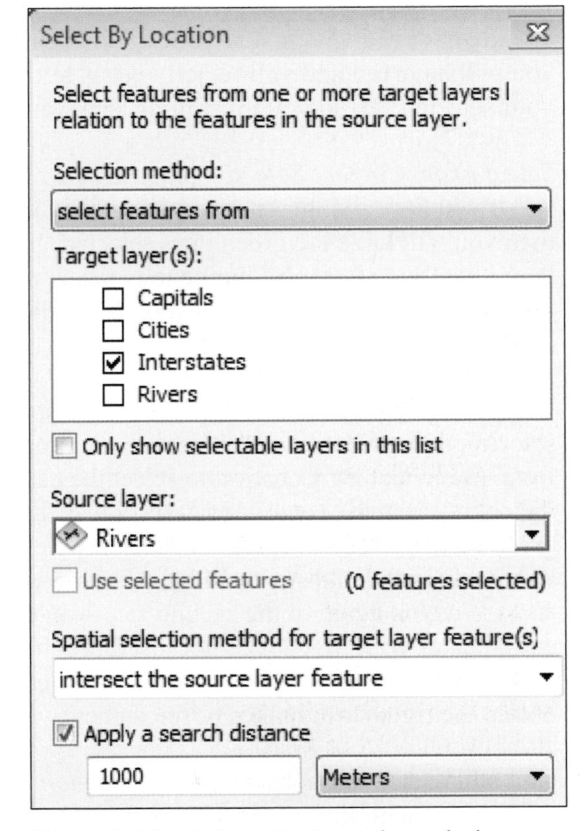

Fig. 5.9. The Select By Location window

If the source layer has a query applied, you must choose whether to honor the selection. For example, if the user previously selected only the Mississippi River and checked the box to use the selected features, then only interstates crossing the Mississippi River would be selected.

The *search distance* option provides a way to manage potential data accuracy issues.

Choosing the selection method

ArcMap has four methods for selecting features (Fig. 5.10). The methods may be specified in two places, from the main Selection menu when doing interactive selection or within the Select By Attributes or Select By Location window.

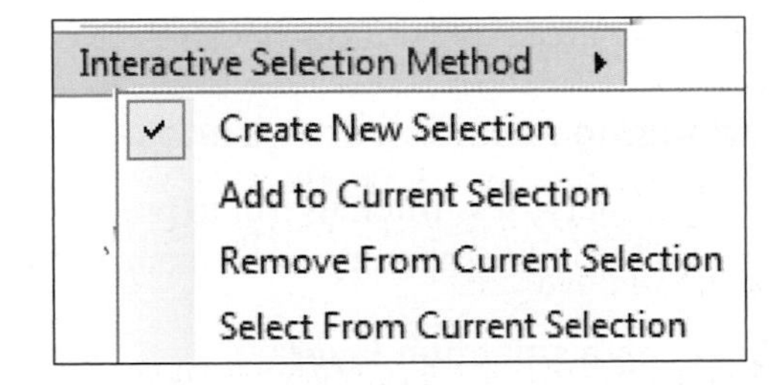

Fig. 5.10. Selection methods

The **selection method** applies to all three types of selections (interactive, selecting by attribute, and selecting by location). These methods give greater flexibility in defining selection sets and enable selection using multiple steps. When performed sequentially, complex expressions involving many criteria are often less confusing than when executed by entering a long expression full of ANDs and ORs and trying to figure out where to put the parentheses.

Create New Selection selects the specified records regardless of what else is already selected, and it replaces any previous selection with the new selected records. For example, if you have selected green jelly beans and then choose to select red jelly beans, then you will only have red jelly beans selected. This method is the default.

Add to Current Selection yields the records that fit the query in addition to any that are currently selected. Thus, if you have green jelly beans already selected and then select the red ones, then you will have both red and green jelly beans selected. Using this method is equivalent to doing a single selection step with a double expression using the OR operator, as when selecting beans where "color" = red OR "color" = green.

Remove From Current Selection takes away the records that meet the selection criteria. If you already have red, green, and yellow jelly beans selected and you "select" green jelly beans, then you will have red and yellow jelly beans left. In this option, what you choose gets deleted from your selected set, so you must think backward on this one. It is equivalent to the NOT operator.

Select From Current Selection chooses the records that fit the criteria from the already selected set. If you have red, green, and yellow jelly beans selected and choose to select the green ones, then you will have the green ones selected. If you choose to select orange ones, then you will end up with *nothing* because orange jelly beans were not in the original selection. Performing several selections in a row with this method is similar to performing a series of AND expressions.

Imagine selecting earthquakes larger than 5.0 magnitude in Mineral County, Nevada (Fig. 5.11). This problem involves selecting by attribute *and* by location, so it occurs in two steps. First, use Select By Location to select the earthquakes within the county, using *Create New Selection* as the method (the green earthquakes in Fig. 5.11). Then use Select By Attributes to choose earthquakes with MAG > 5. However, using *Create New Selection* again in the second step would yield ALL earthquakes with MAG > 5, not just those in Mineral County. You need to specify *Select From Current Selection* in order to obtain the right earthquakes (circled ones).

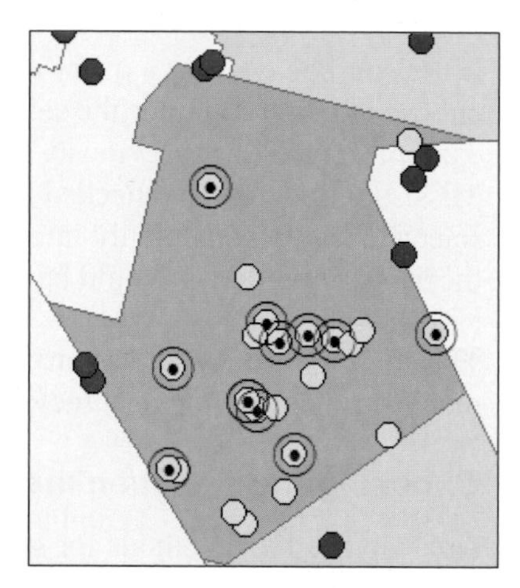

Fig. 5.11. Using selection methods

One other selection option, *Switch Selection,* replaces the selected set with the unselected records. In other words, if you have red, yellow, and green jelly beans available and the red ones are selected, and if you do a switch selection, then you will have green and yellow ones selected.

Managing results from queries

After a query, it is often useful to preserve the selected features for future use. Several options are available, both temporary to permanent.

Creating a selection layer

In Figure 5.12, the user selected the southeastern states and created a layer from the selected features, resulting in the new states selection layer shown above the original states layer.

Creating a selection layer has several advantages. The layer can be given its own symbols and can be displayed separately from the original layer. The layer preserves the selected features for future reference and eliminates the risk of accidentally clearing the selection and having to redo it. Selection layers can be used to input the same set of features to several tools. They are helpful for viewing and recording intermediate results in a complex series of queries. The layer can also be saved as a layer file for use in other map documents.

Fig. 5.12. The South Atlantic states were selected and used to create a layer.

Creating a layer is the best solution when a selected set is intended for use within a single map document (although it can also be saved as a layer file for use in other map documents). It does not create multiple copies of the data, and edits to the source data will be updated automatically. It is the best solution when you need a temporary duplicate of features from your own data or from an organizational database shared by multiple users.

Definition queries

A **definition query** is a property of a layer. It is similar to a selection layer, except that instead of being created from a selected set, the query is performed in the layer properties window. Like selection layers, definition queries refer back to the original database rather than creating a copy of it, and they share the same advantages and disadvantages.

Exporting data

The **export** function saves a new feature class that contains only the selected features, and it is used when a permanent copy of the selection is needed. Because it creates a physical copy of the data, the link to the original feature class is broken. This can be a disadvantage, such as when an organizational database with frequent updates is the source. Changes made to the database will not be reflected in the exported file, which may eventually become out of date. On the other hand, it can be an advantage if the user needs to make changes but does not have write access to the original data. Exports may also be used to develop a new geodatabase based on a subset of existing data, such as developing a Texas geodatabase using feature classes from a US geodatabase.

The Selection window

The Table of Contents has a viewing window that facilitates working with queries, the List By Selection window (Fig. 5.13). It reports which layers currently have selections applied and how many features are selected (Counties and Capitals), which layers are selectable but have no current selection applied (Rivers and Cities), and which layers are not available to interactive selection (States and Interstates). The dim layers are not currently turned on to be visible on the map. This window also allows the user to set selection options and access context menus for various selection options and functions.

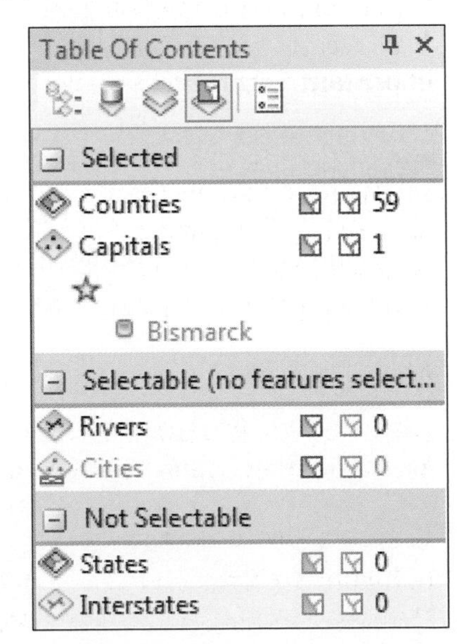

Fig. 5.13. The List By Selection view

Summary

- Queries extract a subset of records or features from a data set based on a set of one or more conditions.
- Attribute queries extract features based on conditions from fields in the attribute table. They are written in SQL and use operators such as =, >, <, AND, and OR. Partial matches to text strings are evaluated with the LIKE operator.
- Spatial queries extract features based on criteria of how two layers are spatially related, using spatial operators to test containment, intersection, or proximity/adjacency.
- Data accuracy and scale issues may affect spatial queries and need to be considered in evaluating the methods and results.
- ArcGIS uses only the selected features in a layer when a function, command, or tool is applied to the layer. If no query has been applied, all features are used. If a query has been applied, only the selected features are used.
- An interactive query uses the Select Features tool and lets the user pick certain features by finding them on the screen and clicking them. Attribute queries are implemented using Select By Attributes, and spatial queries use Select By Location.
- The Selectable Layers option restricts the operation of the Select Features tool to the specified layers. By default all layers are selectable.
- Four selection methods add greater flexibility to queries. These methods include creating a new selection, adding to the current selection, removing from the current selection, and selecting from the current selection.
- Creating a new layer from a set of selected features allows the selection to be stored, displayed, and passed on to tools or commands. Exporting creates a permanent copy.

Important Terms

attribute query	expression	Select By Attributes	Structured Query Language (SQL)
Boolean operators	interactive selection	Select By Location	
containment	intersection	selection method	Venn diagram
definition query	logical operators	spatial operators	
export	proximity	spatial query	

Chapter Review Questions

1. What is a query?

2. Write a valid SQL expression to select cities between 1000 and 10,000 people using a field called POP2000.

3. Write a valid SQL expression to select all counties whose names begin with the letter Q.

4. Let T be a table containing all students attending a community college in New York. Let A be the subset of students living in New Jersey. Let B be the students with a GPA greater than 3.0. The query A AND B yields 200 records. The query A OR B yields 1100 records. The query A NOT B yields 400 records. Construct a Venn diagram for the sets, labeling each section with the number of students. How many students live in New Jersey? How many students have a GPA greater than 3.0?

5. From the information in Question 4, can you determine the number of students attending the community college? If yes, state how many. If not, explain why.

6. What does it mean to set the selectable layers? What is the default setting?

7. Imagine that you have some trail mix composed of peanuts, raisins, almonds, cashews, dried cranberries, and chocolate candies colored red, green, yellow, and orange. Imagine that you apply the following set of "queries" to the trail mix:

   Create new selection all candies
   Add to selection cashews
   Remove from selection red and green candies
   Select from selection all nuts and candies

   What do you have selected now? ______________________________

8. For each of the following queries, state whether the syntax is correct or incorrect. If incorrect, explain why.

   a. [ZONE] = 'COM' AND [ZONE] = 'RES'

   b. [COVTYPE] = 'SPRUCE' AND [CROWNCOV] > 50

   c. [POP2000] > 2000 OR [POP2000] < 9000

   d. [INCOME] < 100000 AND [INCOME] > 50000

9. What is an operator? Describe and give examples of each of the following: arithmetic operators, logical operators, spatial operators, and Boolean operators.

10. List some advantages of creating a new layer from the selected features.

Mastering the Skills

Teaching Tutorial

The following examples provide step-by-step instructions for doing basic tasks and solving basic problems in ArcGIS. The steps you need to do are highlighted with an arrow ➔; follow them carefully. Click on the video number in the VideoIndex to view a demonstration of the steps.

➔ Start ArcMap and open the document ex_5.mxd in the MapDocuments folder.

➔ Use Save As to rename the document and remember to save frequently as you work.

Using interactive selection

First, let's experiment with selecting features with the mouse.

1➔ Choose Bookmarks > Texas from the main menu bar.

1➔ Click on the Select Features by Rectangle tool (on the Tools toolbar), and click on the star that represents Austin, Texas.

SKILL TIP: Learn how to set and use bookmarks with map documents (ArcMap Basics).

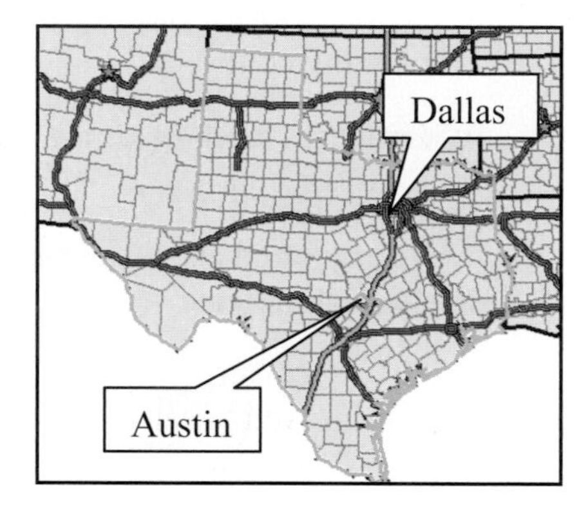

Fig. 5.14. Interactive selection with all layers selectable

Your results may differ slightly, but you should now have several features selected, including Texas, one or more counties near the capital, and an interstate highway (Fig. 5.14). Any feature near the clicked point was added to the selected set regardless of which layer it came from.

1➔ Try clicking several other spots to see what gets selected.

1➔ Click and drag a box around the Dallas area and notice that anything in the box or anything that goes through the box is selected.

You can control which layers are selectable by using the List By Selection window in the Table of Contents.

2➔ Click the List By Selection icon at the top of the Table of Contents.

The upper part of the window shows the layers that currently have selected features, and how many features are selected in each (Fig. 5.15). The lower part shows selectable layers that don't currently have a selection. You can control whether a layer is selectable using the icons in this window.

2➔ Click the open Clear Layer Selection button in the Counties row to clear the counties selection. It moves to the lower panel.

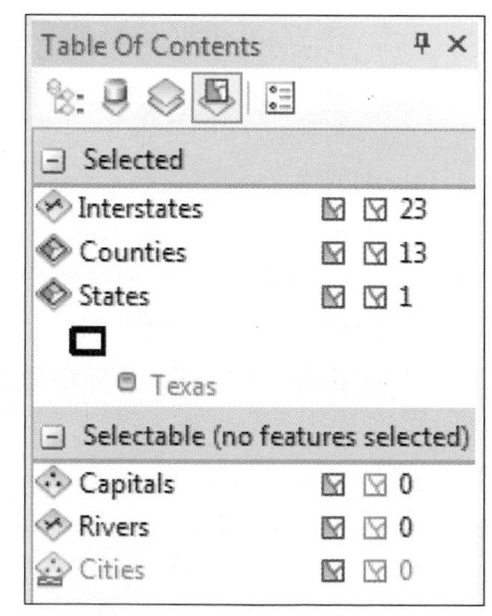

Fig. 5.15. The List By Selection window

2➔ Click the Clear Layer Selection button for the other layers that have selections also.

2➔ Click the Toggle Selectable button to make the Interstates unselectable. Click it again to turn selection for the layer on again.

3➔ Right-click the Counties layer to open a context menu. Examine the choices, and then choose Make This The Only Selectable Layer.

3➔ Use the Select Features by Rectangle tool again to draw a rectangle around the Dallas area. This time, only counties are selected.

3➔ Make Interstates the only selectable layer, and draw a box around Dallas again.

Notice that when a new selection is made the previous selection disappears. This behavior is called the selection method. By default, when a new selection is made, the previous one is discarded. The selection method can be changed.

4➔ Choose Selection > Interactive Selection Method > Add to Current Selection from the main menu bar.

4➔ Click the toggle switch to make the States layer selectable.

4➔ Click somewhere inside the state of Texas. It is added to the selected set.

4➔ Click on Oklahoma, the state north of Texas. It is added as well.

4➔ Click on one of the interstates running west out of Texas. Both the interstate and the state are selected.

The Clear Layer Selection button in the Selection window clears only that layer. You can also clear the selection for all layers.

4➔ Click the Clear Selected Features button on the Tools toolbar, next to the Select Features by Rectangle Tool.

4➔ Choose Selection > Interactive Selection Method > Create New Selection from the main menu bar (returns to the default method).

It's time to experiment some more with interactive selection.

5➔ Choose Bookmarks > Michigan from the main menu.

5➔ Click the Visibility box next to Interstates to turn them off.

5➔ Make Counties the only selectable layer.

6➔ Click one county in Michigan with the Select Features by Rectangle tool.

6➔ Hold down the Shift key and click another county. It is added to the selected set.

6➔ Add three more counties using the Shift-click method.

6➔ Now hold down Shift and click one of the counties that are *already selected.* It will be *removed* from the selected set.

Imagine that you want to efficiently select all the counties in the northern portion of Michigan that lie between Lake Superior and Lake Michigan.

Fig. 5.16. Counties of northern Michigan

7➔ Click and drag a box that goes through as many counties as possible in northern Michigan (Fig. 5.16). Do it several times, if necessary, to get a good result. You will probably not be able to get all the counties, however.

The Interactive Selection Method makes it easier to add to an already selected group.

7➔ Choose Selection > Interactive Selection Method > Add to Current Selection.

7➔ Click on each of the remaining northern Michigan counties. Each one clicked becomes selected with the previous ones, without needing to hold down the Shift key.

7➔ Continue until all of the northern Michigan counties are selected.

8➔ Now add a few southern Michigan counties by drawing a box around them.

8➔ Choose Selection > Interactive Selection Method > Remove from Current Selection.

8➔ Click on or draw a box around the southern Michigan counties to remove them from the selected set.

At this point, you should have all the northern counties selected and none of the southern ones (Fig. 5.16). Recall that the Statistics command shows information about the selected features.

8➔ Right-click the Counties layer and choose Open Attribute Table.

8➔ Right-click the POP2010 field and choose Statistics.

1. How many counties are selected and what is the total number of people in them (in 2010)?

____

8➔ Close the Statistics box and the attribute table.

Now let's get back to testing the different selection methods.

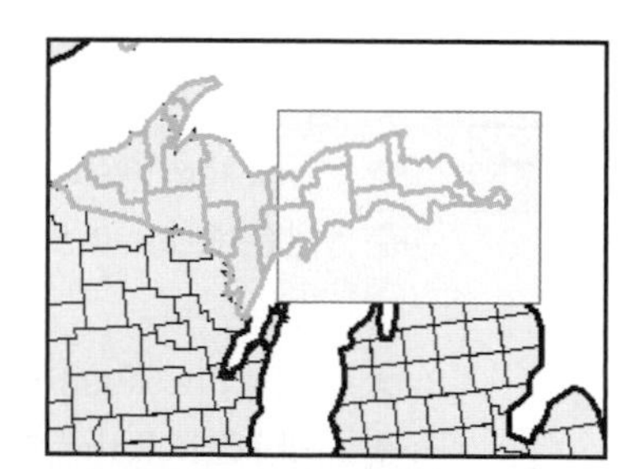
Fig. 5.17. The box

9➔ Choose Selection > Interactive Selection Method > Select from Current Selection.

9➔ Use the Select Features by Rectangle tool to draw a box that encloses the northern peninsula, covering the area shown in Figure 5.17. Be sure to include some of the southern counties in the box.

Notice how the northern counties touching the box were selected, but the southern counties were not because they were not already part of the selected set. Now let's experiment with some other selection tools.

9➔ Set the Interactive Selection method to Create New Selection.

9➔ Click the tool to Clear Selected features on the Tools toolbar.

10➔ Click the Down arrow next to the Select Features by Rectangle tool and examine the other interactive selection tools (Fig. 5.18).

10➔ Choose the Select by Circle tool.

10➔ Click on the state capital, Lansing, and drag to make a circle (Fig. 5.18). Watch the radius measurement in the lower-left corner of the window and try to get it close to 100,000 meters. Release the mouse button when you are satisfied with the circle, and the counties touching the circle will be selected.

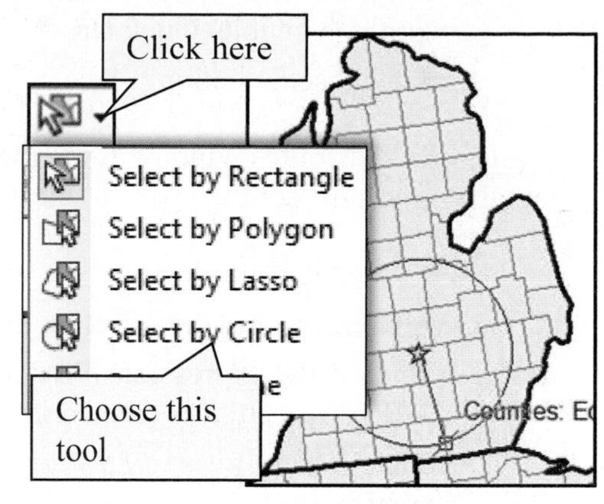

Fig. 5.18. Selecting using a circle

As you drew the circle, you may have noticed that the cursor jumped to align itself with Lansing and various county corners or edges. This behavior is called snapping, and it is often helpful. It made sure, for example, that your circle began exactly on the Lansing point feature. The text that appears indicates which layer is being snapped to, as well as the type of snapping (edge, point, vertex, etc.). The snapping actually made it difficult to control the circle size, but there is another way.

11➔ Click on Lansing again and drag to start the circle.

11➔ With the mouse button still down, press the letter r (for *radius*) on the keyboard. When the box appears you can let go of the mouse button.

11➔ Type the desired radius in the box, 100000 meters, and press Enter.

SKILL TIP: Learn more about snapping (Chapter 12) and how to control it (Editing).

12➔ Choose the Select by Polygon tool.

12➔ Click around the lower peninsula to add the vertices of a polygon that touches all of the southern Michigan counties without including those from other states or the northern peninsula. Double-click to finish the polygon and make the selection.

➔ Take a few minutes to experiment with the other selection tools. When you are satisfied that you know how to use them, clear any selected features and go on with the tutorial.

SKILL TIP: Learn to change the Selection Options, such as the color of highlighted features and whether features must pass through the rectangle/circle or be inside it (Queries).

Selecting by attributes

Chapter 4 demonstrated one way to select by attributes from the Options button in the table. Another method does not require opening the table.

13➔ Choose Bookmarks > USA from the main menu.

13➔ Right-click the Cities layer and choose Properties. Click the General tab.

13➔ Set the scale range to *Show layer at all scales*, and check the *Visible* box to make sure it is visible. Click OK.

13➔ Choose Selection > Select By Attributes from the main menu.

The Select By Attributes window is identical to the one accessed by the Table Options menu, except that you must set the layer to select in the first drop-down box.

14➔ Use the drop-down box to choose Cities as the layer to select.

14➔ Confirm that the selection method is *Create a new selection*.

14➔ Enter the expression "POP2007" > 1000000 (1 million) by double-clicking POP2007 until it is entered in the expression box, clicking the > button, and typing in 1000000.

14➔ Use the Verify button to make sure the expression is correct.

TIP: Most users will find it faster to use the buttons and the windows of the query window rather than trying to type the expressions. If you have trouble making an expression work, then clear the expression box and reenter the expression.

14➔ Click Apply and move the Select By Attributes window to show the Table of Contents, the map, and the window at the same time.

2. How many cities in the United States had more than 1 million people in 2007? ___9___

One advantage of using this window is being able to query another layer from the same window.

15➔ Click the Select By Attributes window again to activate it.

15➔ Change the layer being selected to States. Keep the Create New Selection method.

15➔ Enter the expression to "POP2010" > 5000000 (5 million). Click Apply.

Notice that both cities and states are now selected, even though Create New Selection was the selection method. This happened because States and Cities are separate layers and are treated separately in this query window.

Also note that the states were selected even though they were not a selectable layer. The selectable layer applies only to interactive selection, and it is ignored by the Select By Attributes and Select By Location functions (unless you check the box in the Select By Attributes window to show only selectable layers).

Now let's try a selection with double criteria, all counties that lost population between 1990 and 2000 and have more males than females.

16➔ Clear all selected features using the Clear Selected Features button on the Tools toolbar.

16➔ Click the Select By Attributes box to activate it, if necessary.

16➔ Set the layer to Counties.

16➔ Enter the expression "POP2010" < "POP2000" AND "MALES" > "FEMALES" in the query box. Click Apply.

3. How many counties with more males than females lost population between 2000 and 2010? 209 Where are they mainly located? Rural areas

Now from these we will select those counties that also have fewer than 10,000 people. Be careful to ensure that the new set of counties is pulled only from the currently selected ones.

> 17➔ In the Select By Attributes window, change the selection method to *Select from current selection*.
>
> 17➔ Click the Clear button to erase the old expression.
>
> 17➔ Enter the expression "POP2010" < 10000. Click Apply.

Watch as a few of the counties disappear from the selected set.

4. How many counties remain selected? 177

Springfield seems to be a popular name for cities. Let's see how many there are. Text queries must have quotes around the value being selected.

> 18➔ Clear the selection for the Counties layer.
>
> 18➔ In the Select By Attributes window, change the layer to Cities.
>
> 18➔ Change the Selection method back to *Create a new selection*.
>
> 18➔ Enter the expression "NAME" = 'Springfield'. Click Apply. — 9 cities

When selecting text, using the Get Unique Values field may be valuable and save some typing, but it tends to work best with categorical data. With nominal data like city names, there are so many different choices to scroll through that it is usually faster just to type the desired name.

We might also be curious how many cities end with the word *City*, such as *Oklahoma City*. The LIKE operator is used to search for text within a string instead of an exact match.

> 19➔ Clear the expression box and enter the expression "NAME" LIKE '%City'. Click Apply.
>
> 19➔ Right-click the Cities layer in the Table of Contents and choose Open Table Showing Selected Features to see all the cities that met the query. Make sure that you got what you intended.
>
> 19➔ Close the Table window when finished looking at it.

5. How many city names begin with the word *New*? Be sure to include cities such as New Brunswick, but exclude cities like Newcastle. Check the names in the table. __________

TIP: The Clear Selected Features tool on the Tools toolbar and in the Selection menu clears the selection from ALL layers. To clear the selection for a single layer, use one of the context menus.

> 19➔ Clear all selections and close the Select By Attributes window before going on.

Table of Contents
List by Selection
Shows Count

Selecting by location

Next we will try selecting by location. First we will select all the counties that are adjacent to rivers. Remember, the features you want to select (Counties) are the target layer.

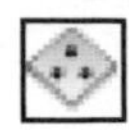

20➔ Click the Visibility icon to turn off the Cities layer.

20➔ Choose Selection > Select By Location, and adjust the window to show both the Selection window and the map.

20➔ In the first drop-down box, choose the selection method to *select features from* (Fig. 5.19).

20➔ Check Counties as the target layer.

20➔ Choose Rivers as the source layer.

20➔ Set the spatial selection method to *intersect the source layer feature.*

20➔ Click Apply and view the map.

6. How many counties in the United States are intersected by rivers? ____________

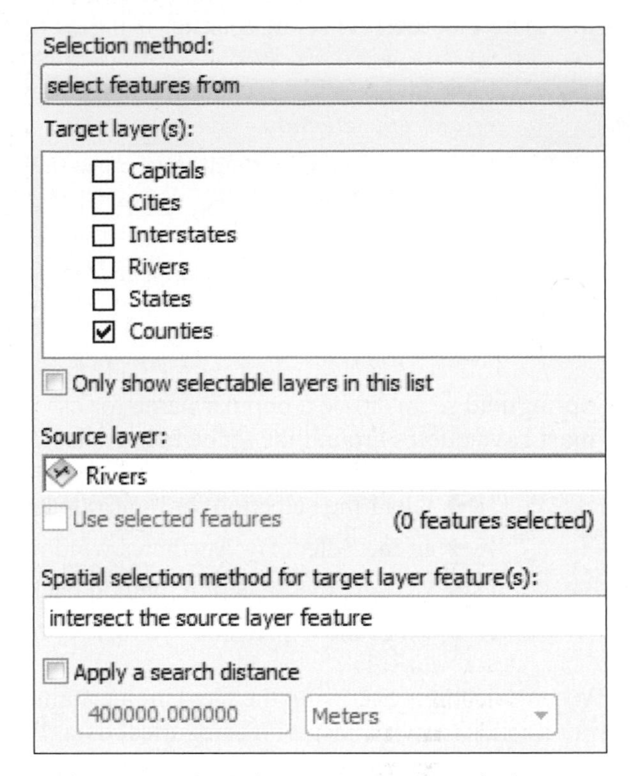

Fig. 5.19. The Select By Location window

Next let's select the counties that contain state capitals.

21➔ Change the source layer to Capitals and the operator to *contain the source layer feature*. Click Apply.

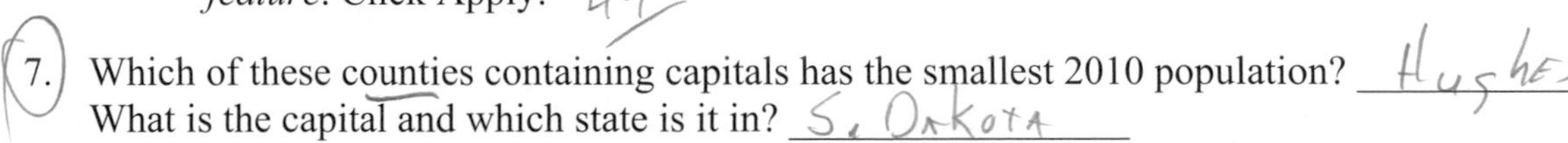

7. Which of these counties containing capitals has the smallest 2010 population? ____________ What is the capital and which state is it in? ____________

Now let's do an example with points and lines by selecting the cities close to interstate highways.

21➔ Click the Visibility icons (now gray) to turn on the Cities and the Interstates.

22➔ In the Select By Location window, change the target layer to Cities (be sure to uncheck Counties as well) and set the Source layer to Interstates.

22➔ Change the spatial selection method to *are within a distance of...(the source layer feature)*.

22➔ Set the search distance to 20 miles.

22➔ Click Apply and view the results. You can zoom in to check the selection, if you like, but then return to the previous extent.

TIP: Layers from a previous query will remain checked in this window until they are unchecked, so get in the habit of making sure only the desired layers are checked.

8. How many US cities are within 20 miles of an interstate highway? ____________

You might wonder what percentage of cities this is, because the map looks full of selected cities. To find out, you need to know how many cities are in the whole layer.

23➔ Right-click the Cities layer and open the attribute table.

23➔ Notice that the bottom of the table says (3321 out of 3557 Selected).

23➔ Use a handheld calculator, or the one on your computer, to calculate the percentage.

9. What percentage of the cities are within 20 miles of an interstate? ____________

TIP: Look carefully at less populous states, like Wyoming and Nevada. Sort the cities by ascending population and examine the CLASS field. Do you think this data set contains all cities in the United States? What impact might the completeness of this data set have on your analysis?

In the previous example, we compared *all* the features in one layer to *all* the features in the other. You can also select features based on a subset, such as selecting the rivers that intersect Texas. We must select Texas first and then perform the Select By Location for the rivers.

24➔ Clear all selected features and close the Table window.

24➔ Turn off the visibility for Cities, Capitals, and Interstates.

24➔ Choose Selection > Interactive Selection Method > Create New Selection on the main menu bar.

24➔ Make States the only selectable layer.

24➔ Click the Select Features by Rectangle tool and select the state of Texas.

25➔ Open the Select By Location window, if necessary.

25➔ Set the target layer to Rivers and the source layer to States.

25➔ To use only Texas, you must check the box to *Use selected features*.

25➔ Set the spatial selection method to *intersect...(the source layer feature).*

25➔ Click OK and view the results.

10. Which rivers intersect Texas? ____

Spatial queries may also be performed within a single layer, using it both as the target and as the criterion layer. The features comprising the criteria must be selected first. For example, select all cities within 400 miles of Kansas City.

26➔ Clear all selected features.

26➔ Turn off the Rivers and turn on the Cities in the Table of Contents.

26➔ Open the Select By Attributes window and make sure the selection method is set to *Create a new selection.*

26➔ Make sure that the box to *Only show selectable layers* is unchecked.

26➔ Select Cities as the layer to select.

26➔ Clear the previous expression and enter a new one: "NAME" = 'Kansas City'.

26➔ Click Apply and close the Select By Attributes window.

27➔ Open the Select By Location window.

27➔ Set both the target and the source layer to Cities.

27➔ Because the target and source layers are the same, you must use the selected set. The box is checked and dimmed, so you cannot change it.

27➔ Change the spatial selection method to *are within a distance of...* and set the distance to 400 miles.

27➔ Click OK and view the map.

Attribute and location queries gain power when combined sequentially. Let's select all the rivers that intersect states in the Mountain subregion. We do this by first selecting the mountain states using an attribute query and then selecting the rivers that intersect these states.

28➔ Clear all selected features. Turn off Cities and turn on Rivers.

28➔ Choose Selection > Select By Attributes.

28➔ Set the layer to States, and enter the expression "SUB_REGION" = 'Mountain'. (This is categorical data, so use the Get Unique Values button.)

28➔ Click OK to close the query window.

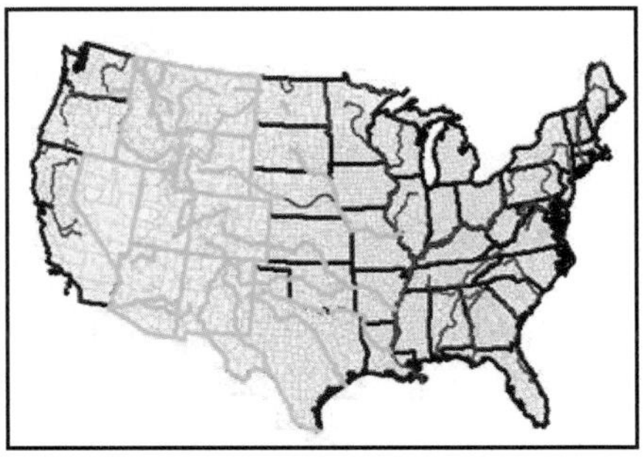

29➔ Open the Select By Location window.

29➔ Change the target layer to Rivers, the source layer to States, and the spatial selection method to *intersect...*

Fig. 5.20. Rivers that cross the Mountain subregion

29➔ Check the *Use selected features* box.

29➔ Make sure that the *Apply a search distance* box is NOT checked.

29➔ Click Apply and view the results (Fig. 5.20).

Notice that many of the rivers extend far to the east to join the Mississippi. Because they do pass through the mountain states, though, they are included. You could try selecting only the rivers that are within the mountain states instead.

29➔ Change the selection method to *are within the source layer feature* and click Apply.

11. Now which rivers are selected? ________________________

Now hardly any rivers are selected. Ideally, you might like to select the *portions* of rivers that fall inside the mountain states. However, Select By Location only works on whole features. With intersect, for example, if even part of a river crosses the state, the entire river is selected. Chapter 7 presents techniques for splitting features when they cross other features.

Combining attribute and location queries can create sophisticated selections. Imagine that a meatpacking company is considering adding the capacity to slaughter bison and wants to survey customers between Pierre and Denver to see if the demand for bison meat justifies the added expense of handling these more difficult animals. The company decides to target the survey to counties that are within 300 miles of both Denver and Pierre and have more than 10,000 people.

30➔ Clear all the selected features and turn on the Capitals layer.

30➔ Right-click the Capitals layer in the Table of Contents and choose Properties. Click the Display tab.

30➔ Check the box to *Show MapTips using the display expression*. Click OK.

30➔ Right-click Capitals in the Table of Contents and choose Make This The Only Selectable Layer.

30➔ Click on the Select Features by Rectangle tool and select Denver in Colorado.

30➔ Open the Select By Location window and set it to select all Counties within 300 miles of the selected features of Capitals (Denver). Click Apply.

31➔ Clear the selection on the Capitals layer only by right-clicking Capitals and choosing Clear Selected Features.

31➔ Choose the Select Features by Rectangle tool, hold down the Shift key, and select Pierre in South Dakota. If you forget to Shift, the county selection will be cleared.

31➔ In the Select By Location window, change the selection method to *Select from the currently selected features in.* Check the *Use selected features* box and click Apply.

31➔ Close the Select By Location window.

TIP: You may encounter a bug in redrawing the selected features. If your selected set does not look like Figure 5.21, try using the Refresh button in the lower-left corner of the map window.

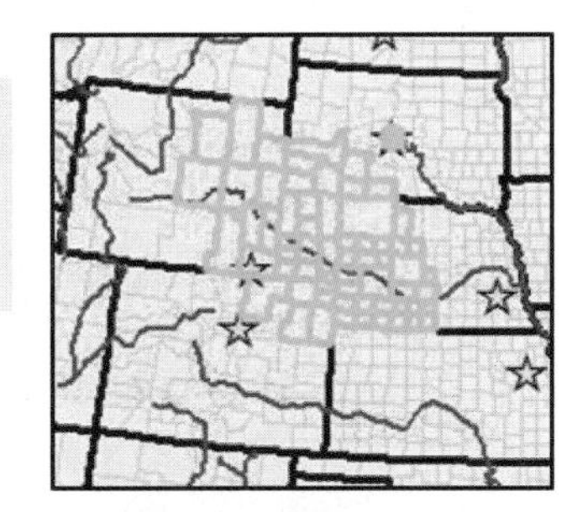

Fig. 5.21. Counties within 300 miles of Pierre and Denver

At this point a patch of counties between Denver and Pierre should show as selected (Fig. 5.21). Now we select from this set the counties with more than 10,000 people.

32➔ Open the Select By Attributes window.

32➔ Set the layer to select to Counties.

32➔ Set the selection method to *Select from current selection.*

32➔ Enter the expression "POP2010" > 10000 and click OK.

32➔ Right-click Capitals and choose Clear Selected Features. The map should look similar to Figure 5.22.

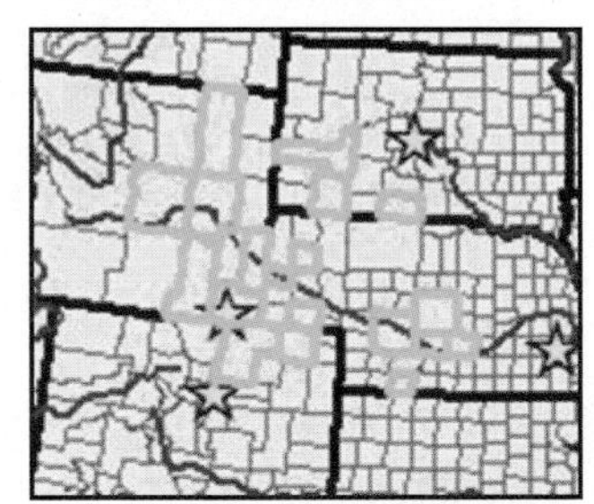

Fig. 5.22. Target counties for the packing survey

33➔ Change the Table of Contents view to List by Drawing Order.

33➔ Right-click the Counties layer and choose Selection > Create Layer from Selected Features.

33➔ Name the new layer Packing Survey. Change the symbol, if necessary, to show up well against the background.

33➔ Clear all selected features.

12. How many counties will be in the survey? __19__ What is the total number of people who live in these counties? __863,967__

Definition queries

A definition query sets a property for a layer that temporarily restricts the layer to features meeting the condition. Imagine that you want to create a map showing only your home state.

34➔ Turn off all layers except States and Counties and zoom to your home state.

34➔ Open the layer properties for Counties and click the Definition Query tab.

34➔ Click the Query Builder button and enter the expression "STATE_NAME" = *'your state name'*.

34➔ After using Verify on the expression, click OK and OK.

TIP: Be sure to Verify when doing definition queries. An error may cause the layer to disappear entirely, which may confuse you.

35➔ Open the properties for the States layer, click the Definition Query tab, and build a query to show only your home state.

The state may appear rotated, especially if it is near the east or west coast. The coordinate system for the data frame is set to US Equidistant Conic, optimized for displaying the entire United States. For a state map, you should choose a coordinate system that is optimized for the state.

36➔ Right-click the Layers data frame name in the Table of Contents and choose Properties. Click the Coordinate System tab.

36➔ Expand the folders for Projected Coordinate Systems > State Plane > NAD 1983 (Meters).

36➔ Choose one of the State Plane zones for your state. If a Central zone is listed, it will likely have the least distortion. However, any of them will work for now.

TIP: Learn more about choosing coordinate systems in Chapter 11.

A definition query will work for a quick map, but for an extended project it is more convenient to create a copy of the data with only the selected feature(s) in it. We'll try that next, but first remove the definition queries on these layers.

37➔ Open the layer properties for States and click on the Definition Query tab. Delete the query expression and click OK.

37➔ Open the layer properties for Counties and delete its definition query also.

Exporting data

A layer with a definition query still refers back to the original feature class. It is a temporary exclusion of the features you don't want. Export creates a permanent subset of a feature class that is stored in its own file. You could use it to build a geodatabase for your own state.

38➔ Open the Catalog tab, if necessary, and expand the mgisdata\usa folder.

38➔ Right-click the Usa folder and choose New > File Geodatabase. Give the geodatabase the name of your state (omitting spaces, if any) and click Enter.

39➔ Turn on the Cities layer.

39➔ Use Select By Attributes to select the cities in your state. (Make sure that the selection method is reset to *Create a new selection*.)

40➔ Right-click the Cities layer and choose Data > Export Data.

40➔ Choose to export the *Selected features* (Fig. 5.23).

40➔ Fill the button to *Use the same coordinate system as the data frame*, so that the new file is stored in State Plane coordinates.

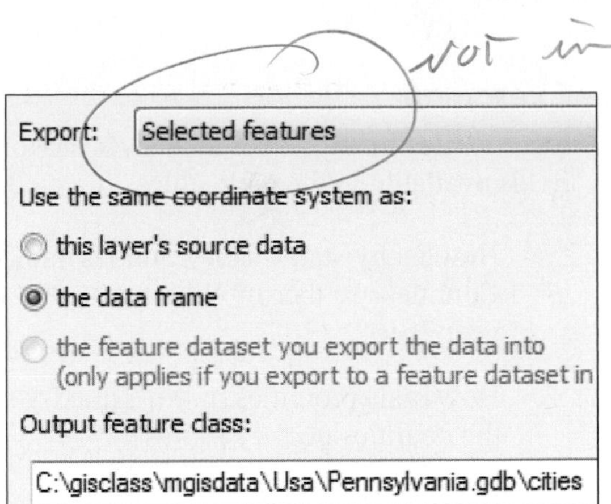

Fig. 5.23. Exporting features

40➔ Click the Browse button. Make sure the *Save as type* is set to *File and Personal Geodatabase feature classes*.

40➔ Navigate inside your new geodatabase and name the feature class cities. Click Save and OK.

40➔ Click Yes to add the new feature class to the map.

40➔ Clear the selected features.

41➔ Repeat Steps 39–40 to select the Counties in your state and export them to your geodatabase, as you did for the cities.

42➔ Repeat Steps 39–40 to select your state from States and export it.

43➔ Save this map document and then open a new blank map document.

43➔ Add the three feature classes from your state's geodatabase to the map and symbolize them as shown for Pennsylvania (Fig. 5.24).

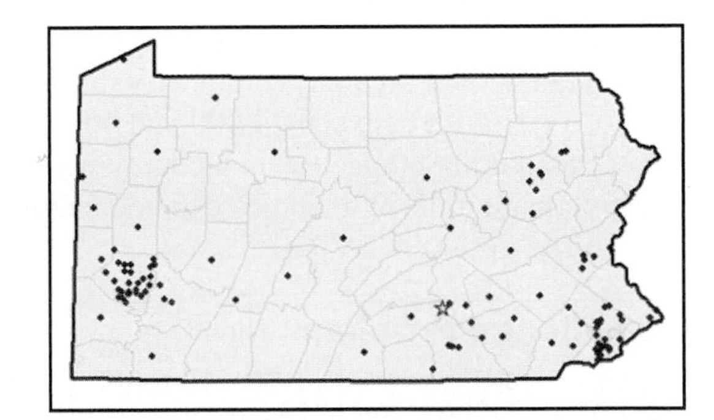

Fig. 5.24. Pennsylvania geodatabase

44➔ Use Select By Attributes on the cities layer to select the capital of your state ("CAPITAL" = 'State').

44➔ Create a layer from the selected feature and name it Capital. Symbolize it with a star.

45➔ Determine how many people live in cities that are within 50 miles of the state capital. __________

This is the end of the tutorial.

➔ Exit ArcMap and save your changes as a map document named for your state.

Exercises

Use the files in ex_5.mxd to answer the following questions. Use the most recent population fields available in the table unless directed otherwise.

1. How many states have counties named for Thomas Jefferson (i.e., how many Jefferson Counties are there)? Which state has the Jefferson County with the most people in the year 2010?
2. How many counties in the United States have more men than women? What percentage of the counties do they represent?
3. How many cities in the United States have more Hispanics than African Americans, a median age greater than 40, and a population between 50,000 and 100,000? List the names of the three largest. (**Hint:** Do your queries in more than one step.)
4. What percentage of counties in the United States have a river intersecting them? What percentage of the US population live in these counties?
5. How many counties have more than 1 million people and contain a state capital? List the states these counties are in.
6. How many cities are within 50 miles of a volcano? (Add the volcano feature class from the Usa\usdata geodatabase.) What is the total number of people living in those cities?
7. How many other volcanoes are there within 300 miles of Crater Lake, a volcano in Oregon? How many of these volcanoes are also within 50 miles of an interstate?
8. How many cities in the West South Central subregion of the United States are less than 200 miles from Oklahoma City? **Capture** a map showing your selected cities.
9. How many capitals are more than 50 miles from a river? Which one has the most people?
10. Imagine that you are looking for a nice place to live. You like being in a small town in a fairly rural area but within easy driving distance of an urban area for fun. You would prefer a city with between 20,000 and 40,000 people, no more than 50 miles from a city with more than 250,000 people, in a county with a population density of fewer than 25 persons per square mile. What cities can you choose from? (**Hint:** Create some intermediate layers as you go.)

Challenge Problem

Congress has awarded FEMA 10 million dollars to help large cities prepare for earthquakes. The cities that qualify for the funding must have more than 500,000 people and be less than 50 miles from one or more earthquakes exceeding 6.0 in magnitude. The bill stipulates that the funding is to be divided among the qualified cities in proportion to their population. Create a feature class that contains only the qualified cities and has a table field listing the amount of funding to be given to each. **Capture** a view of the table showing the city, population, and funding amount.

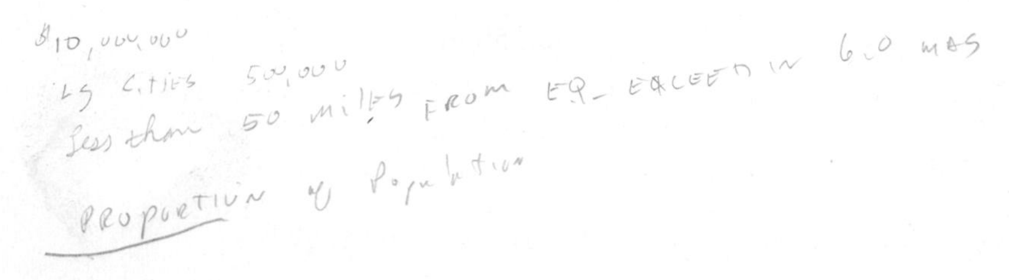

Chapter 6. Spatial Joins

Objectives

- Learning the purpose and capabilities of spatial joins
- Correctly setting up spatial joins based on cardinality and feature type
- Learning to solve problems with spatial joins

Mastering the Concepts

GIS Concepts

What is a spatial join?

Chapter 4 presented attribute joins performed on tables—for example, joining information in an earthquake statistics table to a map of states to create a graduated color map of damages. This join is based on a common field, the state name, and has a cardinality of one-to-one. The join results in combining the two tables as if they were one table. The destination table receives the information from the source table.

A **spatial join** is similar to an attribute join, except that, instead of using a common field to decide which rows in the table match, the *locations* of the spatial features are used. The spatial join uses either a containment criterion (one feature inside the other) or a proximity criterion (one feature close to another).

Like attribute joins, spatial joins designate a source feature class and a destination feature class. Unlike an attribute join, which appends the source attributes to the existing destination table, a spatial join creates a new feature class. It retains the features from the destination layer and appends the attribute information from the source layer (Fig. 6.1). The two original feature classes are unaffected. The destination feature class determines the type of features in the output feature class. If an airports feature class is joined to a cities feature class with cities as the destination, then the output feature class contains cities.

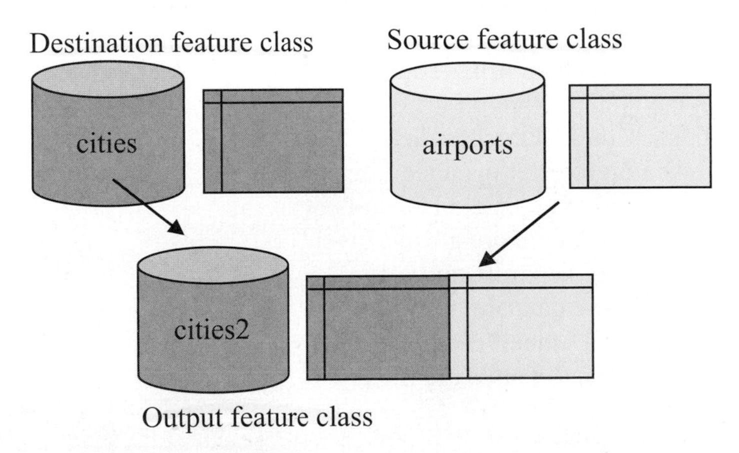

Fig. 6.1. A spatial join keeps features from the destination and appends attributes from the source.

Distance joins

A **distance join** uses a proximity criterion to link one feature and its attributes to another based on whether one feature is closest to another. Figure 6.2a shows the details of the join between airports and cities diagrammed in Figure 6.1. The source feature class, airports, has been joined to the destination feature class, cities. The output feature class contains cities. Each city has been given the attribute information from the airport that lies closest to it, and a new field has been added to record the distance. The attribute table contains two parts, the original data from cities and the joined data from airports. So McNary Field is the closest airport to Adair Village, and Eastern Oregon Regional is the closest airport to Adams.

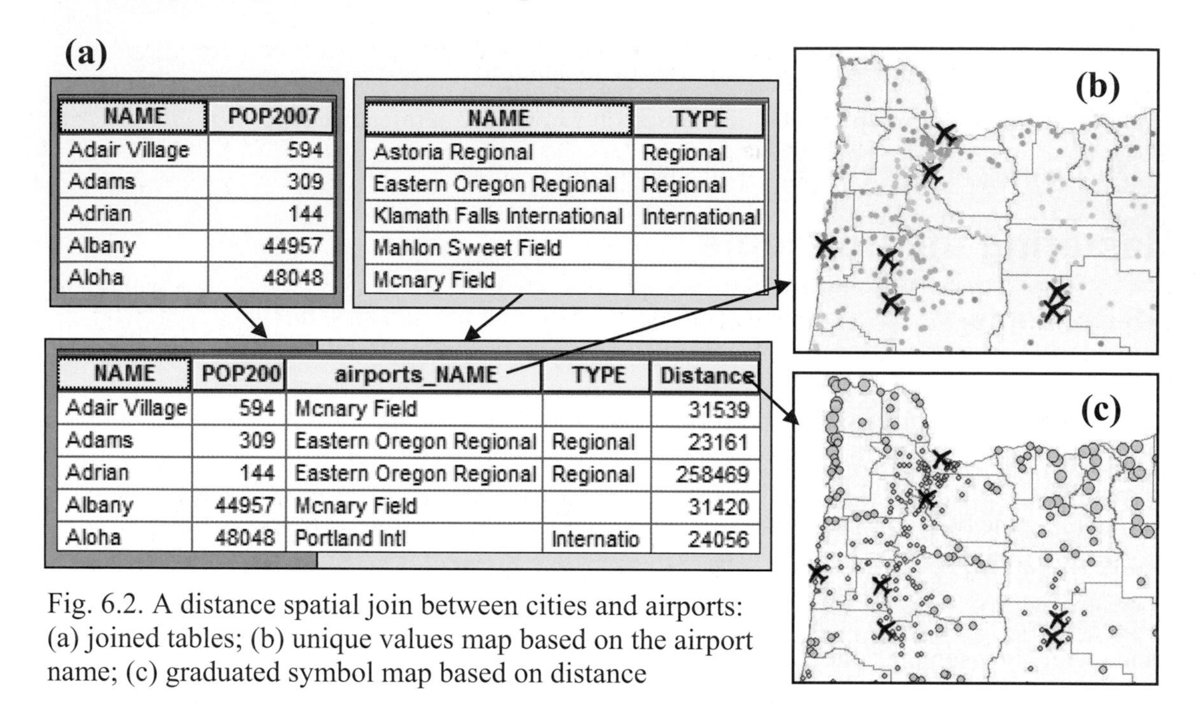

NAME	POP2007
Adair Village	594
Adams	309
Adrian	144
Albany	44957
Aloha	48048

NAME	TYPE
Astoria Regional	Regional
Eastern Oregon Regional	Regional
Klamath Falls International	International
Mahlon Sweet Field	
Mcnary Field	

NAME	POP200	airports_NAME	TYPE	Distance
Adair Village	594	Mcnary Field		31539
Adams	309	Eastern Oregon Regional	Regional	23161
Adrian	144	Eastern Oregon Regional	Regional	258469
Albany	44957	Mcnary Field		31420
Aloha	48048	Portland Intl	Internatio	24056

Fig. 6.2. A distance spatial join between cities and airports: (a) joined tables; (b) unique values map based on the airport name; (c) graduated symbol map based on distance

Two maps have been created from the new joined layer. Figure 6.2b is a unique values map based on the airport name, so each dot gets a color based on the closest airport. The colors indicate which cities are served by each airport, assuming that people will drive to the closest one. The second map is a graduated symbol map based on the distance field, with the larger circles indicating greater distance from the airport (Fig. 6.2c). The units of the distance field are always given in the stored coordinate system units. These data are in the Oregon Statewide Lambert coordinate system, and the units are meters.

Inside joins

In an **inside join**, the records of the feature classes are joined based on whether one feature is inside another (wholly or partly). Figure 6.3 shows a point layer containing septic system locations and a polygon layer showing geological units. Imagine that a community has a porous geological aquifer that provides the city water supplies.

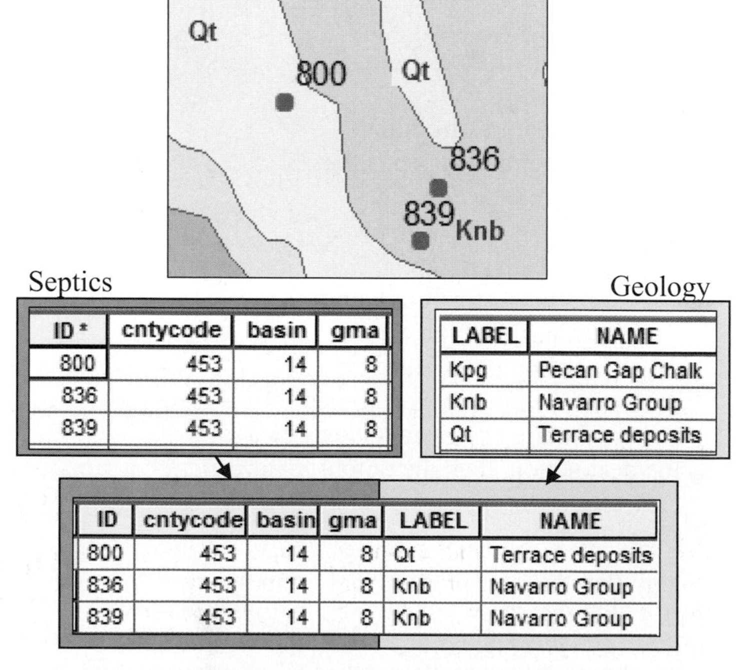

ID *	cntycode	basin	gma
800	453	14	8
836	453	14	8
839	453	14	8

LABEL	NAME
Kpg	Pecan Gap Chalk
Knb	Navarro Group
Qt	Terrace deposits

ID	cntycode	basin	gma	LABEL	NAME
800	453	14	8	Qt	Terrace deposits
836	453	14	8	Knb	Navarro Group
839	453	14	8	Knb	Navarro Group

Fig. 6.3. A spatial join gives each septic system the attributes of the geology polygon within which it lies.

Extensive development outside city limits has caused the city concern about potential contamination of the aquifer by faulty septic systems. Assessing the threat requires identifying the number of septic systems that occur in the outcrop area of the aquifer.

A spatial join solves this problem perfectly. Septics is the destination layer and becomes the output feature class. The geology attributes of the polygon containing the septic system is appended to the output septics table. As shown in Figure 6.3, septic system 800 falls inside the Terrace Deposits (Qt) polygon in the geology layer, and septic systems 836 and 839 fall inside the Navarro Group (Knb) unit. The output table will be helpful in assessing how many septics fall on sensitive geological units.

Spatial joins may be performed on any two spatial data layers. A user can join points to points, polygons to polygons, lines to points, and nearly any combination of the three types of data. The output layer will always have the same feature type as the destination layer.

Cardinality

Cardinality is an important issue for spatial joins, just as it is for attribute joins. Because records in tables are being matched together, the Rule of Joining must be fulfilled in spatial joins also. Each feature record in the destination table must have one and only one matching record in the source table. This condition is met if the cardinality of destination to source is one-to-one or many-to-one. In attribute joins, we had to use a relate if the Rule of Joining was not fulfilled. With spatial joins, we must use a **summarized join** if we encounter a one-to-many relationship.

Recall that the Summarize function calculates statistics for groups of records in a table. It uses one field to divide the records into groups, and then it calculates statistics for other fields for each group. In a summarized join, each feature in the destination layer is matched to many features in the source. Statistics are calculated for that group of features, and the result is appended to the feature record.

In Figure 6.2, imagine reversing the join so that airports is the destination layer and cities is the source layer. Each airport has many cities that are closer to it than to any other airport. Instead of attaching the single closest city, a summarized join finds all of the cities closer to the airport and calculates one or more statistics, for example, the sum of the city populations. Then for each airport we would know the total number of people being served by that airport (i.e., the sum of the populations of the cities that are closer to that airport than to any other). Figure 6.4a shows the output table of the joined layer with a Count field representing the number of cities and the Sum_POP_98 field representing the total people in those cities. So Portland International has 69 cities close to it with a total of 870,598 people. Figure 6.4b shows a proportional symbol map based on the Sum_POP_98 field to represent the total potential population served by the airport.

NAME	TYPE	Count	Sum_POP
Rogue Valley Intern	Interna	22	192645
Astoria Regional	Region	14	34898
Mahlon Sweet Field		23	266623
Klamath Falls Intern	Interna	8	46110
North Bend Muni	Munici	21	99896

(a)

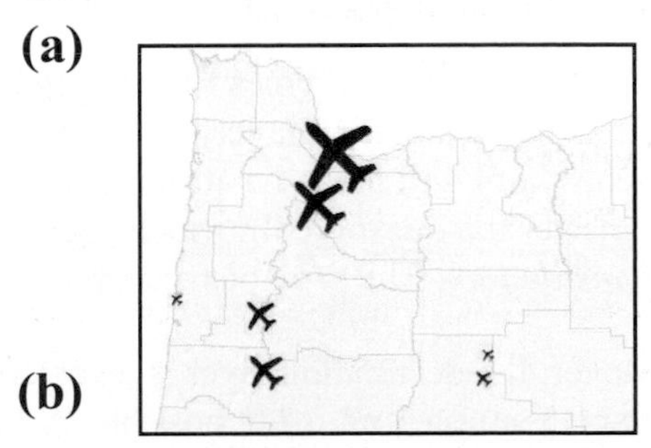

(b)

Fig. 6.4. A summarized spatial join: (a) output table; (b) map based on sum of population served

Types of spatial joins

Spatial joins fall into four main types according to the cardinality of the relationship between the joined layers (simple or summarized) and the choice of spatial criteria (inside or distance). Figure 6.5 shows a matrix of the four possible combinations. A **simple join** may be used whenever the cardinality is one-to-one or many-to-one so that the Rule of Joining is maintained. In a one-to-many relationship, a summarized join must be employed.

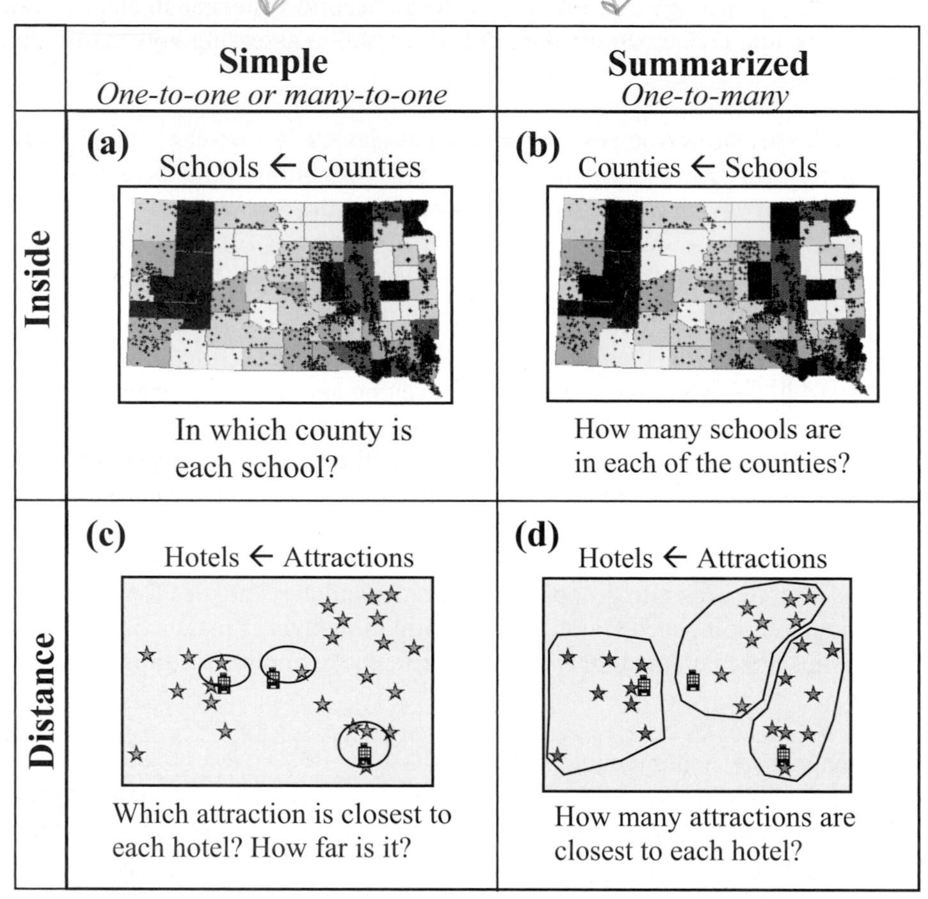

Fig. 6.5. Matrix of spatial joins resulting from different choices of destination table, spatial condition (inside or distance), and cardinality

Consider the simple join types first. In Figure 6.5a, schools and counties are being joined, with schools as the destination layer (some call it the target layer). The output layer contains schools, and each school will have the attributes of the county it falls inside. With the output layer, one could answer the question "In which county does a particular school lie?" This example is a *simple inside join.* The previous example of the septic systems is also a simple inside join.

If, however, the destination layer is reversed, the cardinality of counties to schools is one-to-many, and a simple join is not possible. A summarized join may be employed in this case. The summarized join groups the schools together based on which county they are in and generates a single record of statistics for the group. This single summarized school record can then be appended to the matching county in the destination table. A Count field is always generated, containing the total number of schools in each county. Figure 6.5b shows this as a *summarized inside join.*

Distance joins also come in simple or summarized varieties. Consider evaluating the desirability of several hotels based on their distance to local tourist attractions, using a layer of hotels and a layer of attractions with the yearly number of visitors in each one. What is the cardinality of this relationship? This question seems confusing because there are several hotel sites and many attractions. In a distance join, it is the question that dictates the cardinality. If we ask "Which attraction is closest to the hotel?" in order to find out whether that attraction has a large visitor pool, we have specified a one-to-one criterion because only one attraction can be *closest* to each hotel. A *simple distance join* suffices, and each hotel appears in the output table along with the attributes of the attraction closest to it and the distance between them. In Figure 6.5c, the circles connect each hotel site with its closest attraction.

A different question may be asked: "How many tourist attractions are closer to this hotel site than they are to any other?" In this case, we are interested in evaluating the richness of choice of attractions for each hotel or perhaps the total combined visitor pool from all the closest attractions. In this case, each hotel is connected with many attractions, as shown in Figure 6.5d, and summary statistics for the group, such as the number of attractions and the sum of the visitor pools, may be added to the record for each site. This would be a *summarized distance join.*

Feature geometry and spatial joins

The available join types will depend in part on the geometry of the features being joined. When joining points to points, for example, an inside join type does not apply. Thus, only two options are available when joining points to points, simple distance, or summarized distance. Table 6.1 lists all of the possible geometry combinations and the join types that can be applied to each. (In this table we break the convention of putting the destination first in order to match the descriptions in the spatial join window in ArcGIS.)

Let us examine some other combinations of geometry types. Consider this problem: Many counties in South Dakota do not have a hospital. The state emergency planning office wishes to know the hospital closest to each county and how far away it is. They require a list of counties, each with the closest hospital attached to it. Thus, counties is the destination layer, and hospitals is the source layer. According to Table 6.1, joining points to polygons requires either a summarized inside or a simple distance join. In this case, the simple distance join is the correct choice. After the join, each county has a field indicating the name of the closest hospital (Fig. 6.6) as well as the distance from the county to the hospital. When the hospital is inside the county, the distance of zero is assigned.

cntyhosp

NAME	hospitals_NAME	Distance
Shannon	Battle Mountain National Sanitar	38424
Fall River	Battle Mountain National Sanitar	0
Bennett	Battle Mountain National Sanitar	110128
Pennington	Bennett Clarkson Hospital	0
Lincoln	Canton-Inwood Hospital	0
Yankton	Canton-Inwood Hospital	51493

Fig. 6.6. This table resulted from a distance join of hospitals to counties.

Table 6.1. Join types are available for each combination of feature geometries in a spatial join. The second feature type is the destination layer in each case.

Geometry Type	Join Type	Example
Points to points	Simple distance	Find the hospital closest to each town.
	Summarized distance	Find all the towns closer to one hospital than to any other hospital.
Lines to points	Simple distance	Find the water main closest to the proposed building site.
	Summarized inside	Find the total voltage of all electric lines meeting at a substation.
Polygons to points	Simple inside	Find the soil type that underlies each gas station.
	Simple distance	Find the lake that is closest to each campground.
Points to lines	Simple distance	Find the elementary school that is closest to each residential street.
	Summarized distance	Find the total number of septic systems closer to a particular stream than to any other stream.
Lines to lines	Summarized inside	Find the number of roads that cross each river.
	Simple inside	Give a section of hiking trail the attributes of the road it follows for a short distance.
Polygons to lines	Summarized inside	Give a stream the average erosion index of the soil types it crosses.
	Simple distance	Find the lake closest to a hiking trail or the national park within which a road lies.
Points to polygons	Summarized inside	Find the total number of schools and students in a county.
	Simple distance	Find the town that is closest to a lake. A point inside a polygon is given a distance of zero.
Lines to polygons	Summarized inside	Find the total number of rivers crossing a state.
	Simple distance	Find the carrying capacity of the closest power lines to an industrial site.
Polygons to polygons	Summarized inside	Find the total population of all counties that intersect part of a watershed.
	Simple inside	Find the county within which a lake falls completely.

In Figure 6.7, a unique values map was created based on the hospital name. It is easy to see that, in the northeast part of the state, the purple counties are closest to the hospital in the purple area, the green counties to the hospital in the green area, and so on. The long, hatchet-shaped county in western South Dakota (Pennington County) has several hospitals. Since all three lie inside

Pennington, they all have a distance of zero, and Pennington's closest hospital was arbitrarily assigned based on which hospital was found in the table first. The lighter pink counties in the central part of the state were assigned to a different hospital in Pennington County and appear in a different color. The map in Figure 6.8 shows each county displayed according to its distance to the closest hospital.

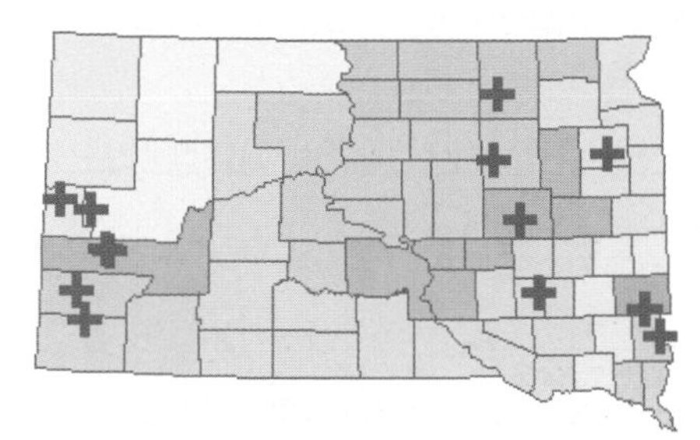

Fig. 6.7. Counties with the same color are closest to the same hospital.

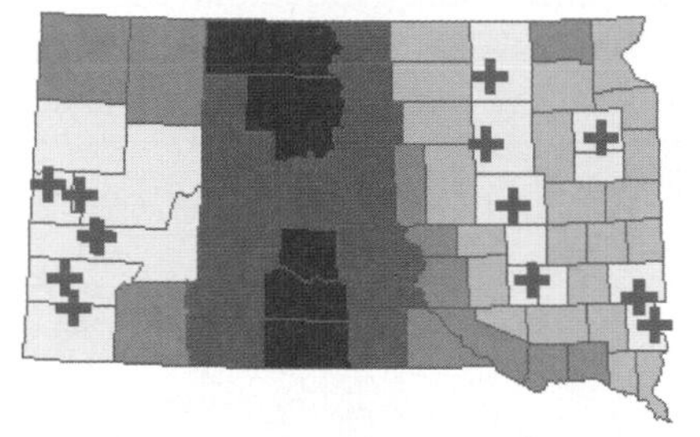

Fig. 6.8. Counties colored according to distance from the closest hospital

Figure 6.9 demonstrates an example of joining points to lines to predict the susceptibility of streams to contamination from septic systems. In this analysis, we make the assumption that, the more septic systems that are close to the stream, the greater the stream's susceptibility to contamination. The point locations represent the centers of one-mile by one-mile sections, and the size of the symbol indicates the number of septic systems in the section. We need to join each stream line to the closest septic systems, so streams is the destination layer. Joining points to lines requires either a simple distance join or a summarized distance join. In this case a summarized distance join, which sums the total septic systems that are closest to each stream, is correct. The map shows the results: the thicker the line symbol of the stream, the more septic systems are closest to that stream and the higher the susceptibility of the stream to contamination.

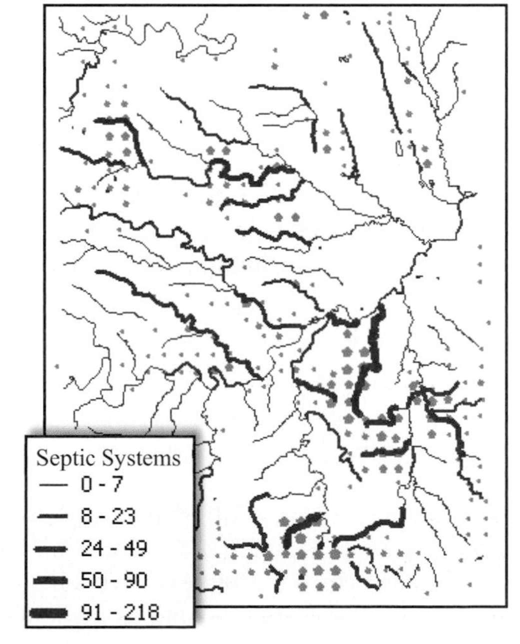

Fig. 6.9. Joining septic systems to streams to evaluate stream susceptibility to contamination

Coordinate systems and distance joins

Distance joins should always be performed on layers with projected coordinate systems. If a join is performed on a layer with a geographic coordinate system (GCS) and units of decimal degrees, two problems arise. First, the distances reported in the table will be in decimal degrees. Degrees cannot be easily converted to miles or kilometers because the conversion factor changes with latitude. Second, the result could be invalid. The distance algorithm relies on the relative *x-y* coordinates of the features to calculate distances and assumes a Cartesian coordinate system to do so. Degrees of latitude and longitude are spherical coordinates, not Cartesian, and the relative distances calculated may be incorrect.

Figure 6.10 shows two cases of a summarized distance join of tourist attractions to hotels. In Figure 6.10a, the join was done with both layers in a GCS. Notice that the three attractions in the northeast corner were assigned to Hotel C based on their distance from it in decimal degrees. In Figure 6.10b, both layers were projected to UTM prior to joining, and the attractions were assigned instead to Hotel B. Since a UTM projection preserves distance within the zone, we know that the second example gives the correct spatial distances between the attractions and the hotels and is the valid result. Notice that the spatial distribution of attractions is elongated in Figure 6.10a, as a result of being displayed in a GCS.

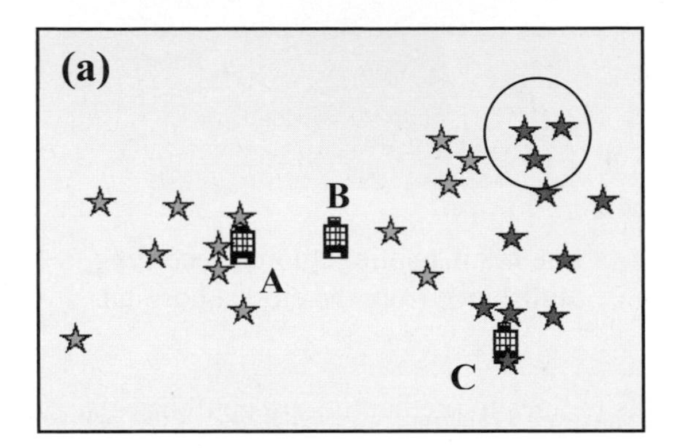

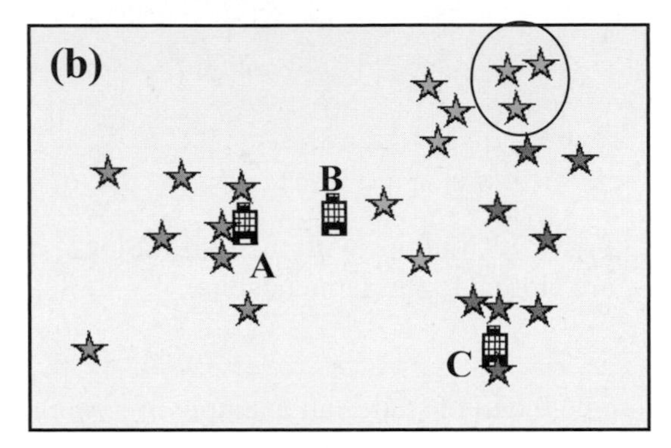

Fig. 6.10. The effect of the coordinate system on a spatial distance join. (a) Three attractions are incorrectly joined to Hotel C (purple stars) when the input feature classes have a GCS. (b) When the feature classes are projected to UTM before joining, the attractions are correctly assigned to Hotel B (green stars).

It is not sufficient to set the data frame to a projected coordinate system; the layers themselves must be projected. Furthermore, the projected coordinate system should be one that preserves distance over the region of analysis, such as UTM, State Plane, Equidistant Conic, and so on. Otherwise, the distance analysis may again be incorrect.

About ArcGIS

Spatial joins are performed on two input feature classes and result in a new feature class that may be stored as a shapefile or geodatabase feature class. The original inputs are unchanged. Spatial joins are initiated using the same method as attribute joins.

Choosing the join type

As in an attribute join, the process begins by deciding which layer is the destination. The user right-clicks the destination layer, chooses to join the data based on spatial location (Fig. 6.11), and enters the desired source layer.

The join menu always offers two ways to decide how to match the fields, taken from the four types shown in Figure 6.5. Consider joining schools to counties, with counties as the destination layer as described in Figure 6.5a. Each county can have more than one school, so it is a one-to-many join. There are two options in the window based on the choices listed in Table 6.1. The user must choose between them based on the desired result (Fig. 6.11).

For this example, Option 1 offers an *inside summarized join,* which gives each county a statistical summary of the attributes of all the schools inside it, such as the total number of schools or the sum of the students. Option 2 specifies that a *simple distance join* should be performed, which gives each county the attributes of the closest school. In this case, the second option is nonsensical. All of the schools are inside the county and would be assigned a distance of zero, so it avails little to find the "closest" one. In this example, the summarized join is the appropriate choice.

2. You are joining: Points to Polygons

Select a join feature class above. You will be given different options based on geometry types of the source feature class and the join feature class.

(◉) Each polygon will be given a summary of the numeric attributes of the points that fall inside it, and a count field showing how many points fall inside it.

How do you want the attributes to be summarized?

[ ] Average [ ] Minimum [ ] Standard Deviation

[✓] Sum [ ] Maximum [ ] Variance

( ) Each polygon will be given all the attributes of the point that is closest to its boundary, and a distance field showing how close the point is (in the units of the target layer).

Fig. 6.11. A one-to-many cardinality is handled by either a summarized join or a distance join.

Figure 6.12 shows the result of the join. The resulting layer is a map of counties with the original county attributes plus a field called Count_ that contains the number of schools in the county. From this field, the graduated color map was created, showing the number of schools in each county. When summarizing, the user can choose from several statistics—minimum, maximum, average, and so on. All numeric attributes in the source table are summarized using the chosen statistics and are placed in the output table. String fields cannot be summed or averaged, so they are not included in the output table.

Setting up a spatial join

Performing the spatial join itself is a simple process. However, determining that a spatial join is required and identifying the destination table and the type of join can challenge beginners. This section presents a series of questions to be answered when setting up a spatial join to help produce the correct result. Too many students of GIS simply try random combinations until they hit on the right one—this process is designed to give the right answer the first time.

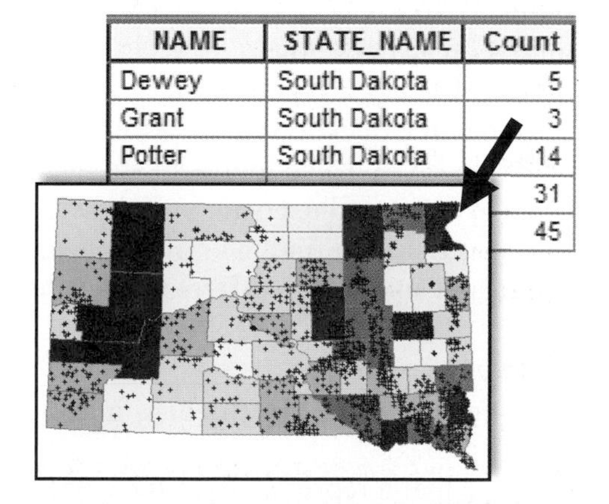

NAME	STATE_NAME	Count
Dewey	South Dakota	5
Grant	South Dakota	3
Potter	South Dakota	14
		31
		45

Fig. 6.12. Using a spatial join of schools to counties, one can create a graduated color map showing the number of schools in each county.

When faced with a suspected spatial join problem, first make a simple sketch of the relationships between the layers to be joined. Then answer this series of questions, designed to lead to the correct solution:

What should the final output layer or table look like?

Which is the destination layer?

Should a distance join or an inside join be used?

What is the cardinality of the join?

Should a simple join or a summarized join be used?

Let us demonstrate this process by using the problem of estimating stream susceptibility to contamination from septic systems, as shown in Figure 6.9. Recall that each point represents a section and stores the total number of septic systems in the section. First sketch the problem and then go on to the questions.

What should the final output layer/table look like? The desired result is a layer of streams. In the attribute table, each stream must be assigned the total number of septic systems that are closer to it than to any other stream. Imagine the fields in the table, with the stream followed by the sum of the number of septic systems. Once the output is envisioned, it is usually easy to see which feature class is the destination layer.

Which is the destination layer? The features in the output layer are always the same features as those in the destination layer. If streams is chosen as the destination, then the output layer will contain streams. If the septics layer is chosen, then the output layer will contain septic points. Imagining our output table again, we see that what we really want is streams, each of which has septic information assigned to it. Thus, streams is the destination layer.

Should a distance join or an inside join be used? Since we're looking for septic systems closest to each stream, this is clearly a distance join.

What is the cardinality of the join? Assignment of cardinality depends on the destination layer, in this case streams, so consider the relationship between streams and the septic systems. Since one stream can have multiple septic points closer to it than to any other stream, this is a one-to-many cardinality.

Should a simple join or a summarized join be used? Since the cardinality is one-to-many, a simple join cannot work. Because the goal is to sum all the septic systems closest to the streams, rather than simply finding the single closest septic to the stream, a summarized join is indicated, with Sum as the statistic.

Now that the questions are answered, the problem setup becomes clear: We need to do a summarized distance join with streams as the destination layer and septics as the source layer. Even experienced users may find that following this suggested procedure when setting up a join makes it easier to find the right approach. In the next few examples, we will apply this process to set up and solve three different spatial join problems.

Problem 1

Number of earthquake deaths in congressional districts

Imagine that a representative from one of the congressional districts in California is sponsoring legislation to provide earthquake emergency planning funds and is looking for support for the bill. She is planning to throw a big party and invite all of the reps from districts with a significant number of earthquake deaths. She asks one staffer, who is a GIS specialist, to draft a list of names for the invitation list. The staffer might approach the problem as follows, using the data in the mgisdata folder.

STOP! Write the answers to the questions and then read on to see if you analyzed the problem correctly.

> What should the output layer/table look like? ______________________________
> Which is the destination layer? ______________________________
> Should a distance join or an inside join be used? ______________________________
> What is the cardinality of this join? ______________________________
> Should a simple join or a summarized join be used? ______________________________

The earthquake table has a state attribute but none for congressional district. The only way to associate the number of earthquakes with districts is to perform a spatial join. The goal is a table of districts with a field containing the number of earthquake deaths for each one. Thus, the districts will be the destination layer. The relationship between districts and earthquakes is potentially one-to-many, so a simple join is out of the question. Since the staffer wants to know the total deaths from quakes in each district, the summarize option will be the best, and the proper statistic is Sum. We will do a summarized inside join, with districts as the destination layer.

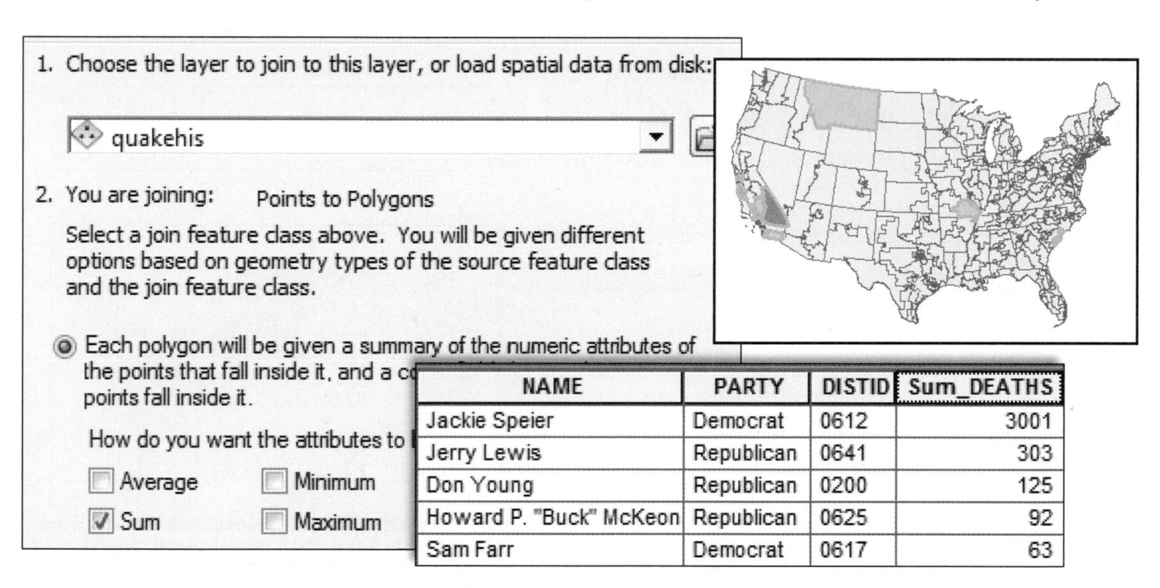

NAME	PARTY	DISTID	Sum_DEATHS
Jackie Speier	Democrat	0612	3001
Jerry Lewis	Republican	0641	303
Don Young	Republican	0200	125
Howard P. "Buck" McKeon	Republican	0625	92
Sam Farr	Democrat	0617	63

Fig. 6.13. Using a summarized spatial join to find the congressional districts that have had more than 10 earthquake-related deaths

Figure 6.13 shows the spatial join window filled out for this problem, as well as the resulting table. After performing the join, the staffer uses an attribute query to select all of the districts that have had 10 or more earthquake-related deaths. The table and the map show the selected records from the query—many of the districts are in California, as one might expect. Finding only 12 districts on the list and knowing that the boss wants a BIG party, the staffer then uses Select By Attributes to find all the districts that have had ANY earthquake deaths. This brings the total up to 29 districts. The guest list should be even bigger, so the staffer decides to use Select By Location to

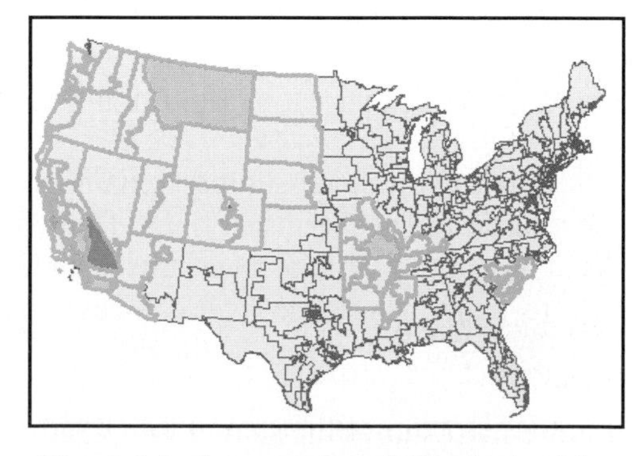

Fig. 6.14. Congressional districts with earthquake deaths or that touch a district with earthquake deaths

also select any districts that touch the districts with earthquake deaths (Fig. 6.14). This brings the guest list to 91, and the staffer can make up the invitations.

Problem 2

Pollution risk of rivers based on county populations

Imagine that Lindsey must do an analysis estimating the risk of pollution to rivers in the United States. She has no detailed information of the sources and types of pollutants, but she can make use of the observation that, generally speaking, large numbers of people are generally correlated with high risks of pollution. She decides to use the population of counties as a proxy for the pollution risk. For each river, she wants to find the total number of people living in counties that intersect the river.

STOP! Write the answers to the questions and then read on to see if you analyzed the problem correctly.

What should the output layer/table look like? ______________________________

Which is the destination layer? ______________________________

Should a distance join or an inside join be used? ______________________________

What is the cardinality of this join? ______________________________

Should a simple join or a summarized join be used? ______________________________

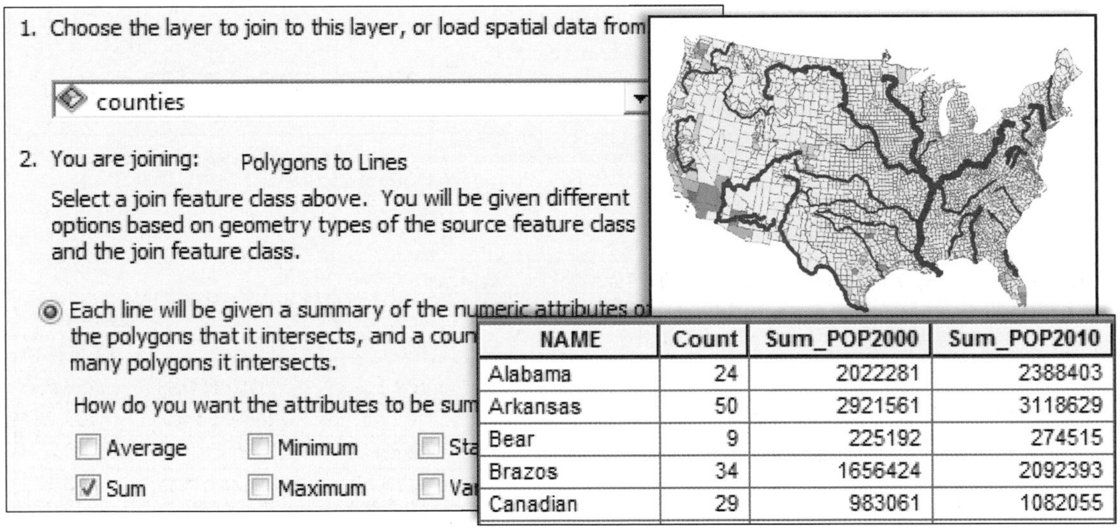

1. Choose the layer to join to this layer, or load spatial data from
counties
2. You are joining: Polygons to Lines
Select a join feature class above. You will be given different options based on geometry types of the source feature class and the join feature class.
Each line will be given a summary of the numeric attributes of the polygons that it intersects, and a coun many polygons it intersects.
How do you want the attributes to be sum
Average Minimum Sta
Sum Maximum Va

NAME	Count	Sum_POP2000	Sum_POP2010
Alabama	24	2022281	2388403
Arkansas	50	2921561	3118629
Bear	9	225192	274515
Brazos	34	1656424	2092393
Canadian	29	983061	1082055

Fig. 6.15. Estimating pollution risk for rivers based on adjacent county populations

The final output layer should contain rivers, with a table listing each river and the total population of counties adjacent to that river. Thus, rivers must be the destination layer. Since the counties must actually touch the river, this is an inside join rather than a distance join. The cardinality is one-to-many, since one river can have many counties touching it. An inside summarized join must be used with rivers as the destination layer, and the SUM statistic should be requested.

Figure 6.15 shows the spatial join window filled out for the join. The output table shows the river name, the number of counties adjacent to it, and the sum of each numeric field, including the POP2000 and POP2010 fields. The figure also shows a graduated symbol map based on the Sum_POP2010 field, in which the width of the river represents the total county population living adjacent to it. Notice that ALL of the numeric fields are summed during the join, not just the POP2000 field. One drawback to summarized spatial joins is that the chosen statistic is applied to all the fields, yielding potentially large attribute tables.

Also note that the field names appear truncated. Because field names are limited to 13 characters, prefixing each with "Sum_" may crop other characters from the end. Notice that the POP2000 field has become Sum_POP200, the POP00_SQMI field has become Sum_POP00_, and so on. The shorter names may cause confusion, and it may be necessary to go back to the original table of the source layer to find out which field is which. The fields will be in the same order as the original. An alias can then be created, if desired, to avoid further confusion.

Problem 3

Closest volcano to a city

In this final example, imagine that a professor who specializes in volcanoes has built a website about them. He would like to put a table on his web site so that schoolchildren all over the United States can enter the name of the city they live in and get an information page about the volcano that is closest to them. Since the professor knows nothing about GIS, his graduate student, Cody, gets to make the table.

STOP! Write the answers to the questions and then read on to see if you analyzed the problem correctly.

What should the output layer/table look like? ______________________________
Which is the destination layer? ______________________________
Should a distance join or an inside join be used? ______________________________
What is the cardinality of this join? ______________________________
Should a simple join or a summarized join be used? ______________________________

1. Choose the layer to join to this layer, or load spatial data from disk:

volcanos

2. You are joining: Points to Points

Select a join feature class above. You will options based on geometry types of the s and the join feature class.

Each point will be given all the attributes of the point in the laye being joined that is closest to it, and a distance field showing how close that point is (in the units of the target layer).

NAME	ST	volc_NAME	TYPE	Distance
Algonquin	IL	Dotsero	Maar	1,505,644
Alhambra	CA	Lavic Lake	Volcanic field	151,255
Alice	TX	Durango Volc	Cinder cones	738,157
Aliquippa	PA	Dotsero	Maar	2,139,036
Aliso Viejo	CA	Lavic Lake	Volcanic field	162,176

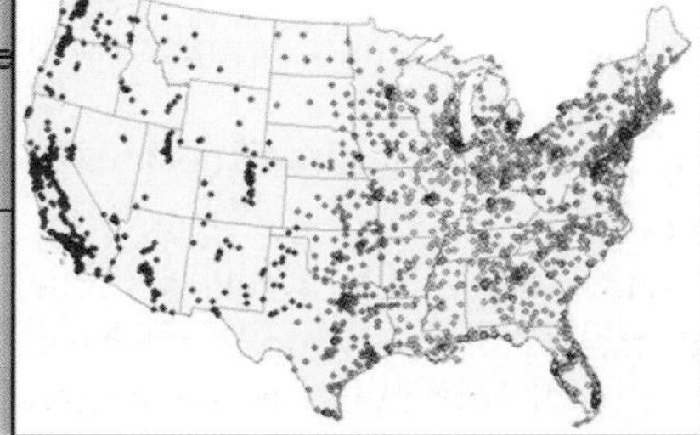

Fig. 6.16. Table showing each city and its closest volcano, with the distance to the volcano in meters: graduated color map is based on the Distance field

The final goal is a cities layer with fields indicating the closest volcano and the distance to it. Thus, Cody realizes, Cities is the destination table, and a distance join to the volcanoes layer will provide the necessary information. Because the closest volcano is the target, this is a simple distance join (only one volcano can be the closest one).

Figure 6.16 shows the spatial join window settings to produce this analysis and the result of the spatial join. The table shows the city name, the state abbreviation, the volcano name, the place where the volcano is located, the volcano elevation and type, and the distance to the volcano.

Notice the field name volcanoes_NAME in the output table. It happens that the Cities layer has a NAME field, and so does the volcanoes layer. The output table cannot have two fields with the same name, so during the join the repeated names are automatically converted to unique names. The Cities name field remains NAME, but the Volcanoes name field becomes NAME_1. In addition, an alias is created, volcanoes_NAME. It is the alias that is being displayed in the table, but if the table is set to show actual field names, then the NAME_1 field will be shown instead.

Also notice that the distance from Boone, Iowa, to the Dotsero volcano in Colorado is over one million. These units are clearly not miles or kilometers. In fact, they are meters, indicating that the coordinate system of the original data is stored with units of meters.

TIP: Always check the original coordinate system after a join to be sure you know the units for the distance values. Also, remember that distance joins should be performed only on projected coordinate systems that preserve distance, or the results will be invalid.

Summary

- A spatial join combines the records of two feature tables based on the location of the features. A new feature class is created by a spatial join.
- An inside join uses the criterion that one feature falls inside another or, in the case of points and lines, on top of each other. A distance join matches the destination layer feature to the record of the closest feature in the source layer. A distance field reporting the distance between the joined features is added to the table.
- A simple join may be used whenever the cardinality of the layers is one-to-one or many-to-one. A summarized join generates summary statistics for all the source features matching the destination features and is used when the cardinality is one-to-many or many-to-many.
- Four types of spatial joins exist based on the combination of the criterion and the cardinality of the relationship: simple inside joins, simple distance joins, summarized inside joins, and summarized distance joins.
- Distance measurements are reported in map units of the input layers.
- Distance joins should only be performed with layers having projected coordinate systems that do not distort distances. Using layers with a geographic coordinate system may yield incorrect results.
- In the joined table, fields with identical names will be renamed in the output file so that all field names are unique, such as NAME to NAME_1. Usually, the source field is renamed.

➢ Use the following series of questions to help you set up a spatial join properly:

What should the output layer/table look like? ______________________________
Which is the destination layer? ______________________________
Should a distance join or an inside join be used? ______________________________
What is the cardinality of this join? ______________________________
Should a simple join or a summarized join be used? ______________________________

Important Terms

cardinality	inside join	simple join	summarized join
distance join	logical consistency	spatial join	

Chapter Review Questions

1. What primary characteristic distinguishes a spatial join from an attribute join?

2. What two options may be used to handle one-to-many relationships in a spatial join?

3. If a polygon feature type is joined to a line layer, with the lines as the destination table, what will the feature type of the output layer be?

4. How many output fields will result if a summarized join is specified and a single statistic (e.g., Sum) is selected?

5. Why should distance joins always be performed on layers with a projected coordinate system? What kind of projection should be used?

6. What happens if the two input layers in a join each have a field with the same name?

For the following spatial join problems, answer the series of questions in the text and then state the type of join that should be used: simple inside, simple distance, summarized inside, or summarized distance.

7. Determine the number of parcels within each of Austin's watersheds.

8. Find the closest school for each house in a realtor's database.

9. Find the land use zoning type associated with each well in Atlanta.

10. Determine the number of counties and the total number of people served by each airport in the United States.

Mastering the Skills

Teaching Tutorial

The following examples provide step-by-step instructions for doing basic tasks and solving basic problems in ArcGIS. The steps you need to do are highlighted with an arrow ➔; follow them carefully. Click on the video number in the VideoIndex to view a demonstration of the steps.

Simple inside joins

➔ Start ArcMap and open the map document ex_6.mxd.

➔ Use Save As to rename the document and remember to save frequently as you work.

Spatial joins produce new feature classes. We only need these for practice and don't want them to become part of our permanent geodatabase. So, for this lesson, we'll create a new personal geodatabase to store the outputs.

1➔ Click the Catalog button or go to the Catalog tab if it is already docked in ArcMap.

1➔ Expand the Folder Connections entry and navigate to the mgisdata\Austin folder.

1➔ Right-click the Austin folder and choose New > Personal Geodatabase.

1➔ While it is highlighted, name the new geodatabase chap6results and click Enter.

TIP: When you see the word **STOP**, pause a moment and set up the problem using the questions presented in the Concepts section. Then read on to see if you analyzed the problem correctly.

- What should the output layer/table look like?
- Which is the destination layer?
- Should a distance join or an inside join be used?
- What is the cardinality of this join?
- Should a simple join or a summarized join be used?

The map currently shows the geology and water wells for Austin, TX. For each well, we would like to know the geological unit associated with it at the surface (i.e., for each well, we want to know the geological unit that the point falls inside). **STOP** and think it through.

The desired output is a table of wells containing a field with the geological unit. Therefore, the Wells layer is the destination. A well can fall into only one geology polygon, making it a one-to-one cardinality. Therefore, a simple inside spatial join is indicated.

2➔ Right-click the Wells layer and choose Joins and Relates > Join from the context menu.

2➔ In the top drop-down box, choose to *Join data from another layer based on spatial location* (Fig. 6.17).

2➔ Choose Geology as the layer to join.

2➔ Choose to join the point to the polygon that *it falls inside*.

2➔ Click the Browse button and change the *Save as type* to *File and Personal Geodatabase feature classes*.

2➔ Navigate inside the Austin\chap6results geodatabase and enter wellgeology as the name of the feature class to be created. Click Save and click OK.

Notice that the new feature class appears at the top of the Table of Contents.

3➔ Turn off the original Wells layer.

3➔ Right-click the wellgeology layer and choose Open Attribute Table.

3➔ Scroll to the right to find the fields from the Geology table.

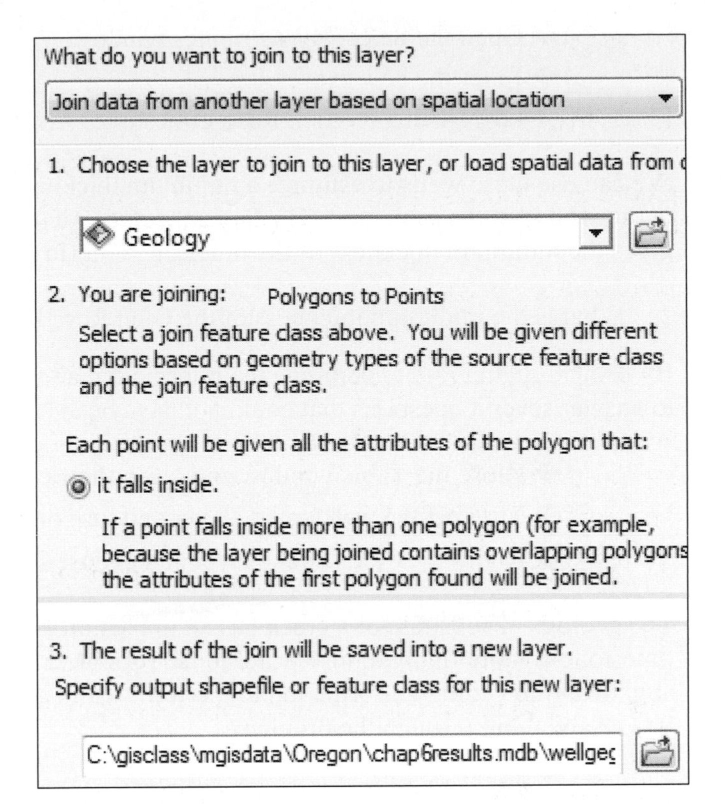
What do you want to join to this layer?
Join data from another layer based on spatial location
1. Choose the layer to join to this layer, or load spatial data from
Geology
2. You are joining: Polygons to Points
Select a join feature class above. You will be given different options based on geometry types of the source feature class and the join feature class.
Each point will be given all the attributes of the polygon that:
it falls inside.
If a point falls inside more than one polygon (for example, because the layer being joined contains overlapping polygons the attributes of the first polygon found will be joined.
3. The result of the join will be saved into a new layer.
Specify output shapefile or feature class for this new layer:
C:\gisclass\mgisdata\Oregon\chap6results.mdb\wellgec

Fig. 6.17. The Join Data window for spatial joins

TIP: The first field from the Geology feature class is named geology_OBJECTID. All fields to the right of it are from the Geology table. Look for the second OBJECTID after any spatial join to find where the appended information begins.

Locate the UNIT_NAME field at the very end of the new table, containing the geological unit. Let's find out how many of the wells are on limestone units.

3➔ Choose Table Options > Select By Attributes and select the wells based on the expression [UNIT_NAME] LIKE '*Limestone*'. Close the window when done.

1. How many wells are situated on limestone? 591 ____________

4➔ Right-click the wellgeology layer and choose Zoom to Layer.

4➔ Expand the legend for the Geology layer and find the Glen Rose Limestone. Where does it outcrop on the map?

4➔ Find the aquifer_code field from the wells part of the table and examine the entries. Look for codes containing "GLR".

The aquifer code field indicates the formation that actually supplies water to the well, at an opening some depth underground. The GLR aquifer codes refer to the Glen Rose Limestone, which is a prominent aquifer (water-bearing rock) in this part of the country. Wells often get water from a different rock formation than is exposed at the surface. Let's find out how many of the wells ON the Glen Rose Limestone actually get water from the Glen Rose Limestone.

4➔ Open the Select By Attributes window.

4➔ Enter the new expression

[UNIT_NAME] = 'Glen Rose Limestone' AND [aquifer_code] LIKE '*GLR*'

We can use these wells to estimate a minimum thickness for the Glen Rose Limestone, since the entire well is in the same unit. The largest well depth (well_depth field) of these selected wells gives the minimum thickness of the limestone unit (in feet).

2. What is the minimum thickness of the Glen Rose Limestone near Austin? 25 feet

By using a spatial join to combine the information about the wells and the geology, we were able to answer several questions that could not have been easily determined from each table alone.

5➔ Close the Table window and clear all selected features.

5➔ Turn off the wellgeology layer and turn on the Creeks layer.

5➔ Right-click the Creeks layer and choose Zoom to Layer.

The geological substrate of a creek has an impact on its connection with groundwater. A creek on limestone is more likely to lose water to the rock. A creek on clay will lose less. Just as we were able to assign geology information to the wells using a spatial join, we can also do it for creeks. **STOP** and think through the problem.

We want a table containing creeks with a field indicating the geological unit that the creek is on. Thus, Creeks is the destination feature class. A simple inside join is called for. We are using a detailed creeks layer in the Plum Creek and Elm Creek watersheds for this analysis.

6➔ Right-click the Creeks layer and choose Joins and Relates > Join.

6➔ Set the source layer to Geology.

6➔ Use the simple join option. (*Each line will be given the attributes of the polygon that it falls completely inside.*)

6➔ Store the output feature class in the chap6results geodatabase and name it creekgeology. Click OK.

7➔ Collapse the Geology layer legend and turn it off.

7➔ Open the Symbology properties of the new creekgeology layer and choose a Categories: Unique values map based on the UNIT_NAME field.

7➔ Click the Symbol heading to change properties for all symbols to a 2-pt.-thick line.

7➔ Choose a color scheme with dark, bold colors that will show up well. Click OK.

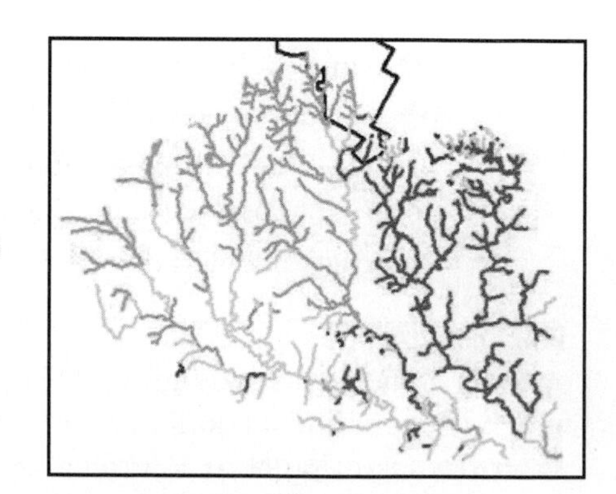

Fig. 6.18. Creeks displayed by geological unit

In the Table of Contents, notice that one of the symbols has a blank, or <Null>, value instead of a unit name.

8➔ Right-click the symbol with the blank and set it to a medium gray color.

8➔ The map should appear similar to Figure 6.18.

8➔ Turn on the Geology layer again.

Notice that all the gray features cross geological contacts. We had assumed that each creek was completely within a single geological unit to do the join, but we can see that it is not true in all cases. When a creek crossed a geological boundary, no unit was joined to it. It received a <Null> value in the attribute table for all of the geology attributes.

8➔ Open the creekgeology attribute table. Examine the fields, particularly the UNIT_NAME field containing the geology.

8➔ Right-click the UNIT_NAME field and choose Sort Ascending to see the <Null> values. Then close the Table window.

One weakness of inside spatial joins is that they join only when features fall completely inside other features. In Chapter 7 we will learn about the Intersect tool, which can perform this analysis more effectively than a spatial join.

A summarized inside join

A simple join is appropriate for a one-to-one or a many-to-one cardinality. A summarized join is needed for a one-to-many cardinality. Think about a watershed, the collection area for surface water. Runoff that comes from a watershed with many people is likely to have more pollutants than runoff from a watershed with fewer people. Let's analyze which watersheds in the Austin area are at greatest risk for polluted runoff.

9➔ Remove the creekgeology and the wellgeology layers from the map document. They are still saved in the chap6results geodatabase if you want them later.

9➔ Turn off the Geology and Creeks layers.

9➔ Turn on the Watersheds and Block Population layers.

9➔ Right-click the Block Population layer and choose Zoom to Layer.

9➔ Open the table for the Block Population layer.

Block Population contains points representing the centroids of block groups, the smallest unit for which the Census Bureau summarizes population data. Each point represents numbers of people and households. We will use these points to determine the number of people in each watershed. **STOP** and think it through.

This time we want to have a list of watersheds with the total number of people in it. Watersheds is the destination layer. One watershed can have many block points in it, so the relationship is one-to-many. A summarized inside join must be used. Since we want to know the total number of people, we need to request the Sum statistic.

10➔ Close the Table window.

10➔ Right-click the Watersheds layer and choose Joins and Relates > Join.

10➔ Set Block Population as the layer to join to.

10➔ Choose the option to summarize the points inside the polygons and check the box for the Sum statistic.

10➔ Name the output feature class watershedpop and put it in the chap6results geodatabase. Click OK.

11➔ Drag the watershedpop layer just below the Block Population layer.

11➔ Open the watershedpop table and scroll all the way to the right.

The Count field tells how many block points were found in the watershed. The fields prefixed with "Sum_" show the totals summed for each of the numeric fields in the source table. Sum_POP2000 tells us the total number of people living in each watershed.

11➔ Close the Table window.

12➔ Open the symbology properties for the watershedpop layer and choose a Quantities: Graduated color map based on the Sum_POP2000 field and the Jenks classification.

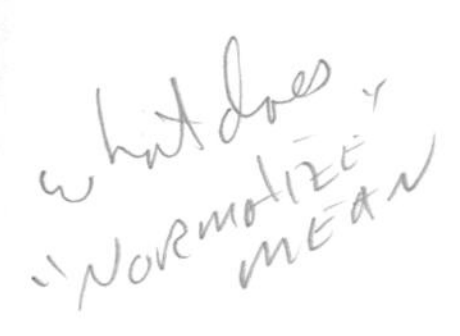

12➔ The population values are influenced by polygon area, so normalize the classification using the Shape_Area field.

12➔ Give it a green-to-red color ramp to indicate hazard potential and click OK.

12➔ Name the layer Watershed Hazard.

Your final map should look similar to Figure 6.19. Note that the watersheds extend beyond the block population data. If this were a real project you would need to address this issue by dropping polygons with incomplete data or expanding the block group data to the full extent of the watersheds.

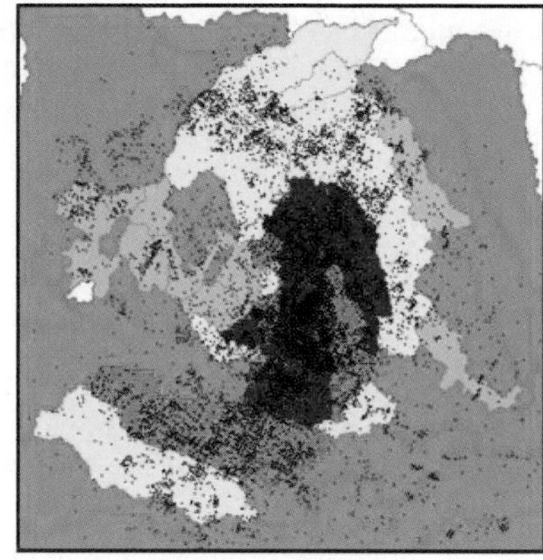

Fig. 6.19. Watershed pollution risk from population

13➔ Turn off the Watershed Hazard, Watersheds, and Block Population layers.

13➔ Turn on the Streets layer. Add the police_districts feature class from the mgisdata\Austin\Austin\Administrative feature dataset.

Imagine that the city has been reviewing its police department staffing levels. To help assess needs, it would be helpful to know the total length of streets for which each district is responsible. You have been given the task of finding this information. **STOP** and think it through.

You want to produce a list of the police districts, each of which has the total street length. Thus, the districts are the destination layer. Each district contains many streets, so the cardinality is one-to-many. A summarized inside join must be used, with Sum as the statistic. First, notice that many streets do not fall in any of the districts. You can reduce the processing time for the spatial join by first selecting only the streets that are within the districts.

14➔ Open the Select By Location window.

14➔ Set the target layer to Streets and the source layer to police_districts.

14➔ Set the spatial selection method to *are within*.

14➔ Click OK.

Now when you do the spatial join, only the selected streets will be used, saving time. You will also have fewer streets with unassigned districts.

15➔ Right-click the police_districts layer and choose Joins and Relates > Join.

15➔ Set the source layer to Streets and choose the summarized option. Check the Sum statistic.

15➔ Place the result in chap6results and name it policestreets.

15➔ Click OK and wait. This join may take several minutes.

16➔ Clear the selection and turn off the Streets layer.

16➔ Open the policestreets table. The original Streets layer contained a field called MILES with the length of the street. Find the Sum_MILES field in the joined table.

16➔ Close the Table window.

17➔ Create a graduated color map of the policestreets layer based on the Sum_MILES field. For miles, it is best to classify the map using a defined interval of 50 miles.

Notice that the downtown districts are smaller and have fewer miles of road than the outlying districts (Fig. 6.20). The next step might be to analyze the total block population for each district to see if they are balanced in that way instead. We will leave that as an exercise, and move on.

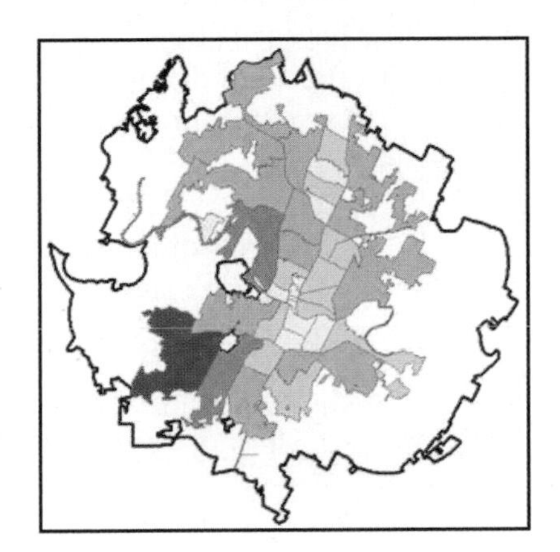

Fig. 6.20. Total road mileage each police district covers

18➔ Turn off the police_districts layer.

18➔ Remove the policestreets layer from the map document.

Simple distance joins

Distance joins combine feature records for features that are closest to each other. Consider for a moment the problem of post office usage. One might assume that, generally, people will go to the closest post office. For each street, it would be nice to know which post office to go to. **STOP** and think it through.

We want a list of streets with a field indicating the closest post office, so Streets is the destination layer. Since only one post office can be "closest," this is a one-to-one relationship. A simple distance join is appropriate.

19➔ Turn on the Streets and Post Office layers.

19➔ Right-click the Streets layer and choose Joins and Relates > Join.

19➔ Set Post Office as the layer to be joined.

19➔ Choose the simple join option, where *Each line will be given the attributes of the point that is closest to it.*

19➔ Name the output file streetpost and save it in chap6results. Click OK.

Now we can analyze the results.

20➔ Open the streetpost attribute table and examine all of the fields. Find the field that contains the names of the different post offices.

20➔ Close the Table window.

3. Which field contains the post office names? FACILITY_N

20➔ Create a unique values map of the streetpost layer based on the FACILITY_N field, so that each street is symbolized by the post office it is closest to. Use a slightly thicker line to see the colors better. The map should look similar to Figure 6.21.

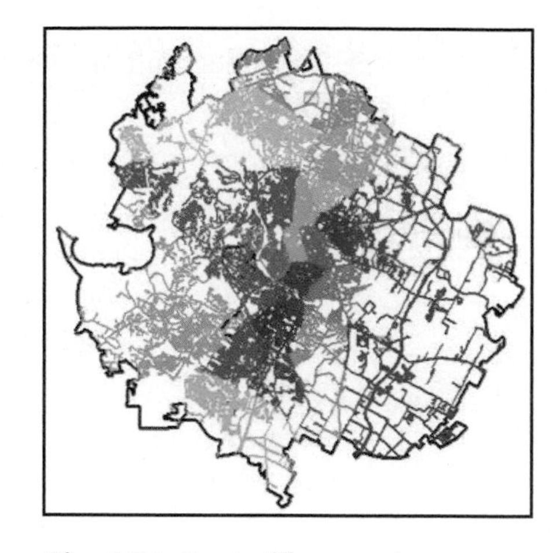

Fig. 6.21. Post office service areas

A distance join also creates a distance field. In this case it shows how far each street lies from the nearest post office (as the bird flies, not driving distance). The units will match whatever the storage units are for the feature class coordinate system.

4. What are the coordinate system and units for the streetpost feature class?

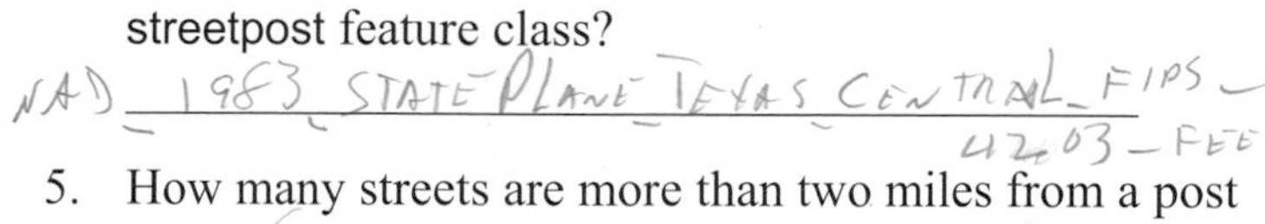

NAD 1983 STATE PLANE TEXAS CENTRAL FIPS
4203 FEET

5. How many streets are more than two miles from a post office (as in Fig. 6.22)? ________________

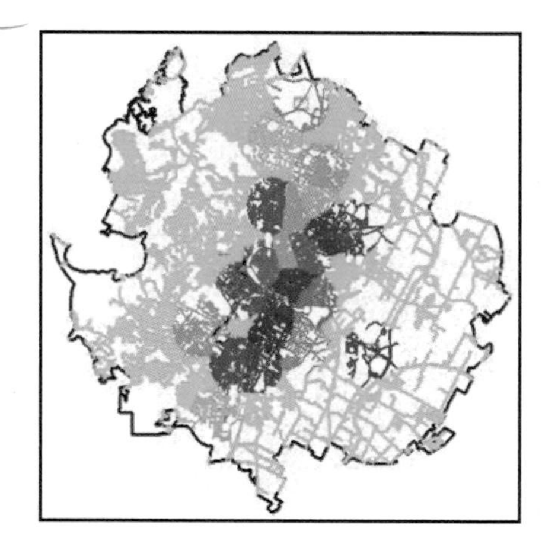

Fig. 6.22. Streets farther than two miles from a post office

21➔ Clear the selected features.

21➔ Create a graduated color map of the streetpost layer based on the Distance field, using a thicker line and a monochromatic color ramp (Fig. 6.23).

22➔ Remove the streetpost layer and turn off the Post Office layer.

22➔ Add the trailheads and restrooms feature classes from the Parks feature dataset in the Austin geodatabase.

Before people start a hike, or after they finish, they like to find a restroom. Imagine that a hiking club wants to analyze the proximity of trailheads to restrooms and make recommendations for adding more restrooms to serve hikers. **STOP** and think through the problem.

The club needs a list of trailheads that includes the distance of each trailhead to the closest restroom. Thus, trail_heads is the destination layer. There can be only one closest restroom to a trailhead, so the cardinality is one-to-one, and a simple distance join will suffice.

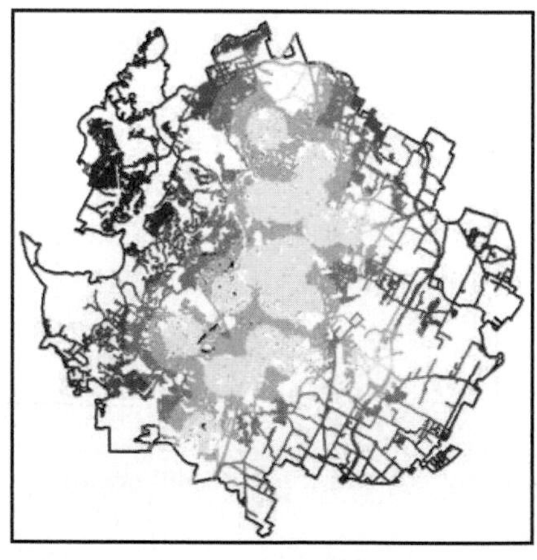

Fig. 6.23. Distance of streets from nearest post office

23➔ Right-click the trail_heads layer and choose Joins and Relates > Join.

23➔ Set restrooms as the layer to join.

23➔ Choose the simple join option to give each point the attributes of the closest point.

23➔ Name the output feature class trailrest and save it in chap6results. Click OK.

The club is most concerned about the trailheads that are more than 1000 feet from a restroom. You will create a map that will help the club identify optimum locations for new restrooms.

24➔ Open the trailrest table and examine the fields. Find the Distance field.

24➔ Use Select By Attributes to select the trailheads that are more than 1000 feet from a restroom. Close the Table window.

24➔ Right-click the trailrest layer and choose Selection > Create Layer From Selected Features. Name it Problem Trails.

24➔ Clear the selected features.

25➔ Add the parks and the trails feature classes from the Parks feature dataset.

25➔ Symbolize the parks in a light green color and the trails as a dark green thick line.

26➔ Click on the restrooms layer symbol to open the Symbol Selector.

26➔ Type restroom in the search box at the top and click the Search button. Choose the symbol you prefer and make it about 16 pt.

26➔ Open the restrooms layer properties, click the General tab, and set the minimum scale to 1:60,000.

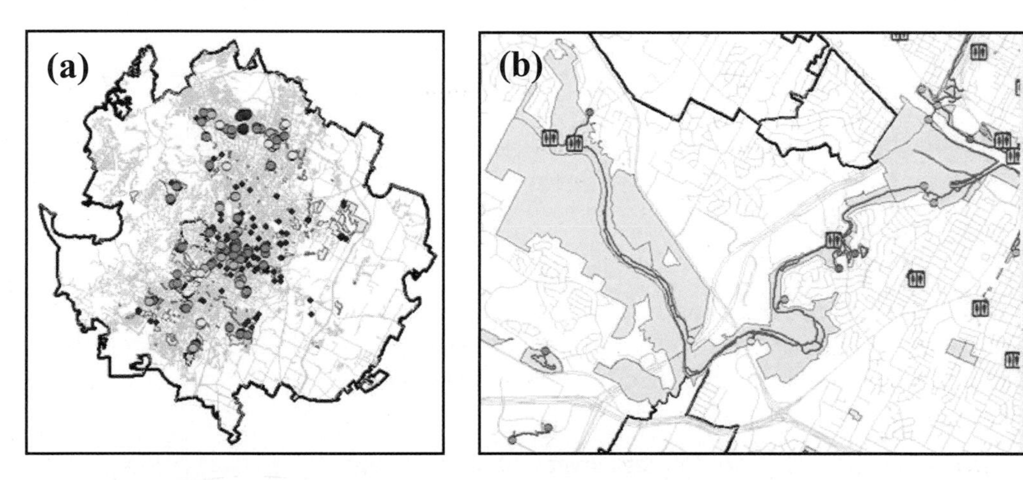

Fig. 6.24. Map showing trailheads more than 1000 feet from a restroom

27➔ Remove the trailrest and trail_heads layers.

27➔ Create a graduated color map of the Problem Trails layer based on distance from the restrooms. The values are skewed, so use the Jenks classification method. Make the symbol slightly larger so you can better see the colors (Fig. 6.24a).

27➔ Zoom in to an area with clusters of Problem Trails to test the map (Fig. 6.24b).

Let's save this work as a group layer file for future use by the club.

28➔ Click on the Problem Trails layer name to highlight it. Hold down the Ctrl-key and click the restrooms, parks, and trails layers.

28➔ Right-click one of the highlighted layers and choose Group.

28➔ Name the new group layer Restroom Map.

28➔ Right-click the Restroom Map group layer and choose Save As Layer File. Save it in the Austin folder as Restroom Map.lyr.

29➔ Turn off the Restroom Map group layer and collapse it.

29➔ Right-click the City Limit layer and zoom to it.

Summarized distance joins

A simple distance join asks "Which feature is closest?" A summarized distance join is needed to address the question "How many features are closer to one location than they are to another?"

Imagine that the Department of Recreation is applying for a major grant to renovate its recreation centers. For guidance, they would like to know the potential population served by each center, making the assumption that most people will go to the closest center. The Block population data will provide the base data set for the analysis. **STOP** and think it through.

We want a list of the recreation centers with a field containing the total number of people who are closest to it. Centers is the destination layer, and the cardinality is one-to-many. A summarized distance join will be used.

30➔ Add the facilities feature class from the Facilities feature dataset in the Austin geodatabase.

30➔ Use Select By Attributes to select recreation centers using the expression "FACILITY_T" = 'RECREATION CENTER'.

30➔ Right-click the facilities layer and choose Selection > Create Layer from Selected Features. Name the layer Rec Centers.

30➔ Remove the facilities layer.

TIP: Creating intermediate selection layers is often a smart thing to do. It provides a visual test of your query, and it saves the selection in case a later mistake requires redoing the analysis. Give each selection layer a descriptive name, to avoid confusion later.

31➔ Right-click the Rec Centers layer and choose Joins and Relates > Join.

31➔ Set Block Population as the layer to join.

31➔ Choose the summarized option. You want to know the total number of people, so check the Sum statistic.

31➔ The default save folder was changed to Austin when you saved the Restroom Map layer file. Click the folder Browse button to change it back.

31➔ Change the Save as type to *File and Personal Geodatabase feature classes*.

31➔ Navigate inside the chap6results geodatabase. Name the output recpop and click Save and OK.

32➔ Open the recpop table and examine the Sum fields.

32➔ Examine the statistics for the Sum_POP2000 field.

6. What are the minimum, maximum, and average numbers of people served by the recreation centers? 907 – 138,275

Clearly there are large discrepancies in the potential usage of each center. What is the spatial distribution of this potential usage?

32➔ Close the Statistics window and the Table window.

32➔ Create a graduated symbol map of the recpop layer based on the Sum_POP2000 field. Use Jenks Natural Breaks for this skewed data set.

The map is interesting (Fig. 6.25). Small centers clustered in the downtown area appear to compete with each other, while the outlying centers draw large groups over larger distances. The department decides to adopt a strategy to turn the downtown cluster into special-interest centers that may draw more people, while expanding standard activities in the outlying centers. Of course, the final plan would take into account many more issues, such as parking, public transportation, and current facilities and usage.

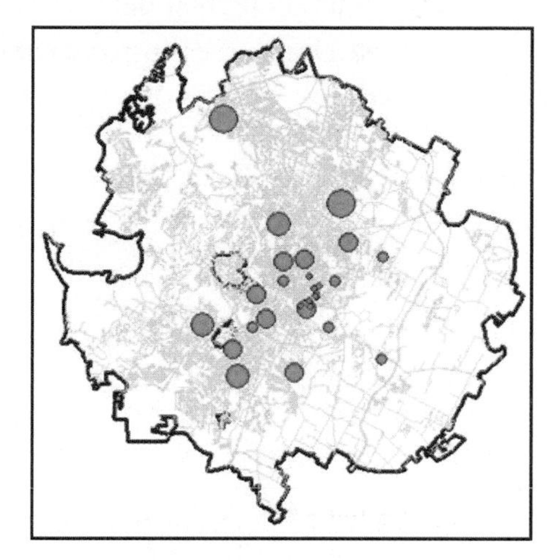

Fig. 6.25. Potential usage for recreation centers

Meanwhile, another group is concerned about the effects of population on stream pollution. The number of people living close to a stream impacts the likelihood for the stream to be affected by pesticides and other pollutants. This group wants to characterize the basic threat to each stream by summing the population closer to that stream than to any other. **STOP** and think it through.

We want a feature class of streams that contains a field with the total population close to it, so streams are the destination layer. The cardinality is one-to-many, so a summarized distance join with the Sum statistic is needed.

33➔ Add the named_creeks feature class from the Environmental feature dataset in the Austin geodatabase.

33➔ Turn off all layers except the City Limit and named_creeks layers.

33➔ Turn on the Block Population layer, but right-click its symbol and change the color to Gray 10%.

Save

34➔ Right-click the named_creeks layer and choose Joins and Relates > Join.

34➔ Set the source layer to Block Population.

34➔ Choose the summarized join option. Fill the button to use the closest features and check the Sum statistic.

34➔ Name the output **creekpop** and save it in chap6results. Click OK.

35➔ Open the creekpop attribute table and examine the Sum_ fields.

35➔ Sort the Sum_POP2000 field in descending order.

7. Which three creeks have the highest risk based on total population?

Little Walnut Creek
Williamson Creek
Walnut Creek

35➔ Rename the creekpop layer **Population Load**.

35➔ Create a graduated color map for the Population Load layer based on the Sum_POP2000 field. The data are fairly skewed, so use Jenks Natural Breaks.

35➔ Use a green-to-red color ramp and thick lines so the colors show well (Fig. 6.26a).

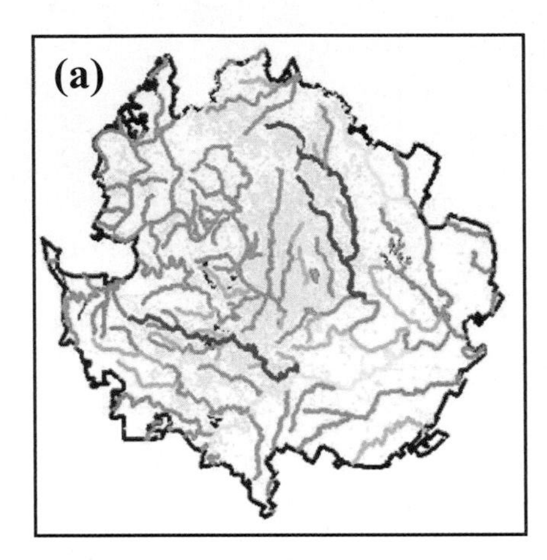

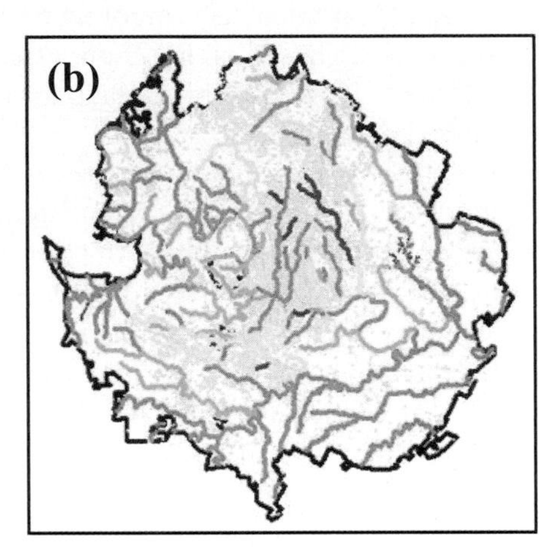

Fig. 6.26. Maps showing pollution hazard of Austin creeks based on (a) population load and (b) hazard index

As you examine the map, it may occur to you that the longer creeks are naturally exposed to more people. You decide to create a hazard index based on the number of people per unit length of stream.

36➔ Click the Table Options menu and choose Add Field.

36➔ Name the field HazIndex and set the field type to Float. Click OK.

36➔ Right-click the empty HazIndex field and choose Field Calculator. Enter the formula [Sum_POP2000] / [Shape_Length] and click OK.

36➔ Examine the HazIndex field and sort in descending order.

8. Which three creeks have the highest risk based on the index?
Barton / Country club / Slaughter

Notice that most of the creeks have an index near or below 1, but the top three are in the range of 3–6. Notice the Shape_Length field also. These high-index creeks have very short lengths, and they are probably scraps of creek remaining after they were clipped to the city boundary. These high index values are therefore probably a fluke and should be eliminated from consideration. To avoid having them skew the map, you will exclude short creeks from the map.

37➔ Close the Table window.

37➔ Right-click the Population Load layer and choose Copy.

Save

37➔ Right-click the Layers data frame name and choose Paste Layer(s). Rename the new layer **Hazard Index**.

37➔ Open the symbol properties for the Hazard Index layer and create a graduated color map based on the HazIndex field.

37➔ Click the Classify button. Use Jenks and click the Exclusion button. Enter the expression [Shape_Length] < 1000 to eliminate the shorter creeks. Click OK and OK.

37➔ Use a green-to-red color ramp and thick lines so that the colors show well (Fig. 6.26b). Click OK to finish the map.

As you can see, there are some differences between the two approaches for evaluating pollution hazards. (Some color differences may also be due to differences in the Jenks breakout of the different values.) Which model do you think might be a more realistic approach?

A hydrologist who specializes in surface water quality would probably question both models, because she knows that surface water runoff is controlled more by elevation than by distance, and that a watershed provides a better unit for summing the population impacts. You already did this using a summarized inside join.

38➔ Turn on the Watershed Hazard layer and open its properties.

38➔ On the Display tab, set the transparency to 50%.

38➔ Turn off the Block Population layer.

38➔ Compare the Population Load and Hazard Index maps visually against the Watershed Hazard map by turning the upper one on and off several times.

Which of the two creek-based methods appears to agree better with the watershed method? Overall, do the three methods agree on which watersheds and creeks bear closer scrutiny?

This analysis brings up an important consideration in GIS analysis. A problem may be approached several ways, using different models based on different assumptions. All three approaches had potential problems. In the watershed-based model, the population data set did not cover the entire watershed area. In the creek-based analysis, the fundamental assumption that distance was an appropriate predictor of influence is flawed, and short creeks resulting from data problems could have impacted the results. Each approach yielded similar, but not the same, results. Three important lessons should be gleaned from this example.

First, *data issues and problems affect nearly all analysis procedures*. You cannot mitigate all issues, but it is important to be aware of what they are, as well as what impacts they might have.

Second, *the fundamental assumptions behind a model should be the best available*. The watershed model had the most realistic assumption about water flow and is the best of the three models (or would be if the population data covered all the watersheds).

Third, *do not push the results of a model beyond its limits*. Even the watershed model neglects important variables. For example, industrial areas produce different amounts and types of pollutants than residential or farming areas do. Stream flow impacts the ability of a stream to dilute pollutants. Although we can map five, or even more, hazard levels, whether the model can realistically separate the hazard into many categories with distinct boundaries is questionable.

In the long run, the best you can say about this analysis is that it highlights potential problem areas that would benefit from further scrutiny. Toward that goal, all three models performed equally well. This observation yields a fourth lesson, that *the most detailed and accurate model is not always necessary to achieve the objective*.

Save before

Data quality issues with spatial joins

Now we will do one more polygon-to-polygon join to demonstrate some issues to consider when joining. We have a census tracts layer for Austin that has a county FIPS code, but no county name. We would like to do a spatial join of counties to the tracts in order to provide a field with the county name. This is a simple inside join with a many-to-one cardinality.

39➔ Turn off all of the layers in the data frame.

39➔ Add the tracts and the counties feature classes from the Administrative feature dataset of the Austin geodatabase.

39➔ Put tracts above counties in the Table of Contents and zoom to the full extent.

40➔ Right-click the tracts layer and choose Joins and Relates > Join.

40➔ Set the join layer to counties.

40➔ Choose the simple join option.

40➔ Name the output tractcounty and place it in the chap6results geodatabase.

41➔ Change the counties symbol to a hollow shade with 2-pt. purple borders, and place it above the new tractcounty layer in the Table of Contents.

41➔ Create a unique values map for the tractcounty layer, using the joined COUNTY field containing the county names (it is near the end of the table).

Notice that the unique values legend has a <Null> entry, and many of the tracts are symbolized with this value. What happened?

42➔ Open the tractcounty table and scroll far to the right where the COUNTY field is.

42➔ Scroll down to examine all the rows in the table.

You can see many <Null> fields in the table. These tracts were not matched to a county at all. This fact is puzzling, because, from the map, the tracts appear inside the counties and should have been matched with one. However, the map shows that the affected tracts all share a boundary with the county (Fig. 6.27).

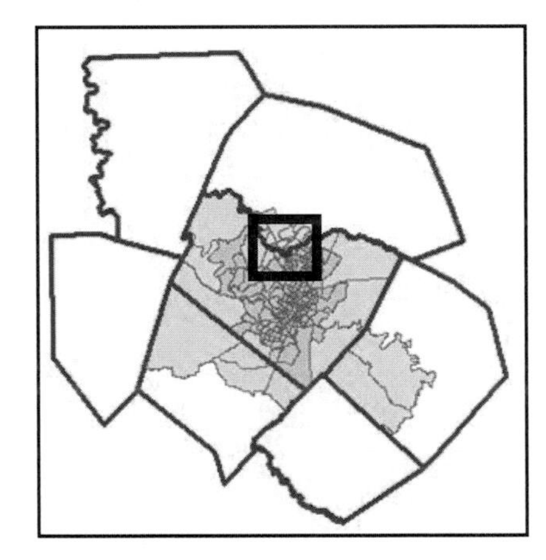

Fig. 6.27. The orange tracts received no match during the join.

42➔ Close the Table window and zoom in to the area shown in the black box in Figure 6.27.

42➔ Switch to the List by Selection view in the Table of Contents and make tractcounty the only selectable layer.

42➔ Click the Select Features by Rectangle tool and click on one of the tracts that share the county border. Click several more.

It becomes apparent that the tract boundary and the county boundary are not exactly the same. In spots, some of the tract lies outside the county (Fig. 6.28). In the real world, a tract always falls inside a county. In this GIS data set, the tracts and counties came from two different sources, and as a result their boundaries do not coincide. To be joined, the tract must fall within the county. Even the small discrepancies seen here are sufficient to prevent the county from being matched to the tract during the join.

Fig. 6.28. Part of this tract lies outside the county.

How well the GIS data features reflect relationships in the real world is called **logical consistency**. In this case the data sets are not logically consistent because the data features with mismatched boundaries do not reflect the real-world boundaries, which are identical.

TIP: Always keep in mind that data accuracy and logical consistency issues may have an impact on an analysis. Be alert for potential problems.

This is the end of the tutorial.

➔ Close ArcMap. You can save your changes.

Exercises

Use the data in the mgisdata\Austin\Austin geodatabase to answer the following questions.

1. Give the name of the watershed in Austin that contains the most wells. How many wells does it contain? Which watersheds have the deepest and shallowest average well depths?

2. How many wells are there in the WATERFRONT zoning code in Austin? Use the O_NAME field from the zoneoverlays feature class. How many of the wells have depths less than 100 feet?

3. Which zoning category (O_NAME) in Austin contains the most wells?

4. What are the minimum, maximum, and average distances from barbeque (bbq) pits to the closest restroom?

5. An elementary school pool fun day is planned. If each school goes to the closest pool to it, which pool will have the most schools attending? Create a map that would be helpful to planners in reassigning schools to less-crowded pools. (**Note:** You must export the elementary school selection to a new feature class before joining, because the distance join does not honor the selected set. This may be a bug.)

6. Examine the table from Exercise 5 closely, and you will find a problem with your initial analysis. What is it? What would you need to change to get a better result?

7. Which post office potentially serves the greatest number of people, based on the block group population data? Which one serves the least? (**Note:** The post office locations are one of the types of information in the facilities feature class.)

Use the data in the mgisdata\Oregon\oregon geodatabase to answer the following questions.

8. Which city in Oregon is farthest from an airport? What is the distance in *kilometers*?

9. Assuming that an airport's service area includes all of the cities that are closer to it than to any other airport, determine how many cities each airport serves. Which airport serves the most cities? How many cities? Which serves the most people? How many people? Why does the Sum_POP2000 field contain negative values?

10. A boating club would like to know which parks in Oregon have the best access to lakes. Find the number of lakes and total lake area in each park. Which park(s) have the most lakes and how many? Which park has the greatest area of lakes? Examine parks with only one lake. The sum of the area fields for these lakes keeps repeating the same few numbers. Explain why.

Challenge Problem

Imagine that a national body recommends the minimal staffing levels for a police district as one officer for every 1000 people plus one officer for every 25 miles of road. The police department has asked you to calculate the minimum recommended staffing levels for each district. Create a table showing the largest 15 to 20 districts with district name, the population, the miles of road, and the number of officers. Place the table on a supporting map layout (screen capture is the easiest method), along with a graduated color map of population and labels for the staff numbers.

Chapter 7. Map Overlay and Geoprocessing

Objectives

- Learning about spatial analysis functions, including overlay, clipping, and buffering
- Using map overlay to analyze multiple spatial criteria
- Understanding differences between spatial joins and overlays
- Geoprocessing with menus, ArcToolbox, and Model Builder

Mastering the Concepts

GIS Concepts

Over the years many procedures have accumulated for characterizing spatial relationships within and between spatial features. Functions can find areas shared by two or more conditions, evaluate distances, extract or erase areas of interest, merge similar features together, examine distance relationships between points, and more. The entire enterprise of GIS, including gathering the data, putting them in digital form, designing the geodatabase, managing the data, and creating maps, is justified by the ability to apply these tools to extract information that might be difficult to obtain any other way. GIS shares many capabilities with CAD systems and database software, but spatial analysis gives GIS unique power to get the most out of map and attribute data.

Chapter 6 discussed spatial joins as one way to analyze spatial relationships. In this chapter we'll learn more tools to solve spatial problems. Often two or more functions will be strung together to solve a specific problem, a practice called **geoprocessing**. We'll begin with a general discussion of spatial functions and then investigate how these functions are implemented within ArcGIS.

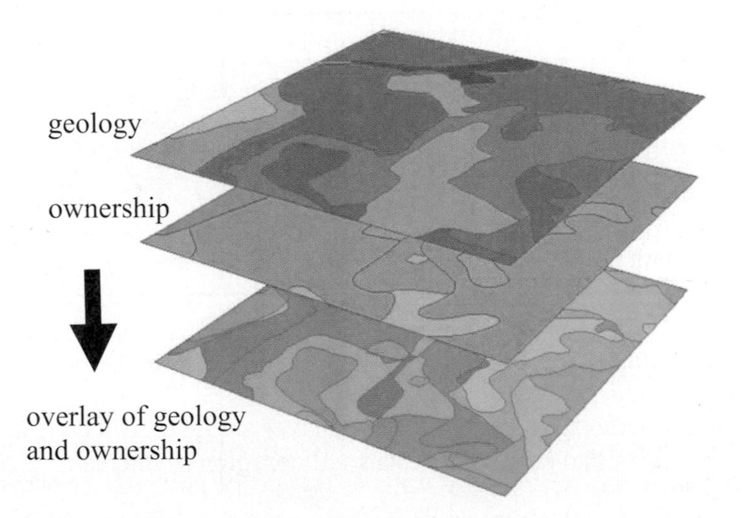

Fig. 7.1. Map overlay combines two feature classes to create a new feature class containing information from both inputs. Both features and attributes may be combined. Here, each new polygon in the output contains both geology and ownership attributes.

Map overlay

Map overlay functions combine two layers to create a new output feature class containing information from both of the inputs. In Figure 7.1, the polygons from a geology layer and from a land ownership layer have been combined to produce a new set of polygons. Each new polygon has been given the attributes of the originals, and each contains information about both geology and ownership for the same feature. The output layer might be useful for selecting areas containing certain gem-bearing rock formations on public land that would be suitable for mineral hunting.

Overlay functions fall into two categories, those that do not combine attributes and those that do. Functions that do not combine attributes, called extraction functions, include **clip** and **erase**. The two functions that combine attributes include **intersect** and **union**.

Extraction functions

Extraction functions separate features of interest from a larger group based on some criteria. Queries are one form of extraction and can be based on attribute data as in Select By Attributes or on spatial criteria as in Select By Location. Queries, however, can only extract whole features from a data set. Two new functions, **clip** and **erase**, are able to extract both whole features and portions of features. They have the ability to truncate features when they cross a boundary, for example, a road crossing a state line. This ability differs from the Select By Location function, which must select (or not select) the entire feature.

A **clip** works like a cookie cutter to truncate the features of one file based on the outline of another. In Figure 7.2, the county roads layer (gray lines) has been clipped by the land use layer (beige) to produce a shapefile of the roads that fall inside the city planning boundaries (red). The features being extracted come from the input layer. The feature class providing the boundary is called the clip layer. The clip layer must always be a polygon feature class, but the input layer may be points, lines, or polygons. Only the outside boundary of the clip layer is used for clipping; internal boundaries, if present, have no effect on the output.

SKILL TIP: To temporarily clip features for display only, use the Data Frame tab of the Data Frame Properties menu (Layouts and Data Frames).

Erase works in the opposite sense of a clip, keeping features that fall outside the erase layer and eliminating those inside. Given the same input and clip layers in Figure 7.2, an erase would keep the roads outside the city (gray) and eliminate the roads inside the city (red). An erase might be used to eliminate privately owned lands from forest protection areas. Erase is only available to users with an ArcGIS Advanced license.

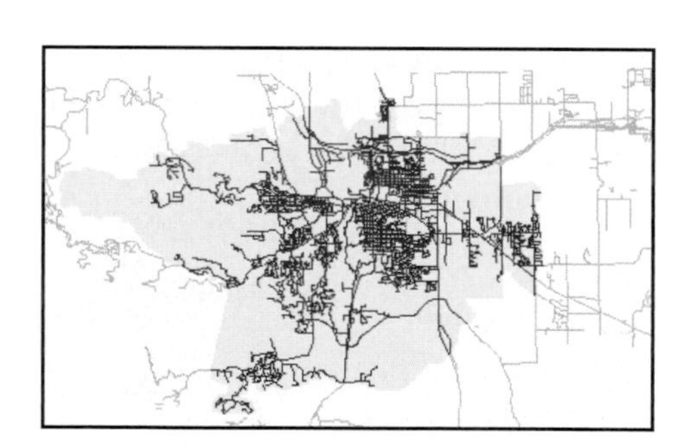

Fig. 7.2. A clip truncates the features of a layer by the boundary of another.

Attributes from both truncated and nontruncated features in the input layer are preserved intact during a clip or an erase operation. However, attributes from the clip or erase layer are ignored. The result of a clip or erase is a copy of the input layer that now occupies a smaller region of interest.

Areas and lengths for features stored in the geodatabase fields Shape_Area and Shape_Length are automatically updated following a clip or any other overlay function. Note, however, that user-defined area or length fields are not automatically updated. A field named MILES or ACRES containing highway miles or tract acres, for example, would contain an incorrect value for any truncated feature. The lengths and areas of user-defined fields must be updated manually using the Calculate Geometry function.

Overlay with attributes

In a clip or an erase, the input features retain their attributes, but the attributes of the clip or erase boundary layer are ignored. The **intersect** and **union** functions combine both features and attributes.

Fig. 7.3. The selected road inside the square crosses three different land use zones. An overlay splits the road so that each segment lies within a single polygon.

The intersect and union functions are related to spatial joins in that they correlate features based on their spatial relationships to each other. However, spatial joins fail when spatial features do not overlap exactly (recall the creeks and geology example from Chapter 6). Consider the map of roads and land use shown in Figure 7.3. The state government has requested a report of the total miles of road falling into each land use category. At first glance, a spatial join might be considered to generate this information; by joining the land use polygons to the roads, one might expect to get a field showing the land use type each road crosses.

However, the single selected road, highlighted in Figure 7.3, crosses three different land use classes. How would a land use type be assigned in this case? If one used the *completely within* join option, no land use class would be assigned for the road because it does not lie entirely inside a polygon. A *summarized join* would do no good because a nominal data type such as land use class can't be averaged.

The ideal solution would split the road into three sections and assign the land use to each section, and this is what overlay functions do, thereby enforcing a one-to-one relationship between features and enabling a perfect correspondence when joining the tables. In Figure 7.3, the original single road is now three sections, and each segment in the output table (Fig. 7.4) is assigned the attributes of the land use polygon containing it.

Fig. 7.4. During an overlay, each new road segment retains its original attributes and receives the land use code and other attributes of the polygon in which it falls.

roadlu

OBJ_ID	ROADNAME	SUFFIX	LU-ID	LU_CODE
3033	2 ST	ST	153	High Density Residential
3034	QUINCY ST	ST	153	High Density Residential
3035	RANGE RD	RD	72	Office/Commercial
3036	RANGE RD	RD	305	Public
3037	RANGE RD	RD	70	Medium Density Residential
3038	SOO SAN DR	DR	72	Office/Commercial

Map overlay with attributes occurs in two forms. **Union** combines two polygon layers, keeping all the areas and merging the attributes for both layers. **Intersect** also merges the attributes but retains only the areas common to both layers.

A **union** creates all possible polygons from the combination of features in both layers. In a union, both input layers must contain polygons. Figure 7.5 shows a union performed on geology and slope class layers to create a landslide hazard map. Landslide risk depends primarily on two factors, the strength of the geological units (sandstones are less susceptible to sliding than shales) and the slope. The lighter areas in the slope class map have lower slopes, and the dark areas have higher slopes. A union produces a new feature class with new polygons, each of which possesses the original attributes of its parents. The new feature class can then be evaluated for the various combinations of slope and geology that may constitute a landslide hazard.

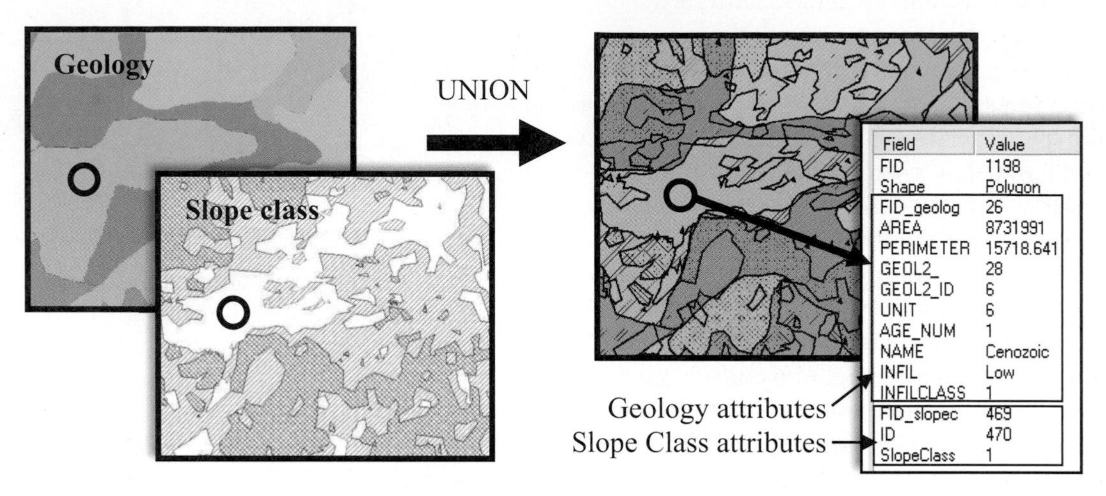

Fig. 7.5. A union creates every possible new polygon from the combined input layers. Each new polygon assumes the attributes from both layers; the new polygon marked with the circle contains attributes from both the geology and slope class layers.

The **intersect** operation is similar to a union, except that it only keeps the polygon areas that are shared by both input layers. This function provides a way to find out where two or more conditions hold simultaneously. Finding the potential habitat for a species with specific environmental requirements constitutes one example of using intersect.

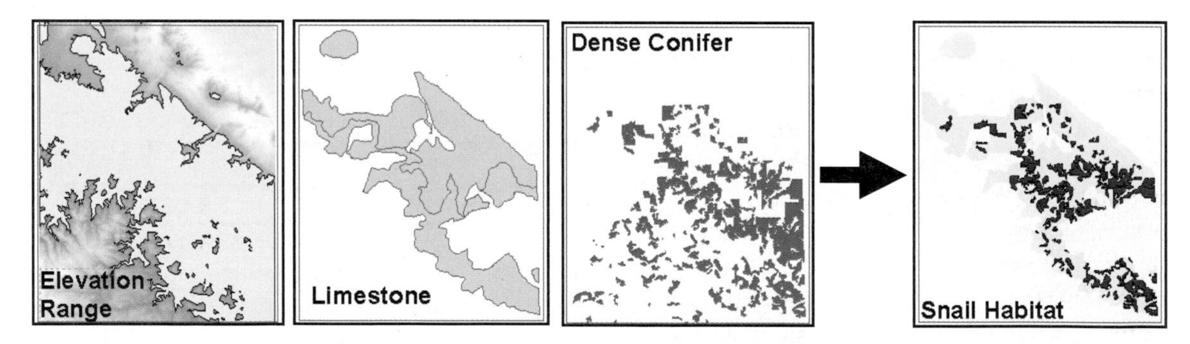

Fig. 7.6. Intersecting elevation, limestone areas, and dense conifer vegetation can help identify areas of potential snail habitat.

Imagine that a rare species of snail inhabits the Black Hills. This snail prefers limey soils in cool and dense coniferous forest and is rarely found above an elevation of 1600 meters or below 1200 meters. By combining a geological map, a vegetation map, and an elevation map, one can identify the likely areas of snail habitat (Fig. 7.6). Such a map would help biologists to create sampling strategies for counting populations, to analyze whether the habitats are interconnected or widely separated, and to make decisions about forest management to protect the snails.

The analysis includes a series of steps using the layers shown in Figure 7.6. First, a digital elevation model (DEM) is queried to create a polygon showing the areas between 1200 and 1600 meters in elevation. Second, a query is applied to the geology features to extract the limestone units as a separate layer. Third, another query separates the dense conifer areas from the other vegetation. Finally, the intersection is performed in two steps, first intersecting the elevation range and limestone layers and then intersecting the result with the dense conifer map. The final output contains the areas common to all three layers—the snail habitat. These habitat polygons contain the full attributes of all three layers, should that information be needed, such as whether the conifers are ponderosa pine or white spruce.

Whereas the union function requires two polygon input layers, the intersect command is more versatile. Both input and overlay layers may contain points, lines, or polygons. The output geometry may vary, but it cannot exceed the dimensionality of the lowest dimension input. In other words, if the inputs are both polygons, the output may be points, lines, or polygons (Fig. 7.7a). If the inputs are both lines, the output may be lines or points (Fig. 7.7b). If the inputs are polygons and lines, the output may be points or lines (Fig. 7.7c). If the input is points, the output must be points.

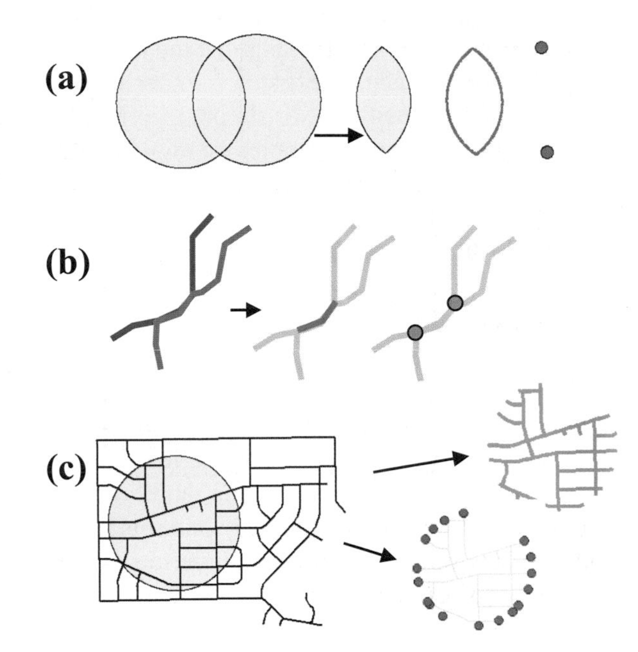

Fig. 7.7. Intersect geometries. (a) Polygons can yield polygons, lines, or points. (b) Lines can yield lines or points. (c) Polygons and lines can yield lines or points.

Comparing overlay functions

Figure 7.8 summarizes the different overlay operations that can be performed using two polygon input layers, including the extraction functions clip and erase, which do not combine attributes, and the intersect and union functions, which do. Clip and erase can also accept point and line feature classes for one input, but the boundary input must contain polygons. The output feature classes will have the same geometry as the input classes.

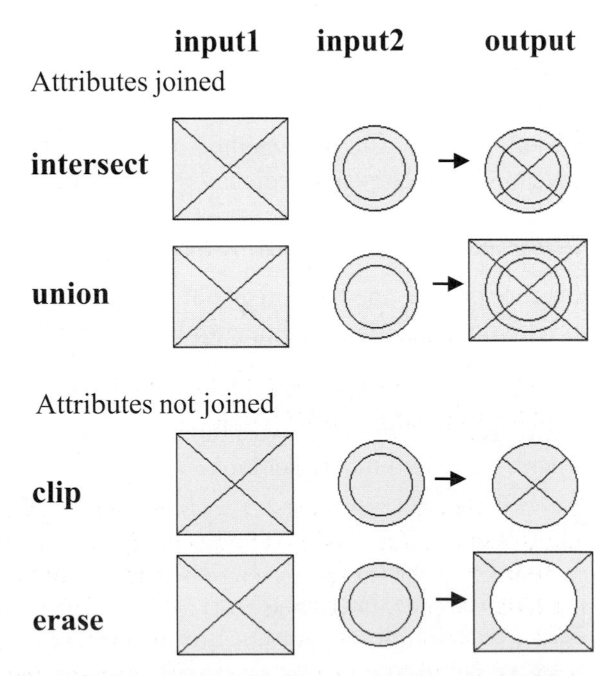

Fig. 7.8. Summary of polygon overlay operations and results

Intersect and union are similar functions, so a word about when to use each may be helpful. The primary use of intersect is to find areas where certain conditions overlap. Typically the input layers occupy different regions, as do the elevation range, limestone, and conifer layers shown in Figure 7.6. The goal is to find the areas common to the inputs. Although the attributes from each input are combined in the final table, the attributes themselves are sometimes of little interest. In the snail habitat problem, the pertinent attributes were already preselected using queries. The main goal was developing a map of the overlaps.

In a union, the primary goal is to combine the tables. Generally the input layers all fully occupy the same region, as the geology and slope class layers do in Figure 7.5. It is expected that the entire map area will remain, but that the new features will contain the attributes of both inputs. After a union, the usual next step is to perform attribute queries to find areas with specific combinations of attributes, or simply to symbolize the combinations to show the areas of interest. In the landslide hazard example, we might have extracted the shale and high slope areas and done an intersect to show where the high-hazard areas exist. The hazard is either there or not there. With a union, though, we are able to analyze different combinations of geology and slope and develop different gradations of hazard instead of a simple yes or no.

Slivers and tolerances

Overlaying layers often produces small extraneous polygons or lines called **slivers** (Fig. 7.9). Sometimes these slivers represent real features. However, more often they arise when overlaying layers share some boundaries, such as voting districts and counties. In theory the shared boundaries should match exactly, but in practice few data sets have been corrected to this level of integrity. These slivers do not represent real combinations of values and are a nuisance when calculating statistics or performing subsequent operations. It is often desirable to prevent slivers when overlaying.

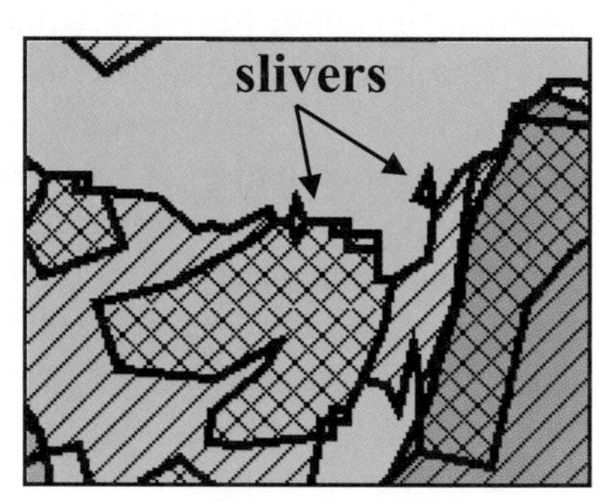

Fig. 7.9. Overlay may produce small sliver polygons.

An **XY tolerance** can be set during an overlay operation. This tolerance specifies the minimum distance between vertices, and it will combine vertices that fall close together into a single vertex. Setting an appropriate tolerance during an overlay will eliminate many slivers. However, the tolerance applies to all vertices, not just slivers. Using a tolerance may combine the vertices of other features and degrade the accuracy of the data set. Caution should be used in setting tolerances that are large enough to correct slivers without shifting other vertices an unacceptable amount. It must also be remembered that outputs from an overlay that uses a tolerance generally have a lower geometric accuracy than the original layers.

Other spatial analysis functions

Overlay is only one type of spatial analysis available in GIS systems. The remainder of this chapter presents additional commonly used functions, including dissolving, buffering, appending, and merging. All of these analysis functions, and many others, are found in ArcToolbox.

Dissolve

A **dissolve** is used to group features together based on whether they share the same value of an attribute field. For example, the road segments in Figure 7.10 have the same street name (Main St), but they are separate features. A dissolve based on the street

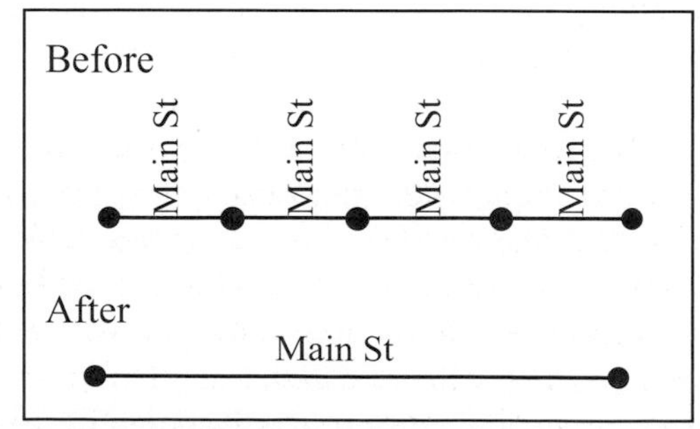

Fig. 7.10. Separate road segments were combined using the dissolve function.

name field would yield a new file in which all streets having that name were one feature. A dissolve can also be used to remove lines between polygon features that share the same value. For example, the map in Figure 7.11 shows ponderosa pine stands of different ages. Dissolving on the cover type field removes the age boundaries and yields polygons based on the cover type (ponderosa) only.

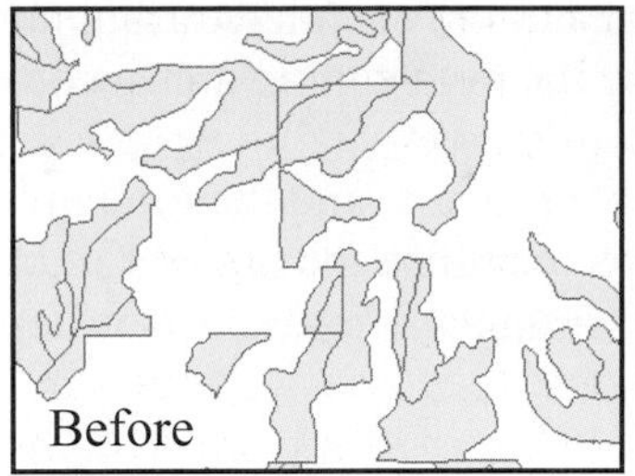

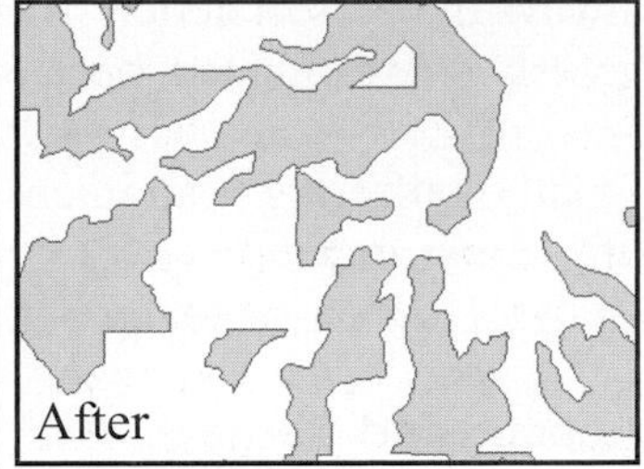

Fig. 7.11. A dissolve removes boundaries between polygons with the same attribute value, in this case, tree species.

By default, a dissolve will produce multifeatures in which spatially distinct lines or polygons may constitute a single feature. Each of the orange "polygons" in Figure 7.11 is actually part of a single polygon multifeature.

When features are dissolved, the output layer is a new file containing the single attribute on which the dissolve was based. However, the user can specify additional fields to summarize the information from the original features. For example, in the dissolve shown in Figure 7.11, one might request the average canopy percentage of each output polygon, based on averaging the canopy percentages of the polygons before the dissolve.

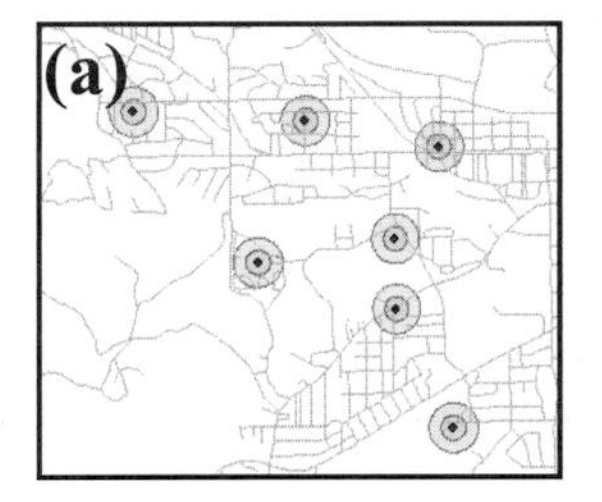

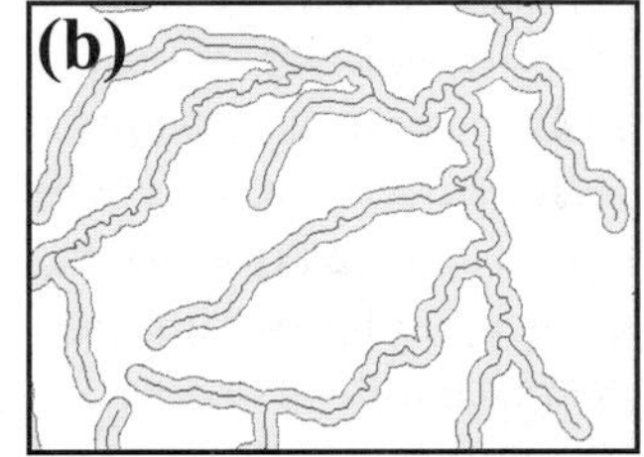

Fig. 7.12. Examples of buffers around (a) wells and (b) roads

Buffer

Buffering is used to identify areas that fall within a certain distance of a set of features. Buffers can be created for points, lines, or polygons (Fig. 7.12). They could be used to find 300-yard drug-free zones around schools or sensitive protected areas within 100 meters of a stream. Negative buffers can be applied to determine setback limits from the edge of a piece of property. Buffers can be created as simple rings or as multiple rings. An attribute can even supply different sizes of buffers for different features—for example, buffering primary roads by 200 meters and secondary roads by 100 meters.

Buffers are created for each individual feature and may overlap. This option might be appropriate in analyzing the distribution of soils in buffers around wells. In many cases, however, it is best to dissolve the buffers, which gets rid of the boundaries between them and removes the overlapping areas to create a single region. Figure 7.13 shows the difference in results obtained when the buffers around roads are not dissolved versus when they are. If the areas of the buffers are of interest, then dissolving is

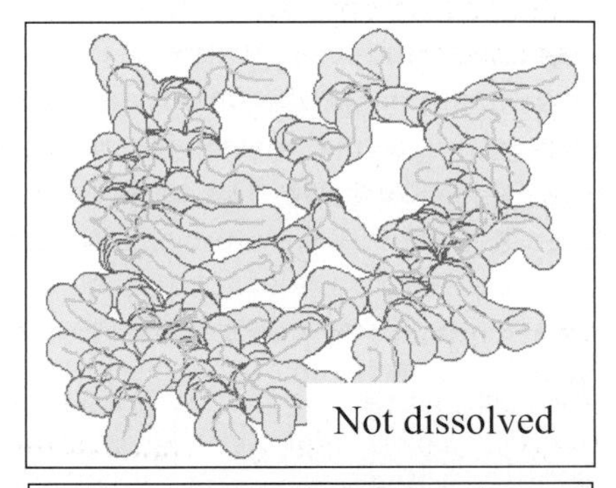

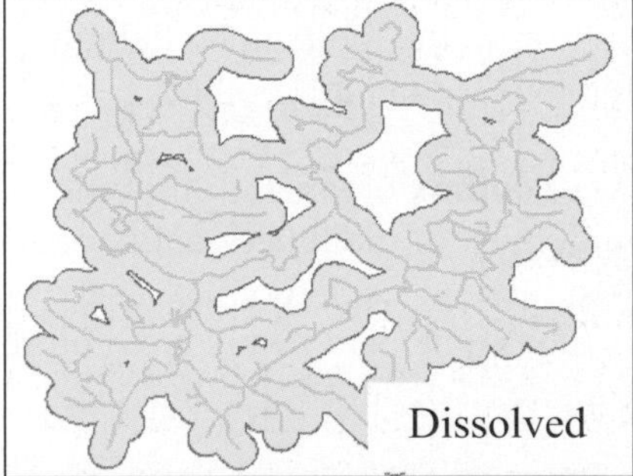

Fig. 7.13. Using the dissolve option to prevent overlapping buffers

always the correct action. If areas were calculated from non-dissolved buffers, they would be grossly overestimated, because the overlapping areas would be counted multiple times.

Finally, buffering is an intensive process, and the time involved increases quickly as the number of features increases. Any steps to reduce the number of buffers created at one time will facilitate analysis, for example, by performing queries before buffering rather than afterwards.

Append and Merge

The **append** function is used to combine the features of two or more layers and place them into an existing target feature class (Fig. 7.14). The appended layers must have the same feature type as the target (i.e., both polygons, both lines, or both points). The two layers must also share the same coordinate system. Overlapping of layers is permitted.

The treatment of attribute tables during an append requires some consideration. If you wish to combine the attribute information from both layers, the attribute fields must have the same definition and must occur in the same order in both tables. If the two tables differ, one can use the NOTEST option. In this case, if a field in the input layer has the same name and data type as the target layer, then the information will be copied into the target. Fields without matching names will not be carried into the target.

Merge is similar to append, except that it creates a new feature class and offers more flexible treatment of attribute tables. Instead of insisting that the two tables match, it allows the user to specify the fields to be included in the output feature class and which table they will come from.

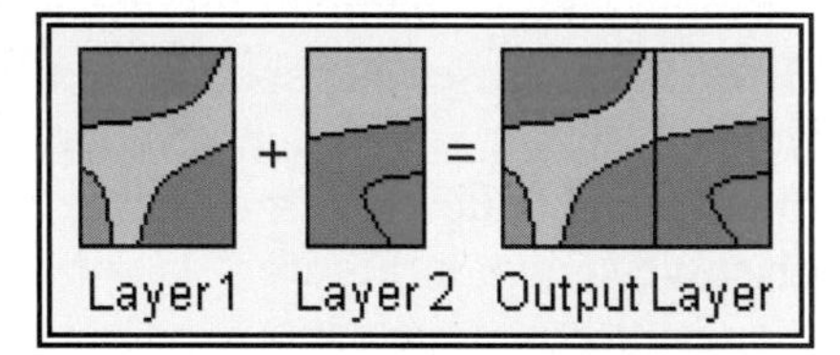

Fig. 7.14. Append combines the features of two adjacent layers.

About ArcGIS

About geoprocessing

GIS analysis involves many functions that operate on data objects, such as feature classes or tables. Geoprocessing applies one or more of these functions in sequence to solve a problem or investigate the properties of data sets. Most ArcGIS tools operate on any allowed data type. In performing a road buffer, for example, one may specify a coverage, a shapefile, or a feature class from a geodatabase. The tools make the necessary allowances to execute the function on the various data types. A few tools operate only on certain data types. An ArcGIS Advanced license, for example, provides the user with a suite of additional tools that work on coverages.

Ways to run tools

Geoprocessing tools may be executed several ways. *Toolbars and menus* provide interactive control of tools. Users may customize menus and toolbars to add frequently used tools that don't appear on the toolbars by default. Users can also create tools and place them on menus, on toolbars, or in ArcToolbox.

ArcToolbox organizes all of the installed tools into one central location. It provides a window with most of the functionality of ArcGIS, containing tools that can be run from ArcMap or from ArcCatalog. Each tool has a set of parameters, or inputs, which must be specified before the tool is run. Some parameters are required, but others are optional, which means that the software supplies a default value that can be changed, if necessary.

ArcMap has a command line that allows the user to type a command and its parameters rather than filling in boxes in a window. Typing requires greater familiarity with the tools, but for experienced users entering commands by typing may be faster and more efficient. Becoming familiar with the command line also paves the way to writing scripts to execute functions. The command line possesses a sophisticated interface to help beginners to correctly enter the commands.

Model Builder provides a graphic canvas to string tools together and execute them in sequence. Models have several advantages. First, they record the steps and parameters used to execute an analysis in case questions on methodology arise later. Second, models can be used to explore the consequences of different parameters on the final outcome. Consider a model that calculates fire danger from precipitation, temperatures, vegetation, and structure locations through a series of tools. With a model, one could redo the analysis based on the current day's precipitation and temperature conditions more quickly than if all the steps were executed interactively, and with less chance of mistakes. Models can be shared with other users, and they permit a less experienced user to reliably repeat an analysis set up by an expert. Models can incorporate decision structures, conditional statements, and iteration functions if needed. Finally, models can be saved as scripts to provide a starting point for creating geoprocessing programs.

A **script** is a program that may contain conditional statements (if/then), iterative loops, and other control structures that permit sophisticated analysis. Scripts may be written in one of several programming languages, although in each case the geoprocessing commands are identical and only the control statements differ. ESRI has adopted the readily available language Python as the recommended scripting platform. A copy of Python is installed with ArcGIS, and all ESRI scripting examples and supplemental tools are written in Python. However, users may elect to use any COM-compliant language such as Jscript, VBScript, or Perl.

Geoprocessing Environments

The operation of tools and commands is impacted by **Environment settings** specified by the user. For example, the default coordinate system setting is *Same as Input,* meaning that the output file has the same coordinate system as the input file. The user could alternatively specify a particular coordinate system, such as UTM Zone 13 NAD 1983. Under this setting, every output feature class would have the UTM coordinate system regardless of its input system.

Figure 7.15 shows some of the environment settings. The Workspace setting tells ArcGIS where to initially look for data and to save output data sets. It also sets a scratch workspace where temporary files are stored. The Output Coordinates setting specifies the coordinate system given to any outputs. If it is set to UTM Zone 13N, then every output data set will be stored in that coordinate system.

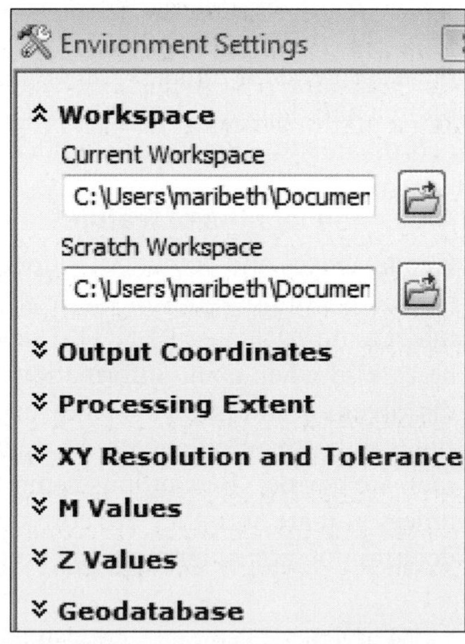

Fig. 7.15. Some Environment settings

Environment settings are hierarchical. Application settings are set using the main menu bar in ArcMap or in ArcCatalog, and they affect every command and tool that is run. However, when you run a tool, you can use an Environments button in the tool to set it differently for that tool as it runs. Models, scripts, and tools can all have their own environment settings, which take precedence over the application settings.

The default settings work fine for most applications. Users may set them when doing so provides more efficient or accurate work. For example, Amy might have to clip 10 different feature classes and project them from a geographic coordinate system to a projected coordinate system. Instead of running both the Clip and Project tools on each layer, she could set the environment coordinate system to the desired projection and then run the clips, reducing the amount of work.

Coordinate systems

Tools will accept any coordinate system or combination of coordinate systems for input, but rules of precedence dictate the coordinate system of the output. Outputs will be projected on the fly, if necessary, according to the following rules:

- If the output is placed in a feature dataset, the coordinate system will always match that of the feature dataset.
- If the coordinate system is set in the Environment settings and the output will not become part of a feature dataset, the Environment settings coordinate system is used.
- If the Environment setting is not set, the default rule applies—that the output will match the coordinate system of the first input to the tool.

When setting the output coordinate system in the Environment settings, you can choose Same as Input (the default), or Same as Display (the current data frame), or set it to match the coordinate system of a particular data set.

Geoprocessing tools are fundamentally spatial in nature and commonly manipulate areas and distances as part of the processing. When geoprocessing, it is best to always use a projected coordinate system, rather than a geographic coordinate system (GCS). As with spatial joins, it is not enough to set the data frame coordinate system to a projection. The tools operate on the saved feature class on disk, and the saved coordinate system is the one being used.

Often a GCS is employed by data providers because they cannot predict what coordinate systems the user will need. When downloading data, it will often be in a GCS, and the user must choose an appropriate coordinate system and project the data to it. Chapter 11 explains this process. When you are setting up a database for a mapping and analysis project, a suitable projection should be chosen based on the scale and the extent of the data being analyzed. It must be kept in mind that distortions present in the map projection, if significant, will affect distance and area measurements. It remains important to choose a suitable map projection with minimal distortion of area and distance. Review the coordinate system properties on the inside front cover and the coordinate system selection guidelines in Chapter 3, if necessary.

Areas and lengths of features

Geodatabases and coverages automatically create and update fields containing the areas and perimeters of polygons or the lengths of lines. Shapefiles do not maintain this information. Users may create AREA or LENGTH fields for shapefiles and calculate the values manually. However, be careful when using information from these fields. If polygons in a shapefile are clipped, dissolved, or intersected or they undergo any other operation that changes their shape, the AREA fields will not automatically be updated. The user must manually update the fields again to ensure they are correct. Be cautious using an AREA or a PERIMETER or a LENGTH field in a shapefile unless you are sure they are correct. Chapter 4 introduced the Calculate Geometry tool for determining areas, lengths, and perimeters in a variety of units.

Using ArcToolbox

ArcToolbox may be used in ArcCatalog or in ArcMap. When a tool is opened (Fig. 7.16), required inputs are marked with a green dot (a). To enter data sets as inputs or outputs, use the Browse button next to the appropriate box (b). If using the tool in ArcMap, you can also select layers in the map from a drop-down list as inputs. Optional inputs are marked (c) and can usually be left to their defaults. Some tools have input parameters that extend beyond the visible pane, and the scroll bar must be used to view and enter them. To get more information about the tool or an input, click the Show/Hide Help button (d) for a brief description of the tool's function. For more detailed help, especially on the input/output data, click the Tool Help button (e).

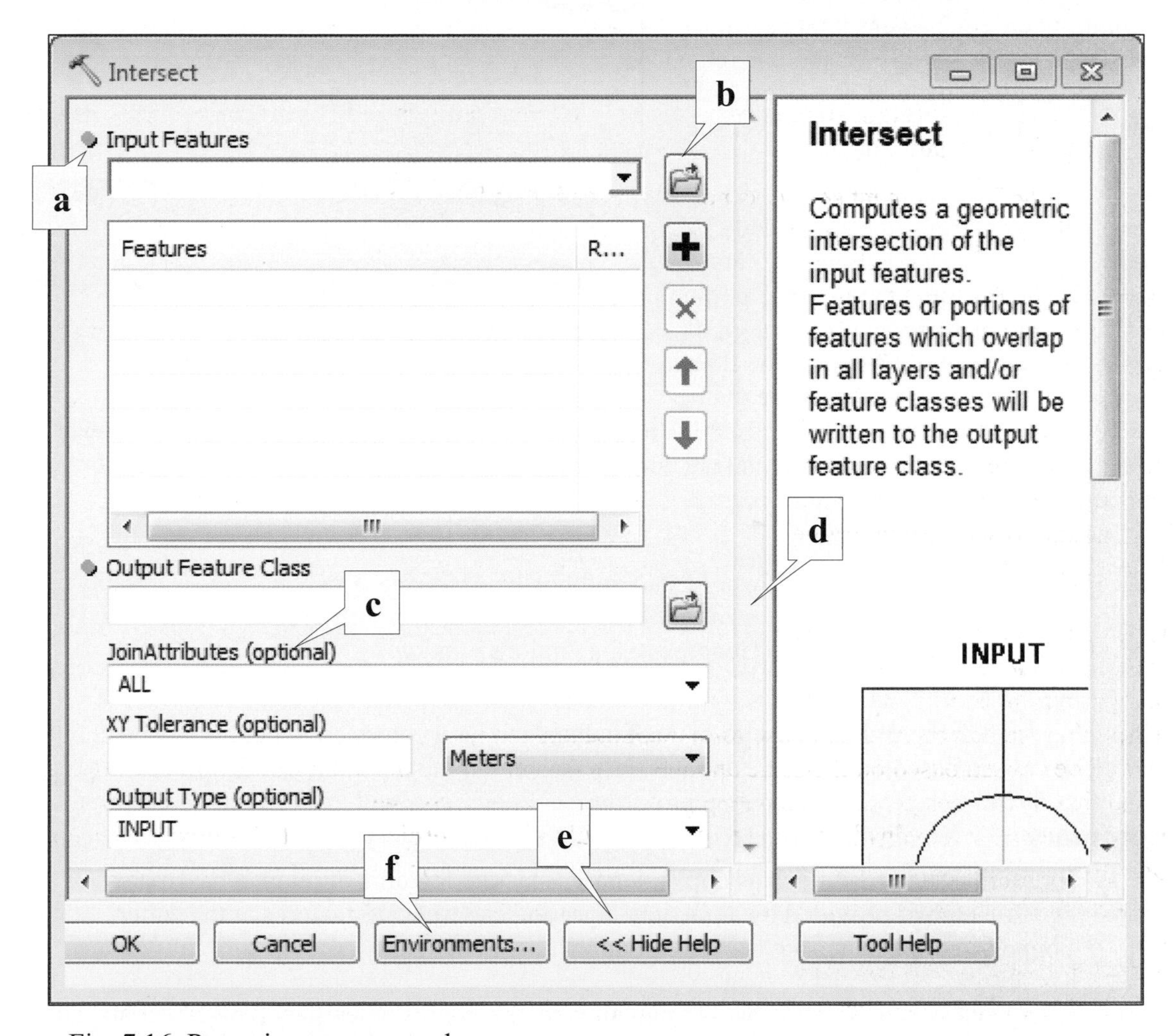

Fig. 7.16. Preparing to run a tool

TIP: When using a tool for the first time, it is strongly recommended that you read the detailed help information. It may contain critical information for making the tool work properly.

The operation of tools may be affected by the Environment settings for the geoprocessing environment. For example, the user can set a default output directory or a default *x-y* extent for the output. These settings can be accessed by clicking on the Environments button at the bottom of the tool (f). When set here, the Environment is set for this one tool, this one time.

The input pane for a tool posts symbols to help you use the tool (Fig. 7.17). A green dot next to an input parameter indicates that it is required and must be entered before the tool will run. A yellow exclamation point indicates that the user should proceed with caution. Placing the cursor over the yellow sign for a moment will display a pop-up message describing the problem; in Figure 7.17, the user is attempting to define a coordinate system for a data set that already has one. A red X indicates an error that will cause the tool to fail. Again, holding the cursor over the X will display the error message.

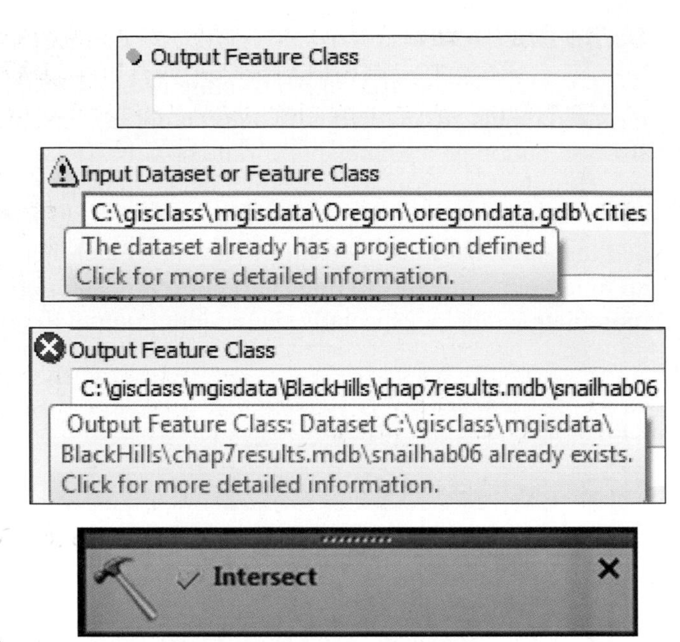

Fig. 7.17. Error handling for tools

By default, a tool runs in background mode so that you can continue working as it runs. When the tool begins running, blue scrolling text will appear at the bottom of the ArcMap or ArcCatalog window. If an error occurs, it will be reported in a dialog box. When the tool completes successfully, a blue completion message will pop up in the lower-right corner of your computer screen (Fig. 7.17). The message will go away automatically, or you can click the X to close it.

TIP: Background processing is slower to start and finish a task, and it may not run as reliably as foreground processing. The author **strongly recommends** that you turn off background processing in the Geoprocessing > Geoprocessing Options window.

Users interested in the full capabilities of ArcToolbox and the geoprocessing environment should consult the extensive sections in the ArcGIS Help (search on the index tab for geoprocessing).

Summary

- Extraction functions, such as clip and erase, separate features of interest based on a "cookie cutter" provided by a polygon layer and can truncate features at the cutter boundaries. Attributes are carried through without change.

- Map overlay resembles a spatial join, but it splits features when they partly overlap. This function enforces a one-to-one relationship between features when their attributes are joined in the output table.

- Map overlay comes in two basic types. A union keeps all the features from both layers. An intersect keeps all the features that are common to both input layers. Attributes from both layers are joined together in the output.

- A dissolve combines features within a data layer if they share the same attribute. This function can be used to convert many street segments into a single line feature or to remove boundaries between parcels with the same zoning.

- Buffers are polygon constructions that enclose the area within a certain distance of features. Buffers may be created for points, lines, or polygons, and they may be constructed as single or multiple rings.

- Append and merge allow feature classes with the same feature type to be combined as a single feature class, such as merging two adjacent quadrangles to make a single file.

- Map overlay and spatial analysis are generally best done using a projected coordinate system, especially if one anticipates determining areas and lengths as part of the analysis.

- Geoprocessing involves stringing together a sequence of commands during spatial analysis. Spatial functions may be executed from the menus, ArcToolbox, the command line, Model Builder, or scripts.

- The Shape_Area, Shape_Length, and Shape_Perimeter fields in geodatabase feature classes are stored and updated automatically. Other area-based or length-based fields, such as Area, Acres, or Road_km, are not automatically updated.

- Lengths and areas of features stored in shapefiles must be updated manually.

Important Terms

append	Environment settings	merge	union
ArcToolbox	erase	Model Builder	XY tolerance
buffer	geoprocessing	parameter	
clip	intersect	script	
dissolve	map overlay	sliver	

Chapter Review Questions

1. How does the clip function differ from Select By Location?
2. What is the most important difference between a spatial join and a map overlay?
3. What are slivers? Explain how they can be prevented.
4. What is a buffer? Why is a dissolve often performed when buffering features?
5. What function would you use to create a map of a study area such that all the features in the map stopped at the study area boundary?
6. What attribute fields will be present in a layer resulting from a dissolve?
7. Why is it usually advantageous to use a projected coordinate system when doing a map overlay?
8. How can you determine the areas of polygons in a geodatabase? In a shapefile?
9. What determines the coordinate system of the output when overlay is used?
10. What is geoprocessing? In what different ways can commands be executed?

Mastering the Skills

Teaching Tutorial

The following examples provide step-by-step instructions for doing basic tasks and solving basic problems in ArcGIS. The steps you need to do are highlighted with an arrow ➔; follow them carefully. Click on the video number in the VideoIndex to view a demonstration of the steps.

To demonstrate map overlays, we will do the problem described earlier in the chapter concerning delineating habitat for the rare Black Hills snail using overlays. This habitat will be defined by three criteria: on a limestone geology unit, in dense coniferous forest, and between the elevations of 1200 and 1600 meters.

➔ Start ArcMap and open the map document ex_7.mxd.

➔ Use Save As to rename the map document. Remember to save it often as you work.

As in Chapter 6, we don't want the new feature classes created during the analysis to become part of the original geodatabase, so we'll create another geodatabase to contain them. We'll also set the geoprocessing Current Workspace as the default location to place output data.

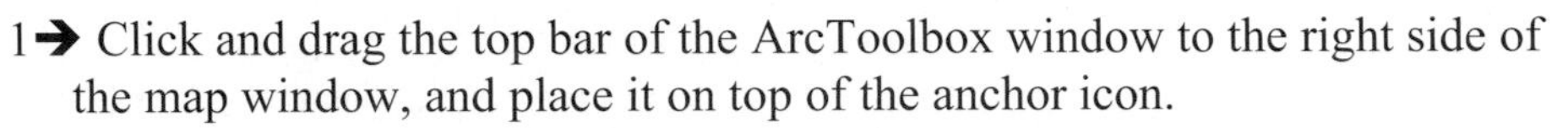

1➔ Click the ArcToolbox under the main menu bar to open it.

1➔ Click and drag the top bar of the ArcToolbox window to the right side of the map window, and place it on top of the anchor icon.

1➔ Click the Pin icon at the top of the Toolbox to pin it to the right side as a tab. When you want a tool, hover over the ArcToolbox tab and the toolbox will open. The rest of the time it is out of the way.

2➔ In ArcToolbox, expand the Data Management > Workspace entries and double-click on the Create Personal GDB tool.

2➔ Click the Browse button for the Location and navigate to the mgisdata folder. Select the BlackHills folder by clicking it once (don't go inside it) and click Add.

2➔ Enter chap7results for the output geodatabase name. Click OK.

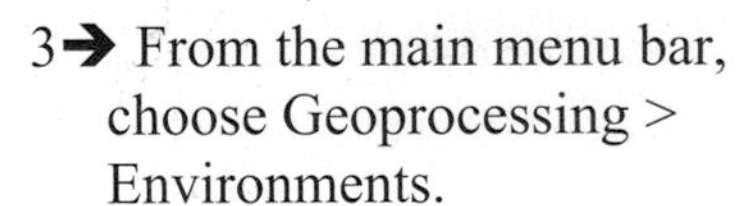

3➔ From the main menu bar, choose Geoprocessing > Environments.

3➔ Expand the Workspace entry and use the Browse button to set the Current Workspace to chap7results (Fig. 7.18).

3➔ Also set the Scratch Workspace to chap7results.

3➔ Expand the Output Coordinates entry and check that the Output Coordinate System is set to Same as Input. Click OK.

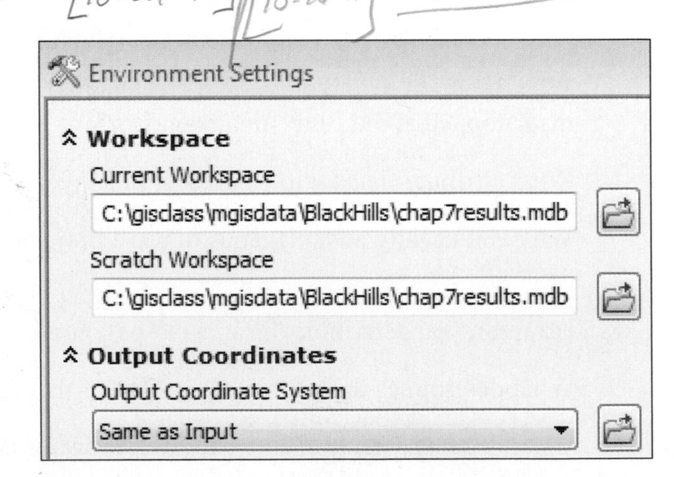

Fig. 7.18. Setting geoprocessing environments

Preparing to overlay

Our first step is to create layers that contain polygons where each condition holds. We will begin with the Geology layer. The limestone areas include the Madison Formation and the Upper Paleozoic units. We will use an attribute query to place these units in a separate layer.

4➔ Choose Selection > Select By Attributes from the main menu bar.

4➔ Set the layer to Geology and enter the expression:

[NAME] = 'Upper Paleozoic' OR [NAME] = 'Madison Limestone'. Click OK.

4➔ Right-click the Geology layer and choose Selection > Create Layer from Selected features.

4➔ The new layer appears at the top of the Table of Contents. Rename it **Limestone**.

TIP: Creating a layer from a selection, although not necessary for processing, is often a good practice. It keeps the selection available if a question about the procedure arises later, or if a tool must be rerun due to an error.

Next, we will select the dense conifers and create a layer from them. We will do the query in two steps. First we will select the ponderosa pine (TPP) and white spruce (TWS). Then we will select the dense areas (class contains C or 5) from the already selected set.

5➔ Turn on the Vegetation layer and turn off the Limestone layer.

5➔ Clear all selected features.

5➔ Open the Select By Attributes window again and set the layer to Vegetation.

5➔ To select only the conifers, enter the expression:

[COV_TYPE] = 'TPP' OR [COV_TYPE] = 'TWS'. Click Apply.

6➔ Change the selection method to *select from current selection.*

6➔ Clear the expression and enter another one that says:

[DENSITY96] = 'C'.

Verify the expression and then click Apply.

6➔ Close the Select By Attributes window.

6➔ Right-click the Vegetation layer and choose Selection > Create Layer from Selected Features.

6➔ Rename the new layer **Dense Conifer**. It should look like the polygons shown in Figure 7.19.

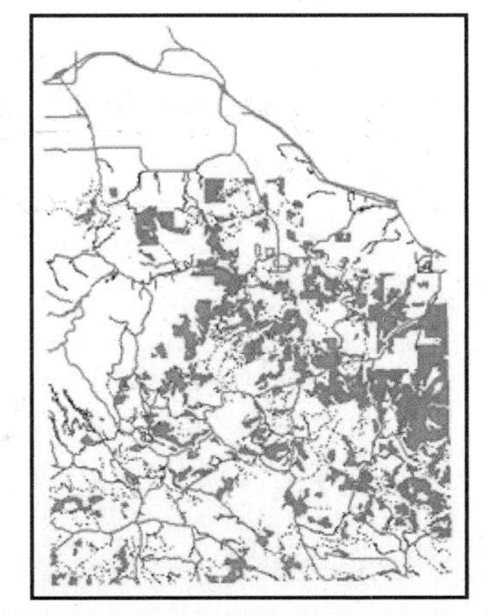

Fig. 7.19. Dense conifer

The elevation range has already been prepared. You are almost ready to intersect.

7➔ Clear the selected features and turn off the Vegetation layer.

7➔ Turn off all layers except the Elevation Range layer, the Limestone layer, and the Dense Conifer layer.

7➔ Zoom in to the middle right of the map and examine the Dense Conifer polygons.

Notice that these polygons have many boundaries inside them because they are divided based on age and density as well as species. The amount of time needed to intersect a layer is proportional to the number of features in the layer. To streamline the intersect, we are going to remove the unnecessary boundaries between these polygons using the Dissolve tool.

8➔ In ArcToolbox, expand the Data Management > Generalization entries and double-click the Dissolve tool to start it (Fig. 7.20).

When using a tool for the first time, it is wise to select Show Help to find out what the tool does and what the parameters mean.

8➔ Click Show Help on the Dissolve tool and read the description.

8➔ Click on the Input Features box. The Help message changes to describe the Input Features parameter.

8➔ Click on and read the descriptions for the other input parameters in the window.

8➔ To bring the tool description back, click on the gray tool area.

8➔ Click on the Tool Help icon at the bottom of the tool for more detailed information. Read about the Dissolve tool.

8➔ Scroll down to the bottom of the description.

8➔ Close the ArcGIS Desktop Help window and click Hide Help on the Dissolve tool.

9➔ Click the drop-down arrow in the Input Features box and set it to Dense Conifer (Fig. 7.20).

9➔ Click the Browse button to place the Output Feature Class in the BlackHills/chap7results geodatabase and name it conifer2.

9➔ Check the box to dissolve on the COV_TYPE field.

9➔ Keep the box unchecked to *Create multipart features*. We want unconnected polygons to remain separate features.

9➔ Notice the other options in the tool. We don't need to change any of them, however.

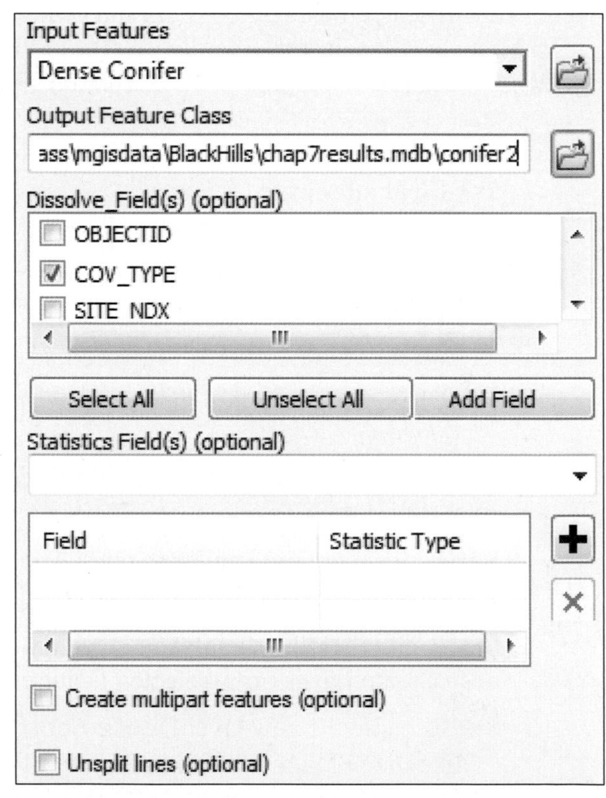

Fig. 7.20. The Dissolve tool

9➔ Click OK. A scrolling banner will appear at the bottom of the ArcMap window to indicate that the tool is running. A blue completion message will pop up in the lower-right corner of the screen when the tool is done.

TIP: When using a tool, be aware that some options may be out of sight, and you must scroll down to see them. Often the defaults are fine, but get in the habit of checking before you run it.

10➔ Examine the output file and note that the intervening boundaries have disappeared.

10➔ Right-click the Dense Conifer layer and choose Remove.

10➔ Rename the conifer2 layer Dense Conifer.

Intersecting polygons

Now we are ready to overlay. We will use Intersect because we want to find the areas common to all three layers. Because the tool can intersect only two layers at a time, we must use it twice.

11➔ Open the ArcToolbox > Analysis Tools > Overlay > Intersect tool (Fig. 7.21).

11➔ Click Show Help and examine the entries.

1. What are the options for the Join Attributes parameter? Which one do you think is best to use? ______________________________

11➔ Click on the drop-down button under Input Features and choose the Dense Conifer layer. It will be added to the list of Features.

11➔ Click the drop-down button again to add the Limestone layer to the list.

11➔ Enter the Output Feature Class, placing it in chap7results and naming it conf_lime.

11➔ Set the Join Attributes option to NO_FID.

11➔ The remaining options can be left with their defaults. Click OK.

11➔ Examine the output. The conifers outside the limestone have disappeared.

12➔ Use the Intersect tool again to intersect conf_lime and the Elevation Range layers, naming the output snailhab and putting it in chap7results.

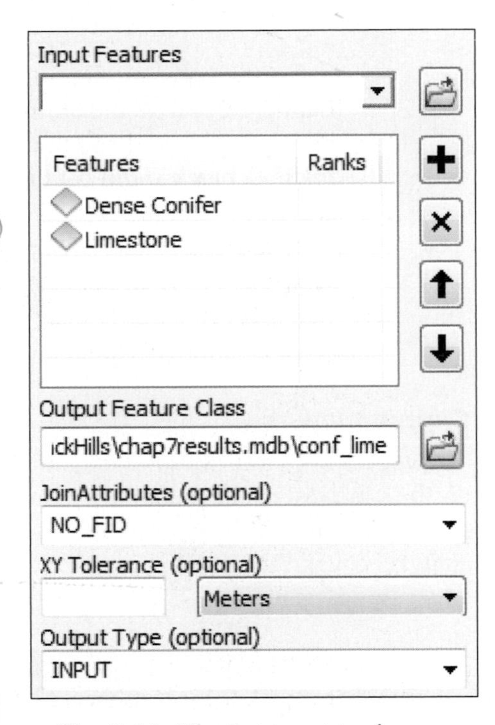

Fig. 7.21. The Intersect tool

> **TIP:** If you have an ArcGIS Advanced license, you can intersect more than two layers at a time.

13➔ Zoom out to the full extent of the data.

13➔ Rename the snailhab layer Snail Habitat.

13➔ Remove the Limestone, Dense Conifer, and conf_lime layers.

13➔ Zoom in to the extent of the Snail Habitat and turn on the Roads layer.

Overlay of lines in polygons

The snails have a three-week breeding season in early June. During this period they seek the open areas offered by roads, and many get crushed. The Forest Service wants to consider closing primitive roads that traverse through snail habitat during the breeding season to lessen the number of crushed snails. They need to assess which roads must be closed. You will make a map with the proposed road closures. This process involves a line-in-polygon intersection.

14➔ Use Select By Attributes to select the Primitive roads [TYPE] = 'PR'. (Make sure you set the method back to *Create a new selection*.)

15➔ Intersect the Roads layer (with the primitive road selection) and the Snail Habitat layer. Name the output **propclose** and put it in chap7results.

15➔ Rename the propclose layer **Proposed Closures**. Clear the selected features.

TIP: Background processing takes longer than foreground processing, and it can cause tools to crash. The author recommends turning off background processing unless it is needed.

16➔ Open Geoprocessing > Geoprocessing Options from the main menu bar.

16➔ Under Background Processing, uncheck the Enable box and click OK.

17➔ Create a map similar to Figure 7.22, with the closure roads highlighted in red and the other roads in black. Turn off the Elevation Range layer. Set the transparency of the Snail Habitat layer to 50% to help bring out the road patterns.

17➔ Save your map document.

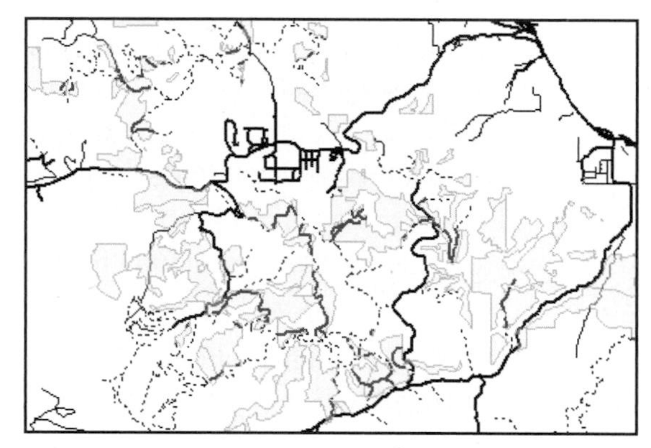

Fig. 7.22. Map showing proposed road closures in snail habitat areas

Clipping layers

The vegetation layer does not extend as far north as the rest of the data, but the map may give readers the false impression that the analysis is valid there. Clipping the Roads layer at the edge of the vegetation layer will prevent any misunderstanding.

Now you know that you need the Clip tool. Let's practice finding tools when you know the name but not where to look for it in the Toolbox.

18➔ On the main menu bar, choose Geoprocessing > Search For Tools.

18➔ Type **clip** in the search box. Several functions appear, including Clip (analysis) and Clip (management). The words in parentheses refer to the area of the Toolbox.

18➔ Ignore the suggestions for now and click on the magnifying glass button.

The list includes some exact matches (clip) and some similar functions, such as clipping rasters, along with a description of each.

18➔ Click the Options button in the search window and choose Search Options.

18➔ Click the General tab and check the box to *Show Pop-up window in search results*.

18➔ Click OK and hover the cursor over the blue tool names to see the descriptions. The Clip (Analysis) appears to be the one to use.

18➔ Click the blue Clip (Analysis) text to open the tool.

TIP: You can pin the Search window to the right tab bar for easy access later.

19➔ Set the Input Features to Roads. Set Vegetation as the Clip Features.

19➔ Store the output in the chap7results geodatabase as **roadclip**. Click OK.

Because you turned off background processing, a window appears to show the tool's progress. Close it when the tool finishes. You can check the box to close it automatically after it finishes. (If it encounters an error, it will leave the box open with the error message.)

19➔ Close the notification box. Turn off the Roads layer.

You can quickly transfer the symbology of the old Roads layer to the clipped roads.

20➔ Open the properties of the roadclip layer and click the Symbology tab.

20➔ Click the Import button and choose to import symbology from a layer or layer file.

20➔ Click the drop-down box to set the layer to Roads. Click OK.

20➔ The road classification is based on the TYPE field. The clipped roads use the same field, so leave the TYPE name in the box. Click OK and OK.

21➔ Rename the roadclip layer **Clipped Roads**.

21➔ Click and drag the Clipped Roads layer below the Snail Habitat layer.

Working with buffers

To prevent snail death on the major roads, which cannot be closed, a Forest Service biologist has suggested thinning the tree stands within 200 meters of the roads to make these areas less attractive to snails and keep them away from roads. We can prepare a map showing the stands to be cleared and determine the total percentage of snail habitat that would be eliminated.

22➔ Use Select By Attributes to select the primary and secondary roads from the Clipped Roads layer, using the expression [TYPE] = 'P' OR [TYPE] = 'S'.

22➔ Create a layer from the selected roads and name it **Major Roads**. Clear the selected features.

TIP: Buffers are time-consuming to create, so it is helpful to reduce the number of features before buffering. We will use the Elevation Range layer to clip the roads before buffering.

23➔ Open the tool ArcToolbox > Analysis > Extract > Clip.

23➔ Set Major Roads as the Input Features and Elevation Range as the Clip Features.

23➔ Save the result in chap7results as **majroadclip**. Click OK to run the tool.

23➔ Turn off the other road layers to examine the new file.

Now buffer the clipped roads.

24➔ Open the tool ArcToolbox > Analysis > Proximity > Buffer.

24➔ Set the Input Features to majroadclip and the Output Feature Class to **roadbuf** in chap7results.

24➔ Set the Linear unit to meters and type **200** in the box.

24➔ Scroll down, if necessary, and set the Dissolve Type to ALL. Click OK.

TIP: By default the Buffer tool does not dissolve boundaries between buffers. This is one case where the default option is not right, and the Dissolve Type must be set to ALL.

Now intersect the road buffers with the snail habitat to determine the potential areas to be cleared.

25➔ Open ArcToolbox > Analysis > Overlay > Intersect.

25➔ Choose Snail Habitat and roadbuf as the Input features.

25➔ Specify the Output file as **proposedthin** in chap7results. Click OK.

26➔ Clean up the map display to show the results. Turn off the roadbuf layer and make the proposedthin layer a symbol that contrasts well with the snail habitat (Fig. 7.23).

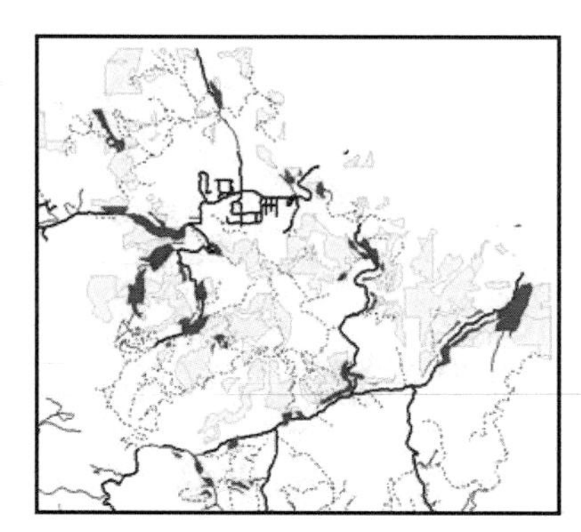

Fig. 7.23. Proposed thinning areas (in pink)

Now the final step is to determine the area of snail habitat that will be eliminated by the thinning. Geodatabases automatically keep track of the areas, lengths, and perimeters of features.

27➔ Open the proposedthin attribute table and examine the Shape_Area and the Shape_Length fields.

These fields were renamed Shape_1_Area and Shape_1_Length during one of the intersections, but they will contain the correct areas and perimeters for the feature class. The units of the Shape_Area field will be in the same coordinates as the map units of the feature class coordinate system, in this case, square meters.

27➔ Use the Statistics function to determine the total area to be thinned.

TIP: To convert square meters to square kilometers, divide by 1 million (1,000,000).

2. What is the total area of the proposed thinning in square kilometers? __________________

3. What is the total area of snail habitat in square kilometers? ________________________

4. What percentage of the habitat would be eliminated by the proposed thinning? ______

TIP: Shapefiles do not automatically maintain area, length, or perimeter fields. Values in these fields may not be correct if geoprocessing actions have been performed since the fields were created. The fields may be updated using the Calculate Geometry tool described in Chapter 4.

Investigating relationships with union

Intersect is valuable for finding areas where known criteria exist simultaneously, such as finding the snail habitat based on known ideal factors of elevation, geology, and vegetation. Union is a powerful tool for investigating relationships between two types of information.

It is well known that rock units have an impact on the soils that are produced from them. It would be interesting to see if the soils then have an impact on the productivity of the forest. We will perform a union on the Geology and Vegetation layers to explore this question.

28➔ Turn off all layers except Geology and Vegetation.

28➔ Open ArcToolbox > Analysis > Overlay > Union.

28➔ Choose Geology and Vegetation as the Input Features.

28➔ Name the output feature class geolveg and put it in the chap7results geodatabase. Click OK.

29➔ Zoom to the full extent and turn off all layers except for geolveg.

First, notice that all areas of both input layers are included in the output, even though the Vegetation layer does not cover as much area as the Geology layer does.

29➔ Click the Identify tool. Click on several of the large polygons at the top and examine the attributes.

29➔ Click on several of the smaller polygons at the bottom.

Field	Value
OBJECTID	135
Shape	Polygon
FID_geology	15
UNIT	5
AGE_NUM	4
NAME	Upper Paleozoic
INFIL	Mod
INFILCLASS	2
FID_vegetation	241
RISDATA	0814130007
DATA	0814130007
OWNER	NFS
COV_TYPE	TPP
SSTAGE96	4B
TREE_SZ96	L
DENSITY96	B
SITE_NDX	65

Fig. 7.24. Attributes from union of geology and vegetation

Most of the smaller polygons occur where both Vegetation and Geology were present. Their attributes are fully stocked with information from both layers (Fig. 7.24). However, the polygons at the top came only from Geology; there were no Vegetation polygons present in this part of the map. The vegetation fields are in the table, but they are full of blanks and zeros, because there was no information to attach.

To investigate the relationship between geology and site productivity, we will Summarize on the Geology NAME field, and request statistics from the SITE_PROD and CROWN_COV fields, both indicators of forest productivity. However, because some of the polygons have no vegetation data, we will exclude the zero values from consideration with a query.

30➔ Close the Identify window and open the geolveg table.

30➔ Open the Table Options > Select By Attributes window and enter the expression [SITE_PROD] > 0. Click Apply and close the query window.

31➔ Right-click the NAME field and choose Summarize.

31➔ Expand the SITE_PROD field and check the boxes for Minimum, Maximum, Average, and Standard Deviation.

31➔ Expand the CROWN_COV field and check the boxes for Minimum, Maximum, Average, and Standard Deviation.

31➔ Name the output table geolproductivity and save it in the chap7results geodatabase. Click OK and add the table to the map.

32➔ Open the geolproductivity table. Find the Average_SITE_PROD field and use Sort Descending. Examine the averages and standard deviations.

32➔ Find the Average_CROWN_COV field and use Sort Descending. Examine the averages and standard deviations.

Although both site productivity and crown cover vary from unit to unit of the geology, for the most part all the averages are within one standard deviation of the others. This analysis does not support the hypothesis that the geological units have a significant impact on forest productivity.

32➔ Close the Table window and clear all selected features.

Working with slivers and tolerances

We will now work on a short exercise to demonstrate the problem of slivers and how to help prevent them using tolerances. We need a different data frame for this part of the tutorial.

33➔ Choose Insert > Data Frame from the main menu bar. Collapse the Sturgis Area data frame to hide the layers you've been working with.

33➔ Add the cd111 and counties feature classes from the mgisdata\Usa\usdata geodatabase.

33➔ Zoom in to the San Francisco Bay Area in northern California (Fig. 7.25).

33➔ Choose Bookmarks > Create Bookmark. Enter the name San Francisco and click OK.

Fig. 7.25. Zoom in to this area.

This area will be our experiment. We don't need to overlay the entire country, so to save processing time, we will use the Environment settings to constrain the output extent to this display window.

34➔ From the main menu bar, choose Geoprocessing > Environments.

34➔ Expand the Processing Extent entry.

34➔ Change the Extent drop-down to read Same as Display. Click OK.

35➔ Open the tool ArcToolbox > Analysis > Overlay > Union.

35➔ Enter cd111 and counties as the input features.

35➔ Name the output union_notol and put it in chap7results. Click OK.

36➔ Zoom to the full extent of the data.

36➔ Notice that only a small group of polygons around San Francisco is included in the output. This is the result of the Processing Extent environment setting.

TIP: Once set to an extent, the Processing Extent setting will continue to use the same range of *x-y* coordinates, so it will not be affected as you zoom in and out.

36➔ Choose Bookmarks > San Francisco, and examine the output.

Notice that many slivers occur (Fig. 7.26). The district and county boundaries coincide in real life, but these two data sets come from different sources, and the boundaries aren't identical. To specify a processing tolerance, it would be helpful to know the typical width of the slivers.

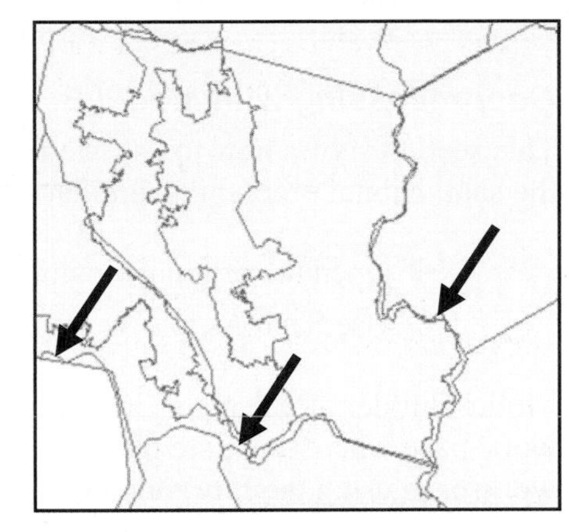

Fig. 7.26. Slivers from the union

37➔ Zoom in to a region with several slivers.

37➔ Click the Measure tool on the Tools toolbar. Set the units to meters.

37➔ Measure across several slivers to estimate their widths. Close the Measure tool.

You will find that a typical width is about 1000 meters. A tolerance is a radius, so we should use about half the measured value. Let's repeat the union and use a tolerance of 500 meters.

38➔ Open the ArcToolbox > Analysis > Overlay > Union tool.

38➔ Set the input layers to cd111 and counties, as before.

38➔ Name the output union500 and place it in the chap7results geodatabase.

38➔ Enter 500 meters as the XY Tolerance. Click OK and examine the output.

38➔ Turn the new layer on and off a few times to compare the boundaries with the union_notol layer. Many slivers are gone in the new layer.

Notice two things about the new layer. First, many slivers are still present. Our initial tolerance was not sufficient to remove them. Second, the district boundaries have lost some of their detail. The XY Tolerance may not have fixed all the slivers, but it already is having an effect on the accuracy of the output. Nevertheless, we will try a larger tolerance.

39➔ Repeat the union of cd111 and counties, but this time use an XY Tolerance of 1000 meters and name the output union1000.

39➔ Turn on just the union_notol and the union1000 layers and compare them by turning the top one on and off a few times.

Most of the slivers are now gone, but the loss of detail in the district boundaries is apparent. The county boundaries don't look much worse, but they were a lower resolution to begin with.

40➔ Try one more union with a 1500-meter tolerance and compare.

Now the generalization of the boundaries is obvious, and a few stubborn slivers still remain. Further generalization and loss of accuracy are not worth getting rid of the few remaining slivers. The 1000-meter tolerance gave the best compromise performance. Before we go on, we must set the Environment settings back to their defaults.

41➔ Choose Geoprocessing > Environments from the main menu bar.

41➔ Expand the Processing Extent entry and set it back to Default. Click OK.

TIP: One challenge of using Environment settings is to remember to reset them after you are finished. Note that the Environment settings are saved with map documents.

Geoprocessing with Model Builder (optional)

This section shows how to use Model Builder to record sequences of geoprocessing steps. Since the snail habitat problem is familiar, we'll use it as our example.

➔ Open the original version of ex_7.mxd without any changes.

➔ Use Save As to save it under a new name and save often as you work.

Model Builder often creates intermediate data sets while it works. You can specify a scratch workspace where these are placed, so that they don't interfere with the original data. You set the workspace using the Environment settings.

42➔ Open Geoprocessing > Environments from the main menu bar.

42➔ Expand the Workspace entry. Set both the Current Workspace and Scratch Workspace to the chap7results geodatabase. Click OK.

Creating and running a model

You store a model inside a geodatabase or in a toolbox. We will create a toolbox for it.

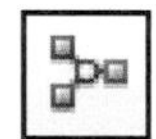

43➔ Click the Model Builder button on the main tool bar.

43➔ Choose Model > Save in the Model Builder window.

43➔ Navigate so that you are in the BlackHills folder.

43➔ Click the New Toolbox button in the Save window.

43➔ Name the toolbox BHtools and click Enter.

43➔ Double-click the new BHtools toolbox to enter it.

43➔ Type SnailHab as the name for the model and click Save.

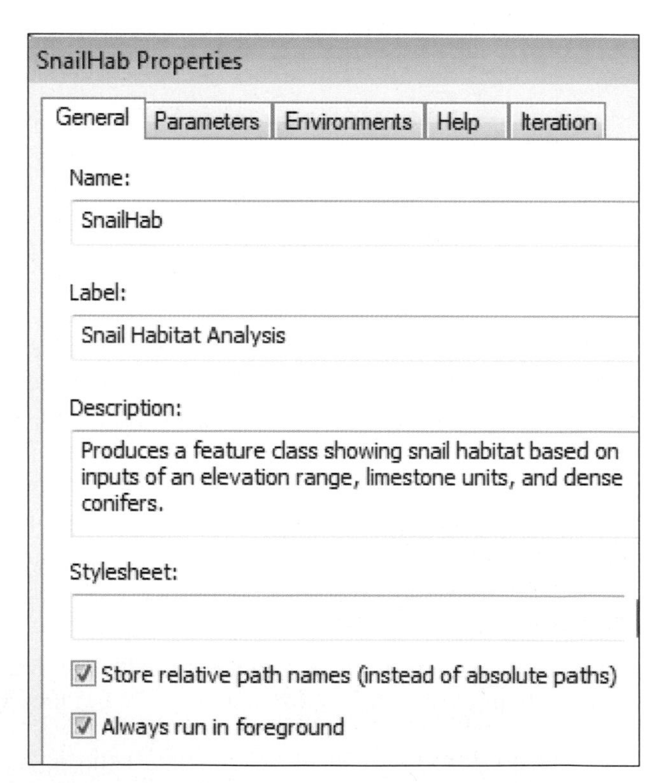

Fig. 7.27. Naming and describing a model

44➔ Choose Model > Model Properties from the Model Builder menu bar.

44➔ Label the model Snail Habitat Analysis and give it a Description (Fig. 7.27).

44➔ Examine the other settings as shown in Figure 7.27 and then click OK.

TIP: Storing relative pathnames for a model makes it work if the model and data together are moved to a different folder. Absolute pathnames only work if the data stays put.

In Model Builder, you paste the desired tool in the window, set arguments for the tool, and run it. Each step can be run separately as the model is built or all together. When building the model, it is helpful to execute the steps one at a time. In the beginning of this tutorial, we used Select By Attributes to select the desired polygons from the geology and vegetation layers. In Model Builder, we use the Make Feature Layer tool.

45➔ Find the ArcToolbox > Data Management Tools > Layers and Table Views > Make Feature Layer tool. Click and drag it to the Model Builder window.

45➔ Right-click the Make Feature Layer tool in the Model Builder window and choose Open.

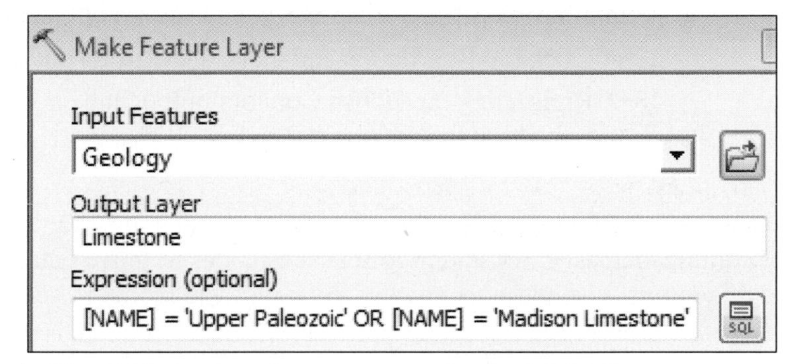

Fig. 7.28. Setting the arguments for the Make Feature Layer tool to create the Limestone layer

45➔ Set the Input Features to Geology and name the Output Layer Limestone, as shown in Figure 7.28.

46➔ Click the SQL button to enter the selection expression: [NAME] = 'Upper Paleozoic' OR [NAME] = 'Madison Limestone'. Verify the expression.

46➔ Click OK to close the SQL window and OK to close the tool.

46➔ The boxes in the model become colored, indicating the tool is ready to run. Choose Model > Run, or click the Run button to execute this part of the model.

46➔ Close the dialog box after the tool completes successfully. A drop shadow behind the model shapes indicates that the tool has been run (Fig. 7.29).

The blue ovals indicate inputs, and the green ovals indicate outputs. Both inputs and outputs have properties that can be set.

46➔ Right-click the green Limestone oval and choose Add to Display. The new layer appears in the Table of Contents and is shown on the map.

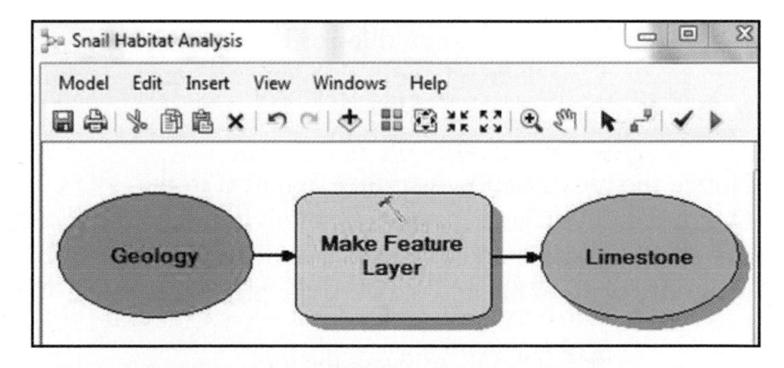

Fig. 7.29. The model after the first step is run

Now we'll use the same tool to select the dense conifers. Because we're using a tool, it is easier to do the selection in one step.

47➔ From ArcToolbox, click and drag another copy of the Make Feature Layer tool onto the model canvas.

47➔ Right-click the tool to open it. Set the Input Features to Vegetation and the Output Layer to Dense Conifer.

47➔ Click the SQL button and carefully enter the expression (with the parentheses):

([COV_TYPE] = 'TPP' OR [COV_TYPE] = 'TWS') AND [DENSITY96] = 'C'.

47➔ Click the Verify button to ensure that you did not make a mistake in the query.

48➔ Click OK to close the SQL window and OK to close the tool window.

48➔ Click the Run button and close the dialog window after it finishes.

48➔ Right-click the Dense Conifers output and choose Add to Display. Make sure that the selection looks correct (see Fig. 7.19).

TIP: Click on a model shape to select it. Use click and drag boxes or the Shift key to select multiple shapes. Click and drag selected shapes to move them around on the model canvas or to resize them. Use these techniques to arrange the model in neat patterns.

The Auto Layout button can be used to quickly arrange the model in a logical pattern. 

49➔ Click and drag the ArcToolbox > Analysis > Overlay > Intersect tool onto the model canvas.

49➔ Right-click the Intersect tool on the model canvas and open it.

49➔ Set the Input Features to Limestone and Dense Conifer. When choosing the layers, be sure to choose the ones with the blue symbol, indicating that they are outputs from the model.

49➔ Name the Output Feature Class conf_lime and store it in the chap7results geodatabase. Click Yes to overwrite the one that exists, if you are asked.

49➔ The tool should look like Figure 7.30. Leave the other arguments set to their defaults.

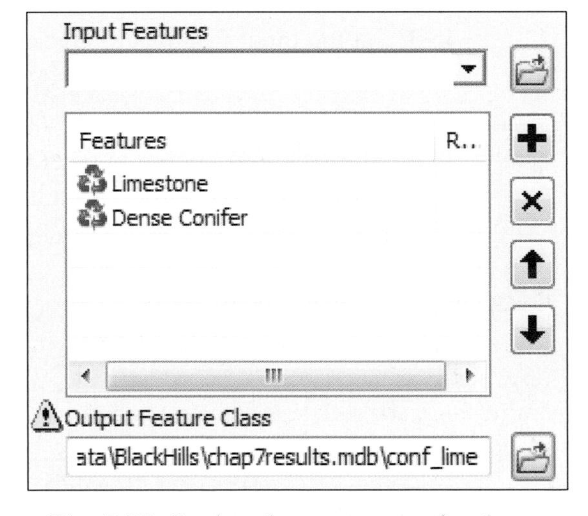

Fig. 7.30. Setting the arguments for the first Intersect tool

Notice the red or yellow warning sign next to the Output Feature Class. Put the cursor on it to read the message that says that this output already exists. Of course it does, since we've done this analysis before. When running models, you generally want to be able to overwrite previous results. We can set this option.

50➔ Click OK to close the tool.

50➔ Choose Geoprocessing > Geoprocessing Options from the main menu bar.

50➔ Check the box to *Overwrite the results of geoprocessing operations*. Click OK.

51➔ Drag another copy of the Intersect tool from ArcToolbox to the model canvas.

51➔ Open the second Intersect tool.

51➔ Select the conf_lime model output from the Input Features drop-down box.

51➔ Choose Elevation Range from the Input Features drop-down box.

51➔ Name the output feature class snailhab2 and place it in chap7results.

51➔ Leave the rest of the arguments to their defaults and click OK.

52➔ Choose Model > Save to save the model in its current form. It should look similar to the model in Figure 7.31.

52➔ Run the model. The two intersections you set up will both be executed.

52➔ Right-click the snailhab2 output to add it to the display and examine it.

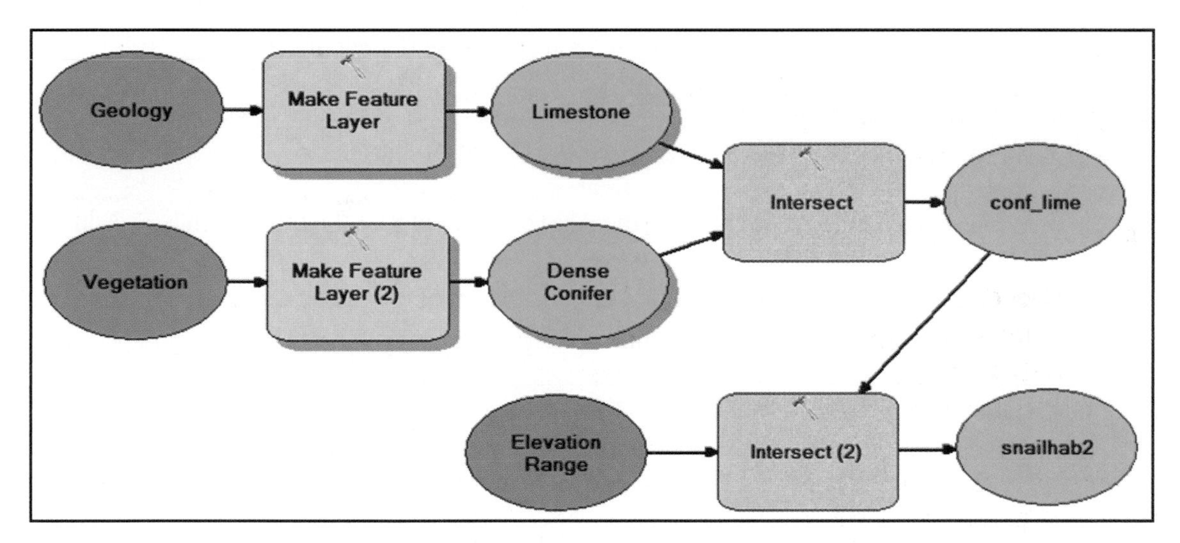

Fig. 7.31. The completed Snail Habitat model

Changing and re-running a model

The saved model provides a record of the inputs and parameters used to derive the output. It can also be edited and run again to try out different solutions. Imagine that a geologist colleague has commented that the Lower Paleozoic geological units also contain limestone and should be included with the other geology layers.

53➔ Right-click the Make Feature Layer input box for the Geology query and open it.

53➔ Click the SQL button and click at the end of the current expression to place the cursor there.

53➔ Add the rest of the expression, OR [NAME] = 'Lower Paleozoic', so that all three units will be selected. Verify the query and then click OK to close the tool.

Notice that the model boxes lose their shadows (except the unchanged Vegetation query), indicating that they need to be run again with the new conditions.

54➔ Right-click the Make Feature Layer tool from the Geology query again and choose Run to run only this step of the model to check your query.

54➔ The previous Limestone output was removed from the display. Right-click its green oval in the model and choose Add to Display to uncheck it. Repeat to add it to the display a second time. Examine it.

55➔ Click the Run button in the Model Builder toolbar to run the rest of the model steps.

55➔ Add the snailhab2 output to the map and examine it to see how it changed.

55➔ Save the new version of the model by choosing Model > Save.

Create a tool from a model

Finally, models can be set up to run as tools, by turning the inputs into **parameters**. This approach is helpful when the same series of steps is to be run with different inputs each time.

56➔ Right-click the blue Vegetation oval and choose Model Parameter. A small P appears to the right of the blue oval, indicating that it is now a parameter.

56➔ Right-click on the snailhab2 green oval and make it a parameter.

56➔ Save the model and close it.

Before we can run the tool, we must add your toolbox to ArcToolbox.

57➔ Open ArcToolbox. Right-click the ArcToolbox entry at the top and choose Add Toolbox.

57➔ Navigate to the BlackHills folder and click on the BHtools toolbox. Click Open.

57➔ Open ArcToolbox again and double-click the BHtools > Snail Habitat Analysis tool.

The model is opened as a tool, with input boxes for the two parameters. Examine the titles above the boxes, Vegetation and snailhab2. The titles are taken from the names assigned to the ovals in the model. Let's rename them so they make more sense to a user.

58➔ Close the tool. Right-click on the model in the toolbox and choose Edit.

58➔ Right-click the Vegetation oval and choose Rename. Call it Input Vegetation Layer. (You can slightly enlarge the box so that the text fits.)

58➔ Rename the snailhab2 oval Output Habitat Layer.

58➔ Save the model and close the model canvas.

58➔ In ArcToolbox, double-click on the model to open it as a tool.

TIP: Double-clicking or choosing Open will open the model as a tool to be run. To open the model canvas, right-click the model and choose Edit.

There is a more recent vegetation feature class in the Sturgis83 geodatabase called veg2006. We'll run the model again using the new data to see if the habitat has changed in the last decade.

59➔ Click the Browse button and find the veg2006 feature class in the Sturgis83 geodatabase. Add it.

59➔ Set the output feature class to snailhab06 and put it in the chap7results geodatabase. Click OK to run the tool.

59➔ Turn off all layers except the Roads, snailhab2, and snailhab06.

59➔ Compare the snailhab2 and snailhab06 layers, noting the change in habitat.

Recall that to create the tool we designated the vegetation input and the habitat output as parameters that could be set each time the tool is run. You can also set individual arguments as parameters. Imagine that we wanted to be able to select different types of vegetation each time we ran the tool.

60➔ Right-click the Snail Habitat tool in ArcToolbox and choose Edit.

60➔ Right-click on the Make Feature Layer box associated with the vegetation query and choose Make Variable > From Parameter > Expression. A new blue input oval appears for the expression.

60➔ Rename the new oval Vegetation Query. Move or resize as needed.

60➔ Right-click the Vegetation Query oval and choose Model Parameter. A P appears next to it (Fig. 7.32).

60➔ Save the model and close it.

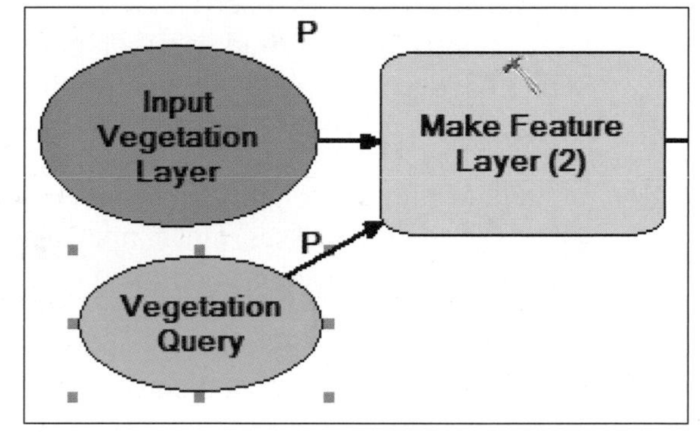

Fig. 7.32. Making the query a parameter

Now let's run the model again. Suppose we want to see the potential habitat if all densely treed areas are considered, not just conifers.

61➔ Open the Snail Habitat model and set the input and output feature classes to Vegetation and snailhab_all.

61➔ Click the SQL button next to the new Vegetation Query input box.

61➔ Clear the existing query and replace it with [DENSITY96] = 'C'.

61➔ Click OK to run the tool again.

61➔ Compare the old snailhab2 and new snailhab_all outputs to see the changes in habitat.

This brief introduction shows only a small part of what Model Builder can do. To learn more, read about Model Builder in the ArcGIS Help files.

TIP: Models can be saved as scripts in the Python programming language. If you learn Python, then you can create full-fledged programs with conditional statements (if/then), looping, and so on.

This is the end of the tutorial.

➔ Close ArcMap. Save your changes.

Exercises

Use the data in the BlackHills folder and Sturgis83 geodatabase to answer the following questions.

1. The Madison Limestone is an important aquifer. Citizens are concerned about gas or chemical trucks overturning on roads and causing contamination. What is the total length of roads crossing the Madison, in kilometers?

2. The infiltration rate of the geological substrate is a strong factor in evaluating the spill hazards for roads. Create a map of all the roads based on the infiltration capacity of the geology, using appropriate symbols to indicate hazard. **Capture** your map.

3. The use of a road is another factor in evaluating spill hazard. Create a field containing a hazard rating for each road based on both infiltration and use. For use, give primary roads a value of 3, secondary roads 2, and primitive roads 1. What is the range of values for the index? Create a map of the roads based on the hazard index. **Capture** the map.

4. Rangeland can cause significant impacts on surface water quality. Using the lulc and wsheds2b feature classes, create a table showing the area of rangeland in each watershed. (Include both herbaceous and mixed rangeland categories.)

5. Areas that are within 300 meters of both streams and primitive roads are the most popular places for fishing. Create a map showing these areas. **Capture** your map.

6. The fishing areas in #5 should be restricted to areas within the national forest (OWNER = 'NFS' in the vegetation feature class). Create a revised map showing the public access fishing areas in green and the private fishing areas in red. **Capture** your map.

7. You need a feature class that shows polygons based only on infiltration and only within the wshds2b area. Symbolize this layer based on infiltration. **Capture** your map.

8. The Forest Service is concerned about septic systems contaminating the Madison Limestone. Create a map showing private land (OWNER = PVT in the Vegetation layer) on the Madison. **Capture** the map.

9. You wish to investigate whether different soil units influence the net productivity of the forest in terms of canopy cover. Find the mean crown cover percent of *only* the forested stands within each soil unit. Which soil unit supports the highest crown cover? Which one supports the lowest? Interpret your findings. (**Hint:** Use the Dissolve tool. The forested stands have a cover type code that begins with T, e.g., TAA, TPP.)

10. The Forest Service is considering adopting a policy to not cut timber within 200 meters of a primary road (TYPE = P). Assuming that cutting timber includes all [SSTAGE96] LIKE '4*', determine the total area in square kilometers lost to harvesting if this policy were adopted. **Capture** a map showing the off-limits areas.

Challenge Problem

Imagine that you have always wanted a vacation home in an aspen stand with a little pond. Create a map showing all aspen stands (COV_TYPE = 'TAA') that lie on low infiltration geology units (INFIL = LOW) and are within 500 meters of a primary or secondary road (TYPE = P or S). **Capture** your map.

Chapter 8. Raster Analysis

Objectives

- Exploring raster analysis through map algebra and Boolean overlays
- Getting familiar with common raster analysis functions
- Understanding the raster data model and the storage of raster in ArcGIS
- Using Spatial Analyst for raster analysis and geoprocessing
- Controlling the analysis environment when using Spatial Analyst

NOTE: This chapter requires Spatial Analyst, an optional extension to ArcGIS. If you do not have it installed on your computer, you will not be able to perform the functions described.

Mastering the Concepts

GIS Concepts

Raster data

The availability of two different data models, raster and vector, provides added flexibility to options for data storage and analysis. (Readers may wish to review the section in Chapter 1 on raster and vector data models.) Neither data model is intrinsically superior. They both have areas in which they excel and areas in which they are at a disadvantage. The GIS professional who has a grasp of both tools holds the keys to developing the most efficient and accurate analysis.

The analysis functions available for rasters number in the hundreds and include very sophisticated applications. This chapter introduces some basic concepts and analysis functions, but the topic spans a much greater breadth of theory and technique than we have time to explore here. The serious student of raster analysis will find plenty of additional material in the software documentation for Spatial Analyst.

Recall that spatial data come in two forms (Fig. 8.1). Discrete data, such as soil types or land use, represent spatial objects and form distinct regions on the map. Continuous data, such as elevation or precipitation, vary smoothly over a range of values and form a surface or field. The vector data model best stores discrete data. Rasters easily store and analyze continuous data.

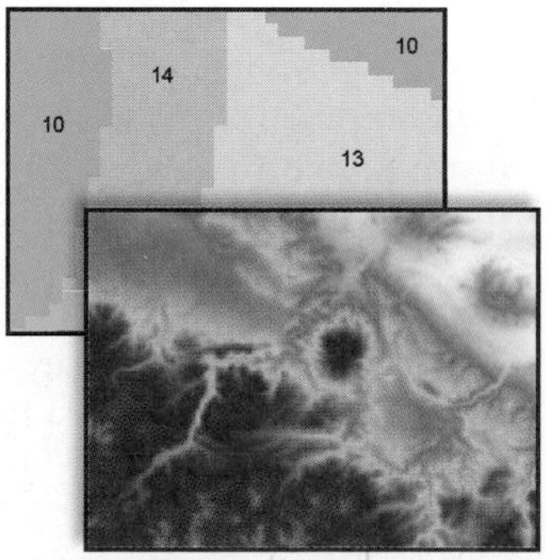

Just like vector data, rasters have a defined coordinate system with a projection and a datum. The raster is georeferenced by specifying the *x-y* coordinates of the center of the upper-left pixel and the cell size (the length of the sides of the pixels). Usually pixels are square. A 30-meter raster has pixels composed of 30×30-meter squares. The cell size is specified in the coordinate system's map units, and meters or feet are typically used.

10
14
10
13

Fig. 8.1. Discrete land use data and a continuous digital elevation model

Although one can store a raster in a GCS, this practice is not common because the distortion gives an unsatisfactory view of the data. Furthermore, many raster analysis functions incorporate distance and angle formulas that will give inaccurate results when applied to a raster in a GCS or in a projection with significant distortion. Therefore, raster data are normally stored in whatever projected coordinate system gives the best and most undistorted view of the data. The main exception occurs for data providers, such as the US Geological Survey, which provide topographic or other data in GCS format. Such data should be projected before it is used.

Raster analysis

Raster processing is fundamentally different from vector processing. Vector analysis manipulates *x-y* coordinates. Raster processing operates on stacks of aligned cells and can be very efficient. This chapter describes the most common functions, but many more are available.

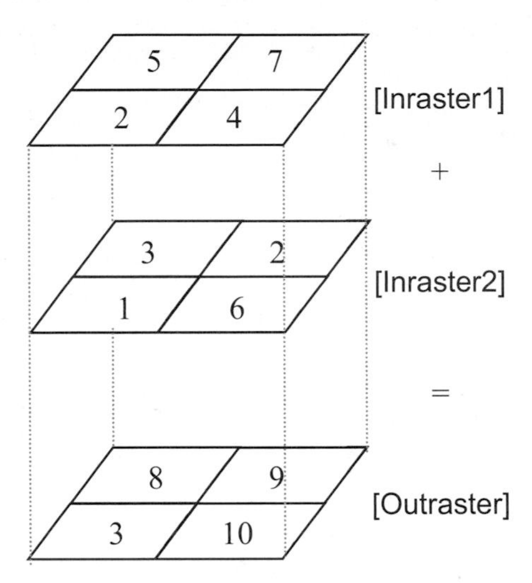

Fig. 8.2. Map algebra addition

Map algebra

A raster is an array of numeric values arranged in rows and columns. Several stacked rasters representing different attributes can be analyzed using cell-by-cell formulas based on mathematical, logical, or Boolean operators. This technique is called **map algebra**. Consider the stack of two input rasters in Figure 8.2. These rasters can be added together cell by cell to produce a new raster with cells containing the sum of the corresponding cells in the inputs. If the cells of the input rasters are not already aligned, then they must be resampled to achieve this.

The map algebra expression for the operation in Figure 8.2 would appear as [Outraster] = [Inraster1] + [Inraster2] with the raster names enclosed in brackets. Imagine a raster of precipitation in inches. The expression [Precip] * 2.54 would produce an output raster with precipitation in centimeters. Expressions can include multiple rasters, functions, and values. A more complex expression to predict erosion potential as a function of rainfall, slope, soil erodability, and vegetative cover might read:

[Precip] * 2 + [Slope] * 4 / ([Erosion] – [Vegcover]).

Map algebra also includes logical functions that can return Boolean rasters, which have values of true (1) or false (0). Figure 8.3 shows the result of entering an expression [Elevation] > 1400; the blue area shows the cells in the elevation raster that meet this criterion.

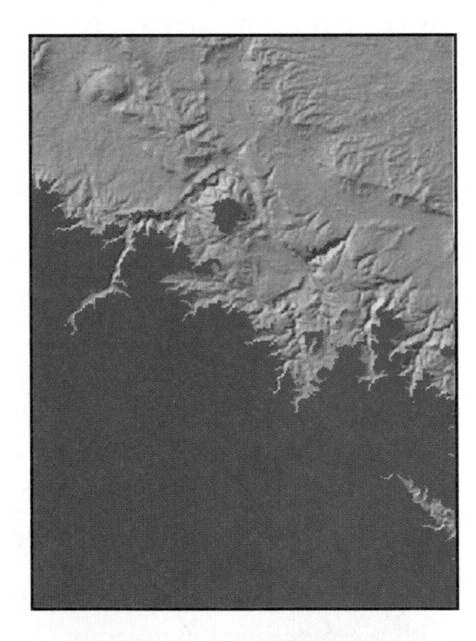

Fig. 8.3. The blue portion of this map shows the pixels where Elevation > 1400 meters.

Boolean overlay

Chapter 7 described a vector process called overlay, used to find areas that meet multiple conditions, as we did for snail habitat. In many cases, raster overlay has advantages over vector overlay. First, it can incorporate conditions from continuous data sets, such as elevation or slope, which cannot be stored as vectors. Second, the process is faster because corresponding cells are

simply added or multiplied together, instead of splicing the existing polygons and creating new ones. Third, raster overlay can process more than two layers at once.

Map overlay with rasters is accomplished using map algebra with Boolean rasters and operators. Chapter 5 introduced Boolean operators, which evaluate two input conditions, A and B, to determine a result that is either true or false. In raster analysis, this function is performed on a cell-by-cell basis (Fig. 8.4). A **Boolean raster** contains two possible values, ones or zeros. A one corresponds to a true result and a zero corresponds to a false result.

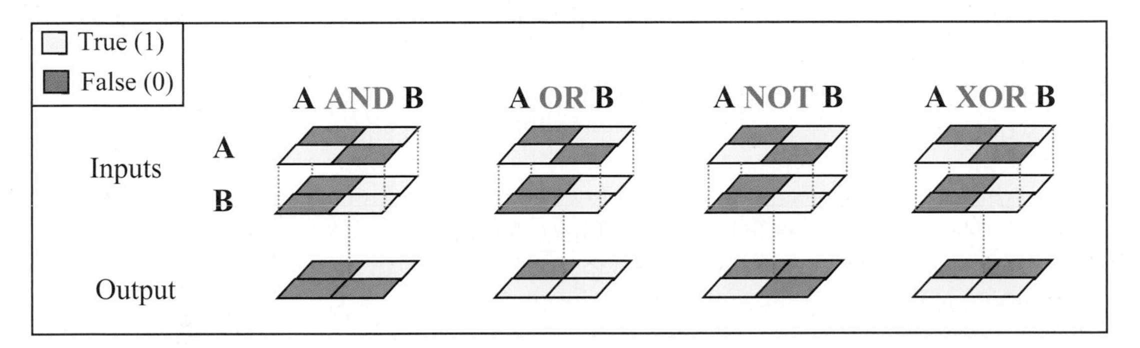

Fig. 8.4. The Boolean operators are applied to rasters. Each input cell is 1 or 0 (true or false). Matching cells in the inputs are evaluated to produce the output result.

Imagine creating a potential habitat map for lodgepole pine trees based on two criteria: areas above 1500 feet in elevation and with more than 15 cm of precipitation. In Figure 8.5a, the Boolean raster shows where the precipitation conditions for lodgepole are favorable. Areas where the precipitation is greater than 15 cm have the value true (1), and areas labeled false (0) have precipitation < 15 cm. Figure 8.5b shows the favorable conditions based on elevation, where the cells containing 1 have elevations >= 1500 meters.

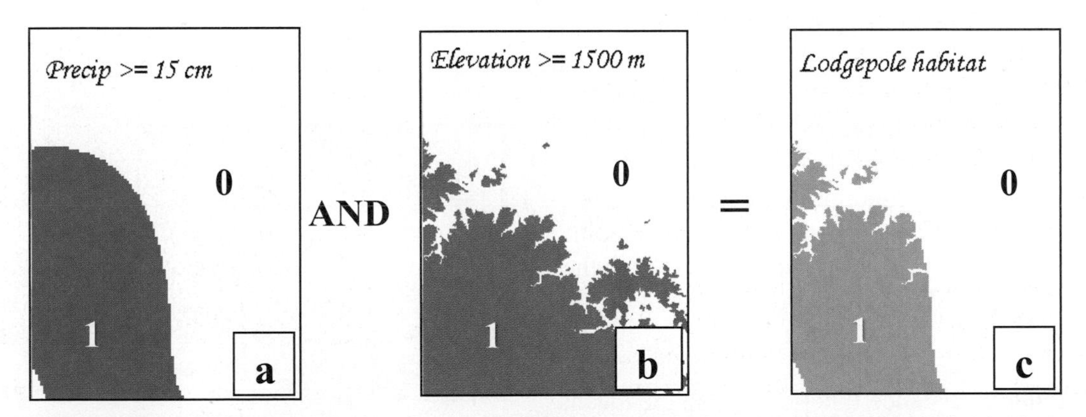

Fig. 8.5. Boolean overlay: (a) a Boolean raster in which the value 1 (true) indicates where precipitation is >= 15 cm; (b) a raster indicating elevations >= 1500 meters; (c) the result of multiplying the first two rasters, showing where both conditions are true

To find the lodgepole habitat, we apply the Boolean operator AND to the input rasters. If both inputs are 1, the output cell is 1 also. If either of the input cells has a 0 or if both of the input cells have a 0, then the output will be 0. Figure 8.5c shows the output raster with the potential lodgepole habitat areas identified as cells with the value 1 (true). The raster AND (sometimes written &) operation is the equivalent of the vector Intersect tool.

AND	Multiplication
1 AND 1 = 1	$1 \times 1 = 1$
1 AND 0 = 0	$1 \times 0 = 0$
0 AND 1 = 0	$0 \times 1 = 0$
0 AND 0 = 0	$0 \times 0 = 0$

Fig. 8.6. Equivalence of the AND and multiplication operators

The Boolean AND operates the same way as multiplication of the two input rasters. Figure 8.6 compares the outputs of AND and multiplication on two Boolean inputs, and they are identical. Thus, either AND or multiplication can be used to find the lodgepole habitat. Since both of these operators are symmetrical, meaning that the order of inputs does not matter, they can be applied to three or more rasters as easily as two.

Map *addition* provides another way to model the habitat—to *rank* the probability of finding lodgepole. In this model, if both elevation and precipitation are ideal, then lodgepole is most likely to occur. If only one is ideal, then it can occur but is not as likely. If neither condition is ideal, then lodgepole is very unlikely. To implement this model, we add the two input maps instead of multiplying them. The final map in Figure 8.7 has three possible values: 0 if neither condition is true, 1 if either condition is true, or 2 if both conditions are true. Thus, 2 indicates the highest probability of finding lodgepole, 1 a moderate probability, and 0 a low probability.

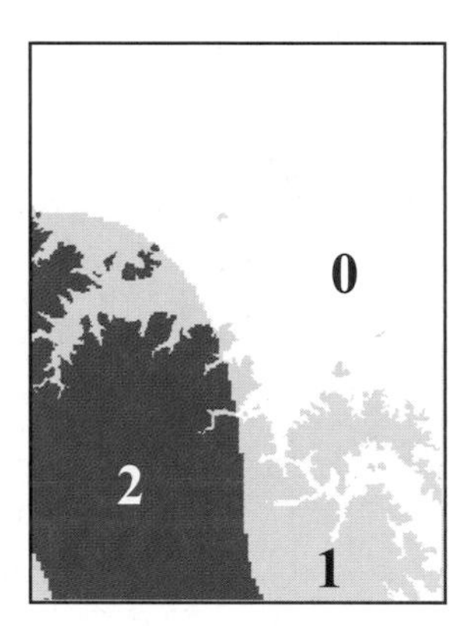

Fig. 8.7. A Boolean addition overlay

The best model to use will depend on which most accurately predicts the actual habitat areas. GIS modeling is always tied to knowledge of the real world. If this project were real, we would proceed to test the model predictions by comparing them to the actual distribution of lodgepole pine. If the model proved accurate, then we could use it to make predictions in other areas.

Distance functions

Different types of distance functions solve different problems. A **Euclidean distance** is the shortest distance between two locations. The Euclidean distance function produces a raster in which each cell represents the shortest distance from a set of specified objects, for example, streams (Fig. 8.8a). A related output is the Euclidean direction function, which records the direction of the closest object. At any given point, these two grids would show the distance from the closest stream and the direction you would have to travel to get to it.

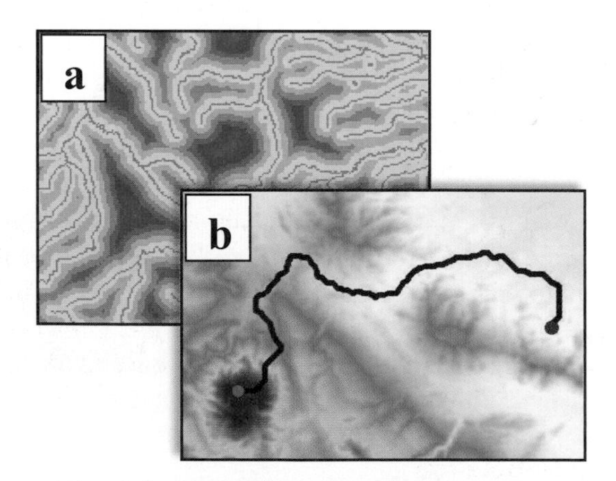

Fig. 8.8. (a) Straight-line distance; (b) least-cost path between two points

The Euclidean distance function is also the means for creating buffers as raster data sets. Once a distance surface is generated, it is a simple matter to identify the distances that are greater than or less than a specified buffer value, to produce a Boolean raster with a 1 value inside the buffer and a 0 value outside.

Cost-distance functions consider more complex and flexible approaches to travel. Consider a mountain with two points on opposite sides. The shortest path to the opposite point goes over the mountain. However, most people would find it easier to go around the base, even though the path is longer, due to the effort (cost) of climbing over the mountain. Slope is one kind of cost. Heavy vegetation, marshes, and off-limits areas are other factors that increase the cost of travel.

In a cost-distance analysis (Fig. 8.8b), a raster is first prepared that estimates the cost of travel over a surface. A single cost factor, such as slope, could be used, or the raster could combine different types of costs. This cost raster is used to develop a cost direction raster, which tells a traveler which directions incur the least cost, and a cost accumulation raster to keep track of the total effort expended as the traveler moves. These rasters can be combined to calculate the least- cost path between two points, indicating the easiest journey rather than the shortest. In Figure 8.8, a slope raster was used to designate the cost, and a least-cost hiking trail was calculated.

Density

A density function is used to create rasters showing the frequency of objects within a given radius. Figure 8.9 shows a point density map of city populations in the United States. A search radius is specified (500 miles), and the function counts the points falling inside a moving circle as it pauses at each location (cell). If an attribute field is specified, the counts are weighted by the attribute. Without the attribute, the function would count the density of cities; with the attribute, it counts the density of population. The counts are given in specified units, as in people per square mile or cities per square kilometer. A large radius produces a smooth map with higher density values (in this map, 0–90 persons/sq km); a smaller radius produces more detail and smaller density magnitudes.

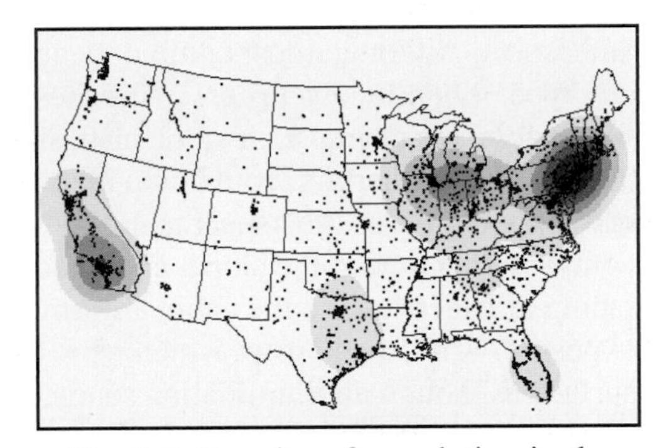

Fig. 8.9. Density of population in the United States

A line density function is similar, except that it counts line occurrences within the circle instead of points, and it weights them by length. It can also weight by a designated field. This function could be used to calculate road density within a national forest. An attribute field might give stronger weights to primary roads and a lesser weight to primitive roads.

A kernel density function works like a point density function, except that it spreads the magnitude of the point attribute over an area defined by a function, such as an inverse distance squared function. A city may be represented by a point with a population attribute, but in reality the people are spread out over an area. The kernel density function provides more realistic modeling of the distribution for such cases.

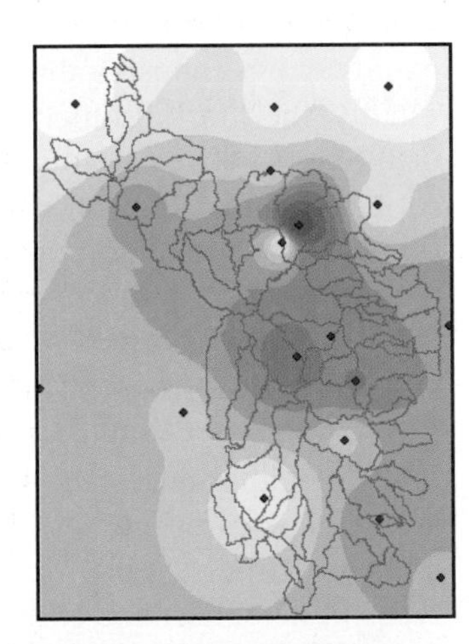

Fig. 8.10. Rainfall raster interpolated from weather station measurements

Interpolation

Interpolation takes measured values at points and distributes them across a raster, estimating the values in between the measurements. A common example involves taking precipitation measurements at weather stations and interpolating the precipitation values in between stations to create a continuous precipitation surface. Figure 8.10 shows the interpolation of the total precipitation measured at the stations over the Black Hills region, with the watersheds superimposed on the image.

Three different interpolation methods are common: inverse distance weighted (IDW), splining, and kriging. Each method uses different assumptions and different means to obtain the result. More information about these three interpolators can be found in the Help documentation for Spatial Analyst. Interpolation is a complex topic and the subject of entire textbooks. GIS professionals should become familiar with the different methods in order to understand the benefits, drawbacks, and assumptions of each one.

Although interpolated and density maps appear similar, the functions work differently. Density functions count occurrences within a radius and divide by the area. Interpolation estimates a value for each raster cell based on nearby measurements.

Surface analysis

Surface analysis includes a set of functions for calculating properties of a surface, such as slope and aspect. Although most commonly applied to elevation data, these functions can be used on any type of continuous raster. Using these functions on a DEM could help find an ideal location for a cabin, for example. A slope map, showing the steepness of the terrain, could help identify sites flat enough to build on. An aspect map, showing the compass direction of the slope, could locate south-facing hillsides to give the cabin a warm southern exposure. A hillshade map, which represents the surface as though an illumination source were shining on it (Fig. 8.11), can help the map reader visualize the landforms. A viewshed analysis, which determines what parts of the surface can be seen from a specified point or points (Fig. 8.11), could help in choosing the site with the best view. Finally, the Cut/Fill function calculates the differences between two input surfaces and is useful for developing estimates of the amount of material that must be added or removed from a building site.

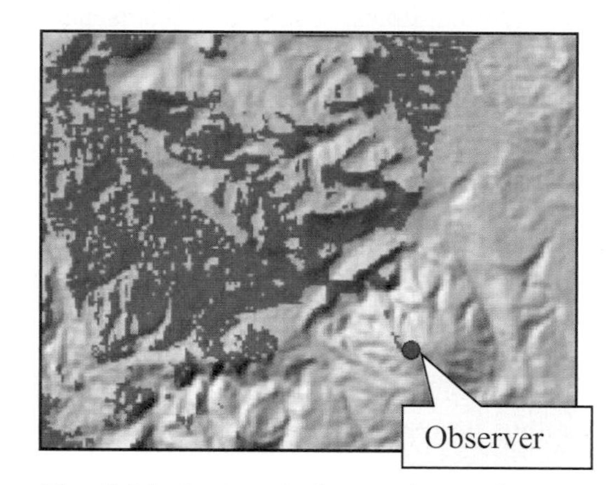

Fig. 8.11. A viewshed map shows the areas visible from a point on a hilltop in red. The background map is a hillshade.

Neighborhood statistics

Neighborhood functions operate on an area of specific size and shape around a cell, known as a neighborhood. A moving window passes over the raster and calculates a statistic from all the cells in the neighborhood, records the result, and moves on. Neighborhoods may have several shapes, including squares, circles, rectangles, and annuli. The statistical functions include minimum, maximum, range, sum, mean, standard deviation, variety, majority, minority, and median. Two types of neighborhood functions are used: block functions and focal functions.

A **block function** moves the window such that it never overlaps the previous window, but it produces a set of side-by-side adjacent windows that are independent of each other. Once the statistic is calculated for the window, every cell in the neighborhood receives the same value before the window moves on. Every cell is used only once.

A **focal function** centers the window on a cell, called the target. It calculates the statistic for the neighborhood and assigns the value to the one target cell. Then it moves to the adjacent cell and repeats the process. Each cell in the raster is evaluated in turn and is used multiple times as the window is evaluated for its neighbors. Consider a 3×3-cell averaging window (Fig. 8.12). For each target cell in the raster, the algorithm finds the average value of the nine cells within the window and puts the average value in the target cell of the output raster. The window then moves to the next cell. An edge cell is still evaluated but will have fewer cells in the neighborhood, such as the cell in the upper-left corner of Figure 8.12. Only four values are averaged, the target and its three neighbors. The "1" cell to its right would have the average of six cells. A **filter** is another name for a focal function.

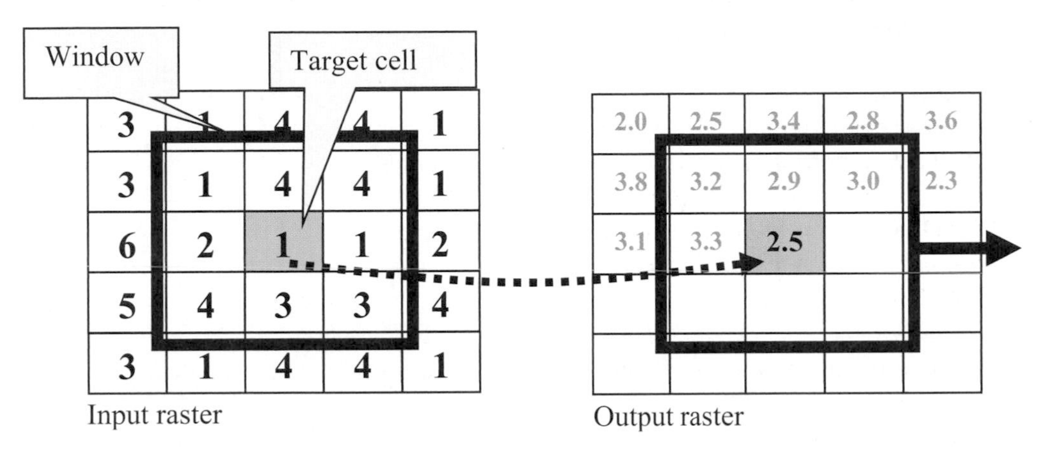

Fig. 8.12. A 3 × 3 averaging window averages the nine cells in the window and places the result in the target cell of the output raster.

Zonal statistics

The **zonal statistics** function is sometimes considered another type of neighborhood function, but the neighborhoods are not a fixed size or shape. Instead, they are defined by another layer, called the zone layer. Zones are regions of a raster or a feature class that share the same attribute. A zone might be a watershed, a parcel, or a soil type. The attribute might be a name, a category, or a unique identifier, such as a parcel-ID or even the table ObjectID.

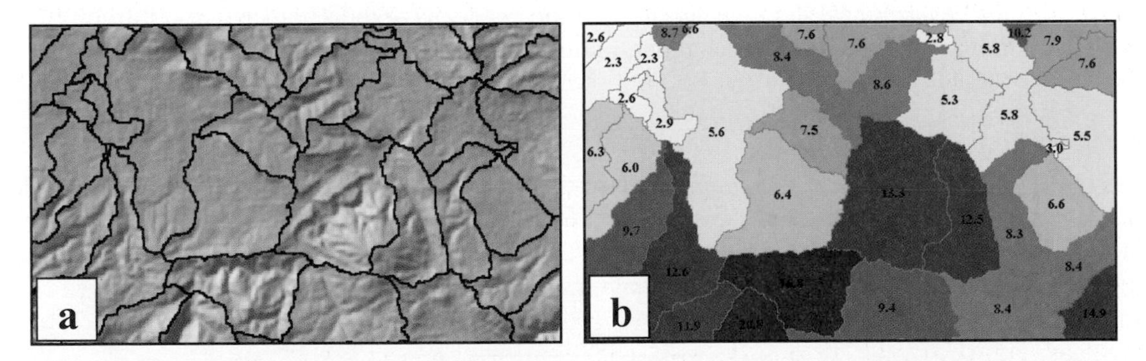

Fig. 8.13. Using zonal statistics to find the average slope of watersheds: (a) the zone raster shown over a hillshade; (b) the watersheds colored by and labeled with the average slope

Within each zone, the function gathers all the cells from an input raster that fall inside a zone and calculates statistics for the zone. For example, if a watershed layer defined the zones and the input raster contained slope, then zonal statistics would calculate the average slope for each watershed. Figure 8.13a shows a set of watersheds displayed on a hillshade map, and you can clearly see which watersheds have more variable topography and a higher average slope. When the zone function is executed, the slope values are averaged for each watershed.

Zonal statistics functions may produce two types of output, a raster or a table. A raster output places the calculated statistical value into every cell in the zone. The table output contains the zone attribute field plus one or more statistics for each zone. This table could be joined to the original watershed polygons using the zone field and used to map the average slope of the watersheds (Fig. 8.13b).

Zones need not be contiguous like the watershed zones, however. The geology map in Figure 8.14 contains many unconnected areas with the same geological unit. The map has only six zones, one for each color.

The zonal statistics function can also assign values from a surface to line or point data, such as finding the average slope along a stream section or attaching an elevation value from a DEM to a layer of wells. Each line or point must have a unique name or ID value that identifies it as a distinct zone.

Fig. 8.14. This geology map has six zones.

Reclassify

The **Reclassify** function changes the values of a raster according to a scheme designed by the user, such as classifying a slope map into three regions of low, medium, or high slope. For a map with only a few values, such as land use codes, the user might specify a new value for each code individually. Perhaps a town has annexed a nearby planning unit that used different land use codes and must convert one set of codes to another; the code 143 becomes 347, and so on. With continuous data covering a range of values, such as slope, the values are typically classified into a discrete number of ranges. Classification uses a tool similar to the one described in Chapter 2 for classifying attribute data.

Resampling

Often rasters must have their values converted to a raster with a different cell size. This process is called **resampling**. Users may resample purposely in order to align rasters with each other or to decrease the resolution of a data set. Resampling must also occur anytime two rasters with different cell sizes are analyzed together. Even if the rasters have the same cell size, the cell rasters might not be aligned with each other, requiring that resampling occur.

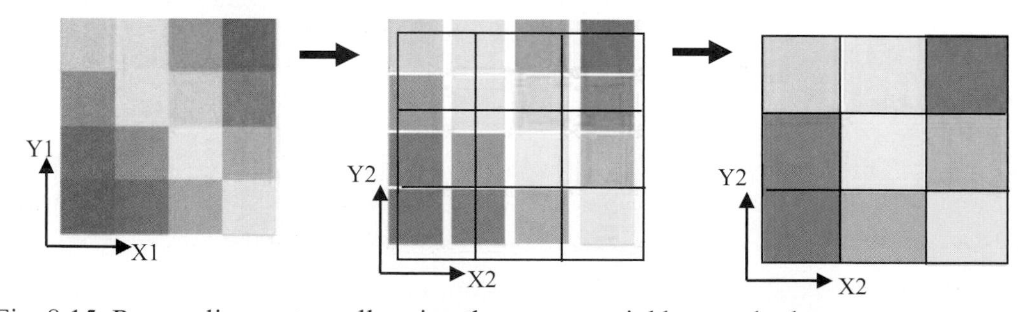

Fig. 8.15. Resampling raster cells using the nearest neighbor method

Figure 8.15 shows an example of resampling. The cell centers on the left are being transformed from coordinate system X1-Y1 to X2-Y2, introducing gaps between the cells. The cell grid for the new coordinate system is shown in the output by the black lines. The old cells are resampled to the new cells using one of three methods. In **nearest neighbor** resampling, the new cell is given the value of the old cell that falls at or closest to the center of the new cell. In **bilinear** resampling, a distance-weighted average is taken from the four nearest cell centers. The **cubic**

convolution resampling method determines a new value by a curve fit through the nearest 16 cell centers. Nearest neighbor is the fastest method and should always be used with categorical raster data because the values do not change. The bilinear method is usually better for continuous rasters such as elevation.

Resampling can degrade the quality of a data set. Best practices dictate that the user retain full awareness of the coordinate systems and cell sizes of all input layers being used for analysis and endeavor to standardize them before processing occurs. Some GIS systems automatically resample without warning. Users must be especially careful with these systems if they are attempting to retain a high degree of spatial integrity in their rasters.

Conversion

Users can easily convert between rasters and vectors, such as turning a geology shapefile into a raster. The user must choose *one* attribute field to populate the values of the raster, such as the geology unit name. Rasters can only store numbers. Numeric fields are converted as is, but text values are assigned arbitrary numeric codes for storage in the raster. Some raster formats place the original text values in a raster table for reference.

You may also convert discrete rasters to polygons as long as the number of polygons does not become excessive. Continuous data, such as DEMs, should never be converted to polygons, since each pixel is different from its neighbors and will become a separate polygon. If too many polygons will result from a conversion, the software may be unable to complete the task.

About ArcGIS

Storing rasters

ArcGIS can read many raster data formats and can read and write some of them. Prior to ArcGIS Version 10, Spatial Analyst could only operate on a format unique to ArcGIS, called a **grid**. Other rasters, such as TIFF files, SID files, and JPEG files, had to be converted to grid format before they could be analyzed, and all raster output generated by Spatial Analyst took the form of grids. At Version 10, Spatial Analyst reads any supported raster format and supports four different output formats: grids, geodatabase rasters, TIFF, and ERDAS Imagine. In this chapter, we will work almost exclusively with geodatabase rasters.

TIP: Rasters should NOT be stored in folders that have pathnames containing spaces, because some functions will not work properly. GIS users are advised to avoid spaces in ALL folder names, no matter what.

TIP: NEVER copy, move, rename, or delete grids using Windows Explorer. Grids have complex storage formats, and Windows cannot manage them properly. Always use ArcCatalog to manage grids.

Rasters come in two forms. Integer rasters store whole numbers in binary format. Floating-point rasters can store decimal values, but they require four times as much space as integer rasters. Thus, it is foolish to use floating-point rasters except when decimal values are required.

When creating output, the raster type will usually be the same as the input. If you add or multiply two integer rasters, the result will also be an integer raster. If one of them is floating-point, the output will be floating-point. Some functions always create floating-point rasters, such as dividing or averaging a raster. (Floating-point rasters may be converted to integer rasters using the INT tool.)

TIP: It is recommended that raster names contain only letters, numbers, the underscore, or a hyphen. Grids have even stricter naming conventions and are limited to 13 characters. Consult ArcGIS Help for more information on supported formats and naming rules.

Raster attribute tables

Rasters can have attribute tables, just as feature layers do. However, since a raster contains only cell values, the attribute tables look different. Each record contains one unique value present in the raster. The table has three fields that are always present: the ObjectID field, containing a unique ID for the table rows; the Value field, showing each unique cell value; and the Count field, indicating how many cells contain that value (Fig. 8.16). This table indicates that the Precambrian rock unit has the largest area in the raster because more cells (64,573) have that value than any other. The set of all cells sharing the same value and having one record in the raster attribute table is called a **zone**.

The attribute table in Figure 8.16 comes from a raster that was converted from a polygon feature class using the rock unit Name field, so the Name is included in the table. The Value field contains numeric codes representing each rock unit, which were arbitrarily assigned at the time the feature class was converted. The Cenozoic unit is represented by the code value 1, the Upper Mesozoic by code 2, and so on. Attribute tables may also include fields added by the user.

geolgrid

OID	Value	Count	NAME
0	1	22607	Cenozoic
1	2	8624	Upper Mesozoic
2	3	38679	Lower Mesozoic
3	4	32925	Upper Paleozoic
4	5	28901	Madison Limestone
5	6	64573	Precambrian
6	7	24983	Lower Paleozoic

Fig. 8.16. The attribute table of a geology raster

Only integer rasters can have attribute tables. The tables of categorical rasters are the most useful, as they document a fairly small set of discrete values. However, an attribute table is automatically created for any raster with fewer than 500 unique values.

NoData values

Rasters may contain a **NoData** value that indicates a null, or missing, value for a cell. Because rasters are always rectangles, the NoData value may be used to mask areas outside a nonrectangular area, as shown by the black cells in Figure 8.17. NoData values are typically displayed with black or transparent color. Some raster formats use 0 instead, which may cause skewing of image statistics. NoData values are usually ignored in processing, providing more consistent results.

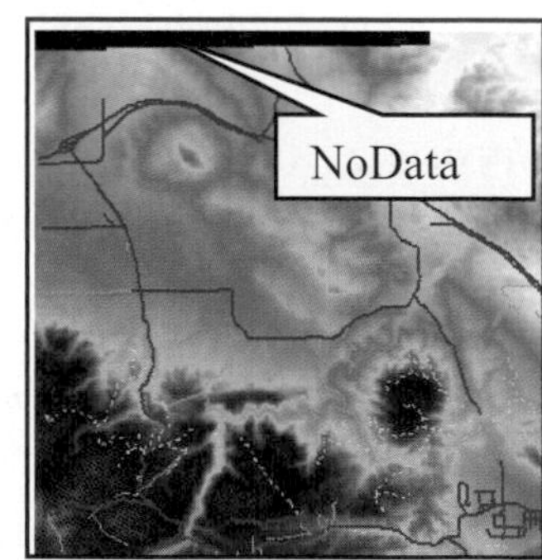

Fig. 8.17. NoData values at the top of a DEM

Using Spatial Analyst

Spatial Analyst is an extension to ArcGIS and must be purchased separately. It contains a large set of tools for ArcToolbox, under the heading Spatial Analyst Tools (Fig. 8.18). The tools are organized by function type. Spatial Analyst tools are fully integrated into the geoprocessing environment. They can be run from the toolbox or the command line, used to build models in Model Builder, and incorporated into scripts.

The tools are well documented in the Help files, in many cases including complete discussions of the concepts behind each tool. Users interested in interpolation, for example, will find extensive information on, and references to, the various interpolation methods (IDW, kriging, and splining). Reading about unfamiliar tools before implementing them is highly recommended.

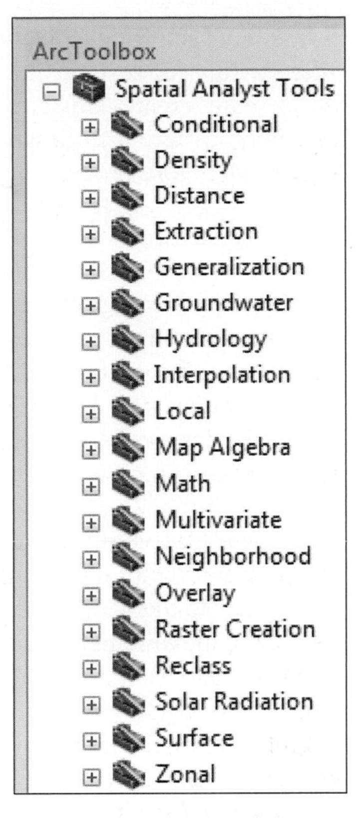

Fig. 8.18. The Spatial Analyst toolbox

Environment settings

The Spatial Analyst tools obey the Geoprocessing Environment settings just like any other tool. Several settings are particularly useful and bear mention.

The Workspace settings specify the default input and output locations for data sets. Spatial Analyst produces many output rasters that may be intermediate steps toward a solution. Specifying a separate folder or geodatabase to hold the outputs helps keep them separated from original data or more permanent results and simplifies cleanup later. Setting the current workspace to this temporary place streamlines the entry of output file names.

The Output Coordinates setting specifies the output coordinate system. It is usually best left to the default, Same as Input. Projecting rasters consumes time and memory and should be done only when necessary, preferably as a conscious step and not as a forgotten background process.

The Processing Extent controls the rectangular area used to preselect features or raster cells to be used in a tool. By default, all areas of the input data sets are used. With rasters, this setting is particularly important because a raster must have a rectangular extent. If two data sets represented by the black squares in Figure 8.19 are input to a tool, the output raster will contain the entire blue area even though the cells outside the black lines will be empty. A good choice for most raster analysis is the Intersection of Inputs option (the yellow area containing all cells with valid values from both inputs). You can also set it to match the display area, or to be the same as a specific data set to which everything else should conform.

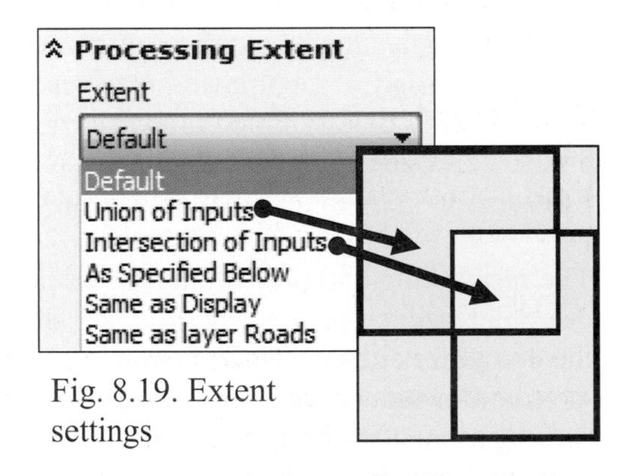

Fig. 8.19. Extent settings

The Cell Size setting controls the output resolution of rasters (Fig. 8.20). It can be set to a specific value or to match an existing data set. The default is the maximum of the inputs. For example, if two tool inputs have a cell size of 50 meters and 30 meters, the output will have 50-meter resolution. The cell size is always measured in coordinate system units and is typically in meters or feet. If a raster is stored in a GCS, however, the units will be degrees.

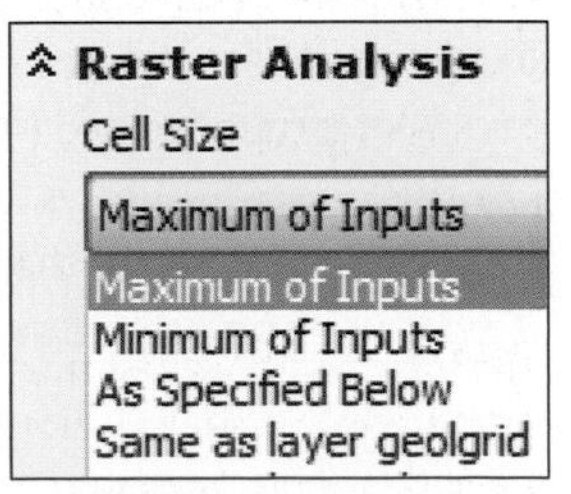

Fig. 8.20. Cell size settings

Some tools ask for a cell size even if the environment is set. In this case, the cell size box will be filled with a suggested value based on the map units and the analysis extent. The suggested size will usually create a raster with a sufficient, but not excessive, number of cells. If the suggested size is far different from what you expect or want, then double-check the extent settings and the map units before proceeding.

The Mask setting is used to specify a feature class or raster that controls which cells will be processed. A mask raster contains NoData cells outside the region of interest, and all output cells there will be given NoData values. If a feature class is used as a mask, then any cells outside the feature class boundary will receive NoData values.

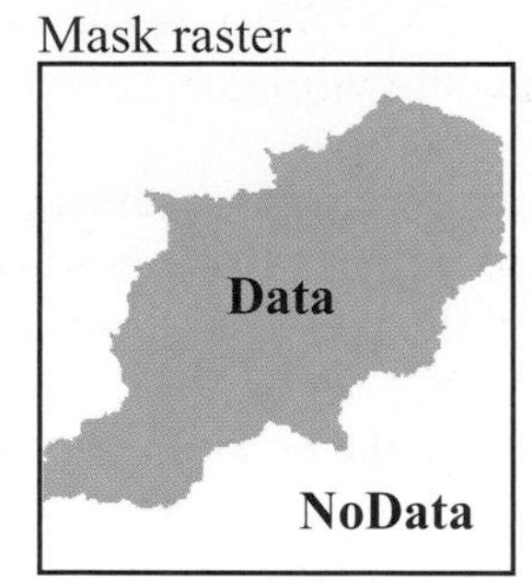

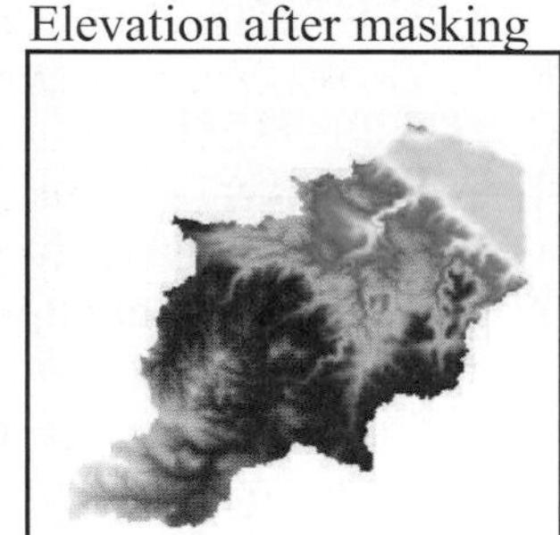

Fig. 8.21. NoData cells in a mask raster are forced to be NoData cells in the output raster.

Using a mask is similar to using the clip function for vector data. In Figure 8.21, areas outside the watershed have NoData instead of elevation after the mask is applied.

TIP: Environment settings persist through a work session unless changed, and they are saved with map documents. They should be checked each time a new work session begins or a new map document is opened.

Coordinate system management

Raster functions treat layers as arrays of numbers, which can be processed as a stack. To perform raster analysis, all of the input rasters must have the same cell size and extent so that the cells in each stack align. If a command receives inputs with different cell sizes or extents, the rasters will automatically be resampled to match each other. By default all rasters are resampled to the most coarse cell size present in the input rasters. When input rasters have different coordinate systems, Spatial Analyst will automatically project the rasters to match.

The resampling and projection process is invisible to the user and seems convenient; however, it does increase the processing time. Resampling or projecting very large rasters may overwhelm the computer and cause a system failure. Since resampling or projecting can also degrade the quality of a data set, it is unsettling to realize that it can happen without the user's knowledge. Storing all rasters to be analyzed together with the same cell size and extent can reduce unintentional resampling.

If multiple coordinate systems are input for analysis, then the same rules discussed in Chapter 7 for geoprocessing are applied, and the output will be given a coordinate system based on this order of precedence: (1) the feature dataset coordinate system if the output is being placed in one, (2) the coordinate system specified by the Environment settings, and (3) the coordinate system of the first input to the tool. An additional rule applies to (3): if one of the inputs is a vector data set, it will use the coordinate system of the first vector feature class.

Summary

- Rasters consist of spatial data stored as individual cells in an array and can represent discrete features or continuous fields of information.
- Rasters have coordinate systems just like other spatial data. Many raster calculations are based on distance and area functions, so it is important to store rasters in a projected coordinate system with minimal distortion.
- Map algebra treats rasters as arrays of numbers that can be added, multiplied, and otherwise manipulated on a cell-by-cell basis. Arithmetic, logical, Boolean, and bitwise operators and functions offer a wide array of options for analyzing rasters and scalars.
- Distance functions calculate costs and paths associated with traveling over a surface.
- Density functions measure the density of objects over an area.
- Raster interpolation estimates a surface based on measurements at isolated locations.
- Statistical measures can be calculated for rasters based on single cells, within a specified neighborhood, or within large, irregular zones.
- Rasters can be reclassified and given new values as needed.
- Resampling occurs automatically when rasters with different cell rasters are analyzed together but can degrade the quality of a data set. Users should attempt to minimize resampling when possible by aligning raster data before analysis.
- Analyst tools can be executed from menus, toolbars, the command line, Model Builder, and scripts. Spatial analyst functions can also be executed by the Raster Calculator along with map algebra expressions.
- The analysis environment controls certain settings for all new rasters created during a session. The settings include the working directory, the extent and cell size of output rasters, and the analysis mask.

TIP: Grids have stricter naming conventions than other ArcGIS files. Raster names must have no more than 13 characters and should contain only letters, numbers, or the underscore character.

TIP: ALWAYS use ArcCatalog to copy, move, rename, or delete grids. Grids have complex storage formats, and Windows cannot manage them properly.

Important Terms

azimuth	Euclidean distance	nearest neighbor	zenith angle
bilinear	filter	neighborhood functions	zonal statistics
block function	focal function	NoData	zone
Boolean overlay	grid	reclassify	
Boolean raster	interpolation	resampling	
cubic convolution	map algebra	surface analysis	

Chapter Review Questions

1. Describe the difference between the terms *grid* and *raster*.
2. Why is it best to store rasters in the same coordinate system in which you plan to use them?
3. You are adding two rasters together in the Raster Calculator using the formula raster1 + raster2. Raster1 is in UTM projection, raster2 is in State Plane, and the geoprocessing Environment setting for the output coordinates is USA Equidistant Conic. By default, what coordinate system will the output raster have?
4. Define a Boolean raster. What values can it have?
5. List and describe the five Environment settings that can be helpful in managing raster analysis. When should these be checked?
6. Imagine that a parcels map is used as the zone layer for a zonal statistics function. Explain how the results would differ if the zone attribute used were the (1) parcel-ID number or (2) the land use designation.
7. You are adding three rasters with cell resolutions of 30 meters, 50 meters, and 90 meters. List the four available options that would determine the cell size of the output raster and state what the output resolution would be in each case.
8. Using Figure 8.4 as a guide, state the results of the four possible combinations of 0 and 1 for the Boolean operators OR, NOT, and XOR. (The results for the AND operator are already shown in Figure 8.6.)
9. Imagine you have Boolean rasters for 10 different criteria regarding the siting of a landfill. If you add the rasters together, what is the potential range of values in the output raster?
10. What is the potential range of values in Question 9 if you multiply the rasters?

Mastering the Skills

Teaching Tutorial

The following examples provide step-by-step instructions for doing basic tasks and solving basic problems in ArcGIS. The steps you need to do are highlighted with an arrow ➔; follow them carefully. Click on the video number in the VideoIndex to view a demonstration of the steps.

➔ Start ArcMap and open ex_8a.mxd in the MapDocuments folder.

➔ Use Save As to rename the document and remember to save frequently as you work.

To become acquainted with raster analysis, we will use rasters to solve the snail habitat problem from Chapter 7 (Fig. 8.22). Spatial Analyst functions are all found in a Spatial Analyst toolbox. Before using the Spatial Analyst extension, however, it must be turned on.

1➔ Choose Customize > Extensions from the main menu bar.

1➔ Check the box to activate the Spatial Analyst extension, if it is not already checked.

1➔ Close the Extensions window.

1➔ Click the List By Drawing Order icon at the top of the Table of Contents.

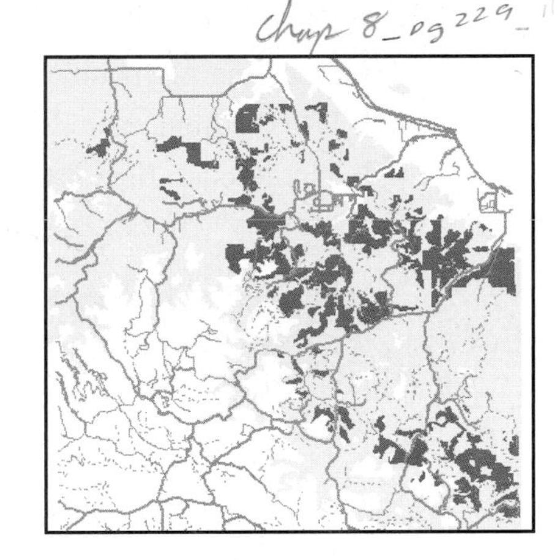

Fig. 8.22. Snail habitat determined by vector analysis

As before, we will create a practice geodatabase to store our results and prevent them from mixing with the original data.

2➔ Open the Catalog tab and expand the Folder Connections > mgisdata folder.

2➔ Right-click on the BlackHills folder and choose New > Personal Geodatabase. Name it chap8results.

It is wise to set some Environment settings before starting. We will be producing many rasters, so it helps to have a default workspace to store them. We should also set a default cell size and extent, so that all outputs have the same area and resolution as our Elevation raster, 30 meters.

3➔ Open Geoprocessing > Environments from the main menu bar.

3➔ Expand the Workspace entry. Set both the Current Workspace and Scratch Workspace to the chap8results geodatabase.

3➔ Expand the Processing Extent entry. Set the Extent drop-down to *Same as layer Elevation*.

3➔ Expand the Raster Analysis entry. Change the Cell Size drop-down to *Same as layer Elevation*. Click OK.

TIP: It is good practice to form a habit of checking the Environment settings at the beginning of a work session. Forgotten settings from previous sessions may cause strange results.

In Chapter 7, we predicted snail habitat from vector layers of geology, vegetation, and elevation. We will now perform the same analysis using three Boolean rasters, each one representing one of the ideal snail conditions. We will then multiply them together to find the habitat. We'll begin by creating the elevation range from the DEM.

Performing a Boolean overlay

The snails prefer elevations from 1200 to 1600 feet. We need to create a Boolean raster in which the cells meeting the elevation criteria have a value of 1 and the rest have a value of 0. The Reclassify tool is used to convert one set of raster values to another.

4➔ Open ArcToolbox and dock it as a tab on the right side of ArcMap.

4➔ Open the ArcToolbox > Spatial Analyst Tools > Reclass > Reclassify tool.

4➔ Use the Input raster drop-down to select the Elevation raster.

4➔ The Reclass field is already set to Value, the field containing the values stored in the raster. This is almost always the field you use when processing rasters.

You can edit the entries here directly (be sure to put spaces around the hyphens if you do). However, it is often easier to use Classify to set up the desired class breaks first, either manually or using one of the classification schemes described in Chapter 2.

4➔ Click the Classify button.

4➔ Change the Classification method to Equal Interval and the number of classes to 3.

4➔ Change the Classification method back to Manual.

5➔ In the Break Values box, change the top number to 1200 and click Enter.

5➔ Change the middle value to 1600.

5➔ Look in the Classification Statistics window and note that the maximum value is 1958. Set the third class break value above it, to 2000.

5➔ Click OK to close the Classification window.

5➔ Enter the 0 and 1 in the New values boxes as shown in Figure 8.23, including 0 for NoData.

5➔ Name the output elevrange and save it in chap8results. Click OK.

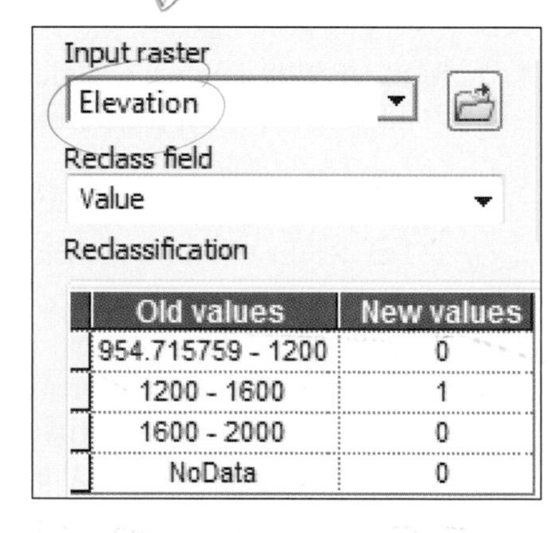

Old values	New values
954.715759 - 1200	0
1200 - 1600	1
1600 - 2000	0
NoData	0

Fig. 8.23. Reclassifying elevation

The new raster should appear as in Figure 8.24 (your colors may be different). This is a Boolean raster with two values: a 1 where elevation is in the desired range (green) and a 0 where it is not.

Fig. 8.24. Boolean elevation range

6➔ Right-click the 0 value symbol for elevrange and set its color to Gray 10%. This helps you visualize the parts of the raster that meet the criteria.

6➔ Turn off the Elevation raster and the Roads layer and turn on the Geology layer.

Next, we will convert the Geology layer to a raster. Rasters can store only one attribute value, so we must specify which field in the geology table will supply the data. We will use the NAME field.

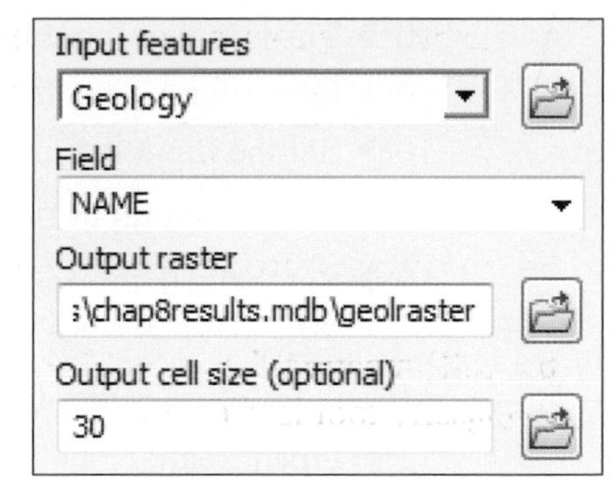

Fig. 8.25. The Feature to Raster tool

7➔ Choose ArcToolbox > Conversion Tools > To Raster > Feature to Raster.

7➔ Select Geology as the input features and set the Field to NAME (Fig. 8.25).

7➔ Name the output geolraster and place it in chap8results.

7➔ Enter 30 for the output cell size. (This tool does not use the Environment setting.) Click OK.

7➔ Turn off the Geology polygon layer.

We converted on a text field (NAME), but a raster can store only numbers. Hence, the legend for geolraster shows seven different numbers and colors, one assigned to each of the unit names.

8➔ Right-click the geolraster layer and choose Open Attribute Table.

In a raster table, the Value field shows the values stored in the cells (Fig. 8.26). A number was assigned to each different name, and the NAME field was added to the table to show what each value represents. The Count field indicates how many cells have that value.

geolraster

Value	Count	NAME
1	62881	Cenozoic
2	23984	Upper Mesozoic
3	107355	Lower Mesozoic
4	91459	Upper Paleozoic
5	80237	Madison Limestone
6	179391	Precambrian
7	69367	Lower Paleozoic

Fig. 8.26. The geolraster table

1. What percentage of this raster consists of Madison Limestone? ________

TIP: To save space and reading effort, future tutorial references to "ArcToolbox > Spatial Analyst Tools > will be written simply as "Spatial Analyst >".

Because we are only interested in the limestone, we must reclassify geolraster to create a Boolean raster containing pixel values of 1 for the Madison and Upper Paleozoic rock units.

8➔ Close the Table window.

Old values	New values
Upper Mesozoic	0
Lower Mesozoic	0
Upper Paleozoic	1
Madison Limestone	1
Precambrian	0
Lower Paleozoic	0
NoData	0

9➔ Open Spatial Analyst > Reclass > Reclassify.

9➔ Set the Input raster to geolraster and the Reclass field to NAME.

9➔ Change the New values to 1 for the Madison Limestone and the Upper Paleozoic units.

9➔ Change the other values, including NoData, to 0.

9➔ Name the output limestone and place it in chap8results. Click OK.

Next, we will create a Boolean raster showing areas of dense conifers.

10➔ Turn on the Vegetation layer and turn off the geolraster layer.

10➔ Open the ArcToolbox > Conversion Tools > To Raster > Feature to Raster tool.

10➔ Set the input features to Vegetation and the value field to DENSITY96.

10➔ Name the output **density** and put it in chap8results.

10➔ Set the cell size to 30 and click OK.

11➔ Turn off the Vegetation layer.

11➔ Open Spatial Analyst > Reclass > Reclassify.

11➔ Set the Input raster to density and the Reclass field to DENSITY96.

11➔ Change the New value to 1 for C, and the rest to 0.

11➔ Name the output **trees** and put it in chap8results.

Old values	New values
0	0
A	0
C	1
B	0
NoData	0

We are ready to overlay. We have three Boolean rasters, each containing 1 where the habitat condition holds and 0 where the condition is absent. When we multiply them, areas that meet all three conditions will be assigned the value 1, and all other cells will be 0.

12➔ Choose Spatial Analyst > Map Algebra > Raster Calculator.

12➔ Double-click the raster names and click the * operator button to enter the expression:

"elevrange" * "limestone" * "trees"

12➔ Name the output **snailhab** and place it in chap8results. Click OK.

13➔ Turn off all layers except for snailhab.

13➔ Zoom to the extent of the snailhab layer and compare it carefully with Figure 8.27 (your colors may be different).

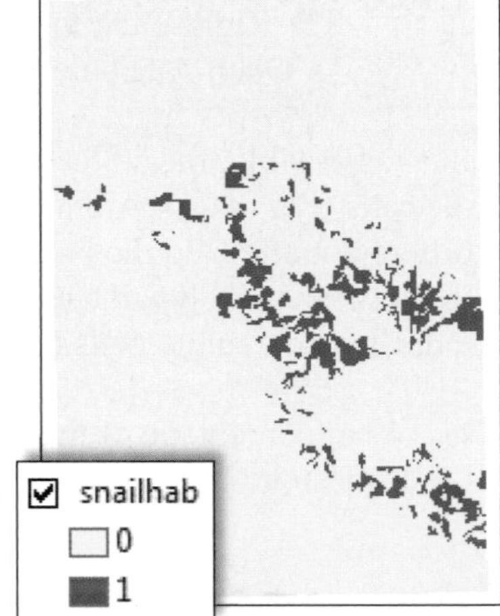

Fig. 8.27. Snail habitat raster

TIP: A bug in the initial release of Version 10.1 gave incorrect results from the Raster Calculator. If your results don't match Fig. 8.27, you can work around the bug using the Spatial Analyst > Math > Times tool. However, it can multiply only two rasters at a time, so you must use it twice: calculate elevrange * limestone to create elevxlime; then calculate elevxlime * trees to create snailhab. Do it now, and check your results before going on. (You may need to change the symbology for snailhab from stretched to unique values, if you don't see a 0 and 1 box in the legend as in Fig. 8.27.) (Ask your system administrator about installing the most recent ArcGIS Desktop service pack to fix the problem.)

In Figure 8.27, the snail habitat has the value 1 and appears purple. It is the raster equivalent of the snail habitat calculated in Chapter 7 by intersecting vector features. With rasters, though, we can choose to <u>add</u> them to get a <u>ranking</u> of suitability, rather than just a simple yes or no.

TIP: If the Raster Calculator is not working, use the Spatial Analyst > Math > Plus tool to add the rasters.

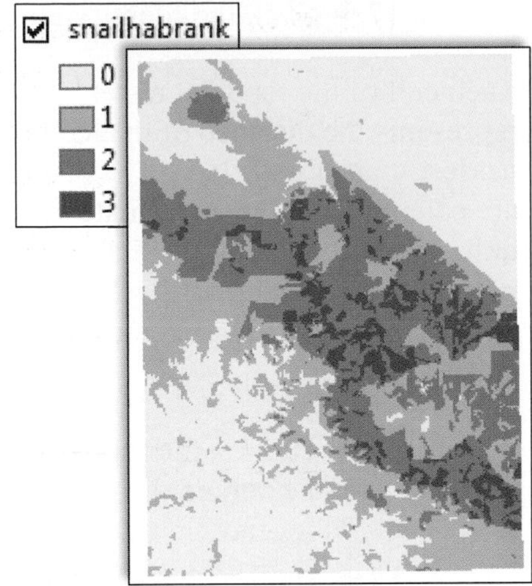

Fig. 8.28. Ranked habitat raster

14➔ Choose Spatial Analyst > Map Algebra > Raster Calculator.

14➔ Double-click the raster names and click the + operator button to enter the expression:
"elevrange" + "limestone" + "trees"

14➔ Name the output snailhabrank and place it in chap8results. Click OK.

14➔ Open the Symbol properties for snailhabrank and change the color scheme to a light to dark green color ramp.

This time the best habitat receives the value of 3, and other areas are ranked by how many of the preferred conditions occur there (Fig. 8.28).

Distance functions and buffers

Vector analysis uses buffers to evaluate proximity to features, such as areas within 200 meters of a road. Distance analysis using rasters is more powerful and flexible, partly because rasters can store continuous distance information, whereas vectors are limited to portraying discrete objects. Let's explore some of the raster distance functions.

15➔ Collapse the legends and turn off all layers except for Roads.

15➔ Open the Spatial Analyst > Distance > Euclidean Distance tool.

15➔ Set the input feature source data to Roads (Fig. 8.29).

15➔ Name the output distance raster roaddist and save it in chap8results.

15➔ Leave the maximum distance blank.

15➔ The output cell size should already be 30 from the Environment setting.

15➔ Give the optional output direction raster the name roaddir and save it in chap8results. Click OK.

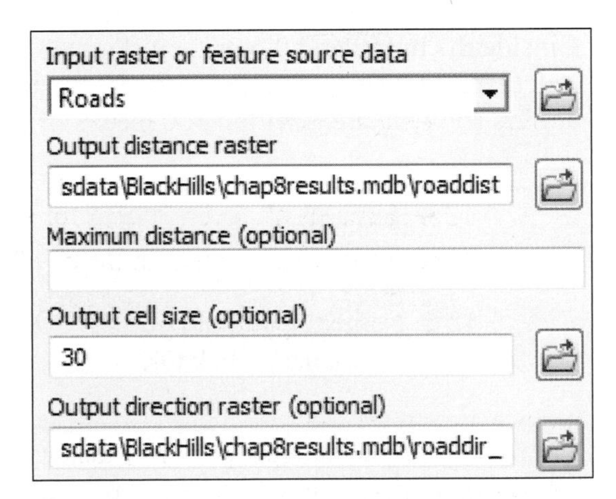

Fig. 8.29. Euclidean Distance tool

16➔ Turn off roaddir for the moment to examine roaddist.

16➔ Open the roaddist layer properties and click the Symbology tab.

16➔ Click Classify and change the method to Jenks. Click OK and OK.

Normally, equal interval is the proper classification for distance rasters, but in this case the roadless area in the upper left is skewing the distance measurements and the symbols.

17➔ Click the Identify tool and click on cells of the raster to see the values they contain.

Each cell of the roaddist raster represents the distance of that cell to the nearest road (Fig. 8.30a). The distances are calculated in the coordinate system units, meters. When you click on a point, the pixel value is the distance you would have to travel to get to the closest road.

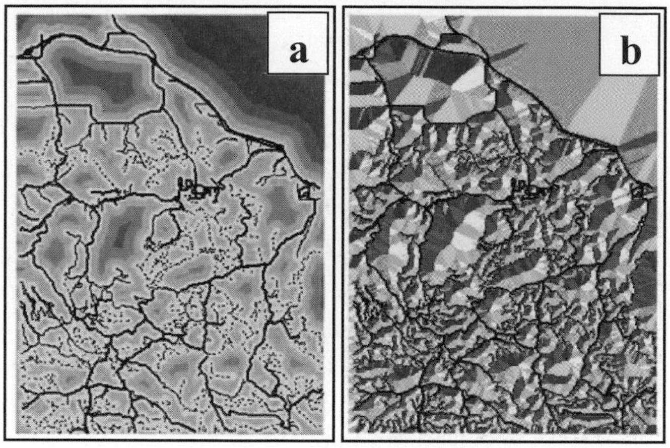

17➔ Turn on the roaddir raster and examine the values in its legend.

17➔ Set the Identify window to identify values from roaddir, and click on several cells. Close the Identify window when finished.

Fig. 8.30. The (a) roaddist and (b) roaddir rasters

The roaddir raster cells record the direction to the closest road (Fig. 8.30b). The raster values are in degrees and range continuously from 0 to 360, with 0 (and 360) indicating north, 90 east, and so on. For any location in the map area, these two rasters tell you how far you are from the closest road and what direction you would have to travel to get there.

Chapter 7 introduced polygon buffers. A distance raster can be used to create raster buffers in the form of a Boolean raster with 1 inside the buffer and 0 outside it (Fig. 8.31). The Raster Calculator is a quick way to produce it. Imagine we want buffers showing areas within 300 meters of the roads.

18➔ Open Spatial Analyst > Map Algebra >Raster Calculator.

18➔ Enter the expression "roaddist" <= 300.

18➔ Name the output `roadbuf300` and save it in chap8results. Click OK.

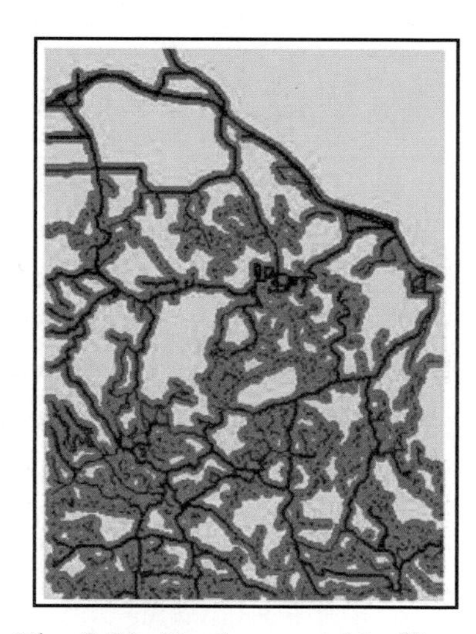

Fig. 8.31. Boolean road buffers

TIP: The attribute table of a discrete raster tells you how many cells have a given value. Each cell has a given area defined by the cell size. You can use this information to calculate areas of rasters, where: Area = # cells × cell size × cell size.

2. Using the cell size (30 meters) and attribute table of roadbuf300, calculate the total area of the buffers in square kilometers. ______________________

We can also create multiple ring buffers with rasters. As an example, consider the snail habitat again. Because snails tend to get crushed on roads, the best habitat is further away from roads. Biologists have determined that the snails rarely travel more than 400 meters during their lifetimes. We can rank the habitat based on distance by using 200-meter ring buffers.

19➔ Open Spatial Analyst > Reclass > Reclassify.

19➔ Set the input raster to roaddist and the Reclass field to Value.

19➔ Click the Classify button. Set the number of classes to four.

19➔ In the Break Values box, click the first entry and change it to 200. Change the next two to 400 and 600.

19➔ Set the last break value above the maximum, 9400.

19➔ Click OK to close the Classify window.

19➔ Make sure that the New values in the Reclassify tool go from 1 to 4, as shown here.

Old values	New values
0 - 200	1
200 - 400	2
400 - 600	3
600 - 9400	4
NoData	NoData

19➔ Name the output ringbuf200 and save it in chap8results. Click OK.

Now we have a raster with 1 signifying the least desirable distance and 4 the best. Let us multiply the ring buffers with the snail habitat raster. Recall that habitat has a value of 1 and non-habitat a value of 0, so multiplying by ringbuf200 will yield a new raster with values from 1 to 4, but only in the snail habitat.

20➔ Use the Raster Calculator or the Times tool to evaluate the expression:

"snailhab" * "ringbuf200"

20➔ Name the output disthabitat and save it in chap8results.

20➔ Turn off all layers except disthabitat and roads.

20➔ Change the color scheme of disthabitat so that 1 is red, 2 orange, 3 yellow, and 4 green (Fig. 8.32).

This map might raise new concerns, because very little of the habitat lies in the safer regions away from the roads.

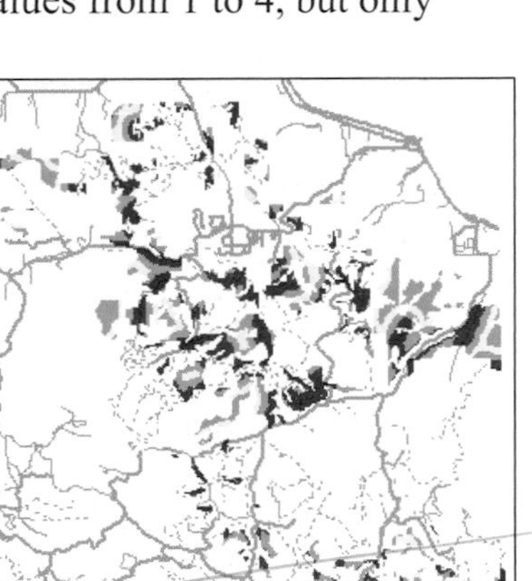

Fig. 8.32. Snail habitat ranked by distance to roads

It is important to note that the classified display of a distance raster (Fig. 8.30a) looks similar to buffers, but do not be fooled. A distance raster contains continuous distances, whereas a buffer raster contains discrete values of 0 or 1. You cannot create buffers by changing the symbology of a distance raster. Even though it may appear to give you buffers, the values are still distances.

21➔ Compare the legends and rasters of the ringbuf200, roadbuf300, and roaddist layers.

Nor can you create buffers using the maximum distance feature in the Euclidean Distance tool. This is a common misconception.

22➔ Open Spatial Analyst > Distance > Euclidean Distance.

22➔ Set the input features to Roads, the output to mindist, and the maximum distance to 300. Don't specify a direction raster; you don't need it. Click OK.

22➔ Turn off all layers except mindist and roadbuf300. Zoom in and compare them carefully. Compare their legends in the Table of Contents as well.

Superficially, the mindist raster looks like buffers, because only the cells close to roads have values. However, the values are distances. The true Boolean roadbuf300 has 1 in the buffer cells.

To turn mindist into buffers, you would still need to reclassify it. So setting the maximum distance does not save any effort, and the output will not work correctly for Boolean overlay.

Topographic functions

Topographic functions form an important subset of raster tools that are probably used more than any other. They are most often used with elevation data. In this section, we will use them to find low-slope, sunny areas as prospective sites to build a winter ski cabin.

23➔ Save the map document and open a new, blank map.

23➔ Save the map with a new name, and save frequently as you work.

23➔ Add the raster mgisdata\BlackHills\Sturgis83\topo30m to the map.

23➔ Set the topo30m symbology to this stretched color elevation ramp.

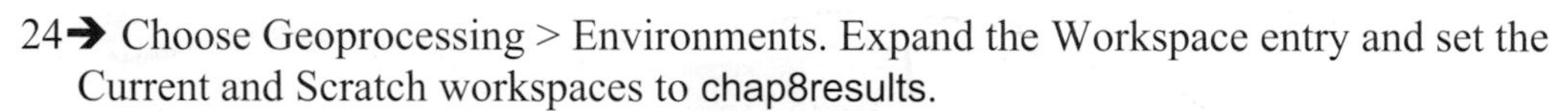

You have opened a new document, so check the Environment settings.

24➔ Choose Geoprocessing > Environments. Expand the Workspace entry and set the Current and Scratch workspaces to chap8results.

24➔ Expand the Processing Extent entry and set the Extent to *Same as layer topo30m*.

24➔ Expand the Raster Analysis entry and set the cell size to *Same as layer topo30m* (which is 30 meters). Click OK.

The slope function calculates, in layperson's terms, how steep the surface is. In mathematical terms it calculates the rate of change of the surface.

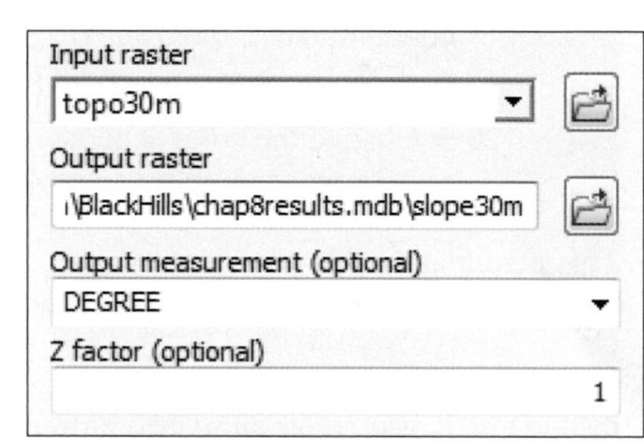

Fig. 8.33. The Slope tool

25➔ Open Spatial Analyst > Surface > Slope.

25➔ Set the input surface to topo30m.

25➔ Name the output slope30m and save it in chap8results.

25➔ Keep the defaults for the output measure (DEGREE) and the Z factor (1) (Fig. 8.33). Click OK.

25➔ Turn off the topo30m and examine slope30m.

3. What is the maximum slope of this raster? ______________

TIP: From this point onward, only the output file name will be given. Make sure that all output is saved in chap8results unless otherwise indicated.

Let's find the areas that are flat enough to build a cabin, with slope less than 10 degrees.

26➔ Open Spatial Analyst > Map Algebra > Raster Calculator.

26➔ Enter the expression "slope30m" < 10 and name the output lowslope. Click OK.

26➔ Set the 0 values of lowslope to light gray and the 1 value to a nice color.

To find sunny areas, we need an aspect raster. The aspect is the direction a slope faces—measured in degrees from north, with true north being 0 (and 360) degrees, east 90 degrees, and so on.

27➔ Turn off the slope30m and lowslope rasters.

27➔ Choose Spatial Analyst > Surface Analysis > Aspect.

27➔ Set the input raster to topo30m, and name the output **aspect30m**. Click OK.

4. The aspect map is automatically colored and labeled according to compass direction. Which color indicates a northern aspect? __________ A southern aspect? __________

Let's find the areas with a warm southerly exposure, where the aspect is between 130 and 230 degrees (50 degrees either side of south).

28➔ Open Spatial Analyst > Map Algebra > Raster Calculator.

28➔ Carefully enter this expression, including the parentheses:

("aspect30m" > 130) & ("aspect30m" < 230)

28➔ Name the output **southerly**.

TIP: The parentheses are required to force the calculator to evaluate the two conditions (aspect > 130) and (aspect < 230) before it executes the Boolean AND (&).

Now we do a Boolean overlay to find potential cabin sites.

29➔ Use the Raster Calculator or the Times tool to evaluate the expression:

"lowslope" * "southerly"

29➔ Name the output **cabinsites** and click OK.

29➔ Compare your results with Figure 8.34.

29➔ Collapse all legends, and turn off all layers.

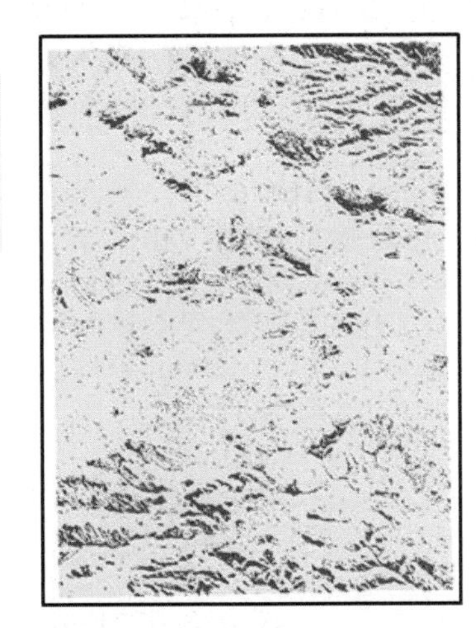

Fig. 8.34. Cabin sites

Two other surface functions are worth investigating. The hillshade function creates a raster that mimics the illumination of the surface from a light source at a specified direction (**azimuth**) and altitude (**zenith angle)**, both measured in degrees. The result looks similar to what a person might see when flying over the surface.

30➔ Choose Spatial Analyst > Surface > Hillshade.

30➔ Set the input raster to topo30m, and name the output **hillshd30m**. Leave the other settings to their defaults. Click OK.

Compare the elevation and hillshade rasters, and notice how the hillshade brings out fine topographic details. It makes an excellent base map.

The Viewshed function determines what areas are visible from a set of points. It is used for applications such as placing fire watchtowers or determining whether timber clear-cut areas are visible from a lookout. We will use it to determine the visibility offered by three fire towers.

31➔ Add the towers feature class from the BlackHills\Sturgis83 geodatabase.

31➔ Open Spatial Analyst > Surface > Viewshed.

31➔ Set the input surface to topo30m, and the observer features to towers. Name the output towerview and click OK.

31➔ Save the map document.

5. Examine the attribute table of towerview. What values does the raster actually store? Use the tool Help to determine what the values mean.

TIP: Notice that, as you accumulate rasters, you have to wait while each draws. For best performance, get in the habit of turning off or removing any rasters not currently in use.

Neighborhood functions

A neighborhood function examines a target window in a raster and calculates statistics from cells in the window. Consider the idea of habitat variety.

32➔ Add the landcover raster from the BlackHills\Sturgis83 geodatabase.

32➔ Open the Symbology properties and change the map type to Unique Values. Change the Value field to Covertype and click OK.

Some areas are homogeneous with only one or two landcover units, but other areas are more variable. We can analyze the variety of landcover types in different areas using neighborhood functions. Let us calculate the variety of landcover types within a 10×10-cell moving rectangle.

33➔ Open Spatial Analyst > Neighborhood > Focal Statistics.

33➔ Set the input raster to landcover and name the output lcvarfoc (Fig. 8.35).

33➔ Choose a rectangle neighborhood. Enter 10 cell units for both height and width.

33➔ Select VARIETY for the statistic. Click OK.

33➔ These are ordinal data, so change the display color scheme to a dichromatic color ramp—yellow to blue works well.

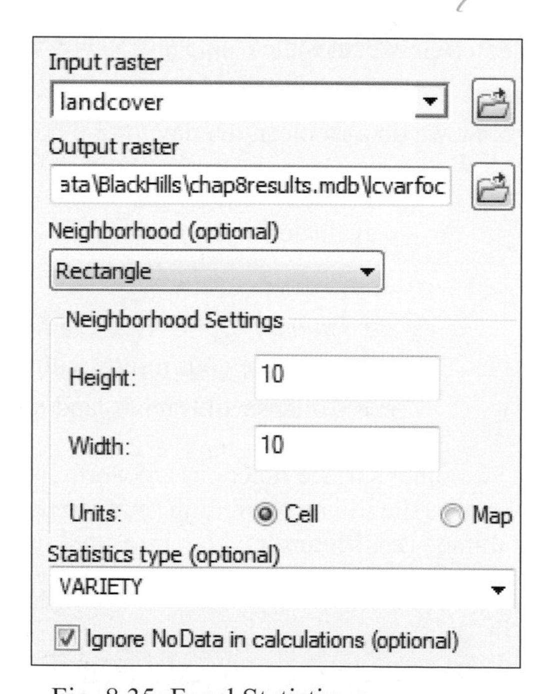

Fig. 8.35. Focal Statistics

A value of 1 in the lcvarfoc raster indicates that only one landcover type is found within a 10×10-cell (300×300-meter) area; it is homogeneous. A value of 9 indicates that nine different types were found. This raster gives an objective estimate of landscape heterogeneity.

A block statistics function works much in the same way except that the target window, instead of moving cell by cell, moves over to a completely new area each time.

34➔ Open Spatial Analyst > Neighborhood > Block Statistics.

34➔ Use the same inputs as for the Focal Statistics tool, but name the output **lcvarblock**. Run the tool.

34➔ Change the display color scheme to the same one as lcvarfoc. Compare the two rasters. Then turn them both off and collapse the legends.

The neighborhood functions can use different statistics. The focal majority function, for example, does a good job of simplifying a raster for conversion to features. This landcover data set was generated from raster satellite data, but perhaps you would like a vector version. In its original form it would make a complex set of polygons with many tiny spots and stringers. It would be best to simplify it. The Majority Statistic tool examines the target window and selects the landcover value that occurs most frequently. We will use the focal version to get a smoother result. The larger the window, the greater the generalization.

35➔ Open Spatial Analyst > Neighborhood > Focal Statistics.

35➔ Set the input raster to landcover and name the output **major5**.

35➔ Select a 5×5-cell Rectangle window and the MAJORITY statistic. Click OK.

When simplifying, a smaller rectangle done two or three times usually works better than a larger rectangle, so run the majority filter once more.

36➔ Open Spatial Analyst > Neighborhood > Focal Statistics.

36➔ Set the input raster to major5 and name the output **major5x2**.

36➔ Select a 5×5-cell Rectangle window and the MAJORITY statistic. Click OK.

To compare the final result, it would help if the colors matched. The landcover raster had a colormap stored with it, but it was not transferred to the new raster.

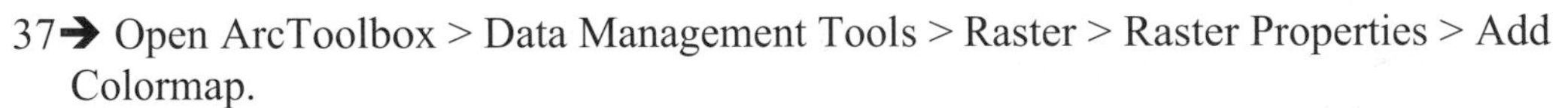

37➔ Open ArcToolbox > Data Management Tools > Raster > Raster Properties > Add Colormap.

37➔ Set the input raster to major5x2 and the input template to landcover. Click OK.

37➔ Remove the major5x2 raster and add it again from chap8results to see the new colors. Turn off the major5 layer.

37➔ Compare major5x2 (Fig. 8.36b) with landcover (Fig. 8.36a).

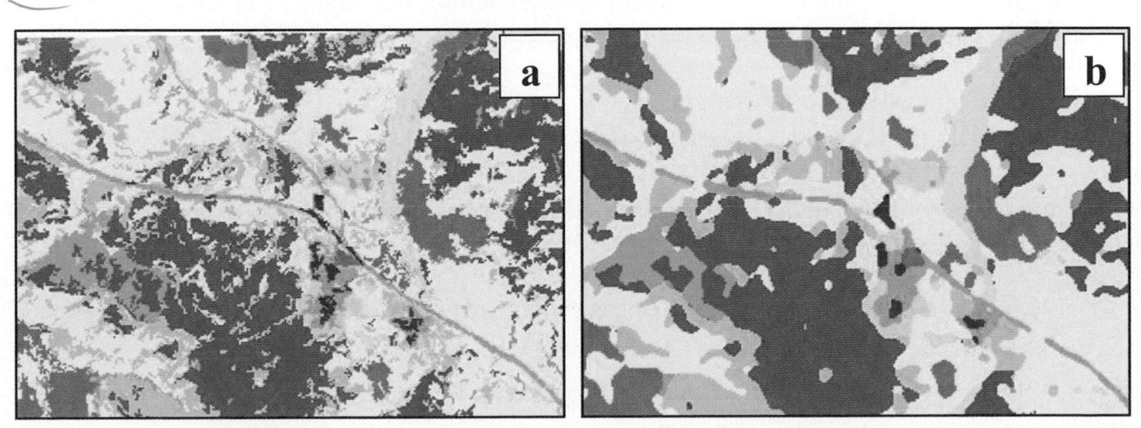

Fig. 8.36. (a) Original land cover and (b) land cover after two passes of a 5×5 majority filter

Notice the simpler appearance of the new map. It will convert to polygons more easily. However, keep in mind that you have modified the original resolution and accuracy of the data and made them more generalized. Keep this in mind when interpreting results based on the new data.

Converting rasters to features

Since we've done the simplification, let's go ahead and convert the landcover to polygons.

38➔ Open ArcToolbox > Conversion Tools > From Raster > Raster to Polygon.

38➔ Set the input raster to major5x2 and the field to Value.

38➔ Name the output landcoverpoly and check the box to *Simplify polygons*. (Otherwise, the polygon edges will follow the cell boundaries.) Click OK.

The field you specified, Value, is transferred to the polygon table as an attribute named *grid_code*. Unfortunately, the text field containing the names of the landcover classes, which was present in the original landcover raster, was omitted when the neighborhood majority tool was run. However, we can copy the field from the original landcover table using a join.

39➔ Open the landcoverpoly table and examine the grid_code field.

39➔ Choose Table Options > Add Field and add a text field called Covertype with a length of 25 characters.

39➔ Choose Table Options > Joins and Relates > Join.

39➔ Make sure that the first drop-down reads *Join attributes from a table.*

39➔ Set the first key field to grid_code, the join table to landcover, and the second key field to Value. Click OK.

40➔ Right-click the empty Covertype field and choose Field Calculator.

40➔ The names are odd, but enter the second Covertype field name from the landcover image (appears as [r_1.img.vat.Covertype]) in the expression box. Click OK.

40➔ Choose Table Options > Joins and Relates > Remove Join(s) > Remove All Joins.

40➔ Examine the newly populated Covertype field and then close the Table window.

40➔ Symbolize landcoverpoly using a unique values map based on the Covertype field.

The snailhab raster from the first analysis will serve to demonstrate another potential issue when converting rasters to polygons.

Old values	New values
0	NoData
1	1
NoData	NoData

41➔ Collapse and turn off all layers.

41➔ Add the snailhab raster from chap8results.

Notice that the raster has 0 and 1 values. Both will be converted to polygons, even though we only want the polygons with 1 value. We must use Reclassify to convert the 0 values to NoData.

42➔ Choose Spatial Analyst > Reclass > Reclassify.

42➔ Set the Input raster to snailhab and the Reclass field to Value.

42➔ In the New values box, change the value for 0 to NoData. Change the 1 value to 1.

42➔ Name the output snailhab2 and save it in chap8results. Click OK.

42➔ Turn off snailhab to better see the new raster.

43➔ Open ArcToolbox > Conversion Tools > From Raster > Raster to Polygon.

43➔ Set the input raster to snailhab2, and set the field to Value.

43➔ Name the output snailhab2poly and check the Simplify box. Click OK.

43➔ Examine the polygons. Then save the map document.

Interpolation and zonal statistics

Interpolation calculates a raster surface from point measurements. For example, climate stations in the Black Hills record precipitation. In between the stations, we don't know what the precipitation is, but we can estimate it at each cell to create a precipitation raster.

44➔ Open the ex_8b.mxd map document. Use Save As to save it under a new name.

44➔ Open the Climate Stations attribute table and examine the fields. This data set contains monthly precipitation values in centimeters for 1997, with annual totals.

44➔ Close the attribute table.

This data set covers a greater extent than the others, and keeping the default cell size at 30 will generate rasters with large storage requirements. Nor do we need that much resolution at this scale. We will increase the default resolution.

45➔ Choose Geoprocessing > Environments from the main menu.

45➔ Expand the Workspace entry and set the Current and Scratch workspace to chap8results again.

45➔ Expand the Raster Analysis entry, set the cell size to As Specified Below, and enter 200 underneath it. Click OK.

We will create a raster with an estimate of annual precipitation from the Sum field.

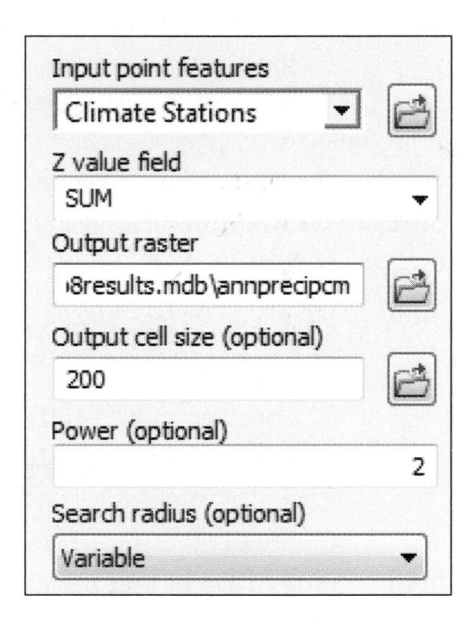

Fig. 8.37. The IDW tool

46➔ Open Spatial Analyst > Interpolation > IDW.

46➔ Set the input points to Climate Stations and the Z value field to SUM.

46➔Name the output annprecipcm. Keep the defaults on everything else (Fig. 8.37). Click OK.

46➔ Examine the new raster.

6. Do you think the precipitation raster is an integer or a floating-point raster? __________ Examine its properties to check your answer.

TIP: Using interpolation functions properly requires understanding the settings and using them appropriately. Spend some time with the Help and/or textbooks when ready to learn more.

Although we needed all the stations to get the best possible interpolation, the estimate worsens farther from the stations. We also are only interested in areas inside the watersheds. We can extract part of a raster using a mask. A mask works like a clip, keeping the areas of interest and assigning NoData to the others. Either rasters or polygon features may be used as masks.

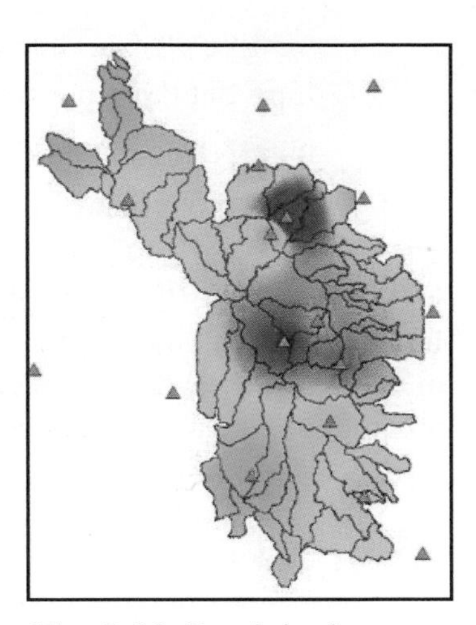

Fig. 8.38. Precipitation raster after Extract by Mask

47➔ Open Spatial Analyst > Extraction > Extract by Mask.

47➔ Set the input raster to annprecipcm.

47➔ Set the mask data to the feature class Watersheds.

47➔ Name the output **precipmask** and click OK.

48➔ Turn off annprecipcm.

48➔ Symbolize the precipmask raster using the stretched method with a green-blue color ramp (Fig. 8.38).

The Raster Analysis Environment settings include a mask setting. You can specify a mask, and every new raster created will have NoData in the space outside the mask area.

49➔ Choose Geoprocessing > Environments and expand the Raster Analysis entry.

49➔ Set the mask to Watersheds and click OK.

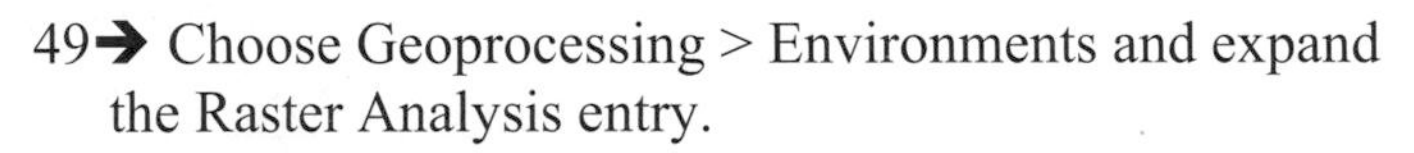

Now we can use this new raster to estimate the annual volume of precipitation per cell. The depth of received water in the cell in meters is precipmask* 0.01. The area of each cell is 200 × 200 meters. Therefore, V(m^3) = precipmask (cm) * 0.01 m/cm × 200 m × 200 m.

49➔ Choose Spatial Analyst > Map Algebra > Raster Calculator and enter the expression:

"precipmask" * 0.01 * 200 * 200

49➔ Name the output **precipvol** and click OK.

The volume per cell ranges from about 16,000 to 35,000 m^3. Next, we would like to know the total volume of water that falls annually in each watershed. The precipvol raster contains the volume that falls in each cell. If we add all the cells in the watershed, we get the total volume of water in the watershed.

Zonal statistics will sum the cell volumes in one raster (precipvol) based on zones defined by another raster or feature class (Watersheds). In the zone feature class, we must choose an attribute field that holds the unique values that define the zones. We will use the field SHEDNAME.

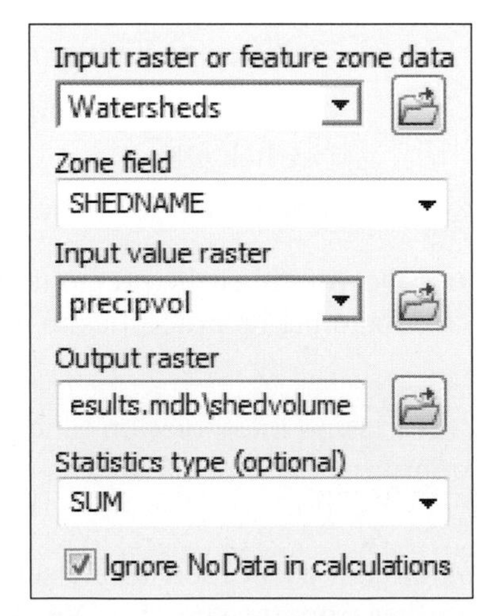

Fig. 8.39. Zonal Statistics

50➔ Choose Spatial Analyst > Zonal > Zonal Statistics (Fig. 8.39).

50➔ Set the input zone data to Watersheds and the Zone field to SHEDNAME.

50➔ Set the input value raster to precipvol.

50➔ Name the output **shedvolume** and set the statistic to SUM. Click OK.

50➔ Open the symbol properties of shedvolume and give it a light to dark blue color ramp (Fig. 8.40).

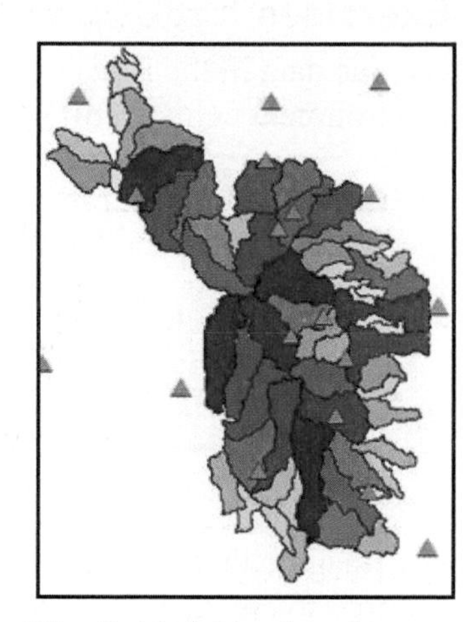

Fig. 8.40. Map showing volume of water in each watershed

In the output raster, each cell in the watershed contains the summed volume over the watershed. The Zonal Statistics as Table tool provides an alternative way to calculate and view the totals.

51➔ Choose Spatial Analyst > Zonal > Zonal Statistics as Table.

51➔ Fill out the first three boxes as before (Fig. 8.39).

51➔ Name the output table **shedvoltable**.

51➔ Instead of the single SUM statistic, choose ALL. Click OK.

51➔ Open the shedvoltable table and examine the statistics for the watersheds.

The output table lists more information, including the cell count, the area, and statistics about the water volumes (min, max, etc.). This table could be joined back to the Watersheds layer if needed.

7. What is the name of the watershed that has the largest volume of water? What is the volume?

____

TIP: You can also use zonal statistics on zones defined by points or lines. You could calculate the elevation of a set of summits, for example, or the average slope of individual streams.

This is the end of the tutorial.

➔ Close ArcMap. You can save your changes.

Exercises

Use the data in the BlackHills\Sturgis83 geodatabase to answer the following questions. Set the Environment cell size to 30 meters for all problems and make sure any masks are off.

1. Create a Boolean raster showing areas where the density of vegetation is open (DENSITY 96 is 0 or A) and the land is owned by the Forest Service (OWNER is NFS). **Capture** the map.
2. Create a Boolean raster showing the areas that are within 1000 meters of a primary or secondary road (TYPE = P or S). **Capture** your map.
3. Create a Boolean raster showing the areas that have an average slope of less than 10 degrees over a 300×300-meter area. (**Hint:** Use Block Statistics.) **Capture** your map.
4. Imagine that you are looking for a good landfill site. Use the rasters produced in Exercises 1 to 3 to create a Boolean raster showing the areas where all three conditions are met. What is the total area of these potential sites? **Capture** the map.
5. The site must be at least 1 sq km in area to be considered for the landfill. Create a helpful map of the sites meeting this criterion, labeled with the site area in square kilometers and including helpful information, such as streams and roads, for the final selection. How many potential sites have the minimum area? (**Hint:** Convert the sites to polygons.)
6. Which geological unit has the highest average slope? Which has the lowest? What is the average slope for each one?
7. The canopy raster contains the forest canopy percentage. Create a map of wshds2c showing the average canopy percentage over each watershed. **Capture** your map.
8. Prime harvestable timber in the Black Hills has SSTAGE96 = 4C or 5 and is more than 200 meters from a stream. How much good timber is left, in square kilometers? Create a map showing the timber and streams. **Capture** the map.
9. Which of the summits has the highest elevation? Which has the lowest?
10. Create a raster with an integer canopy index that ranges from 1 to 5. Create another raster with an integer slope index that ranges from 1 to 5. From these, produce a raster showing an erosion potential index based on canopy and slope. **Capture** the map. (**Hint:** Use Reclassify.)

Challenge Problem

Imagine that you would like to build a cabin somewhere in the Sturgis area. Develop a map showing potential land on which to build based on the following criteria:

-Must be on private land (vegetation, OWNER = PVT)
-Must have a slope less than 20 degrees
-Must have a southerly aspect (between 90 and 270 degrees)
-Must be within 500 meters of an existing road
-Must be more than 100 meters from a stream
-The land use category in landcover must contain the term *Forest*.

Create and capture a map showing your potential sites with the roads shown in light gray. What is the total area in square kilometers of potential land you found?__________

Chapter 9. Network Analysis

Objectives

- Learning the function and terminology of networks
- Understanding the different properties of transportation and utility networks
- Performing tracing analysis on networks
- Understanding how networks are constructed

Mastering the Concepts

GIS Concepts

About networks

A network consists of a system of paths traveled by a variety of things, such as traffic, water, sewage, or electricity. Common examples of networks include roads, utility lines, airline routes, and streams. Modeling the behavior of a "commodity" flowing through the network can help answer questions such as these: How long would it take a chemical spill in this watershed to reach the city water supply? If this transformer blows, which parts of the city will be out of power? If a problem develops with a sewer line, which shutoff valve can we use to halt flow during repairs while affecting the fewest possible people?

All feature classes participating in a network must take the role of either an **edge** or a **junction** (Fig. 9.1). An edge feature represents a path along the network. Examples of edges include roads, pipelines, and electric cables. Edges must come from a line feature class. A junction represents a point in a network where edges meet or where an object such as a valve exists. A street intersection, for example, is represented by a junction. T-valves or straight-line connectors would constitute junctions in a pipeline network. Junctions can come from two sources: they can be automatically constructed during network building wherever edges meet, or they can be specific objects that come from a point feature class, such as valves.

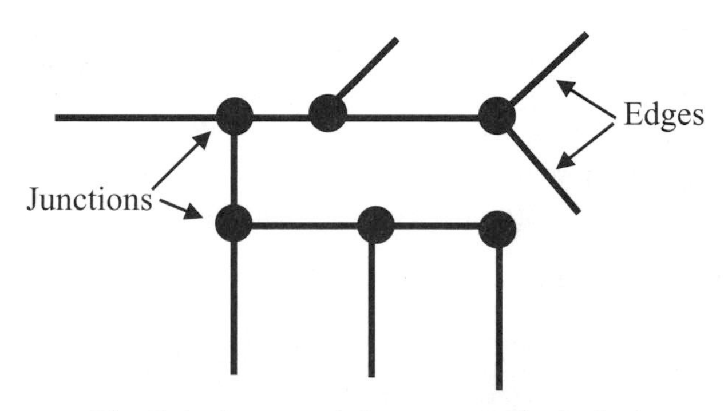

Fig. 9.1. A network is composed of edges and junctions.

Network features can take one of two states: enabled or disabled. An enabled edge or junction allows material to flow through it. A disabled edge or junction blocks travel, whether of water, electricity, or cars. The feature state is initially assigned during construction of the network; however, it may be temporarily modified during analysis, such as disabling a street edge to represent road construction that blocks travel. The states of the network features are stored in the attribute table of the feature class in a field named Enable.

Network storage in a data model has two aspects (Fig. 9.2). The **geometric network** consists of the actual points and lines in the feature classes that participate in the network. These features are stored as the familiar feature classes containing points and lines. Certain fields are added to the attribute tables of the feature classes to help track the participation of each feature in the network. The **logical network** includes a set of tables that store information about the construction and operation of the network elements. Users do not usually have direct access to the logical network. The files are created at the time the network is built from the feature classes. The geometric network and the logical network are interconnected, and changes in the geometric network require changes to the logical network. During editing, the relationships between the geometric and the logical networks are maintained automatically.

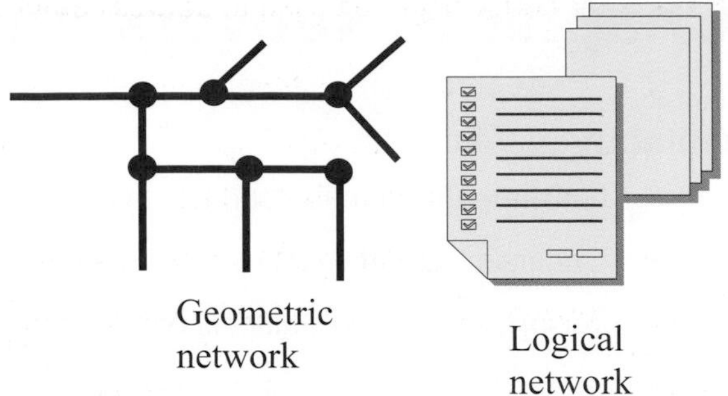

Fig. 9.2. The geometric network consists of features in a feature class. The logical network contains relationships, connections, and behavior as tabular data.

Types of networks

Networks fall into one of two basic categories: transportation networks or utility networks. The material in a transportation network (cars) has few constraints on travel; the drivers go where they choose (Fig. 9.3). The vehicles can usually travel in either direction along an edge and through intersections, represented by junctions.

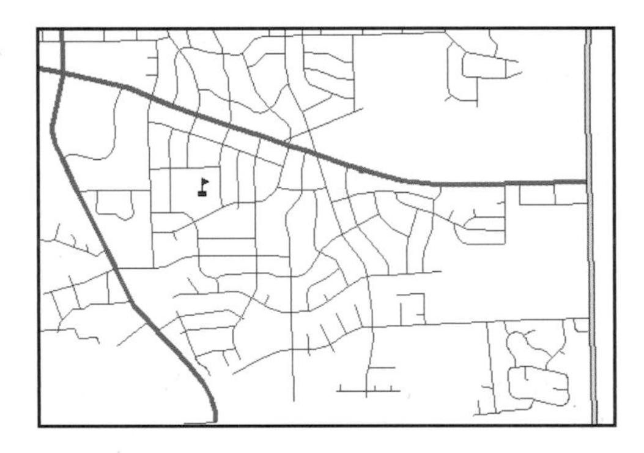

Fig. 9.3. In a transportation network, the drivers choose the way to travel.

Utility networks have established directions of flow determined by the topology of the connections and the location of sources or sinks (Fig. 9.4). A **source** point provides material to the network and "pushes" the material away from itself. A power plant is a source for an electric grid; a water treatment plant might act as a source for a water line network. A **sink** represents a location where material is used or leaves the network. Sinks draw the material toward themselves. Electric meters or water meters on a building might be considered sinks, as would a sewage treatment facility discharging to a lake.

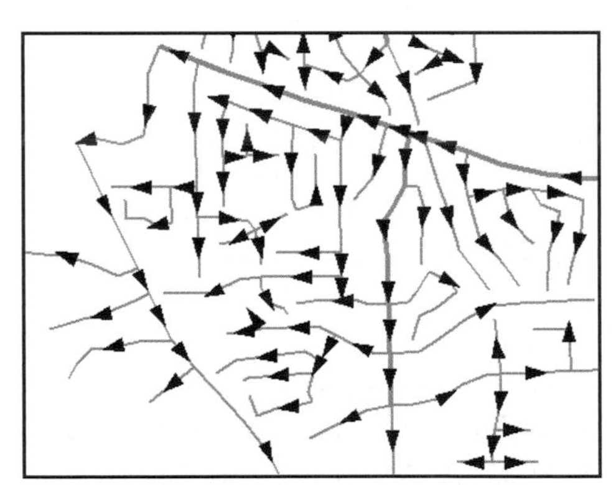

Fig. 9.4. In a utility network, the flow of the material is determined by the network configuration.

Transportation networks and utility networks have different types of analysis functions available for use in modeling flow. A transportation network could be used to solve problems such as the following:

- What is the best path to travel to 16 delivery locations?
- What is the likely service area of a fire station based on travel time?
- What is the shortest path from point A to point B?

Utility networks might be used to address other kinds of problems:

- If this valve fails, which customers will be affected?
- If I have to close this pipe for repairs, can I reroute water through another path to minimize service disruption?
- How will contamination at one location propagate through the network?
- Which sewer lines serve only residential customers?

Utility network flow must be established before the network can be analyzed. You establish flow by specifying the sources or the sinks in the network and solving for the flow direction. A utility network edge may have three possible states: it may be unassigned if network flow has not yet been established; it may have a determinate flow, meaning that a single flow direction was established; or it may have an indeterminate flow, meaning that the flow direction is ambiguous and cannot be uniquely determined. The third case generally results from a topological error in the network or from an error in specifying the sources and the sinks.

Network analysis

Network analysis problems are tackled using a program called a **solver**. Many types of solvers are possible. Figure 9.5 shows the set of solvers that comes with ArcMap. Some general features of using solvers are discussed in this section.

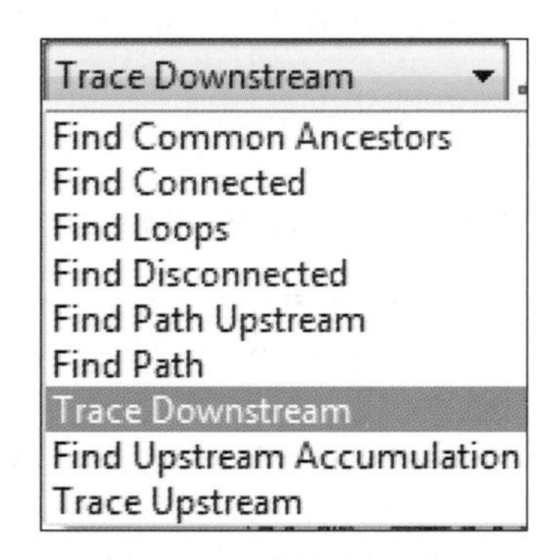

Fig. 9.5. Tracing solvers

A **flag** pinpoints a location of interest, such as the starting and the endpoints of a journey, the delivery stops on a route, or a location from which to trace up- or downstream. Flags may be placed along edges or at junctions (Fig. 9.6).

A **barrier** represents a temporary blockage or outage in the network, such as a damaged bridge, a closed valve, or a broken power line. As with flags, barriers may be placed on either edges or junctions (Fig. 9.6). When placed on a feature, barriers temporarily override the normal "Enabled" state of the feature, as stored in the layer's attribute file.

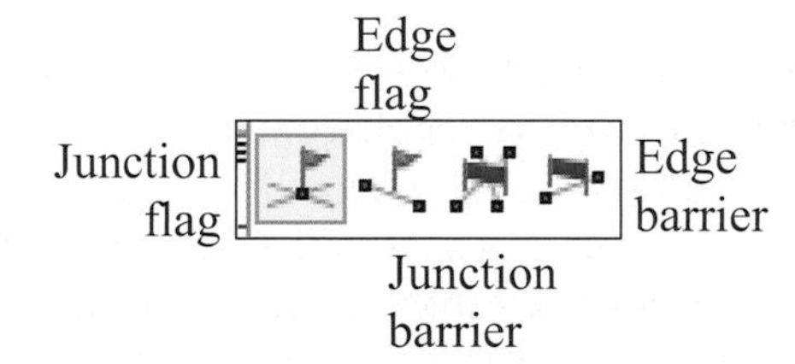

Fig. 9.6. Flags mark locations of interest. Barriers stop flow.

Network weights can be stored for each edge or junction as attributes and specify the "cost" of traversing the network. Common edge weights include distance along the edge or travel time as dictated by the street length and speed limit. Common junction weights include a loss of pressure at a pipe T-junction or a waiting time imposed at a traffic light. Solvers can take these weights into account when finding solutions, such as finding the path with the least wait time at intersections rather than finding the shortest path between two points.

The output of a network tracer usually takes the form of a set of edges and junctions along the traced path. Network analysis offers several options concerning the form of this output. The path solution may be returned either as a graphic or as a set of selected features. If returning a selection, one may choose to return only edges, only junctions, or both.

Generic trace solvers

The solvers in this section may be applied to either utility or transportation networks. They carry out their work without regard to the direction of flow through the edges and junctions.

Finding paths

This solver traces the least-cost path between two or more flags placed on the network. It can find a way to travel from one location to another or to plan a path to most efficiently visit a series of locations such as delivery stops. When finding paths, the solver keeps track of the "cost" of traveling the path. By default the cost is the number of edges traversed. Figure 9.7a shows the lowest-cost path, in terms of the number of edges traversed, for driving from a student's house to a school. The flags are shown as green boxes at the locations of interest; in this case, edge flags are used.

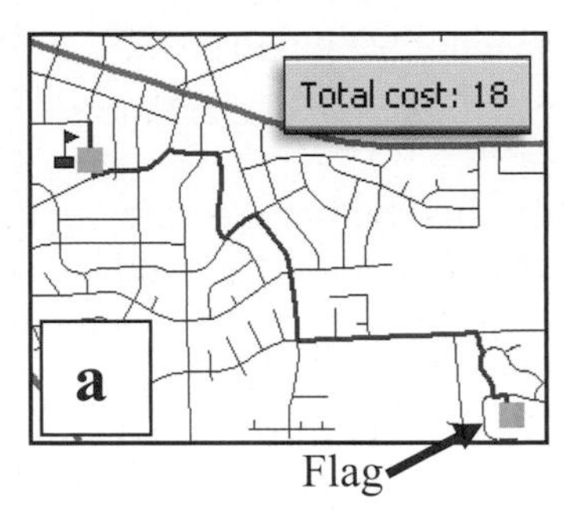

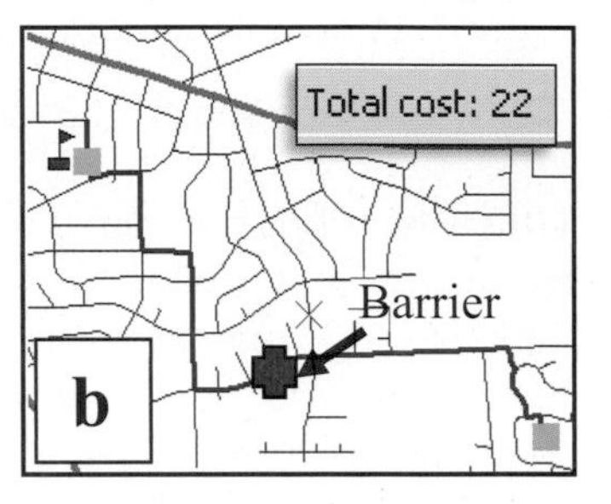

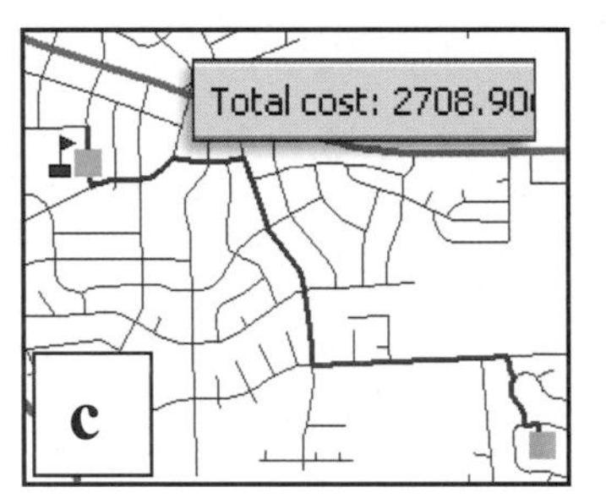

Fig. 9.7. Finding paths:
(a) finding the shortest path with cost in edges;
(b) finding the shortest path with a barrier;
(c) finding the shortest path using a distance weight (meters)

The user can modify solver behavior by specifying barriers to travel, such as road construction. Figure 9.7b shows the new path found when a barrier is placed along the original one. The cost has increased to 22 edges, but the new path is still the least-cost path, given the presence of the barrier.

Users can also specify a weight, such as distance or travel time. Without weights, the solver simply counts the number of edges traversed. However, the lowest number of edges does not always mean the shortest path; in some cases the shortest path could have more edges. Finding the shortest path requires a distance weight. In Figure 9.7c, the distance along the edges has been chosen as the weight. A slightly different path results than when simply counting edges. The cost shows up in the distance units—in this case, meters.

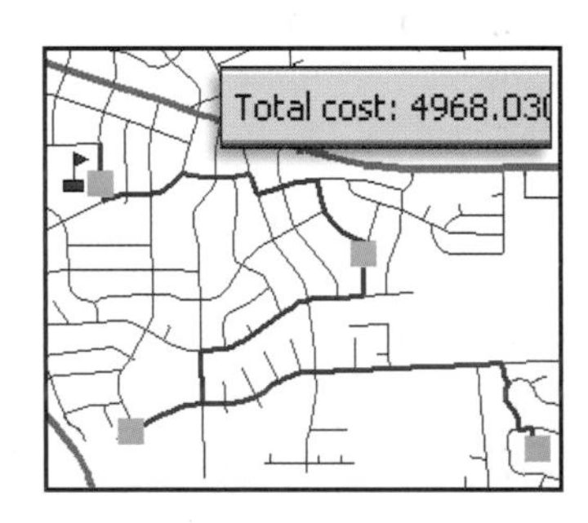

Fig. 9.8. Finding the best path to pick up a child's friends on the way to school

Finally, the user can specify multiple flags in order to find the least-cost route for making several stops in sequence. Figure 9.8 shows a path to pick up a child's friends and take them to school. The Find Path tracer simply visits the stops in the order given. Other solvers must be used to solve the classic "traveling salesman" problem: given a set of locations to visit, devise the most efficient sequence and path to follow.

Find Connected/Find Disconnected

The Find Connected solver traces along the features and highlights those that are connected to the flag(s) placed on the network. Such a solver might be used to find all of the water lines that are supplied by a single water intake gallery along a river (Fig. 9.9). Such information might be useful in gauging the impact of a chemical spill near the intake and determining which customers must be warned to purchase drinking water until the contamination is removed.

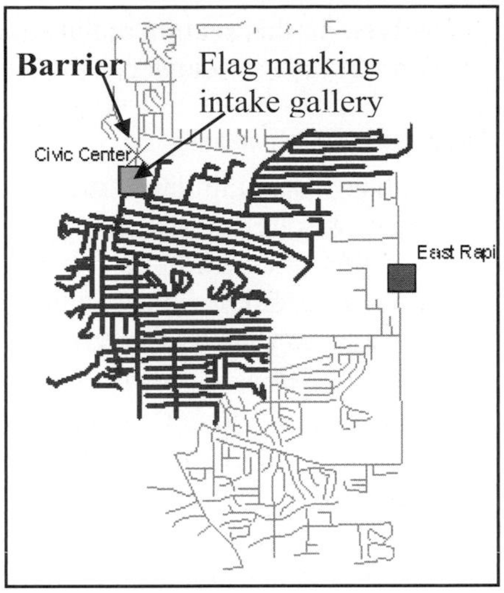

Fig. 9.9. Find Connected returns all parts of the network connected to one or more points of interest.

Barriers placed at specific locations affect the results of this solver. Any feature with a barrier between it and the flag location will be considered disconnected from the network (Fig. 9.9).

Find Disconnected performs the opposite function of Find Connected; it locates portions of the network that can't be reached from the flagged locations. This solver is particularly useful for diagnosing problems when constructing networks. Sometimes a failure in snapping two line ends together or neglecting to split a line at an intersection will result in part of the network being disconnected. The solver finds these problem areas so that they can be fixed.

The Find Connected and Find Disconnected solvers are most useful for utility networks, but they can also be applied to transportation networks. A road network is designed to facilitate connecting all locations for easy travel. Thus, Find Connected usually results in the selection of the entire network (not a very useful result most of the time). However, one useful application might include finding which parts of the city might be seriously affected by construction placed at a critical location. Find Disconnected also helps locate network errors, as previously noted.

Finding loops

A loop is a section of a network that forms a continuous loop, or ring. In a transportation network, loops are generally the rule, as most streets eventually connect to other streets and can be traversed in a circular fashion. This solver finds little utility in transportation networks. In utility networks, loops indicate a topological error rather than a legitimate situation. A water pipe that forms a loop would cause problems when the emerging flow reentered the loop. The Find Loops tracer, then, has its primary use as a tool for detecting and fixing loops. A flag must be specified as the place to begin tracing, and the solver finds all loops that lie downstream of the flag.

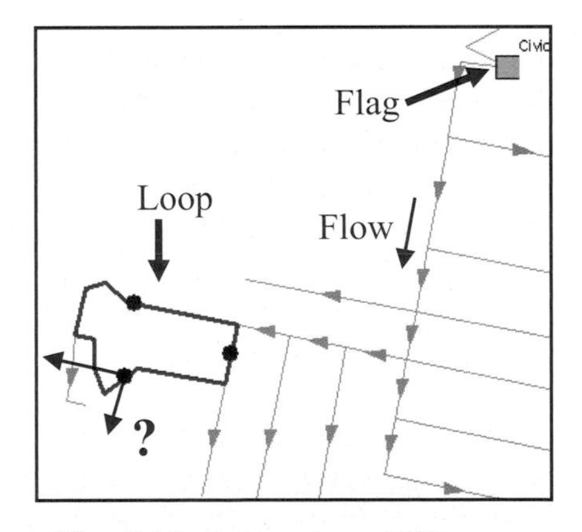

Fig. 9.10. A loop in a utility network cannot flow.

Figure 9.10 shows an example of a loop in a utility network. The flag has been placed at the upper right of the picture. The arrows indicate the direction of flow along the network. The loop found by the tracer is marked with small circles, which indicate that flow in the loop is indeterminate; because of the loop, the flow direction cannot be established from the two possible solutions. To fix the loop, one end must be disconnected from the circle, or an edge or a junction must be disabled.

Utility trace solvers

The solvers in this section are used only for utility networks. They rely on an established flow direction in order to carry out the analysis.

Find Path Upstream

This solver works similarly to the Find Path solver previously discussed, except that movement along the edges is always constrained to be in the upstream direction. The tool can locate a path between a single location and its source (Fig. 9.11). If more than one flag has been specified, the paths of each flag are traced upward.

Fig. 9.11. Find Path Upstream traces the flow from a location to its source.

Find Upstream Accumulation

This solver is a variation of the Find Path Upstream tool, but it keeps track of the weights on each edge and junction and applies them to determine a total accumulation of resources along the path.

Trace Upstream/Downstream

Trace Upstream and Trace Downstream provide a way to follow the flow from a specified location. Trace Upstream is essentially the same solver as Find Path Upstream. The downstream trace finds all the edges that lie downstream of a flagged location or locations. Such a solver might be used to find the service areas that would be disrupted by a downed power line or a water line break (Fig. 9.12).

Fig. 9.12. Trace Downstream could find the service area disrupted by a broken pipe.

Find Common Ancestors

This solver takes a group of flags and finds any upstream edges and junctions that are shared by all the individuals in the group. The tool can be used by utility companies to help locate failures in the system based on reports by customers. Suppose that a utility company receives 50 calls in one evening from customers complaining about a loss of water pressure. By entering the locations of these reports as flags, the company can isolate possible locations of the suspected break by finding the set of pipes that is common to all the calls. The break will most likely be found near the lower end of the common ancestors (Fig. 9.13).

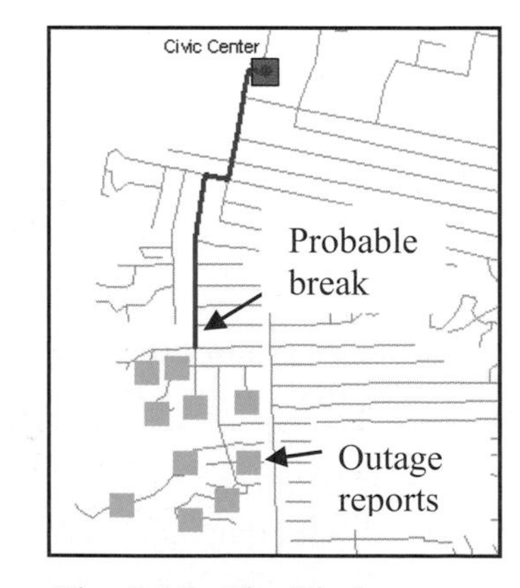

Fig. 9.13. The Find Common Ancestors solver can locate service failures.

About ArcGIS

Geodatabases are used in ArcGIS to create and store networks. Networks are topological data structures that can be built from existing simple feature classes. A feature class containing roads, for example, can be used to create a transportation network. Multiple layers can be combined to form more complex networks. A water supply network might include feature classes representing pumping stations, pipes, valves, and meters.

Networks may only be created in feature datasets. All feature classes participating in a network must belong to the same feature dataset; however, the dataset can contain other feature classes that do not participate in the network. A Transportation feature dataset containing roads and rail lines could have a roads network that did not include the rails.

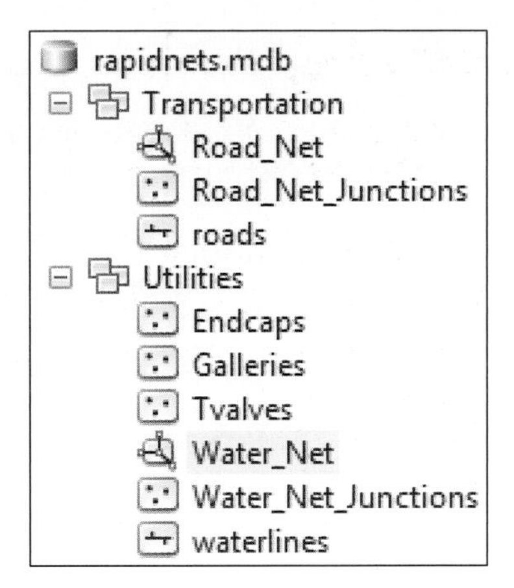

Fig. 9.14. Networks are built from feature classes in feature datasets.

Figure 9.14 shows a geodatabase containing two networks. The Transportation feature dataset contains a single line feature class, roads, plus two feature classes that constitute the network, Road_Net and Road_Net_Junctions. The Utilities feature dataset contains four feature classes plus the Water_Net and Water_Net_Junctions. It is not obvious from looking at the geodatabase which feature classes in the feature dataset participate in the network; however, this information can be obtained by examining the properties of the network in ArcCatalog.

The Utility Network Analyst toolbar

A **trace** is an algorithm that solves network problems, and ArcMap comes with a set of simple traces that follow network connections (see Fig. 9.5). More advanced and complex network analysis tools are also available for users who purchase the Network Analyst extension. Knowledgeable users may employ the special customization languages provided with ArcGIS to write new solvers. ArcMap solvers require input in four forms: the network itself, the weights of elements along the network, the locations of interest, and any barriers to flow.

ArcMap contains a special toolbar used to direct and control the analysis of networks (Fig. 9.15). The main features of this toolbar are as follows:

- The Network drop-down list controls which network is being analyzed. One network may be analyzed at a time.
- The Flow menu controls the display of symbols representing flow directions in a utility network. The user can turn the symbols on and off and change the symbols being used.
- The Establish Flow button next to the Flow menu is used during editing to establish the initial flow direction in a network, and it requires an ArcGIS Standard or Advanced license. ArcGIS Basic can perform analysis on networks for which this task is already complete.
- The Analysis menu manages the analysis by placing flags or barriers, clearing both flags and barriers, or clearing previous results in preparation for a new analysis. The Trace Task menu specifies which kind of analysis will be performed.
- The drop-down flag tool controls which type of flag or barrier to add to the map.
- The Solve button must be clicked to perform the trace. All of the trace tasks require that at least one flag be added to the map before any analysis can occur. If no flags are present, the Solve button will be dimmed.

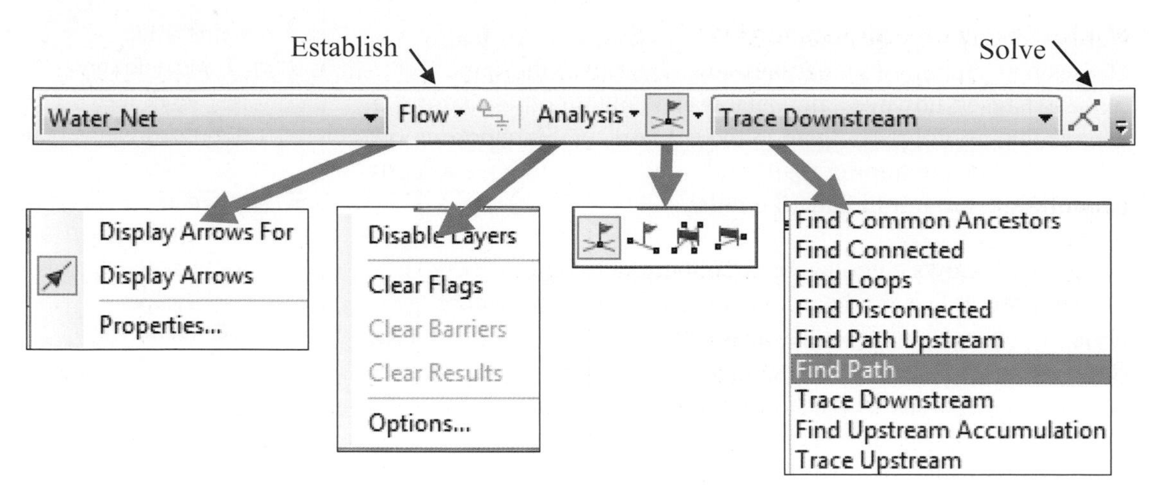

Fig. 9.15. The Utility Network Analyst toolbar contains all the functions needed to analyze networks in ArcMap.

The overall process of analyzing a network includes specifying the network to be used, choosing the tracing task, adding flags or barriers, and solving. Once a solution is obtained, it may be cleared before going on.

The Analysis > Options menu choice opens a window with four tabs. These options control the analysis by specifying which features will be traced, whether the features will be weighted, and the format of the results. The output of a trace may take the form of a drawing or a set of selected features (Fig. 9.16). The features returned can include all features matching the criteria of the trace task or simply those stopping the trace. The user can also choose whether to return edges or junctions or both.

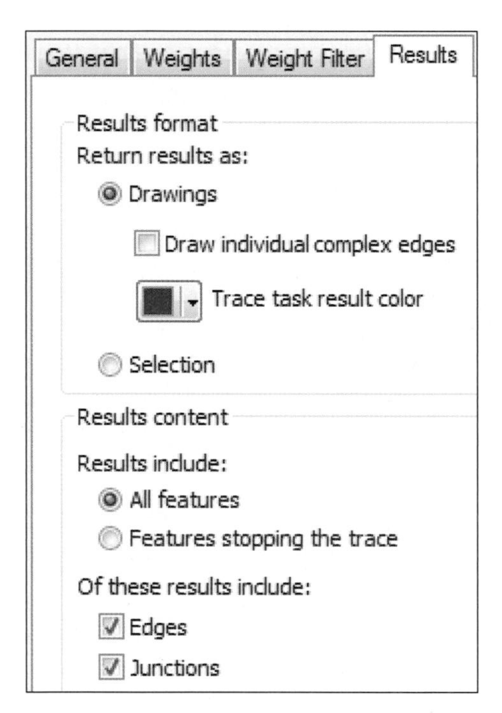

Fig. 9.16. Choosing the format of the output results

Building networks

Building valid networks requires ingenuity and attention to detail, especially if the network is to be used for sophisticated modeling. The planning stage is important, and the builder must consider a number of issues in designing the network, including using simple or complex edges, assigning weights, using domains or subtypes, and establishing connectivity rules.

SKILL TIP: An ArcGIS Standard or Advanced license is required in order to create networks. If you have the license and want to try creating a network, learn how in the Skills Reference (Files and Geodatabases).

Simple edges or complex edges?

Network edges occur in two different styles, simple edges and complex edges (Fig. 9.17). A simple edge can have junctions only at its endpoints. A complex edge can have one or more junctions in its middle.

Complex edges offer advantages when keeping track of long features. For example, a water main along a street has junctions composed of T-valves that branch off to laterals carrying water to each individual house. Using simple edges, the main would have to be broken into multiple pieces between each T-valve. However, for purposes of queries and maintenance, the water main is more logically and easily treated as a single feature. When clicking on the water main, it would be better to select the entire feature, not just one tiny segment between two laterals. By using a complex edge network, the water main can remain a single feature while allowing the junctions of the laterals to connect to it.

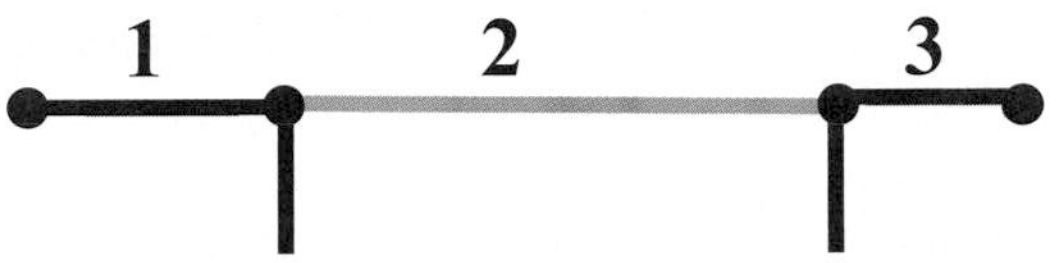

1
2
3
Simple edges end at junctions. Each segment between the junctions is a separate line feature. This water main has three segments with two laterals joining it.
1
Complex edges may have junctions in the middle of the line feature. This water main is a single feature with two laterals joining it in the middle.

Fig. 9.17. Networks with simple edges versus networks with complex edges

Junctions also can be implemented as simple or complex features. A complex junction can be built from a set of simple junctions and simple edges. For example, a dam in a stream network might be represented as a complex junction, which itself is composed of edges and junctions representing flow gates, generators, turbines, and more. The geodatabase model comes equipped with simple junctions, simple edges, and complex edges. Complex junctions must be created using advanced modeling tools, and their construction is beyond the scope of this book.

Network weights

When creating a network, the user can include certain attributes that establish weights for each edge and junction. Such weights allow for more realistic modeling of flow. For example, in analyzing possible routes from a fire station to a fire, the shortest route is not necessarily the fastest. A longer route with less traffic, fewer traffic lights, and a higher speed limit might serve better than the shortest route. A more realistic travel network could include a travel time attribute for the road features and point junctions representing traffic lights with different average delays.

When creating the network, these attributes would be specified as weights, to take into account costs associated with traversing the edges or passing through the junctions. Weights can only be established during network creation.

Summary

- Networks are used to model the flow of objects or materials along an interconnected system of edges and junctions.

- Transportation networks are distinguished by the fact that the material can independently determine how it will flow. In a utility network, the direction of flow is determined by the network topology and the location of sources and sinks.

- Programs called solvers may be applied to analyzing flow through a network. Solvers typically use three types of input: the network itself, flags indicating points of interest such as delivery locations, and barriers that prevent flow.

- Network weights are stored as attributes of edges. Weights can be used during analysis to determine, for example, the shortest path or the shortest travel time. They can also be used to accumulate the costs of moving along the network.

- Networks are stored in feature datasets, and they may include some or all of the feature classes in the dataset.

- Network topology, connectivity, and relationships are stored in data tables and collectively represent the logical network. The actual line and point features represent the geometric network.

- Network topology is created in ArcCatalog. Once established, it is updated and maintained during editing. Building networks requires an ArcGIS Standard or Advanced license.

- Networks may be composed of simple edges, which always end at junctions, or complex edges, which can contain junctions in the middle of the feature.

TIP: Weights are set up at the time the network is created. If no weights are listed in the drop-down boxes, then the network contains no weights, and none can be used.

TIP: Some tracing tools will not work if the flags or the barriers are of different types. For an analysis, try to always use edge flags and edge barriers together or junction flags and junction barriers together.

Important Terms

barrier	geometric network	network weights	source
edge	junction	sink	trace
flag	logical network	solver	

Chapter Review Questions

You may need to consult the Skills Reference section to answer some of these questions.

1. What kinds of feature classes can be used to create networks?
2. Explain the difference between a geometric network and a logical network.
3. What characteristic distinguishes a transportation network from a utility network?
4. Networks are composed of what two spatial elements? What is the role of each?
5. What are sources and sinks? In what type of network are they found?
6. What purposes do flags and barriers serve?
7. What are loops? Are they desirable?
8. By default, does the Find Path solver always find the shortest path? Explain.
9. What is the purpose of using network weights?
10. How do simple edges and complex edges differ?

Mastering the Skills

Teaching Tutorial

The following examples provide step-by-step instructions for doing basic tasks and solving basic problems in ArcGIS. The steps you need to do are highlighted with an arrow ➔; follow them carefully. Click on the video number in the VideoIndex to view a demonstration of the steps.

➔ Start ArcMap with the ex_9.mxd map document in mgisdata\MapDocuments.

➔ Use Save As to rename the document and remember to save frequently as you work.

1➔ Open the Catalog tab and expand the Folder Connections, if necessary.

1➔ Navigate to the mgisdata\Rapidcity folder.

1➔ Expand the rapidnets geodatabase.

1➔ Expand the Transportation feature dataset.

1➔ Right-click the Road_Net network and choose Properties.

1. What, if any, weights have been created for this network?____________________

1➔ Close the Transportation network properties.

1➔ Click the ArcCatalog tab, expand the Utilities feature dataset, and open the properties for the Water_Net network.

2. Which feature classes participate in the network, and what roles do they play?

____

3. What weights does the Utilities network have? ____________________________

1➔ Close the Utilities network properties.

Finding paths

We will begin by finding some simple paths through this transportation network.

TIP: Be sure to use the *Utility* Network Analyst toolbar, not the Network Analyst toolbar. Network Analyst is an extension that must be purchased.

2➔ Choose Customize > Toolbars from the main menu bar and select the Utility Network Analyst toolbar.

2➔ Make sure that the Network is set to Road_Net in the Utility Network Analyst toolbar.

2➔ Set the trace task to Find Path.

2➔ Click on the Junction Flag tool.

2➔ Click on the ends of two roads on opposite ends of town, as shown in Figure 9.18.

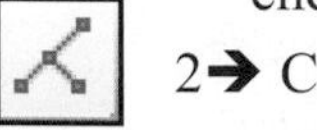

2➔ Click the Solve button.

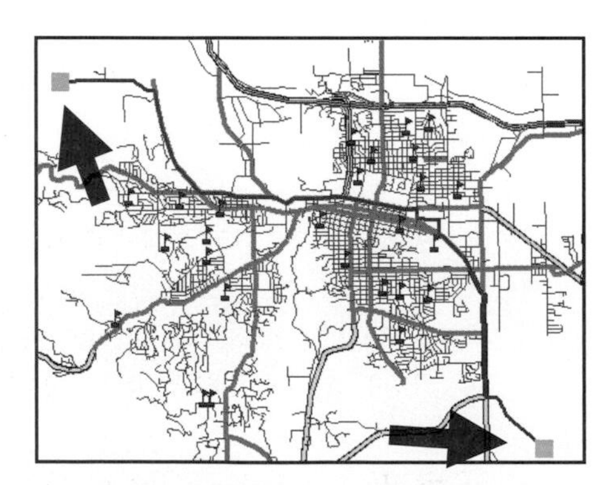

Fig. 9.18. Finding a path between two flags

TIP: Flags and barriers should match types. For best results, always use junction flags and junction barriers together or edge flags and edge barriers together.

The path solution appears as a bold red line connecting the flags. This path meets the criterion that it follows the minimum number of edges through the network. Since this is a transportation network, traffic can travel in either direction. Now let's try another one.

3➔ Choose Analysis > Clear Flags from the Utility Network Analyst menu.

3➔ Choose Analysis > Clear Results from the Utility Network Analyst menu.

3➔ Click the Junction Flag tool and add two flags at any locations you choose.

TIP: If the flag does not snap to the right location, clear the flags and start again. Zooming in will make it easier to snap to the right spot.

3➔ Click Solve.

4➔ Clear the flags and the results again using the Analysis menu.

4➔ Choose the Edge Flag tool, try adding a pair of edge flags, and then solve.

4➔ Clear the flags and the results.

TIP: Flags and barriers will be snapped to the closest feature within the snap tolerance set from the General tab in the Analysis > Options menu on the Utility Network Analyst toolbar.

Now let's explore the use of barriers to help find the best way to take a child to school. Imagine that you live on Harney Dr and that your child goes to Southwest Middle School.

5➔ Choose Bookmarks > School Trips 1 from the main menu bar.

5➔ Click the Find tool on the Tools toolbar.

5➔ Type Harney as the text for which to search, and set the layer to Roads. Click Find.

5➔ Hold down the Ctrl-key and click on each Harney Dr entry in turn to select them all. Don't select Harney Pl or the others.

5➔ Right-click on one of the entries and choose Select. Harney Dr will be selected.

5➔ Use the Find tool to locate Southwest and select the school in the Schools layer.

5➔ Close the Find window.

6➔ Zoom in to the extent of Harney Dr to more easily mark the right spot with the flag.

6➔ Click the Edge Flag tool.

6➔ Place a flag near the middle of Harney Dr, and use the Zoom to Previous Extent button to return to the original extent.

7➔ Place another edge flag on Park Dr next to the school, zooming as necessary.

7➔ Zoom back to see both flags, and make sure the trace task is set to Find Path.

7➔ Click Solve.

Well, that was probably the answer you expected. Perhaps you are wondering, however, how far it is from home to school.

8➔ Click the Solve button again without moving the mouse off it, and this time keep an eye on the lower-left corner of the map window to see the cost.

4. What is the cost of this trip and what are the units of the cost? ________________________

It would be more interesting to learn the distance. To find distances, you must use a weight.

8➔ Choose Analysis > Options from the Utility Network Analyst menu bar.

8➔ Click the Weights tab.

8➔ Set both the along and the against edge weights to Distance. Click OK.

8➔ Clear the results (but not the flags).

8➔ Click Solve again and watch for the cost.

5. What is the cost and what are the units? ________________________________

TIP: When specifying weights in a Transportation network, it is critical to specify a weight both FOR and AGAINST the direction of travel. If only one weight is chosen, then cost will accumulate only when travel is in the same direction as the line, and the total will be too small.

On the first day of school, you discover that an enormous traffic jam builds up at the intersection south of the school. You wonder if there is an alternate route through the tangle of residential streets north of the school and what the street names are. First, though, let's turn on Map Tips.

9➔ Right-click the Roads layer and choose Properties.

9➔ Click the Display tab. Check the *Show MapTips* box.

9➔ Make sure that the Display Expression Field is set to ROADNAME. Click OK.

With Map Tips on, use the cursor to hover over a street to find out its name. (In addition, the layer is set to label the streets at scales greater than 1:20,000.)

10➔ Clear the results (but not the flags).

10➔ Zoom in to the middle school to see it better and to see the road labels appear.

10➔ Click the Edge Barrier tool and place a barrier on Corral Dr between Park Dr and Sheridan Lake Rd.

10➔ Zoom to the previous extent when done.

10➔ Click Solve.

6. What is the distance of this alternate route? ___________________________

To make it easier to find the way tomorrow morning, make a list of the streets involved in this path. One way to get a list is to return the path as a selection instead of as a drawing.

11➔ Choose Selection > Interactive Selection Method from the main toolbar, and make sure that it is set to Create New Selection.

11➔ Switch to the List by Selection view in the Table of Contents. Click the toggle switches to make all layers selectable.

11➔ Set the Table of Contents view back to List By Drawing Order.

12➔ Choose Analysis > Options from the Utility Network Analyst toolbar.

12➔ Click the Results tab.

12➔ Change the results format to Selection, and uncheck the box for returning junctions. The edges will contain the street names. Click OK.

12➔ Clear the results (but not the flags or the barriers).

12➔ Solve again.

TIP: If no features are selected when solving a trace, check that the selection method is set to Create New Selection and that the network is one of the selectable layers.

13➔ Open the Roads attribute table.

13➔ Click the button to show only the selected records.

7. How many (different) street names are included on this path? ______________________

13➔ Set the results format back to Drawings in the Analysis > Options menu, and check the box to return junctions once more. Click OK to close the window.

13➔ Close the attribute table.

Finally, your other child attends Meadowbrook Elementary School. Normally she walks to school, but you want to plan a route to take both children on rainy days. In addition, you have agreed to pick up her friend who lives on Player Dr. Find a new path including all of these elements. The Middle School starts earlier than the others, so you plan to stop there first, pick up the child on Player Dr next, and then arrive at Meadowbrook. You need to enter the flags in the order of your planned stops (Fig. 9.19).

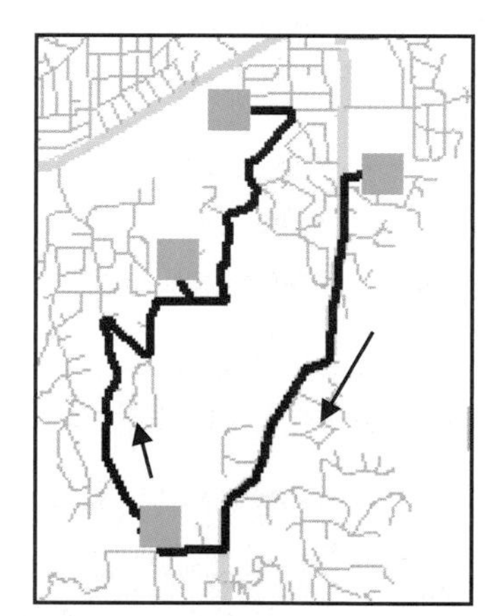

Fig. 9.19. Traveling a sequence of stops

14➔ Use Clear Selected Features from the Selection menu on the main menu bar to clear the trace.

14➔ Clear the barriers and the flags.

14➔ Place an edge flag on Harney Dr.

14➔ Place an edge flag next to Southwest Middle School.

15➔ Locate Player Dr using the Find button or Map Tips, and place an edge flag on it.

15➔ Place an edge flag next to Meadowbrook Elementary School (west and slightly north of Harney Dr).

15➔ Solve for the path. When finished, clear the results and the flags.

Advanced path problems

Combinations of selections, barriers, and other techniques can tackle more advanced path problems. To start, assume that you live on Red Dale Dr and work at the South Dakota School of Mines and Technology (SDSM&T). You decide to map a route to work.

16➔ Find Red Dale Dr and place an edge flag on the west end.

16➔ Zoom to the extent of the Roads layer.

16➔ Find SDSM&T (search for Mines), zoom in, and place an edge flag on E. St. Joseph St next to the school.

16➔ Zoom to the extent of the Roads layer again.

16➔ Solve for the path.

Now imagine that a doctor has placed you on a strict, low-cholesterol diet. Driving home from work in the evening, you find it nearly impossible to drive by a restaurant without stopping for something wonderfully greasy to eat. Try to find a route home that avoids all the restaurants in town. To begin, select all the road segments that are close to restaurants.

17➔ Choose Selection > Select by Location from the main menu bar.

17➔ Select roads that are within 100 meters of a restaurant.

17➔ Close the window after completing the selection.

Next, we will set the trace analysis options so that the selected roads act as barriers to travel.

18➔ Choose Analysis > Options from the Utility Network Analyst toolbar.

18➔ Click the General tab and fill in the button to *Trace on: Unselected features*.

18➔ Click OK.

18➔ Clear the previous trace results.

18➔ Solve for the path (Fig. 9.20).

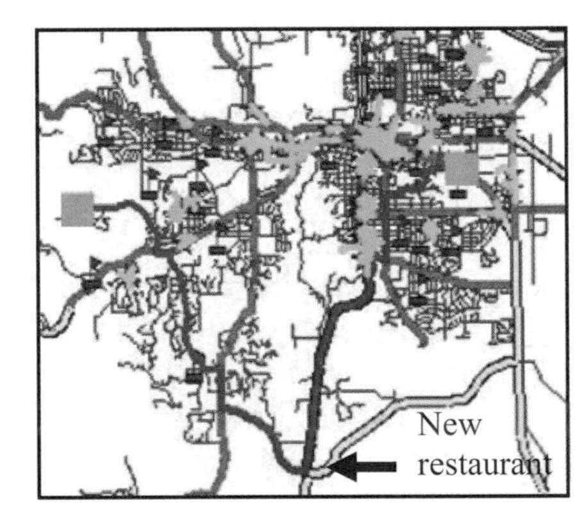

Fig. 9.20. Avoiding restaurants on the way home from work

Unfortunately, you discovered that last week a new restaurant with juicy burgers and crisp fries opened at the intersection of Catron Blvd and Highway 16. This new restaurant is not in the database, so enter barriers by hand.

19➔ Zoom in to the intersection of Catron Blvd and Highway 16 (Fig. 9.21).

19➔ Add two edge barriers to block off the intersection.

19➔ Zoom back to the previous extent.

19➔ Clear the previous result and solve for the path again.

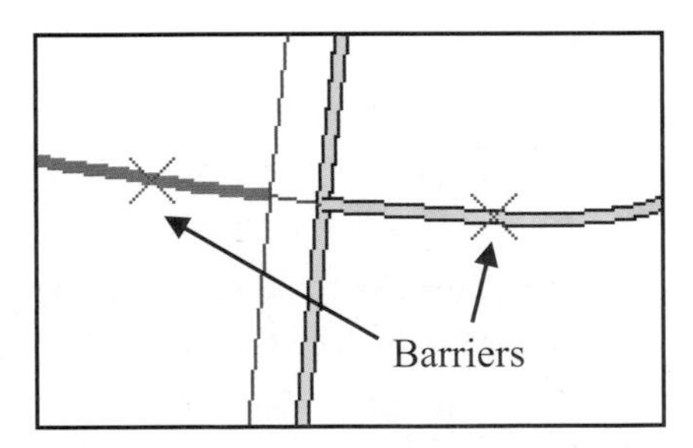

Fig. 9.21. Placing barriers on Catron Blvd

Looks like you can still get home, although the path is a little tortuous!

20➔ Clear the flags, the barriers, and the results.

20➔ Clear the selected set of roads.

20➔ Turn off the Restaurants and Schools layers.

Finding transportation network errors

Errors in topology are often created during the process of digitizing and creating networks. Once the network is built, the tracing tools can help you locate and fix these problems. One useful tool is the Find Disconnected solver. In a road network, it should be possible to get to any road. A disconnected road is usually the result of a failure to snap the end of a new road to the existing network or to split a road at a junction.

21➔ Zoom to the extent of the Road Network group layer.

21➔ Choose the Junction Flag tool and click to add a flag anywhere on the network.

21➔ Set the trace task to Find Disconnected and click the Solve button.

At first glance, no trace result appears. This network is in good shape. However, look closely at the north-south extension of the yellow interstate, I-190, for a small red glint.

22➔ Zoom in to this area for a closer look (Fig. 9.22).

22➔ Zoom in to the triangle with the two small red spots.

22➔ Zoom in to one of the spots until the red street fragment is clearly visible.

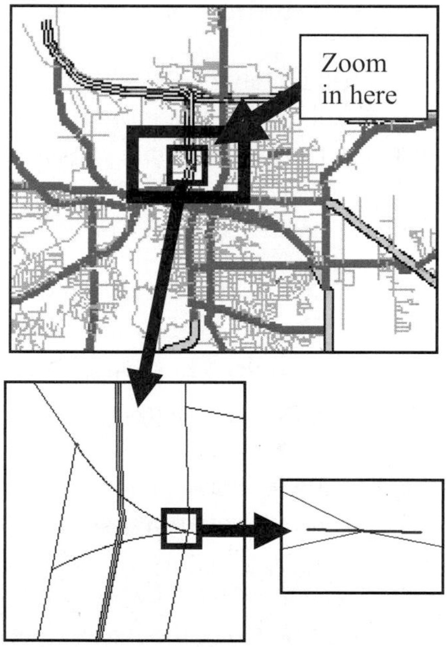

Fig. 9.22. Finding a disconnected road segment

Here's the problem. A tiny sliver of road, which is probably a processing error, lies across the intersection. You could use editing (Chapter 12) to delete the slivers. For now, we will simply go on; these small extra segments will not affect the behavior of the network.

We can also use the Find Loops on transportation networks.

23➔ Zoom back to the extent of the Road Network group layer.

23➔ Change the trace task to Find Loops.

23➔ Click the Solve button.

This time nearly the entire network is returned as the result. We can expect this because typically road networks DO let people drive in loops. (How many times have you driven around the block?) However, the roads that are not part of the result do have a common feature—they are all "dead end" streets. Of course, some of them end merely because they go outside the city limits. Generally, finding connections and loops is not very useful with transportation networks.

Tracing on utility networks

Although finding driving routes is amusing, utility networks also have interesting tracers. Now we will take a look at a portion of a hypothetical water utility system for east Rapid City.

24➔ Clear the flags and the results.

24➔ Turn off the Road Network group layer and collapse it using the minus box.

24➔ Turn on the Water Network group layer.

24➔ Zoom to the Water Network group layer.

24➔ Change the network to Water_Net in the Utility Network Analyst toolbar.

24➔ Examine the network.

This network contains more elements than the roads network. The water mains themselves form the edges. In addition, there are galleries, end caps on the pipes, and T-valves at the pipe junctions. All of these features participate in the network.

25➔ Right-click the Galleries layer and choose Open Attribute Table.

The attribute table contains three fields. The Name field simply indicates the name of the gallery. The Enabled field is a part of the network setup. The value True indicates that the galleries are turned on and functioning as part of the network. The AncillaryRole field says Source, indicating that the galleries are water sources for the network. They are underground chambers that pull water in from Rapid Creek, filtering and treating it before it enters the water lines.

25➔ Close the attribute table.

26➔ Choose Bookmarks > Pipes from the main menu bar.

26➔ Choose Flow > Display Arrows from the Utility Network Analyst toolbar.

Arrows appear at the center of each edge, indicating the direction of flow. Here's how to edit the symbols used for portraying flow.

26➔ Choose Flow > Properties from the Utility Network Analyst toolbar.

26➔ Click the Arrow Symbol tab.

There are three different conditions of flow possible in the network. Uninitialized flow means that the flow direction has not yet been established for the network. Indeterminate flow means that flow has been initialized but cannot be determined for that section. Determinate flow indicates that flow has been established.

26➔ Click on each flow type to examine the symbol used to display it.

8. What is the symbol for Indeterminate flow? ________________________

27➔ Click the symbol for Determinate flow. The Symbol Selector appears.

27➔ Change the symbol color to blue. Click Apply. The map arrows change color.

27➔ Change the symbol for Indeterminate flow to a 12-point pink square. Click Apply.

Like other features in ArcMap, a display scale can be set so that the networks are not cluttered with arrows at small scales.

28➔ Click the Scale tab (still in the Flow Display Properties).

28➔ Fill the button for *Don't show arrows when zoomed out beyond* and enter 30,000. Click OK.

28➔ Zoom to the extent of the Water Network group layer to check the scaling. The arrows should disappear.

28➔ Return to the previous extent.

Notice that the arrows take a few moments to draw. Turn them off for now.

29➔ Choose Flow > Display Arrows again to turn the symbols off.

Let's explore this network more by using some of the tracing tools. Let's start by seeing which parts of the network are connected to each gallery.

29➔ Zoom to the extent of the Water Network group layer to show the entire network.

29➔ Choose the Junction Flag tool and click to add a flag at the Civic Center gallery.

29➔ Set the trace task to Find Connected. Click the Solve button.

Now it is easy to see which areas of town are served by the Civic Center gallery. We can use Find Disconnected to highlight the other pipes that are not connected to this gallery.

30➔ Clear the results and change the trace task to Find Disconnected.

30➔ Click the Solve button.

Next, let's use the Find Loops tracer to see if there are any loops in this network.

31➔ Clear the results.

31➔ Add another junction flag to the other gallery so that both subnets are examined for loops.

31➔ Change the trace task to Find Loops and Solve.

31➔ Zoom in to the single red loop result that appears (Fig. 9.23).

31➔ Choose Flow > Display Arrows from the Utility Network Analyst toolbar.

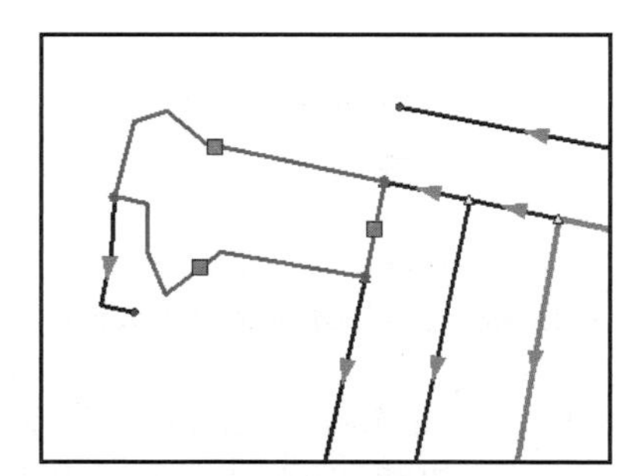

Fig. 9.23. Flow in a loop is indeterminate.

Note that all of the lines except the loop have blue determinate flow arrows. The loop has the pink square indeterminate flow symbols. The water could go around the loop in either direction. Fixing this would require editing the loop to break the connection at one point.

32➔ Clear the flags and the results.

32➔ Zoom to the extent of the Water Network group layer.

One useful benefit of using network models is the ability to predict what areas of town will experience service disruptions due to repairs and outages. The city has been planning to repair a leaking pipe on St. Patrick St between 6th and 7th Ave. They will need to turn off the flow to the pipe for several hours, and they want to notify the customers who will be affected.

33➔ Turn on the Road Network group.

33➔ Use Select by Attributes to select the roads where [STREET] = 'ST PATRICK'. Click Apply rather than OK to keep the selection window open.

33➔ Change the Selection method to Add to Current Selection.

33➔ Select the roads where [STREET] = '6' OR [STREET] = '7'.

33➔ Close the Selection window.

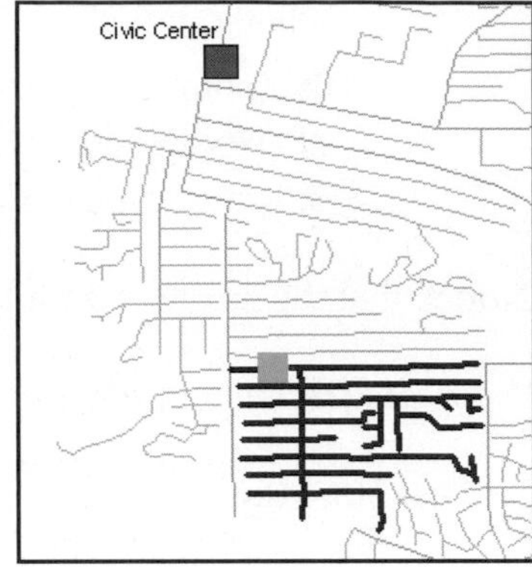

Fig. 9.24. Finding a service disruption due to repairs on St. Patrick St

34➔ Zoom in and add an edge flag to the segment of the water line that follows St. Patrick St between 6th and 7th. Then return to the previous extent.

34➔ Change the trace task to Trace Downstream.

34➔ Click the Solve button.

Now you know where to notify customers (Fig. 9.24). If this network were complete down to the building connections for each parcel, then the utilities company would be able to automatically compile a mailing list to send to the affected customers.

The Trace Upstream tool performs the opposite task, finding the path that water has traveled to get to a particular location.

35➔ Clear the flags and the results.

35➔ Clear the selected features.

35➔ Use the Find tool to locate and select the streets N. and S. Grand Vista Ct. They are on the middle of the west side of the network (Fig. 9.25).

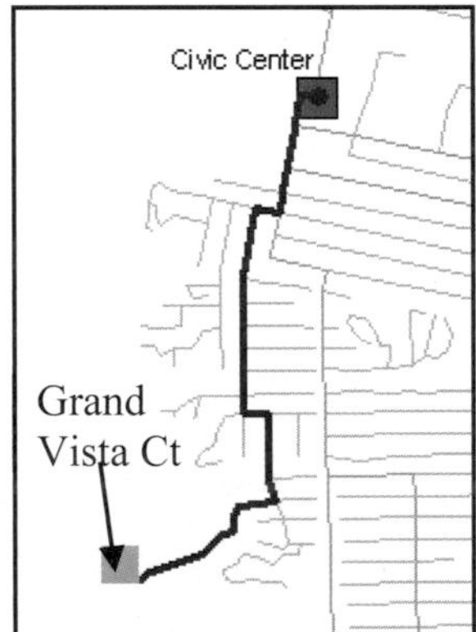

Fig. 9.25. How water gets to Grand Vista Ct

Tracing can also be done with accumulation, for example, to find the total distance water travels to get to a particular location.

36➔ Zoom in to the location of these two streets.

36➔ Place a junction flag on the end of the water main serving these streets, and return to the previous extent.

36➔ Change the task to Find Upstream Accumulation.

36➔ Click Solve and watch for the cost in the lower-left corner of the window.

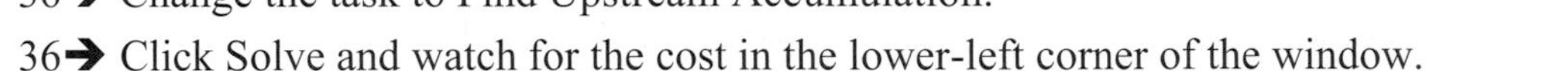

9. What is the cost in edges to get from the Civic Center to Grand Vista Ct? __________

37➔ Choose Analysis > Options from the Utility Network Analyst toolbar.

37➔ Click the Weights tab and set both the along and against weights to Distance. Click OK.

37➔ Click the Solve button again.

10. How far does the water travel to get there? _________________

There is no downstream equivalent to tracing with accumulation. However, we can use the selection result option to select the downstream features from a point and determine the total length of pipe beyond that point. First, we must make sure that Waterlines is a selectable layer.

38➔ Switch to the List By Selection option on the Table of Contents.

38➔ Check the selection toggle to make sure that Waterlines is selectable.

38➔ Switch back to View By Drawing Order.

39➔ Clear the flags and results.

39➔ Place a junction flag at the East Rapid gallery.

39➔ Choose Analysis > Options from the Utility Network Analyst menu.

39➔ Click the Results tab and change the result format to Selection. Click OK.

39➔ Change the trace task to Trace Downstream and Solve.

Now that the water lines are selected, it is easy to get their total length.

40➔ Open the attribute table for the Waterlines layer.

40➔ Right-click the Shape_Length field and choose Statistics.

11. What is the total length in kilometers of pipe served by this gallery? _______________

40➔ Close the Statistics window and the attribute table.

40➔ Clear the selected set.

This is the end of the tutorial.

➔ Close ArcMap. You do not need to save your changes.

SKILL TIP: Learn how to create a geometric network if you have an ArcGIS Standard or Advanced license (Files and Geodatabases).

Exercises

1. Find the shortest path from Hidden Timbers Dr to Stevens High School. **Capture** a map showing the route.

2. What is the distance covered by the route in Exercise 1?

3. What is the shortest alternate route if you want to avoid the before-school congestion on Park Dr by Southwest Middle School? **Capture** the map. How long is the new route?

4. The cafeteria at Stevens High School makes daily lunches for several elementary schools in town. They are delivered in this order: Pinedale, South Canyon, Upper Rapid, Canyon Lake, Meadowbrook, and Cleghorn. Find the most efficient route to make the deliveries. **Capture** the map.

5. You are trucking a load of Canadian hemp from a distributorship on Olde Orchard Rd to a cowboy rope-making factory on Stacy St. Because of a bizarre technicality in the drug laws, you can be arrested if the truck is stopped within 250 meters of a school. Find a path to the destination that avoids this risk. **Capture** the map.

6. A construction worker accidentally breaks a water line at the intersection of E. St. Patrick St and Ivy Ave. Create a map showing the area of disrupted service. **Capture** the map.

7. Of the two water mains leaving the Civic Center gallery, which serves a greater length of pipes? What is the total pipe length for each main (north and south)?

8. What is the total length of water *mains* served by the East Rapid gallery versus the Civic Center gallery? (Do not include laterals or other water line types.)

9. The water main that ends at 5th St and Cathedral Dr/Fairmont Blvd passes through how many T-valves?

10. Joe did Exercise 9 with Selection as the result format and got 26 *selected* features. Mary did the same problem and got 51 *selected* features. Explain why they got different numbers. What is the cost in each case?

Challenge Problem

Find a route that goes from Alta Vista Dr to the east end of E. Knollwood St and that travels ONLY on local streets. (**Hint:** You cannot ignore the role of junctions in this problem.) **Capture** your map.

Chapter 10. Geocoding

Objectives

- Understanding the geocoding process
- Knowing the different styles of geocoding
- Setting up an address locator service
- Geocoding using several styles of locators

Mastering the Concepts

GIS Concepts

What is geocoding?

As everyone knows, a street address tells a postal worker where to deliver mail. It contains a type of spatial information. However, the postal worker requires additional knowledge to get the letter to the house. Unlike a latitude-longitude point, which uniquely locates a house on the globe, a street address requires information about the city and state, the location of the street, and the sequence of house numbers. A newly hired postal worker delivering a piece of mail in a strange city would need a good map in addition to the address.

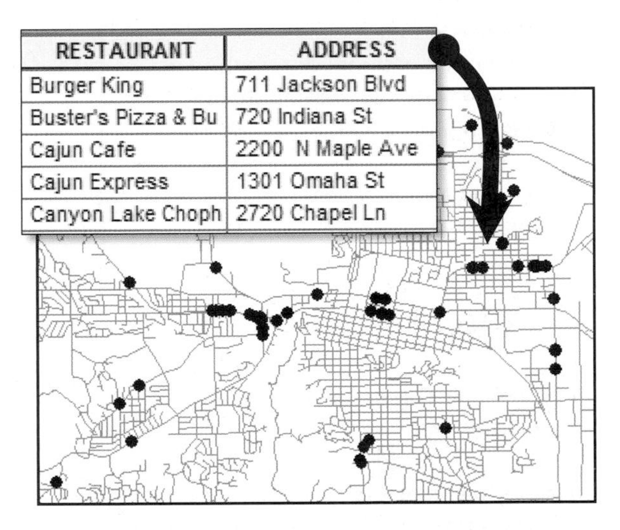

RESTAURANT
ADDRESS
Burger King
711 Jackson Blvd
Buster's Pizza & Bu
720 Indiana St
Cajun Cafe
2200 N Maple Ave
Cajun Express
1301 Omaha St
Canyon Lake Choph
2720 Chapel Ln

Fig. 10.1. Geocoding can convert restaurant addresses in a table to points on a map.

Likewise, geocoding combines map information with street addresses in order to locate a point uniquely. It enables someone to convert a list of addresses into points on a map (Fig. 10.1). Geocoding has many uses.

911 addressing

One critical use being developed by many governmental agencies concerns finding street addresses given over the phone during an emergency call in order to quickly route response vehicles to the location. Routing of vehicles is accomplished by providing a map or a global positioning system (GPS) location in vehicles equipped with GPS units. Because lives are at stake, this application requires highly accurate address information, and it is a challenging application to implement.

Driving directions

Many people have gone to one of the popular web sites, such as MapQuest or Google Maps, to find directions from a starting address to a destination. The results include step-by-step driving directions and a map showing the route. Other people may have tried the GPS systems installed in cars, such as OnStar or Magellan, which guide people to their destination from their current position as identified by GPS. Both of these applications are built upon geocoding technology.

Crime analysis

Police records of crime incidents include the address, and these locations can be converted to points on a map through geocoding. Analysts could use these data to help identify "hot spots" in order to help plan patrol routes for officers or to target efforts to develop neighborhood watch programs. Police departments can glean information about spatial patterns for different types of crime or target the time of day that particular crimes tend to occur. Such information can be critical in developing strategies to reduce crime and to protect residents and businesses.

Marketing

Many companies collect address information about their customers as a regular part of conducting business. Converting addresses to locations and analyzing their distribution might help a company provide better service by opening new stores in underserved areas. Mail-order businesses can locate areas with low sales for targeted marketing efforts. Sales and service companies can better plan where to place agents to serve their clients. Recently, a study helped a university in South Dakota to understand the national distribution of its students, and it is helping them to plan marketing strategies to increase national enrollment (Fig. 10.2).
A whole host of business applications is available once information about customer locations is known, and many companies now routinely use GIS in their operations.

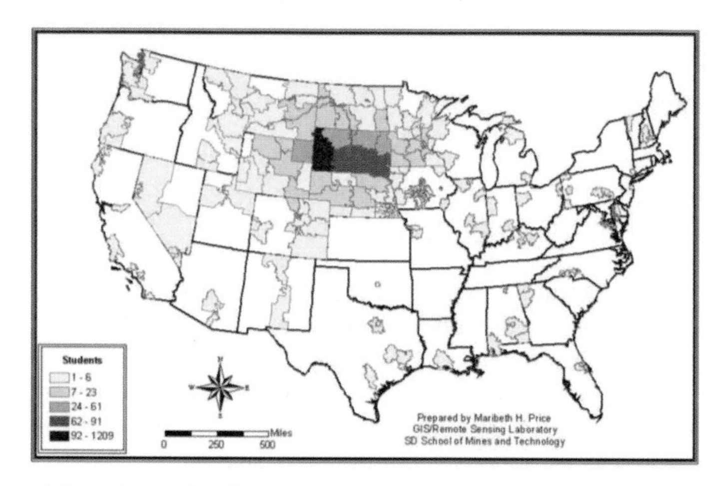

Fig. 10.2. Distribution of students attending the South Dakota School of Mines and Technology in 2004

Public Land Survey System locations

Many parts of the United States, particularly out west, utilize the Public Land Survey System, or PLSS, as a means of locating points of interest. In the days before GPS units, field workers often located themselves on a topographic map and marked field notes or sample locations by means of the township, range, section, and section quarters, rather than units of latitude or longitude. Property descriptions also use this format, and they are still common terms in western agriculture.

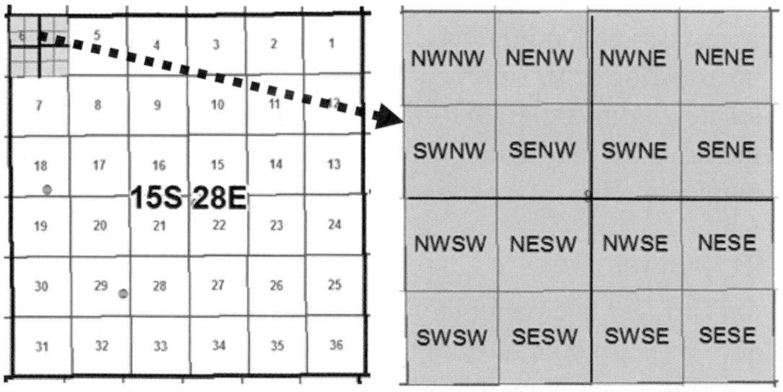

Fig. 10.3. A township, sections, and quarter-quarter sections

The basic unit of PLSS is the township, a 6-mile by 6-mile square divided into 36 one-mile square sections (Fig. 10.3). The township is referenced by its position relative to a surveyed east-west baseline and north-south meridian, defined for the state or region. Figure 10.3 shows a township that is the 15th one south of the baseline and the 28th one east of the meridian for Oregon. Notice that the sections are numbered in alternate rows: 1–6 from east to west on the first row and then 7–12 from west to east on the second. The section in the upper-left corner has been divided once into quarters (quarter-sections NW, NE, SW, and SE), and those quarters divided into quarters again (quarter-quarter sections also designated NW, NE, SW, SE). The full designation for the upper-left quarter-quarter section is T15S, R28E, S6, NWNW. The quarter-quarter section designation comes first, then the quarter-section.

A GIS practitioner may need to convert PLSS locations from old index cards or field notebooks or legal descriptions to points in a geodatabase. Geocoding provides a quick and straightforward method to convert descriptions to points once the descriptions are entered in a spreadsheet.

However, the geometric accuracy of these methods is much lower than even the cheapest GPS. A section is a mile on the side, a quarter section is a quarter-mile, and a quarter-quarter section, the smallest unit typically used, is a 16th of a mile, or 330 feet. Geocoding finds the right polygon unit and places a point at its center. So if you locate a fossil or a rock sample to the quarter-quarter section, you still have a large area to hunt through for its source (40 acres!).

TIP: If you are not already familiar with PLSS history and terminology, a practical review can be found at Wikipedia.org.

How does geocoding work?

Just as a stranger requires a map to find an address in a city, the computer requires map data in order to turn an address into a location. This spatial data is called the **reference layer**. Geocoding typically uses a street layer composed of polylines with specific attributes indicating each street's name and the address ranges present on that street. However, other types of geocoding can use point or polygon reference data, such as matching store franchises to points representing cities or matching a list of owner addresses to parcel polygons.

Although humans automatically recognize a variety of address styles, a computer requires specialized software to interpret addresses. An address might be written 123 Maple Street, or 123 Maple St, or 123 N. Maple St, or even 123 Maple, and the postmaster would have no difficulty finding the house. To computers, however, each of these addresses is different. Geocoding programs must interpret different styles of addresses and manage typographical and other errors, such as someone entering Maple Rd instead of Maple St. They accomplish this task by dividing the address into different components and then scoring how well they match the reference components. A high score (> 80) indicates a good match, a medium score (60–80) indicates a tentative match, and a low score (< 60) suggests a poor match.

To interpret an address, a geocoder begins by parsing the address into individual fragments of information. The address "235 W. Main Str" would be divided into four pieces: the address number (235), the direction prefix (W.), the street name (Main), and the street type designation (Str). After separating the components, each fragment is standardized so that it has the best chance to match the information in the reference layer. All letters may be capitalized (MAIN), periods removed (W), and common abbreviations reduced to a single standard (ST). This process is called **address standardization**. This step is necessary because a computer, confronted with the problem of comparing (Str) in the address table to (ST) in the reference layer, will conclude that the two do not match. A human would hardly notice the difference, but computers can be tiresomely literal. An address may include these components:

Prefix direction. A direction indicator that comes before the street name, such as *E* Maple St or *North* Main St. Some towns and cities use prefixes for directions; others use suffix directions (e.g., Main St North). This part may have any of the standard directions, N, E, S, W, NE, SW, SE, and NW. The complete spellings North, East, South, and West are also recognized.

Prefix type. An indicator of the road type that comes before the street name, such as *Highway* 85. In this case, 85 is the street "name" and Highway, or Hwy, is the prefix type.

Number. The address number, such as *123* Maple St

Street name. The name of the street, such as 123 *Maple* St or *Mount Rushmore* Road

Street type. A type indicator that comes after the street name, such as Main *St* or Maple *Ave*. The software recognizes a variety of street types (Street, Lane, Place, Road, etc.) and understands most common abbreviations, such as St, St., Rd, Av, and Ave.

Suffix direction. A direction indicator that comes after the street name, such as Florman St *E* or 87th St *North*. This part may have the same values as the prefix direction.

Zip code. The standard U.S. zip code or, optionally, the Zip+4 nine-digit codes

Table 10.1. Examples of address standardization.

Address	**Prefix dir**	**Prefix type**	**Number**	**Street**	**Type**	**Suffix dir**
123 Maple Road			123	MAPLE	RD	
15 Center St East			15	CENTER	ST	E
314 Hwy 85 N		HWY	314	85		N
234 E. St. Patrick St	E		234	SAINT PATRICK	ST	

Table 10.1 shows some examples of addresses and their standardizations. Note that, in addition to separating the address into components, the text is converted to capital letters to facilitate comparisons in case-sensitive databases. The complete words in the directions and types are also converted to standard abbreviations, so that Road becomes RD and the periods are omitted.

Once the address is standardized, the components are compared to attributes in a spatial data layer, such as a line shapefile containing the city streets. This reference layer must contain certain attribute fields to compare with the standardized components. The score for each component match is averaged to give an overall match score for the address, in terms of percentage. A score of 100 indicates a perfect match. A zero indicates that none of the components match. Usually, a score of 80 or better is considered a good match, and a point will be created for the address. When the score is lower than 60, the geocoding software looks for other possible candidates. Typically, candidates have scores above 30. The user can examine the candidates individually to find out if any of the candidates might provide a reasonable match.

Figure 10.4 shows an example of a street feature class used for geocoding. Notice that most of the fields described are present, although the names are not the same. Also notice that, instead of a single address, the street has two fields (LEFT_FRADD and LEFT_TOAD) that describe the range of addresses that fall along the left side of the street, and another pair of fields for addresses on the right side. This reference layer can place an address on either the left or right of the street. The FRADD field is the From address and contains the lower range value; the TOADD field is the To address and contains the higher range value.

Streets

PREFIX_DIR	PRE_TYPE	STREET_NAM	STREET_TYP	SUFFIX_DIR	LEFT_FRADD	LEFT_TOADD	RIGHT_FRAD	RIGHT_TOAD
		NORTHLAND	DR		3301	3345	0	0
		CEDAR BEND	DR		1605	1617	1610	1616
		NORTHLAND	DR		3051	3225	0	0
		BRECOURT MANOR	WAY		7621	7637	7620	7632
		CARENTAN	DR		7701	7709	7700	7708
N	IH	35	SVRD	NB	0	0	3213	3223
N	IH	35	SVRD	NB	0	0	3201	3211
		E 32ND ST TO N IH 35	RAMP		0	0	0	0
		DECKER	LN		10714	10806	10715	10807
		DESSAU	RD		10206	10214	0	0
		DECKER	LN		10808	11016	10809	11017

Fig. 10.4. The attribute table of a shapefile that could be used as the basis for geocoding

To create a point feature based on an address and street reference layer, the following logic is used. The number of the address is tested to see if it falls within the range of addresses on that street section. If it does, the point locating the address is placed along the street in proportion to its value between the range endpoints. In Figure 10.5, for example, Meadowlark St has an address range of 700 to 799. If matching the address 750 Meadowlark St, the point feature would be placed halfway between the street ends because 750 is halfway between 700 and 799. Likewise, 725 Meadowlark St would be placed a quarter of the distance along the street. This method of placing points is called the **One Range** method because a single address range is used for each street.

For more advanced geocoding, a **Dual Ranges** method can be used. In this method each street has four attributes defining its address ranges, a Left From and Left To address range for the left side of the street and a Right From and Right To address range for the right side. When using the Dual Ranges method, the addresses will be located on the left or the right side of the street, as dictated by the address ranges. In addition, the user can specify an offset distance such that the actual point is placed slightly to the side of the street instead of directly on top.

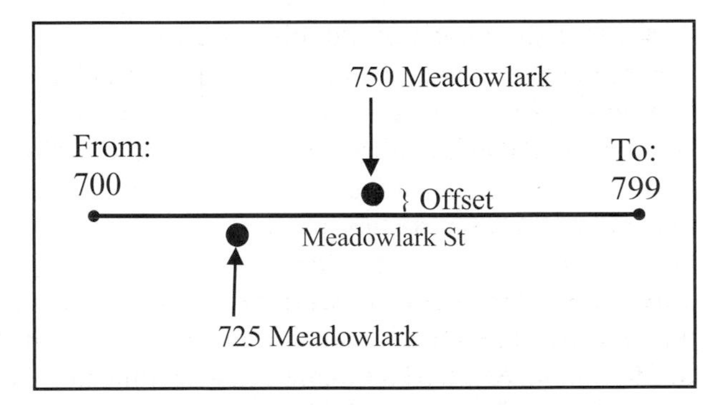

Fig. 10.5. Placing a located point along the street based on its proportional distance from the ends

The assumption that a house or other structure is located along a street segment at a distance proportional to its address number does not always work as well as it should. Parcel sizes may vary along the street, or some address numbers may be skipped entirely. This problem occurs most severely in rural areas where the spacing of addresses may be very irregular. If Meadowlark

St in Figure 10.5 represents a mile-long section line and all the houses are clumped together at one end, then the actual location and geocoded location may be quite different. Geocoded locations should always be considered approximations.

One also encounters great differences in the accuracy of the reference data used for geocoding. Large national data sets, such as TIGER, are generally less accurate and up to date than locally produced data intended for city governments and 911 applications. High-growth areas may experience difficulty keeping reference layers up to date. Users need to be aware of these potential pitfalls when geocoding. Creating and maintaining reference layers is a time-consuming and expensive process. Fortunately, the advent of cloud-based geoprocessing services has provided access to some national and international geocoding services through ArcGIS Online. These resources may not be as accurate as high-quality local data, but they are free and cover large areas in a standard way.

One Range and Dual Ranges are examples of different geocoding styles. The style chosen will depend on the way in which the streets layer has been set up for geocoding. Other styles are also available, as described in the next section. The user must ensure that the attributes in the reference layer support the desired geocoding style.

Although geocoding is most often associated with placing addresses, it applies to many types of data. Points can be located by city and state, by zip code alone, or even by township/range information. These different styles of geocoding have two common themes: they convert attributes to point locations using a reference layer, and they allow the user to link data that are similar but are not necessarily an exact match.

Available geocoding styles

Geocoding takes a variety of approaches, from finding a specific house location on one side of the street to simply finding a city and country on a world map. The available options depend on the attributes of the reference layer, the setup in the address table, and the type of matching to be done. A **geocoding style** specifies what type of matching will be done and defines a set of required attributes in the reference layer. To choose an appropriate style, the user must be familiar with the organization of the reference data.

Single Field

A single field locator uses one field to match to the reference layer, which typically contains points or polygons, although lines could also be used if desired. Any single field, such as the zip code, a state name, or a place name, can be used as the key to match the locations. This method expands geocoding from addresses to *any* type of information. For example, a table might contain information about petroleum wells keyed to specific well IDs, such as “TenRun Hole No. 6.” Such IDs may change slightly from database to database, depending on who typed them in (Ten Run Hole #6 or Ten Run Num 6). Performing an attribute join to link the table with a point layer of well locations would yield many missing values because the matches are not exact for each ID. Geocoding would work better because it can link close matches, not just exact matches. Single field locators are also used for township-range locations.

U.S. Alphanumeric Ranges

Many people are familiar with using grid zones to locate streets on a map by consulting a list of streets, finding its zone (such as H7), locating the zone by means of letters and numbers on the map margin, and scanning the streets in the zone to find the one in question. This geocoding style uses this approach to reduce the search time while geocoding. Each address is attributed with its zone identification, and the geocoding algorithm need only search inside that zone. The required fields are the grid zone, the street name, and the From and To ranges for both sides of the street.

U.S. Cities with State

This style accommodates national scale matching of location based on the city and state names only. Cities may be matched to points or polygon features. This style could create a map showing the locations of customers across the United States. The required fields are the city name and a state name or state abbreviation.

U.S. Single House

This style uses points or polygon reference data set up so that each address corresponds to one point or polygon (Fig. 10.6). For example, it could be used with points representing buildings or polygons representing parcels. The required fields include street number and street name. The user may also include prefix directions, suffix directions, and other address components.

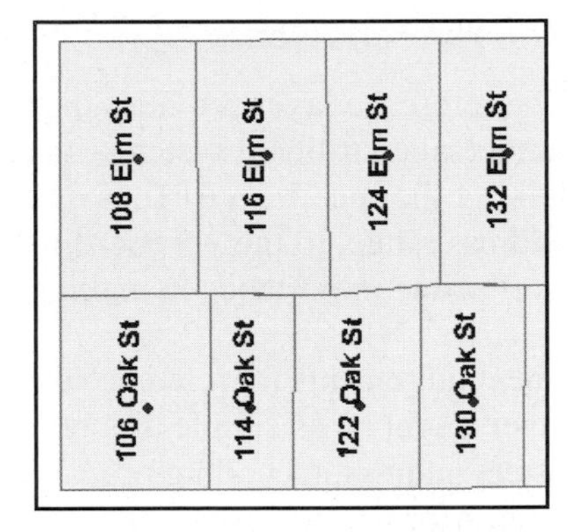

Fig. 10.6. The US Single House style codes to point or polygon layers, such as these parcels.

U.S. Streets–One Range

This style uses one range of addresses along a street rather than having ranges for both the left and right sides of the street. The required fields are the street name, the From address, and the To address. Prefixes, suffixes, and street types may also be used if desired.

U.S. Streets–Dual Ranges

This style relies on a reference layer with dual ranges of addresses, one for each side of the street. The required fields include the street name, a From and To address for the left side of the street, and a From and To address for the right side. Prefixes, suffixes, and street types may also be used if desired.

U.S. Hyphenated

This style follows conventions in locations, such as Queens, New York, where housing locations are indicated by a cross-street designation followed by the number and the street of the house. For example, the address 72-40 38th Ave specifies a location at 40 38th Ave near the cross street 72nd.

World Cities with Country

This style allows matching of a city and a country on a world map. The reference layer may contain cities as points or as polygons. The required fields include the city name and the country name or abbreviation.

Zip 5-Digit and Zip+4

The Zip style matches zip codes to points or polygons in a zip code reference layer. The only field required is the 5-digit zip code. The Zip+4 style works the same way but has an additional separate field containing the 4-digit extension.

With Zone

Many of the above styles offer a zone option. A zone is an additional piece of information in the reference layer, such as state, city, or zip code, which is added to the other required fields for that style. The addition of the zone allows matching over larger areas. For example, the US Streets style can only find addresses within a single city because only the street address is included; there is no city name to distinguish between Main St in Rapid City and Main St in Sioux Falls. To match street addresses across all of South Dakota requires a city name or a zip code as a zone, provided that such a field exists in the reference layer and the address table.

The reference data

The reference layer provides the key link that allows the interpretation of an address to a more universal coordinate system—the Cartesian *x-y* coordinates of a map projection. It does this by linking the specific attributes of a spatial feature stored in a table, such as the street name or the address range, to the *x-y* coordinates stored for the feature. As such, the validity of the attributes is of prime importance in making this translation.

The requirements for reference data vary widely with the type of geocoding being done. A reference layer might be as simple as a point layer of cities with states or zip codes in the attribute table. Street address geocoding probably has the most stringent requirements because it needs address ranges for one or both sides of a street. Such information is tedious to gather and requires significant preparation and maintenance time to ensure that it remains up to date and has minimal errors. U.S. Census data, or TIGER data, are the most common starting point for assembling reference layers, although they require a fair amount of conversion, assembly, and error checking before they are ready to use. Reference data are also available from a number of commercial vendors, with considerable variability in completeness, quality, and price.

About ArcGIS

Before starting, the user must have access to an address locator. In some cases, the administrator provides the locator with a geodatabase. Recently, online geocoding services have become available and can be used for free. However, if nothing is available that fits the need, the user must create a locator using suitable reference data.

Setting up an address locator

Before geocoding, the user must create an **address locator** that specifies the style, the reference layer, and the fields used. It takes a snapshot of the reference data and stores it within the locator. Locators may be copied and moved to new locations without losing the source data. Once an address locator is created, it can be used many times.

A locator may contain several sources of reference data. The Primary table contains the main reference layer with the features and associated address component fields. Every locator must have a Primary table. Two additional tables may be specified to aid geocoding (Fig. 10.7). An Alternate Name table provides nicknames or dual names for features. For example, HWY 192 may have another local name, such as Hightstown Road. The table will ensure that addresses are coded regardless of which name is used. An Alias table defines an actual address for an informal place name, such as the White House, so that users can find the location either by place name or by the actual address, such as 1600 Pennsylvania Ave.

rc_alternate

NAME	ALTERNATE
79	SW Connector
8th Street	Mt. Rushmore Rd
79	Sturgis Rd
190	Hwy 16
44	Omaha St

rc_alias

ALIAS	ADDRESS
Rushmore Mall	2100 N Maple Av
City Hall	300 6th St
County Courthouse	315 St Joseph St
Public Library	610 Qunicy St
Rushmore Plaza	444 Mt Rushmore Rd

Fig. 10.7. An Alternate Name table (top) and Alias table (bottom)

Address locators are created using ArcCatalog. First, the geocoding style must be designated. Figure 10.8 shows the settings for creating a Dual Ranges locator. A single streets reference layer is assigned and must be designated the Primary table. Supporting information in an Alias table may also be added at this time. The fields preceded by an asterisk (*) are required; the others are optional.

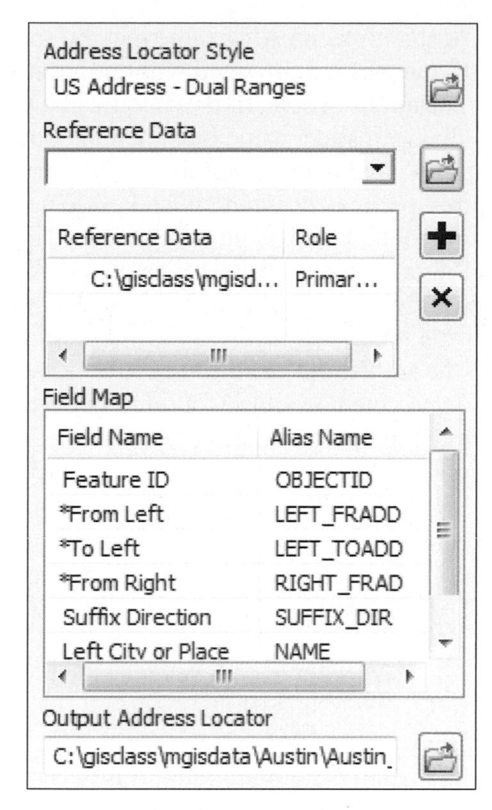

Fig. 10.8. Defining the reference layer for a new address locator

Next, the user must enter the fields from the reference data that correspond to the address components needed for the geocoding style. The fields shown in the setup will vary depending on the style of locator requested. This Dual Ranges style requires many different fields. The Single Field style requires only one.

Finally, the user specifies a name and location to store the locator. Locators may be stored in a folder or inside a geodatabase.

Geocoding

Geocoding starts with a table of addresses to be matched. The table may come from a dBase file, a geodatabase, or even Excel. At minimum it must have an address field and any other required fields for the style of locator being used. Geocoding may be performed live in ArcMap or using the Geocoding tools from ArcToolbox.

First, the user specifies the fields in the address table that will be used for geocoding (Fig. 10.9). If the common names are used for the table, as shown, then they will be recognized and included by default, so the user only has to change them if they are incorrect.

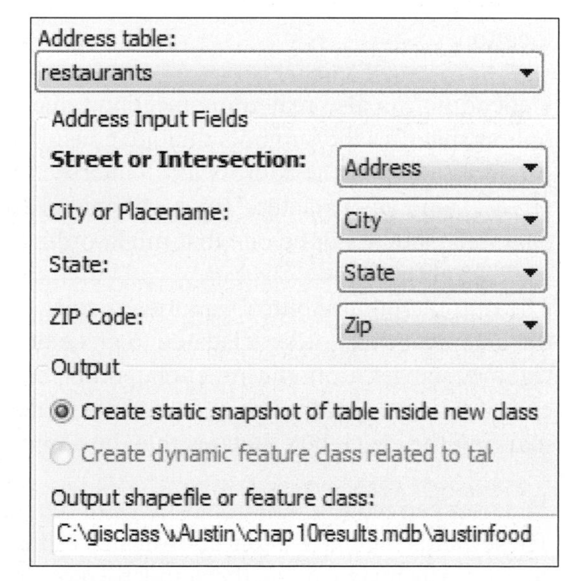

Fig. 10.9. Starting to geocode

The user must specify a shapefile or feature class to contain the output points and table. Usually, the default static option is used.

Geocoding options

Next, the user has the chance to set various geocoding options that control how the locator decides on candidates and how it structures the output (Fig. 10.10). These options may also be adjusted later if necessary. Most users will be happy using the defaults at first.

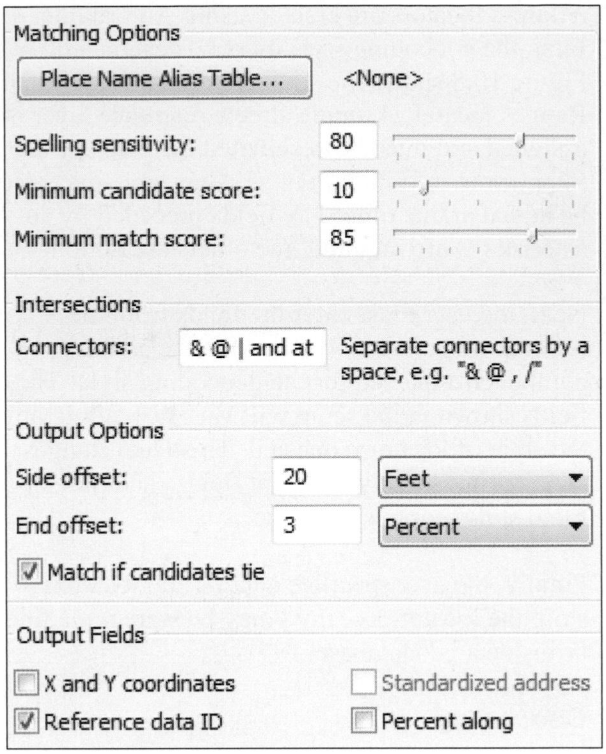

At this time, an Alias table may be specified if one was not provided with the locator. The **spelling sensitivity** specifies how closely a street name must match to be a valid candidate. With this option, small misspellings or typing mistakes do not prevent a match from being made. Thus, Pierce St would still be recognized even if it were spelled Peirce St.

The **minimum candidate score** is used to determine how closely a reference feature must match the address to be considered a possible candidate for a match. Generally, a low value is desirable to aid in finding candidates for mistyped addresses, but if the user finds that too many candidates are being generated, the score can be raised.

Fig. 10.10. Geocoding options

The **minimum match score** is the lowest score that will constitute a match. In applications in which an occasional mismatch is not important, it is acceptable to lower the default in order to match as many records as possible. To a company analyzing customer locations, a few wrong addresses are not likely to significantly affect the analysis. In other cases, the minimum score might be raised. To an emergency management team, a poorly matched address may be a matter of life and death. In this case, poor matches should be identified and individually considered by the dispatcher to avoid sending the team out to a wrong location.

Geocoding can also match intersections, such as the corner of 5th St and St. Joseph St. This option specifies the symbols that are considered to be connectors so that the address locator can properly interpret the address as an intersection rather than a regular address. The default values are **&**, |, and @ characters, but additional characters may be used. However, the separation character should not be one that might ordinarily appear in an address.

The output options control parameters related to the output file. First, for the Dual Ranges style, the user can set an offset distance to place the locations slightly off the street to the left or to the right. A setback from the street endpoints may also be specified so that no point ends up at the end of a street. Finally, when an address matches more than one street with the same score, marking the check box dictates that the address will be matched to the first street found.

The user can request additional fields in the output file for more information about the locations. These fields include the *x-y* coordinates of the point, the reference data ID (the feature-ID of the street it was matched to), the standardized address, and the percentage distance along the street where the point was placed.

After modifying the options, actual geocoding begins.

Matching addresses

In the first pass through the address table, the locator matches all of the addresses it can automatically. When finished, it issues a report of how many addresses were matched, how many were tied, and how many remain unmatched. If all addresses were matched (rare), you can close the window and the geocoding is done. Usually, though, the next step is to review and match addresses using the Interactive Rematch window (Fig. 10.11).

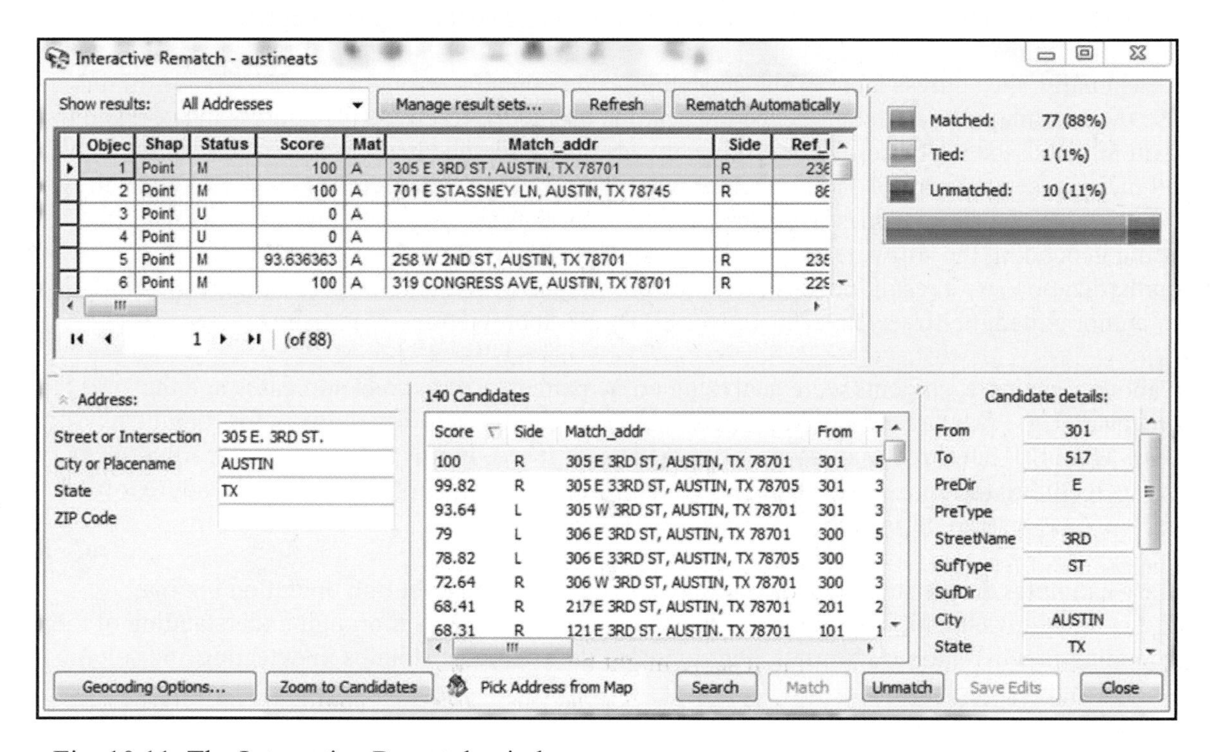

Fig. 10.11. The Interactive Rematch window

The top part of the window shows the records from the address table with additional fields added by the geocoding, such as the matched status (U for unmatched or M for matched), the score, and the address from the reference layer to which it was matched. A Show Results filter allows you to see all the addresses, just the unmatched ones, just the tied ones, and so on.

At this point you have two choices. You can modify the geocoding options from this window and then rematch automatically, or you can review each unmatched address individually and try to find matches that way. Before deciding, it is helpful to examine the unmatched addresses to see what types of problems are occurring.

You might notice that the unmatched addresses have multiple references to place names, such as City Hall or Capitol Building. In that case, it would be wise to create or locate an Alias table and specify it using the geocoding options. Then you would rematch automatically.

You might see that the address table is fully of sloppy typing mistakes, and that candidates are being generated that are mostly acceptable but do not quite meet the minimum match score. In that case, you would probably use the geocoding options to lower the spelling sensitivity and minimum match score and rematch automatically.

If you don't see any obvious repeated issues, then interactive matching is the next step. Each unmatched address is examined. Suspected typing mistakes can be corrected on the fly followed by a new search for candidates. Tied candidates or those with lower scores can be evaluated individually to see if a match is acceptable. If no candidates appear, you can try lowering the spelling sensitivity to generate more. However, it is not always possible to match every address. The address may be located beyond the extent of the locator reference layer, or errors in either the address table or reference layer may be present. Once you have matched all the addresses that can be matched, the Interactive Rematch window is closed and geocoding is complete.

Occasionally, the address table or the geocoding layer contains errors or inconsistencies. In this case the user may decide to halt geocoding, fix the problems, re-create the address locator, and try again to match the addresses. For example, the original reference layer prepared for this textbook had many streets named for saints, such as Saint Patrick St. In the reference layer as well as the address table, the street names were abbreviated to St. Patrick, St. Anne, and so on. However, during geocoding the software automatically converted *St* to *Saint* during the address standardization. As a result, dozens of addresses remained unmatched. To streamline matching, the author edited the streets layer to replace all the St. abbreviations with Saint.

In another example, students were analyzing crime patterns using street addresses and the StreetMap USA product. The police reports used the standard address prefixes for Rapid City (*West* Main St), but StreetMap uses suffixes (Main St *West*). In this case, the easiest solution involved editing the spreadsheet containing the addresses so as to place them in the address file with suffixes instead of prefixes.

These examples demonstrate that geocoding is not a simple push-button operation but may require some investigation and ingenuity on the part of the user. A thorough understanding of the geocoding process, address formats, and reference data being used helps in detecting and solving barriers to matching.

Summary

- Geocoding converts information in a table to points on a map, using reference feature classes that contain the relevant information tied to feature locations.
- Although geocoding most often refers to addresses, other types of information, such as zip codes, city-state pairs, or township-range, are also possible data on which to match table records to features in a reference layer.
- Geocoding requires an address locator that specifies the style of information to be used and stipulates the fields in the reference layer to be used for matching. Locators may come ready-made or may be created by a user for a specific purpose.
- National and international locators are available as cloud-based geoprocessing services on ArcGIS Online.
- Address geocoding breaks down addresses into standard components, compares these to the attributes in the reference layer, and scores how closely the components match possible candidates in the reference layer.
- Once a match is found, the address location is placed along the street at a distance proportional to the address value between the street address ranges.

- Geocoding styles vary on a number of factors, such as whether zip codes are used and whether the address ranges are given for the street as a whole (One Range) or for both the left and the right sides of the street (Dual Ranges).

- The process begins by matching as many addresses as possible. Then the user may change the geocoding options and try again, or interactively review individual addresses and attempt to match them. The process may be repeated until as many addresses as possible have been matched.

- Locators store a snapshot copy of the reference layer. If changes are made to the reference layer, a new locator should be created.

TIP: If you edit the features or attributes of a reference feature class, you must rebuild or re-create the address locator.

Important Terms

address locator	geocoding style	One Range
address standardization	minimum candidate score	reference layer
Dual Ranges	minimum match score	spelling sensitivity

Chapter Review Questions

1. What two data objects are required for geocoding?
2. Describe the difference between One Range and Dual Ranges geocoding.
3. If the 1200 block of Maple Street is 1000 ft long, how many feet from the 1200 corner would the address 1275 Maple be located during geocoding?
4. What is a street type? Give examples.
5. Which components are present in the address 1298 Bear Valley Road NW?
6. How does automatic matching differ from interactive matching?
7. What is an address locator and what does it include? If you don't have one, where might you look?
8. What is an Alias table and what is its purpose?
9. What is an Alternate Name table and what is its purpose?
10. Discuss the factors that would impact the geometric accuracy of a feature class derived from geocoding.

Mastering the Skills

Teaching Tutorial

The following examples provide step-by-step instructions for doing basic tasks and solving basic problems in ArcGIS. The steps you need to do are highlighted with an arrow ➔; follow them carefully. Click on the video number in the VideoIndex to view a demonstration of the steps.

> ➔ Start ArcMap and open ex_10a.mxd in the MapDocuments folder.
>
> ➔ Use Save As to rename the document and remember to save frequently as you work.

Setting up an address locator

For our first example, we will geocode restaurants in Austin, Texas. The addresses were obtained from an Internet search, copied or typed into a spreadsheet, and converted to a geodatabase table. However, before we can start, we need to decide what style of geocoding we will use, and we need to create an address locator.

> 1➔ Open the Streets attribute table and examine the fields.

1. For each of the following address components, write down the name of the field in the table that contains the information.

Prefix direction	____________
Prefix type	____________
Street numbers	____________ and ____________
	____________ and ____________
Street name	____________
Street type	____________
Suffix direction	____________
City	____________
State	____________
Zip code	____________

2. Based on the fields just listed, which geocoding style is best to use? ____________

> 1➔ Open the restaurants table and examine the fields and records. (If the table is not visible, switch to the List by Source option in the Table of Contents.)

3. Which field contains the restaurant addresses? ____________

4. Are there any intersections for addresses? ______ Which connector symbol is used? ______

> 1➔ Close the Table window.
>
> 1➔ Open the Properties of the Streets layer and examine the Source tab contents.

5. What is the source layer for Streets and where is it located? ____________

Now we have all the information we need to set up an address locator using the Streets as the reference layer.

2➔ Close the Properties window for Streets.

2➔ Open the Catalog tab in ArcMap.

2➔ Open the Folder Connections, if necessary. Right-click the mgisdata\Austin\Austin geodatabase and choose New > Address Locator.

2➔ Click the Browse button for the Style and choose US Address – Dual Ranges (Fig. 10.12). Click OK.

2➔ Choose Streets as the Reference Data. Ignore the warnings for a moment.

2➔ Click in the Role field for Streets and use the drop-down to change it to Primary Table, if it does not already say it.

Address Locator Style

US Address - Dual Ranges

Reference Data

Reference Data	Role
Streets	Primary Table

Fig. 10.12. Set the style and reference layer.

Next, we specify the fields to be used for the address components. Note that the required fields are preceded by an asterisk (*).

3➔ For the Feature-ID field name, click in the box under Alias Name, and use the drop-down to select the OBJECTID field (Fig. 10.13).

3➔ For the From Left and To Left components, select the LEFT_FRADD and LEFT_TOADD fields.

3➔ For the From Right and To Right components, choose the RIGHT_FRAD and RIGHT_TOAD fields.

3➔ For the Prefix Direction and Prefix Type components, choose the PREFIX_DIR and PRE_TYPE fields.

3➔ For the Street Name component, choose STREET_NAM.

3➔ For the Suffix Type and Suffix Direction components, choose the STREET_TYPE and SUFFIX_DIR fields.

4➔ For both the Left City and Right City components, choose the NAME field.

4➔ For both the Left and Right ZIP code components, choose ZIPCODE.

4➔ For both the Left and Right State components, choose STATE.

4➔ Leave the remaining components set to the <None> value.

4➔ Store the locator in the Austin geodatabase and name it Austin_Streets.

4➔ Click OK. Wait for the tool to finish; it may take a few minutes.

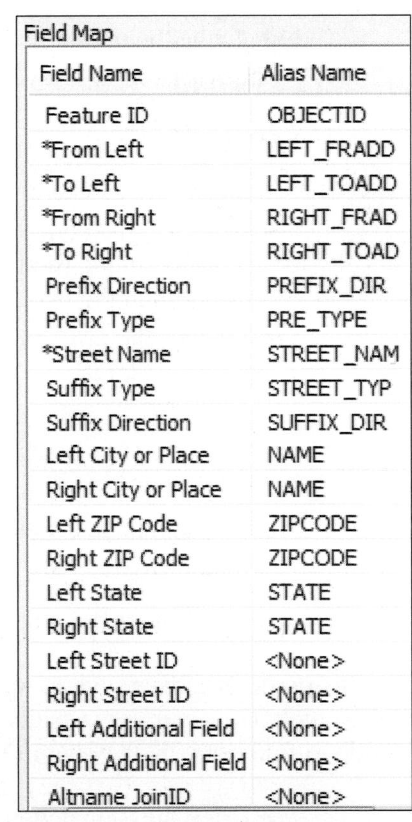

Field Map

Field Name	Alias Name
Feature ID	OBJECTID
*From Left	LEFT_FRADD
*To Left	LEFT_TOADD
*From Right	RIGHT_FRAD
*To Right	RIGHT_TOAD
Prefix Direction	PREFIX_DIR
Prefix Type	PRE_TYPE
*Street Name	STREET_NAM
Suffix Type	STREET_TYP
Suffix Direction	SUFFIX_DIR
Left City or Place	NAME
Right City or Place	NAME
Left ZIP Code	ZIPCODE
Right ZIP Code	ZIPCODE
Left State	STATE
Right State	STATE
Left Street ID	<None>
Right Street ID	<None>
Left Additional Field	<None>
Right Additional Field	<None>
Altname JoinID	<None>

Fig. 10.13. Setting the locator fields

Geocoding

Now we can geocode the restaurants. As always, we will create a practice geodatabase to hold the results of this chapter's work.

5➔ Open the Catalog tab and navigate to the Austin folder. Right-click it and choose New > Personal Geodatabase. Name it chap10results.

6➔ Right-click the Restaurants table in the Table of Contents and choose Geocode Addresses.

6➔ Choose the Austin_Streets locator from the list and click OK.

TIP: If the desired locator is not already listed, click Add, navigate to the place where it is stored, and select it.

6➔ Check the address input fields. The field names in the table are recognized by the software and added automatically, but check them to be sure (Fig. 10.14).

6➔ Name the output feature class austinfood and save it in chap10results.

6➔ Click the Geocoding Options button and examine the options.

6➔ Note that the connector symbol you identified in the Restaurants table, &, is present in the list.

6➔ Change the *Minimum match score* to 80 and click OK.

6➔ Click OK to start geocoding.

6➔ Examine the progress bar. When it is finished, you will find that you have some unmatched addresses.

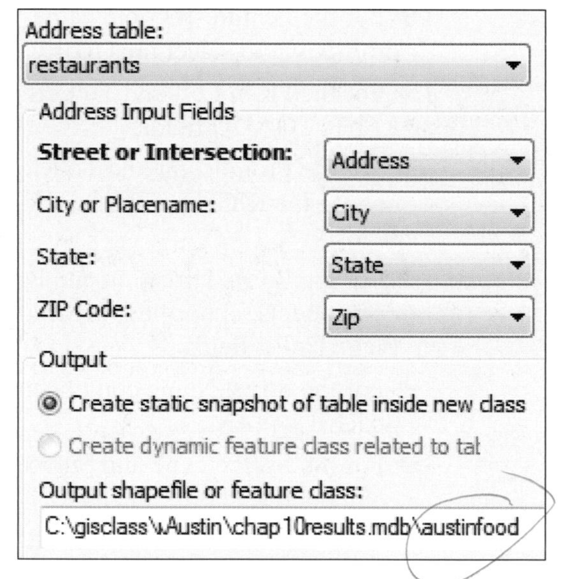

Fig. 10.14. Starting to geocode

6. How many restaurants were matched? __78__ Tied? __2__ Unmatched? __8__ (Your results may differ slightly.)

At this point, you have several ways to proceed. You could change the geocoding options and try to rematch. However, it is usually instructive to examine the unmatched addresses in interactive mode to find out why they won't match.

7➔ Click the Rematch button.

7➔ Examine the Rematch window. Note that it is currently set to Show All Addresses, both matched and unmatched.

7➔ Examine the fields in the upper part of the window, showing the table from the geocoding output feature class.

Figure 10.15 shows the upper table, with some of the fields resized so that you can see more of them than in your window. The Status field contains an M for Matched or U for Unmatched. The match score is shown in the Score field. The Match_addr field shows the final matched address. In this table, the first two fields were matched and the second two were not matched.

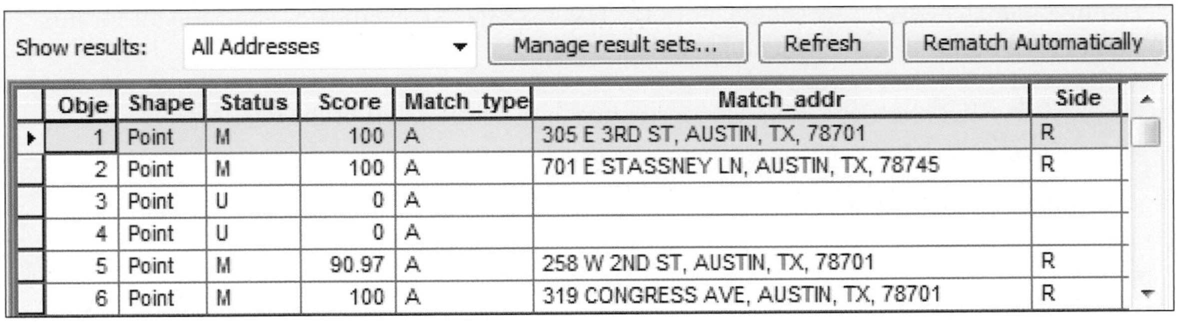

Show results: All Addresses | Manage result sets... | Refresh | Rematch Automatically

	Obje	Shape	Status	Score	Match_type	Match_addr	Side
▸	1	Point	M	100	A	305 E 3RD ST, AUSTIN, TX, 78701	R
	2	Point	M	100	A	701 E STASSNEY LN, AUSTIN, TX, 78745	R
	3	Point	U	0	A		
	4	Point	U	0	A		
	5	Point	M	90.97	A	258 W 2ND ST, AUSTIN, TX, 78701	R
	6	Point	M	100	A	319 CONGRESS AVE, AUSTIN, TX, 78701	R

Fig. 10.15. All Addresses results from the first geocoding attempt, with field widths adjusted

Examine the lower half of the window. The area to the left shows the address for the currently highlighted field. A list in the middle shows potential candidates, sorted by score. The area to the right shows the details for the currently highlighted candidate. Take a moment to examine the various candidates and scores, to get a feel for how the locator scores and matches the addresses.

8➔ Change the Show Results box from *All Addresses* to *Matched Addresses with Candidates Tied.*

8➔ Examine the lower panel for the first entry, 201 W 3RD ST, seeing the actual address on the lower left, and the candidates on the right.

The address in the table did not specify E or W 3rd Street, so there are two equally likely candidates. A little detective work might help.

8➔ Scroll the upper table to the right until you can see the restaurant name.

The restaurant's name is Catina Laredo (probably a misspelling of Cantina). You could whip out your Austin telephone directory (or yellowpages.com) and look up the restaurant, discovering that it is in fact on West 3rd Street.

8➔ Click on the 201 W 3RD ST candidate to highlight it, and click Match.

8➔ If you see more tied addresses, examine them and choose the top candidate. Click Match for each one if you can find a match.

9➔ Change the Show Results box to Unmatched Addresses.

9➔ Resize the first few fields to make them smaller.

9➔ The Match_addr field is empty and very wide, preventing you from seeing the addresses. Right-click the Match_addr field heading and choose Turn Field Off.

9➔ Highlight the first row in the table by clicking on the open gray box to its left.

TIP: You can adjust the size of each window section, or the size of the fields within a section, as needed. Place the cursor on the right side of the lower-left Address window, if necessary, and resize it so that you can see the entire address.

The first unmatched address should be 13343 HWY 183N. No candidates appear at all for this one. Let's try adjusting the geocoding options.

9➔ Click the Geocoding Options button at the bottom left of the window.

9➔ Relax the spelling sensitivity to 60 and click OK.

9➔ Click Search to look for more candidates. No new candidates appear.

9➔ Set the spelling sensitivity back to the default of 80.

Sometimes the commonly written address does not match the reference layer. The next step is to search the reference layer manually to see if this road is being portrayed differently.

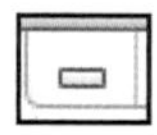

10➔ Minimize (do not close!) the Interactive Rematch window.

10➔ Click the Find tool on the Standard toolbar and enter the text to search as 183.

10➔ To refine the search, choose to search only in the Streets layer and only in the STREET_NAM field (since that is what you used in the locator). Click Find.

The search shows that there are many different ways that this reference layer stores HWY 183. This problem is not uncommon in reference layers. It is hard to be consistent when assembling data sets—with troublesome results, as you can see.

11➔ Click a few of the entries to flash them and see where they are located.

11➔ Click the first entry, scroll to the end, hold down the Shift key, and click the last record, to highlight them all.

11➔ Right-click on one of the entries and choose Select, to select them all.

With all of the segments highlighted, it appears that most designate the same road, although a few streets form spurs off the main road. You note these points as issues with the reference layer that may need fixing later. In the meantime, there are one or two tricks to try to get a match right now.

TIP: If you closed the Interactive Rematch window by mistake, or if you want to rematch a geocoded layer at any time, you can open the Interactive Rematch window by right-clicking the layer in the Table of Contents and choosing Data > Review/Rematch Addresses.

12➔ Close the Find window and clear the selection.

12➔ Restore the Interactive Rematch window.

12➔ In the Address section, edit the address field, placing a space between 183 and N.

12➔ Click Search.

12➔ If any reasonable candidates appear, click Match. If not, leave it unmatched and go on.

Address:

Street or Intersection	13343 HWY 183 N
City or Placename	AUSTIN
State	TX
ZIP Code	

TIP: Between the initial release of Version 10.1 and the first service pack, the behavior of the geocoding changed substantially. Your results may not correspond exactly with the instructions. You may find new unmatched addresses or fewer than described below. For an unmatched address, you may find different candidates or no candidates at all. Don't worry about it; just be flexible and try to match all the unmatched addresses that you can.

13➔ Highlight the row of the next unmatched candidate, 3107 S. IH35.

A couple of candidates may appear, but with low scores. Keeping in mind the issue with HWY 183, let's try a couple of alternate road names to see if better candidates appear. IH35 probably refers to Interstate-35. The roads look reasonable, but the address ranges are not satisfactory.

13➔ Edit the address, trying variations such as `IH 35` or `I 35`. Try removing the period after S. Click Search after each try.

13➔ It might find the correct road, but not the correct address range. Try editing the City to AUSTIN instead of ROUNDROCK. Click Search.

13➔ Now the top candidate looks reasonable. Highlight it and click Match.

TIP: If you have room on your computer screen, move the Interactive Rematch window off the map. Potential candidates are highlighted on the map so you can examine them spatially if needed. The highlighted candidate is shown in yellow, the rest in blue.

14➔ Click on the next unmatched address, 500 NORTH I-35. Note that some candidates refer to IH 35. This designation seems to be standard for this Austin dataset.

14➔ Edit the address, replacing I-35 with `IH 35`, and click Search. Now the top candidate is a good match with a score over 90. Click Match.

15➔ Click on the next unmatched address, 605 West St. The top candidate is barely below the minimum match score of 80, and it looks reasonable, so click Match.

15➔ Click on the next unmatched address, 2900 N RM 620. The top candidate appears acceptable. RM is a mistyping of FM, perhaps. Click the top candidate and Match.

15➔ Click the next, 3506 RR 620. Match the top candidate if any appear.

16➔ Click the next, 4894 HWY 290 WEST. Match the top candidate if any appear.

16➔ Click the next, 5001 EAST BEN WHITE. Match the top candidate if any appear.

16➔ Click the next, 4715 HWY 290 WEST. Match the top candidate if any appear.

17➔ Click the next, 6500 BEE CAVES ROAD. Match the top candidate if any appear.

At this point you might have about 86 matched or tied and 2 unmatched, although your values may differ depending on the service pack you have. (Service Pack 1 yielded 84 matched and 4 unmatched.) In any case, you are finished.

18➔ Click Close to finish rematching.

18➔ Examine the geocoded austinfood layer.

18➔ Zoom in to the cluster of restaurants in downtown Austin and examine the restaurants. Identify a few if you wish.

18➔ Return to the Full extent of the map.

18➔ Save your map document.

Geocoding with the Single House style

Instead of using street features, points or polygons can be used if they represent houses or parcels. A small section of Austin will serve to demonstrate, as these reference layers can be very large.

Imagine that a political group has come to City Hall with a set of 100 signatures regarding a neighborhood issue. You have been asked to verify whether the addresses on the petition are real.

19➔ Right-click the Crestview data frame name and choose Activate.

19➔ Zoom in to a block of houses and examine the parcels with the address points.

19➔ Return to the previous extent.

19➔ Turn off the Address Points layer.

20➔ Open the Parcels attribute table and examine the fields. Note the address component fields, similar to the ones we used in the Streets layer.

20➔ Open the Address Points attribute table and examine the fields.

20➔ Open the petition table and examine the fields.

20➔ Close the Table window.

We could use either the Address Points layer or the Parcels layer for this style of geocoding, but we will use the Parcels.

21➔ Open the Catalog tab, right-click on the Austin geodatabase, and choose New > Address Locator.

21➔ Click the Browse button for the Address Locator Style and choose the US Address – Single House style. Click OK.

21➔ Set the Reference Data to Parcels and make sure that it is the Primary Table.

22➔ Enter the reference fields (Fig. 10.16).

22➔ Name the locator Austin_Parcels and save it in the Austin geodatabase.

22➔ Click OK and wait for the locator to build.

Field Name	Alias Name
Feature ID	OBJECTID_1
*House Number	ADDRESS
Side	LEFT_RIGHT
Prefix Direction	PREFIX_DIR
Prefix Type	<None>
*Street Name	STREET_NAM
Suffix Type	STREET_TYP
Suffix Direction	SUFFIX_DIR
City or Place	CITY
ZIP Code	ZIPCODE
State	STATE
Longitude	<None>
Latitude	<None>
Street ID	<None>
Additional Field	<None>
Altname JoinID	<None>

Fig. 10.16. Single House

It's time to geocode to the parcels.

23➔ Right-click the petition table and choose Geocode Addresses.

23➔ Highlight the Austin_Parcels locator and click OK.

23➔ Set the Street or Intersection field to FULL_NAME.

23➔ Leave the City, State, and ZIP Code fields set to the defaults found in the table.

23➔ Name the output **petitionparcels** and save it in the chap10results geodatabase.

23➔ Click OK. Examine the results when it finishes.

You should have about 94 matched and 6 unmatched petition names.

24➔ Click Rematch.

24➔ Change Show Results to Unmatched Addresses.

24➔ Examine each of the first five unmatched addresses. None of them come up with good candidates, so leave them unmatched.

25➔ The last one, 1006 CHESAPEEK DR, has no candidates, but it might be a misspelling of CHESAPEAKE.

25➔ Edit the address and search for new candidates. Match the top one.

You should now have 95 matched addresses and 5 unmatched. The unmatched ones are either not real addresses or lie outside this neighborhood area. Just to be sure, though, you will check the reference layer.

26➔ Close the Interactive Rematch window, completing the geocoding session.

26➔ Notice the geocoded points that have been placed inside the parcel polygons.

26➔ Open the attribute table for Parcels and sort on the STREET_NAM field.

26➔ Scroll down and look for the missing street names.

Oertli is there, although the address number does not match any of the parcels. The other streets are not listed at all. You can be sure that these signatures on the petition are invalid.

27➔ Open the Geocoding Result: petitionparcels table.

27➔ Use Table Options > Select by Attributes to select the unmatched addresses using the expression [Status] = 'U'.

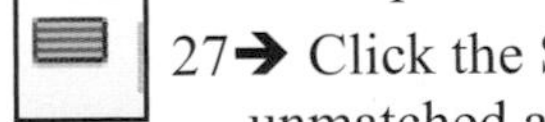

27➔ Click the Show selected records button in the table and scroll right to find the unmatched addresses.

Address
212 OERTLI LN
8617 N LAMAR BLVD
8311 SWALLOW CT
8409 GOLDFINCH CT
8413 GOLDFINCH CT

You could export or screen-capture this list for the report you will prepare for your supervisor, indicating that these five signatures should not be counted.

27➔ Close the Table window and save the map document.

Geocoding U.S. cities

Next, we will practice our geocoding skills by matching a table of precipitation data for U.S. cities to a feature class of U.S. cities. When finished, we'll make a map of monthly normal precipitation totals for the United States. Now let's prepare to do our geocoding and examine the table we'll be using for the "addresses."

28➔ Open a new, blank map document and set the home geodatabase to the mgisdata\Usa\usdata.

28➔ Add the cities and states feature classes from the usdata geodatabase. Also add the table normal_precip.

29➔ Zoom in to the conterminous United States.

29➔ Open the normal_precip table and examine the fields. The precipitation values are given in inches for each month; a yearly total is also given.

7. Which style of geocoding will we use this time? ______________________________

29➔ Open the attribute table of the cities layer, which will be the reference layer.

8. Which field will we use for the city name? _________________ Which field do we use for the state abbreviation? ___________________

29➔ Close the Table window.

30➔ Open the Catalog tab and navigate to the mgisdata\Usa folder.

30➔ Right-click the usdata geodatabase and choose New > Address Locator.

30➔ Click the Address Locator Style browse button and select the General – City State Country style.

30➔ Specify cities as the reference layer. Make sure it is the Primary Table.

30➔ Select the NAME field for the City Name component and the ST for the State component.

30➔ Name the locator US_City_State and save it in the usdata geodatabase.

30➔ Click OK to create the locator.

All right, let's give it a try.

31➔ Right-click the normal_precip table and choose Geocode Addresses.

31➔ In the Choose an Address Locator window, select the US_City_State locator just created. Click OK.

31➔ In the Geocode Addresses window, specify CITY for the City field and STATE for the State field.

31➔ Put the output in the chap10results geodatabase and call it cityprecip.

31➔ Click OK to start geocoding.

31➔ Examine the results window.

9. How many matched/tied records are there? __________ How many unmatched? __________

That is not a bad effort. Let's look at the unmatched records.

32➔ Click Rematch and change the Show Results to Unmatched Addresses.

32➔ Highlight the first row, and examine the Address entry on the lower left of the window. It says NOME, AK. No candidates have appeared.

You've probably heard of Nome, Alaska, and think that the spelling is correct. Perhaps this smaller city is just not within the cities feature class?

32➔ Open the Find tool on the Standard toolbar.

32➔ Set it to search only in Cities, and type Nome as the Find text. Make sure it is searching in all fields, and click Find.

It did not find Nome, so apparently this city was not in the feature class.

TIP: If you are not certain Find is working, search for a major city, like Boston, to check.

33➔ Leave the Find window open in case you need it again, moving it aside to see the Interactive Rematch window.

33➔ Leave NOME unmatched and highlight the next unmatched record.

33➔ The next unmatched record is WINSLOW, AZ. Use Find to search for it in the cities layer. Nothing is found, so leave it unmatched.

34➔ The next several cities are not in the reference layer. Check each one to be sure and then leave them unmatched, until you get to MINNEAPOLIS-ST.PAUL.

When Find turns up nothing for Minneapolis-St. Paul, a major city, we are immediately suspicious. It must be in the cities layer.

35➔ In the Find window, search just for Minneapolis.

Now we see the problem. The precipitation station is named for the combined city of Minneapolis-St. Paul, but in Cities they are listed separately.

Address:	
City Names	MINNEAPOLIS
State	MN

35➔ Edit the address in the City Names address box, changing it simply to MINNEAPOLIS.

35➔ Click Search. The right candidate appears. Match it.

35➔ Click the next unmatched record, OMAHA EPPLEY.

Omaha, Nebraska, you have heard of. Eppley is actually the airport. In this case, the climate station is named not just for the city, but also for its location at the airport. It can be matched.

36➔ Edit the address to read just OMAHA, click Search, and match it.

36➔ The next unmatched one, ATLANTIC CITY AP, appears to have the same issue. Edit the address and match it to Atlantic City, NJ.

36➔ The next unmatched one is a second climate station for Atlantic City. Edit the address and match it.

Some large cities have more than one climate station, like Atlantic City and New York. Simply match all the climate stations to the same city. A point will be created for each station, stacked on top of each other.

All of the remaining unmatched records have one of the issues we've already discussed.

37➔ Keep moving through the table, examining each address in turn. Match them if you can, or leave them unmatched if you cannot.

37➔ Also examine the tied candidates and check that they have been assigned the best match.

TIP: The initial release of Version 10.1 caused inappropriate ties to occur when matching some cities. For example, a climate station in Little Rock, AK, was matched to North Little Rock instead of Little Rock (both had scores of 100).

When finished, you should have about 222 matched stations and 4 unmatched ones.

37➔ Click Close to close the Interactive Rematch window. Also close the Find window.

Once matched, the data can be used to analyze the distribution of rainfall across the United States.

38➔ Turn off the Cities layer to see the precipitation locations better.

38➔ Create a graduated symbol map of the Geocoding Result layer, showing the annual precipitation (field ANN) in symbols that increase in size with precipitation (Fig. 10.17).

38➔ Save the map document.

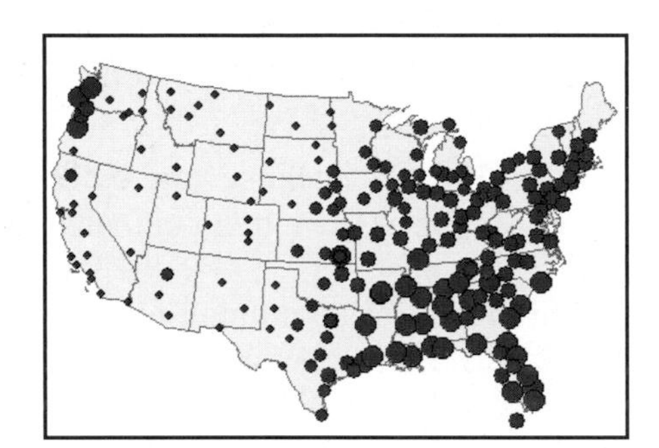

Fig. 10.17. Annual precipitation of the matched cities

10. Can you think of two reasons the geocoding method works better to accomplish this task, as opposed to joining the precipitation data to the Cities layer based on a common CITY_NAME attribute?

Geocoding township-range data

Geocoding provides an effective way to convert township-range-section locations to points on a map. Imagine that you have some rock samples that a retired professor collected from Grant County, Oregon, located by township, section, and quarter-quarter section. You want to produce a map showing the locations of the samples. Someone has already typed the locality information into a spreadsheet, as shown in Figure 10.18.

STATE	MERID	TSHP	RANGE	SECTION	QQSEC	QSEC
OR	33	T0150S	0270E	023	SE	NW
OR	33	T0170S	0260E	016	NW	SE
OR	33	T0160S	0270E	010	SE	SW
OR	33	T0150S	0290E	018	SW	SE
OR	33	T0170S	0280E	029	NE	SE

Fig. 10.18. Rock sample localities

39➔ Open the map document ex_10b.mxd and save it under a new name.

39➔ Zoom in to one of the gray squares, a township. Township labels and sections will appear.

39➔ Zoom in to one of the red sections. The quarter-quarter section outlines and labels appear.

39➔ Zoom to the full extent of the map.

The townships, sections, and quarter-quarter section feature classes have been included just as they come from the BLM web site, except that they have been converted from shapefiles to geodatabase feature classes.

40➔ Open the QQsection table. Notice the lndkey, sectn, and qqsection fields and examine their contents carefully for the first row in the table.

Figure 10.19 shows the relevant attributes of the QQsection table. The lndkey field contains a concatenated value representing the state, zone, and township. The first row refers to the NE corner of the NE quarter of Section 19 of Township 17S-26E. Because these fields are formatted with additional zeros to keep the codes a consistent width, your data tables must include the zeros, as has been done in Figure 10.18. Note that the spreadsheet columns must be specifically formatted as text to retain the leading zeros.

The geocoding style that you must use is the General – Single Field style. Because you have only one field to match on, you must contrive a way that both your location data and your reference data have a single unique field that identifies the polygon and sets up the match. You can get this by concatenating the strings into a single key field, in both the reference layer and the spreadsheet.

lndkey	qsection	sectn	qqsection
OR33T0170S0260E	NE	019	NENE
OR33T0170S0260E	NW	020	NWNW
OR33T0170S0260E	NW	020	NENW
OR33T0170S0260E	NE	020	NWNE
OR33T0170S0260E	NE	020	NENE

Fig. 10.19. Relevant QQsection attributes

40➔ Choose Table Options > Add Field.

40➔ Type KEY_QQSEC as the field name, and make it a text field with a length of 25. Click OK.

40➔ Right-click the empty KEY_QQSEC field and choose Field Calculator.

40➔ Enter the expression [lndkey] & [sectn] & [qqsection]. In this window, the & symbol is a string concatenation operator. Click OK.

The KEY_QQSEC field in the first row should now read OR33T0170S0260E019NENE, including all the information needed to uniquely identify the polygon. (You may need to expand the field width to see it all.) Next, you must edit the spreadsheet to create matching identifiers.

TIP: If you don't have access to Excel™, simply use the rocksampleskey.xls spreadsheet, which already has the key added for you, and skip to Step 50.

41➔ Close the Table window.

41➔ Start Excel™, and open the mgisdata\Oregon\rocksamples.xls spreadsheet.

41➔ Create a new column in the spreadsheet named KEY.

41➔ In the top row, enter the formula =concatenate(B2,C2,D2,E2,F2,G2,H2).

41➔ Copy the formula down to the other rows.

ArcMap cannot read spreadsheets with formulas, so you must convert the formulas to text.

42→ Select the cells containing the formulas and click Ctrl-C to copy them.

42→ Right-click inside the highlighted cells and choose Paste Special.

42→ Fill the button to Paste Values and click OK.

42→ Save the spreadsheet and close it. (ArcMap will not open the file while it is open in Excel™).

Now you build the address locator.

43→ In ArcMap, open the Catalog tab.

43→ Right-click on the mgisdata\Oregon folder and choose New > Address Locator.

43→ Click the Browse button for the Style and choose General – Single Field.

43→ Select QQsection as the reference data and make sure it is set as the Primary Table.

43→ Set the Key field to KEY_QQSEC.

43→ Name the locator QQsection_locator and store it in the mgisdata\Oregon folder.

43→ Click OK to start building the locator.

Now you can geocode the rock sample locations.

44→ Click Add Data and navigate to the mgisdata\Oregon folder.

44→ Double-click the rocksamples.xlsx file to open it.

44→ Select Sheet1$ and click Add.

44→ Open the table to make sure it looks all right, and then close it.

45→ Right-click Sheet1$ and choose Geocode Addresses.

45→ Select the QQsection_locator and click OK.

45→ Set the Key field to KEY.

45→ Name the output samplelocalities. and save it in the GeolProject geodatabase. Click OK.

All of the records should be matched, so you shouldn't need to do any rematching.

46→ Click Close to close the matching window.

46→ Change the samplelocalities layer symbol to a large green circle so it shows up better against the geological map (Fig. 10.20).

46→ Zoom in to one of the townships containing sample points.

46→ Zoom in until you can see the quarter-quarter sections, and the sample points placed in the center of one.

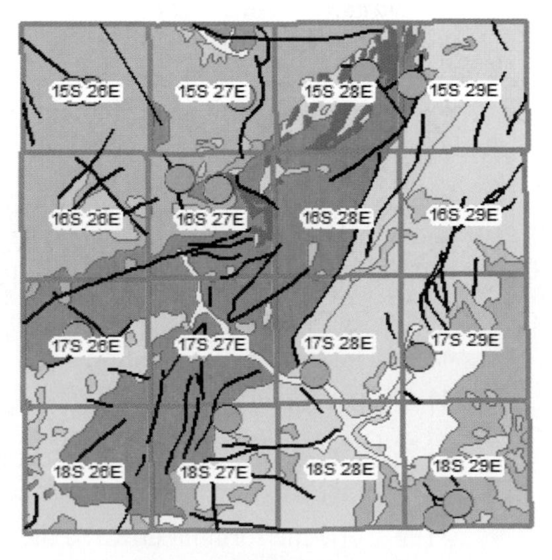

Fig. 10.20. Rock sample localities

46➔ Return to the full extent of the map.

This geocoding technique can also be applied to locating just townships, sections, or quarter-sections, depending on the level of detail in the data you need to geocode. Simply modify the key field and use the appropriate reference layer.

TIP: GIS files of PLSS units are freely downloadable as shapefiles from the site http://www.geocommunicator.gov/geocomm/lsis_home/home/index.shtm.

This is the end of the tutorial.

➔ Exit ArcMap and save the map document.

Exercises

1. Search on the Internet for 10 to 15 hotels in Austin, Texas. Type the names and addresses into a spreadsheet and geocode them. **Capture** a map showing the hotels.

2. Geocode the Austin hotels using the online geocoding service. Compare them with the results of Exercise 1. **Capture** a map showing both sets of hotels.

3. In the mgisdata\Rapidcity folder, you will find a table of gas stations called gas_loc.dbf and an Alias table called rc_alias.dbf. Create a One Range locator using the roads feature class in the rapidnets geodatabase, and geocode the gas stations. **Capture** a map showing the stations.

4. In the mgisdata\Usa\usdata geodatabase, you will find a table called customers. These data come from the test marketing of a catalog for natural landscaping to minimize watering of plants (xeriscaping). Create a map showing the location of the customers. **Capture** your map.

Challenge Problem

Search the Internet for a data set that interests you, is geocodable, and contains at least 25 data points. Create a spreadsheet that can be read by ArcMap. Then geocode it. With your answer, include the source of the data, the style of locator used, and the service or reference layer used.

Chapter 11. Coordinate Systems

Objectives

- Learning the basic properties and uses of coordinate systems
- Understanding the difference between geographic and projected coordinate systems
- Understanding different projections and the distortions they cause
- Choosing appropriate projections for a geodatabase
- Managing and troubleshooting coordinate systems of feature classes and images

Mastering the Concepts

GIS Concepts

About coordinate systems

A successful GIS system depends in large part on managing coordinate systems correctly, and a person's skill (or lack of it) can make the difference between usable and trashed data. Coordinate systems are also a vast and complex topic. Experience working with coordinate systems and regular review of these concepts will ultimately provide a sound basis for effectively working with coordinates.

In order to display and analyze maps on the screen, a GIS system uses **coordinate pairs**, which specify the location and shape of a particular feature. A point is described by a single *x-y* coordinate pair located in space; a line is a series of *x-y* coordinates. A **coordinate space** is an agreed-upon range of coordinates used to portray the features.

Imagine that a surveyor is mapping an industrial site using typical surveying instruments that measure angles, directions, and distances. He may begin at a fence corner at the lower right of the complex and assigns that point the coordinates (0, 0). This is the **origin** of the coordinate space (Fig. 11.1). The surveyor then determines that the southwest corner of building A is 175 meters east and 200 meters north of the origin, and he assigns that corner the coordinates (175, 200). The surveyor uses meters to record the distances and determine the coordinates, thereby establishing meters as the **map units**. The combination of origin and map units becomes the coordinate space of this map. In this case the coordinate space is arbitrary and chosen for convenience.

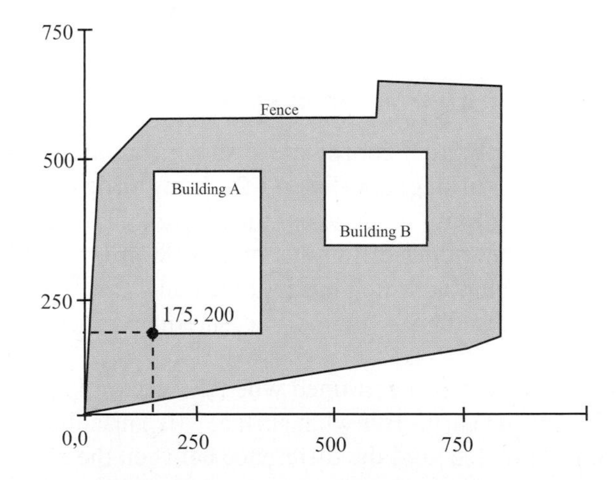

Fig. 11.1. An arbitrary coordinate space used for surveying a site

Now imagine that the surveyor's colleague is surveying another site across town. She also chose the corner of her site as the origin, but she is measuring distances in feet. The two surveyors take their drawings back to the office and decide to create a map showing the location of both sites, so they each transfer the coordinates into a GIS. When they plot the sites, however, the two different sites show up in exactly the same spot, even though in real life they are several miles apart. This problem occurs because both surveyors used an arbitrary (0, 0) origin unrelated to the other site. Moreover, the woman's buildings look about three times as large as the man's buildings because she is measuring in feet instead of meters. To plot these together the two surveyors would need to define a single common origin and convert both sites to the same map units.

To avoid problems such as this, geographers and map creators usually employ a standard **coordinate system**, a coordinate space whose characteristics are defined and established according to cartographic standards. The surveyors could have used a global positioning system (GPS) instrument to establish the *x-y* coordinates of their origins in a common coordinate system, such as Universal Transverse Mercator (UTM), and then entered their surveyed points relative to that base. Then their maps would have been located correctly with respect to each other.

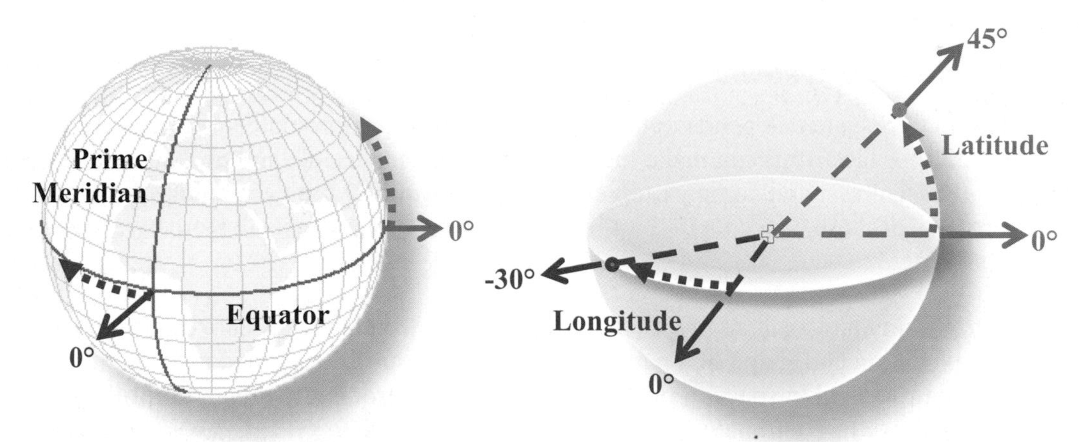

Fig. 11.2. Longitude measures E-W angles in the plane of the equator; latitude measures N-S angles above or below the plane of the equator.

Geographic coordinate systems

One commonly used coordinate system is based on measuring angles from the center of the earth and has units of **degrees** (Fig. 11.2). **Longitude** measures horizontal angles east or west of the **Prime Meridian**, which is the line where longitude equals 0 that passes through Greenwich, England. Longitudes fall in the range –180 to +180. **Latitude** measures vertical angles above or below the equator, which has latitude equal to 0, and range from –90 at the South Pole to +90 at the North Pole. This system of measurement is called a **geographic coordinate system**, or **GCS**.

A GCS location is determined when the vector defined by a latitude-longitude pair intersects the surface of the earth. But what surface? Because the earth's shape is irregular, its surface can only be approximated, and the difference between the true shape and the approximate shape is a source of positional errors. For this reason, different approximations have been developed to meet specific mapping needs, each one representing a slightly different shape for the earth. Such an approximation is called a **datum**.

How is a datum defined? The earth's rotation causes it to bulge at the equator, so the first task is to model the earth's surface using a smooth figure called a **spheroid**, an oblong sphere with a major and a minor axis. (The bulge in Figure 11.3 is highly exaggerated.) The chosen spheroid becomes a part of the datum definition. (Some people use the term **ellipsoid** rather than spheroid.)

Different spheroids have been used in the past because estimates of the earth's major and minor axes have varied over time. Clarke 1886 was a common spheroid used in North America, but it is being replaced by more recent and accurate satellite-measured spheroids.

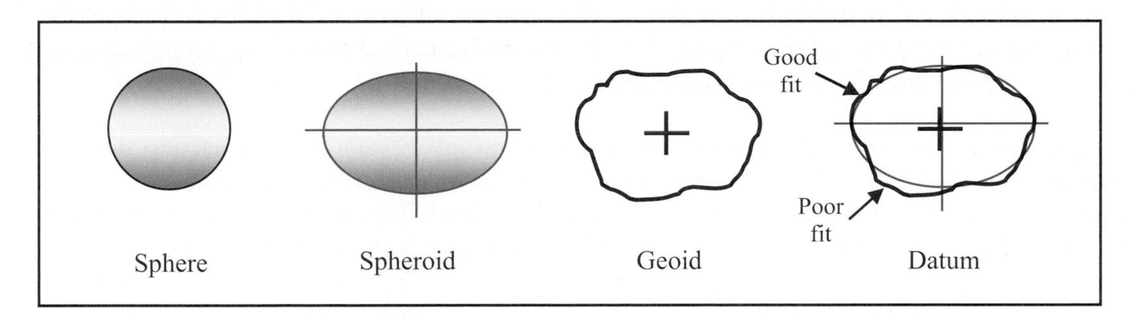

Fig. 11.3. Relationship among the spheroid, geoid, and datum

However, a spheroid does not adequately represent the shape of the earth either, so a second correction is made. The closest approximation to the earth's shape is the **geoid** (Fig. 11.3), an irregular, equipotential surface based on gravity. Sea level varies from place to place due to rotational, topographic, and compositional differences in the earth's mantle, and the geoid is often described as what the shape of the earth would be if there were only seas and no continents. The geoid can be modeled mathematically, but the equations are too computationally intensive to serve well as the basis for mapping. Instead, the spheroid is shifted relative to the geoid until a best-fit solution is obtained (Fig. 11.3). A datum, then, consists of the chosen spheroid and the location of its center relative to the geoid center.

geoid

A datum that finds the best fit overall is called an *earth-centered datum*. The World Geodetic Survey of 1984 is an earth-centered datum based on the WGS 1984 spheroid and is commonly used for international data sets and web maps. It is the default datum for most GPS units.

Another approach is to find the best fit of the spheroid to the geoid for a particular region, such as North America, to produce a *local datum*. The improved fit in one place, however, often means a poorer fit elsewhere, so a local datum should be used only in the regions for which it was developed. The North American Datum of 1927 (NAD 1927) is based on the Clarke 1866 spheroid. The North American Datum of 1983 (NAD 1983) is based on the more accurate GRS 1980 spheroid. Both datums use a network of surveyed benchmarks to further customize the fit and provide more accurate positions; this approach is often employed for local datums. Coordinates stored using NAD 1927 and NAD 1983 can differ by up to several hundred meters, which, although not significant for a national map, can cause serious registration issues for larger-scale maps. Figure 11.4 shows the offset between an image stored using NAD 1983 and roads stored using NAD 1927.

Fig. 11.4. Offset of roads due to different datums

Adjustments of NAD 1983, based on better and better surveying of the benchmarks, take place periodically and may be used for very high-precision surveying applications when differences on the order of centimeters are important. Such adjustments have names such as CORS96, HARN, and NSRS2007. However, unless you have been told that you are using one of these specialized NAD adjustments, it is best to choose the "vanilla" datum, NAD 1983.

NAD 1983
NAD 1983 (2011)
NAD 1983 (CORS96)
NAD 1983 (CSRS)
NAD 1983 (NSRS2007)
NAD 1983 HARN

The datum forms the base definition of a geographic coordinate system (GCS) used to store locations in degrees of latitude and longitude. The same place will have slightly different values of latitude and longitude depending on which datum was used. The GCS typically takes its name from its defining datum and, to many, the terms *datum* and *GCS* are interchangeable.

At times it is necessary to convert from one datum to another; the process is called datum **transformation**. It does not always use exact mathematical formulas and may require localized estimates and fitting. Often, multiple methods are available. The transformation itself may introduce new errors in coordinate locations, on the order of several meters, so it is not desirable to transform GIS data back and forth repeatedly and multiply these errors. Usually, one picks the datum to be used for a project (or organization) and sticks with it, converting data once at the time they are brought into the geodatabase, using the best available transformation.

Map projections

A GCS is a three-dimensional coordinate system, but maps need to be flat. Therefore, cartography requires converting the locations on a globe to locations on a piece of paper. This process is called **projection**. One starts with a defined GCS and applies a set of mathematical formulas that convert degrees of latitude and longitude into planar *x-y* coordinates on the paper. During this conversion the three-dimensional GCS becomes a two-dimensional representation, and degrees are converted into feet or meters. The chosen earth shape (the GCS) influences exactly where locations end up on the paper, so the GCS is inherited by the projection and becomes a part of its definition.

You can imagine the projection process visually as a clear globe with the graticules of the earth printed in black. If you place a light bulb at the center of the earth and wrap a piece of paper around the equator in the shape of a cylinder (Fig. 11.5), the graticules will be cast as shadows on the paper. Tracing them gives you a map, which you can unroll and lay flat.

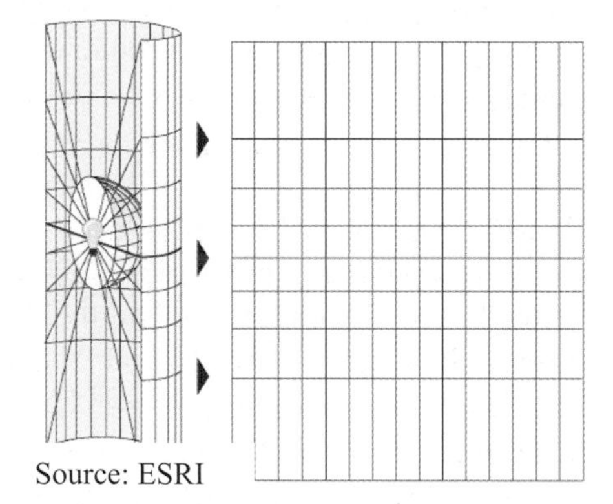
Source: ESRI

Fig. 11.5. Projecting a map from 3D to planar coordinates

Projections are grouped into three major classes, depending on the shape of the surface onto which the GCS locations are projected. A **cylindrical** projection uses a cylindrical surface that lies tangent to (touches) the earth at the equator along a great circle (Fig. 11.6). Rotating the cylinder sideways and making it tangent to the earth along a line of longitude produces a **transverse** projection. An **oblique** projection places the tangent at an angle. In each case, distortion is absent along the tangent and increases away from it. Cylindrical projections typically preserve shape or direction at the expense of area and distance.

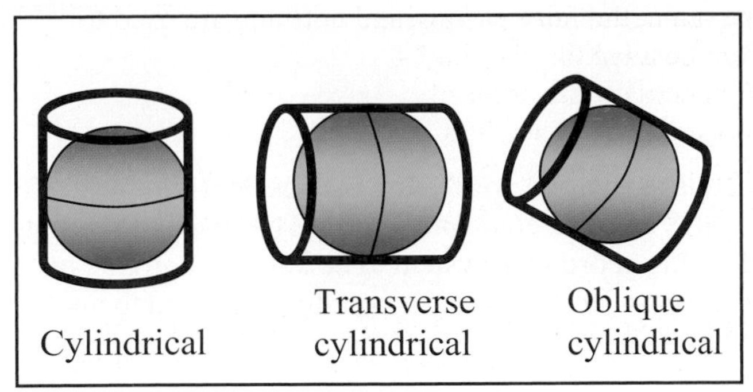

Fig. 11.6. Cylindrical projections

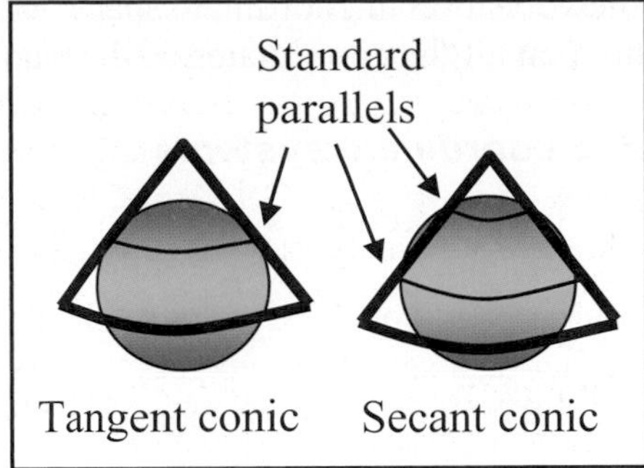

Fig. 11.7. Conic projections

A **conic** projection is based on setting a cone on the sphere (Fig. 11.7). If the cone is tangent to the globe along a line of latitude, then the projection is a **tangent** projection. The line of tangency is called the standard parallel. One can also place the cone *through* the sphere so that it touches in two places—a **secant** projection. Such projections have two standard parallels. (Cylindrical projections may also be tangent or secant.) Distortion increases away from the parallels. Conic projections typically preserve area or distance at the expense of direction and shape. One can also place a plane tangent or secant to the sphere, producing an **azimuthal** projection (Fig. 11.8). Other names for this method are **stereographic** and **orthographic** projection. Typically, these projections are used for displaying the earth's poles, and for that reason they are sometimes called polar projections. Distortion is minimal at the point or circle of intersection and increases away from it. Distortions associated with azimuthal projections vary with the projection method, but they generally preserve area or distance rather than direction or shape.

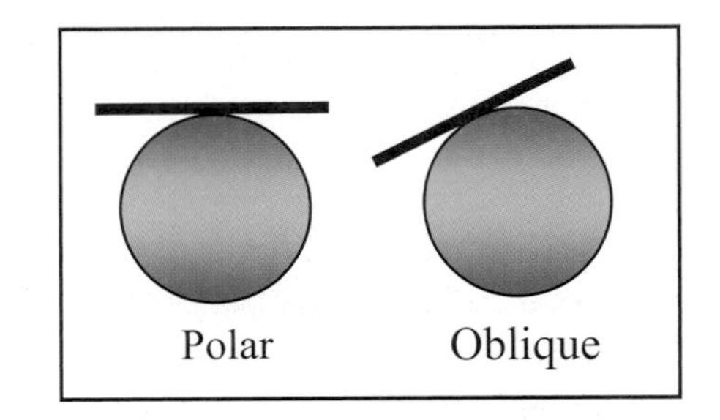

Fig. 11.8. Azimuthal projections

Map projections have certain properties that are described using projection **parameters** (Fig. 11.9). One parameter is the line of longitude, which constitutes the **central meridian**, or $x = 0$ line on the map. The central meridian is chosen based on its position near the center of the area being mapped. A map of England would have a central meridian close to 0 degrees. A map of the United States would have a central meridian of about –100 degrees, because that meridian is close to the center of the continental United States. The **latitude of origin**, also sometimes called the **reference latitude**, is the $y = 0$ line. The equator is often used as the reference latitude. A map projection will also have one or more **standard parallels** in its parameters—the latitudes that define the lines of tangency or secancy. Maps may also have a **false northing** or **false easting** (Fig. 11.9). These are arbitrary numbers added to the x-y coordinates in order to translate the map

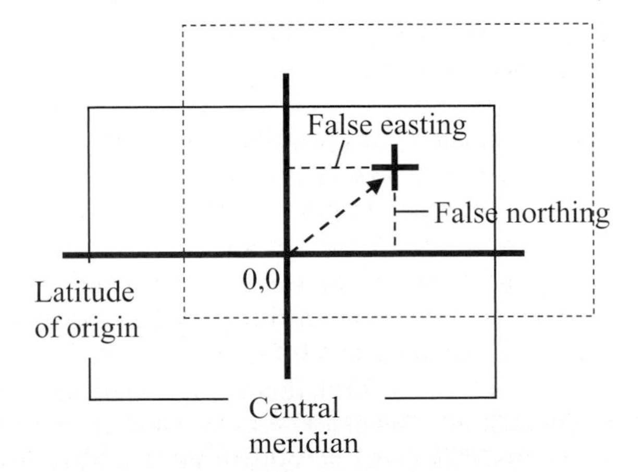

Fig. 11.9. Some parameters of map projections

to a new location in coordinate space. Most often, the false easting and northing are used to ensure that all the coordinates of the map are positive.

Raster coordinate systems

Rasters have coordinate systems and projections just as vector feature classes do. (Figure 11.10) shows a South Dakota elevation raster in a GCS and in a conic projection. The shape of the state is different, and you need to know the underlying coordinate system to be able to align this raster with other data sets. Like vectors, rasters can be projected from one coordinate system to another, although the algorithms are more complicated.

Assigning coordinate systems to rasters

Many rasters are provided with a coordinate system at the outset. However, sometimes a user must assign the coordinate system. A raster that has been assigned a real-world coordinate system is said to be **georeferenced**. Two cases commonly occur.

Case I involves a raster for which the projection, datum, cell size, and *x-y* coordinate of the upper-left corner are known, but the information has not been stored in accessible form. Some image formats store georeferencing information inside a header file or in the image itself. Other formats use a **world file**, which is a six-line text file containing the parameters needed to georeference the raster. If the georeferencing information is missing, the user can create a world file with the appropriate parameters. A description of the contents of the world file can be found in the ArcGIS Help.

TIP: A world file has the first and third characters of the image file extension plus a "w" on the end. A .tif file would have a .tfw world file. A .bil file would have a .blw world file.

Case II involves a raster for which the coordinate system parameters are completely unknown. Such files are commonly generated by downloading an image from the Internet or by scanning a map. Images derived from aerial photography often fall into this category as well. The coordinate system is in pixel units from the screen or the scanner rather than real-world coordinates in feet or meters. To georeference this raster, the user must establish a list of **ground control points** that link the pixel *x-y* values to real-world *x-y* values in a coordinate system. From this list, a set of equations can be developed to convert the pixel coordinates to real-world coordinates. To finalize the georeferencing, the user may either save the information to a world file or **rectify** the image to create a new, permanent image.

Several different methods of georeferencing are used. A first-order (**affine**) transformation can translate, rotate, and skew the image to match the points. A second-order or third-order transformation allows the image to stretch or contract in some places relative to others in order to fit the points. This process is often called **rubbersheeting**, likening the image to a sheet of rubber that can be stretched and pulled to get a good fit. After rectification there may still be offsets in the two sets of ground control points. The algorithm calculates a root mean square error (**RMSE**) between the two sets of points, which serves as a measure of the accuracy of the transformation. The RMS

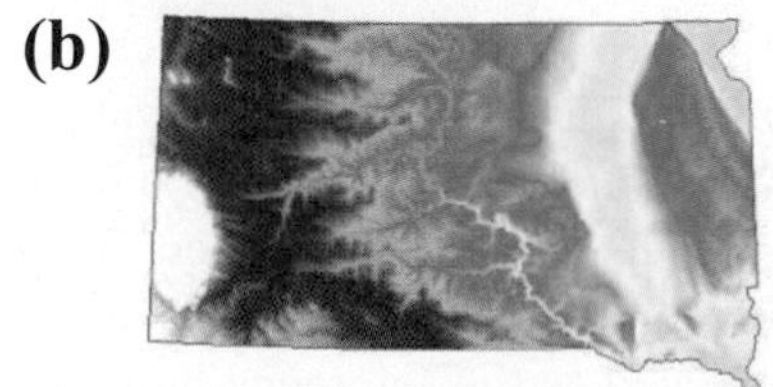

Fig. 11.10. An elevation raster in (a) GCS and (b) State Plane coordinate systems

error is usually reported in map units of the new coordinate system. The transformation RMS error should be included with the geometric accuracy report of the metadata for the image.

The ground control points are established by selecting points on the image that can be matched to another image or feature class that already has a real-world coordinate system. Road intersections, building corners, or other distinctive features are typically used. At least three pairs of points must be chosen for an affine transformation, but more points generally result in a more accurate transformation (lower RMSE). Rubbersheeting requires additional control points. Once a set of points is defined, the geometric transformation is applied to convert the image to the same coordinate system as the map.

During rectification, the raster is being converted from one grid of cells to another with potentially different sizes and spacing. The old cell grid in many cases will not match the new cell grid, and so the cells must be resampled, as discussed in Chapter 8. The key point to remember is that rasters containing categorical data should be resampled with the nearest neighbor method, and continuous rasters should be resampled with the bilinear method.

Projecting rasters

Projecting rasters works much in the same way as projecting vectors, but there are additional considerations. A GCS elevation raster of South Dakota (Fig. 11.10a) has a different shape than the raster stored in a State Plane coordinate system (Fig. 11.10b). Just looking at the shapes in Figure 11.10 suggests that the GCS pixels must be squeezed together to produce the State Plane map. The GCS map units are in degrees, and the cell size is 30 arc seconds, or 0.008333 degrees, with 964 columns and 422 rows. A single degree of latitude at the equator is 111.3 km. The height of the cell is

$$0.00833 \text{ degrees} \times 111.3 \text{ km/degree} = 0.927 \text{ km},$$

or about 900 meters. The cell is not square because 30 arc seconds of latitude is not equal to 30 arc seconds of longitude except at the equator—degrees of longitude become smaller with increasing latitude. (At this latitude the cells represent an area of about 600 × 900 meters.)

To project the GCS raster to State Plane, the user must specify a new cell size in the State Plane map units (meters). The old cell grid in degrees will be converted to a new cell grid in meters. Choosing the right size cell is important. Since the resolution in ground units is no better than 900 meters, 900 should be the minimum cell size specified. In this case, a cell size of 1000 meters was chosen; many people prefer to have nicely rounded cell sizes. As in rectification and georeferencing, resampling will need to occur to convert the cells to the new grid spacing.

Projecting rasters is more complex and computer intensive than projecting vector features, in which each *x-y* coordinate is simply converted to its new location through the projection equations. For rasters, each cell center must be recalculated and resampled, possibly degrading the image and introducing artifacts. Projecting the same raster multiple times to different coordinate systems should be avoided if possible. For best performance and quality when viewing multiple coordinate systems, the data frame should be set to match the coordinate system of the raster(s) rather than the vector data.

Common projection systems

Several projections deserve special description because they are so commonly used. One, the GCS, is not actually a projection but is often treated as one. UTM is a family of coordinate systems utilized worldwide. State Plane is another such family used within the United States, and other countries have similar national or state systems used for large-scale maps.

Geographic coordinate systems

As a three-dimensional coordinate system, a GCS cannot, strictly speaking, be displayed on a flat map or screen. However, because data sets may be stored as unprojected data in a GCS, and users would like to be able to see them, a GIS must have some method of displaying GCS data. A simple equirectangular projection is applied, which essentially treats the degree as a planar distance measurement rather than an angle (Fig. 11.11). However, distortions are introduced by this practice. Notice on the globe that the latitude lines get increasingly shorter at higher latitudes, going to zero at the poles, yet in the planar display, these lines are all represented as the same length. The further north or south an area is from the equator, the more the map gets distorted. Furthermore, analysis functions also treat a GCS data set as if it were planar. If the functions use area or distance, then the results may be incorrect.

TIP: It is inappropriate to use a GCS to portray maps or do spatial analysis, and doing so may cause incorrect results.

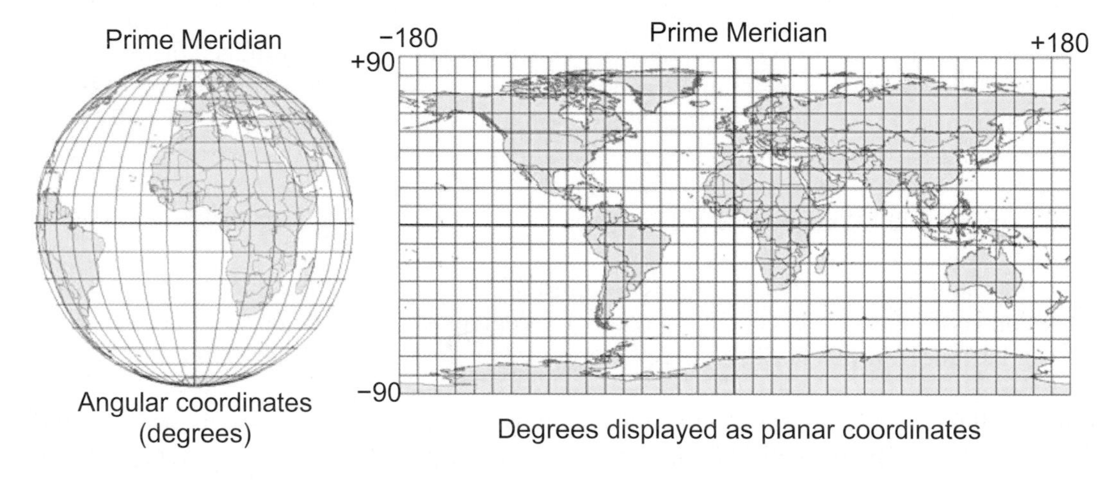

Fig. 11.11. A GCS is projected flat for display in a GIS, causing distortion.

The UTM system

The **UTM** system is based on, as the name suggests, a secant transverse cylindrical projection. Two lines of tangency of the cylinder to the sphere fall about 180 km to each side of the central meridian (Fig. 11.12a). A zone is defined to include 3 degrees on each side of the central meridian between the 80S and 84N latitude lines. Since distortion is absent along the lines of tangency and increases with distance from them, the distortion inside the zone is minimal. To map different locations, the cylinder is rotated about the sphere in 6-degree intervals, producing 60 zones around the world. Areas within a single zone have negligible distortion in all four map properties of shape, direction, distance, and area.

Each UTM zone has its own central meridian and splits at the equator into a north and south component. In order to eliminate negative *x-y* coordinates, southern zones have a false northing of 10 million meters applied, and both northern and southern zones have a false easting of 500,000 meters. UTM is convenient because users need only know the zone number and hemisphere.

In the United States and in ArcGIS, each zone is indicated by a number and the designation N for the northern hemisphere or S for the southern hemisphere, for example, UTM Zone 14N. (The worldwide UTM system uses letters C to X to represent different latitudinal zones.) Figure 11.12b shows the UTM zones of the conterminous United States.

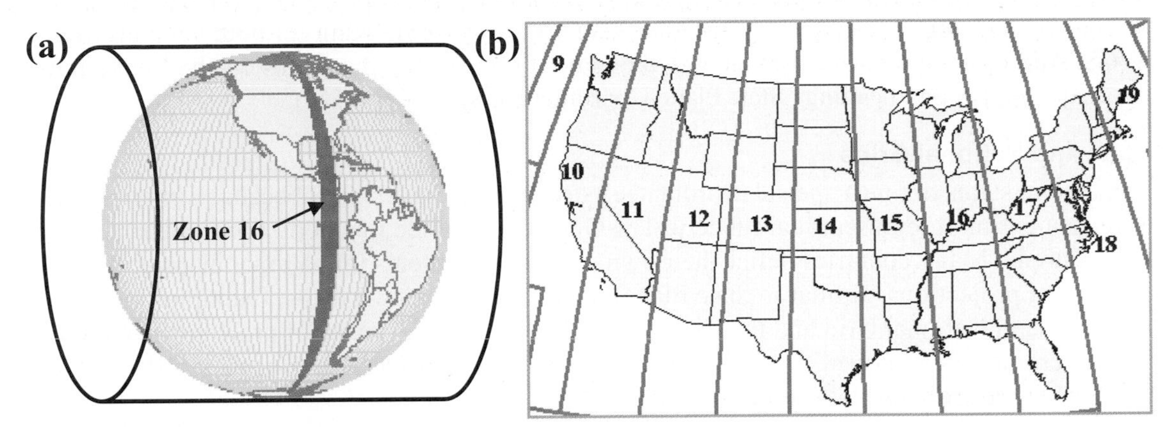

Fig. 11.12 (a) The UTM projection system has 60 north-south zones, each 6 degrees wide. (b) This map shows the UTM zones of the conterminous United States.

The UTM system is used extensively for US Geological Survey 1:100,000 and 1:24,000 scale publications, for county maps, for maps of states within single zones, and for project maps covering small areas. If the region of interest lies on the boundary between two zones, the user can choose the zone containing the larger area to represent the map. This choice may be acceptable if only a small area extends out of the zone. If not, State Plane zones may provide a better alternative.

The State Plane Coordinate System

The **State Plane** Coordinate System (SPCS) includes an assortment of coordinate systems developed in the 1930s by the U.S. Coast and Geodetic Survey for large-scale mapping in the United States. To maintain the desired level of accuracy, many states are broken into zones with different parameters used to minimize distortion in each zone (Fig. 11.13). Like the UTM system, an SPCS is identified by its name and zone, and distortion is negligible within a single zone.

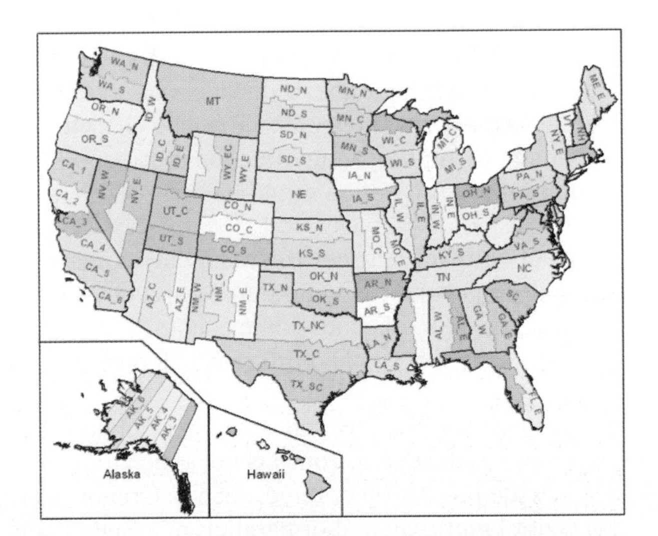

Fig. 11.13. State Plane zones of the United States

The SPCS uses three different projections: Lambert Conformal Conic, Transverse Mercator, and Oblique Mercator. Zones oriented north to south use a Transverse Mercator projection, as it provides the least distortion. Zones oriented east to west use a Lambert Conformal Conic projection. A state may use more than one projection depending on the size and shape of its zones. Alaska has 10 zones and uses all three projections. Zones are formally identified using a FIPS code composed of a two-digit state code and a two-digit zone code (e.g., FIPS 0407 for Zone 7 of California). Users may refer to them informally by state name and zone, for example, California State Plane Zone 7.

The original SPCS was based on NAD27 and used units of feet. Improvements in surveying necessitated the revision of the SPCS in 1983. At that time, it adopted the NAD83 datum and metric units. Certain zones were also revised and in some cases eliminated. Even though the official units changed from feet to meters, users may specify either unit for both versions of SPCS. Additional information about both the State Plane and UTM systems can be found in the ArcMap Help (index headings State Plane Coordinate System and UTM).

State and National Grids

Some large states define a special coordinate system for statewide maps because a single State Plane zone cannot represent it accurately. The Oregon Statewide Lambert coordinate system is one example. Many countries define their own coordinate systems for national maps or their own systems of projections (similar to State Plane) for local and regional maps. Examples include the New Zealand National Grid and the Canada CSRS98 system. A variety of projections, datums, and spheroids are employed. ArcMap maintains a collection of these coordinate system definitions for use with state and country data.

Choosing projections

At the beginning of every GIS project, one of the first tasks is to establish the coordinate system to be used, as best practices dictate that all data sets in a project share the same projection and datum. Unprojected coordinate systems should be avoided because spatial analysis using a GCS may yield incorrect results. Likewise, the choice of projection impacts which map properties are preserved, and distortions caused by projections may also cause incorrect analysis results. Here are some guidelines for choosing projections for a project. Preference should be given to one of the standard coordinate systems for the sake of interoperability, but if none of the existing systems provide the right properties, a user can define his or her own.

Maps of the world. Distortions cannot be avoided in small-scale maps, so they must be managed. Consider the applications. If distances or areas are to be analyzed as part of the project, then an equidistant or equal area map should be chosen. If navigation or wind/water currents are important, then preserving direction is critical. If portraying attractive maps is the main goal, then a compromise projection might be best.

Maps of countries or continents. Many countries or regions have one or more projections defined for general mapping purposes, such as USA Contiguous Equidistant Conic or Europe Albers Equal Area Conic. Some countries have their own defined mapping systems. Like world maps, the application determines which properties of the map are important to preserve.

Maps of states. Smaller states within a single UTM or State Plane zone can use them. Mapping a large state with several zones is not as accurate as using a single zone map, and some states have a statewide projection defined, such as Oregon Statewide Lambert. If none exists, you can modify the central meridian and/or parallels of a State Plane zone to work better for the entire state. A good rule of thumb is that the central meridian should bisect the map extent, and the parallels for secant conic maps should divide the north-south extent into thirds.

Local and regional maps. Any region that fits inside a single UTM or State Plane zone can use either. The preferred system would place the region to be mapped closer to the center of the zone where the distortion is least. However, other factors may take precedence, such as the existing coordinate system of your data sources, the decision of the agency that oversees the project, or your preference for using feet or meters as the map units.

Sometimes the region to be mapped falls close to a UTM or State Plane zone boundary or spans two zones. In this case, a customized coordinate system is desirable. Typically, one starts with a UTM or State Plane zone and adjusts the central meridian and/or the standard parallels to minimize the distortion. The disadvantage is that nearly all external data must be reprojected to be included in the data set from its original source.

Regional maps in other countries. Many countries, such as Canada and Australia, have defined their own mapping systems similar to the State Plane system in the United States. Some research can be used to determine the typical projection systems and datums that are commonly used in other countries.

About ArcGIS

Labeling coordinate systems

ArcMap is able to display data together even when they have different projections. This capability relies on every data set having a coordinate system label, which records the characteristics of the coordinate system, including the datum and projection (Fig. 11.14). This label serves two purposes: it documents this information about the coordinate system for the users, and it helps GIS properly display and manage the data. Making sure that all data sets have correct coordinate system definitions is a critical step in building a GIS database. Shapefiles store the labels in a separate file with a .prj extension. Geodatabases store it within the feature class. Rasters must also have coordinate system labels, storing them in various places depending on the raster format. In all cases, the coordinate system label is stored and copied as part of the data set. A coordinate system may also be stored as a file with a .prj extension. These files may be created, saved, copied, and used to define the spatial reference for new data sets.

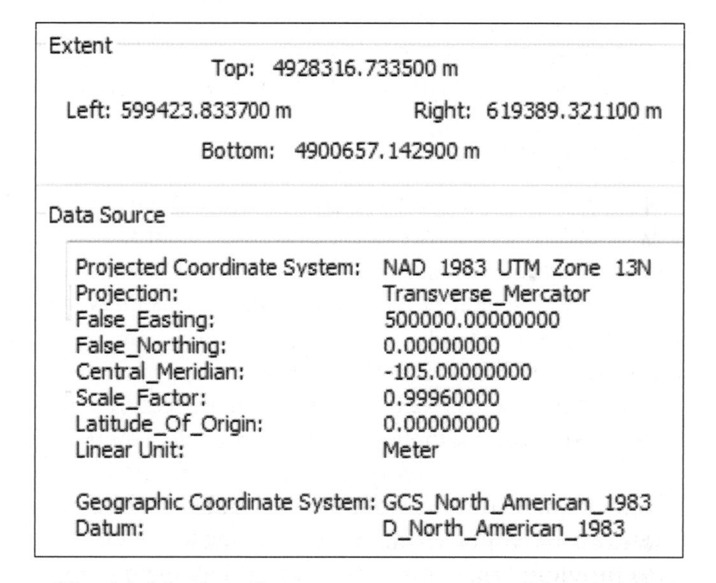

Fig. 11.14. Coordinate system extent and label

Data sets may contain unprojected or projected data. If unprojected, the coordinates are stored in decimal degrees of latitude and longitude. The feature class coordinate system label would include simply the GCS and its underlying datum, such as GCS North American 1983. The stored *x-y* coordinates of a data set on the disk would fall in the range –180 to +180 degrees for *x* and –90 to +90 degrees for *y*. The map units would be degrees.

Projected data have had a map projection applied to them, converting the GCS latitude-longitude values to feet or meters in a planar coordinate system. The map units become meters or feet. Typically, the *x-y* values in a projected coordinate system encompass millions or billions of meters. The feature class coordinate system definition includes the GCS/datum plus a description of the projection and its parameters, such as the central meridian, reference latitude, standard parallels, and false easting and northing.

The coordinate system is part of a larger description called the **spatial reference**, which includes the coordinate system, the X/Y domain, and the resolution. The **X/Y domain** is the range of allowable *x-y* values that can be stored in a feature class. Storing a world map in degrees requires a small domain, since degree values range at most between –180 and +180. Storing a world map in feet requires a large domain, because there are more than 115 million feet in the circumference of the earth. The **resolution** represents the underlying accuracy of the values; a resolution of 0.001 meter means that coordinate values are stored to the nearest thousandth of a meter.

The **extent** of the data layer is the range of *x-y* coordinates of the features actually in the feature class. You can view the extent of a data layer in ArcMap using the Source tab of the layer properties, as shown in Figure 11.14. This information can help you decide if the layer is projected or unprojected. Values between +180 and –180 indicate an unprojected GCS coordinate system with map units in degrees, whereas large values indicate a projected coordinate system with units of feet or meters.

A note on terminology

Many people use the terms *coordinate system* and *projection* interchangeably to refer to the spatial reference (the complete label) of a data set. Using the terms this way is not strictly correct, as a coordinate system is only part of the spatial reference, and a projection simply is a mathematical conversion from a 3D to a 2D coordinate system. However, it is a common practice, and usually the intended meaning is clear in context.

Missing coordinate system labels

Occasionally, a user will encounter a GIS feature class or raster without a coordinate system label, and the coordinate system appears as "Unknown" when viewed in ArcMap or ArcCatalog. To use the data effectively, the user must figure out the real coordinate system of the data and create a label for it. Finding the coordinate system can be a challenge because it usually cannot be determined by looking at the data set itself. The user must find documentation supplied by the data provider, such as a Web page or document at the site where the data were downloaded or a separate text or document supplied with the data. The user might have to contact the data supplier in person to determine this information. As a last resort, the user might have to guess one of the more commonly used coordinate systems and test whether it is correct.

Once the coordinate system is known, the user may proceed to create the coordinate system label by using the **Define Projection tool** in ArcToolbox or by using the XY Coordinate System tab from the feature class properties in ArcCatalog. The Define Projection tool labels a data set with a coordinate system and allows the user to specify all the coordinate system properties, including the GCS, datum, central meridian, and so on. Once defined, the data set is ready for use in ArcMap. The same tool is used to define a label for rasters.

On-the-fly projection

A data frame may be assigned any desired coordinate system, and then ArcMap will project all data in the frame to match. To do this, ArcMap must know the coordinate system of each data set. If the data set coordinate system definition is Unknown, then ArcMap will issue a warning message and will display the data with whatever coordinates it has. Sometimes it will fortuitously match the other data in the frame, but other times it will appear totally out of place. If it matches, then you know that the Unknown layer has a coordinate system that is the same as, or at least very similar to, the data frame, and you can create a coordinate system label for it.

The importance of having a correct coordinate system label on a data set cannot be overstated. Imagine that someone gave you the number 37 and asked you to convert it to meters. You would naturally want to know the units of the number 37; otherwise, you would not be able to choose the correct equation. Likewise, ArcMap needs to know the coordinate system of a data set in order to choose the appropriate equations to project it to match the data frame. If you give a feature class the wrong coordinate system label, then ArcMap will choose the wrong equations, and the features will end up in the wrong place.

ArcMap can adjust differences in datums on the fly, but the user has to select an appropriate transformation method, which many users lack the advanced knowledge to do, and the transformation can cause new positional errors up to several meters. Therefore, ArcMap does not automatically perform datum transformations. The user is warned if a data set is added with a different GCS than the data already in the frame, and he or she can then select a transformation if desired. For national- or world-scale data, the differences between datums is small compared to the accuracy of the data, so no transformation is necessary. For large-scale data in situations where high accuracy is needed, the user should consult an expert on transformation methods. There is one exception: ArcMap automatically applies a transformation called NADCON if the two datums are NAD 1927 and NAD 1983.

Projecting data

Usually, GIS projects choose a single coordinate system and convert all data to match it. Many organizations establish an official projection for their area of interest and use it consistently. Outside data from different coordinate systems are always converted to the official one at the outset, using the **Project tool**.

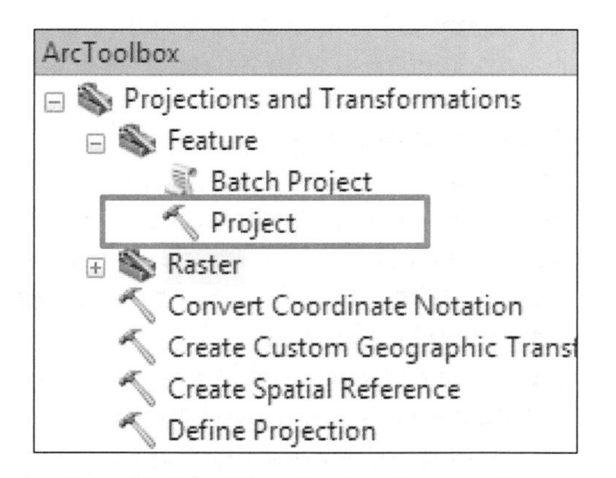

Fig. 11.15. The Projection tools

The Project tool is found under the Data Management section of the Toolbox (Fig. 11.15). It converts the *x*-*y* coordinate values in a feature class to a different coordinate system and saves them in a new feature class with a new coordinate system label. The original data remain unchanged. The original data set must have a correctly labeled coordinate system before it is projected. If the label is Unknown, the Project tool cannot work. The Project Raster tool is used to project rasters.

Troubleshooting coordinate system problems

A coordinate system problem occurs when two map layers should map on top of each other—but don't. They might be off by a few hundred meters or by thousands of miles. Most of these problems result from the mislabeling of a data set with the wrong coordinate system. The usual remedy is to determine the correct coordinate system and correct the label.

ArcMap can be used to troubleshoot such problems. In Figure 11.16, the user is examining the properties of a feature class. The extent shows *x-y* values in the range –103.6 to –103.5 for *x* and 44.3 to 44.4 for *y*. These values are clearly degrees. However, the label underneath lists NAD 1983 UTM Zone 13N as the coordinate system, and the *x-y* values are listed with meters as the map units. The label does not match the *x-y* extent values. So which is correct?

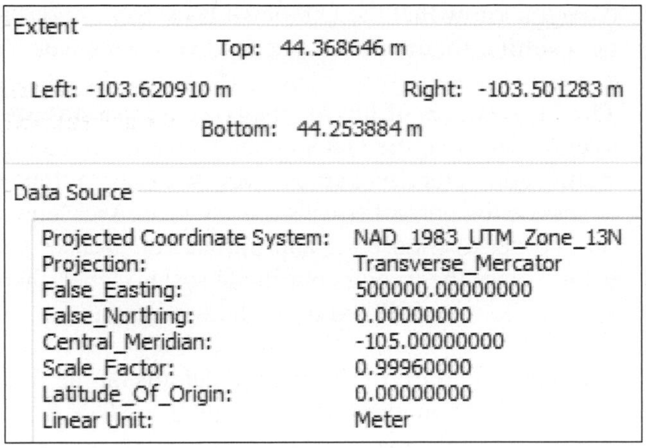

Fig. 11.16. An incorrectly labeled coordinate system

The *x-y* values are always correct because they are obtained from the values stored in the file. This data set really is stored in degrees, and the label is incorrect. Imagine adding this feature class to a data frame set to the UTM coordinate system. ArcMap reads the label and mistakenly concludes that the *x-y* values are already in UTM meters and don't need to be projected on the fly. The *x-y* values are plotted as degrees in the UTM coordinate space, so they don't match the other data and appear in a completely different location.

To fix the problem, you must know the correct coordinate system and use Define Projection to modify the label. You will need another data set in the same area that is known to be correct so that you can compare the result and test the new label. If the data set aligns with the known one, then you set the label correctly. If it is still misaligned, then a different label is needed.

Troubleshooting Steps

- ➔ Add a data set known to be correct to ArcMap first to set the data frame coordinate system. Record the coordinate system of the reference data.
- ➔ Add the data set causing the problems. Examine its coordinate system and extent, and make sure that the label and extent appear to match. If they don't, or if the data set has an Unknown or Undefined coordinate system, you must try to figure out what it is.
- ➔ Examine the extent and consider what you know about the source of the data. If the coordinates are in degrees, they are in a GCS. If not, make your best guess as to the coordinate system and datum using any clues or resources you have available.
- ➔ Use the Define Projection tool to assign your guess to the problem data set. Rasters may need to be removed and added again afterward to see the difference.
- ➔ If your guess is correct, the data sets should align. If not, try a different guess.

Data in a GCS are relatively easy to correct because the choice of possible datums is limited. Projected data are virtually impossible to guess unless they are in one of the common projections or you have some outside information to narrow down the possibilities.

Define Projection versus Project

Why would a data set be mislabeled with the wrong coordinate system? The usual culprit is a user who does not understand the tools used to manage coordinate systems, and who makes the mistake of using Define Projection when he or she should use the Project tool.

Imagine Frank has a shapefile with GCS coordinates and wants to convert it to a UTM projection to match his other data. He carelessly selects the Define Projection tool from ArcToolbox and sets the coordinate system to UTM. However, the tool merely puts a UTM label on the shapefile without changing the coordinates inside. So now the shapefile is incorrectly labeled UTM and the coordinates inside are still in a GCS. (This situation is analogous to placing an albacore tuna label on a can of cat food, with disastrous results to your dinner recipe.) When the data are added to ArcMap, it will choose the wrong equations to convert the data to the data frame of the coordinate system, and the feature class will no longer appear where it should.

WARNING: Many people get confused about the functions of the Define Projection tool and the Project tool. This confusion can ruin data sets, so **pay attention** to the next two paragraphs and then be careful to select the correct tool for the task at hand.

The **Project tool** acts on the *x-y* coordinates of a layer and converts them to a different coordinate system, producing a new feature class and leaving the original feature class unchanged. The new file has a new coordinate system and label. You use the Project tool if you have a layer in one coordinate system and want to permanently convert it to a different one. The Project tool should be used only on layers with correct coordinate system definitions.

The **Define Projection** tool only changes the coordinate system label of the feature class, without affecting the coordinates inside. It should only be used on a data set that has an Unknown coordinate system or on a data set that was previously mislabeled and needs to be fixed.

Summary

- All GIS data have coordinate systems, which define the units and axes used to represent map features as *x-y* coordinates. The complete definition is called the spatial reference.

- A geographic coordinate system, or GCS, is based on angular measures of latitude and longitude with units of degrees. A GCS is based on a datum.

- A datum includes a spheroid and its location relative to the earth's geoid. It is used to reduce map errors introduced by differences between the spheroid and the earth's actual surface. Common datums in North America include the NAD 1927, NAD 1983, and WGS 1984.

- Data for a project should be stored in a common GCS. Transforming from one datum to another should be minimized because the process may introduce new positional errors.

- Map projections are mathematical equations that convert degrees from a GCS into planar *x-y* coordinates of meters and feet so that the map may be portrayed on a flat sheet of paper. All map projections introduce distortions of area, distance, direction, and shape.

- Rasters also have coordinate systems. Assigning coordinate systems to rasters may involve creating world files for them or georeferencing them based on control points.

- UTM and State Plane are common projection systems in the United States. Each is a family of projections designed to minimize distortion over the area covered (a zone).

- Every data set used in ArcGIS should have the correct coordinate system defined and stored with the map features. ArcGIS uses these definitions to display data correctly.

- Most coordinate system problems result from missing or incorrect labels on data sets. The Define Projection tool is used to update a data set with an undefined or incorrect label.

- Data may be permanently projected from one coordinate system to another using the Project tool or Project Raster tool in ArcToolbox.

TIP: The ArcGIS Help section also has some good material about projections.

Important Terms

affine transformation	geographic coordinate system (GCS)	Project tool	reference latitude
central meridian	geoid	projection,	resolution
coordinate pair	georeferenced	azimuthal	RMSE
coordinate space	ground control points	conic	rubbersheeting
coordinate system	latitude	cylindrical	spatial reference
datum	latitude of origin	oblique	spheroid
Define Projection tool	longitude	orthographic	standard parallel
degrees	map units	secant	State Plane
ellipsoid	origin	stereographic	Transformation
extent	parameters	tangent	UTM
false easting	Prime Meridian	transverse	world file
false northing		rectify	X/Y domain

Chapter Review Questions

1. If a data set's features have *x* coordinates between –180 and +180, what is the coordinate system likely to be? In what units are the coordinates?

2. What are the *x-y* coordinates of a map's origin? ___________ What is the *x* coordinate along the central meridian?_______________________

3. What is the difference between a spheroid and a geoid?

4. Examine Figures 11.5 through 11.7, and explain why conic projections usually conserve area and distance but cylindrical projections typically preserve direction.

5. What extra step is performed when projecting rasters that is not needed when projecting vector data? What happens during this step?

6. What is the difference between a central meridian and the Prime Meridian?

7. You have a shapefile with an Unknown coordinate system, but a file on the web site says that the coordinate system is UTM Zone 13 NAD 1983. What is your next step?

8. True or False: A shapefile of the United States with a GCS coordinate system would have an *x-y* extent that contains entirely positive values. ________ Explain your answer.

9. You have a shapefile with a UTM Zone 10 NAD 1983 coordinate system, and you want to bring it into your state database, which uses the Oregon Statewide Lambert coordinate system. What is your next step?

10. You need to create a map for the entire state of Idaho. What options do you have? How could you get the most accurate map possible?

Mastering the Skills

Teaching Tutorial

The following examples provide step-by-step instructions for doing basic tasks and solving basic problems in ArcGIS. The steps you need to do are highlighted with an arrow ➔; follow them carefully. Click on the video number in the VideoIndex to view a demonstration of the steps.

Displaying coordinate systems

➔ Open ArcMap and choose to start with a new, empty map.

➔ Use Save As to save the document with the name of your choice.

1➔ Use the Add Data button to add the country and latlong shapefiles from the mgisdata\World directory.

1➔ Change the data frame name from Layers to World.

1➔ Right-click the country layer, choose Properties, and then click the Source tab.

1. What is the coordinate system for this feature class? ________________________

Observe the values in the top of the window, which show the extent of *x-y* values present in the shapefile. The extent is the range of *x-y* values of features stored in the data set.

2➔ Close the Layer Properties window.

2➔ Observe the coordinates shown in the lower part of the window and confirm that they are in degrees. The data frame units default to the units of the first data set loaded.

2➔ Zoom in to the tip of Florida in the United States.

2➔ Use the cursor to hover over the most southeastern tip of Florida and observe the *x-y* coordinates (Fig. 11.17).

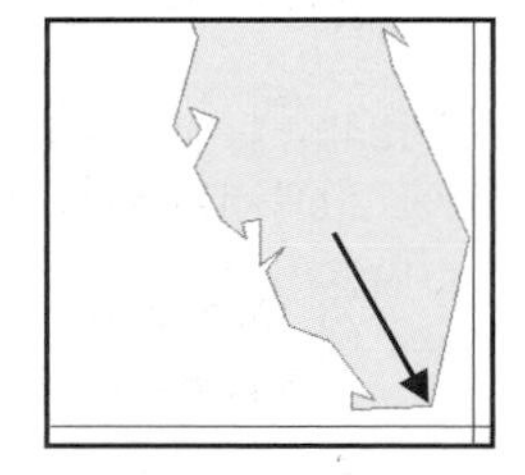

Fig. 11.17. Hover here.

2. What are the coordinates of the SE tip of Florida? __________

3➔ Right-click the World data frame name and choose Properties.

3➔ Click the General tab.

3. What are the map units of this frame? ________What are the display units? __________

The map units are dimmed because you can't set them—they are based on the data frame coordinate system. The display units can be set to any units desired by the user.

3➔ Set the display units to miles and click OK.

4. What are the coordinates for Florida's tip now? ______________________

The *x* value tells you the distance Florida lies from the central meridian of the map (in this case the Prime Meridian that runs through Greenwich), and the *y* value tells you its distance from the reference latitude (the equator).

4➔ Click the Full Extent button to see the whole world again.

4➔ Open the World data frame properties.

4➔ Click the Coordinate System tab and notice what it says.

The data frame coordinate system defaulted to match the first data layer added to it. However, the data frame can display maps in any coordinate system, either geographic or projected.

4➔ In the folder tree, scroll back to the top and collapse the Geographic Coordinate Systems folder.

4➔Expand the Projected Coordinate Systems folder and examine its contents.

4➔Expand the World folder. Click on the Mercator (world) projection and click OK.

TIP: As you work with the *data frame* coordinate systems in ArcMap, keep in mind that they can be different from the coordinate systems of the *feature classes* you are working with.

5➔ Zoom in to the tip of Florida again. The display units are still miles because you set them. What are the units of this projection?

5➔ Open the data frame properties and click the General tab.

5. What are the map units for this Mercator projection? _______________

5➔ Close the data frame properties window.

5➔ Zoom to the previous extent and use the cursor and coordinate box to locate the coordinate origin, where both x and y equal zero (approximately—look for where the coordinates change from positive to negative).

Mercator is a cylindrical projection. The central meridian where $x = 0$ corresponds with the Prime Meridian that runs through Greenwich, England. The $y = 0$ line is the equator and is called the latitude of origin. Coordinates west of the central meridian and south of the reference latitude have negative values. Florida is west of the central meridian; its x coordinates are negative.

6. Which continent has primarily negative x AND negative y coordinates in this projection? ____________ Which one has primarily positive x and y coordinates? ______________

6➔ Use the Insert menu to add a new data frame to the map. Name it USA.

6➔ Add the states feature class from the mgisdata\Usa\usdata geodatabase.

6➔ Zoom in to the lower 48 states.

This feature class is stored in a projected coordinate system, North America Equidistant Conic, and the data frame defaults to it when the feature class is loaded.

7➔ Open the Properties for the USA data frame and click the Coordinate System tab. Read the properties of the current coordinate system.

7. What longitude is the central meridian? _________What is the latitude of origin? _________

8. Examine the standard parallels and the latitude of origin, and predict whether any areas of the United States have negative y coordinates in this projection. ______ Why or why not?

The coordinate system label indicates that this map has no false easting or northing applied. It has two standard parallels where the projection is free of distortion.

9. Is this Equidistant Conic projection a tangent or secant projection? ______________________
   How can you tell? ___________________________

   7➔ Close the USA data frame properties window.

Changing the projection parameters changes the appearance of the map. Let's try an example.

   8➔ Right-click the World frame and choose Activate from the menu.

   8➔ Choose Add Data, and add the states feature class from the usdata geodatabase.

   8➔ A Geographic Coordinate Systems Warning window should appear. Read it.

   8➔Click the hyperlink *About the geographic coordinate systems warning*. Read it.

10. Why did this warning appear?

11. Sometimes ArcMap automatically chooses a datum transformation for you. Will it do so in this case? Why or why not?

   9➔ Close the Help window, if necessary, and click the Transformations button.

   9➔ In the *Convert from:* box (Fig. 11.18), select GCS North American 1983 (the states GCS). Leave the *Into:* box set to GCS_WGS_1984.

   9➔ Click the *Using:* drop-down and examine the long list of transformation choices available.

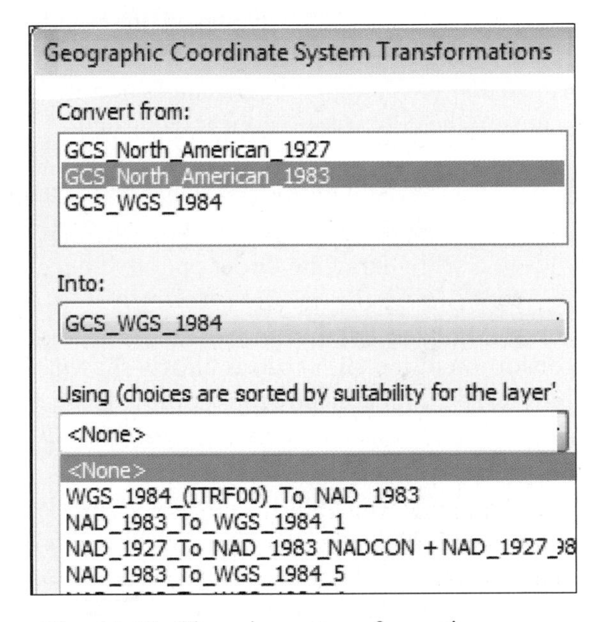

Fig. 11.18. Choosing a transformation

Converting between datums is more complex than changing projections and requires knowledge beyond that of the typical user. This complexity is one reason it is advisable to assemble your data sets in a single GCS so that the appearance of this window is rare.

In this case, however, we are looking at world data. The difference between datums is never more than a few hundred meters, which would not be noticeable at this scale and is smaller than the geometric accuracy of the data, anyway. Therefore, we will not apply a transformation.

   9➔ Click Cancel.

   9➔ Check the box *Don't warn me again in this session* and click Close.

Understanding map distortion

The circumference of the world at the equator is ~25,000 miles. The circumference at the poles is zero. Latitudes in between have intermediate lengths. Let's look again at the WGS 1984 GCS.

   10➔ Open the World data frame properties and select the Coordinate System tab.

10➔ Collapse all the folders until only the main four are shown; then expand the Layers folder.

Favorites
Geographic Coordinate Systems
Projected Coordinate Systems
Layers
GCS_WGS_1984
North_America_Equidistant_Conic

The Layers folder shows coordinate systems of layers in the map document. GCS_WGS_1984 would also be found in the Geographic Coordinate Systems folder, but it is easier to select it here.

10➔ Expand the + sign next to GCS_WGS_1984 to see the layers with this GCS.

10➔ Click the GCS_WGS_1984 coordinate system to select it and click OK.

When a GCS data set is displayed, the angular units are automatically drawn in an equirectangular projection, as if the degrees were a planar x-y coordinate system from $x = -180$ to $+180$ and $y = -90$ to $+90$. Distances become increasingly distorted away from the equator, causing the countries to appear elongated.

11➔ Add the circles shapefile from the mgisdata\World folder.

11➔ Drag the circles above the latlong layer in the Table of Contents (switch to the List by Drawing Order option, if necessary).

Each circle has a radius of 5 degrees and would appear identical on a globe, but here the distance distortion of the GCS map is clearly shown. The areas and distances, if determined from these circles, would be incorrect over most of the globe. *Any GIS analysis based on distance or area measurements would be wrong when data are stored in a GCS.*

12➔ Open the World data frame properties. Move the window so that you can leave it open and still see the map.

12➔ Collapse/expand the folders to select the Projected Coordinate Systems > World > Mercator (world) projection. Click Apply.

Area and distance are also distorted in this projection, except at the equator where the sphere touches the paper. However, Mercator does preserve direction and shape. The circles remain circles, and the longitude meridians point north.

13➔ Change the coordinate system to Sinusoidal (world). Click Apply.

Sinusoidal is an equal-area projection, meaning that it preserves area and, to some extent, distance. However, it does so at the expense of shape and direction. Notice the squashed look of the United States and that the longitude lines converge toward the poles. The circles have equivalent areas now, but some are no longer circles. The distortion gets worse away from the equator and the central meridian.

14➔ Change the coordinate system of the frame to Robinson (world). Click Apply.

Robinson is called a compromise projection because it minimizes distortion in all four properties but preserves none. Distortion is unavoidable when using small-scale world and national maps. There are excellent projection systems for large-scale maps. Two common systems in the United States are the Universal Transverse Mercator (UTM) system and the State Plane Coordinate System (SPCS).

15→ Collapse the World folder and expand the UTM folder. Navigate to UTM > WGS 1984 > Northern Hemisphere > WGS 1984 Zone 35N and select it. Click OK.

15→ Turn off the latlong layer.

15→ Add the shapefile utmzone35 from the mgisdata\World folder. Make the symbol hollow with an orange outline.

The narrow strip of UTM Zone 35 runs along the tangent of a transverse cylindrical projection. Within the zone, the circles remain circles and all four properties of area, distance, shape, and direction are preserved (except in Antarctica, which is not in Zone 35N).

TIP: You can use Figure 11.12 or 11.13 to select UTM or State Plane zones for a particular map area. If you need a more detailed map, feature classes of UTM zones and State Plane zones are in the World folder and usdata geodatabase, respectively, and can be viewed in ArcMap.

12. Which UTM zone should be used for the state of Nevada? ______________________

Troubleshooting projection problems

ArcMap relies on correct coordinate system labels to perform on-the-fly projection.

16→ Click the New Map File button to start a new, empty map. Save the map changes.

16→ Add the usfsrds feature class from mgisdata\BlackHills\Sturgis83.mdb.

16→ Open the Properties for usfsrds, click the Source tab, and write the coordinate system and datum of the layer here: ______________________________

16→ Also examine the Extent information, which shows the range of *x-y* coordinates as stored in the file. Close the Properties window.

17→ Add the majroads feature class from mgisdata\Usa\usdata geodatabase and write its coordinate system here: ______________________________ Close the window.

Even though the two coordinate systems are different, the maps overlay each other correctly because each has an accurate label. There are some offsets, but these are to be expected for two sets of roads derived from different-scale maps (originally 1:24,000 and 1:250,000).

13. State the data frame coordinate system WITHOUT looking at its properties. ____ How could you know?

18→ Remove the majroads layer from the map.

18→ Add the topo30m raster from the BlackHills\Sturgis83 geodatabase and give it an attractive color ramp.

Your supervisor has requested that you add benchmark locations to this geodatabase. A colleague has offered to send them to you as four quadrangles. You accept, not realizing that this colleague is not as well trained as he should be.

Your task is to make sure that all four shapefiles are in your geodatabase's coordinate system of NAD 1983 UTM Zone 13N, and then you will merge them into a single feature class.

19→ Add the benchmarks1 shapefile from the BlackHills folder.

The first problem surfaces with a warning that this shapefile has an unknown spatial reference. It lacks a label identifying its coordinate system, and you must give it one. You call your colleague to find out what they use but learn that he is on vacation for two weeks. You decide to see if you can work it out yourself.

19➔ Click OK on the warning window. The layer appears in the Table of Contents but you can't find any points on the map. Maybe the symbol is too small?

19➔ Change the symbol for the benchmark1 layer to Triangle 3. Still nothing.

19➔ Right-click the benchmarks1 layer and choose Zoom to Layer.

Well, here are the benchmarks, but they don't appear in the same coordinate space as the other data. This error generally results from an improper or missing coordinate system label.

20➔ Open the properties for benchmarks1 and click the Source tab.

20➔ Examine the coordinate system and the extent carefully (Fig. 11.19).

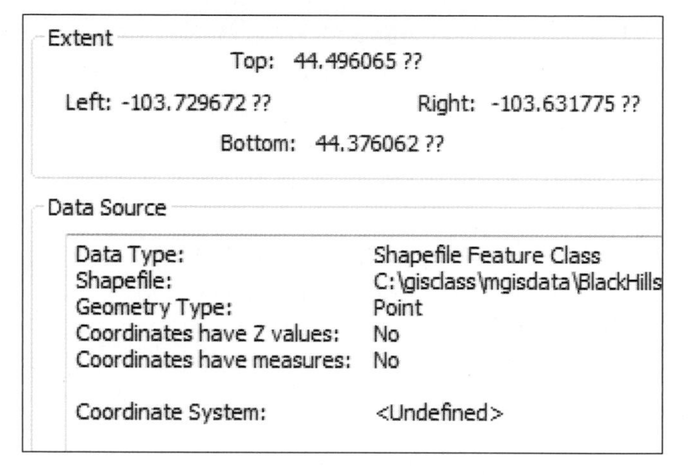

Fig. 11.19. An unlabeled coordinate system

This coordinate system is undefined, as you already know. The Extent shows the range of *x-y* coordinates as small numbers, less than 180, which is a sure sign that you are dealing with an unprojected coordinate system in degrees, or a GCS.

So you know to assign a GCS label, but you also need to assign a datum. There is no way to know for certain, but NAD83 is the most common datum for US data, so you will try it first.

21➔ Close the Layer Properties window.

21➔ Open ArcToolbox > Data Management > Projections and Transformations > Define Projection.

21➔ Set the input dataset to benchmarks1.

21➔ Click the coordinate system browse button to open the Spatial Reference Properties.

21➔ Expand Geographic Coordinate Systems > North America and select NAD 1983.

21➔ Click the Add to Favorites button, as this GCS is very common.

21➔ Make sure that NAD 1983 is still highlighted and click OK.

21➔ Click OK to run the Define Projection tool.

22➔ Click the Full Extent button to return to the view of your study area.

Now the benchmarks appear on the map near roads and peaks. You think that you have solved the problem; however, to be absolutely sure, you will check the benchmarks against a reliable source.

22➔ Click the Add Data drop-down button and select Add Data From ArcGIS Online.

22➔ Search for topo, and add the USA Topo Maps map service (not the LPK file).

22➔ Zoom in to several benchmarks and confirm that they lie on or very close to the benchmarks on the topo map (a black X or triangle labeled with the elevation).

22➔ Click the Full Extent button to return to the larger view.

Oops—the zoom now goes to the full extent of the map service rather than the Sturgis area. However, we can fix this using a property of the data frame that controls how the full extent is defined.

23➔ Open the Layers data frame properties and click the Data Frame tab.

23➔ In the middle panel under Extent Used By Full Extent Command, fill the Other button and click Specify Extent.

23➔ Fill the button for Outline of Features and select the usfsrds layer from the drop-down list. Click OK and OK.

23➔ Click the Full Extent button again. Now it shows the area you want.

23➔ Turn off the USA Topo Maps layer for now, but keep it in the map document.

So benchmarks1 is now correctly labeled, but it is in a GCS. You want all the files in your geodatabase to be in NAD 1983 UTM Zone 13N. To convert it, you must use the Project tool.

24➔ Open the ArcToolbox > Data Management > Projections and Transformations > Feature > Project tool.

24➔ Set the input dataset to benchmarks1.

24➔ Set the output as a shapefile named bench1utm.shp in the BlackHills folder.

24➔ Click the Browse button to set the spatial reference.

24➔ Expand the Layers folder.

24➔ Select the NAD_1983_UTM_ Zone_13N coordinate system.

24➔ Click the Add to Favorites button and click OK.

24➔ Confirm that the Project tool arguments appear as in Figure 11.20, and click OK.

Input Dataset or Feature Class
benchmarks1
Input Coordinate System (optional)
GCS_North_American_1983
Output Dataset or Feature Class
C:\gisclass\mgisdata\BlackHills\bench1utm.shp
Output Coordinate System
NAD_1983_UTM_Zone_13N
Geographic Transformation (optional)

Fig. 11.20. Projecting the benchmarks

TIP: We used the coordinate system listed in the Layers folder to ensure that the benchmark's new coordinate system matched the data we already had.

25➔ Remove benchmarks1 and give the new bench1utm layer the Triangle 3 symbol.

25➔ Add the benchmarks2 shapefile from the BlackHills folder and give it the Triangle 3 symbol with a different color.

25➔ Open the properties for benchmarks2 and click the Source tab.

This shapefile is also missing the spatial reference, but in this case the benchmarks appear in the right places. Looking at the Extent, you can see that the values are far too large to be degrees, so you are dealing with a projected coordinate system. But which one is it?

Your clue is that the benchmarks are in the right places. If a data set has no label, then ArcMap does not attempt to project it on the fly but simply plots the *x-y* coordinates in the file. Thus, you know that the file coordinate system matches that of the data frame, NAD 1983 UTM Zone 13N, and that is what needs to go on the label.

26➔ Close the Layer Properties window.

26➔ Open ArcToolbox > Data Management > Projections and Transformations > Define Projection.

26➔ Set the input dataset to benchmarks2 and choose NAD_1983_UTM_Zone_13N from the Favorites folder. Click OK and OK to run the tool.

26➔ Zoom in to check a couple benchmarks against the USA Topo Maps layer, and then turn the topo layer off and zoom back to the full extent of the map.

Now the benchmarks2 coordinate system is labeled, and it already matches our geodatabase coordinate system, so we don't need to project it. It's time for the next shapefile.

27➔ Add the benchmarks3 shapefile from the BlackHills folder.

27➔ Give it the Triangle 3 symbol in a third color, but no benchmarks appear.

27➔ Open its layer properties and examine the Source tab (Fig. 11.20).

This one is interesting. According to the Data Source, the coordinate system is NAD 1983 UTM Zone 13N, but the Extent values are small and appear to be in degrees (Fig. 11.21). We have a disagreement between the extent and the label. Which one is correct?

TIP: The Extent *x-y* values show the values actually in the file and are the best guide to the true coordinate system.

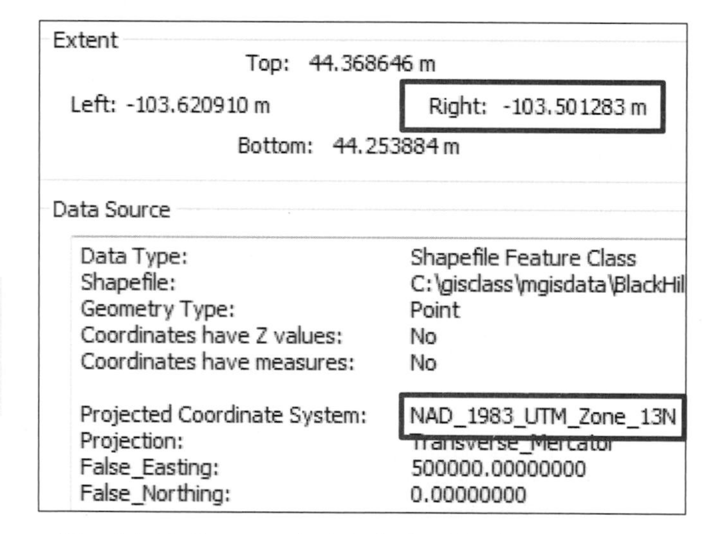

Fig. 11.21. Incorrectly labeled GCS shapefile

This data set exemplifies what happens when a user does not understand the difference between the Project and Define Projection tools. Your GIS colleague started with a correctly labeled GCS shapefile in degrees. He wanted the data in UTM. Instead of using the Project tool to convert from GCS to UTM, he used Define Projection, which only changed the label. As a result, the GCS data set is now incorrectly labeled as a UTM data set. To fix it, you must re-label it with a GCS label.

28➔ Close the Layer Properties window.

28➔ Open the Define Projection tool.

28➔ Set the input data set to benchmarks3. A warning appears because it already has a coordinate system label. Ignore it—you know that the label must be corrected.

28➔ Click the Browse button and select the NAD 1983 coordinate system from the Favorites folder. Click OK and OK.

28➔ The benchmarks appear in their proper places. Check some against USA Topo Maps, and then zoom back to the full extent.

Now they are correctly located, but the benchmarks3 shapefile is still in a GCS, and it needs to be projected to UTM (as the original user should have done).

29➔ Open the Project tool.

29➔ Set the input to benchmarks3. Save it in the BlackHills folder as bench3utm.shp.

29➔ Browse and select NAD 1983 UTM Zone 13N from Favorites. Click OK and OK.

29➔ Remove benchmarks3 and give bench3utm the Triangle 3 symbol.

30➔ Add the benchmarks4 shapefile from the BlackHills folder and give it the Triangle 3 symbol in a fourth color. You get a warning, but the benchmarks appear on the map.

30➔ Open the properties for benchmarks4 and click the Source tab. Examine it.

It is another undefined coordinate system, but it appears to be in UTM because the benchmarks show up in the right spots. You just need to set the label as before.

31➔ Close the Layer Properties window.

31➔ Open the Define Projection tool and set the input data set to benchmarks4.

31➔ Browse and select NAD 1983 UTM Zone 13N. Click OK and OK.

31➔ Zoom in and compare several benchmarks to the USA Topo Maps layer.

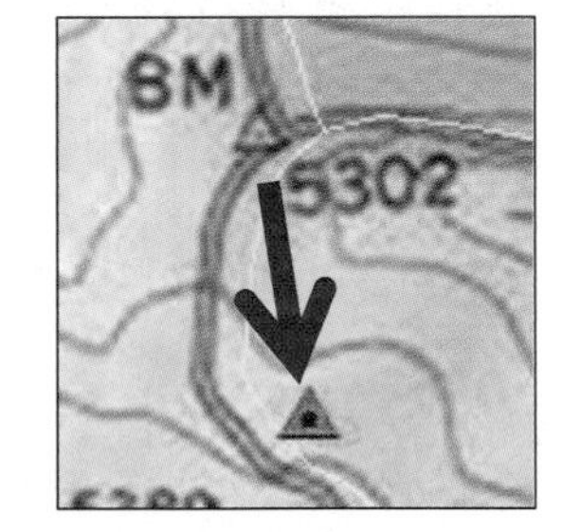

This time you notice something odd. These benchmarks appear several hundred meters south of the benchmarks in the topo map. Something is still wrong. An offset of this magnitude is most likely to be a datum mismatch. You surmise that perhaps benchmarks4 was in the NAD 1927 datum, and so the label you assigned was not correct. To test your guess, you will reassign the label.

32➔ Open the Define Projection tool again and set the input data set to benchmarks4.

32➔ Browse in the Projected Coordinate Systems > UTM > NAD 1927 folder and select the NAD 1927 UTM Zone 13N coordinate system. Click OK and OK.

32➔ Compare the benchmark locations to the USA Topo Maps layer again.

You were right. With the NAD 1927 UTM label assigned, the benchmarks now appear where they should. However, you want all your data in the geodatabase to be in the same coordinate system, so you still need to project this file from NAD 1927 UTM to NAD 1983 UTM.

33➔ Open the Project tool and set the input to benchmarks4.

33➔ Save the output in the BlackHills folder as bench4utm.shp.

33➔ Browse and select NAD 1983 UTM Zone 13N from Favorites. Click OK.

Notice that this time a transformation method is entered to convert the NAD 1927 datum to NAD 1983. The best one is chosen for you, NADCON.

33➔ Click OK to run the tool.

33➔ Remove benchmarks4 and give bench4utm the Triangle 3 symbol.

33➔ Zoom to the full extent and turn off USA Topo Maps.

You are ready to merge the four shapefiles, now all in NAD 1983 UTM Zone 13, to a single new feature class in the Sturgis geodatabase.

34➔ Open ArcToolbox > Data Management > General > Merge.

34➔ Use the drop-down under Input Datasets four times to add the four layers: bench1utm, benchmarks2, bench3utm, and bench4utm. The order does not matter.

34➔ Save the output as benchmarks in the Features dataset of the Sturgis83 geodatabase. Click OK.

34➔ Turn off the four original benchmark layers and give the new benchmarks layer the Triangle 1 symbol (Fig. 11.22).

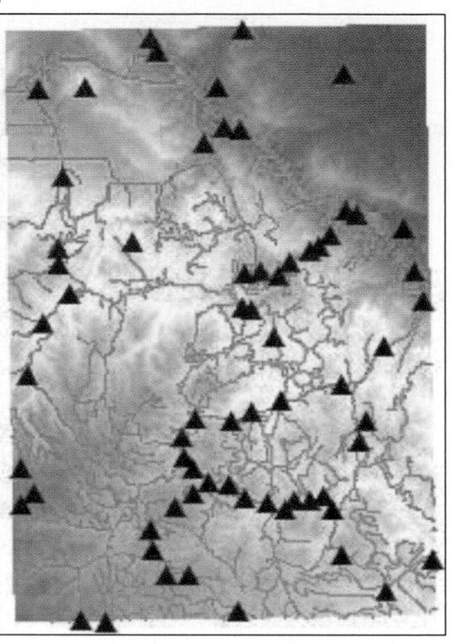

Fig. 11.22. Final benchmarks

Creating a custom coordinate system

Sometimes you don't have a predefined coordinate system for the area you are working on and need to define a custom one. Imagine creating a geodatabase for Turkey to display the general features with no need for equal areas or equal distances. A search of the available coordinate systems in ArcMap yields none specific to Turkey. You begin by choosing a suitable projection.

35➔ Start a new, empty map. Save changes to the previous map document.

35➔ Add the country shapefile from the mgisdata\World folder.

36➔ Use Select by Attributes to select Turkey.

36➔ Right-click the country layer and choose Selection > Create Layer From Selected Features.

36➔The layer appears in the Table of Contents. Rename it Turkey and zoom in.

37➔ Open the Layers data frame properties and click the Coordinate System tab.

37➔ Type turkey in the search box and click the Search button.

37➔ Expand the Projected Coordinate Systems > Gauss Kruger folder and the Turkey subfolder. Seven different zones are suggested.

37➔ Expand the National Grids folder and the Turkey subfolder. A number of coordinate systems are in here. Click several of them just to see their parameters.

37➔ Close the data frame properties window.

The predefined coordinate systems in ArcMap are all based on the Gauss Kruger projection (which is another name for Transverse Mercator). We know that this projection is best for north-south-oriented areas, but Turkey is oriented east-west and is too wide for a single TM zone. Our best bet will be to design our own coordinate system based on a conic projection, with a central meridian in the center of Turkey, and standard parallels that cut it into approximate thirds.

38➔ Find the approximate central longitude and range of latitudes of Turkey using the cursor and reading the values from the location display.

Central longitude ________ Southern latitude ________ Northern latitude_______

Armed with this information, we can now define a custom coordinate system, which we will base on the Lambert Conformal Conic projection.

39➔ Open the data frame properties and click the Coordinate System tab.

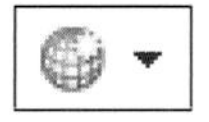

39➔ Click the Add Coordinate System drop-down button and select New > Projected Coordinate System.

39➔ Enter Turkey Lambert Conformal Conic WGS84 as the name (Fig. 11.23).

39➔ Choose Lambert_Conformal_Conic from the drop-down list for the projection name.

39➔ Carefully change the Central_Meridian to 35 degrees.

39➔ Set Standard_Parallel_1 to 37.5 degrees. Set Standard Parallel_2 to 39.5 degrees.

39➔ The Geographic Coordinate System defaulted to the data frame GCS, WGS 1984. Leave it.

39➔ Click OK and OK.

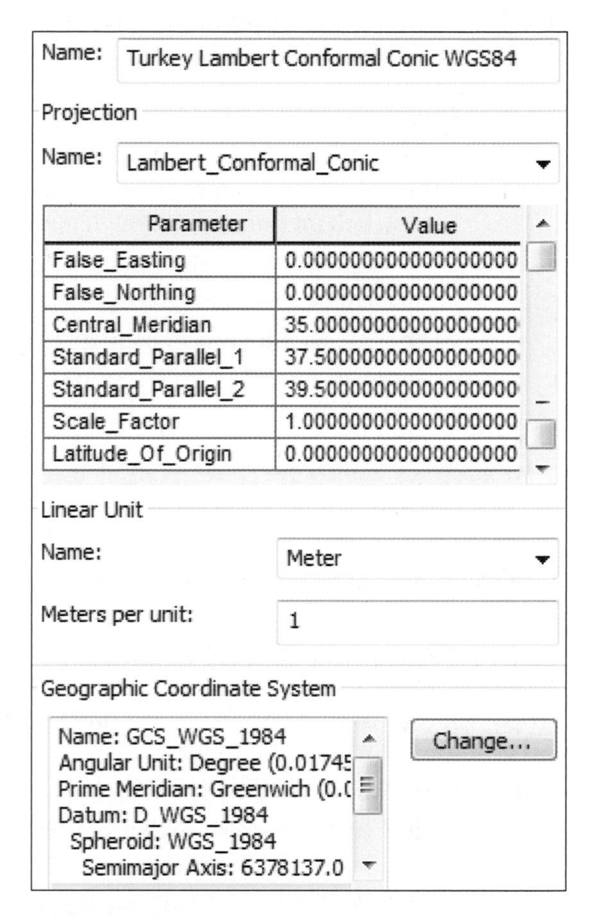

Fig. 11.23. A custom coordinate system

The map looks fine. Now that we are sure that this coordinate system is acceptable, we can save the projection file to use again.

40➔ Open the data frame properties again.

40➔ Right-click the new coordinate system in the Custom folder and choose Save As.

40➔ Navigate to the World folder and save the .prj file using the suggested name.

40➔ Close the data frame properties.

TIP: To use the .prj file, click the Add Coordinate System drop-down and choose Import. You can open it in Notepad if you want to see what it looks like—it is a text file.

Georeferencing a raster

In this section, you will download an image from the Internet and georeference it. If you don't have Internet access or if you have trouble obtaining an image, use the nwsradar_sample.gif image stored in the mgisdata\Usa folder. Your Web browser instructions might be slightly different if you are not using the Internet Explorer Web browser.

41➔ Open your Web browser and go to http://www.nws.noaa.gov/radar_tab.php.

When georeferencing, you want to use a coordinate system that is as close as possible to the one in the image. Try to identify this one. Use the inside front cover of the book to make comparisons, if you wish.

14. On what type of projection does the map appear to be based? _______________

41➔ Click on Full resolution version.

41➔ Right-click the enlarged image and choose Save Picture As. (If a polygon appears instead of a menu, right-click again inside the polygon to bring up the menu.)

41➔ Navigate to your mgisdata\Usa folder and save the image as a GIF file. Name it nwsradar.gif.

42➔ Open a new map document and add the states feature class from the mgisdata\Usa\usdata geodatabase.

To georeference the image, you must set the data frame to the coordinate system the image uses, in this case, a GCS. You will assume it is NAD 1983, the most common US datum.

42➔ Change the data frame coordinate system to NAD 1983 from the Favorites folder.

42➔ Zoom to the full extent to see the states layer again, if necessary.

42➔ Switch back to the browser for a moment and compare the shapes of the states in ArcMap and in the image. They appear similar.

42➔ Close the browser window and return to ArcMap.

43➔ Zoom in to the conterminous states so that the image and the map window are showing a similar map extent. This will facilitate finding control points.

43➔ Change the symbol for the states so that it is hollow with a high-contrast outline color, such as hot pink. It needs to show up well against the image.

43➔ Choose Customize > Toolbars > Georeferencing from the main menu bar.

43➔ Click on Add Data and add the nwsradar.gif image from the Usa folder. Click OK if you get a warning or are asked to build pyramids.

The image does not appear yet because it is in a different coordinate system than the map. You begin by bringing the image into the same map extent as the states, which will make it easier to match control points.

44➔ On the Georeferencing toolbar, choose Georeferencing > Fit to Display. The ArcMap session should look similar to Figure 11.24.

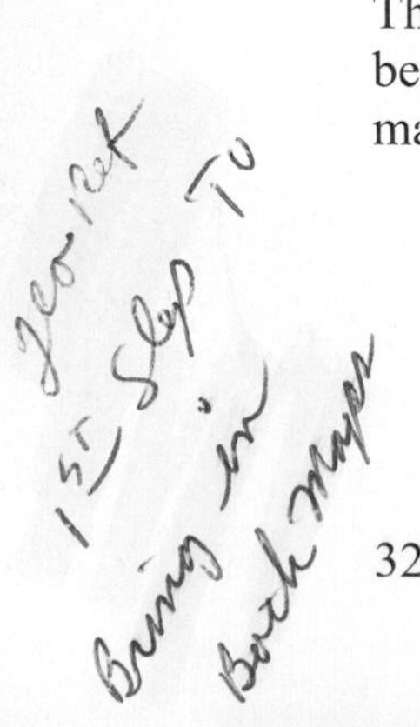

The points work best if they are spread around the map. You will first add two at diagonal corners and then two near the other corners. If the state corners or boundaries are obscured by the radar colors on your map, choose another point that is close by and clearly visible.

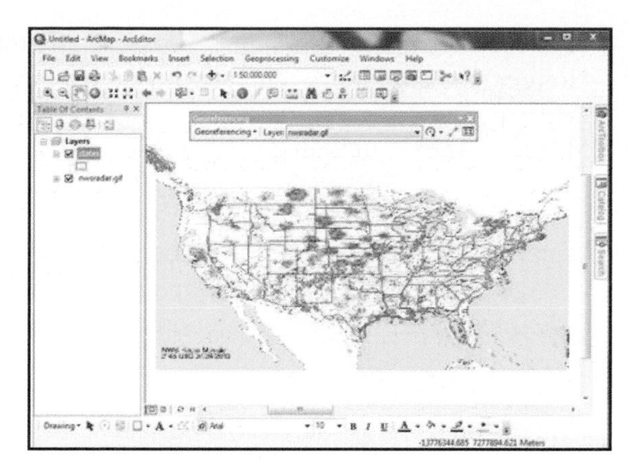

Fig. 11.24. Ready to begin georeferencing

44➔ Locate the southwestern corner of California and *mentally* match the corner of the state outline on the image to the corner of the pink state outline on the map.

44➔ Use the Zoom In tool to zoom in to an area containing the two matching points.

44➔ Click the Add Control Points tool on the Georeferencing toolbar.

44➔ Carefully click the corner on the image first and then click the corresponding corner on the states map. ArcMap will help by snapping to the vertex of the state polygon. This is your first control point.

44➔ Notice that the map updated after you added the control point so that the corner of California is better aligned.

TIP: The pairs of control points must always be added in the same order. Decide if you will click the image point first and then the map or vice versa. Then stick with your choice! For the rest of this exercise, you will always click the image first, since you did the first one that way.

45➔ Zoom to the previous extent, and then zoom in to the New England region.

45➔ Mentally locate a matching point on the Maine state boundary.

45➔ Switch back to the Add Control Points tool.

45➔ Click the location on the image first and then click the corresponding location on the states map. Notice that your control points are being numbered.

The match will already look much better, indicating that you chose an appropriate coordinate system. Continue switching between the Zoom and Control Point tools to add the next points, even if they are almost identical. You need at least three points for an affine transformation. Be careful not to snap to a state vertex when adding an image point.

46➔ Add a control point on the Florida peninsula. Click the image first!

46➔ Add a control point in the Pacific Northwest. Click the image first!

Examine the map and locate any areas where the state boundaries are still poorly aligned. These areas should receive additional control points. If you don't see any at the national scale, zoom in to a couple of locations in the middle of the country to check.

47➔ Zoom to the state corner between Idaho, Wyoming, and Utah. Add a point, if necessary (image first).

47➔ Zoom to the western corner of Kentucky. Add a point, if necessary (image first).

For many maps, you would continue to add more control points in misaligned areas until you no longer saw any improvement when new points were added.

48➔ Click on the View Link Table button to see the points you've added.

The points are listed in the order that you added them (Fig. 11.25), and yours will differ from the figure. The X and Y Source fields show the coordinates of each control point in the original image units (pixels). The X Map and Y Map columns show the point coordinates in map units (degrees). The 1st Order Polynomial (Affine) Transformation method is being used.

Note the residuals for each point and the total RMS error, reported in map units. In Figure 11.25, the residuals range between 0.01 and 0.03 degree, and the RMS error is about 0.02 degree. At the equator that error would be about 2 km, a reasonable value for a national map. Yours should be in the same range if you entered the control points carefully.

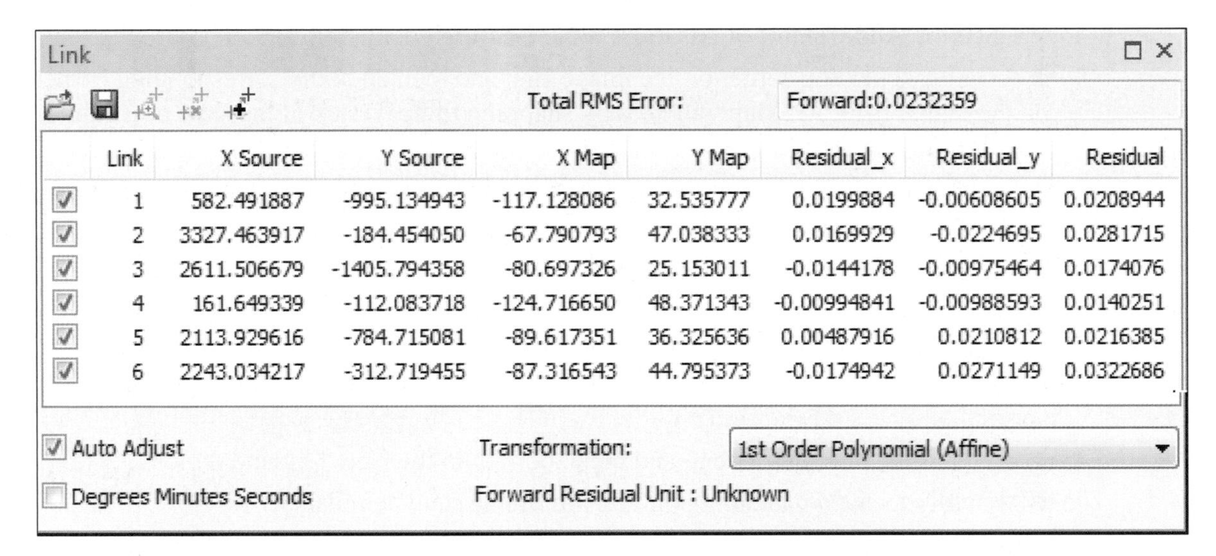

Link	X Source	Y Source	X Map	Y Map	Residual_x	Residual_y	Residual
1	582.491887	-995.134943	-117.128086	32.535777	0.0199884	-0.00608605	0.0208944
2	3327.463917	-184.454050	-67.790793	47.038333	0.0169929	-0.0224695	0.0281715
3	2611.506679	-1405.794358	-80.697326	25.153011	-0.0144178	-0.00975464	0.0174076
4	161.649339	-112.083718	-124.716650	48.371343	-0.00994841	-0.00988593	0.0140251
5	2113.929616	-784.715081	-89.617351	36.325636	0.00487916	0.0210812	0.0216385
6	2243.034217	-312.719455	-87.316543	44.795373	-0.0174942	0.0271149	0.0322686

Fig. 11.25. The Link Table shows the ground control points that have been added.

TIP: A few additional buttons on the link table are handy if you need to (from left to right) load a saved set of points, save a set of points, zoom to a point, delete a point, or add a new empty point.

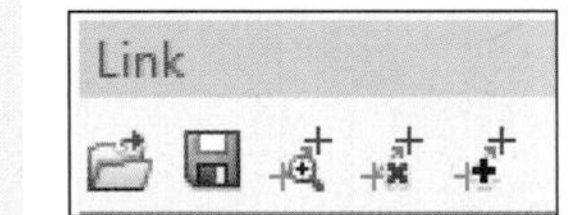

48➔ On the Link Table, click the Transformation button. It shows the methods currently available with the number of control points you have.

TIP: If the RMS error is large and does not improve when more points are added, try a second-order transformation. Second-order transformations require more control points.

48➔ Once you are happy with the registration, close the Link Table. If you achieve an RMS error less than 0.02 degree, then you have done well.

48➔ To save the new coordinate system information with the image, click Georeferencing > Update Georeferencing.

At this point the image is georeferenced, and the control point information has been saved with it as a world file. The next time you open the image, it will appear in the correct location.

SKILL TIP: Learn how to rectify an image, or permanently save it in the georeferenced coordinate system (Coordinate Systems).

Projecting a raster

You want to place the georeferenced image in the same coordinate system as the other United States data, so we must project it. First, however, estimate the expected resolution of the projected raster. Looking at the layer properties Source tab would tell you that the cell size is 0.017979 degree. One degree of latitude is about 111 km, so the cell size in meters should be 0.017979 × 111,000 = 1995.7, or about 2000 meters.

49➔ Start with a new, empty map. Don't save changes to the previous map.

49➔ Open the ArcToolbox > Data Management > Projections and Transformations > Raster > Project Raster tool.

49➔ Click the Browse button, navigate to the Usa folder, and set the Input raster to nwsradar.gif. Note that it has the GCS NAD 1983 coordinate system now.

49➔ Name the output raster usradar and put it in the mgisdata\Usa\usdata geodatabase.

50➔ Click the Browse button to open the Spatial Reference Properties window.

50➔ Click the Add Coordinate System drop-down button and choose Import (to ensure that the raster is projected to the same coordinate system as the other US data).

50➔ Select one of the data sets in the usdata geodatabase and click Add. Note that the new coordinate system will be North America Equidistant Conic.

50➔ No green dot appears by the Geographic Transformation, so no change of datum is required. Leave it blank.

50➔ Set the resampling technique to NEAREST, since the image contains discrete data.

50➔ Set the output cell size to 2000 meters.

50➔ Click OK to start projecting.

TIP: The cell size suggested in the tool is an exact conversion from the old to new cell size, but often the user rounds it to an even number.

51➔ Add the states feature class from usdata and give it a hollow symbol to ensure that the projection worked correctly (Fig. 11.26).

This is the end of the tutorial.

➔ Close ArcMap.

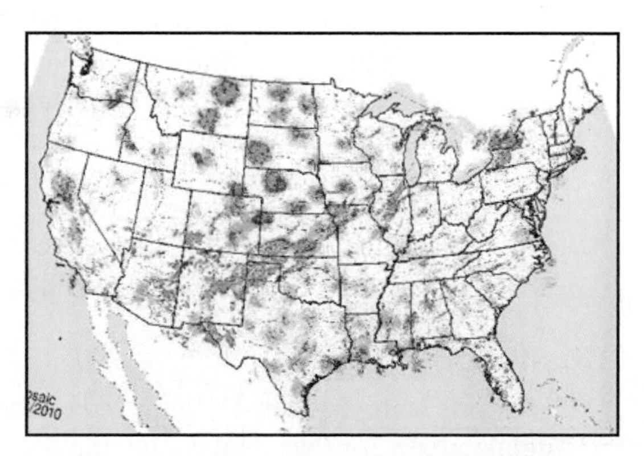

Fig. 11.26. The projected radar image

Exercises

1. Examine the coordinate system for the streets feature class in the Austin geodatabase. What is the name of the coordinate system? Is it projected or unprojected? What are the map units?

2. What is the name of the projection (not the coordinate system) used by the feature classes in the Oregon geodatabase? What are the central meridian and the standard parallel(s)? Does it use the equator for the latitude of origin?

3. Open the map document ex_11.mxd in the mgisdata\MapDocuments folder. What is the coordinate system of the data frame in this map? _______ What are the map units of the frame? _______ What are the display units of the frame? ________ What are the map units of the feature class coordinate systems? __________

4. Choose a good projection/coordinate system from the predefined coordinate systems in ArcGIS for maps of the following areas. Explain your choice for each. (**Hint:** Use the UTM zone and State Plane zone feature classes to help locate the area relative to a zone, if needed.)

   Humboldt County, California

   Grafton County, New Hampshire

   State of Nevada

   State of New Jersey

   England

   Antarctica, true distances required

5. You are working on a statewide Illinois project and decide to define a custom coordinate system. Start with one of the State Plane zones and modify it slightly to make it better for the whole state. Explain your approach and **Capture** the window showing the custom coordinate system description you created.

6. You are working on a statewide Colorado project and decide to define a custom coordinate system. Start with one of the State Plane zones and modify it slightly to make it better for the whole state. Explain your approach and **Capture** the window showing the custom coordinate system description you created.

7. The Austin folder contains two shapefiles showing dog off-leash areas as points and polygons. Both have coordinate problems. Describe the problem for each; then fix them and create a map showing both the points and the polygons with a backdrop of the major transportation arteries in Austin. (**Hint:** Open ArcMap and load arteries first, in order to be able to compare the other data sets against it.) **Capture** the map.

8. Find a map image on the Internet, save it as a JPEG, and georeference it. Display it with at least one other data set from the mgisdata folder.

Challenge Problem

Pick one state, besides Oregon, and choose the best coordinate system for it. Create a geodatabase for that state containing the same feature classes as the usdata geodatabase, in the projection you've chosen. (**Hint:** Use the Clip tool to crop linear features at the state boundary.) Find an image of the state on the Internet and include that in the geodatabase also, correctly georeferenced and projected to match the feature classes.

Chapter 12. Basic Editing

Objectives

- Understanding topological errors
- Using snapping to ensure topological integrity of features
- Adding features to map layers using basic editing functions
- Using the sketching tools and context menus to precisely position features
- Entering and editing attribute data

Mastering the Concepts

GIS Concepts

Editing can update existing feature classes or create new ones. If a housing subdivision is added to a city, the new roads must be added to the city's roads feature class. Likewise, new parcels, sewer lines, and other infrastructure need to be added to the city database to ensure that it is up to date. Entirely new feature classes can be created, for example, if the city planning department decides to create a map of garbage collection zones where none existed before. Parcel ownership attributes must be updated when parcels are sold.

When editing, care must be taken to create and maintain topological integrity between features. If two parcels share a common boundary, then the boundaries should match exactly. Line features that connect in the real world, such as streets or water lines, should connect in the feature class. Lines that cross each other should intersect at a node. Lines and polygon boundaries should not cross over themselves. These basic rules must be observed to ensure the logical consistency of features so that they properly represent the relationships of their real-world counterparts.

Topological data models permit the user to test, locate, and fix topological errors. However, topology is even more important when editing spaghetti models because the user must manage it without help. Basic editing includes two capabilities that aid in maintaining topological integrity: snapping and coincident boundary editing.

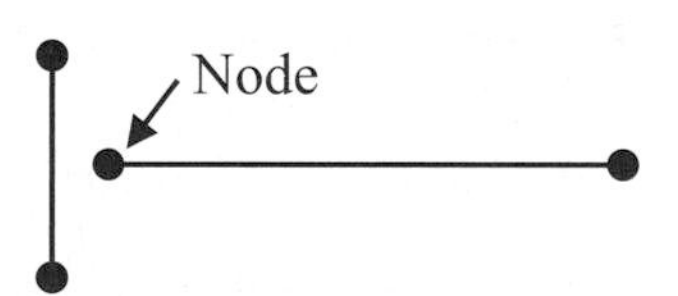

A dangle—the two lines fail to connect.

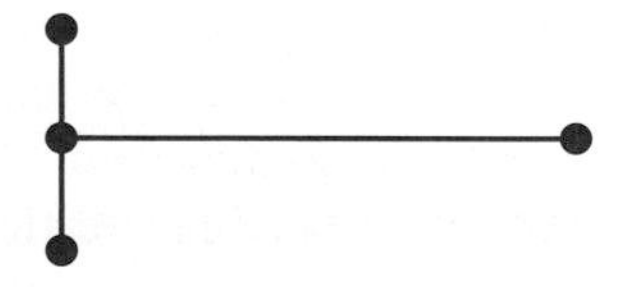

Correct topology—the horizontal line intersects the vertical one, creating three lines.

Fig. 12.1. Topological relationships between lines

Snapping features

When creating line features that connect to each other, such as roads, one must take care that the features connect properly (Fig. 12.1). A line that fails to connect is called a **dangle**. Although you may not be able to see the gap between the lines, the gap will exist unless the endpoints of each line (nodes) have exactly the same coordinate values. Even though the map may look correct, certain functions, such as tracing networks or locating intersections, will not work properly. It is impossible to intersect lines properly by simple digitizing.

Snapping ensures that the nodes of lines and the vertices of polygons match. When snapping is turned on, it affects features being added or modified. If you place the cursor within a specified distance of an existing node or vertex (Fig. 12.2), then the new feature is automatically snapped to the existing one—that is, the coordinates are matched at that one point. This distance is called the **snap tolerance**.

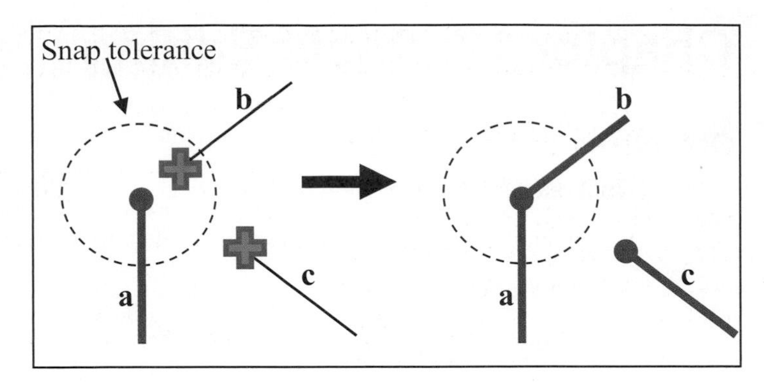

Fig. 12.2. Snapping. Line "b" snapped to "a" because the cursor was inside the snap tolerance. Line "c" did not.

Care must be taken in specifying the snap tolerance. If it is too small, then you will have difficulty making features snap. If it is too large, then you may snap to objects when it isn't appropriate. Snap tolerances are set in either pixels on the screen or meters in the map units. A snap tolerance of 10 pixels indicates that features will be snapped if the cursor moves within 10 screen pixels of an existing feature. When set in pixels, the snap tolerance remains the same as the user zooms in and out, which is convenient. Setting the snap tolerance in map units is useful when using a digitizing tablet, or if the user is trying to maintain a particular level of geometric accuracy independent of zoom scale. Four types of snapping can be used.

Point snapping is only used for point feature classes and snaps to an existing point.

End snapping only allows a new vertex to be snapped to the endpoints of an existing line. End snapping can ensure that new streams connect to the ends of existing streams, and only to their ends. End snapping only applies to line features.

Vertex snapping allows the endpoints of the new line to be snapped to any vertex in an existing line or polygon. It can ensure that adjacent parcels connect only at existing corners.

Edge snapping constrains the feature being added to meet the edge of an existing line or polygon feature. In this case, the vertex being added could be placed anywhere on or between vertices. Edge snapping can ensure that a street ends exactly on another street.

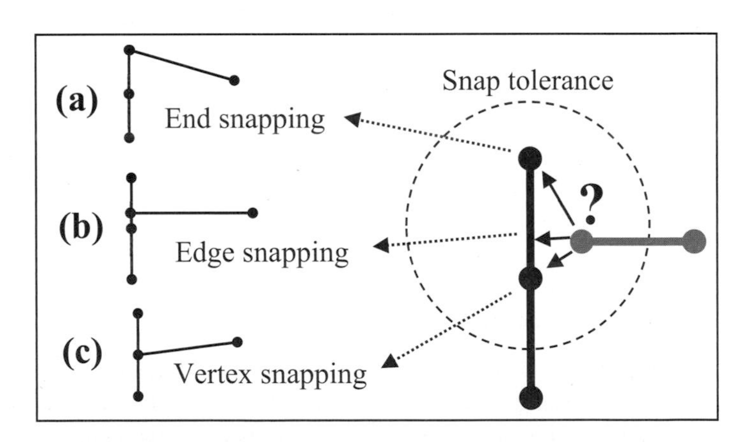

Fig. 12.3. Where will it snap? The snap type dictates where the new horizontal line will end.

Figure 12.3 illustrates how snapping changes depending on the snapping type. The vertical line already exists in the layer, and the horizontal line is being added. The dashed circle shows the snap tolerance. If end snapping is turned on, then the new line will snap to the endpoint of the existing line (Fig. 12.3a). If edge snapping is turned on, then the line can be snapped anywhere between the end and the vertex (Fig. 12.3b). If vertex snapping is set, then the new line will connect to the closest vertex or end (Fig. 12.3c).

If more than one kind of snapping is turned on, then the most inclusive one takes precedence. To ensure snapping only to ends, therefore, both edge and vertex snapping must be turned off.

Creating adjacent polygons

Two adjacent polygons should share the same boundary. In shapefiles and geodatabases the boundary gets stored twice, once for each polygon. However, the vertices should exactly match in both features. If this condition holds, the polygons are said to share a coincident boundary (Fig. 12.4). If they fail to match, gaps or overlaps are generated, which constitute topological errors and may cause problems during display or analysis.

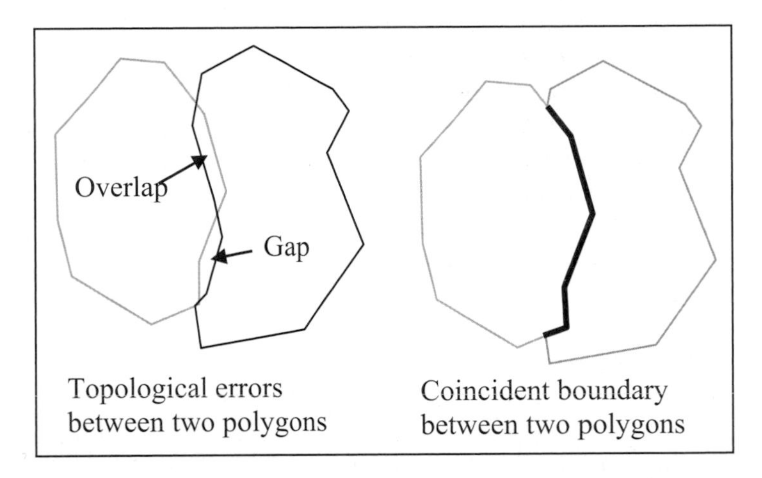

Fig. 12.4. Topological relationships between polygons

About ArcGIS

The editing process begins by opening the Editor toolbar, starting an edit session, and identifying which folder or geodatabase will be available for editing. Only one folder or geodatabase (called a workspace) can be edited at a time, but all the feature classes in the folder or geodatabase can be edited together.

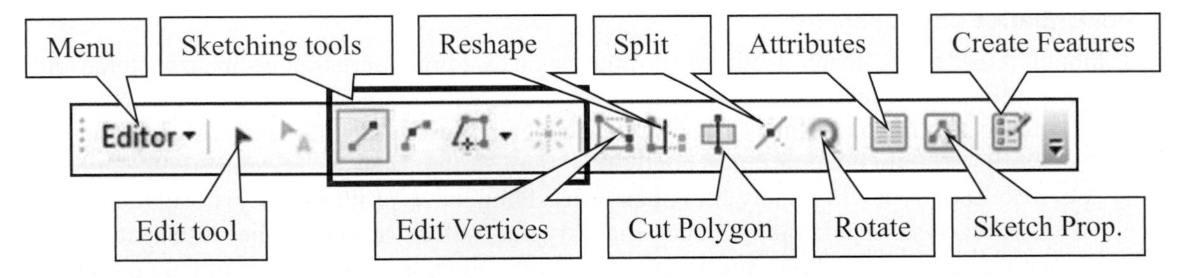

Fig. 12.5. The Editor toolbar contains many editing options and functions.

The Editor toolbar

The Editor toolbar provides access and shortcuts to many editing functions (Fig. 12.5). The first button on the left opens the Editor menu, used for turning editing on and off, setting options, and performing certain editing tasks. The Edit tool selects one or more features for editing. It is similar to the Select Features tools on the Standard toolbar but should always be used for editing.

Next come the Sketching tools, an assortment of methods to add vertices to a feature. The default is a straight line segment between vertices, but one can also create curves, right angles, midpoints, and other specialized vertices. The Sketching tools provide a lot of flexibility in creating features. The next four buttons provide access to common tasks, including editing the vertices of an existing feature to change its shape, splitting a line into two features, cutting a polygon in two, and rotating a feature.

The final three buttons launch editing windows. The Attributes window provides a convenient place for editing feature attributes. The Sketch Properties window is used to examine and, if desired, edit *x-y* coordinates of sketches. The Create Features window contains and manages editing templates, which are used to create new features.

What can you edit?

ArcGIS offers different levels of editing capability depending on the type of license. ArcGIS Basic can edit shapefiles and personal or file geodatabases. ArcGIS Standard and ArcGIS Advanced can also edit SDE databases, geometric networks, and planar topology. All of these levels are accessed through the same ArcMap interface.

ArcMap can edit several layers at once, as long as they are all in the same folder or geodatabase (workspace). This capability makes it easier to view and edit related layers simultaneously or to copy features from one layer to another.

TIP: Coverages cannot be edited in ArcMap. Use the ArcEdit program in Workstation ArcInfo instead, or convert the coverages to shapefiles or geodatabases for editing.

Editing and coordinate systems

ArcMap can edit layers with different coordinate systems from the data frame. The edits will automatically be converted to the coordinate system of the layer being edited. Consider editing a shapefile of roads stored in a GCS while displaying the roads with a digital orthophotoquad (DOQ) in a UTM projection. The roads coordinates will be converted into decimal degrees before they are put into the shapefile.

Performance and reliability

Although ArcMap has many useful capabilities, such as editing across coordinate systems and editing multiple files simultaneously, one must be somewhat cautious in taking advantage of these benefits. Editing coordinate features can be a complicated process, and software does not always work perfectly under arduous conditions. If an editing scenario is demanding extensive system resources by having many files open or by doing many projection transformations during an edit session, system performance and reliability may suffer. Moreover, because spatial data files are complex constructions, the price of an editing glitch may be the loss of data integrity; an ill-timed system glitch might not only lose recent changes but also corrupt the entire data set. Thus, you are encouraged to use the following procedures while editing:

- **Always have a backup copy of the file being edited,** stored elsewhere on the disk or on a different medium. This precaution is especially important when working on data sets that would be expensive or impossible to replace, should something go wrong.
- Shapefiles are the simplest and most robust type of data file to edit and the hardest to damage during editing.
- Geodatabases offer special capabilities and tools for helping ensure that data are entered and maintained correctly.
- Limit the number of open files. A prolonged editing session should be conducted in its own map document and include only the layers that are needed.
- Avoid extensive editing in map documents whose primary purpose is cartography or analysis, and in which you have invested hours of effort. The author has found that

editing sometimes corrupts map documents. They look fine, but tools will crash when run. The map documents then need to be re-created, potentially wasting hours of effort.

- Save edits frequently in case of a system glitch. ArcMap does NOT automatically save edits at regular time intervals. However, remember that saving also clears the "Undo" memory of edits. After edits are saved, they can no longer be undone.

Feature templates

Editing utilizes feature templates. A **feature template** stores the information needed to edit a particular layer in the map document. It includes the feature class in which the edits will be stored, the attributes that are assigned to new features, and the default tool that is used to create them. The template shown in Figure 12.6a is named Commercial. It saves features into a polygon layer called buildings. By default, the plain Polygon construction tool is used. The fields contain default values that are added to each feature, including its designation as a Commercial building.

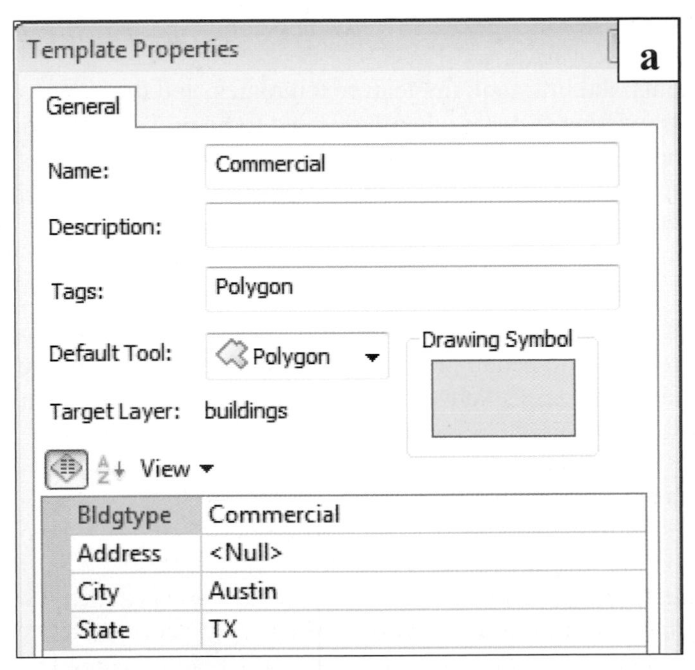

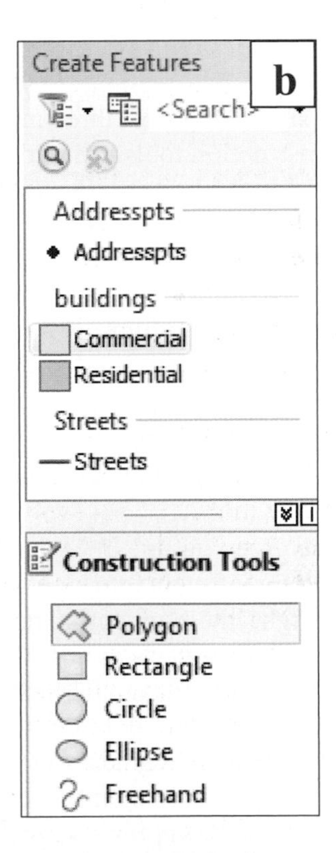

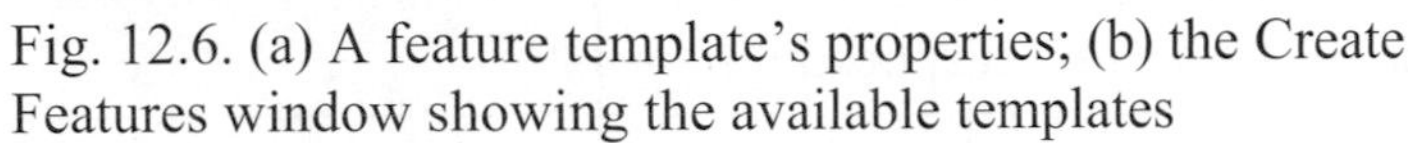
Fig. 12.6. (a) A feature template's properties; (b) the Create Features window showing the available templates

Multiple templates may be constructed for each feature class. As shown in Figure 12.6b, the buildings layer has another template called Residential. It is identical to the Commercial template except that the building type field, Bldgtype, contains "Residential". If the user is digitizing a new commercial building, she clicks on the Commercial template. When the building is added, the default attributes "Commercial", "Austin", and "TX" are automatically placed in the attribute fields. If she wishes to add a residential building, she clicks the Residential template, she digitizes the building, and the attributes are entered. Using templates can save a great deal of typing to enter attributes. Only the address field, which is unique to each feature, must be edited manually.

Templates are accessed and managed in the Create Features window (Fig. 12.6b). The upper panel lists each available template, and the user selects the desired template from this panel when ready to create a new feature.

The lower panel shows the available construction tools for that feature type, with the default one defined by the template at the top. Each feature class type has its own set of construction tools used to create various shapes or perform special tasks, such as creating an ellipse or a rectangle. The two buttons at the top of the Create Features window control the grouping and sorting of the templates (left) and the creation and management of templates (right).

Templates should be automatically created for each active layer in a feature class when editing is initiated. However, the user can also create and edit templates as needed. Moreover, templates can be saved and reused, and so they also contain descriptions and tags that can help in searching for stored templates. The Search window provides access to stored templates.

How editing works

An editing session must be initiated before any changes to a file can be made. This requirement protects the user from accidentally making changes to a file without realizing it. Also, because ArcMap can edit within only one directory or one geodatabase at a time, opening the session establishes the workspace being edited.

Basic editing uses three main components: the Edit tool, the feature templates, and the construction tools. The Edit tool selects existing features when they need to be moved, rotated, deleted, and so on. It is analogous to the Select Features tool on the Standard toolbar, but it should always be used for editing. The templates control the features being added and in which layer they go. The construction tools control the characteristics of the feature being created.

Creating sketches

The construction tools are used to create a **sketch**, a provisional feature that is being worked on, analogous to an artist sketching a figure lightly in pencil prior to inking in the final lines of the picture. This sketch is not actually added to the target layer until it is "finished," which is usually done by double-clicking the last vertex.

The green outline in Figure 12.7 shows a nearly completed polygon sketch, in which green vertices are separated by straight line segments. The latest vertex, just added, is shown in red. New vertices are added to a sketch using one of the Sketching tools on the Feature Construction or Editor toolbar.

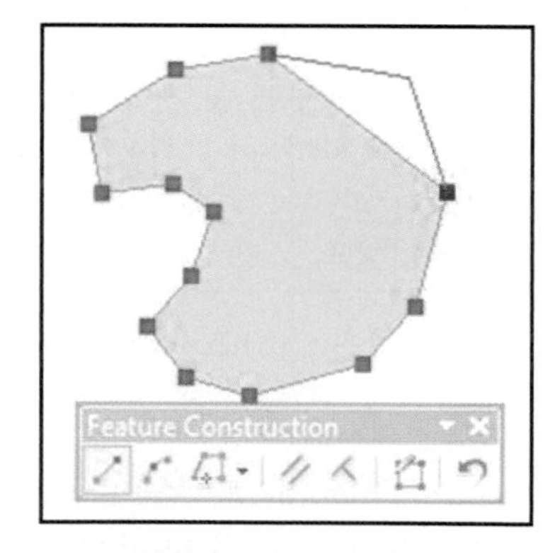

Fig. 12.7. A polygon sketch

The Straight Segment sketching tool is the default. It appears as the first button on the left on the Editor toolbar. It merely creates a straight line segment between the last and the new vertex. It was used to create the polygon in Figure 12.7. Another tool that is commonly used and placed on the Editor toolbar is the End Point Arc Segment tool, used to create a smooth curve between two endpoints.

Additional Sketching tools are accessed by clicking the drop-down menu to the right of the Straight Segment and End Point Arc Segment tools (Fig. 12.8). The Right Angle tool, for example, forces the sketch corners to be square, making it much easier to create buildings. Other sketching tools can create curves, make parallel or perpendicular lines, trace along existing features, create midpoints or tangents, and do

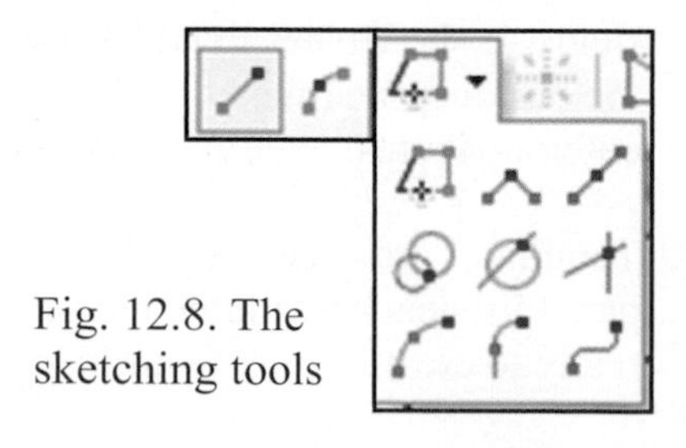
Fig. 12.8. The sketching tools

other specialized tasks. In this chapter we learn to use two of the most common sketching tools, the default Straight Segment tool and the Right Angle tool.

SKILL TIP: The functions and instructions for all of the Sketching tools may be found in the Skills Reference section (Editing).

While the user is sketching, a mini-toolbar known as the Feature Construction menu floats near the cursor on the screen. It provides quick access to different sketching tools and a choice of ways to add the next vertex. If you find the floating Feature Construction menu annoying, you can turn it off from the Editor toolbar using the Editor > Options menu.

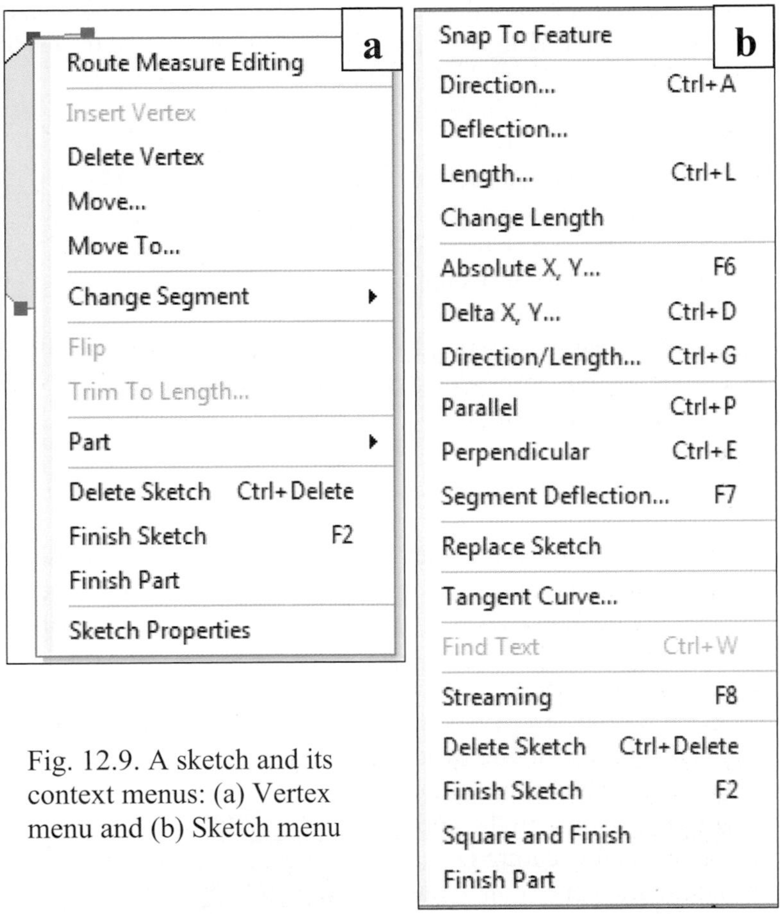

Vertex and Sketch menus

During editing you can access two context menus to aid in certain tasks or change edit settings (Fig. 12.9). The **Vertex menu** appears when you right-click on the sketch (Fig. 12.9a). This menu provides functions for adding, deleting, or moving vertices. The **Sketch menu** appears when you right-click off the sketch (Fig. 12.9b). This menu provides functions for specifying exact angles, lengths, distances, and more.

Fig. 12.9. A sketch and its context menus: (a) Vertex menu and (b) Sketch menu

The Snapping toolbar

A small toolbar to control snapping allows easy access to snapping tools in the middle of a sketch. The four basic kinds of snapping (point, end, vertex, and edge from left to right in Figure 12.10) are controlled with the click of a button. A menu extends the snapping types to intersection snapping (where two lines intersect at a common node), midpoint snapping at the midpoint of a line, and tangent snapping to a curve. The snap tolerance can also be controlled here.

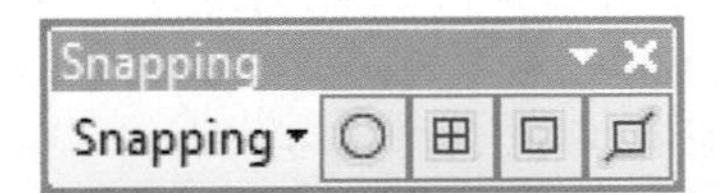

Fig. 12.10. The Snapping toolbar

Adding adjacent polygons

A special construction tool is used for adding adjacent polygons. The Polygon construction tool is first used to enter a complete polygon with no neighbors. To add an adjacent polygon, you must switch to the Auto Complete Polygon tool and digitize only the new part of the polygon. The editor ensures that the polygons share a coincident boundary and are free from topological errors, including gaps and overlaps (Fig. 12.11).

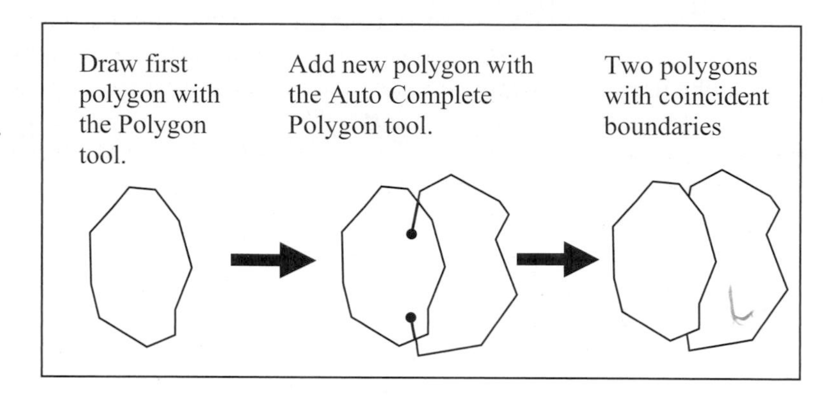

Fig. 12.11. Creating a coincident boundary between two adjacent polygons with the AutoComplete task

Another way to add adjacent polygons uses the Cut Polygons construction tool. With this method, an outer polygon is constructed around an area encompassing two or more polygons, and then the internal boundaries are added by cutting the polygons along the intended coincident boundary.

Editing attributes

Editing features often includes editing their attributes. The user can modify attributes in two ways. In the first way, edits are typed directly into an attribute table, as discussed in Chapter 4. The second method uses the Attributes window to edit the values of single or multiple features.

The Attributes window is accessed through a button on the Editor toolbar (Fig. 12.12). The upper panel shows the currently selected records; if no features or records are selected, then both areas will be blank. The lower panel shows the attributes of the highlighted feature in the upper panel. When the feature is clicked in the upper panel, the feature flashes on the screen to indicate which one it is, and the attributes in the lower panel may be edited.

The window also allows editing of attributes for multiple features. If the top row containing the feature class (Streets) is clicked, then the attribute values go blank. Further updates to the attributes are applied to all of the selected features.

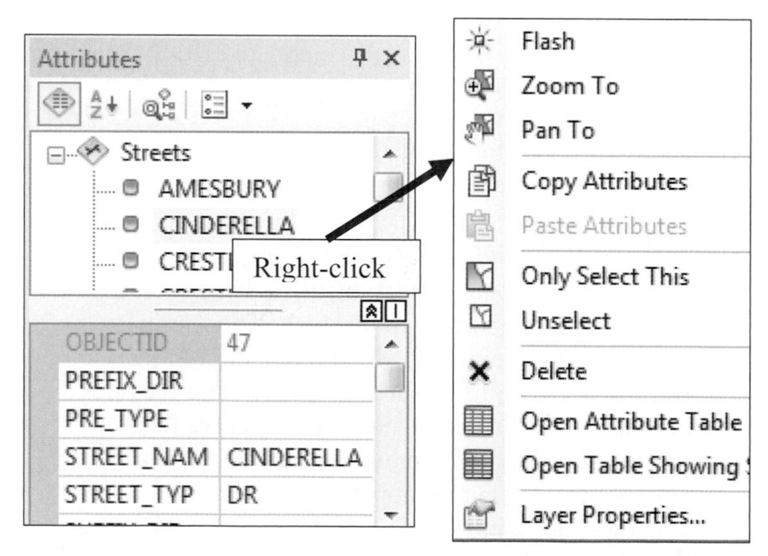

Fig. 12.12. The Attributes window can display and edit many records at once.

Summary

- Editing creates and maintains features in a feature class. Care must be taken to ensure topological integrity when editing.
- Snapping ensures that features connect properly with each other by automatically matching the coordinates of features within a certain distance of each other (the snap tolerance). Features may be snapped to points, endpoints, vertices, or edges.
- When adding adjacent polygons, you must ensure that they have coincident boundaries by using snapping, the Auto Complete Polygon tool, or the Cut Polygons tool.
- Editing sessions are initiated and controlled using the Editor toolbar. You may edit multiple layers of data, but only one geodatabase or folder may be edited at a time.
- Edits are stored in the same coordinate system as the layer, regardless of the coordinate system of the data frame. However, editing is streamlined when the layer and frame coordinate systems match so that no reprojection need take place.
- Basic editing uses the Edit tool, feature templates, and the construction tools. The Edit tool selects features to edit.
- Feature templates define all the properties needed to create new features, including the feature class to place them in, the default construction tool to use, and the attributes with any default values in them.
- Multiple templates may be created for one feature class, enabling different attributes values to be given to different classes of features.
- The construction tools create preliminary figures called sketches. Finishing the sketch creates the new feature in the feature class.
- Different Sketching tools provide a variation of functions for placing vertices. Context menus provide additional tools for specialized functions.

Important Terms

dangle	feature template	Sketch menu	Vertex menu
edge snapping	point snapping	snap tolerance	vertex snapping
end snapping	sketch	snapping	

Chapter Review Questions

1. List the main types of topological errors.
2. Which type of snapping (end, vertex, or edge) would work best in the following situations? (a) digitizing streams, (b) digitizing parcels, (c) digitizing streets, (d) digitizing traffic lights at intersections, (e) digitizing stream gages on streams
3. How are editing functions and options accessed in ArcMap?
4. If a data layer is in UTM coordinates and the data frame is set to State Plane coordinates, in which coordinate system will the edits be stored?
5. True or False? Because you can Undo edits, it is not necessary to make a backup copy of any data set before editing. Explain your answer.
6. When you create a new feature, how do you specify which layer it belongs to?
7. What are dangles and how are they prevented?
8. What is the purpose of the AutoComplete task? What errors does it prevent?
9. How many context menus can you access while sketching? How do you control which one you get?
10. How can feature templates save the user from typing in attribute values?

Mastering the Skills

Teaching Tutorial

The following examples provide step-by-step instructions for doing basic tasks and solving basic problems in ArcGIS. The steps you need to do are highlighted with an arrow ➔; follow them carefully. Click on the video number in the VideoIndex to view a demonstration of the steps.

➔ Start ArcMap and open the map document ex_12a.mxd in the MapDocuments folder.

➔ Use Save As to rename the document and remember to save frequently as you work.

Creating new feature classes

You are working on a project for Austin regarding the outcrop of an important aquifer, the Edwards Limestone. You will digitize from a USGS map that is already scanned and georeferenced, using the same coordinate system as the rest of the Austin data. Before starting, you need to create empty feature classes that will contain your new digitized features.

1➔ Open the Catalog tab. Right-click the mgisdata\Austin folder and choose New > Personal Geodatabase. Name it Edwards.

1➔ Open ArcToolbox > Data Management Tools > Feature Class > Create Feature Class.

1➔ Click the Browse button for Feature Class Location (Fig. 12.13). Navigate to the mgisdata\Austin folder and select the Edwards geodatabase. Click Add.

1➔ Name the new feature class karstfeatures.

2➔ Set the Geometry Type to POINT.

2➔ Leave the Template Feature Class blank.

2➔ Scroll down, if necessary, to locate the Coordinate System box. Click the Browse button.

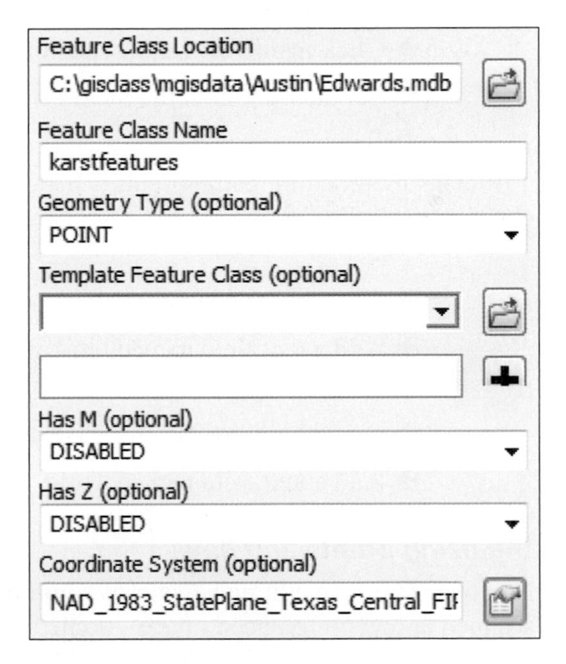

Fig. 12.13. Creating a feature class

2➔ Click the Add Coordinate System drop-down button and choose Import.

2➔ Navigate to mgisdata\Austin\Austin geodatabase and select the Environmental feature dataset. Click Add.

1. We used Import in order to guarantee that the coordinate system of our new feature class matched our other data. What coordinate system is it? ____________________

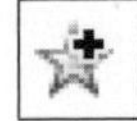

2➔ Click the newly added coordinate system and click the Favorites button. Click OK to close the Spatial Reference Properties window.

2➔ Click OK to finish creating karstfeatures.

The new feature class is automatically added to the map document. Next we'll create two more feature classes for faults and geology.

3➔ Use the Create Feature Class tool again to create a line feature class named faults in the Edwards geodatabase. Make sure that the feature geometry is set to POLYLINE and that you select the right coordinate system from the Favorites folder.

4➔ Create a polygon feature class named geology. Make sure that the feature geometry is set to POLYGON and that you select the right coordinate system from the Favorites folder.

We also need to set up some fields for these feature classes.

5➔ Open the Catalog tab, find the Edwards geodatabase, and expand it.

5➔ Right-click the faults feature class and choose Properties. Click the Fields tab.

5➔ Click inside the first empty Field Name box and type Linetype to name the field.

5➔ Click inside the Data Type box for Linetype and choose Text for the Data Type.

5➔ Set the Length in the Field Properties box to 10. Click OK.

TIP: This method of creating fields is better when multiple fields need to be added to a feature class, instead of using the Table Options method we've used previously.

6➔ Use the Catalog tab to open the properties for the Edwards/geology feature class.

6➔ Add a text field named Unit with a length of 6. Click OK.

7➔ Open the properties for the Edwards/karstfeatures feature class.

7➔ Add a text field named Type with a length of 15. Click OK.

Digitizing points and lines

Now we set up the editing session. We will be digitizing features by tracing them from the scanned and georeferenced image—sometimes called "heads up" digitizing.

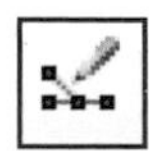

8➔ If the Editor toolbar is not visible, click the Editor button.

8➔ From the Editor toolbar, choose Editor > Start Editing.

8➔ Choose the geodatabase to edit by selecting the Edwards geodatabase from the lower part of the window. Click OK.

8➔ If no Create Features window appears automatically, choose Editor > Editing Windows > Create Features.

TIP: In Edit mode, some other functions are disabled, such as adding fields, to help prevent damage to data sets. It is wise to be in Edit mode only when you are actually editing.

Examine the Create Features window. It shows a **feature template** for each feature class. A template is a collection of edit settings for a layer. Below the template panel lies a Construction Tools panel, showing the tools available for that template.

8➔ Click on the karstfeatures template. Notice that two construction tools are available, *Point* and *Point at end of line*, and that *Point* is currently active.

8➔ Click on the geology template. A longer list of different tools appears.

TIP: If you cannot see the entire list of tools, place the cursor on the line between the upper and lower parts of the window and move it up to enlarge the lower window.

First we are going to digitize some springs, shown on the geological map as blue dots. Look for a group of six springs near the SW corner of the map (Fig. 12.14).

9➔ Zoom in to the group of springs.

9➔ Click the karstfeatures template again.

9➔ Click inside one of the blue dots to add a point.

9➔ Continue adding points until you have digitized all of the springs in view.

Notice that the last one clicked is highlighted, indicating that it is selected.

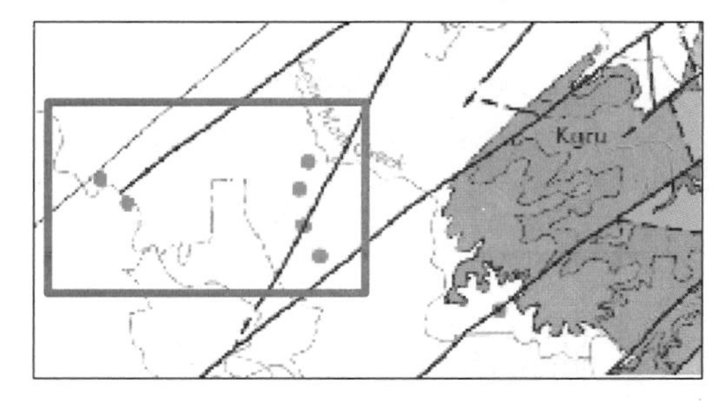

Fig. 12.14. Adding the springs

9➔ Click the Delete key. It will delete the highlighted point, the last one entered.

9➔ Click on the spring again to reenter the point.

We need to enter attributes for the springs as well.

10➔ Right-click the karstfeatures layer in the Table of Contents and choose Open Attribute Table.

10➔ Click inside the highlighted box for the Type field and type `spring`.

10➔ Click the next box up and type `spring` in it also.

10➔ To select a feature from the table, click on the gray area to the left of a row with an empty Type box to select the feature.

The Attributes window is a handier way to enter attributes while editing.

11➔ Click the Attributes button on the Editor toolbar.

TIP: If the Attributes window does not appear, look at the bottom of the Create Features window for an additional tab. To open it, click the tab and drag it away from the Create Features window.

11➔ The top panel has one feature, the currently selected one, labeled with its ObjectID.

11➔ In the lower panel, find the Type field. Click in the box that currently says <Null> and type `spring`. Click Tab to enter it. Watch it update in the Table window.

You may imagine that it will become tedious to type "spring" for every single feature. But there's a better way.

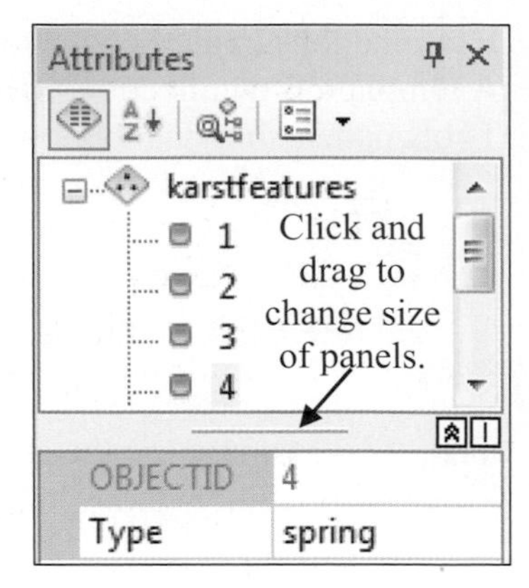

Fig. 12.15. The Attributes window

12➔ Dock the Attributes window under the Table of Contents.

12➔ Adjust the boundaries between the Table of Contents and the two Attributes panels so that you can see several rows of space in the upper panel and the two fields in the lower panel (Fig. 12.15).

12➔ Also resize and position the Table window so that you can see it.

13➔ Choose the Edit tool from the Editor toolbar.

13➔ Click and drag a rectangle around all of the springs to select them. The Attributes window now shows all of the springs (Fig. 12.15; only four are visible).

13➔ Click the karstfeatures heading at the top of the list of selected features.

13➔ Type `spring` in the Type field at the bottom and click Tab. The value is entered in all rows of the table.

TIP: Highlight the feature class name at the top of the Attributes window to make the field edits apply to all of the selected records.

14➔ Right-click one of the features in the top panel of the Attributes window and examine the entries of the context menu.

14➔ Close the Table window.

14➔ On the Editor toolbar, choose Editor > Save Edits.

14➔ Clear the selected features using the usual button on the Standard toolbar.

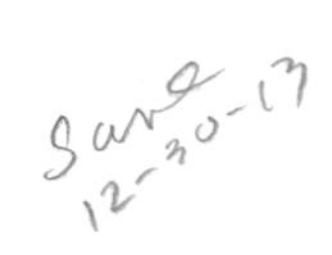

One of the advantages of templates is that you can easily switch between layers.

15➔ Find the fault that travels diagonally across the map through one of the springs. Pan/Zoom so that you can see the entire fault, both solid and dashed ends.

15➔ Click the faults template. Notice that *Line* is the default construction tool.

15➔ Click to place a vertex at the lower end of the solid part of the line.

15➔ Click another point on the line some distance from the first.

15➔ Keep adding vertices until you reach the end of the solid part. Double-click there to complete the sketch. A highlighted line feature appears.

TIP: If you are not satisfied with the fault, click Delete (while it is still highlighted) to get rid of it and try again. If it is no longer highlighted, use the Edit tool to select it.

16➔ In the Attributes window, click in the Linetype box and enter `solid`. Click Tab.

16➔ Move the cursor toward the lower end of the fault just added, as if you were creating the next fault, but don't click yet.

Notice that, as you get close to the end of the line, some text and a symbol appear, called Snap Tips. They show what the next vertex will snap to.

16➔ Move the cursor along the fault and watch as the Snap Tips change.

The text "Faults: Endpoint" appears at the ends of the existing fault. The text "Faults: Vertex" appears when you run over a vertex. The text "Faults: Edge" appears in between vertices. Note the symbols that go with each type.

16➔ Place the cursor on the lower end of the fault, making sure that "Faults: Endpoint" is being displayed. Click to enter the first vertex for the new fault.

16➔ Add several more vertices along the dashed line. Double-click at the end to finish the sketch.

16➔ Enter dashed in the Linetype box of the Attributes window and click Tab or Enter.

You probably can't see the first fault now because its thin symbol does not show up well against the black lines in the image.

17➔ In the Table of Contents, change the symbol for faults to a 2-pt. green line.

17➔ Add the upper dashed fault, taking care to snap to the end of the solid fault.

17➔ Enter dashed in the Attributes window.

TIP: The floating mini-toolbar can be annoying. To turn it off, choose Editor > Options, click the General tab, and uncheck the box to *Show feature construction toolbar*.

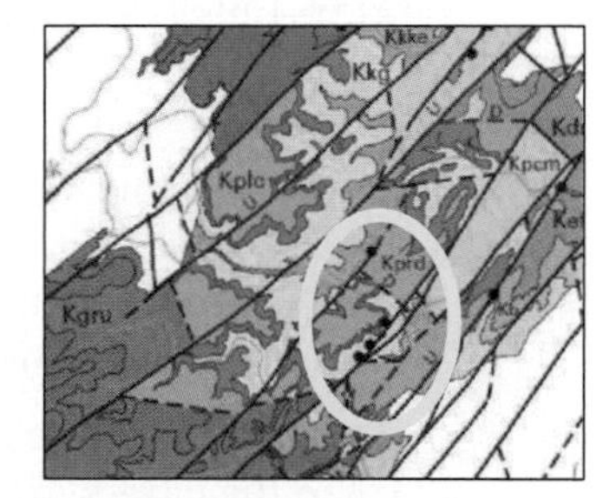

18➔ Zoom to the full extent of the map. Find some black dots east of the six springs and zoom to them. These are more karst features—caves or sinkholes.

18➔ Enter the black points using the karstfeatures template.

18➔ Click the Edit tool and draw a rectangle around them to select them.

18➔ Highlight the top karstfeatures entry in the Attributes window and enter the text cave/sink into the Type box in the Attributes window.

TIP: To change the text used to label the records in the Attributes window, change the Display Expression on the Display tab of the layer properties.

19➔ Just for a little more practice, click on the faults template.

19➔ Digitize two or three of the short dashed faults in this view. Remember to assign their attributes as dashed.

19➔ Choose Editor > Save Edits from the Editor toolbar.

19➔ Click the Save button on the Standard toolbar to save the map document.

TIP: Saving the map document does NOT save the edits, and vice versa.

Working with snapping

Snapping ensures that all the features that are supposed to touch actually do. It helps prevent topological errors between features. Let's see how it works.

20➔ Choose Editor > Snapping > Options.

20➔ Make sure that the snap tolerance is set to the default 10 pixels.

20➔ Examine the Snap Tips check boxes, which control the snapping guides. Click OK.

TIP: Use the Show Tips check box to control whether Snap Tips appear and what they include.

20➔ Choose Editor > Snapping > Snapping Toolbar.

20➔ Choose Snapping from the Snapping toolbar and examine the menu choices. (If it does not appear, try choosing it again from Customize on the main toolbar).

The Snapping toolbar gives quick access to snap settings. You can turn snapping off entirely in this menu. The toolbar buttons turn each type of snapping on/off.

21➔ Dock the Snapping toolbar in an accessible place.

21➔ Look above the Kprd text in the blue polygon and find the fault that starts dashed at the upper end and changes to solid at the lower end (A in Fig. 12.16).

21➔ Click the faults template and digitize the dashed upper end, moving from NE to SW. Stop where the dash ends and double-click to finish the sketch.

21➔ Enter dashed in the Attributes window.

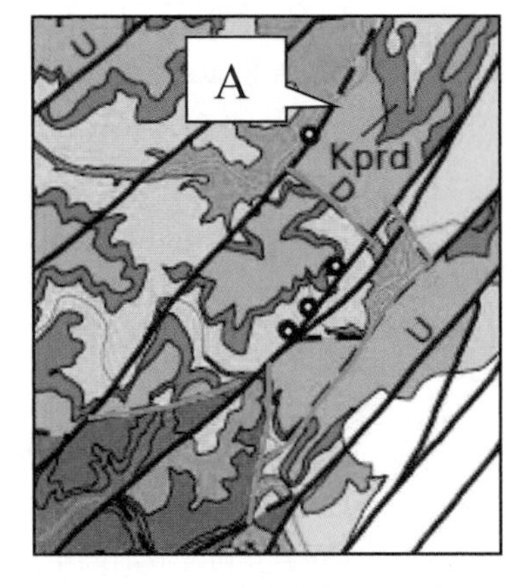

Fig. 12.16. Digitizing faults

In the next section, you want to snap only to the end of the existing fault—not to its edge.

22➔ On the Snapping toolbar, click both the vertex and edge snapping symbols to the off position.

22➔ Begin the next solid section of the fault. As you get to the cave symbol, however, the point snapping is a problem.

22➔ Turn off Point snapping on the Snapping toolbar (even in the middle of a sketch!).

22➔ Continue digitizing the fault. Snap to the ends of any faults that touch it, if they have been digitized already.

22➔ When you run against the edge of the map, don't finish the sketch yet.

22➔ Switch to the Pan tool and move the view to show the rest of the fault.

22➔ Click on the faults template again, and your sketch resumes where you left off.

22➔ Continue digitizing, and finish the sketch when you get to the end of the fault.

22➔ Enter solid for the attribute.

23➔ Click on the edge of the image to start the next fault below the first.

23➔ Alternate between sketching and panning to digitize the entire solid part of this fault, ending where it appears to split into two segments by the caves. Enter solid.

23→ Start another fault on the end, taking the upper branch and digitizing up to where the fault splits again near the text "Buda". Enter solid.

24→ North and northeast of "Buda" you will see additional caves and springs.

24→ Change to the karstfeatures template and add the springs (blue circles).

24→ Use the Edit tool to select them all and assign them the spring attribute.

25→ Click the karstfeatures template and add the caves (black circles).

25→ Change to the Edit tool, select them all, and assign them the cave/sink attribute.

25→ Choose Editor > Save Edits. Clear any selected features.

Using the sketch context menus

Now that you can perform some simple editing functions, let's look at ways in which the Sketch menu enhances editing, such as correcting mistakes during a sketch.

26→ Zoom to the full extent of the map. Find a clear area and zoom to it. The next editing is just practice.

26→ Click on the faults template.

26→ Enter some vertices along a random line. Notice that the last vertex is red and the other vertices are green.

26→ Right-click on top of the last red vertex to open the Vertex menu.

26→ Choose Finish Sketch.

Choosing Finish Sketch does the same thing as double-clicking at the end of the sketch.

27→ Start another line and add some vertices.

27→ Right-click the last red vertex to open the Vertex menu and choose Delete Vertex.

Notice that the last vertex is deleted and the second-to-last one is now red. This menu can delete any of the previous vertices.

27→ Right-click on any of the green vertices and choose Delete Vertex.

27→ Right-click another vertex and examine the menu again.

27→ Click on the window bar at the top to close the menu without choosing anything.

Notice that all this time we are still in the middle of creating this sketch.

28→ Right-click on the sketch *between* two vertices.

28→ Choose Insert Vertex from the context menu.

Remember, clicking ON the sketch gives the Vertex menu. A different menu appears if you click OFF the sketch.

28→ Right-click OFF the sketch, somewhere on the map. The Sketch menu appears.

Some of the entries, such as Finish Sketch and Delete Sketch, are found in both menus. If you make a mistake as we continue, you can delete the sketch and start over at any time.

28➔ Choose Delete Sketch from the menu.

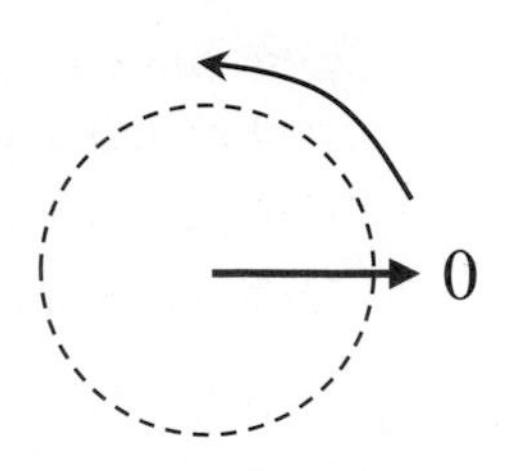

Fig. 12.17. Angles are measured in degrees from horizontal.

29➔ Start a new sketch with several vertices.

29➔ Right-click off the sketch to open the Sketch menu.

29➔ Choose the second entry, Direction.

29➔ Type 90 in the Direction box and press Enter.

As the mouse moves, the cursor is constrained along a vertical line, forcing the next segment to follow the specified angle. Angles are measured counterclockwise from the east (Fig. 12.17).

29➔ Click along the line to enter a vertex. The new segment is exactly vertical.

29➔ Double-click to finish the sketch.

2. What angle would you enter to create a line going S25°W? ______________________

Next, let's create a segment of a specific length.

30➔ Start a sketch and add several vertices.

30➔ Right-click off the sketch and choose Length from the Sketch menu.

30➔ Type 5000 into the box and press Enter.

3. What do you think are the units of these distance values? ______________________

30➔ Move the cursor around, noting how the last segment remains the same length.

30➔ Click somewhere to enter the vertex.

TIP: If the last segment goes off the page or is too small to see, try entering a different length. Simply right-click off the sketch again and type in a different length value.

The Sketch menu also combines the length and direction functions to enter a line with a particular bearing and direction. We will use this feature to enter the next segment of the same sketch.

31➔ Right-click off the sketch and choose Direction/Length.

31➔ Type 45 for the direction and 5000 for the length (you can use different values, if necessary, to keep the sketch on the screen). Click Enter when both boxes are filled.

31➔ Note that you didn't have to click to enter the vertex. Double-click to finish.

SKILL TIP: Learn to use the other functions on the Vertex and Sketch menus (Editing).

Digitizing polygons

Now we will learn how to create polygons. First we will practice with the construction tools.

32➔ Use the Edit tool to select the practice lines and delete them.

32➔ Click on the geology template. The Polygon construction tool is the default.

32➔ Start a polygon by adding several vertices. Notice how the shape remains closed, with the last vertex always connected to the first.

32➔ Double-click the last vertex to finish the sketch.

32➔ Try adding several more polygons that don't touch each other, for practice.

Next we will create some adjacent polygons. It is important to use the Auto Complete Polygon tool so that the polygons have a coincident boundary.

33➔ Turn on Vertex snapping by clicking the toggle on the Snapping toolbar.

33➔ Change the construction tool to Auto Complete Polygon.

33➔ Click INSIDE an existing polygon to start the sketch (Fig. 12.18).

33➔ Snap to a vertex of the existing polygon on the way out, for neater polygons.

33➔ Add more vertices to define the new polygon.

33➔ Add the last vertex INSIDE the existing polygon and double-click to finish.

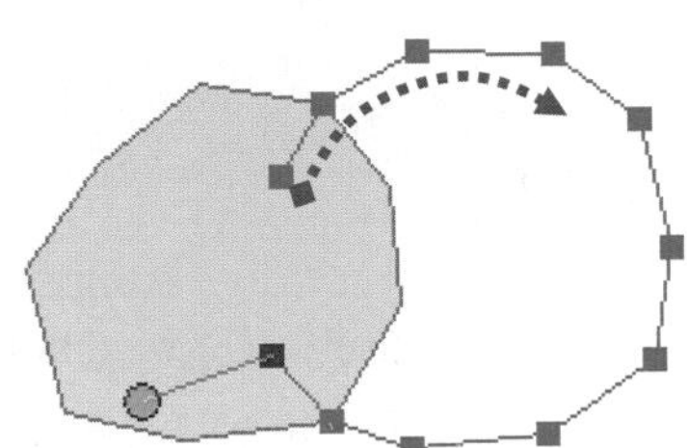

Fig. 12.18. Adding an adjacent polygon

TIP: The Auto Complete tool will not construct a polygon unless the space defined by the existing polygon(s) and the sketch is completely enclosed.

34➔ Begin a new sketch inside one of the polygons, digitize the boundary of a new polygon, and end inside the other polygon (Fig. 12.19a). Finish the sketch.

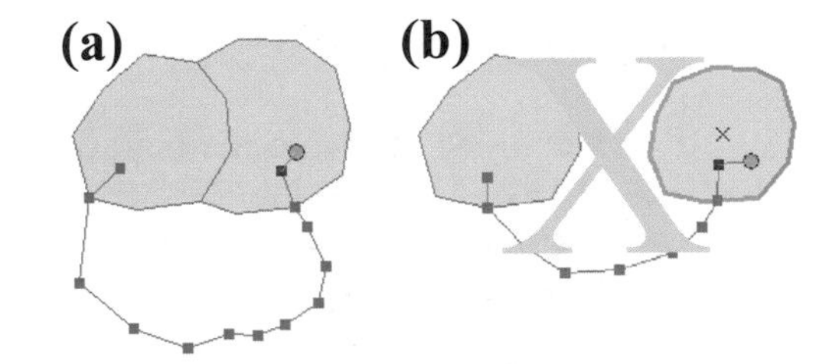

Fig. 12.19. Creating adjacent polygons: (a) works; (b) does not work because the area is not closed.

Splitting provides another way to create adjacent polygons with coincident boundaries.

35➔ Select one of the polygons with the Edit tool (Fig. 12.20).

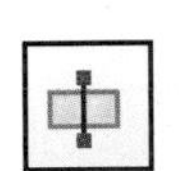

35➔ Click the Cut Polygons tool on the Editor toolbar.

35➔ Start a sketch OUTSIDE the selected polygon and add a line that goes through it.

35➔ Add the last vertex OUTSIDE the polygon and double-click to finish the sketch.

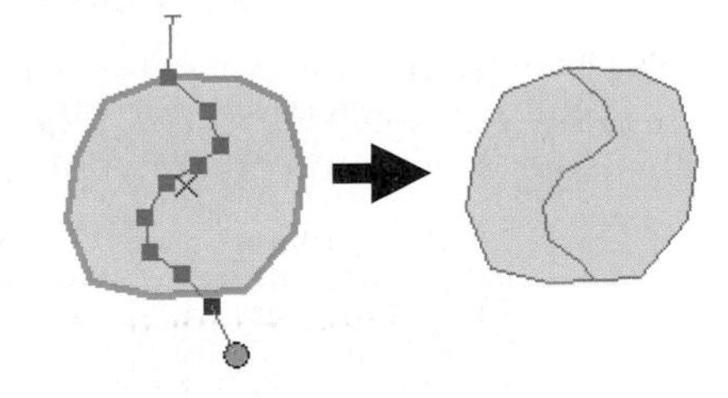

Fig. 12.20. Cutting a polygon

Cutting works on multiple polygons, too.

35➔ Check that the two polygons are still selected. If not, select them (Fig. 12.21).

35➔ Sketch another line through both and finish the sketch.

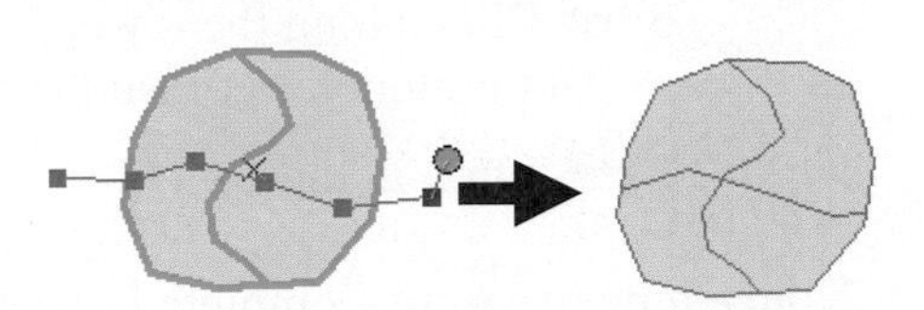

Fig. 12.21. Cutting multiple polygons

Now you have the basic skills to do some real polygon editing. First, let's clean up.

36➔ Use the Edit tool to select and delete the practice polygons.

36➔ Zoom to the full extent of the map.

36➔ Pan and zoom until you can see the polygons shown in Figure 12.22.

TIP: Sometimes features do not disappear when you press Delete. If this happens, click the Edit tool on the Editor toolbar and press Delete again. The Delete key does not work with some of the cursor tools, but it always works if the Edit tool is active.

Now we will try digitizing some geology polygons on the SW end of the outcrop area. Treat the faults as geological contacts that define boundaries of the polygons.

37➔ Click the geology template.

37➔ Start digitizing polygon A in Figure 12.22. Be reasonably careful, but it doesn't have to be perfect. Double-click to finish.

TIP: For best results, place vertices at sharp corners. Use fewer vertices on straight lines and more on curves.

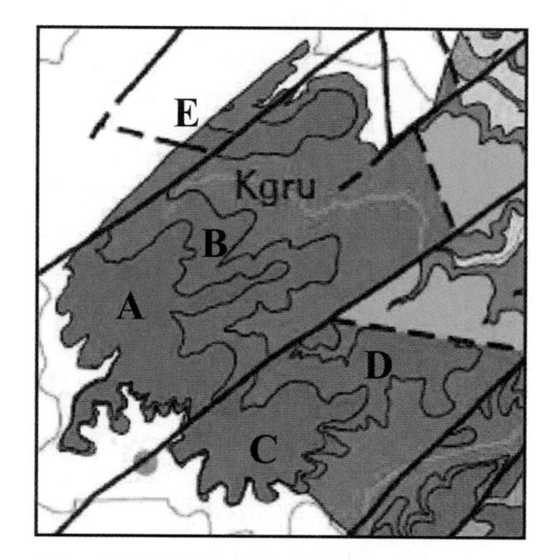

Fig. 12.22. The first polygons

38➔ Type Kkd into the Unit field in the Attributes window.

38➔ In the Table of Contents, change the geology symbol to 10% Cross Hatch.

38➔ Open the geology layer properties and click the Labels tab. Turn the labels on and change the Label Field to Unit. Set the font to Arial 12-pt. Bold. Click OK.

Polygon B is adjacent to polygon A. To ensure that they have a common boundary, we must use the Auto Complete Polygon tool.

39➔ Turn on Edge snapping by clicking the toggle button in the Snapping toolbar.

39➔ Click on the Auto Complete Polygon tool in the Construction Tools panel.

39➔ Start by snapping to the upper-right vertex of polygon A.

39➔ Click along the fault until the purple area ends; then follow the geological contact between the purple and brown polygons to the lower fault.

39➔ Click along the lower fault. End by snapping to the vertex of polygon A. Double-click to finish the polygon.

39➔ Click in the Unit box in the Attributes window and type Kkbn.

Notice how the Auto Complete Polygon tool used the boundary from polygon A, ensuring that the two polygons shared an identical boundary and that you did not have to digitize it twice.

40➔ Select the Auto Complete Polygon tool again.

40➔ Start at the lower-left corner of the Kkd polygon A and snap to the vertex.

40➔ Digitize counterclockwise around pink polygon C.

40➔ At the end, snap to an edge or a vertex of polygon B. Double-click to finish.

40➔ Enter the unit name, Kkd, in the Attributes window.

41➔ Start polygon D in the lower-middle part of polygon C, snapping to its vertex.

41➔ Start trying to digitize this narrow slice of polygon D.

Polygon D is close to polygon C in places, and you may find that it is snapping when you don't want to. Two things might help: zooming in and reducing the snap tolerance.

41➔ Right-click near the sketch and choose Delete Sketch from the context menu.

41➔ Zoom in so that polygon D fills the screen.

41➔ From the Editor toolbar, choose Editor > Snapping > Options. Change the Tolerance from 10 pixels to 5 pixels and click OK.

42➔ Click the Auto Complete Polygon tool and start digitizing polygon D again, snapping to the vertex of polygon C as before.

42➔ Don't let the sketch cross into polygon C—keep the unit narrow and thin along the existing polygon.

42➔ When ending, be sure to snap to an edge or a vertex of polygon B.

42➔ Enter the Unit name Kkbn in the Attributes window.

Cutting polygons is sometimes easier than using Auto Complete.

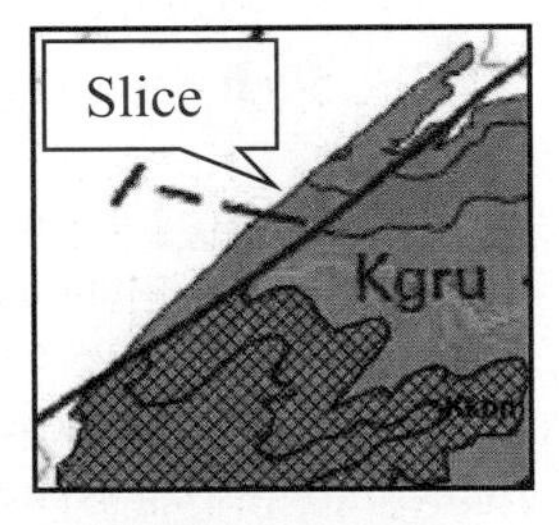

43➔ Pan so that you can see the long, thin slice above the fault.

43➔ Click the Auto Complete Polygon tool. Snap to the upper-right corner of Polygon B, and enter vertices along the fault, then up and around to enclose the entire slice. End by snapping to the edge or vertex of Polygon A.

44➔ While the slice is highlighted, click the Cut Polygons tool from the Editor toolbar.

44➔ Digitize across the slice along the fault, starting and ending outside the slice. The slice is cut in two.

44➔ Digitize across the still-selected slice again, following the geological contact between the purple and brown units.

45➔ Click the Edit tool and select the two brown parts of the slice (hold the Shift key down to select both).

45➔ Click the top geology entry in the Attributes window and enter the unit for both polygons, Kgru.

45➔ Select the middle purple unit, and enter its attribute, Kkbn.

46➔ Switch to the Auto Complete Polygon tool and digitize the purple polygon below the slice, starting and ending along the slice edge.

46➔ Enter its attribute, Kkbn.

47➔ Pan or Zoom, if needed, to see the entire large Kgru brown polygon.

47➔ Use Auto Complete Polygon to enter it, being sure to snap it to existing polygon vertices at both ends. Enter its attribute, Kgru.

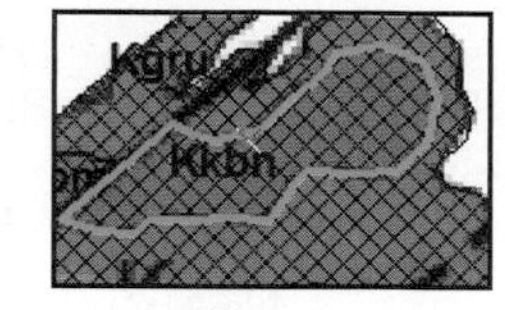

Now we have a problem. The new polygon has completely enclosed the purple Kkbn polygon, and now they overlap each other. We should fix it.

48➔ Click the Edit tool and select the purple Kkbn polygon.

48➔ Chose Editor > Clip from the Editor toolbar.

48➔ Fill the button to *Discard the area that intersects* and click OK.

49➔ Click the Auto Complete Polygon tool and digitize the small purple polygon that remains, snapping it to the edge of the polygon above it. Enter Kkbn.

49➔ Close off the last little section of brown polygon along the dashed fault. Enter Kgru.

50➔ Zoom to the new polygons (Fig. 12.23).

50➔ Choose Editor > Stop Editing. You will be prompted to save your edits. Say yes!

50➔ Save the map document.

When digitizing adjacent polygons, the order is important. New polygons must always enclose a space completely using a single line. In this exercise you were instructed to create them in a suitable order, but you must learn to look ahead and plan. Experience is the best teacher, so don't be afraid to practice, delete mistakes, and try again.

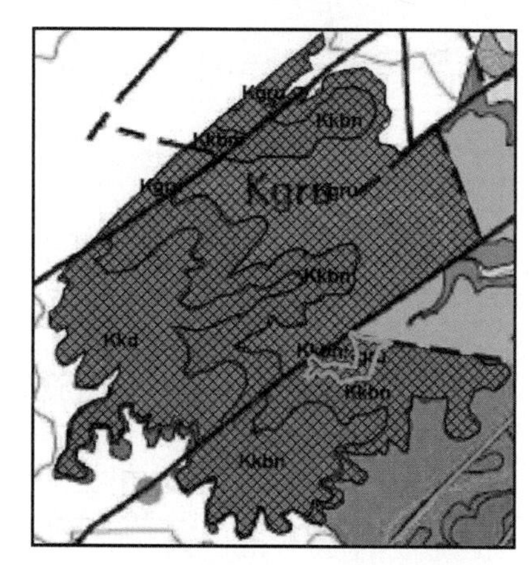

Fig. 12.23. The first polygons

Adding buildings

Now that you have some experience creating features, we will investigate some additional tools while digitizing in a city geodatabase.

➔ Open the ex_12b.mxd map document and use Save As to save it with a new name.

Let's look at another way to create a new, empty feature class for building footprints.

51➔ Open the Catalog tab and navigate to mgisdata\Austin.

51➔ Right-click the Crestview geodatabase and choose New > Feature Class.

51➔ Enter buildings for the name of the feature class. No alias is needed.

51➔ Leave the Type set to Polygon Features and click Next.

51➔ For the coordinate system, click the NAD 1983 State Plane Texas Central coordinate system from the Favorites folder. Click Next.

51➔ Accept the default XY Tolerance and resolution, and click Next.

52➔ Click in the first available blank field box and type Bldgtype. Make it a text field with a length of 25 characters.

52➔ Click in the next empty field box and create a text field called Address with 50 characters.

53➔ Click in the next empty field box to create a text field called City with 25 characters.

53➔ Click in the Default Value box above the Length box properties and enter Austin.

53➔ Click in the next available field box to create a text field called State. Give it a length of 2 characters and a default of TX.

53➔ Click Finish.

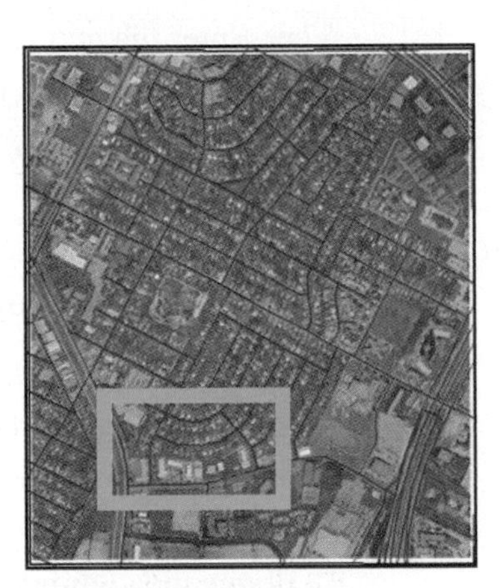

Fig. 12.24. Zoom here.

You will see the use of the Default Values in a moment.

54➔ Zoom in to the block of larger buildings in the southern part of the image, shown by the green box in Figure 12.24.

54➔ Choose Editor > Start Editing.

Only one geodatabase was used in this map document, so you don't have to choose one to edit.

SKILL TIP: An editing template should be created for every layer in the editing workspace. If one is missing, you can create it yourself (Data Management: Editing).

55➔ Right-click the buildings template in the Create Features window and choose Properties.

55➔ Examine the template settings (Fig. 12.25). In particular, note the Default Tool, the Drawing Symbol, and the attribute fields.

General

Name: buildings

Description:

Tags: Polygon

Default Tool: Polygon

Drawing Symbol

Target Layer: buildings

View

Field	Value
Bldgtype	<Null>
Address	<Null>
City	Austin
State	TX

Fig. 12.25. Editing template properties

The City and State fields already have values, Austin and TX, because you set default values when you created the fields. The Drawing Symbol is taken from the symbol assigned to the layer in the Table of Contents.

55➔ Click Cancel to exit without making any changes.

56➔ Open the Attributes window and dock it below the Table of Contents if necessary.

56➔ Click the buildings template. Start with a corner of the first large building on the left, and digitize polygon vertices around it. Double-click to finish.

56➔ In the Attributes window, enter **Commercial** for the Bldgtype attribute. Notice that Austin and TX are already entered!

56➔ Digitize the next building to the right, and enter its attribute as **Commercial**.

Default values are nice, but templates can make it even easier.

57➔ Right-click the buildings template and choose Properties.

57➔ Change the template name to **Commercial**.

57➔ Enter the text **Commercial** in the Bldgtype field. Click OK.

57➔ Digitize the next building to the right. When you finish it, look in the Attributes window. "Commercial" is already entered!

58➔ Right-click the Commercial template and choose Copy.

58➔ Right-click Copy of Commercial and choose Properties.

58➔ Change the template name to **Residential**.

58➔ Enter **Residential** in the Bldgtype field and click OK.

59➔ Click the Residential template and digitize one of the smaller residential buildings above the commercial buildings.

59➔ Open the symbol properties for the buildings layer from the Table of Contents.

59➔ Create a unique values map for buildings based on the Bldgtype field, so that the commercial and residential buildings are shown with different colors.

Note that the template colors reflect the map colors as well. You can switch templates to enter the appropriate kind of building and save a lot of typing.

59➔ Save the map document so it will remember the templates.

Buildings have square corners. The Right Angle tool makes them square by helping the next vertex to form a right angle from the previous segment.

60➔ Zoom/pan, if necessary, to put several undigitized commercial buildings in view.

60➔ Click the Commercial template.

60➔ Enter two vertices defining one edge of the building (Fig. 12.26).

Fig. 12.26. Using the Right Angle sketch tool

60➔ Click the first drop-down button on the Editor toolbar. A set of alternate tools appears.

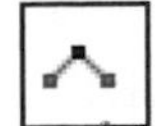

60➔ Find the Right Angle tool and click it.

60➔ Enter the vertex on the third corner of the building.

60➔ Add the remaining vertices on the building corners until you reach the last one. Instead of adding it, right-click the spot and choose Square and Finish from the menu.

TIP: For best results, use the longest, clearest side of the building for the first edge.

61➔ Continue digitizing the commercial buildings until you reach the end of the block.

TIP: If you click on a zoom tool, or on another template, it defaults back to the Straight Segment tool, and you will need to select the Right Angle tool again from the Editor toolbar.

62➔ Zoom in a little more, if needed, and switch to the Residential template.

62➔ Select the Right Angle tool again from the Editor toolbar. Digitize a house and finish the sketch.

62➔ Finish digitizing the rest of the houses in this block (south of the street only).

62➔ Choose Editor > Save Edits.

Now we must assign the addresses to the buildings. The Attributes window is useful here.

63➔ Turn on the Addresspts layer and open its label properties.

63➔ Check to label the features, use the FULL_NAME field, and make the symbols 10-pt. text. Click Symbol and Edit Symbol and the Mask tab to set a halo.

63➔Right-click the buildings layer and choose Selection > Make This The Only Selectable Layer.

64➔ Click the Edit tool and select the two houses (and one outbuilding) with addresses on Isabelle Dr.

64➔ In the Attributes window (Fig. 12.27), click the buildings entry at the top so that edits are applied to all three selected buildings.

64➔ Click the Caps Lock key. Type XX ISABELLE DR in the Address field.

That put the street in all three attributes. Now all we need to do is edit the house number. The XX helps us select it quickly.

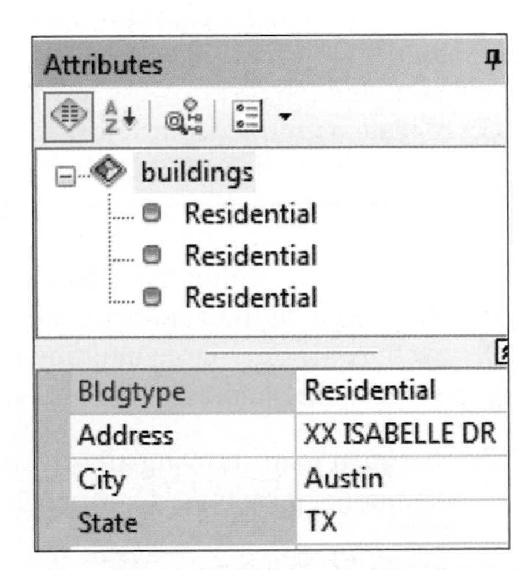

Fig. 12.27. Adding the addresses

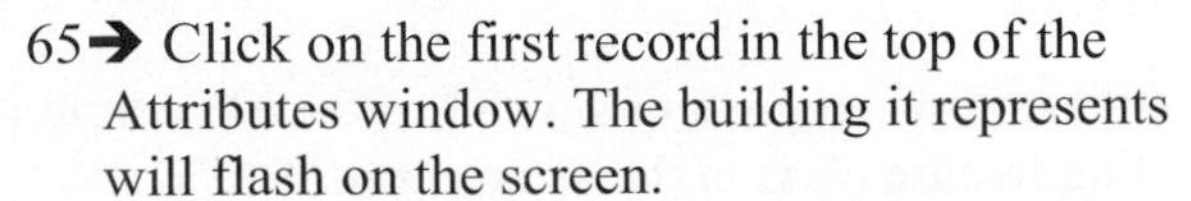

65➔ Click on the first record in the top of the Attributes window. The building it represents will flash on the screen.

65➔ Click several times, if need be, until you see which house number it needs.

65➔ Double-click the XX in the Address field and add the house number.

65➔ Continue the same steps with the other two addresses until all three have numbers.

66➔ Use the Edit tool to select another group of residential buildings on ESTHER DR.

66➔ Use the same technique to enter the addresses.

4. Can you think of a way to automatically assign the addresses to the buildings? What would be the disadvantage(s) of using it?

67➔ Open the attribute table for buildings and examine the fields. Close the table.

67➔ Save your edits, and save the map document.

67➔ Choose Stop Editing from the Editor toolbar.

This is the end of the tutorial.

➔ Close ArcMap.

Exercises

1. Digitize all of the remaining caves/sinks and springs in the Edwards outcrop area map, placing them in the karstfeatures feature class. Remember to assign the attributes.
2. The Travis-Hays County line runs NW-SE across the Edwards outcrop area map, cutting it roughly in half. Digitize all of the faults SW of the county line, placing them in the faults feature class. Remember to assign the attributes.
3. Create a map showing only the Edwards (State Map) layer outline, the karst features, and the faults. Assign different symbols to cave/sink and spring features, and to the dashed and solid fault lines. **Capture** the map.
4. Digitize the geological units shown inside the yellow line in Figure 12.28. The area is just NE of the polygons you digitized in the tutorial. Assign each polygon the appropriate attribute. Be sure to use the Auto Complete Polygon tool or the Cut Polygons tool, and plan ahead.
5. Create a unique values map of your geology polygons based on the Unit field. **Capture** a map showing only these polygons.
6. Digitize the buildings in the full block north of Esther Dr. Also include the houses on that section of Isabelle Dr, as well as the very large building behind them. Finish entering the addresses for all of the buildings.

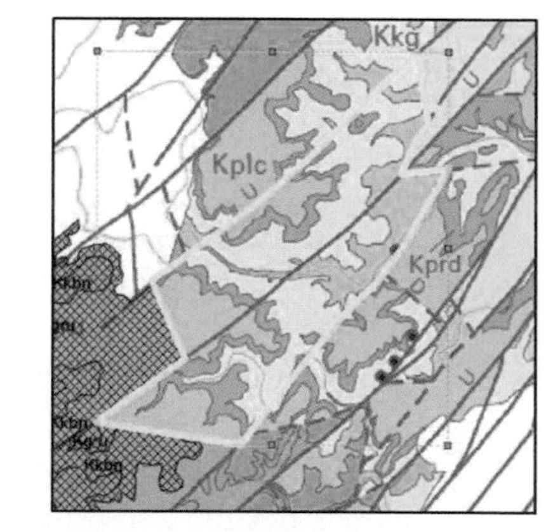

Fig. 12.28. Digitize this area.

7. Create a map showing all of your digitized buildings, labeled with their addresses and symbolized by type. **Capture** the map.

Challenge Problem

The Skills Reference section (Editing) contains instructions for using all of the Sketching tools and Sketch menu items. Create a practice line feature class in the same coordinate system as the Austin data. Experiment with each tool and menu item to get familiar with what it does and how it works. Submit a set of features created with each tool, labeled with the tool name, as a layout.

Chapter 13. Editing and Topology

Objectives

- Understanding topology and learning to preserve topological rules during editing
- Learning techniques for digitizing complex groups of polygons
- Using additional editing tools to help create and preserve topological relationships: merge, union, intersect, clip, modify, and reshape
- Editing shared lines and polygon boundaries with map topology
- Creating and editing with planar topology

Mastering the Concepts

GIS Concepts

Chapter 12 presented some basic techniques for creating new features in a feature class. In this chapter, we examine additional ways to form and modify features with an emphasis on maintaining correct topology during editing.

Topology errors

Topology concerns how the features are spatially related to each other. The four types of spatial relationships introduced in Chapter 1 included adjacency, connectivity, overlap, and intersection. One major goal during editing, other than simply getting features into a feature class, involves ensuring the **logical consistency** of features, in other words, making sure that features are free of geometric errors and that their topological relationships are adequate for the purpose intended. Some of the topological errors to avoid during editing are shown in Figure 13.1.

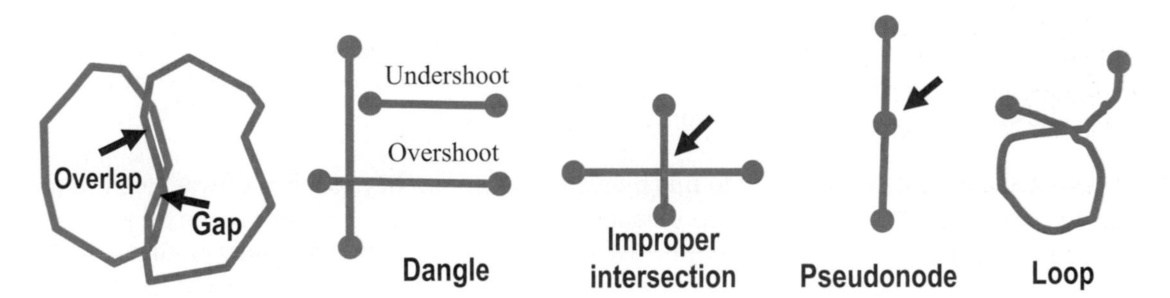

Fig. 13.1. Topological errors can occur when editing is not carefully performed.

Basic topology rules

(1) Adjacent polygons must share a coincident boundary that is exactly the same for both. There are no gaps or overlaps between adjacent polygons.

(2) Lines should always end on other lines. Failure of two lines to meet is called a **dangle**. Dangles can occur when one line is not quite long enough to meet the other (**undershoot**) or when one line crosses too far over the other (**overshoot**).

(3) Lines that intersect should each have a node (endpoint) at the intersection. Lines crossing without nodes are termed improper intersections.

(4) Nodes should exist only where three or more lines intersect. A node where only two lines meet is called a **pseudonode**.

(5) Lines or polygon boundaries should not cross over themselves and form loops.

(6) There should not be duplicate copies of any points, lines, or polygons.

Perfectly legitimate exceptions can occur to topological rules. For example, a dead-end street by definition does not meet another road and must be a dangle. A highway overpass can cross over a street without intersecting it. Polygons showing areas sprayed with pesticide on different dates might have gaps or overlaps between them.

The effects of topological errors can vary. In some cases they are merely a visual nuisance. At times they can be a legal liability, such as an overlap between two parcels. In other cases they contribute to undesirable outcomes, like the slivers than can be formed during geoprocessing operations. In the worst case, they can cause a failure of the data set for its purpose—a street network containing improper intersections will not properly route traffic. Each user must evaluate the purpose of a data set and establish the topological rules needed. In some cases these rules will include more than the six listed previously. However, these six should be considered a basic level of integrity to which every data set should conform (not including unavoidable exceptions).

Topological integrity can be established at the outset by careful editing of features to minimize the occurrence of errors. Most GIS software includes tools for ensuring basic topological integrity during editing, such as snapping and the Auto Complete Polygon tool discussed in Chapter 12.

Topology rules and tools

Even careful editing cannot avoid all topological errors. Moreover, one does not always have control over the creation of a data set; one may be stuck with data that were poorly digitized without attention to topology. A GIS that is capable of storing topology will generally also possess tools that can be used to find errors and fix them.

Many topological errors are difficult to find by mere inspection; they require software algorithms to test and evaluate features at the *x-y* coordinate level. These tools are generally run on feature classes after editing is complete. The errors that are found are fixed by additional editing, which can create new errors. So the process is repeated until all errors have been found and fixed. These routines are extremely important in developing a logical consistency report for the feature class metadata. This report describes the procedures used and tests applied to a data set to ensure that it meets the topological rules we have listed.

Some GIS systems contain advanced capabilities for defining topological rules and ensuring that they are met. The ArcGIS geodatabase model is one of these. In Chapter 9 you were introduced to network topology and some of the solvers that evaluate connectivity. Feature datasets within geodatabases can also contain **planar topology**, which expands the use of rules to model a variety of topological relationships, not just within a single layer but between layers as well.

One class of rules applies to features within a single feature class. The types of rules available vary according to feature geometry. Each of the six topological rules cited above are represented by rules such as Must Not Have Gaps, Must Not Have Overlaps, Must Not Have Dangles, and Must Not Have Pseudonodes. Figure 13.2 shows a planar topology in ArcGIS that uses these rules to test the topology of geology and faults.

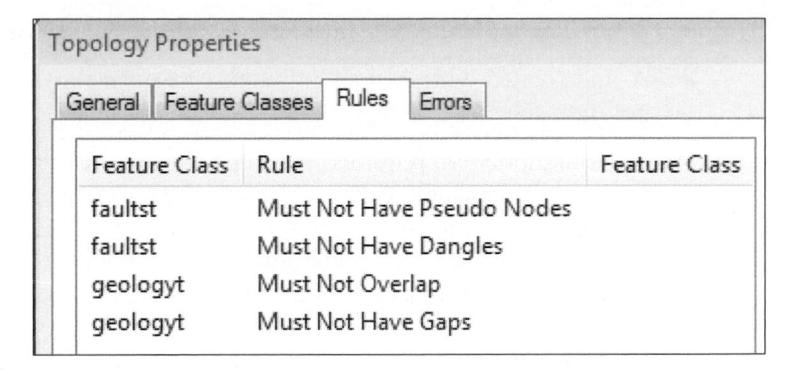

Fig. 13.2. Topology rules in a planar topology of geological units and faults

Other rules specify relationships *between* feature classes, such as the Must Cover Each Other rule that describes the relationship between counties and states in which every bit of the state area must belong to a county and no part of a county should lie outside the state.

Figure 13.3 shows typical topological errors occurring between layers. The boundary of the Pine Ridge Indian reservation (stippled area) should match the boundary between Shannon and Bennett counties, but it does not. Neither does it match the South Dakota–Nebraska boundary.

Creating planar topology makes it easier to locate and eliminate boundary errors by using special Topology tools while editing. This approach is helpful because many errors are too small to see at normal viewing scales—in Figure 13.3 the county boundaries appear to match the state boundaries but might not actually do so. Such errors typically occur because data come from different sources or because some people do not understand or use snapping effectively.

Using planar topology requires an ArcGIS Standard or Advanced license. The tutorial for this chapter includes a demonstration of using topology to test for pseudonodes, dangles, gaps, and overlaps for users who have the appropriate license. However, it is nothing more than a brief introduction to an extensive and complex topic. Users interested in learning about topology should read the appropriate ArcGIS Help files.

Users with an ArcGIS Basic license have much more limited tools for working with topology and must take responsibility during editing to keep features as correct as possible.

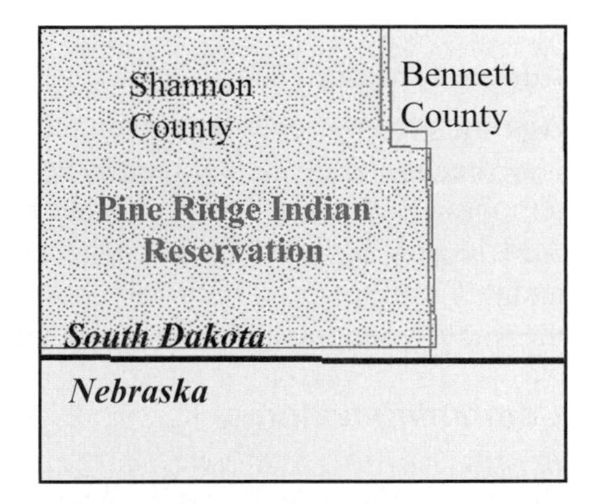

Fig. 13.3. Topological errors between layers

Editing complex polygon shapes

Some map types, such as geology and soils, have a complex interweaving of different units. It takes some forethought and planning to efficiently create polygons with correct topology. Experience digitizing such maps is the most effective way to learn appropriate techniques and avoid common mistakes. When planning, the key constraints to remember are

- A line created with the Auto Complete Polygon tool must completely enclose a region.
- A line created with the Cut Polygons tool must start and end outside of an existing polygon selection.
- An island polygon requires creating a separate feature, which will then overlap the larger polygon underneath. The overlap must be corrected afterwards by clipping.

Consider the geology map shown in Figure 13.4. It would be easy to get "painted into a corner" and not be able to finish the sequence of polygons. For example, if one digitized polygons 2 and 5 first, polygon 3 would be a hole that would be a nuisance to fill. An island polygon, such as 13, cannot use the Auto Complete Polygon tool at all because it lies completely inside another polygon. It must be added after the polygons around it, which causes it to overlap the existing polygon underneath, violating a topology rule. The following are two ways to approach the editing. Both methods should help avoid gaps and overlaps and result in topologically correct polygons.

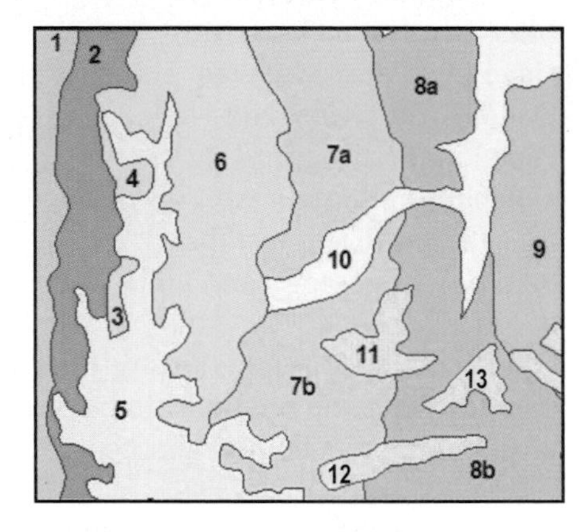

Fig. 13.4. Complex editing

Divide and conquer

Divide and conquer starts by digitizing the outline of a large area, such as the square outline of the map. The Cut Polygons tool is then used to carve out polygons from the larger region. One might start with polygons 1, 2, and 6 (creating a fake boundary between 6 and 7b that goes through 5). Then create 7 as a single unit, 8 as a single unit, and 9. Finally, use Create New Feature to create 5, 10, 11 to 13, and the remaining islands on top of the larger units. After creating each island, the Clip command (discussed in the next section) is used to eliminate the overlap with the larger polygon underneath. Occasionally with this method one is forced to create a superfluous boundary within a polygon, and then merge the two pieces afterwards.

Adding territory

Adding territory relies mainly on the Auto Complete Polygon tool. One might start with Create New Feature to create polygon 1. Switch to Auto Complete Polygon and add 2 and 3. Add 4, temporarily closing the narrow neck between it and 6. Add 5 and 6. Go back and merge 4 and 6. Add 7a and 7b, leaving a hole where 10 is. Add 11 and 12 using Auto Complete. Add 8a, 8b, and 9. Finally, close off 10. Go back to Create New Feature and create 13, making sure to clip out the underlying polygon area beneath it.

Combining features

During editing, especially when editing complex shapes, creating a single feature at a time is often not sufficient. We mentioned that creating island polygons generates an overlap with the polygon around it. Occasionally it is easier to create a polygon in two or more pieces, after which they must be combined. During editing several useful techniques can aid in maintaining correct topology. The editing functions are similar to the geoprocessing tools, such as Clip and Intersect, but they operate on selected features, rather than on entire layers. Moreover, the attributes are treated differently. Instead of combining all the attribute fields from both inputs, the output feature simply has the same attribute fields as the layer being edited.

When using these combinations, the user should be sensitive to how attributes of the features are handled. In the case of a merge, a union, or an intersect, the resulting features will be given the attributes of one of the original features. In a geodatabase, the attributes of the feature selected FIRST will be copied to the output feature. For shapefiles and coverages, the feature with the lowest feature-id will be copied. If the feature attributes matter, then the user must pay attention to which attributes are being copied and must correct any attributes that were not copied as desired.

Merge

A merge takes two or more features and combines them into a single feature (Fig. 13.5). If the features are adjacent, then the boundaries between them are removed. If the features are separate, then a multipart feature will be created. This function might be used frequently to update a parcels map when an owner has purchased two adjacent parcels and combined them into one. It is also useful when one finds it easier to digitize a feature in two or more pieces and then combine them afterwards.

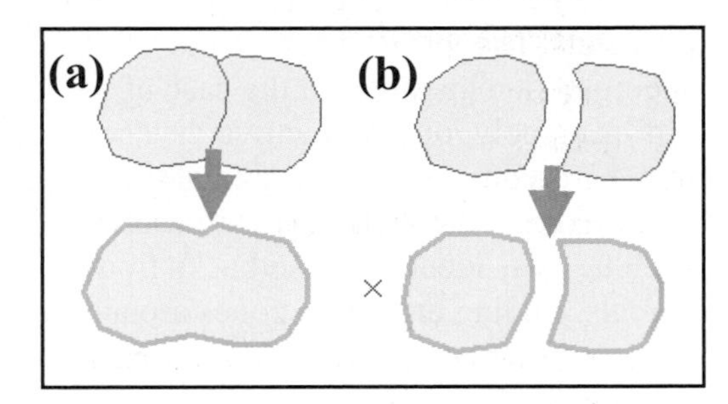

Fig. 13.5. (a) Merging adjacent polygons to create a new feature; (b) merging separate polygons to form a multipart feature

Union

A union performs the same operation as a merge, except that the original features remain unchanged and a new feature is created in addition to the originals.

Intersect

An intersection creates a new feature from an area common to both original features (Fig. 13.6). A new feature is created, and the original ones are maintained. If two features are selected and an intersection is performed, the resulting new polygon will consist only of the areas shared by the original polygons. The new feature is created in addition to the original features, which are not changed or deleted. This function might be used to identify repeat infestations of pine bark beetle attacks. Two features representing infestations in two different years could be intersected to reveal the area attacked twice.

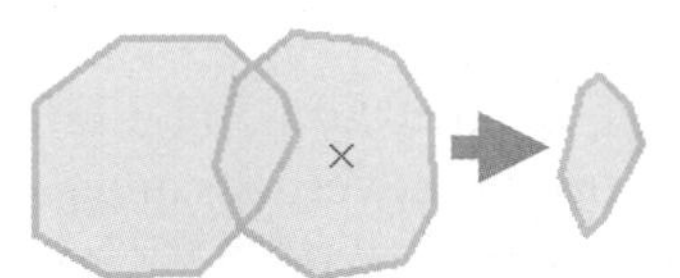

Fig. 13.6. Intersection of two polygons to create a third polygon

Clip

A clip behaves in a cookie-cutter fashion. If one feature lies over another, the underlying feature will be cut along the boundaries of the overlying feature (Fig. 13.7a). Two options may be specified, preserving the area common to both features (Fig. 13.7b) or discarding the area common to both features and retaining what is outside the clip polygon (Fig. 13.7c). In either case, the feature used for clipping is retained unchanged, and the feature underneath is modified. As shown in Figure 13.7c, clipping is one way to create a "donut" polygon. Clipping is useful anytime a polygon must have internal boundaries, such as a residential area with a park in the middle. It is also useful for creating an island polygon inside another. Clipping is crucial to enforcing the "must not overlap" rule by removing sections of a polygon that lie underneath another polygon.

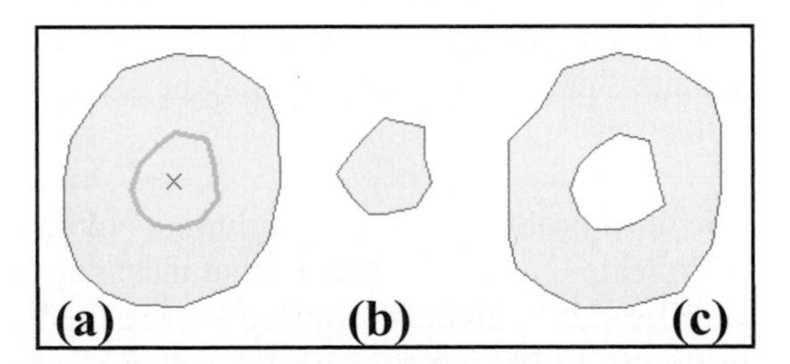

Fig. 13.7. Clipping: (a) ready to clip the outer polygon with the selected polygon; (b) option to preserve the common area; (c) option to discard the common area

Buffering features

A buffer delineates the area within a specified distance of a feature and can be created from points, lines, or polygons (Fig. 13.8a–c). The output may be lines or polygons. The buffer distance must be specified in map units. In the case of buffering polygons, a negative distance may be used to reduce the size of the feature (Fig. 13.8d). Buffers are useful for such tasks as identifying setbacks from parcels, finding drug-free zones around schools, or creating road widths from a set of centerlines.

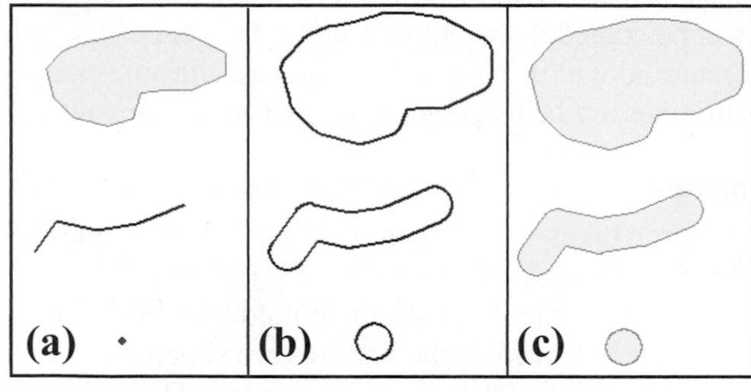

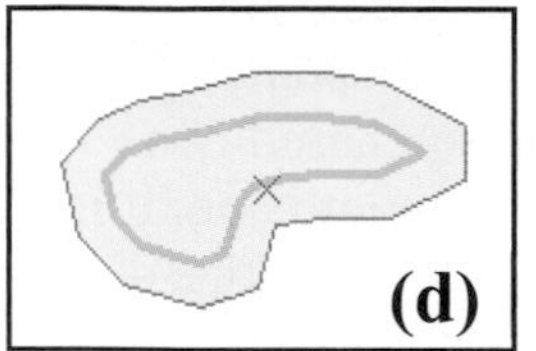

Fig. 13.8. Buffering: (a) the original features; (b) buffer output as lines; (c) buffer output as polygons; (d) a negative buffer

Stream digitizing

Two methods can be used to place vertices for lines and polygons. So far you have used the point-by-point method in which each vertex is placed with a click (Fig. 13.9a). **Stream digitizing** offers another approach. The user chooses a tolerance that controls how far apart each vertex should fall. The first vertex is added with a click. Then the user smoothly moves the mouse pointer over the line to be digitized. Each time the tolerance distance is traversed, a new vertex is added automatically (Fig. 13.9b). Stream digitizing can be faster than point-by-point digitizing, but it does have some drawbacks.

First, in point-by-point digitizing the user adds only as many vertices as needed. Straight edges require fewer vertices and curved edges need more closely spaced vertices. In stream digitizing, one tolerance is used for both. Thus, straight lines often have more vertices than needed and may not be as straight, and curved sections may not have enough vertices. As a result, the features may be less accurately captured. Furthermore, the file size is often larger due to the extra vertices. The choice of stream tolerance is critical for good results.

Second, when creating a line point by point, the user can place a vertex exactly on the inflection points of the feature to capture the shape with greatest accuracy (Fig. 13.9a). In stream digitizing, the vertex falls when it reaches the set tolerance and may be offset from the actual inflection.

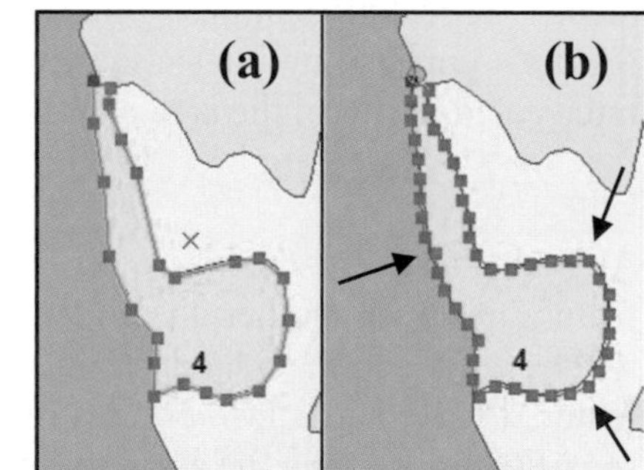

Fig. 13.9. A polygon captured with (a) point-by-point and (b) stream digitizing

The third problem is that a slight movement off the line will create unwanted vertices that must be cleaned up later. A steady hand is required for stream digitizing. In Figure 13.9b, the sketch shows three places where the curve deviates from the feature being digitized (arrows).

Most users find that they develop a preference for one kind over the other. The style of the features being digitized affects the choice as well. Feature classes with many long, straight lines and sharp corners will be much more efficiently digitized using the point-by-point method. Feature classes with many smooth curves can benefit from stream digitizing. It is also possible to switch back and forth between styles while digitizing a single feature, using stream for curved sections and point-by-point for straighter sections.

About ArcGIS

We now turn to a discussion of some additional tools and techniques for editing in ArcMap.

Changing existing features

Once a polygon or line has been created, it can be changed by eliminating unwanted vertices, adding vertices, or moving existing vertices. Other modifications include flipping lines, trimming them, extending them, and performing a variety of other editing functions. These functions are useful for cleaning up errors from stream digitizing.

Modifying features

The Editor toolbar contains an option to **modify** existing features by adding, deleting, or moving vertices. The Edit Vertices tool causes the sketch of the feature to appear. When the edits are complete, the feature is updated to the current form of the sketch. Figure 13.10 shows how an original, circular polygon is modified by moving vertices in the sketch.

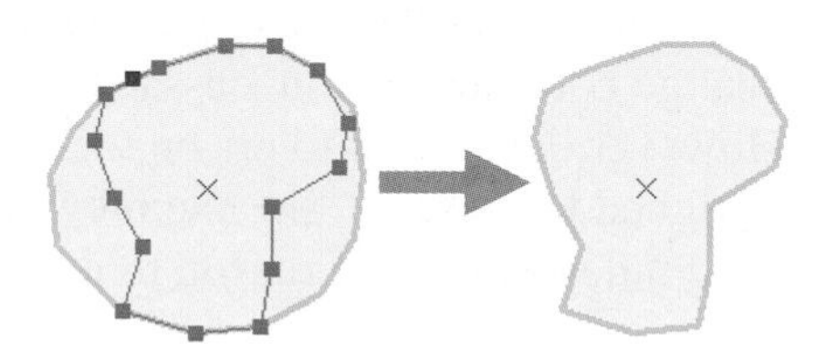

Fig. 13.10. Modifying a feature

Reshaping features

The Reshape Feature tool reenters a feature or a portion of a feature by using a new sketch to define the revised shape. The sketch must start and finish on the original feature, so it is helpful to have vertex or edge snapping on. When the sketch is finished, the original feature is modified to follow the sketch. Figure 13.11 shows how to **reshape** parts of lines and polygons.

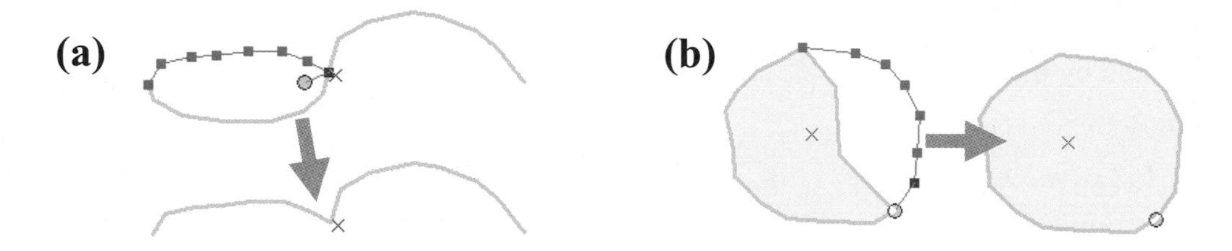

Fig. 13.11. Reshaping features: (a) reshaping part of a line; (b) reshaping a polygon

Flipping lines

As we found in the chapter on geocoding, lines can have a direction. Lines begin at the "from" node and end at the "to" node. Normally the direction of a line generates little concern, and either direction works just as well. In some instances, however, the direction matters. In geocoding, it matters because it defines which way the addresses increase along the street. A streams feature class used for network analysis uses the line direction to encode the flow of water. A similar situation would hold for constructing a network of water pipes or sewers in which the direction of flow must be constrained and recorded.

The direction of lines may be determined by drawing them with arrow-ended symbols. Once drawn, you can see the direction of flow and flip the direction, if necessary (Fig. 13.12).

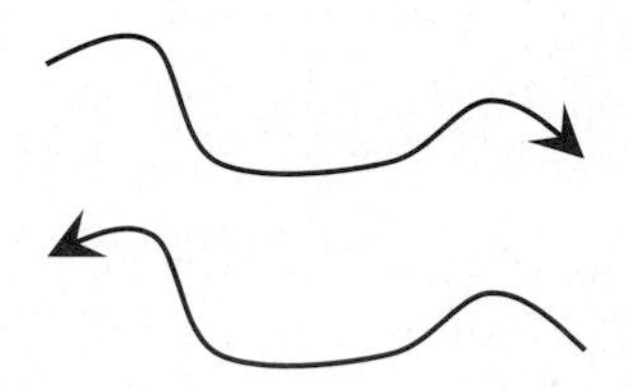

Fig. 13.12. A flipped line

Editing with map topology

Building planar topology with rules for a feature dataset and using it during editing requires an ArcGIS Standard or Advanced license. However, users with ArcGIS Basic licenses or users who are editing shapefiles can use a function called **map topology** to edit features with shared edges or vertices. Map topology creates temporary relationships between features so that they can be edited together. Its purpose is to preserve existing coincident boundaries and nodes.

Topological editing uses a **cluster tolerance** to help enforce snapping and coincident boundaries. When two vertices of features being edited fall closer together than the cluster tolerance, the vertices are made coincident. Setting the cluster tolerance requires great care. If too large, it will change coordinates unnecessarily. If too small, it does not help prevent topological errors. However, it is better to err on the side of too small than too large. The default tolerance is designed to preserve the accuracy of features rather than do extensive corrections and should be used unless you have reason to do otherwise.

The user creates map topology by selecting which feature classes will participate in it. The topology is created on the fly for the set of features currently in the data view. The endpoints of lines are called **nodes**, and lines or polygon boundaries are called **edges**. The user selects the node or edge to be edited using the Topology Edit tool, which is similar to the regular Edit tool, except that it selects shared edges or nodes for editing. When the Topology Edit tool is used to select a feature or part of a feature, all features sharing that boundary are also affected by the edits.

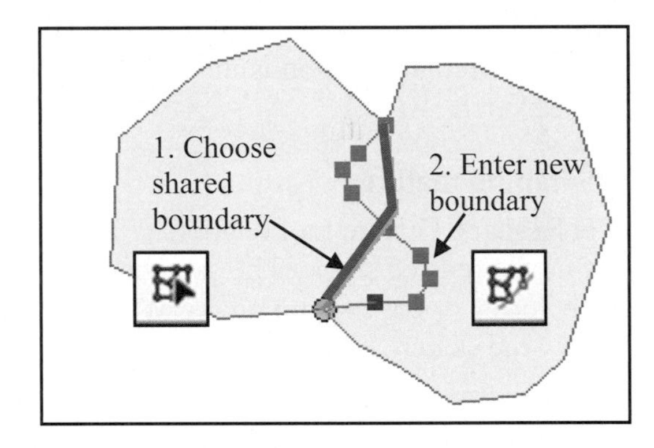

Fig. 13.13. Reshaping a polygon

Figure 13.13 shows the use of the Topology Edit tool and the Reshape Edge tool to change the boundary between two polygons. First, the shared boundary is selected using the Topology Edit tool. The selection color of the Topology Edit tool is purple to distinguish the selection from those made with the Edit tool. The Reshape Edge tool is then used to draw the new boundary between the polygons. When the sketch is finished, the new boundary replaces the old one, and the change is applied to both polygons.

The Topology Edit tool is also convenient when working with connected line features, such as roads. If one road node is moved, the roads attached to the node move also. In Figure 13.14, the road node was selected with the Topology Edit tool and then moved downward. The attached vertices of the other lines are also moved when the sketch is finished.

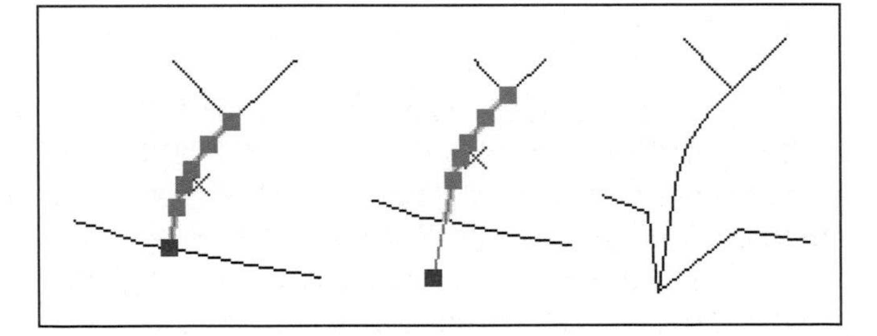

Fig. 13.14. When editing shared nodes, the node is moved, and the attached roads follow.

The Topology Edit tool and other topology editing tools should always be used for editing a map topology. They can be applied to moving features, reshaping them, and modifying them. However, the map topology must be created prior to using the topology tools.

Editing with planar topology

With an ArcGIS Standard or Advanced license, feature classes within a feature dataset may be assigned to participate in a planar topology. The topology is created in ArcCatalog, and specific rules are invoked concerning the desired topological relationships. Once the rules are defined, **validation** is performed to assess adherence to the rules. Errors are found and marked during the validation process. Afterwards, the user can view the errors and use specialized tools to fix them.

The available rules vary depending on what type of feature (points, lines, or polygons) they govern and whether they apply to a single feature class or to relationships between two feature classes. Figure 13.15 shows the rules that could be applied to lines (above) or polygons (below).

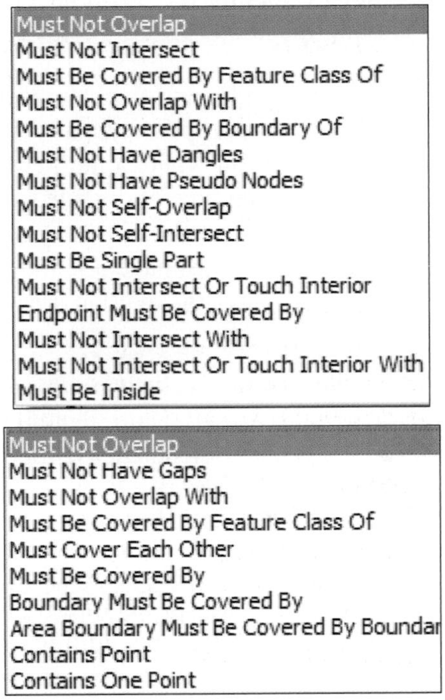

Must Not Overlap
Must Not Intersect
Must Be Covered By Feature Class Of
Must Not Overlap With
Must Be Covered By Boundary Of
Must Not Have Dangles
Must Not Have Pseudo Nodes
Must Not Self-Overlap
Must Not Self-Intersect
Must Be Single Part
Must Not Intersect Or Touch Interior
Endpoint Must Be Covered By
Must Not Intersect With
Must Not Intersect Or Touch Interior With
Must Be Inside

Must Not Overlap
Must Not Have Gaps
Must Not Overlap With
Must Be Covered By Feature Class Of
Must Cover Each Other
Must Be Covered By
Boundary Must Be Covered By
Area Boundary Must Be Covered By Boundar
Contains Point
Contains One Point

Fig. 13.15. Rules for lines (above) and polygons (below)

The topology is assigned a cluster tolerance when it is created. The cluster tolerance is the minimum distance allowed between vertices. During validation, if two or more vertices lie within the cluster tolerance, they will be snapped together.

By default, the cluster tolerance is the same as the feature class XY tolerance, or 0.001 meter. This value is designed to preserve coordinate accuracy. If the default is accepted, then virtually no vertices will be snapped together. Yet it is often desirable to set the cluster tolerance to a larger value, because then the topology can fix some errors automatically. The cluster tolerance should never be less than the XY tolerance.

Choosing the cluster tolerance can be tricky. Both the resolution of the data set and the size of the typical errors must be considered. If the tolerance is too small, then all edits must be manually done. If too large, it can degrade the data set accuracy. Imagine a road data set collected by GPS from a truck, with vertices collected every 25 meters. As lines are stopped and started at intersections, an offset of 3–5 meters typically occurs. These data are then placed in a topology. If the default cluster tolerance of 0.001 meter is employed, then none of the ends will be snapped, and each intersection must be edited manually. However, if the cluster tolerance is set to 6 meters, then most of the road intersections will be snapped automatically, leaving far fewer errors to edit. Since vertices average 25 meters apart, the cluster tolerance will not collapse actual measurements together.

During validation, several processes occur. First, cracking places an extra vertex on edges that fall within the cluster tolerance of another edge, vertex, or end, and then clustering collapses the vertices together. It is important to realize that the validation step can move the vertices of features from their original locations. Users can specify ranks for the different feature classes. Lower-ranked features will be moved to match higher-ranked features. If features have equal ranks, then both sets of vertices may be moved.

TIP: Because validation can change locations of features, it is vitally important that a backup copy of the data be established before validating, in case of an error specifying the cluster tolerance or rules.

The validation produces a topology feature class that contains point, line, and polygon errors, which may be drawn and symbolized during an editing session. Figure 13.16 shows some topology errors for the Edwards aquifer geology and faults. The pink boxes indicate point errors; the pink line, a line error; and the red outline a polygon error.

The user examines and fixes the identified errors singly or in batches, as appropriate. Unavoidable errors, such as the dangles that occur where a fault ends, are marked as exceptions. Each type of error has different possible fixes that can be applied. Dangles may be snapped, extended, or trimmed. A gap must have a feature created to fill it. An overlap must have the overlap area removed.

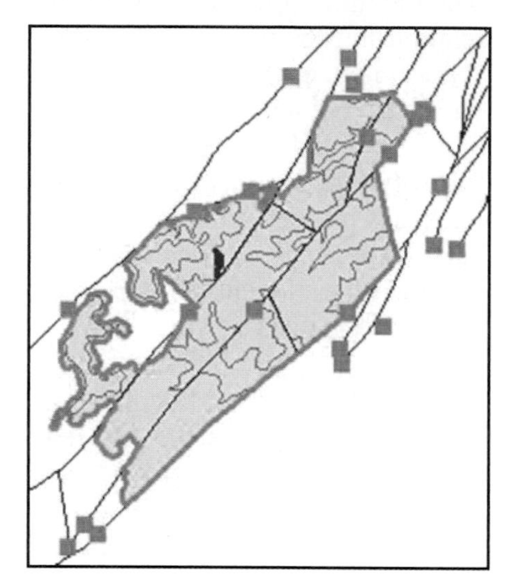

Fig. 13.16. Topology errors

TIP: There is an excellent summary of topology rules and error fixes in ArcGIS Help. Search for "topology error fixes" and select "Geodatabase topology rules and topology error fixes."

After all errors are found and fixed, the user must validate again. Often fixing one set of errors produces new ones. The process of validating, editing, and validating again must be repeated until no new errors are found.

Summary

- Careful editing helps create and maintain topological integrity between features. Topological rules establish requirements for feature adjacency, connectivity, overlap, and intersections.
- Planar topology establishes rules about the spatial relationships within and between layers, and it can be used to help locate and eliminate errors.
- Strategic planning helps when digitizing complexly arranged polygon feature classes.
- Features may be combined to create new ones using merge, union, intersect, and clip. Care must be taken to ensure that attributes are correctly copied during these transactions.
- New features can be created by buffering points, lines, or polygons.
- Stream digitizing can help create smooth curved boundaries, but it has some drawbacks.
- Modifying and reshaping are two ways to change existing features.
- Temporary topological relationships, called map topology, may be used when editing with an ArcGIS Basic license, allowing features that share vertices or boundaries to be edited simultaneously with the Topology Edit tool.

- Planar topology may be created for features in a feature dataset in a geodatabase if one has an ArcGIS Standard or Advanced license.

- Validation tests features against the assigned rules and automatically repairs some errors using a cluster tolerance. Unfixed errors may be visually inspected and repaired during editing.

Important Terms

cluster tolerance	map topology	planar topology	undershoot
dangle	modify	pseudonode	validation
edge	node	reshape	
logical consistency	overshoot	stream digitizing	

Chapter Review Questions

1. For each of the six topology rules listed, explain which of the four spatial relationships they deal with: adjacency, connectivity, overlap, intersection.
2. What is meant by the term *logical consistency*?
3. What two approaches are helpful in digitizing complexly related polygons?
4. What is the difference between modifying a feature and reshaping a feature?
5. What is the difference between a union and a merge?
6. What determines the direction of a newly created line?
7. Examine Figure 13.17. In (a), the polygons were created one after the other using Create New Feature. In (b), a clip was performed after digitizing the second polygon. They look identical, but they are not. What is the difference?

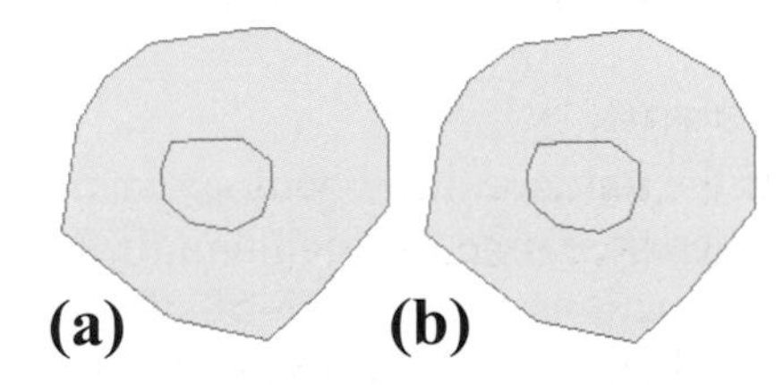

Fig. 13.17. What's the difference?

8. What is a cluster tolerance? What does it do?
9. What are the differences between using map topology and using planar topology during editing? How are they similar?
10. What is validation?

Mastering the Skills

Teaching Tutorial

The following examples provide step-by-step instructions for doing basic tasks and solving basic problems in ArcGIS. The steps you need to do are highlighted with an arrow ➔; follow them carefully. Click on the video number in the VideoIndex to view a demonstration of the steps.

1➔ Start ArcMap and open ex_13a.mxd in the mgisdata\MapDocuments folder.

1➔ Add the geology layer from the mgisdata\Austin\Edwards geodatabase created in the preceding chapter.

1➔ Use Save As to rename the document and remember to save frequently as you work.

Editing complex polygon topology

Digitizing separated buildings as polygons offers no special challenges, but interlocking polygons require experience and planning to avoid dead ends and ensure proper topology.

2➔ Give the geology layer a unique values symbology based on the Unit field.

2➔ Change the symbols so that Kkd is bright pink, Kkbn is purple, and Kgru is brown, as they are in the scanned map.

2➔ Use the layer properties Display tab to make the geology layer 30% transparent.

2➔ Give the geology layer 10-pt. bold labels based on the Unit field.

3➔ Choose Start Editing from the Editor toolbar.

3➔ Choose the Edwards geodatabase as the workspace to edit.

3➔ Open the Attributes window and dock it beneath the Table of Contents.

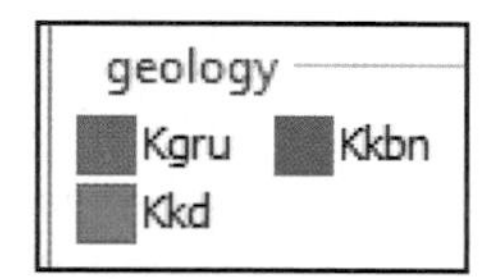

If the unique values geology map is made before starting to edit, then three edit templates should be created, one for each unit. If they were not created, then create them now by adapting the instructions in Steps 57–58 of Chapter 12.

3➔ Zoom in to the pink and purple nested polygons near A in Figure 13.18, so that they fill the screen, even if you can see pixels.

Nested polygons create a challenge because creating them can produce overlaps. These aren't fully nested because they share a boundary, so the Cut Polygons tool will work.

4➔ Click the Kkbn template and digitize around the outer purple polygon.

5➔ On the Editor toolbar, click the Cut Polygons tool. The first polygon must still be selected.

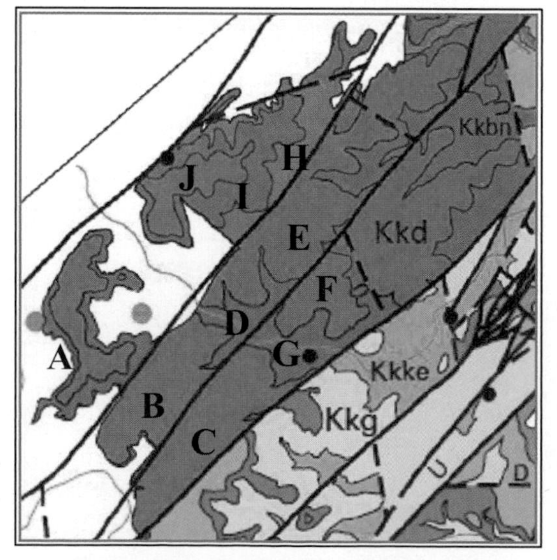

Fig. 13.18. Polygon labels for editing

5→ Starting outside of the shared upper boundary (#1 in Figure 13.19), digitize around the pink polygon, ending outside again at #2. Double-click to finish.

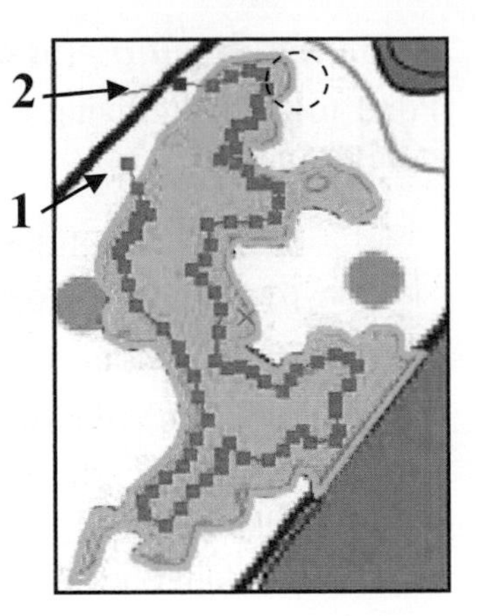

Fig. 13.19. Cutting to create an inner polygon

Now both polygons are selected.

5→ In the Attributes window, click one of the polygons to flash it. Make sure that the Unit code says Kbn for the outer polygon, and type Kkd for the inner one.

The floating toolbar and solid polygon that may appear during digitizing are annoying. You can turn these off.

6→ Choose Editor > Options and click the General tab.

6→ Uncheck the box to *Use symbolized feature during editing*, and uncheck the box to *Show feature construction toolbar*. Click OK.

Notice the tiny pink island in the purple polygon. It is almost too small to digitize, but it illustrates an important point. It is entirely enclosed by the purple polygon, so we cannot use the Cut Polygons tool as we did last time. We must create a new feature.

7→ Click the Kkd template and digitize around the pink island polygon. Finish the sketch. The new polygon should be highlighted.

7→ Turn off the edwardsoverview map image for a moment.

7→ Click the Edit tool and click inside the new polygon, and drag it away from the other polygons. Notice that the purple unit underlies it.

7→ Choose Edit > Undo Move from the main menu bar.

The new feature rests upon the old one like the top layer of a cake, creating an overlap. We must use Clip to delete the overlapping area.

8→ While the new polygon is still selected, choose Editor > Clip from the Editor toolbar.

8→ Make sure that the button is filled to *Discard the area that intersects*, and click OK.

8→ Drag the new polygon away from the others again, noting that there is now a hole.

8→ Choose Edit > Undo Move from the main menu bar.

8→ Turn the image back on again.

8→ Choose Bookmarks > Chapter 13 from the main menu bar.

Although you might be tempted to digitize the next group of nested polygons north of the fault, an experienced editor would realize that it would produce a problem connecting the two groups. You usually want to continue using Auto Complete rather than starting new polygons.

9→ Zoom in to polygons B and C, shown in Figure 13.18.

9→ Click the Kkd template and select the Auto Complete Polygon construction tool from the lower panel of the Create Features window.

9→ Digitize polygon B, taking care to snap to the vertices of the adjacent polygon.

9→ Choose Editor > Save Edits. Also save the map document.

Stream digitizing and section editing

Point-to-point digitizing gives the best control, but some people prefer stream digitizing. Let's try it on the next polygon.

10➔ Choose Editor > Options and click the General tab.

10➔ Set the Stream tolerance to 100 map units to start with and click OK.

10➔ Click the Kkd template and choose the Auto Complete Polygon tool.

10➔ Enter the first vertex of polygon C, snapping to the lower-left corner of polygon B.

10➔ Right-click off the sketch to open the Sketch menu. Choose Streaming.

10➔ Enter the next vertex to start streaming. Then move the cursor, without clicking, along the boundary of the polygon. Notice how closely spaced the vertices are.

10➔ Right-click and choose Delete Sketch.

The initial stream tolerance was too small, producing more vertices than needed and increasing the file size. Let's try a larger tolerance.

1. If you planned to create topology for the geology, what size should the cluster tolerance be relative to the stream tolerance? ____

11➔ Use Editor > Options to set the Stream tolerance to 300 map units.

11➔ Click on the vertex of polygon B to start. Streaming is still on, so you will begin streaming after that first click. Go all the way around, even if you mess up.

11➔ End by snapping to the lower-right vertex of polygon B. Double-click to finish.

Chances are that your polygon has wobbles in it. Stream digitizing takes some practice. Notice how many vertices were entered. They serve an important purpose on the curve edges but are more than needed on straight edges along the faults. You can switch back and forth between stream and point modes during a single sketch using the F8 function key.

12➔ Click the Delete key to delete the streamed polygon.

12➔ Click on the starting vertex again, and then click the F8 key to turn streaming off.

12➔ Digitize the straight edge along the fault, using only the vertices you need.

12➔ At the corner where the contact leads away from the fault, pause and press F8.

12➔ Click once to start streaming, and then move the cursor along the boundary, pausing when you get to the lower fault.

12➔ Press F8 to stop streaming. Enter the straight boundary along the fault.

12➔ Press F8 to start streaming again and digitize the contact between the faults.

12➔ When you get to the upper fault, press F8 to stop streaming. Digitize the remaining vertices and finish the sketch.

TIP: If your computer has no function keys, you must turn streaming on and off by right-clicking to open the Sketch menu. Streaming works wonderfully with a tablet pen instead of a mouse.

Chances are that, even with this new method, you will see spots where your streaming wandered from the actual boundary. This is a good place to learn about the Edit Vertices tool.

13➔ Make sure that the new polygon is still selected.

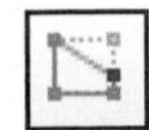

13➔ Click the Edit Vertices tool from the Editor menu.

13➔ The sketch returns, showing the individual vertices.

13➔ Hover over a vertex. The cursor changes to a little box, allowing you to click and drag the vertex to a new location.

13➔ To add a vertex, right-click a segment between vertices and choose Insert Vertex from the context menu.

13➔ To delete a vertex, right-click it and choose Delete Vertex.

13➔ Edit the vertices until you are satisfied. Then right-click on the sketch and choose Finish Sketch.

Streaming is faster but not as easy to control. Let's try another one.

TIP: If you forget that streaming is on and mess up the start of a polygon, just right-click and choose Delete Sketch so you can start over.

14➔ Pan, if necessary, so that you can see all of purple polygon D.

14➔ Click the Kkbn template. Choose the Auto Complete Polygon tool.

14➔ Start on the lower corner of the Kkd polygon just finished, on the fault.

14➔ Snap the first vertex and enter more vertices along the fault (with streaming off).

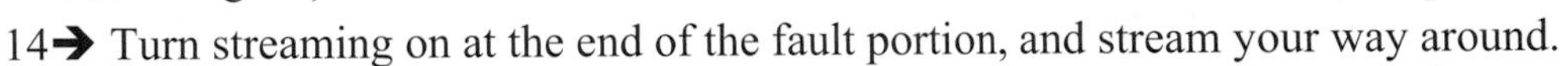

14➔ Turn streaming on at the end of the fault portion, and stream your way around.

14➔ At the end, click the mouse to snap to the corner vertex of polygon B.

TIP: When using Auto Complete with streaming, always click once on the snap point at the end to ensure that the polygon actually closes.

Sometimes you stream well around most of the polygon but have a poor section that you would like to fix without having to redo the entire polygon or edit each vertex. The Reshape Feature tool is perfect for this task. Consider the southern end of polygon C. Imagine that you would like to reenter just the portion between the faults.

15➔ Click the Edit tool on the Editor toolbar and select polygon C.

15➔ Pan/zoom in to the section to be edited so that you can see it clearly.

15➔ Choose the Reshape Feature tool from the Editor toolbar.

15➔ Click on the upper end of the section, being sure to snap to the vertex of the polygon.

15➔ Click F8 to turn streaming off so that you can be careful where you place the section vertices.

It is important to have snapping on to start the new section, because Reshape will not work if you don't start and end exactly on the existing section. However, snapping will probably interfere with where we place the new vertices, as some will probably be close to the old ones.

16➔ If the Snapping toolbar is not already open, choose Editor > Snapping > Snapping Toolbar.

16➔ Click the toggle switches for vertex and edge snapping to turn them off.

16➔ Enter the new section of the polygon, pausing before you place the last vertex (Fig. 13.20).

16➔ Click the toggle switch to turn vertex snapping back on.

16➔ Enter the last vertex of the new section, snapping it to an existing vertex. Finish the sketch.

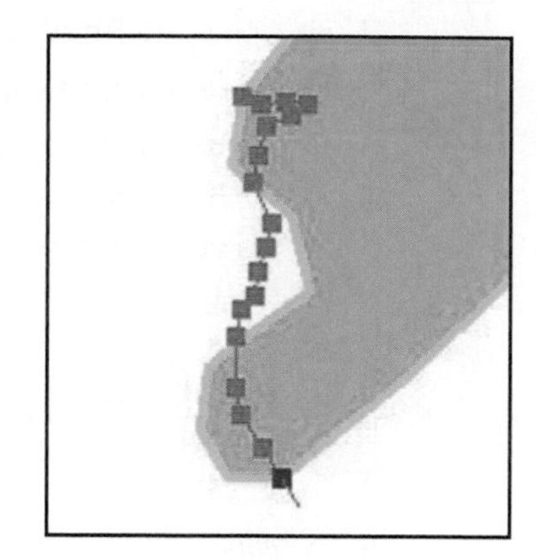

Fig. 13.20. Reshaping a section of polygon C

17➔ Pan to the other end of polygon C that was digitized using streaming.

17➔ Reshape this boundary also, for practice.

17➔ Save your edits.

Both the Edit Vertices and the Reshape Feature tools are handy for editing part of an existing feature. They are useful for fixing problems with streaming but can be used at other times also.

From this point on, you can choose whether you continue to use streaming. You can use it all the time, only for some sections, or not at all, depending on your preference.

Digitizing by adding territory

We have been using an "adding territory" approach to digitizing using the Auto Complete Polygon tool. We will continue this method for a little longer, even though the polygons are getting more complicated. Since we will be using the Auto Complete Polygon tool nearly all the time, let's change the default construction tool for the templates.

18➔ Right-click the Kkd template and choose Properties. Change the Default Tool to Auto Complete Polygon (not Auto Complete Freehand). Click OK.

18➔ Change the default tool for the other two templates as well.

18➔ Save the map document, so the changes to the templates are saved.

19➔ Pan or zoom, as necessary, to see all of polygon E (Fig. 13.21).

Look at polygon E. It is large, so take a moment to follow the boundary around so that you know where to start and stop. We will use the dashed fault on the upper end as a boundary.

19➔ Use template Kkd and the Auto Complete Polygon tool to digitize polygon E.

19➔ Use Edit Vertices or Reshape Feature, if necessary, to fix any glaring errors.

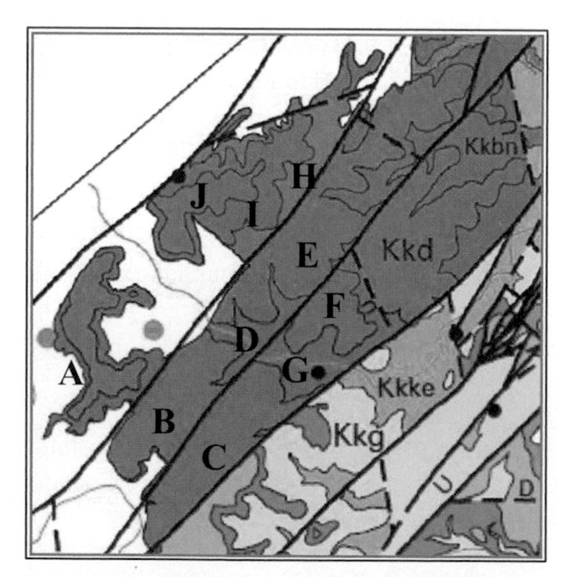

Fig. 13.21. Complex polygon topology

20➔ Pan or zoom to see polygons F and G, as necessary.

20➔ Click the Kkd template and digitize the pink polygon F.

21➔ Click the Kkbn template and digitize the purple polygon G. On the top right edge, ignore the fault this time and include the two little triangles in with the rest.

22➔ Pan/zoom to see polygons H–J.

22➔ Click the Kgru template and digitize polygon H. Note that it stops against polygon I at the upper dashed fault, so you don't need to go all the way around on top.

23➔ Click the Kkbn template and digitize polygon I. Go all the way around, leaving the pink island of Kkd (polygon J) enclosed in the middle.

24➔ Choose the Cut Polygons tool from the Editor toolbar. Start outside, cross the shared boundary, and digitize all the way around polygon J to the outside again.

24➔ In the Attributes window, find the inside polygon and change the Unit field to Kkd.

The last ribbon of brown Kgru is so narrow that it would be easier to digitize it with snapping off.

25➔ Click the Kgru template and enter the first vertex, snapping to the Kkbn polygon.

25➔ Click the toggle switches to turn off vertex and edge snapping.

25➔ Digitize the boundary of the Kgru polygon, stopping before the last vertex.

25➔ Click the toggle switch on the Snapping toolbar to turn vertex snapping on again.

25➔ Enter the last vertex, snapping it to the Kkbn polygon.

We have been treating the faults as geological contacts because it makes the digitizing easier. However, geologists generally do not want faults to separate polygons with like units. We can use the Merge command to eliminate false boundaries created by faults.

26➔ Choose Bookmarks > Chapter 13 from the main menu bar.

26➔ Click the Edit tool on the Editor toolbar.

26➔ Select the two pink Kkd polygons at the southern end (B and C).

26➔ Choose Editor > Merge from the Editor toolbar.

A window appears with both polygons listed. You need to choose one polygon as the host. The host will absorb the other one, and all attributes in the host will become the attributes of the absorbed polygon. In this case they are equivalent, so it does not matter which you choose.

26➔ Click one polygon in the list, and then the other. The flash tells which is which.

26➔ Click one of them to highlight it, and click OK.

27➔ Select the two adjacent purple Kkbn polygons (D and G) and merge them.

28➔ Select the two adjacent pink Kkd polygons (E and F) and merge them.

28➔ Save your edits.

Digitizing by divide and conquer

Instead of the "adding territory" approach, some people prefer to "divide and conquer." In this method we digitize the boundary of an area containing many polygons and then cut it up.

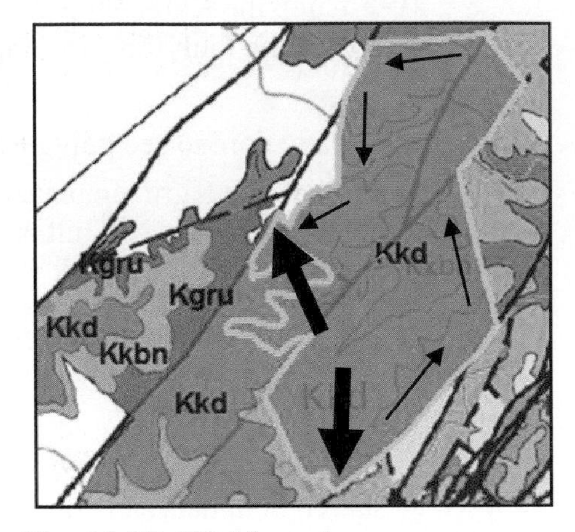

Fig. 13.22. Divide and conquer area

29➔ Zoom to the area shown in Figure 13.22.

29➔ Click the Kkd template.

29➔ Use the Auto Complete Polygon tool to create the polygon highlighted in Figure 13.22, starting at the lower arrow, following the boundary along the faults, and closing it off against the brown Kgru polygon at the upper arrow.

30➔ Click the Cut Polygons tool on the Editor toolbar.

30➔ Cut out the lowest purple polygon. Be sure to start and end outside of the "conquer" area.

30➔ After it is done, use the Attributes window to figure out which of the two selected polygons is the purple one and enter Kkbn in the Unit field.

31➔ Choose the large pink Kkd polygon next to it as the next piece to carve out. Remember to snap to or start outside of the polygon that is being cut.

31➔ Make sure the attributes are correct.

TIP: The polygon to be cut must always be selected before you start cutting. You can turn the geology layer off and on if you need to see the colors underneath.

32➔ Continue carving up the "conquer" area. There are various ways to go about it. If you need to create extra polygons to get a shape cut out, do it and merge them afterwards. Try to find the most efficient way that requires the fewest merges.

TIP: If you end up with an island that does not share a boundary, use the Polygon construction tool to create it as a top "layer cake" and then use Clip to remove the overlap.

33➔ When you are done carving, merge any polygons that require it.

34➔ Zoom to Bookmarks > Chapter 13.

34➔ Turn off the edwardsoverview image so that you can see the polygons clearly. They should look similar to the polygons in Figure 13.23.

34➔ Save your edits, and save the map document.

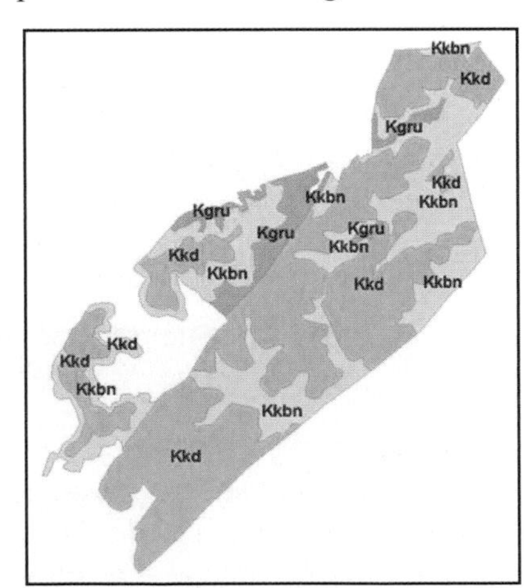

Fig. 13.23. The final polygons

Using map topology to edit polygons

So far we have demonstrated some techniques for creating polygons with logically consistent relationships with a minimum of effort—certainly less effort than it takes to fix topology errors later. However, creating polygons with correct topology is only the first part of the battle. One must also be careful that future editing does not introduce new topology errors. For this goal, editing with a map topology provides the answer.

35➔ Choose Editor > More Editing Tools > Topology.

The Topology toolbar provides functions for editing topology. First we must create a map topology, a temporary topology created on the fly.

35➔ Click the Map Topology tool on the Topology toolbar.

35➔ Check the box to have geology participate in the topology.

35➔ Expand the Options button so you can examine the Cluster Tolerance. Accept the default value and click OK.

The regular Edit tool selects an entire feature for editing. The Topology Edit tool selects a shared edge or node for editing.

36➔ Turn the edwardsoverview image on.

36➔ Zoom in to the nested polygons (H, I, and J) in the central part of the map.

36➔ Find an edge between the nested polygons that could have been more accurately digitized.

36➔ Zoom in closer to the section to be reshaped.

First we will see what happens when we edit adjacent polygons <u>incorrectly</u> using the Edit tool.

37➔ Click the Edit tool on the Editor toolbar and select one of the polygons next to the edge to be reshaped.

37➔ Click the Reshape Feature tool on the Editor toolbar.

37➔ Click to enter the first vertex of the section to be reshaped, snapping to a vertex.

37➔ Turn off vertex snapping and digitize the rest of the section. Turn snapping on again to enter the last vertex. Double-click it to finish the sketch.

Chances are that you introduced a visible gap between the two polygons. You may have also created overlaps, although they can't be seen yet.

37➔ Click the Edit tool again, hold down the Shift key, and select the other polygon adjacent to the reshaped edge.

With both polygons selected, you can clearly see two boundaries: the original boundary of the polygons and the reshaped edge (Fig. 13.24). The Edit tool selected a single polygon feature, and as a result you edited its boundary alone, introducing gaps and overlaps.

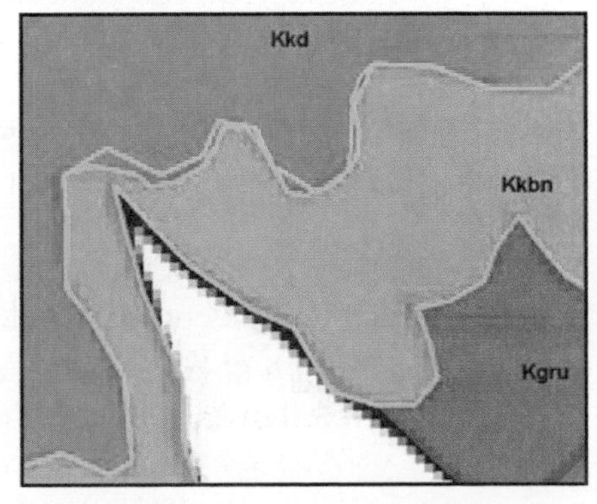

Fig. 13.24. Overlaps and gaps introduced by improper editing

37➔ Choose Edit > Undo Reshape from the main menu bar.

Now let's do it properly.

38➔ Click the button on the Standard toolbar to clear the selected features.

38➔ Click the Topology Edit tool on the Topology toolbar.

38➔ Click on the boundary between the two polygons. It will turn purple instead of blue to indicate that it is a selected shared boundary.

To reshape this time, you will use a different tool. It is similar to the Reshape Feature tool, but it works on edges.

38➔ Click the Reshape Edge tool on the Topology toolbar.

38➔ Click the first vertex of the section to be reshaped, as before.

38➔ Turn off snapping, enter the rest of the section, and turn snapping on to add and double-click the last vertex.

39➔ Click the Edit tool and select both polygons—no gaps and overlaps this time. Clear the selected features.

39➔ Find another section that needs reshaping.

39➔ Select it with the Topology Edit tool and reshape it with the Reshape Edge tool. Double-click to finish the sketch.

The Modify Edge tool lets you edit individual vertices along a shared boundary.

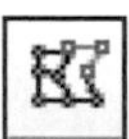

40➔ With the boundary selected, click the Modify Edge tool on the Topology toolbar.

40➔ Edit a few vertices; then right-click on the sketch and choose Finish Sketch.

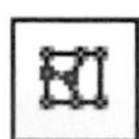

41➔ Click the Shared Features button on the Topology toolbar. It shows you which polygons share the selected edge.

41➔ Close the Shared Features window and click off the polygons to clear the edge selection.

41➔ Choose Bookmarks > Chapter 13.

41➔ Save your edits and stop editing. Also save the map document.

Using map topology to edit lines

Editing roads presents a different set of challenges.

➔ Open the ex_13b.mxd map document from the Map Documents folder.

➔ Save it with a new name and remember to save often as you work.

42➔ Choose Editor > Start Editing.

42➔ Zoom to the area outlined in Figure 13.25.

Fig. 13.25. Zoom to these roads.

42➔ Use the layer properties for Streets to give them 10-pt. green labels using the STREET_NAM field.

42➔ Zoom in to BRENDA St.

42➔ Turn off all snapping except for end snapping.

Notice that the intersections and centerlines are offset from the center of the street. We will try the wrong way first to demonstrate what happens.

43➔ Right-click the Streets layer in the Table of Contents and choose Selection > Make This The Only Selectable Layer.

43➔ Click the Edit tool and select the portion of BRENDA between MIRANDA and ESTHER.

43➔ Click the Edit Vertices tool on the Editor toolbar.

43➔ Move the vertex at the intersection of BRENDA and ESTHER to the center of the intersection. Right-click and finish the sketch.

Now the BRENDA segment is in the centerline, but the three other segments remain where they were. We've lost the connection between BRENDA and the other roads. We would have to edit the other three segments to match them up again. Map topology makes this editing easier.

43➔ Choose Edit > Undo Modify Feature from the main menu bar.

43➔ Clear the selected features.

44➔ Open the Topology toolbar, if necessary, and position it in a convenient spot.

44➔ Click the Map Topology button on the Topology toolbar.

44➔ Choose Streets and Parcels as the layers to participate, and accept the default Cluster Tolerance under Options. Click OK.

45➔ Click the Topology Edit tool and click in the intersection of the roads to select the junction. If you get an edge, try again.

45➔ Move the Edit tool cursor on top of the junction. The cursor will change to a move symbol. Move the junction to the center of the intersection.

With map topology, all four lines stay attached to the node, which is far easier than editing them separately. We will turn end snapping off so it doesn't snap to the existing location.

45➔ Click the toggle to turn off end snapping on the Snapping toolbar.

45➔ Select the junction at the intersection of BRENDA and DEBORAH and center it.

45➔ Select and center the intersection of BRENDA and HUNTLAND.

45➔ Pan west to ISABELLE St and center its junctions as well.

Even with the junctions centered, ISABELLE does not follow the street centerline. Its vertices need editing.

46➔ Select the lower edge of ISABELLE with the Topology Edit tool.

46➔ Click the Modify Edge tool from the Topology toolbar.

46➔ Edit the vertices to center ISABELLE. You will need to insert a few vertices as well as moving existing ones. Right-click and choose Finish Sketch when done.

That worked, but it was a bit tedious. Reshaping can provide a quicker alternative.

47➔ Pan/zoom to MIRANDA between ISABELLE and BRENDA. Turn on end snapping.

47➔ Select the edge of MIRANDA with the Topology Edit tool.

47➔ Click the Reshape Edge tool on the Topology toolbar.

47➔ Click on the intersection vertex to start, enter the vertices of the new MIRANDA, and be sure to snap and double-click to the vertex at the other end.

This is a good opportunity to show off the capabilities of another Sketching tool.

48➔ Still using the Reshape Edge tool, snap a vertex to one end of MIRANDA.

48➔ On the Editor toolbar, click the Sketching tools drop-down button and select the Arc Segment tool.

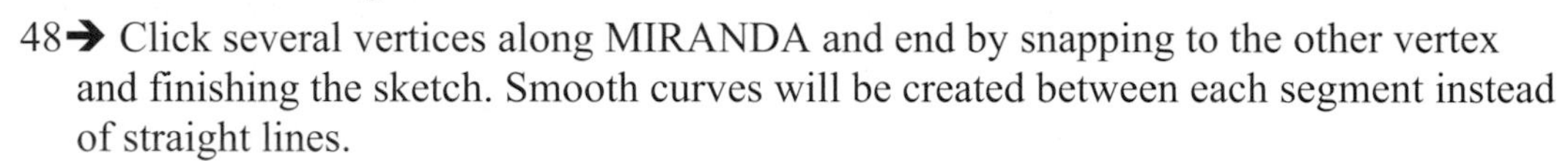

48➔ Click several vertices along MIRANDA and end by snapping to the other vertex and finishing the sketch. Smooth curves will be created between each segment instead of straight lines.

49➔ Zoom out slightly, if needed, to view the entire block of ESTHER.

49➔ Select ESTHER with the Topology Edit tool and reshape it using the Reshape Edge and Arc Segment tools.

49➔ Save your edits and your map document, and stop editing.

TIP: To have attached features proportionally stretched when moving nodes, check the box on the General tab of the Editing > Options menu. This option provides easier adjustments but causes greater changes to coordinates.

The next part of the tutorial requires an ArcGIS Standard or Advanced license. If you do not have one, you can stop the tutorial here and go on to the exercises.

Editing with planar topology (optional)

In this tutorial we will see how planar topology can help locate and fix topology errors. Imagine that someone has partially digitized the faults and geology of the Edwards aquifer. To create topology you must put them in a feature dataset.

50➔ Click the New Map File button to start a new, blank map.

50➔ Open the Catalog tab and navigate to your mgisdata\Austin folder.

50➔ Right-click the Austin geodatabase and choose New > Feature Dataset.

50➔ Name it Edwards and click Next.

50➔ Import the coordinate system from one of the feature datasets in the Austin geodatabase. Click Next.

50➔ Do not specify a vertical coordinate system. Click Next.

50➔ Accept the default tolerance and resolution values. Click Finish.

51➔ In the Catalog tab, right-click the new Edwards feature dataset and choose Import > Feature Class (Multiple).

51➔ Navigate to the mgisdata\Austin\Topology folder and select both shapefiles. Click Add and OK to start importing.

TIP: You can't create a map topology when the participating feature classes are open in ArcMap.

Now we need to build the topology.

52➔ In the Catalog tab, right-click the Edwards feature dataset and choose New > Topology. Click Next.

52➔ Accept the default name, Edwards_Topology, and the default cluster tolerance. Click Next.

52➔ Check both boxes so that faultst and geologyt will participate. Click Next.

52➔ The geologyt lines are narrower on the image and therefore more accurate, so rank them 1 and the faultst 2. Click Next.

Next we do the rules. We want to ensure that the geology polygons have no gaps or overlaps. The faults were digitized in pieces that we would like to join into one if it is continuous, so we don't want to allow pseudonodes. We also want to make sure that faults close together meet, so we don't want dangles.

Feature Class	Rule
faultst	Must Not Have Pseudo Nodes
faultst	Must Not Have Dangles
geologyt	Must Not Overlap
geologyt	Must Not Have Gaps

Fig. 13.26. The topology rules

53➔ Choose Add Rule. Select the faultst layer and set the first rule as *Must Not Have Pseudo Nodes*. Click OK.

53➔ Choose Add Rule again. Select the faultst layer and choose the second rule, *Must Not Have Dangles*. Click OK.

53➔ Add a rule that geologyt *Must Not Overlap*.

53➔ Add a rule that geologyt *Must Not Have Gaps*.

When finished, the rules window should look like Figure 13.26.

53➔ Click Next and Finish. Click Yes when asked to validate the new topology.

54➔ When it is finished, add the entire Edwards feature dataset to the map.

54➔ Choose Editor > Start Editing.

54➔ Choose Editor > More Editing Tools > Topology to open the Topology toolbar if necessary.

We will work on the fault errors first. We can look only at the errors we are working with.

55➔ Turn off the geologyt layer for now.

55➔ Right-click the Edwards_Topology layer and choose Properties. Click the Symbology tab.

55➔ Uncheck all boxes except Point Errors. Highlight the Point Errors entry, and choose to *Symbolize by error type*. Click OK.

The pseudonodes fall into two groups. Some separate a fault with a solid line from one with a dashed line and need to be retained and marked as exceptions. The others connect two faults of the same type and can be eliminated.

56➔ Create a unique values map of faultst based on the Linetype field, using dashed and solid symbols for the respective types.

56➔ Zoom in to the blue pseudonode in location A as shown in Figure 13.27. This pseudonode connects a dashed and a solid fault and needs to be retained.

57➔ Click the Fix Topology Error tool on the Topology toolbar.

57➔ Click on the pseudonode. It will turn black, indicating that it is selected.

57➔ Right-click on the selected error and choose Mark as Exception.

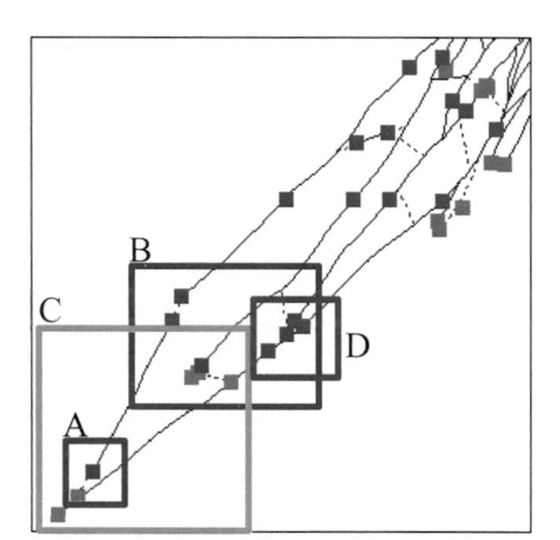

Fig. 13.27. Editing locations

58➔ Zoom to the full extent of the map and then zoom in to the next group of pseudonodes (B).

58➔ Click the Fix Topology Error tool. Hold down the Shift key and select each of the four pseudonodes that connect different fault types.

58➔ Right-click one of the selected pseudonodes and choose Mark as Exception.

58➔ Find the pseudonode that connects two dashed faults at an angle. They are the same type but are different faults, so mark it as an exception also.

58➔ Leave the pseudonodes joining similar fault types alone for now.

59➔ Continue zooming in to each group of pseudonodes, marking the ones that connect two fault types as exceptions, until all are marked.

59➔ Zoom to the full extent when done.

Now that the exceptions are marked, we can automatically fix the others by merging the two like faults on either side. The Error Inspector lets you search for and fix errors of a specific type.

60➔ Click the Error Inspector button on the Topology toolbar. It is convenient to dock the window to the lower edge of the ArcMap window.

60➔ Set the drop-down to Show: Faultst – Must Not Have Pseudo Nodes.

60➔ Uncheck the box for *Visible Extent only*. Click Search Now.

60➔ Click the top row. Then scroll down to the bottom, hold the Shift key, and click the bottom row, so that all are selected. Right-click one of the rows and choose Merge to Largest. All pseudonodes should now be gone.

For the dangles, some simply indicate the end of the fault, and these will be exceptions. Others are errors that need to be fixed.

61➔ Zoom to the five dangles at the southwestern end of the fault map (C in Figure 13.27).

61➔ Click the Fix Topology Error tool on the Topology toolbar.

61➔ Click on the three dangles that are just the ends of faults and select them all.

61➔ Right-click on one error and examine the menu choices (Fig. 13.28). Then choose Mark as Exception.

61➔ Zoom to the remaining two dangles in this view.

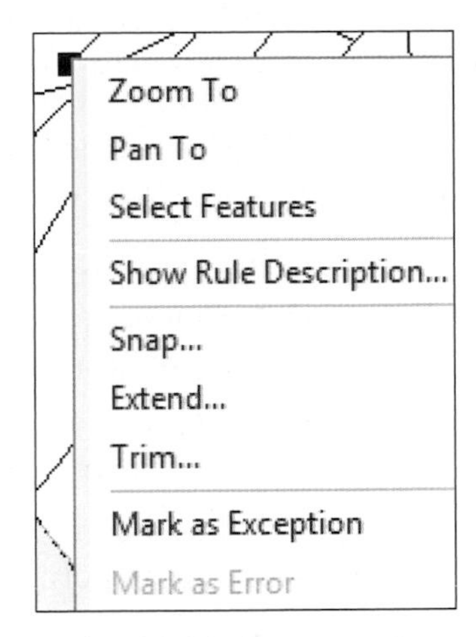

Fig. 13.28. Suggested fixes for dangles

62➔ Use the Fix Topology Error tool to select the lower of the dangles that appears to need snapping to the adjacent fault.

62➔ Right-click it and choose Snap. Type 500 into the box and click Enter.

The other one might be the end of a fault, or it might need to be connected. We should look at the original map to be sure.

63➔ Add the edwardsoverview.png image from the Austin folder.

63➔ We should connect the fault. Turn off the image until the next time we need it.

63➔ Select the dangle with the Fix Topology Error tool and right-click it. Choose Extend and enter 1000 in the box that appears.

64➔ Zoom in to the next group of five dangles (D in Figure 3.27). These all appear to be exceptions.

64➔ Click and drag with the Fix Topology Error tool to make a rectangle to select all five. Unfortunately, more than dangles are selected.

We can set the selectable errors to ensure that we get only what we want.

65➔ Click away from the selected errors to clear the selection.

65➔ Right-click the Edwards_Topology layer and choose Properties.

65➔ Click the Selection tab. Uncheck all the boxes except for *Select Errors* and *Must Not Have Dangles*. Click OK.

65➔ Select the dangles with the Fix Topology Error tool rectangle again. This time only the five dangles should be selected.

65➔ Right-click one and choose Mark as Exception.

66➔ Continue the process with the rest of the dangles, using one of the error fixes or marking them as exceptions. Consult the image if you are not sure which to do.

TIP: If a Snap or Extend does not work, try entering a larger tolerance in the box. If two dangles need to be snapped together, select both and snap instead of trying to do them one at a time.

67➔ Save your edits and zoom to the extent of geologyt.

67➔ Turn geologyt back on and make it 50% transparent.

67➔ Open the Properties for Edwards_Topology again and click the Symbology tab.

67➔ Check the boxes to display the area and line errors as well as the point errors.

67➔ Highlight the Area Errors entry and choose to *Symbolize by error type*.

67➔ Click the Selection tab. Turn on the check boxes for *Must Not Overlap* and *Must Not Have Gaps*. Click OK.

68➔ Use the Fix Topology Error tool to select the pink line that goes all the way around one set of polygons. It will turn black to indicate that it is selected.

68➔ A glance at the Error Inspector panel reveals the type of error.

The rule is Must Not Have Gaps. The gaps occur because the outer polygons are not touching other polygons. This arrangement can't be avoided—the map has to end somewhere. So this outer boundary, and the one around the other polygon group, must be marked as exceptions.

68➔ Right-click the selected black boundary and choose Mark as Exception.

68➔ Select the other outer boundary and mark it as an exception also.

Now we can concentrate on the smaller errors.

69➔ In the Error Inspector, set the Show box to geologyt – Must Not Overlap and click Search Now. Five errors appear.

69➔ Right-click the first entry and click Zoom To. The selected feature is highlighted with a black boundary.

This one appears to be an island error, where the digitizer forgot to clip. The Create Feature fix will create a new polygon and remove the overlap.

69➔ Right-click the error and choose Create Feature. It is fixed and disappears from the Error Inspector.

69➔ In the Error Inspector window, right-click the top entry and choose Zoom To.

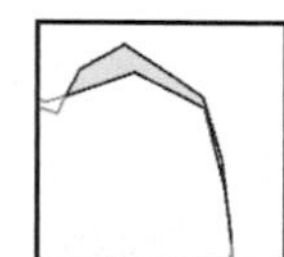

69➔ Right-click the error and examine the menu choices for fixing it.

Subtract will eliminate the overlapping area entirely. Create New Feature will add another one on top. The proper solution is to merge the overlap with one polygon or the other.

69➔ Turn on the image to see which edge best represents the boundary. Zoom out a little if you need to see more to decide.

69➔ Right-click the error and choose Merge. Flash the two polygons in the Merge window and choose to merge the sliver with the one that looks like the best choice.

70➔ Continue working in the Error Inspector until all of the Overlap errors are fixed.

71➔ Change the Show box in the Error Inspector to geologyt – Must Not Have Gaps.

71➔ Uncheck the box to search in the *Visible Extent only*.

71➔ Click Search Now. You will find four errors.

Gaps have no polygon covering them. The usual fix is to create a new polygon to fill in the space and then merge it with one of the adjacent polygons.

72➔ Right-click the first error and zoom to it.

72➔ Right-click the error again and examine the choices. Then choose Create Feature to fill in the space.

72➔ Use the Edit tool to select the filled space and one of the adjacent polygons. Choose Editor > Merge.

73➔ Fix the other three gap errors using the Error Inspector. You may need to click Search Now again in the Error Inspector window after using the Edit tool.

74➔ Zoom to the full extent of the map.

74➔ Save your edits.

74➔ Click the Validate Topology in Current Extent button from the Topology toolbar.

A few new point errors appear, which often happens after validating. You are not really done with topology until the validation step produces no new errors. You can fix the new errors now if you wish, or do them later. It may take several iterations.

This is the end of the tutorial.

➔ Save your edits and stop editing. Save the map document.

Exercises

1. The Travis-Hays county line cuts the Edwards geology map approximately in half. Digitize all of the geology south of the county line. **Capture** a unique values map of the polygons.

2. Finish adjusting the streets in the streets feature class to match the aerial photo.

3. Create a layout showing your adjusted roads, with the aerial photography in the background, and create a PDF map from it.

4. If you have a Standard or Advanced license, construct planar topology for the geology and faults feature classes that you digitized, using the same rules from the tutorial. Validate the topology and fix any errors that you find.

Challenge Problem

Find a map image on the Internet showing a topic or region that interests you. Georeference the image if necessary, create one to three feature classes, and digitize features from the map. **Capture** a map showing the image and your digitized features.

Chapter 14. Geodatabases

Objectives

- Understanding the geodatabase model
- Creating geodatabases and importing layers
- Setting up a feature dataset
- Setting up and using attribute domains and split/merge policies
- Creating subtypes and using them during editing
- Creating and editing annotation stored in geodatabases

Mastering the Concepts

About ArcGIS

As we have learned, the ArcGIS software has a long history of development, and coverages were the first data model used. Later, shapefiles were developed for use in the ArcView package. The geodatabase model arrived with the release of ArcGIS 8. The new model offers several advantages over both coverages and shapefiles. Like coverages, it can store topological relationships between features for added functionality in modeling real-world phenomena, but the geodatabase is simpler in construction and more robust in general usage. Like shapefiles, the features of a geodatabase are simply constructed and difficult to corrupt, but they provide many more benefits than shapefiles. Another advantage of geodatabases over shapefiles is the automatic tracking of lengths and areas for features. Geodatabases also provide behavior and validation rules that control how features behave, how they can connect, and what attributes are allowed. A user can also set up planar topology for features and construct topological rules, such as specifying that a state boundary must always match county boundaries or that counties must never cross a state boundary. We've used feature classes extensively in this text, but this chapter describes the feature dataset and the geodatabase capabilities available to users with an ArcGIS Basic license. More advanced and powerful features are available to those with an ArcGIS Standard or Advanced license; some of these are mentioned as well.

About geodatabases

The geodatabase model is constructed on the architecture of standard database systems and can be implemented with several commercially available database software packages. Three types of geodatabases may be created.

Personal geodatabases are designed for use by individuals or small workgroups and are stored in a single Microsoft Access file. This file is limited to 2 GB in size and works only in the Windows operating system.

File geodatabases are also designed for individuals and small groups, but each data set is stored as a separate file within a system folder, and each file can be up to 1 terabyte in size. File geodatabases can be accessed by different operating systems and are best for cross-platform operations, such as a company that supports Windows, Linux, and Unix computers.

SDE geodatabases, also known as *enterprise geodatabases*, store GIS data within one of several commercial relational database management systems (RDBMS), such as Oracle®, SQLServer®, and IBM Informix®. SDE geodatabases are designed to meet security and management needs for large data sets accessed by many simultaneous users. Such complex systems are often called organizational databases, or Enterprise GIS. For example, at least three groups of users might need to edit parcel records in a large city database: the tax department, the city surveyors, and the deeds office. Instead of maintaining three sets of data, an ArcSDE geodatabase allows users to "check out" certain portions of data for editing and merges the changes back into the central database. Potential conflicts caused by simultaneous editing by different users are avoided. An additional software package called ArcGIS for Server is required to translate geographic data into a consistent and customized format in the large database and share it with large groups.

Organizational databases have distinct advantages when the spatial data sets are large and/or are accessed by many different people. For example, such a system might be developed for the planning department of a large city, which may have dozens of people in several departments needing access to the information, both for reading and for editing. An enterprise setup allows many people simultaneous access to data sets, provides security and permission handling, and manages cases of multiple simultaneous editing of the data. It also makes it possible to set up version tracking for added data security and protection.

An additional advantage of ArcGIS for Server is that it allows the administrator to share geographic data and geoprocessing tools over the Internet. It can publish data services like those found in ArcGIS Online and make data or geoprocessing tools available through web pages or mobile devices. It is not easy or cheap to set up an enterprise geodatabase, but it does offer cost benefits. For example, it may be that 100% of the employees in your organization need to use your GIS data, but 80% of them could perform their tasks using a web interface with a few simple tools. Offering this capability on a web page means that you don't need to buy nearly as many ArcGIS Desktop licenses. Many people are discovering productivity perks by moving some of their work to tablets or smartphones; now this option is available for GIS data and tools as well.

However, the additional work in setting up an enterprise geodatabase only makes sense when these issues are present. In this chapter we will work exclusively with personal or file geodatabases, but most of the commands and functions are the same for the organizational version.

A geodatabase may contain a variety of objects (Fig. 14.1). Feature classes are used to store the spatial information. We've used these extensively in this book already. Geodatabases may also contain rasters, tables, layers, relationships, geometric networks, and feature datasets. A feature dataset is a collection of related feature classes with the same coordinate system. The oregondata database in Figure 14.1 contains two feature datasets: Transportation and Water. The Water feature dataset has been expanded to show the feature classes, including rivers and waterbodies (lakes). The geodatabase also contains standalone feature classes such as cities and counties, as well as rasters like gtopo1km. In this case, the standalone feature classes and rasters share the same coordinate system, but they do not have to.

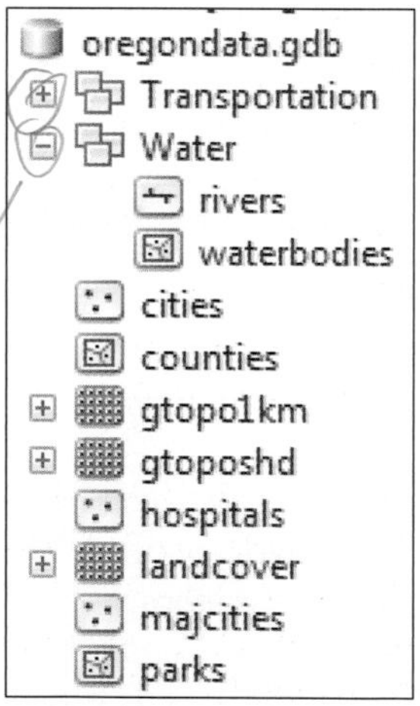

oregondata.gdb
Transportation
Water
rivers
waterbodies
cities
counties
gtopo1km
gtoposhd
hospitals
landcover
majcities
parks

Fig. 14.1. A geodatabase

Designing and creating geodatabases

Good databases do not happen by accident but are the result of careful planning. A project database is often accumulated on the fly with no terrible repercussions, but a long-term repository of data for an organization benefits from forethought. The designer needs to list the feature classes to be included and the attributes that they will have. She must decide whether feature datasets are needed and how they will be organized, and whether any of them will have topology or networks. He must decide on coordinate systems, data formats, and even what individual feature classes and files will be named.

It is even important to think through the contents of various attribute fields and decide on content. Will street names be all uppercase or a mix of upper- and lowercase? How many different categories of buildings are needed? At every step, decisions must be enforced to ensure that fields and names are consistent and easily understood.

Developing a written plan is a must for a good geodatabase. Writing the plan facilitates the organization itself, and the written plan serves as a guide to the creator, who may have forgotten a particular design decision several months later and needs to look it up, as well as to other workers responsible for implementing the design.

However, no database plan is ever a static, finished thing. As the database evolves, the plan can evolve as well, and it should be updated accordingly.

TIP: Designing and creating good geodatabases involve many issues far beyond the scope of this book. Users planning to work with geodatabases should read more about them.

Geodatabases must be created as empty shells to which feature classes and other database objects are later added. Feature classes may be loaded from a variety of sources. Coverages and shapefiles may be imported into the geodatabase, either as standalone feature classes or as part of a feature dataset. INFO and dBase tables may be loaded. Users may also create empty feature classes and use editing to add features, as we did in Chapters 12 and 13. Rasters, tables, annotation, and relationships can also be stored inside a geodatabase. Sometimes new feature classes are added to a geodatabase as a result of the operation of a tool or command.

The design and structure of a geodatabase, including the datasets and feature classes, the fields in the tables, the standalone tables, the relationship, and other objects, are together termed the database **schema**. The description of all the objects in a geodatabase can be stored and saved without any actual data and then reused to generate multiple geodatabases with the same structure. This approach offers many advantages for developing large, complex databases with a well-designed and tested schema. A city planning department that is relatively new to GIS might choose to use a published schema and populate the database with the local information rather than trying to design the geodatabase from scratch. Schemas for various industries and applications may be downloaded free from the ESRI web site.

Many databases are developed by importing shapefiles or coverages of existing data. This process is usually straightforward. The entire data set, including features, attribute tables, and coordinate systems, is transferred into the database. In the case of coverages, the attribute fields must be converted from the coverage field types to the geodatabase field types. This conversion is handled automatically, although the user can customize the interpretation of certain fields if desired. The user can also request a coordinate system reprojection as part of the transfer process.

Unlike layer files, which store a link to source data, the geodatabase stores the feature data itself. Importing a shapefile creates a copy of the features inside the database, leaving the original features unchanged. The user then has two copies of the data.

Creating feature datasets

A feature dataset is a collection of feature classes that are related to each other in some way (Fig. 14.2). For example, a transportation feature dataset might contain feature classes for roads, railways, airways, and interstates. These features could be used to construct a geometric network so that the movement of people or goods along the network could be modeled, as discussed in Chapter 9. Similar models could be constructed for utilities by using feature classes representing water mains, water sources, sewer lines, and treatment plants. Feature datasets may also contain planar topological relationships between the feature classes, allowing for easier identification and correction of errors, such as gaps or overlaps between polygons. (See Chapter 13 for a description of planar topology.)

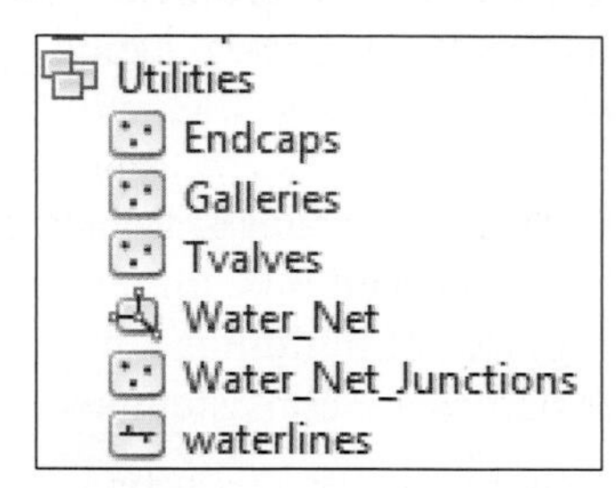

Fig. 14.2. A feature dataset contains related layers with a common coordinate system.

About the spatial reference

The feature classes in a feature dataset must all share the same spatial reference, which includes the coordinate system, the X/Y domain, and the resolution. The coordinate system was discussed in Chapter 11.

The **X/Y domain** is the range of allowable *x-y* values that can be stored in a feature class. Integers are used because the computer can process them several orders of magnitude faster than floating-point values; in the case of SDE databases, they can also be compressed for more efficient storage space. The software converts the integers to real numbers on the fly when it needs them (e.g., when reporting the current *x-y* cursor location in ArcMap).

The **resolution** refers to an underlying coordinate grid to which values are snapped. The units are the same as those for the defined coordinate system. A resolution of 0.001 meter in a projected coordinate system stores values to the nearest thousandth of a meter. A resolution of 0.000001 in a GCS stores values to the nearest millionth of a degree.

The user chooses the coordinate system for a feature dataset, and the software suggests appropriate values for the domain and resolution. The default resolution and domain are determined based on the coordinate system chosen by the user and will be fine for most applications. Users requiring very high resolution, for example, in laying out a surveying grid, can increase the resolution at the expense of a smaller domain.

Default values

One advantage of using geodatabases lies in setting up default values for attributes. For example, if a user is digitizing roads, residential streets (LOCAL) are usually the most common type and typically have two lanes and a speed limit of 25 mph. The user can set default values for each of these fields (Fig. 14.3). Then when each road is added, the road type, lanes, and speed limit are automatically set to LOCAL, 2, and 25, respectively. The information needs editing only if the new road differs from the default. A user could even

Field Properties

Alias	ROADTYPE
Allow NULL values	Yes
Default Value	LOC
Domain	RoadType
Length	3

Fig. 14.3. Default values for attributes save time during data entry.

set the defaults one way, digitize all the local roads, and then close the layer, change the defaults, open the layer again, and digitize the next group. This approach saves editing time and helps reduce attribute errors.

The introduction of editing templates in ArcGIS 10 provides another way to work with default values. Different templates can be created for each type of feature, with different default values stored in the template.

Domains

One great advantage of geodatabases is their tools for helping maintain correct attributes. An attribute **domain** constrains the values that may be entered for a particular attribute. Such constraints help ensure correct data entry and keep out incorrect values. For example, imagine an attribute field named PipeDiam that contains the diameters of water pipes. Water pipes don't come in an infinite variety of sizes; in fact, a town might use only 1-inch, 3-inch, 6-inch, and 12-inch pipes. Setting up a domain for the PipeDiam field provides a "pick list" of choices for the field and prevents any user from mistakenly entering a 4-inch or 5-inch pipe. These constraints can enforce data entry rules in situations when more than one person is entering data and not everyone knows or can remember all the rules.

Domains come in two types. The first type is a **range domain**, which specifies the lowest and highest possible values but lets the values take any number in between. Range domains are applicable only to numeric data. For example, a range domain could be applied to a student Grade Point Average (GPA) field. A GPA ranges between 0 and 4.0. A range domain would allow any value between 0 and 4.0 but would exclude negative values or values greater than 4.0. The second type is a **coded domain**, which allows only certain values taken from a list. The PipeDiam domain mentioned previously is a coded domain because it allows only four specific values from a list: 1, 3, 6, or 12.

Domains are created and maintained in ArcCatalog as a property of the geodatabase rather than as a property of a single feature class or attribute field. Therefore, a domain can be used more than once, and it can be used in multiple feature classes. For example, a range domain called Percent, which allows values from 0 to 100, could be used in many feature classes whenever percentage values were being stored in an attribute.

Figure 14.4 shows how the properties of a percentage domain would be set up in ArcCatalog. The field type is specified as short integer. The domain type is range, and the minimum value of 0 and the maximum value of 100 are entered. A domain can also have split and merge policies assigned to it, which we will cover in the following section.

Domain Properties:

Field Type	Short Integer
Domain Type	Range
Minimum value	0
Maximum value	100
Split policy	Default Value
Merge policy	Default Value

Fig. 14.4. Properties of a range domain

The field type of the domain must always match the field type of the attribute to which it will be assigned. (See Table 4.1 if you need a review of field types in ArcGIS.) Thus, a short integer domain can be used only with a short integer attribute field and not with a long integer or float field. Users must pay careful attention to the fields for which the domain is intended in order to select the appropriate field type for the domain. If a database had some fields that stored decimal percentages and others that stored integer percentages, then two different Percent domains would need to be created, one with a Float field type and one with a Short field type.

An example of the PipeDiam domain is shown in Figure 14.5. The field type is short integer, and the domain type is set to coded values. The codes themselves are typed in below. The code is the value actually stored in the attribute field. The description appears in legends and tables during an ArcMap session, providing easily understood information to the people working with the attributes. However, the descriptions do need to fit well inside the menus and legends, so it is best to keep them reasonably short.

Domain Properties

Field Type	Short Integer
Domain Type	Coded Values
Split policy	Duplicate
Merge policy	Default Value

Coded Values:

Code	Description
1	1-in
3	3-in
6	6-in
12	12-in

Fig. 14.5. Properties of a coded domain

Coded domains provide a valuable way to combine the ease of numeric codes with the luxury of understandable text information for people. Land use planning, for example, often uses numeric codes to indicate various zoning types, such as 32 = Residential and 14 = Commercial. Numeric codes are advantageous because they are less susceptible to typing errors and take less space to store. However, it is difficult for people to remember the meanings of many codes. With a coded domain, the zoning codes may be stored as integers, but the information presented to the user takes the form of the textual description (Fig. 14.6).

Coded Values:

Code	Description
12	Heavy Industrial
14	Commercial
16	Low Density Residential

Fig. 14.6. Using domains to link numeric codes to understandable descriptions

Split and merge policies

Domains offer a way to control the updating of attributes during a split or a merge operation. The default policy of making copies or entering blanks may be overridden by alternate and more useful actions. Split and merge policies are associated with each attribute domain. Setting up these policies correctly can save much time during editing.

The split policy

The **split policy** is applied when a feature is split in two. The policy choices include Default, Duplicate, and Geometry Ratio. If Default is chosen, both split feature attributes will contain the default value for that field. If Duplicate is used, then both new features retain the value of the original feature. If Geometry Ratio is used, then the new value is assigned based on the relative size of the original and split features.

Examine an example of splitting a parcel (Fig. 14.7). The Parcel_ID field is nominal data, so it is not practical to assign a domain. Instead, the normal rules for splitting and merging features without domains would apply. At least one of the new parcels will need a new number.

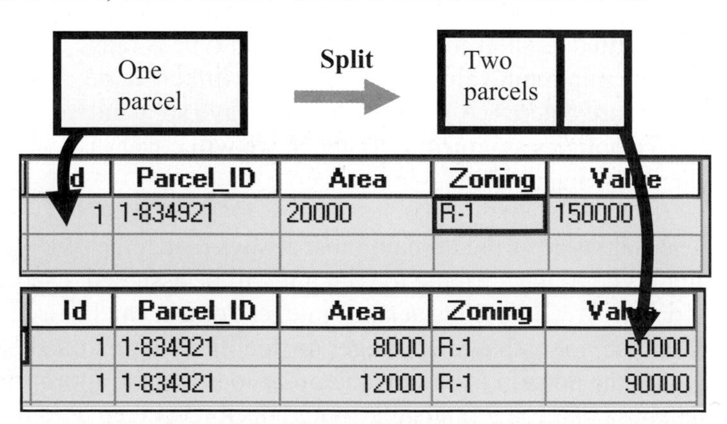

Id	Parcel_ID	Area	Zoning	Value
1	1-834921	20000	R-1	150000

Id	Parcel_ID	Area	Zoning	Value
1	1-834921	8000	R-1	60000
1	1-834921	12000	R-1	90000

Fig. 14.7. The attributes of two parcels created by splitting one parcel according to the split policy

The Area and Value fields use the Geometry Ratio policy. The proportion of areas in the two new polygons is 60-40; thus the smaller parcel is assigned 40% of the original area and value, and the larger parcel is assigned 60% of the original area and value. Finally, the Default policy is applied to the Zoning field, again assigning the most common value of R-1.

In deciding whether to use split and merge policies, one must consider the number of attributes involved and the amount of editing that a data layer receives. Setting up and applying domains to attributes takes some time and thought. Unless the layers are being frequently edited, the effort may not be worth it. When appropriately used, however, domains can save a substantial amount of time and prevent many incorrect data entries.

TIP: A user can cause a field to receive a blank during a split or merge by setting the policy to Default and leaving the default value blank.

The merge policy

A **merge policy** in ArcGIS is not evaluated during editing. It exists so that programmers writing advanced applications can take advantage of them, but they will not help the average user. Thus, only a brief sketch of merging policies is given.

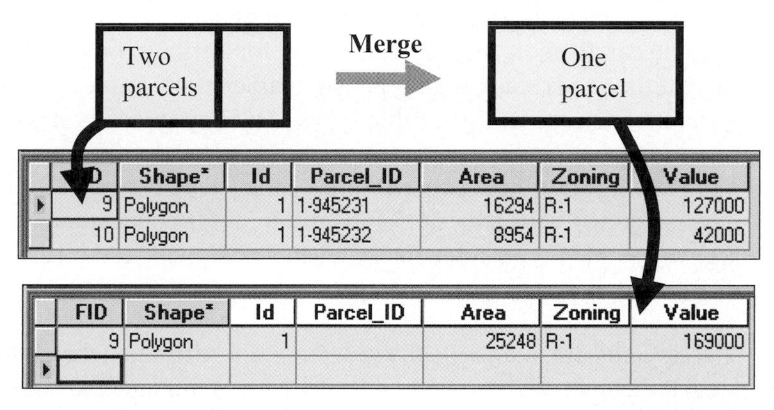

D	Shape*	Id	Parcel_ID	Area	Zoning	Value
9	Polygon	1	1-945231	16294	R-1	127000
10	Polygon	1	1-945232	8954	R-1	42000

FID	Shape*	Id	Parcel_ID	Area	Zoning	Value
9	Polygon	1		25248	R-1	169000

Fig. 14.8. Attributes from two parcels are updated according to the merge policy when they are combined into one.

By default in coverages or in shapefiles, the merged parcel will contain the attributes of the first feature found in the database. In geodatabases, it will contain the attributes of the first feature selected before merging. The merge policy overrides this haphazard method by specifying a different action for each field.

Three different policies may be assigned: Default, Weighted Average, or Sum. Default works the same way as for the split policy. Weighted Average assigns the new value based on the relative areas of the two features, and Sum assigns the sum of the two input features. In Figure 14.8, the Zoning field was assigned the Default policy. The Area and Value fields would logically use the Sum policy.

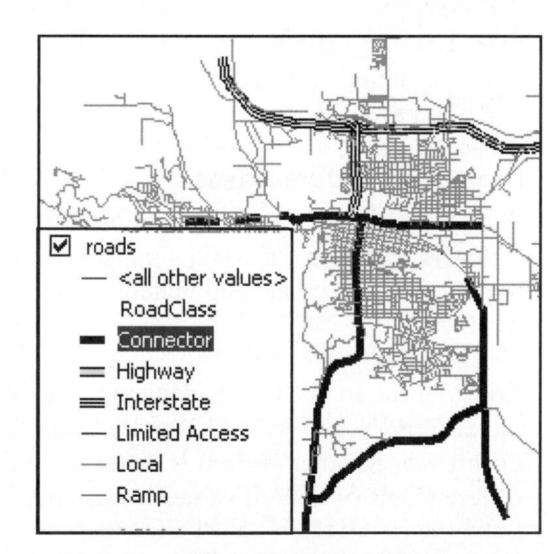

Fig. 14.9. Subtypes are formal classes of features within a layer.

Subtypes

Subtypes are another way to facilitate the entry and validation of attribute data. A subtype must be based on categorical data, and each feature in the class must belong to one of the categories. Roads, for example, may belong to one of several types, such as interstates, highways, connectors, or local roads (Fig. 14.9). On the surface, subtypes are not that different from a categorical data field. However, subtypes have additional requirements and capabilities.

A subtype must be based on a short integer field that contains a numeric code representing each category (1 for interstates, 2 for highways, etc.). Like a domain, the subtype also has a text description that lets the user see what the code stands for in plain words. The subtype essentially defines its own coded domain for the feature class.

Subtypes provide more functionality for a feature class than a simple categorical field with a domain, however. You can think of them as sub-feature classes within a feature class. First, when a data layer is loaded into a map, the subtypes are automatically displayed with different symbols. When edit templates are created, each subtype has its own template (Fig. 14.10).

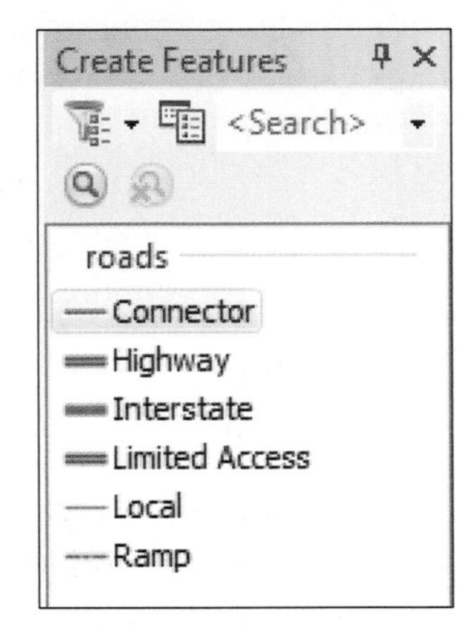

Fig. 14.10. Subtype editing templates

One of the most advantageous properties of subtypes is that each subtype can have its own default values for attribute fields. In the case of the roads, each road type has characteristic values that are different from those of the other types. An alley typically has one lane, a local street has two lanes, a connector has four lanes, and an interstate has four lanes. Speed limits can also be assigned for each type, such as 15 for alleys, 25 for streets, 35 for connectors, 65 for highways, and 75 for interstates. When a subtype is established, different default values can be specified for each of the types.

The defaults are automatically entered in the attribute table when new features are created. To add local streets, the user would select the Local template and begin adding streets. As they were digitized, each street would immediately receive the Local default values: two lanes, 25 mph speed limit, and so on. To enter a connector, the user would select the Connector template, and subsequently added streets would receive the default values for connectors. Subtypes streamline both data entry and the assignment of attributes to features. Occasionally, features may deviate from the default values, such as a connector having six lanes, but such exceptions may be corrected as needed.

Organizing feature classes

Subtypes are an intermediate alternative between placing different groups of features into different feature classes and simply having a field that distinguishes them. Because subtypes require effort to create and manage, one must consider carefully the design question of whether to use them.

Consider the roads. If one plans to do any network analysis it would clearly be a mistake to save the different road types in separate feature classes. However, the different types have distinct differences in other attributes (lanes, speed limits) that make it attractive to be able to assign different sets of default values to each road type. Roads are a good candidate for subtypes.

Next consider census boundary designations: blocks, block-groups, tracts, and so on. These certainly seem like different categories of census polygons. However, they cover each other in overlapping sets, like a layer of jigsaw puzzles with ever larger pieces. Trying to map or display them together in the same feature class would be a mess. Census areas definitely need to be in separate feature classes.

Finally, consider vegetation polygons. Vegetation type is certainly categorical data, but there are so many different types of vegetation, generally, that one would need dozens of different categories, making it very tedious to construct the subtypes. Moreover, it is unlikely that other fields in the feature class would have values dependent on the vegetation type alone, so that there is little opportunity to make use of defaults. Vegetation is not a good candidate for subtypes.

In general, it is better to keep similar features together in a single feature class. For some reason, beginners seem to proliferate feature classes, for example, creating a geology map with separate feature classes for each geological unit or each quadrangle. This strategy of separation is often unnecessary. Queries make it easy to separate a subset of features from a feature class when it is needed. If one wants a map showing only the volcanic rocks, a definition query or a layer selection can accommodate that wish. The Select By Location or Clip functions solve the issue of focusing on a smaller region.

When features are stored separately, it is a much harder task to integrate them. It is a nuisance, for example, to have to load and symbolize 20 feature classes of different geological units to see a single geological map. The Append and Merge tools are valuable but tend to have unsatisfactory results unless all the inputs have identical attribute tables; otherwise, attributes may be lost when bringing the classes together. In addition, gaps and overlaps will be generated between adjacent units and will take time and effort to eliminate.

In short, it is easy to extract the features you want when they are all stored together, but it is problematic and time consuming to put separated features together. When in doubt, keep them together. You can easily separate them later if needed. Subtypes provide one way to be able to keep features together while still permitting them to have different defaults and behaviors.

Geodatabase annotation

In Chapters 1 through 3 we learned about dynamic labels and creating annotation that exists as simple text within a map document. **Annotation** can also be stored in a geodatabase, in which case it can be used in multiple map documents. Annotation feature classes must be modified using the Editor, which is why a discussion of feature annotation has waited until this chapter.

roadsAnno

Status *	TextString	FontName	FontSize	Bold	Italic	Underline	VerticalAlign
Placed	IVY AV	Arial	8	No	N		
Placed	BIRCH AV	Arial	8	No	N		
Placed	MICHIGAN AV	Arial	8	No	N		
Placed	HAWTHORNE AV	Arial	8	No	N		
Placed	E IOWA ST	Arial	8	No	N		
Placed	HOEFER AV	Arial	8	No	N		
Placed	IDAHO ST	Arial	8	No	N		

Fig. 14.11. Annotation feature class in a geodatabase

Annotation is another type of feature class, like a line or a point feature class, except that it stores labels. Like any other feature class, it has an attribute table, and the table stores properties of the annotation, such as the text string, the font, the size, and the formatting (Fig. 14.11).

Annotation may be created as standard annotation, where each text string is a standalone feature not connected to anything else. Alternatively, it may be created as feature-linked annotation, in which the text string is linked to the feature that it belongs to. With feature-linked annotation, deleting the feature deletes the annotation, and vice versa. Feature-linked annotation requires an ArcGIS Standard Advanced license, but standard annotion can be edited with ArcGIS Basic.

Creating annotation begins by creating dynamic labels with the desired font size and other properties. When the labels are converted to annotation, they are stored in a geodatabase. Labels that overlap are stored in the attribute table as unplaced annotation, and they can be accessed and placed at another time.

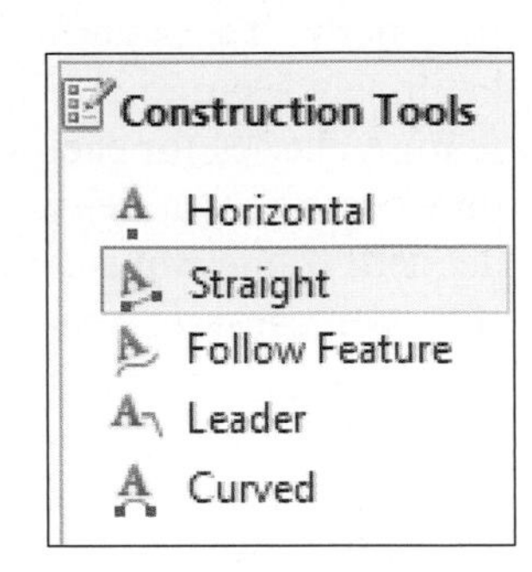

Fig. 14.12. Annotation construction tools

After the labels are created, editing can modify the location, style, and font attributes for each label. Editing annotation is not that different from editing other types of features. Annotation has editing templates and a variety of construction tools (Fig. 14.12). Annotation can be placed horizontally or on a straight line at an angle. It can be placed along a curve or set to follow along a feature.

Annotation is always given a **reference scale**, which is the scale at which the labels appear at their assigned size. The user specifies the reference scale at the time that the annotation is created. If the reference scale is 1:24,000, for example, and the annotation size is 10 points, then it only appears as 10-pt. text when the map is shown at 1:24,000. If the user zooms in, the labels will increase in size; if the user zooms out, the labels will decrease in size. Before creating the annotation, users should view the map at the desired reference scale to ensure that the font and size selected for the labels are acceptable.

Summary

- A geodatabase is the native data model developed for ArcGIS. It uses commercially available standard database formats and technology.

- Geodatabases offer significant advantages over shapefiles and coverages, including simpler and more robust implementation, the storing of topology, geometric networks, and behavior and validation rules.

- Geodatabases are created as empty containers and then filled with feature classes, feature datasets, layers, tables, relationships, and other objects.

- Feature datasets contain related feature classes with the same spatial reference. The X/Y domain specifies the range of *x-y* coordinates allowed in the dataset, and the resolution specifies the underlying grid to which features are snapped.

- Attribute domains define allowed values for attributes in geodatabases. Attribute domains may cover ranges of numeric values, or they may have specific coded text or numeric values.

- Domains also have split and merge policies, which define the assignment of attributes when features are merged or split.

- Subtypes allow features in a layer to be classified into categorical groups. Each group may have its own set of default attribute values.

- Annotation stored in a geodatabase is available for use in many map documents. Standard annotation is simply a type of feature, such as a point or a polygon. Feature-linked annotation is connected to its feature and is deleted when the feature is, or vice versa.

Important Terms

annotation	merge policy	resolution	subtypes
coded domain	range domain	schema	X/Y domain
domain	reference scale	split policy	

Chapter Review Questions

You may need to consult the Skills Reference section to answer some of these questions.

1. List three advantages of the geodatabase model.
2. What is the function of a geodatabase schema?
3. How does a personal geodatabase differ from a file geodatabase and an SDE geodatabase?
4. What is the difference between a feature class and a feature dataset?
5. Why is a resolution of 0.001 appropriate for a parcels feature class stored in UTM but not for one stored in a GCS?
6. Can you have a geodatabase that contains one feature dataset in UTM Zone 13 and another feature dataset in South Dakota State Plane? Is it good practice to do so?
7. You have a forest feature class with polygons showing individual stands of trees. It contains the three attribute fields TreeSpecies, CanopyCover, and Acres. TreeSpecies contains the dominant type of tree (e.g., ponderosa pine, aspen). CanopyCover gives the percentage of the stand covered by tree crowns. The Acres field contains the area of the stand. For each of these attributes, list the most appropriate split policy and merge policy.
8. For each of the following zoning attributes—Zoning, StreetAddress, Value, and PercentImperviousArea—state whether it would be suitable for a domain. If yes, then say whether a coded or range domain would be more appropriate. In each case, explain your reasoning.
9. Domains are set up as properties of a geodatabase rather than as properties of feature classes. Why is this arrangement an advantage?
10. How does one establish a default value for an attribute in a feature class? In a shapefile?

Mastering the Skills

Teaching Tutorial

The following examples provide step-by-step instructions for doing basic tasks and solving basic problems in ArcGIS. The steps you need to do are highlighted with an arrow ➔; follow them carefully. Click on the video number in the VideoIndex to view a demonstration of the steps.

We will build a geodatabase of Rapid City by importing existing data layers from the mgisdata\Rapidcity\archive folder. We will start with ArcCatalog, since we have a lot of data management to do before we need to use ArcMap.

TIP: ArcCatalog is similar to the Catalog tab, but if you find you are confused, review the Skills Reference section (ArcCatalog Basics).

1➔ Start ArcCatalog and navigate to the mgisdata\Rapidcity folder.

1➔ Right-click the Rapidcity folder and choose New > File Geodatabase.

1➔ Type in rcdata as the name of the geodatabase and press Enter.

1. What coordinate system do the feature classes in the archive folder use?

____

The new geodatabase will be in UTM NAD 1983. As we work, we will be careful that each coordinate system is converted during the import process.

Creating feature datasets

Most of the feature classes will go in feature datasets, which are required to have a predefined coordinate system, and which we will carefully create using NAD 1983 UTM Zone 13N. Feature classes imported to these feature datasets will automatically be converted to match.

2➔ Right-click the rcdata geodatabase and choose New > Feature Dataset.

2➔ Name it Admin and click Next.

2➔ For the coordinate system, expand Projected Coordinate Systems > UTM > NAD 1983 and select UTM Zone 13N.

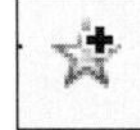

2➔ While the UTM Zone 13 coordinate system is still highlighted, click the Add to Favorites button, so that it will be easily available next time. Click Next.

2➔ Do not set a vertical coordinate system; click Next.

2➔ Accept the default tolerances, resolution, and domain, and click Finish.

3➔ Use the same procedure to create another feature dataset in rcdata named Environmental. (This time choose NAD 1983 UTM Zone 13 from the Favorites.)

4➔ Create a third feature dataset named Transportation.

5➔ Create a fourth feature dataset named Watersystem.

Adding coverages to a feature dataset

We will begin by importing two coverages, the city boundary and the land use data, to the Admin feature dataset. Recall that coverages contain multiple feature classes, so we must specify polygons as the class to import.

6➔ Right-click the new Admin feature dataset and choose Import > Feature Class (single) tool (Fig. 14.13).

6➔ Click the Browse button for the Input Features, navigate to the mgisdata\Rapidcity\archive folder, and double-click the landuse coverage to see its feature classes.

6➔ Select the polygon feature class and click Add.

6➔ Enter landuse for the Output Feature Class Name.

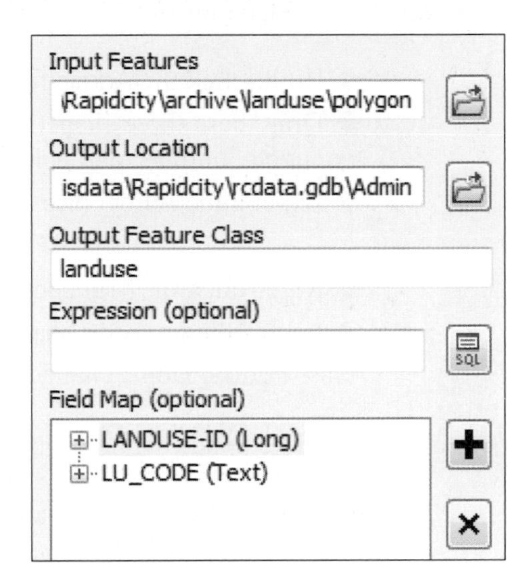

Fig. 14.13. Importing a feature class

The Field Map section lists the attributes in the feature class and controls how they are imported. Fields may be removed or added, or the order can be changed. With coverages, some fields are part of the coverage storage structure and contain no useful information after the conversion. It is best to delete them.

7➔ Click on the AREA field to highlight it, and then click the X button to delete it. (It will not be deleted from the original coverage.)

7➔ Also delete the PERIMETER and LANDUSE# fields from the map.

7➔ Leave the rest of the options set to their defaults.

7➔ The window should appear as in Figure 14.13. Click OK.

7➔ Use the Preview tab to examine the newly imported polygons.

Everything looks in order. Next we import the citybnd coverage.

8➔ Right-click the Admin feature dataset and choose Import > Feature Class (single).

8➔ Set the input feature class to the polygons of citybnd in the Archive folder.

8➔ Set the output name to citybnd.

8➔ Highlight and delete the AREA, PERIMETER, and CITYBND# fields.

8➔ Click OK to start importing.

Often, users have multiple feature classes to import into the same geodatabase. Another tool allows multiple imports to be set up and run in batch mode.

9➔ Right-click the Environmental feature dataset and choose Import > Feature Class (multiple).

9➔ Expand the archive folder in the Catalog Tree. Click on the watersheds shapefile and drag it over to the Input Features box on the tool. The feature class is placed in the list to process.

9➔ Click and drag the stategeol shapefile to the Input Features box. Click OK.

9➔ Expand the Environmental dataset and examine the new feature classes.

Notice that, when using the tool for importing multiple feature classes, you cannot set options individually for each one, as you could with the single input tool. We continue importing.

10➔ Right-click the Admin feature dataset and choose Import > Feature Class (multiple).

10➔ Drag these three shapefiles from the archive folder to the Input Features box: connects, parcels, and buildings. Click OK.

10➔ Expand the Admin dataset and check the new feature classes.

11➔ Right-click the Transportation dataset and choose Import > Feature Class (single).

11➔ Import the rc_roads shapefile from the archive folder. Name the output feature class roads. Don't change the Field Map settings. Click OK.

We also would like to add schools, but the name of the feature class gives us pause.

12➔ Click the sdschools shapefile in the archive folder and click the Preview tab above the display panel, if it is not already clicked.

As suspected, this shapefile contains schools for the whole state, although we want only the ones in the Rapid City area. We need to use the Clip tool this time. The geology map has a good boundary for clipping what we need.

12➔ Open ArcToolbox > Analysis Tools > Extract > Clip.

12➔ Click and drag the sdschools shapefile into the Input Features box on the tool.

12➔ Click and drag the stategeol shapefile into the Clip Features box.

12➔ Name the output schools and place it in the Admin feature dataset.

12➔ Accept the default XY Tolerance and click OK.

Finally, we plan to have a feature class to contain water and sewer lines. This feature class does not yet exist—we will digitize it. We must create the empty feature class.

13➔ Right-click the Watersystem feature dataset and choose New > Feature Class.

13➔ Name the feature class waterlines and set the type to Line Features. Click Next.

13➔ Accept the default configuration key word options. Click Next.

13➔ We are not going to create the fields for this feature class yet. Click Finish.

Creating attribute domains

During the planning phase for this geodatabase, it was determined that it should have domains. A table was developed, listing the domains and their properties (Table 14.1). With the planning done, we are ready to start adding the domains to the geodatabase.

TIP: ALWAYS, ALWAYS click Apply after EACH domain is added, before adding the next one. If you make an error in one domain but add more before clicking Apply, then it may not be clear which one has the error. You'll have to delete them all and start again.

14➔ In ArcCatalog, right-click the rcdata geodatabase and choose Properties. Click the Domains tab.

14➔ Type the domain name, RoadType, into the first empty box, and enter the description Type of road (Fig. 14.14).

14➔ Set the Field Type to Text.

14➔ Set the Domain Type to Coded Values.

14➔ Set the split policy to Duplicate.

14➔ Set the merge policy to Default Value.

15➔ Type the three-letter codes into the Code column and enter each description in the Description column: INT = Interstate; HWY = Highway; RMP = Ramp; CON = Connector; and LOC = Local.

15➔ Click Apply.

Domain Name	Description
RoadType	Type of road

Domain Properties:

Field Type	Text
Domain Type	Coded Values
Split policy	Duplicate
Merge policy	Default Value

Coded Values:

Code	Description
INT	Interstate
HWY	Highway
RMP	Ramp
CON	Connector
LOC	Local

Fig. 14.14. Setting up the RoadType domain

Table 14.1 summarizes the characteristics of all the domains to be added. Use it for reference, if needed, as you continue adding the domains in Steps 16–23.

Table 14.1. Domains for the geodatabase

Domain	Description	Field	Type	Values
RoadType	Type of road	Text	Coded	INT, HWY, RMP, CON, LOC
Directions	Cardinal directions	Text	Coded	N, S, E, W, NE, NW, SE, SW
StreetAbbrev	Common street abbreviations	Text	Coded	AVE, ST, RD, DR, BLVD, CT, LN, WAY, CIR, PL
Diameter	Pipe diameter in inches	Short	Coded	1, 3, 6, 12, 24
Materials	Pipe construction material	Text	Coded	COP, PVC, LEA, TIL, CON, OTH
Flow	Flow range in gpm	Short	Range	1–2000
LengthFloat	Generic length for split policy	Float	Range	0–9999999
Use Duplicate and Default Value for the split and merge policies for all domains except LengthFloat.				

2. Why do you suppose the Materials domain uses three-letter codes? ___________________

16➔ Click in the next open Domain Name box and enter the name Direction.

16➔ Type Cardinal directions for the Description.

16➔ Set the field type to Text, and the split and merge policies to Duplicate and Default Value.

16➔ Enter the eight codes from Table 14.1. Be sure to also enter the Description for each one using the same values, such as N.

16➔ Click Apply.

Code	Description
N	N
S	S
E	E

TIP: You must *always* enter a description, even if it will be the same as the code. If the description is left blank, then the values will appear blank when you try to view or edit them. ArcMap always displays the descriptions, not the codes.

17➔ Create the StreetAbbrev coded domain from the information in Table 14.1. Enter the street codes in the Code boxes, also entering the full-word equivalent for each one.

17➔ Click Apply.

Code	Description
AVE	Avenue
ST	Street
RD	Road

18➔ Create the Diameter coded domain from the information in Table 14.1. Enter the number for the Code (it is a Short Integer field type, but in the Description enter 1-in, 3-in, and so on).

18➔ Click Apply.

Code	Description
1	1-in
3	3-in
6	6-in

19➔ Create the Materials domain from the information in Table 14.1. Enter the three-letter code values for Code, but for Description enter the full-length words: copper, pvc, and so on.

19➔ Click Apply.

Code	Description
COP	copper
PVC	pvc
LEA	lead
TIL	tile
CON	concrete
OTH	other

The last two domains are range domains, so we won't need to type any codes—just the minimum and maximum range values.

20➔ Create the Flow domain. Make sure it uses the Short Integer type.

20➔ Set the minimum value to 1 and the maximum value to 2000, and set the policies as Duplicate and Default Value, as for the others.

20➔ Click Apply.

This last domain is intended to provide suitable split and merge policies for feature lengths, such that if you split a line each piece will receive a proportional value of the original length. Thus, it will use different policies than the other domains. The Range is irrelevant, but it needs to be large enough to hold any reasonable length measure.

21➔ Create the LengthFloat domain. Make sure it uses the Float type.

21➔ Set the minimum value to 0 and the maximum value to 999999999.

21➔ Give it Geometry Ratio for the split policy, and Sum Values for the merge policy.

21➔ Click Apply.

21➔ Click OK to close the Database Properties window.

3. What would you do if you wanted to apply a similar length domain to fields with a Double field type? ____

TIP: Domain policies and codes may be edited after the domain is created. However, the field type and domain type, once set, cannot be altered. To change them, you must delete the domain and create a new one. To delete a domain, select it and press the Delete key.

Assigning domains and default values

The roads feature class already exists, and we can assign domains to the appropriate fields.

22➔ Open the Properties for the roads feature class in the Transportation feature dataset.

22➔ Click the Fields tab.

22➔ Click in the little gray box to the left of the PREFIX field to highlight it.

22➔ Click in the Domain drop-down box below and select the Direction domain.

The Direction domain actually applies to two fields in this table, the PREFIX and SUFFIX2 fields. Domains can be reused in multiple feature classes.

22➔ Highlight the SUFFIX2 field and set the domain to Direction.

22➔ Highlight the SUFFIX field and set the domain to StreetAbbrev. Click Apply.

Next, the road type field does not yet exist in this feature class, so we must create it. We will also assign a default value, since local residential streets outnumber the other types.

23➔ In the next available field name box, type ROADTYPE. Make it a Text field. Give it a length of 3 and set the domain to Roadtype.

23➔ In the Default Value box, enter LOC, the code for local roads. Click OK.

Later in the lesson we will see how these domains function during editing.

Creating subtypes

The waterlines feature class will have four subtypes: water mains, water laterals, sewer mains, and sewer laterals. Subtypes must be based on a short integer field, so we must create this field in the feature class in addition to the other fields. We will also assign a domain to each field, if one is available for it.

24➔ Open the Properties for waterlines in the Watersystem feature dataset.

24➔ Click the Fields tab.

24➔ Type Linetype in the first available field name box and set the Data Type to Short Integer. Click Apply. This will be the subtype field.

25➔ Type PipeDiam in the next box and set the type to Short Integer. Click in the Domain box and choose the Diameter domain. Click Apply.

25➔ Type PipeMaterial in the next box and set the type to Text. Set the length to 3 and the domain to Materials. Click Apply.

26➔ Type InstallDate in the next box and set the type to Date. Click Apply.

26➔ Type MaxFlow in the next box and set the type to Short Integer. Set the domain to Flow. Click Apply.

26➔ Type Feet in the next box and set the type to Float. Set the domain to LengthFloat. Click Apply and OK.

TIP: Note that the list of domains varied for the fields. The domain field type must match the attribute field type. A short integer domain cannot be applied to a long integer attribute.

Before we create the subtypes, we must know the default values to be assigned. If domains are used, then the defaults must list the actual code value, not the description (6 rather than 6-in, for example). For some fields, such as the length in feet or the install date, we cannot predict a typical value ahead of time, so there is no point in setting a default for it.

27➔ Open the properties for the waterlines feature class.

27➔ Click the Subtypes tab in the Feature Class Properties window (Fig. 14.15).

27➔ Set the Subtype Field to Linetype.

27➔ In the first Subtype box, enter 1 for the code and Water Main for the description.

27➔ Highlight the top Water Main row to edit the boxes below.

28➔ Under Default Values and Domains, enter the value 6 in the Default Value field for PipeDiam.

28➔ Enter COP in the Default Value field for PipeMaterial.

28➔ Enter 700 in the Default Value field for MaxFlow.

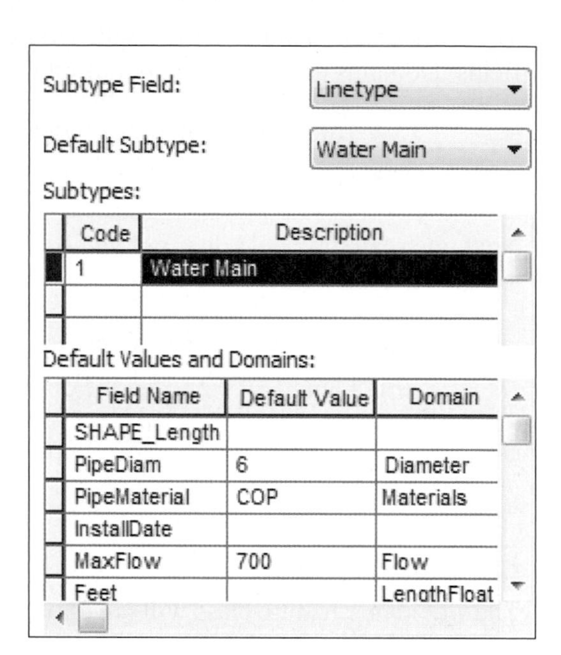

Fig. 14.15. Entering the first subtype

28➔ Make sure that the window appears as in Figure 14.15 and click Apply.

Table 14.2. Default values for each subtype

Attribute	Water main	Water lateral	Sewer main	Sewer lateral
PipeDiam	6	3	12	6
PipeMaterial	COP	PVC	CON	PVC
MaxFlow	700	125	1500	230

29➔ Enter the Code 2 and the description Water Lateral in the second row.

29➔ Click on the second row to highlight it.

29➔ Enter the default values for water laterals as listed in Table 14.2. You don't need to assign domains; they were assigned for the Water Main subtype. Click Apply.

30➔ Enter subtype 3, Sewer Main, and then enter the default values. Click Apply.

30➔ Enter subtype 4, Sewer Lateral, and the default values. Click Apply.

30➔ Make sure that the Default Subtype drop-down box at the top is set to Water Main. The final list of subtypes should appear as shown to the right.

30➔ Click OK to close the Feature Class Properties window.

Code	Description
1	Water Main
2	Water Lateral
3	Sewer Main
4	Sewer Lateral

Using domains while editing

Now we are ready to see how domains and subtypes work during an editing session, which is when they are most noticeable.

31➔ Close ArcCatalog.

31➔ Start ArcMap with a new, blank map. Set the default geodatabase to the mgisdata\Rapidcity\rcdata geodatabase.

31➔ Add the roads layer from the Transportation feature dataset.

31➔ Open the Editor toolbar, if necessary, and choose to Start Editing.

32➔ Use Selection > Select By Attributes to select all the roads where:

[STREET] = 'MOUNT RUSHMORE'.
Close the window when done selecting.

32➔ Click the Attributes button on the Editor toolbar. Dock the window in a convenient location and resize it, if needed, to see the field values.

32➔ Click one of the entries in the list and examine its attributes.

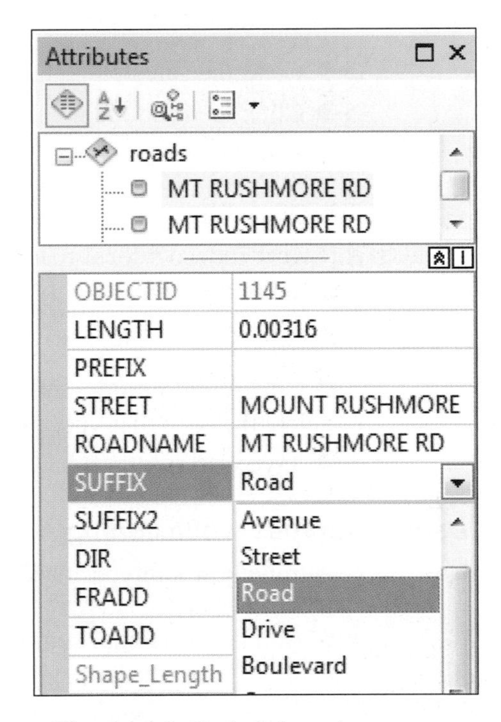

Fig. 14.16. Coded domains appear as drop-down lists during editing.

First, notice that the SUFFIX field entries are listed as "Road" rather than RD, which is the code actually in the field. This resulted from giving this field the StreetAbbrev domain.

33➔ Click in the space next to the SUFFIX field.

A coded value domain has a drop-down list showing the codes for that attribute (Fig. 14.16). It prevents the user from entering an inappropriate value such as AV instead of AVE, and it allows the user to see a real word rather than a cryptic code.

The ROADTYPE field was newly added to the feature class and thus contains no values at all. Hence, the <Null> value appears in the box. We can set it to the appropriate value now.

33➔ Click the roads heading at the top, so that an edit will apply to all selected records.

33➔ Select the Connector road type, setting it for all the selected records.

It would be reasonable to assume that roads designated as "Circle" or "Court" are local roads. We can select them and assign the road type.

34➔ Open Select By Attributes from the main menu bar and clear the expression.

34➔ Start entering the expression [SUFFIX] = and then click Get Unique Values. Notice that both code and description are shown.

34➔ Continue adding the full expression [SUFFIX] = 'CIR' OR [SUFFIX] = 'CT' by double-clicking the appropriate values in the Get box. Notice that only the code is placed in the expression. Click Apply and close the selection window.

34➔ Click the top entry in the Attributes window so that edits apply to all the selected features. Set the ROADTYPE to Local.

Unfortunately, this task must be completed to assign a road type to all of the roads. This is why designing your database ahead of time is so important. Adding fields after the features exist creates extra work. We'll leave this task for another time.

Recall that we assigned a default value, Local, to the ROADTYPE field. Our reasoning was that most of the roads in a city are local ones, so whenever we add a local road we do not have to enter the type.

35➔ Clear the selected features.

35➔ Click the roads template in the Create Features window, and add a line somewhere.

35➔ Scroll down and examine the ROADTYPE field in the Attributes window.

Most of the fields are <Null> for this new road, but the ROADTYPE is automatically set to Local. If this were indeed a local road, we would not have to enter the value. If it were not, we would. Thus, defaults are useful only when a particular value occurs often enough that it saves time overall.

35➔ Click the Edit tool on the Editor toolbar. Press the Delete key on the keyboard to delete the road that you just added.

35➔ Save your edits.

Using subtypes while editing

Now let's examine how subtypes make our work easier.

36➔ Add the buildings, connects, and parcels feature classes from the Admin feature dataset. Turn off the roads.

36➔ Zoom to the extent of the parcels layer, and then zoom out a little bit.

36➔ Add the waterlines feature class from the Watersystem feature dataset.

This feature class is empty, so no features appear on the map, but each subtype appears with its own symbol in the Table of Contents.

Linetype
Sewer Lateral
Sewer Main
Water Lateral
Water Main

37➔ Modify the waterlines symbols so that the mains are thick lines and the laterals are thinner lines. Make the water lines blue and the sewer lines orange.

No editing templates have been created for waterlines yet. Let's do it now.

38➔ In the Create Features window, choose the Organize Templates button.

38➔ Click New Template, and make sure waterlines and connects are checked.

38➔ Click Finish and close the Organize Feature Templates window.

Now let's digitize some water lines.

39➔ Open the Snapping toolbar, if necessary, using Editor > Snapping > Snapping Toolbar. Turn off all types of snapping.

39➔ Click the Water Main template and add a water main that runs north of the block. The precise location is not important. See Figure 14.17.

39➔ Examine the attributes in the Attributes window.

The defaults that we set for a water main, a 6-inch copper pipe with a flow rate of 700 gpm, are already entered in the attribute fields, saving us the trouble.

39➔ Click the Sewer Main template and add a sewer main above the water main.

39➔ Examine the Attributes window. A different set of defaults was entered for the sewer, just as we set up.

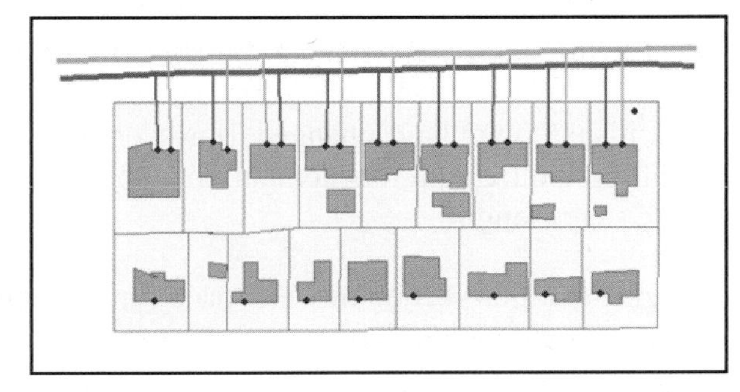

Fig. 14.17. The water and sewer lines

40➔ Click the Water Lateral template and turn on edge and point snapping.

40➔ Add a lateral from the connection point on each house to the water main, being sure to snap from the point to the water main edge. Note the default attributes.

41➔ Click the connects template and add another connection point to each house, next to the first one, snapping to the edge of the building.

41➔ Click the Sewer Lateral template and add a lateral from each second connection point to the main sewer line. Note the default attributes.

42➔ Open the attribute table for waterlines.

Look at all the information entered in the table, even though we never typed a word. You may have thought that the subtypes were a hassle to create, but now perhaps you begin to see how they are worth the effort! Now we just need to assign the attributes that did not have defaults.

42➔ Clear the selected features, if any.

42➔ Right-click the Feet field and choose Field Calculator. The map units are meters, so enter the expression [SHAPE_Length] * 3.281 and click OK.

42➔ Right-click the InstallDate field and choose Field Calculator. Enter the expression "3/17/1997" including the double-quotes. Click OK.

42➔ Close the Table window and save your edits.

We still have not seen the split/merge policies at work.

43➔ Select the water main with the Edit tool and examine its attributes (Fig. 14.18). Your lengths will differ.

43➔ Click the Split tool on the Editor toolbar, and click on the water main about one-third of the way from its left end.

43➔ Select the short end with the Edit tool and view the attributes.

43➔ Select the long end with the Edit tool and view the attributes.

OBJECTID	1
SHAPE_Length	221.0599
Linetype	Water Main
PipeDiam	6-in
PipeMaterial	copper
InstallDate	3/17/1997
MaxFlow	700
Feet	725.2975

Fig. 14.18. Water main attributes before splitting

Most of the fields were copied from the original, using the assigned Duplicate split rule, which makes sense. If the pipe was 6-inch copper before it was split, it should not have changed. However, the Feet field was assigned the Geometry Ratio split rule from the LengthFloat domain, and the length was automatically updated using the proportion to the shape length.

Imagine that this section of water main ruptured and was repaired using a bigger PVC pipe. You are updating the record.

44➔ In the Attributes window, change the PipeMaterial to pvc.

44➔ Click the InstallDate box and click the Today link at the bottom of the calendar.

44➔ You intend to enter 850 for the new flow rate, but pretend to make a mistake and type 8500 instead.

44➔ Clear the selected features and save your edits.

Now let's see how to take advantage of the range domain error checking. It is done by means of a separate step called validation.

45➔ Use the Edit tool to click and drag a rectangle that selects all of the water lines but none of the other features.

45➔ Choose Editor > Validate Features from the Editor toolbar.

If errors are found, it reports the number in a dialog window and places the incorrect records in the selected set, ready for editing. Only attributes with domains will be tested. In this case it found your incorrect entry of 8500, and the offending water main remained selected.

45➔ Change the flow rate to 850 in the Attributes window.

45➔ Stop editing and save your edits.

Editing annotation (optional)

Annotation can be stored in a feature class within a geodatabase, which makes it available for use in many map documents. We will create road name annotation for part of the roads layer.

46➔ Turn on the roads layer.

46➔ Zoom out from the parcels to the area shown in Figure 14.19.

46➔ Turn off all layers except the roads.

46➔ Choose Bookmarks > Create Bookmark from the main menu bar. Enter the name Anno and click OK. Now you can return to this extent whenever you wish.

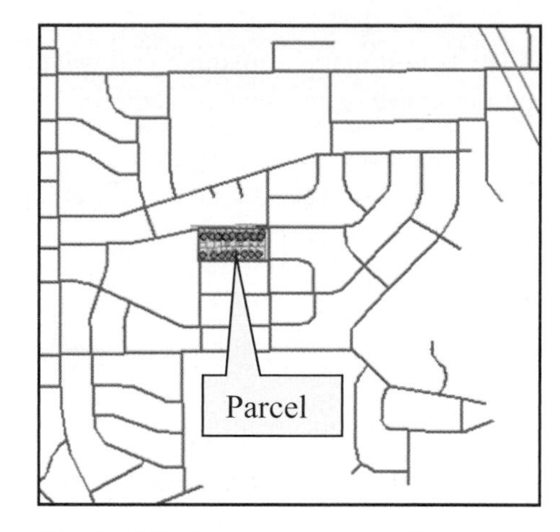

Fig. 14.19. Area to zoom to for editing annotation

Creating annotation begins by creating dynamic labels with the desired properties.

47➔ Open the Labels tab of the roads layer properties.

47➔ Create labels using the ROADNAME field with 8-pt. Arial font. Click OK.

47➔ Set the map scale to exactly 1:10,000.

This will be your **reference scale**, the scale at which the annotation appears at the 8-pt. size. You want to make sure that the labels appear as you wish at this scale before you create the annotation. It looks fine, so let's continue.

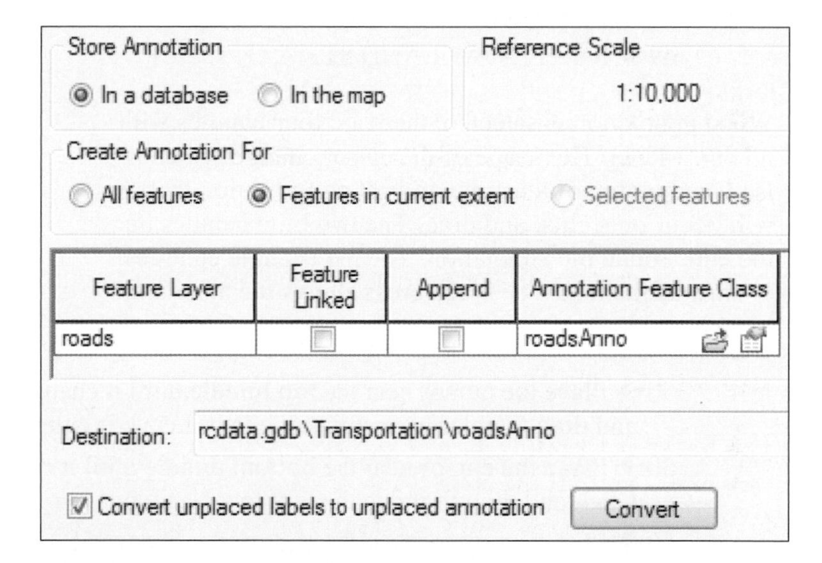

Fig. 14.20. Creating an annotation feature class

47➔ Right-click the Layers data frame name and choose Convert Labels to Annotation.

47➔ Choose to store the annotation in a database. Notice the reference scale (Fig. 14.20).

47➔ Fill the button to create annotation for features in the current extent.

47➔ Uncheck the box to create Feature Linked annotation.

47➔ Notice the Destination box; the annotation will be saved in the Transportation feature dataset of the rcdata geodatabase with the name roadsAnno.

47➔ Click Convert. The roadsAnno feature class is added to the data frame.

Fig. 14.21. Annotation

The annotation is stored as a separate feature class in the Transportation feature dataset, as shown in Figure 14.21.

TIP: If you made a mistake and want to start over, delete the annotation feature class from the geodatabase using the Catalog tab.

First, let's examine the effect of the reference scale.

48➔ Zoom in to a smaller area and watch the annotation get larger. Zoom out and watch it get smaller. Then return to 1:10,000. You can zoom in a little more, if you wish, to see and work with the annotation better.

49➔ Start Editing.

49➔ An editing template for roadsAnno should appear in the Create Features window, but if not, create one using the Organize Templates button.

Selecting annotation for editing requires a different tool than feature editing.

49➔ Click the Edit Annotation tool on the Editor toolbar.

49➔ Select E OAKLAND ST.

When annotation is selected, there are four handles with it (Fig. 14.22). The shape of the cursor changes when it is placed on top of a handle, indicating the action that will occur on a click and drag. The two blue handles on the ends rotate the annotation. The red triangle enlarges or shrinks the text. The black cross allows the annotation to be moved.

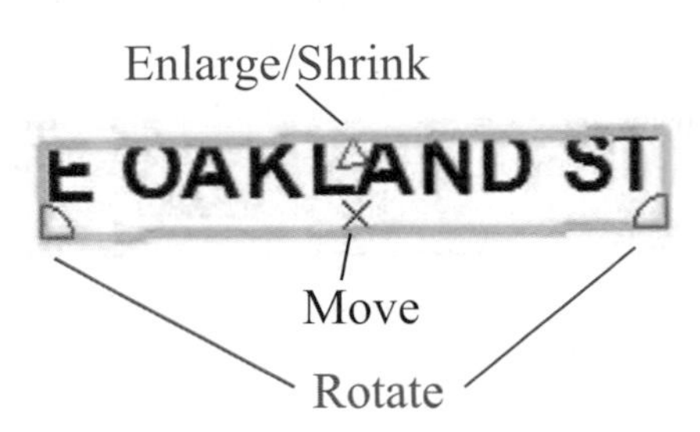

Fig. 14.22. Action points on selected annotation

50➔ Place the cursor near the top middle until it changes to a two-ended arrow. Click and drag to enlarge or shrink the text. Leave it large.

50➔ Place the cursor near the bottom middle until it changes to a four-ended arrow. Click and drag to move the text to a clear spot.

50➔ Place the cursor near the end until it changes to a rotate tool. Click and drag to rotate the text.

50➔ Shrink the text back to its approximate original size and move it back to its starting location.

Unplaced labels are put in the table, with the designation "Unplaced" in a Status field. The Overflow window does not open automatically, but you can open it to place the annotation.

51➔ Choose Editor > Editing Windows > Unplaced Annotation.

51➔ Set the Show drop-down box to roadsAnno and click Search Now.

51➔ Right-click the row WALNUT DR, and click Place Annotation. It appears on the map, selected.

51➔ Adjust the location or angle, if necessary, using the Edit Annotation tool.

51➔ Right-click the next row and place the annotation. Adjust the location.

51➔ Close the Unplaced Annotation window.

We will not continue placing annotation right now, but anytime you want to continue placing the annotation you can open this window and search for unplaced labels.

The Attributes window has additional tabs and options when working with Annotation.

52➔ Open the Attributes window and dock it in a convenient place.

52➔ Select the E OAKLAND ST annotation with the Edit Annotation tool.

52➔ Click the Annotation tab in the lower panel of the Attributes window and examine the settings. You can set the font properties and retype the text, if necessary.

52➔ Examine the Attributes tab. Each piece of annotation can have all its properties set individually.

52➔ Return to the Annotation tab. Click the drop-down box, choose Default for the symbol, and click Apply. The E OAKLAND ST text will go back to original size.

53➔ The Edit Annotation tool should still be active. Hold down the Shift key and click on another nearby piece of annotation to select it also. Now both annotation features appear in the Attributes window.

53➔ Click on the roadsAnno entry at the top of the Attributes window.

TIP: You may need to enlarge the lower panel of the Attributes window to see the font size and formatting buttons, or undock the entire window.

53➔ Click the B button for boldface type and click Apply. Both annotation items are updated to bold font.

53➔ Click B again to return to normal font and click Apply. Leave the Attributes window open so you can use it as needed.

Now we will create new annotation for some streets in the southeast corner. These probably have unplaced labels waiting, but we will create new ones, just for the practice.

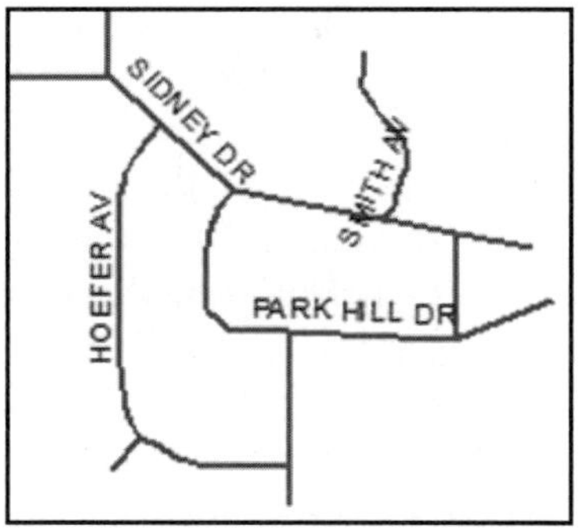

SIDNEY DR
SMITH AV
HOEFER AV
PARK HILL DR

Fig. 14.23. Street names in the SE corner of the map

54➔ Click the roadsAnno template, noticing that the default construction tool is Straight. The Annotation Construction window appears.

54➔ Type SIDNEY DR in the open area of the Annotation Construction window.

54➔ Click on the center of SIDNEY DR, shown in Figure 14.23.

54➔ Move the cursor to rotate the text to the desired angle and click again to place it.

54➔ If you don't like where it was put, press the Delete key to get rid of it and then place it again. (You don't need to type the text again.)

Observe an unlabeled street running north into PARK HILL DR. You can use Identify to determine its name.

55➔ Click the Identify tool and click the unlabeled street. Leave the Identify window open but tuck it out of the way.

55➔ Click the roadsAnno template again and type CHURCHILL DR in the Annotation Construction window.

55➔ Place the annotation on the unlabeled street.

55➔ Save your edits so far, and save the map document.

TIP: MapTips would be a great way to quickly find out unlabeled street names.

56➔ Click the Edit Annotation tool on the Editor toolbar.

56➔ Click on E MEADE ST to select it. Move it to the center of the block.

56➔ Right-click E MEADE ST and choose Copy.

56➔ Right-click anywhere and choose Paste. Choose to paste it into roadsAnno.

A copy of E MEADE ST will be placed on top of the original.

56➔ Click and drag the copy to the eastern part of the street below E ST FRANCIS ST. Rotate and adjust its position until you are satisfied.

Flipping annotation, like flipping a line, makes it go in the opposite direction.

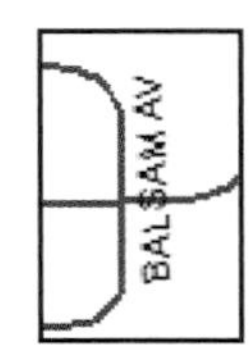

57➔ Select BALSAM AV with the Edit Annotation tool.

57➔ Right-click it and choose Flip Annotation. Now it matches the other nearby annotation for easier reading.

57➔ Select E TALLENT ST and move it west of HAWTHORNE AV.

Placing curved annotation is slightly more complicated, but not much.

58➔ In the Create Features window, click the Curved construction tool

58➔ Type E TALLENT ST in the Construction box.

58➔ Find the curved eastern end of E TALLENT ST and click underneath on the left, the middle, and the end to enter a curve that follows the street. Double-click to end the sketch.

58➔ If you don't like the result, delete it and try again.

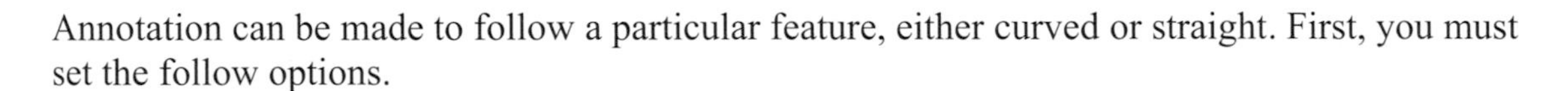

Annotation can be made to follow a particular feature, either curved or straight. First, you must set the follow options.

59➔ Select ROBBINS DR with the Edit Annotation tool.

59➔ Right-click the annotation and choose Follow > Follow Feature Options.

59➔ Choose Curved and set the offset to 15 map units. Click OK.

You only need to set the options the first time. Now place the annotation.

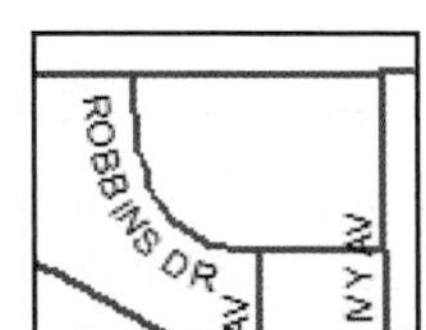

59➔ Right-click on the curved section of road that you want ROBBINS DR to follow and choose Follow This Feature. The text is moved to the point you clicked and rotated to follow the street.

59➔ Place the cursor on the text until you see the four-ended arrow. Click and drag the text above and below the street, and then leave it below.

59➔ Click and drag the text along the street, noticing how it hugs the feature. When you are satisfied with the placement, release the mouse button.

You can also create new annotation that follows a feature.

60➔ Click the roadsAnno template and select the Follow Feature construction tool.

60➔ Type City Bus Route in the Text box.

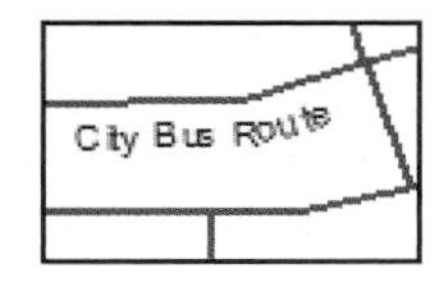

60➔ Click along the inclined segment of E Meade St to select the feature on which to place the annotation.

60➔ Move the mouse until you are satisfied with the position and then click again to place the annotation.

60➔ Click the Annotation tab of the Attributes window and change the font to italic. Click Apply.

61➔ Click the Edit Annotation tool and select MAYWOOD DR.

61➔ Press the O key on the keyboard to open the Follow Feature Options.

61➔ Fill the button for Curved and click OK.

61➔ Right-click on the center of MAYWOOD DR and choose Follow This Feature.

61➔ While the annotation is still selected, place the cursor on it and click and drag it above, below, or along the road to get the best placement and curvature. Release the mouse button when you are satisfied with the placement.

62➔ Select the Horizontal construction tool.

62➔ Type Robbinsdale School in the Text box.

62➔ Click in the center of the large, open area west of IVY AV to place the text.

62➔ In the Attributes window, change the font to 10-pt. Bold Arial and make it red. Click Apply.

You can stack annotation for better placement.

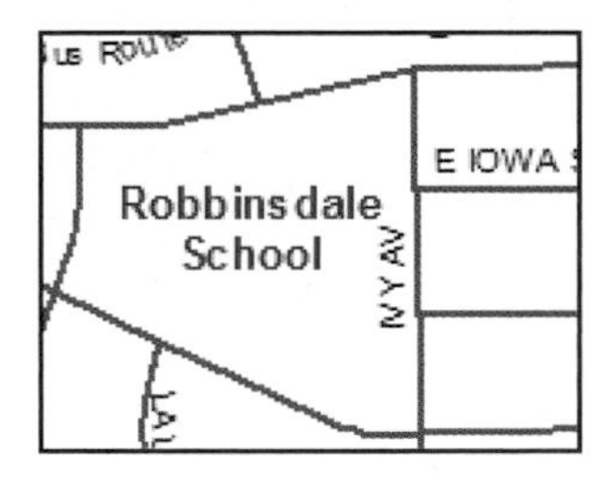

63➔ Click the Edit Annotation tool and click on the Robbinsdale School annotation to select it.

63➔ Right-click the selection and choose Stack.

63➔ Click and drag the annotation to a good position.

This tutorial demonstrated only a few common annotation editing tasks. You can learn more using ArcGIS Help.

63➔ Save your edits and stop editing. Also save the map document.

➔ Exit ArcMap.

This is the end of the tutorial.

Exercises

1. Consider a buildings feature class such as the one you started in the Crestview geodatabase for Austin. Make a list of likely attributes that such a feature class might have.

2. Take the list of buildings attributes and decide which ones could have domains and what types of domains and values they should have. Construct a table similar to Table 14.1 showing these domains.

3. Divide the buildings into categories that would make suitable subtypes. Consider the different default values that each subtype would have, and construct a table similar to Table 14.2.

4. Using the information assembled in Exercises 1 through 3, create a buildings feature class in the Crestview geodatabase. Set up the geodatabase domains, create the fields for the feature class, and assign default values.

5. If you have an ArcGIS Standard license, create subtypes for the buildings with the default values you determined in Exercise 3. If you do not, create editing templates instead of subtypes, with the default values stored in the templates. Consult the Help and learn how to save and load editing templates.

6. Digitize approximately 50 buildings from the Crestview area, distributed among your subtypes. Capture a map showing the buildings and a screenshot of the attribute table showing the various fields of information.

7. Finish editing the roadsAnno annotation in the rcdata geodatabase to make an aesthetic and clearly legible set of labels. Capture a map showing the streets and annotation.

8. Create geodatabase annotation for the streets in the Crestview geodatabase and edit them for optimum placement and legibility. Capture a map showing the streets and annotation.

Challenge Problem

Create a geodatabase with at least four feature classes and at least one feature dataset. The data you use may come from any source but should comprise a logical collection of related data. **Capture** the layout of your geodatabase in ArcCatalog and create a map layout showing your geodatabase data. On the layout, include the coordinate system chosen for your database and a graphic showing its contents (as in Fig. 14.1).

Chapter 15. Metadata

Objectives

- ➢ Knowing about the dominant metadata standards and their relationship to each other
- ➢ Gaining a basic understanding of data quality issues
- ➢ Understanding the layout and main sections of CSDGM and ArcGIS/NAP metadata
- ➢ Learning to use the metadata editor and templates to create metadata

Mastering the Concepts

GIS Concepts

Providing metadata is a fundamental part of creating GIS data sets. It is a professional obligation when data will be distributed to other users or the general public. Best management practices dictate that metadata should also be prepared for in-house data so that critical information about the source and processing of a data set will not be lost. Figure 15.1 shows metadata developed for the radar image georeferenced in Chapter 11, which you will create in this chapter.

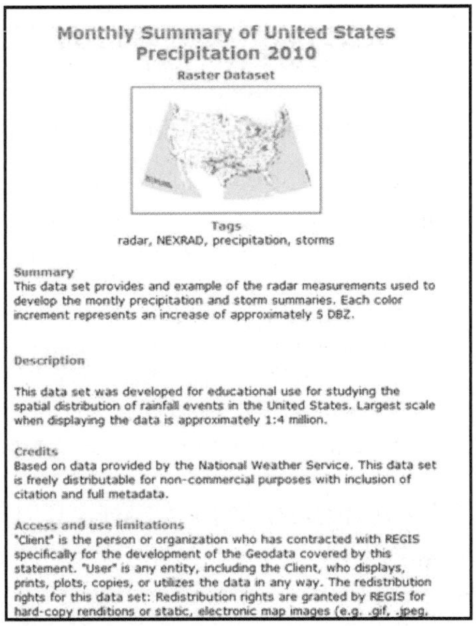

Monthly Summary of United States Precipitation 2010

Raster Dataset

Tags

radar, NEXRAD, precipitation, storms

Summary

This data set provides and example of the radar measurements used to develop the montly precipitation and storm summaries. Each color increment represents an increase of approximately 5 DBZ.

Description

This data set was developed for educational use for studying the spatial distribution of rainfall events in the United States. Largest scale when displaying the data is approximately 1:4 million.

Credits

Based on data provided by the National Weather Service. This data set is freely distributable for non-commercial purposes with inclusion of citation and full metadata.

Access and use limitations

"Client" is the person or organization who has contracted with REGIS specifically for the development of the Geodata covered by this statement. "User" is any entity, including the Client, who displays, prints, plots, copies, or utilizes the data in any way. The redistribution rights for this data set: Redistribution rights are granted by REGIS for hard-copy renditions or static, electronic map images (e.g. .gif, .jpeg,

Fig. 15.1. Metadata example

Preparing a 100% complete set of metadata is a daunting task. However, even partial metadata is better than none. This chapter provides a basic introduction to the most important sections of metadata and how they can be developed.

The metadata standards

To facilitate sharing of metadata, which cross over different GIS formats and programs, the Federal Geographic Data Committee (**FGDC**) and the International Organization for Standardization (**ISO**) have worked to develop metadata standards that specify the layout and information provided. The FGDC "is an interagency committee that promotes the coordinated development, use, sharing, and dissemination of geospatial data on a national basis" (www.fgdc.gov). Its web site contains a wealth of useful information not only on metadata but on geospatial data in general. The metadata standard is described in its entirety in publications offered by the FGDC, and these reports serve as the basis for many metadata questions and answers. The FGDC also serves as a liaison to the international body of standards.

A **metadata standard** lays out the specific information fields that should be included and defines which ones are optional and which are mandatory. It also specifies the exact format and organization of the information so that programs and metadata interpreters can always find the information in the same place and so that the metadata are transportable from system to system.

Currently, metadata standards in the United States are in a state of flux. For many years, users in the United States used a national standard prepared by the FGDC that was commonly known as FGDC metadata, or the Content Standard for Digital Geographic Metadata (**CSDGM**). It has been in use since 1994. Companies that produce data for federal agencies are required to provide them in this format, and many data contracts specify that metadata be compliant with this standard. Much effort has been expended in development, software, and training for CSDGM, and much of the data that you download and use today will use this content standard.

However, the international community has been using a more generic standard, called the ISO 19139 standard, for nearly as long. It is not particularly suited for geographic data, so another standard, the ISO 19115, has been established for geographic data. Part of this standard specifies that a community of users may develop and publish a **profile**, a customized version of ISO 19115 that is better suited to the needs of that community, as long as they follow rules that ensure that the profile will be maintained and will remain compatible with the original 19115 standard.

The United States and Canada geographic data committees agreed to jointly develop a profile that is now known as the North American Profile of ISO 19115, or NAP. It has been formally adopted by Canada. The FGDC has formally endorsed the ISO 19115 standard, and federal agencies are being encouraged to move to the ISO standard as they are able, although this process may take years.

The practical upshot of this situation is that users are likely to encounter both CSDGM and NAP metadata for some time and must be familiar with both standards (as if metadata weren't difficult and confusing enough before). Much of the information contained in the two standards is similar, but the organization is different, and the cross-walk is not perfect or simple.

The following publications are especially recommended for users who need to learn more about metadata standards and implementation. All but the last one refer to the CSDGM standard. Users are also encouraged to explore the FGDC web site http://www.fgdc.gov/metadata for updated information and new publications (http://www.fgdc.gov/metadata/metadata-publications-list) as they become available.

Geospatial Metadata Fact Sheet (2011). Brief explanation of what metadata is and how it is organized. Includes a discussion of the new NAP standard and why it is being implemented.

Geospatial Metadata Quick Guide (2005). The famous "Don't Duck Metadata" publication, which describes some of the basic types of information that should be included.

Top Ten Metadata Errors (2006). A short, fun read.

CSDGM Essential Metadata Elements (2008). The guide for those who have resigned themselves to producing metadata and want to achieve a minimum standard.

Content Standard for Digital Geospatial Metadata Workbook (2000). Too exhaustive for a beginner but an excellent reference for the serious metadata enthusiast. Contains some good examples that beginners might find helpful.

Preparing for International Metadata (2010). A discussion of the development and layout of the North American Profile of the ISO 19115.

Institutionalize Metadata Before It Institutionalizes You (2005) by Linda Wayne. Useful discussion of templates and incorporating metadata creation into the organizational workflow.

Some of the information in metadata is easy to understand and document, such as who produced the data. Many of the terms used in the standards and literature presuppose an understanding of geospatial data topics, such as logical consistency or completeness. Users will develop expertise in these areas as they travel the path to becoming geospatial professionals. Merely trying to understand these terms as you try to produce metadata will make you a better creator of data as well. Some sections of metadata are not applicable to a particular data set and can be ignored. As you gain experience reading and creating metadata, you will understand more and more about geospatial data, where it comes from, what it is good for, and what happens to it during its lifetime. You will also gain experience in evaluating the quality of data and deciding whether it is good enough for a particular project or application.

Data quality issues

One fundamental purpose of metadata is providing sufficient information for a user to decide whether a data set is of sufficient quality for a particular application. Assessing the quality of a data set is not a simple judgment of good or bad but a complex analysis of whether a particular data set is fit for an intended purpose. A data set that is inadequate for one application might be completely suitable for another. The data quality information in metadata is provided so that a potential user can evaluate the data set in light of what he or she wants to do with it. During this process, six key issues must be considered: (1) lineage, (2) positional accuracy, (3) attribute accuracy, (4) logical consistency, (5) completeness, and (6) temporal accuracy. The following sections discuss these issues and provide examples of how a user might document the information in the metadata.

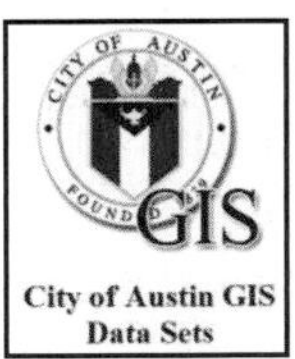

Lineage

Lineage is concerned with documenting the original **source** of the data set (such as the City of Austin web site that provided most of the Austin data for this text) and recording the operations and transformations that have occurred between the source and the final product. Questions to be answered about the data include who collected the data, how and why they were collected, what was the original scale or accuracy, how they were converted into digital form, and what operations have been performed on since.

The metadata include sections on the **process steps**, which list the sequence of operations that have been performed in producing the data. In particular, the impacts of any of the processing steps on the attribute or positional accuracy need to be recorded. For example, if the lines are generalized to the nearest 500 meters to prepare a detailed states map for a faster-drawing national map, then that information should be recorded. If the feature class is the result of intersecting two layers with an XY tolerance of 25 meters, then that information, with the potential impact on the feature accuracy, must be noted.

➢ You digitize published maps of aquifer outcrop areas from two adjacent quadrangles generated by the US Geological Survey. You merge the two quadrangles and match the edges between them. In the metadata, you include the publication information and a complete citation for both maps. You record a process step for digitizing the maps, including the RMS errors reported for registering the maps on the digitizer. You report a second process for edge-matching and merging the two maps, including the RMS error associated with that step.

Positional accuracy

Positional accuracy is an assessment of how closely features in the data set correspond to their actual locations in the real world. Impacts on positional accuracy occur at several stages in the development of a data set: (1) errors present in the source data or in the original data collection, (2) errors associated with transformations or georeferencing, and (3) errors associated with subsequent processing.

(1) Source data errors are generated in many ways, but the most common are by digitizing a paper map or scan, by surveying, by photogrammetry or remote sensing, or by GPS measurement. Each of these methods has different sources of error that impact the positional accuracy.

At minimum, one assumes that maps are accurate to within one line width, or approximately 0.5 mm. This value can be converted to ground units using the scale, as demonstrated in Chapter 1. A 1:24,000 scale map has an effective resolution of 12 meters, a 1:100,000 scale map has an effective resolution of 50 meters, and so on. If a map is digitized, its accuracy can be estimated from the scale. This estimate serves as a minimum, however, for it neglects potential inaccuracies in assembling the map data. Some phenomena or objects can be less precisely located than others. A road is easy to locate, but a soil map has an inherent uncertainty in placing the boundary between two soil types that grade into one another.

Surveying and photogrammetric methods follow rigorous standards during development and are usually subjected to quality assurance and control steps along the way. In general, base data sources, such as topographic maps and derived digital products, can be assumed to conform to the National Map Accuracy Standards. These standards require testing by comparing at least 20 checkpoints composed of well-defined and locatable points to corresponding locations on a reference map of higher accuracy. No more than 10% of the points can be off by more than 1 in 10,000 units. In practice, few maps are actually tested due to the time and cost constraints associated with mapping projects, although the procedures developed to create the maps have been, and therefore they provide some assurance that most of the products meet the standard.

The accuracy of remotely sensed information varies greatly due to the differences in resolution between satellites and the differences in geometric correction that have been applied. GPS units also vary widely in their accuracy.

(2) A paper map or a scanned digital image must be georeferenced, transforming the digitizer, scanner, or screen coordinates into a real-world coordinate system (see Chapter 11). The RMS error associated with this transformation should be recorded and included with the metadata.

(3) Various geoprocessing operations, such as generalizing lines or intersecting feature classes, can degrade positional accuracy. Processing tolerances, such as the XY tolerance used when intersecting to remove slivers or a cluster tolerance applied when correcting topology errors, provide an estimate of the impact. These tolerances should be recorded when listing the process steps in the metadata.

- You digitize the power lines on a 1:24,000 scale topographic map. You infer the source map accuracy to be 12 meters. The RMS error reported when you registered the paper map on the digitizer was 4 meters. Report both of these in the metadata.

- You go out in an ATV and collect GPS data for the boundaries of prairie dog towns. Previous testing indicated that your GPS unit has an accuracy of 10 meters, and you estimate that you can drive within 5 meters of the town boundary without damaging the burrows. Report the accuracy as 15 meters.

Attribute accuracy

Attribute accuracy assesses how well the values in the attribute fields correspond to the true values in the real world. Does the label on Rapid Creek represent its real name? Does the 40% crown cover value of a tree stand really reflect the actual density of the crown foliage? Because there are so many different kinds of attributes, assessing accuracy is complex, variable, and unguided by any sort of standard methodology. However, some of the issues that should be considered are as follows.

For categorical data, attention should be given to the classification scheme. How detailed is it? Are the crown-cover categories divided into 10 groups of 10% each or into three groups? Do the land use categories represent all possible land use types? Do the geological units provide sufficient detail to locate potential gold horizons or are they simply grouped by age or gross rock type?

Issues related to all data types include the following: What is the expected rate of error in assigning values to the field? How often is something misclassified or wrongly recorded? How much heterogeneity exists within the measurement unit? The sand percentage in a soil polygon, for example, cannot be expected to be uniform over the entire area. Have any attempts been made to assess the variability within a unit?

If any information on attribute accuracy is known or if any tests have been conducted to assess the accuracy, these should be reported in the metadata. If nothing is known, then it is acceptable to put Unknown. Here are some examples.

- A land use map has been generated from a Landsat satellite image. You have followed industry practice and tested the classification against a set of ground data points and developed the overall accuracy, producer's accuracy, and user's accuracy for the classification. Include these three values in the attribute accuracy report.

- Your survey agency has surveyed all the counties in the state on their approval of the governor's performance and is reporting an approval rating for each county. The survey method has an accuracy of +/− 3 percentage points. Include this accuracy in the metadata.

- You have digitized a geological map from a plate in a master's thesis. The author did not include any accuracy assessment, and you have no funds to evaluate the accuracy of the map. Put Unknown in the attribute accuracy report in the metadata but include the scale of the original map.

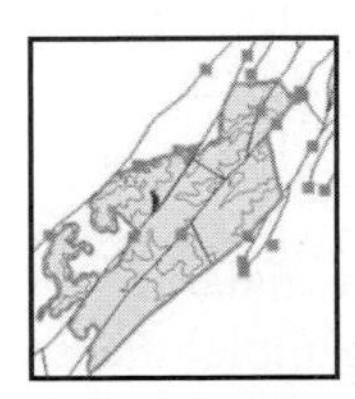

Logical consistency

Logical consistency is a measure of how well the features in the data model correspond to their counterparts in the real world and of how well the data model is able to model the relationships in the real world. This is another complex and difficult area that is rarely tested in full. However, it is usually possible and desirable to test the internal logical consistency of a data set and how well it fulfills the topological rules that have been defined for it. These topological rules were described in Chapter 13. Polygons should not have gaps or overlaps, lines should intersect at nodes and not cross each other, and so on. The logical consistency report can include the topological rules that were tested and the results of the test.

- You develop a planar topology for a Transportation feature dataset and check for dangles, self-intersection, and improper intersections. All identified errors were resolved. Include a statement of the topological rules that were tested, and report that the feature dataset does not contain violations of those rules in the logical consistency report.

- You digitized a geological map as a shapefile using snapping and Auto Complete Polygon, and you are reasonably sure that most of the polygons have no gaps or overlaps. However, you don't have an ArcGIS Standard license and cannot test the topology. Report that snapping and coincident boundary protocols were followed during digitizing and that the data layer is free or nearly free of gaps and overlaps.

Completeness

The **completeness** of a data set refers to how well it has captured every possible instance of the objects in the data set. Completeness for polygons is usually easy to assess; do the polygons completely cover the area of interest or not? Lines and points can be less certain. Did the department of transportation digitize every known road in its GPS survey, or could it have missed some? Did it digitize only public roads, or are private roads also included? Did it include gravel and dirt tracks or only paved roads? Does the table of oil and gas wells in the state include every known well, or have some historical records been lost or misplaced prior to the assembly of the data?

- You assembled a database of septic systems by identifying houses on digital orthophotos taken five years ago. You realize that you may have missed systems for houses that were built after the photos were taken or for houses under heavy tree cover that were not visible in the photo. State these limitations in the completeness report.

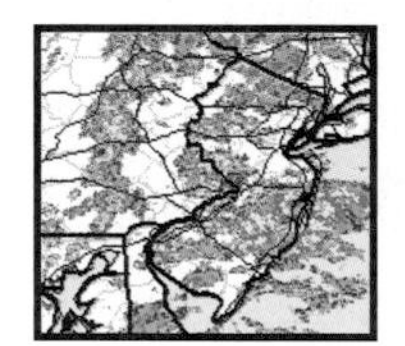

Temporal accuracy

Temporal accuracy is the time period for which a data set is considered valid. Some geospatial data, such as a map of greenness condition or a NEXRAD radar rain map, represent ephemeral conditions valid for a few minutes to days. Other data sets represent phenomena that change more slowly, and data sets might be considered valid for several years. Population, for example, is usually referenced to a census year or census update. Land use data change slowly as cities grow and new developments spring up. Some data can be considered fairly stable. Geology, major road networks, rivers, and international boundaries change infrequently on the scale of human affairs.

For data that will change significantly over the lifetime of the data set, it is important to designate how frequently the data provider intends to update the information. Some data are undergoing continual revision and are considered to be in progress, with updates planned at irregular or regular intervals. Census data, for example, are gathered every 10 years, and estimated updates are available every 4 to 5 years for some portions of the data set. However, a lot of GIS data is simply produced once, and no updates are ever planned.

Two sections in the metadata refer to the time period, that of the source and that of the data product being documented. These two are often the same but can be different. For example, the city planning office might have digitized the original platting records that were updated manually until 1995 to produce its parcels feature class (the source), but continued updates bring the feature class time period to within the past year (the product).

- Your office performed a GPS survey during 2004 and 2005 to locate and classify the existing off-road trails in the national forest, both official and unofficial. No funds are provided to locate and map new trails created in the future. The temporal period of both the source and the product is 2004–2005, and no updates are planned.

- Your office maintains the county parcel records that were digitized from plat maps by a consultant in 2001. Every sale or modification is reported to the county as required by law through the office of deeds and land titles. You process the updates and make changes to the internal database as they come in. You push the updates to the public version of the database on your web site every three months. In this case, the time period of the source is 2001, the time period of the data set is the date it was last pushed to the public site, and the update frequency is quarterly.

Metadata format

FGDC metadata are usually stored either as a simple text file or as a text file in eXtensible Markup Language (**XML**). XML is similar to the more familiar HTML used to develop web pages. In addition to the basic information fields, it also contains formatting tags that tell a program how to read and display the information.

The CSDGM metadata content standard divides the information into seven main sections (Fig. 15.2). The North American Profile of ISO 19115 has its own section organization. In some cases they correspond fairly closely with the CSDGM sections, but not in others. The following descriptions refer to the CSDGM sections, with the NAP sections that contain similar information noted in parentheses.

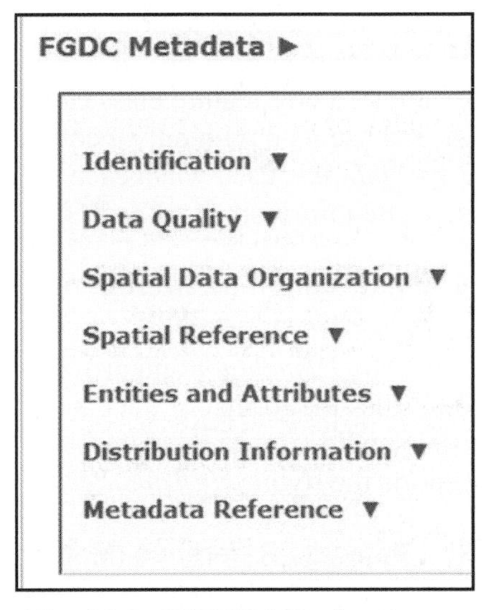

Fig. 15.2. CSDGM sections

1. Identification (Identification, Maintenance)

This section contains basic information about the original source of the data: who created them, what they contain, and any restrictions that have been placed on their use. The *Abstract, Purpose,* and *Supplemental Information* communicate what the data set contains and notifies the user of any important information at the top. The *Currentness Reference* indicates the time period that the data represent, and the *Status* says how frequently they are updated. The *Data Set Credit* and *Citation Details* provide information on who created the data product and how it should be cited in the literature. The *Spatial Domain* and *Keywords* help anyone searching for data to determine the location and the content of the data. The *Access Constraints* and *Use Constraints* specify who can use the data and whether they can be given to others and usually contain standard liability and release statements to protect the data provider from lawsuits.

2. Data Quality (Quality)

This section provides information on the important data quality issues discussed in the previous section. Sections are included on the *Logical Consistency* and the *Completeness* of the data set as a whole. There is a section to report on the *Attribute Accuracy* and the *Positional Accuracy.* The original scale and accuracy of the *Source* data must be given, including the temporal accuracy; the accuracy of the source may be different (better or worse) than the final data set. A complete *Source Citation* should also be included. Finally, there is a *Process Step* section that records the processing steps between the source and final product and the impact of these steps on the accuracy of the data.

3. Data Organization (Spatial Representation)

This section contains information on the format and organization of the data, such as whether they are vector or raster data. ArcCatalog fills out this section automatically.

4. Spatial Reference (Reference System)

This section contains the complete **spatial reference** information on the coordinate system, including the GCS used, the map projection and its parameters, and the map units. ArcCatalog fills out this section automatically.

5. Entity Attribute (Content)

This section describes the attribute fields present in a vector data set or the Value field in a raster data set. This section is critical for giving data users information, such as what the land use codes represent or whether the AREA field is in square kilometers or square miles. The user can choose one of two methods to describe the attributes. The *Overview Description* can be used when all of the fields are similar or self-explanatory and can be described together in a paragraph or two. For example, a city feature class containing decadal population values since 1800 is easy to explain. A land use map with only a couple of fields might also fall into this category. The *Detailed Description* is used when each field requires a different explanation. The metadata creator enters separate information for each field.

6. Distribution (Distribution)

This section describes how the data are made available, by whom, and under what conditions. It states the *Distributor* and provides information for contacting that person or organization. It describes the format of the data (DVD, CD, FTP download) and the *Standard Process* by which a user should request the data, such as the ordering information or download instructions. It also reports the *Available Time Period* when the distributor intends to make the data available.

7. Metadata Reference (Metadata, Maintenance)

This section is metadata about the metadata, including who developed them, in what format they are stored, and if any restrictions are placed on the distribution of the metadata (usually, the metadata are freely distributable even if the data set is not). Some of this section is filled out by ArcCatalog.

Standards identify certain mandatory elements for compliant metadata. There are several categories: mandatory, mandatory if applicable, and optional. Table 15.1 provides a brief overview of mandatory elements in NAP and FGDC; users should strive to cover these sections at a minimum. Some metadata editors provide guidance on required versus optional fields.

Acquiring metadata

Many projects utilize data sets that come from another organization, which, if it follows industry practice, has already supplied metadata for the data set. In this case, implementing the metadata becomes easy. The user need only ensure that the metadata are properly stored with the data set in ArcCatalog and that any subsequent processing or modification after acquisition has been recorded in the Process Step section.

For example, if the user has downloaded wetlands from the National Wetland Inventory and extracted the features in a particular county for a project, he or she only needs to record the extraction. This information would go into the Completeness Report and the Process Step sections of the metadata. The rest of the metadata need not change. In fact, it should not change;

Table 15.1. Comparison of mandatory elements

North American Profile	FGDC
Metadata Information Date stamp Contact Metadata Standard File identifier Language Character set Parent Identifier Hierarchy Level Locale	Metadata Information Date Contact Metadata standard Metadata version
Identification Information Citation Title Date Responsible Party Status Language Topic Category Extent	Identification Citation Title Publication Date Originator Description Status Time Period of Content Spatial Domain Keywords Access Constraints Use Constraints

it is important to retain the originators of the data set so that they receive proper credit and so that the prior information about the data is not lost.

However, this process may be complicated by the fact that not all organizations store the metadata as part of the data set, as ArcGIS does. The metadata may exist in a completely different document close to, but not with, the data they represent. They might be somewhere on the web site where the data are downloaded, in another folder on their source DVD, or occasionally even further afield. In many cases the separate file will be in a standard FGDC format (text or XML) and can be easily imported to ArcGIS. Sometimes, however, the format is different or not quite correct, and the user will have a more difficult time implementing the metadata. After working with metadata for a while, you will learn to recognize the formats that are suitable and that can be imported into ArcGIS with a minimum of fuss.

Metadata development

For data sets developed by you or your organization, the metadata process must start at the beginning. Although entering the metadata may be one of the final steps in producing the data set, the process begins far earlier, when the data are first being assembled. Critical information available at the start of the process and throughout its duration must be recorded so that it can be added to the metadata when the time comes.

Imagine that you have scanned and are digitizing an historical map of the Lewis and Clark journey. You will need to record information about the map itself: its author, year of publication, citation information, scale, medium on which it was drafted, and so on. Is there any information on the positional accuracy of the features on the map? For what time period were the source data relevant? Then, as you go through the digitizing process, information must be recorded along the way. What was the RMS error of the georeferencing step? What kinds of techniques or tests were used to ensure topological integrity? What are the definitions and content of the attribute fields added to the map? What do you know about the attribute accuracy and the completeness?

If you wait until the end, then critical information may be lost. Thus, it is good practice to develop a form on which notes can be recorded during the creation of the data, organized so that the information can be easily translated to the metadata at the end. You can record these notes on paper or electronically. Even if all of the information recorded does not make it into the metadata, the notes will provide a valuable resource for your organization. Table 15.2 shows a sample of a form that might be used. A copy of this form is stored in the mgisdata\Metadata folder for your use.

Table 15.2. Sample notes template for recording information about data layers

Project:		Feature Layer:	
Compiler:	Start Date:		End Date:
___Download	___From Client	___ CD/DVD	__Other:
Original Scale:		Original Format:	
Original Coordinate System:			
Originator Information (name/agency, contact info, and/or URL):			
Publication Reference (title, author, date, publication, version, edition, series, etc.):			
Horizontal Accuracy:		Vertical Accuracy:	
Field Collection Notes or Other Notes:			
Processing Steps and RMS When Applicable:		Notes on Attributes/Units/Definitions:	

About ArcGIS

CSDGM metadata were commonly produced for many years and the CSDGM standard was used by ArcCatalog for editing metadata in ArcGIS 9.3.1 and earlier versions. Other editing programs were also available, but the ArcGIS editor was convenient because the metadata were stored along with the data set, and were copied and renamed with it, so they were never lost. In fact, ArcGIS 9.3.1 had two metadata editors, one for FGDC metadata and one for ISO 19115, with a translator to convert it (somewhat imperfectly) to ISO 19139 if needed. No capability existed to produce NAP metadata at 9.3.1. On the other hand, ArcGIS 9.3.1 was adept at accepting the classic text-based format of FGDC metadata and applying it to data sets.

The problem with 9.3.1 was that the metadata storage and editing capabilities were heavily dependent on the standard being used, making it difficult to meet the needs of all users in North America and internationally. In ArcGIS 10, the major goals were to support several standards simultaneously, as well as multiple profiles. This strategy would help make ArcGIS flexible to changing standards, and it would enable all users to use the same editor. This goal was accomplished by storing metadata information in an internal ArcGIS format, then translating it to meet a specific standard only when needed.

Users who are not required to produce data accompanied by a specific metadata format can simply stay within the ArcGIS metadata framework. Two different levels of metadata support are provided for these users.

A very simple and streamlined set of descriptive information, called the **item description**, can be filled out within five minutes. This approach helps solve a long-standing problem—metadata were so complex and difficult that many people never bothered documenting a data set at all. The item description is easy to understand and to edit, and although it falls far short of a complete metadata record, it is much better than nothing. An item description can be created for any item type in ArcGIS, not just shapefiles and feature classes but also models, tables, scripts, and more. This minimum Item Description is required for every data set shared on ArcGIS Online.

Users desiring greater levels of documentation can access and edit additional metadata information. They must specify a Metadata Style to work with (Fig. 15.3). The style is based on one of the metadata standards, and it controls which fields you will see when editing metadata. It also defines how the metadata are validated and exported to an XML file if a standard-compliant metadata file is needed.

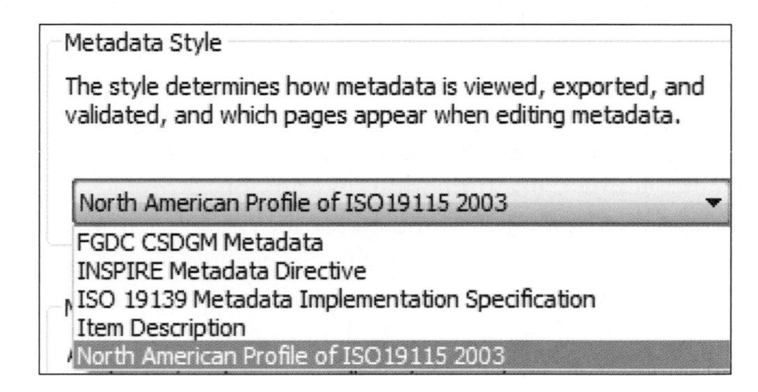

Fig. 15.3. Metadata styles

Finally, users who are tasked with creating full-fledged metadata conforming to a standard must edit the ArcGIS metadata and then convert them to a separate XML file to meet their obligations. Metadata created in ArcGIS 9.3 using the FGDC or the ISO editor can be viewed in ArcGIS 10 but must be upgraded to be edited. This upgrade will be offered if ArcCatalog detects the presence of these metadata formats.

In this chapter we will cover the basic editing of ArcGIS metadata items using the North American Profile of ISO 19115, more as an introduction to understanding metadata and some of their key components rather than as a serious effort to train a metadata producer.

TIP: The relationship among FDGC metadata, ArcGIS metadata, ISO metadata, and other styles is complex. You can ignore this complexity if you are simply creating new metadata, but if you must work with data sets containing FGDC metadata developed prior to ArcGIS 10, you will need more guidance. See the ArcGIS Help under Geodata > Data Types > Metadata.

The Metadata Editor

In Chapter 1 you learned how to view the item description in the Catalog tab. To move beyond the simple description, you must specify a metadata style. Then the Metadata toolbar (Fig. 15.4) allows you to view, edit, import, export, and validate metadata (compare them to the requirements of the standard) using that style.

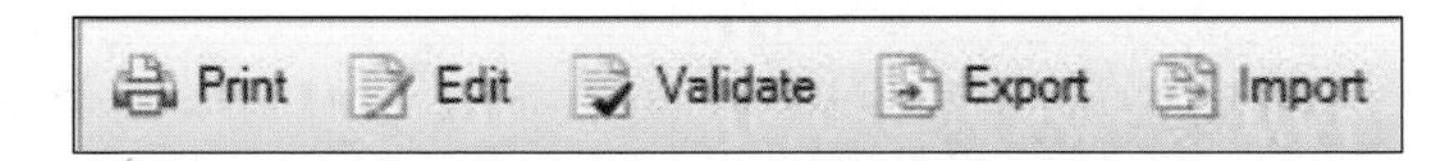

Fig. 15.4. The Metadata toolbar in ArcCatalog

Clicking the Edit button takes you to the metadata editor (Fig. 15.5). It is easy to use, but its organization may take some time to learn. There are three main sections: an Overview of the data set, a Metadata section containing information about the metadata themselves, and a Resource section containing information about the data set. There is no set order; the sections can be visited in any order and at any time. Each section has multiple items to be filled out. In some cases they already exist, and the user simply fills them in. In other sections, users must create containers first, which are then filled with content.

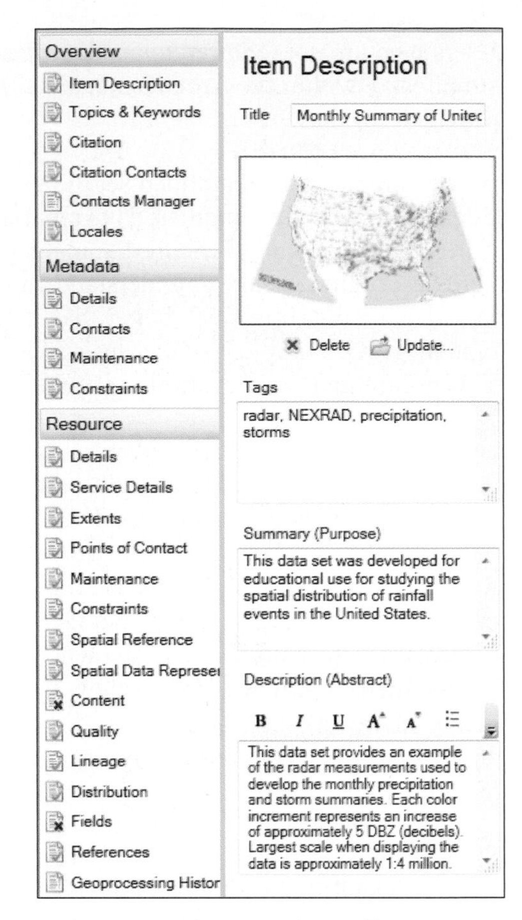

Fig. 15.5. The metadata editor

Certain metadata fields are required by the content standard in order to produce minimally compliant metadata. These boxes are filled in red to assist you in completing the information. In addition, error messages at the top of the editor let you known when an item is missing. Clicking the hyperlink should open the relevant box if it is not currently visible (although this feature worked unreliably in the first release of 10.1).

Some sections allow the user to add multiple items in an information category. For example, in developing a data set you might have used three different data sources and need to document all three (Fig. 15.6). Documenting at least one Source is required information, and the entry boxes appear automatically. The Down arrows expand a section so that you can see its contents, and the Up arrows hide it again. You can type a general description above. You can also add items, such as a full-fledged citation, by clicking the + New Citation text. To add a second source, you would click the + New Source text to provide another container. The small red X symbol lets you delete a container if you decide you don't need it after all.

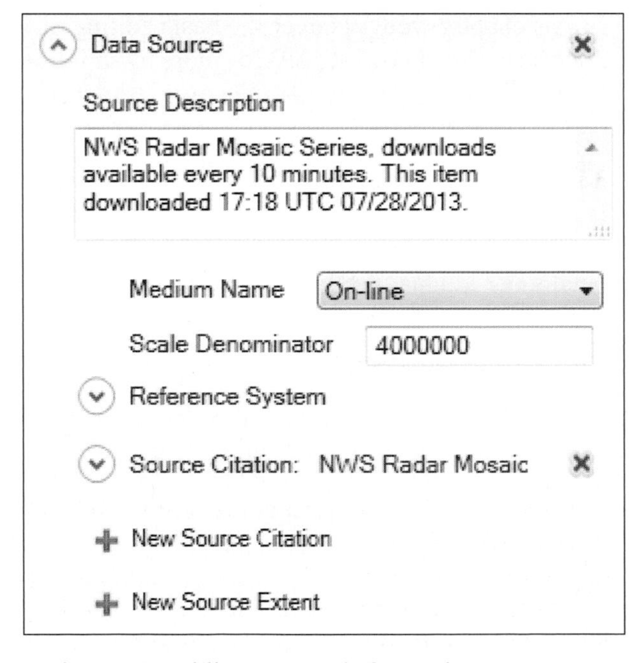

Fig. 15.6. Adding source information

One advantage of the ArcGIS metadata editing format is that, in many cases, containers are generated only when the user has metadata to put in it. The old FGDC editor had places to enter all types of data quality reports, such as logical consistency, horizontal accuracy, vertical accuracy, and completeness, which in some cases were not applicable to a data set or the user had no useful information to impart. In the new editor, a report entry is created only if you need it. You add the report, specify the type, and fill in the relevant details. Thus, the final metadata that you see contain only useful information, not rows of blank entries.

Metadata templates

Every metadata set has an impressive amount of information to be entered, but much of this information is repeated verbatim in many different data sets produced by the same organization. This fact makes the use of templates very attractive. A **metadata template** contains the information that will be common to every data set produced by an organization or for a particular project. The contact information and access and use constraints, for example, are metadata fields that are likely to be consistent within an organization.

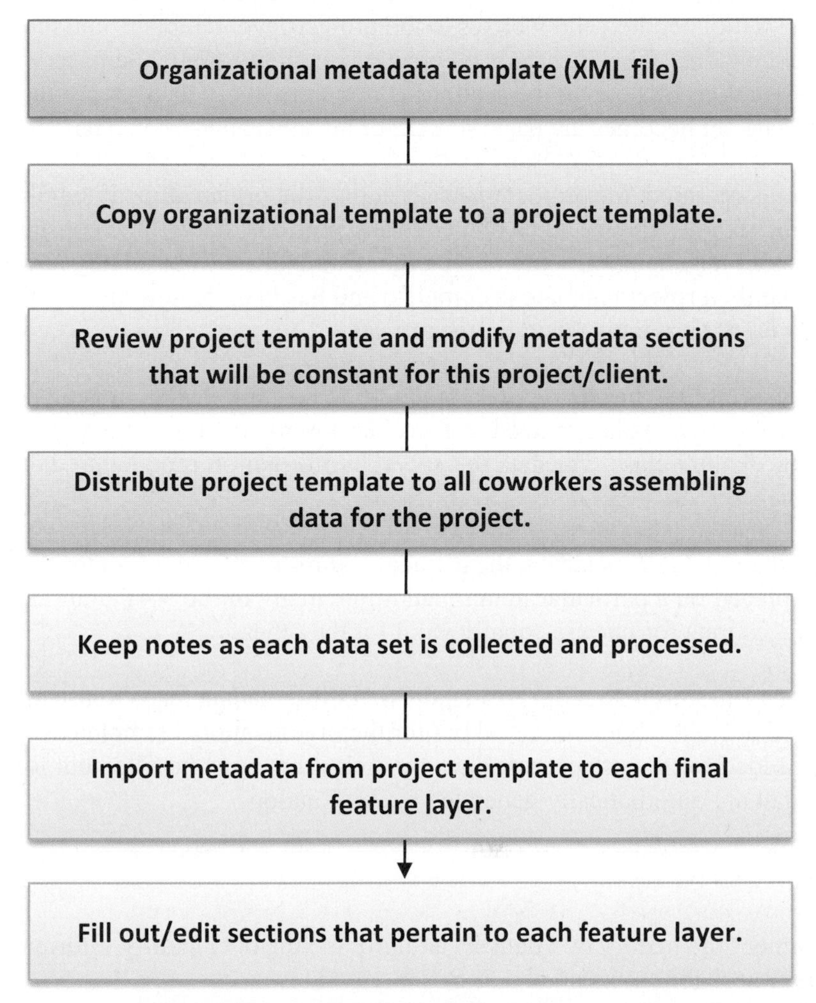

Fig. 15.7. A possible organizational process for using metadata templates

Metadata templates benefit from a hierarchical approach (Fig. 15.7). Some information common to every data set for the organization can serve as the starting point. This template is further edited to include common information for a particular project, for example, the contact information for the project manager. The usual practice for a company might be to make all data available only to clients, but if a certain project is funded as an educational effort, then the usual access constraints might be edited to make the data freely redistributable. Templates might have three levels of specificity, as follows.

Organizational level: Contains fields that are the same for every data set produced. Examples are the organizational contact, the access and use constraints, and the metadata distribution constraints.

Project level: Contains fields that are the same for every data set produced for a particular project. Some of the organizational-level fields might be edited for this level. Examples are the Purpose, the Data Set Credit, the organizational contact, and some of the key words.

Data set level: Fields that must be added for each data set, such as the Summary and Description, the citation information, and the accuracy reports

The overall process might look something like Figure 15.7. The blue box refers to the organizational-level template, the yellow boxes to the project-level template, and the green boxes to the data set-level metadata. The user starts by creating an empty XML file to serve as the metadata template. He then fills out all the sections common to every data set produced by the organization. This data set becomes the basic source for all subsequent metadata.

When a new project is initiated, the project manager copies the organizational template to a project template. She reviews the metadata and edits any portions that will be constant for the project. This might include changing some of the organizational fields or adding new information to the template. Once the project template is complete and has been reviewed, it is distributed to all of the workers who are producing data for the project.

As each data set is assembled, the workers take notes on the source and processing. When the data set is ready to have its metadata created, the worker imports the project template into the metadata for the new feature class. The data set–specific information is added to the metadata to make them complete.

For guidance in filling out NAP metadata, the metadata editor itself provides tags that are visible when the cursor is placed on a particular item. In addition, many of the NAP and CSDGM items are similar, and suggestions for one are equally valid for the other.

The end of this chapter has both ArcGIS/NAP and CSDGM metadata maps to use as a resource. The maps suggest what fields should be filled out for the organizational template, which ones might be edited or added for the project template, the fields that must be filled out for each data set, and the fields that are automatically generated by ArcCatalog.

Summary

- Metadata document the history of a data set and provide the data quality information needed to ascertain its fitness for an intended purpose.

- Metadata standards and formats have been established by the International Organization for Standardization (ISO) and the Federal Geographic Data Committee (FGDC).

- United States metadata standards are currently in flux as users migrate from the current FGDC CSDGM standard to a new international standard, the North American Profile of ISO 19115. International users are more likely to use the ISO 19139 or 19115 standard.

- Six key data quality issues are lineage, positional accuracy, attribute accuracy, logical consistency, completeness, and temporal accuracy.

- CSDGM metadata are organized into seven main sections, including Identification, Data Quality, Data Organization, Spatial Reference, Entity Attributes, Distribution, and Metadata Reference. NAP metadata have a different organization, but many of the sections are similar.
- ArcCatalog automatically fills out some sections of metadata.
- Establishing templates can reduce the work needed to prepare metadata.

Important Terms

attribute accuracy	item description	positional accuracy	temporal accuracy
completeness	lineage	process step	XML
CSDGM	logical consistency	profile	
FGDC	metadata standard	source	
ISO	metadata template	spatial reference	

Chapter Review Questions

1. What is the mission of the FGDC?
2. What is the difference between the source and the product when preparing metadata?
3. How much metadata is required to share data on ArcGIS Online?
4. Why have formal guidelines for testing attribute accuracy not been established?
5. What aspects of logical consistency are most frequently tested and reported?
6. Contrast the temporal accuracy of a rivers feature class with that of a congressional districts feature class.
7. Why will US users need to be familiar with two metadata standards for the near future?
8. Which section of metadata would you consult to determine how to order the data and how much they cost?
9. Which section of metadata would you consult to determine if the ROADLENGTH field of a roads feature class is in kilometers or miles?
10. You download a data set from the web and discover that it has no metadata when viewed in ArcCatalog. What is your next step?

Mastering the Skills

Teaching Tutorial

The following examples provide step-by-step instructions for doing basic tasks and solving basic problems in ArcGIS. The steps you need to do are highlighted with an arrow ➔; follow them carefully. Click on the video number in the VideoIndex to view a demonstration of the steps.

In this tutorial, we will first examine some metadata examples to see what good metadata look like. We will then create metadata for the weather map that we georeferenced in Chapter 11. Metadata are one of the tasks for which it is easier to use ArcCatalog instead of the Catalog tab in ArcMap.

➔ Start ArcCatalog.

TIP: ArcCatalog is similar to the Catalog tab, but if you find you are confused, review the Skills Reference section Basics: ArcCatalog Basics.

Examining metadata

Most of the feature classes in the usdata geodatabase came from the ESRI Data and Maps DVD that is distributed with the ArcGIS software. The ESRI metadata provide excellent examples to learn from.

1➔ Navigate to the mgisdata\Usa folder and expand the usdata geodatabase.

1➔ Click the quakehis layer to highlight it, and click the Description tab.

1➔ Skim the information here, making sure you scroll to the end.

By default, ArcCatalog shows only the basic descriptive information here. To view and work with more extensive metadata, the setting must be changed.

2➔ Choose Customize > ArcCatalog Options from the main toolbar.

2➔ Click the Metadata tab.

2➔ Change the metadata style to *North American Profile of ISO 19115 2003*. Click OK.

2➔ Click the Preview tab and then click the Description tab to update the view.

2➔ Scroll down again and notice how much more information there is, including two sections for ArcGIS Metadata and FGDC Metadata.

Styles may be based on a standard such as ISO 19115. ArcGIS stores the metadata information internally in its own format. The metadata style controls which information items are displayed and edited. The metadata can be exported to the standard format if needed.

TIP: The large-font blue headings are section headings and may be collapsed (side arrow) or expanded (down arrow). Use these headings to hide or view the information as directed in the next steps.

ArcGIS Metadata ►
Topics and Keywords ▼
Citation ▼

3➔ Start by collapsing all the headings under ArcGIS Metadata, if necessary, so that you can see the organization of the 18 sections.

3➔ Expand the Topics and Keywords section and skim through the information.

3➔ Close it when you are finished, by clicking the Topics and Keywords heading again or by using the Hide hyperlink at the bottom of the section.

TIP: After you are finished examining each section in the following steps, close it again.

3➔ Examine the Citation section.

1. What was the publication date for this data set? ______________

3➔ Examine the Citation Contacts section.

2. Who originated this data set?________________ Who is the publisher? ___________

4➔ Examine the Resource Details section.

4➔ Examine the Resource Points of Contact to find out whom you might contact with questions about this data set.

4➔ Examine the Resource Maintenance section to find out how often this data set is updated.

Let's look more closely at where this data set came from.

5➔ Expand the Lineage section and look at the process steps

3. How many process steps are listed? _______ When did they occur? ________________

4. What source was used to compile this data set? ________________________

5. Can you find the original scale (scale denominator) of this data set? _________________

6➔ Examine the Spatial Reference section to see the coordinate system.

6➔ Examine the Spatial Data Properties.

7➔ Expand the Data Quality section. There are five data quality reports for this data set.

7➔ Examine the names and content of the five reports, comparing them to the data quality issues discussed in this chapter.

8➔ Examine the Distribution section.

6. How does one obtain a copy of these data? ___________________________________

8➔ Examine the Metadata Details and Metadata Contacts section.

7. Who produced the metadata? ___________________ What standard was used?____________

8➔ Examine the Fields section.

8. What is the earthquake depth unit of measurement? _____________

Clearly, it takes a lot of effort to document data sets to this level of detail. Not all data sets are this well documented, but all data sets need at least minimal information. As you continue to learn about metadata, you can go back to the data sets in this usdata geodatabase for good examples of how metadata are organized and what types of information go into each section.

Creating a project template

You have been provided with an organizational template from a fictional company called Re-GIS, Inc., which distributes weather and climate data sets by FTP download, usually at a cost. Imagine that you are the project manager for a new initiative to provide a freely distributable monthly summary of precipitation based on National Weather Service radar data. You will create a project template for use with the data sets to be developed for this project.

9➔ In ArcCatalog, navigate to the mgisdata\Metadata directory and click on the REGIS_OrgTemplate XML file to highlight it.

9➔ Click the Description tab, if necessary, and examine the various sections of ArcGIS metadata.

9➔ Scroll through the metadata, noting the type of information that is already filled out.

Since this is only a template, many of the sections are still blank. The spatial information, for example, is only applicable to spatial data and is updated automatically by ArcCatalog. Therefore, it is not present in this XML template.

The organizational template contains information that will be the same for all data produced by the company. Your project template will include this information, as well as information that will be the same for all data sets produced for the project. The sections unique to each data set will be left blank, to be filled in as each data set is created.

10➔ Right-click the REGIS_OrgTemplate XML file and choose Copy.

10➔ Right-click the Metadata folder entry and choose Paste.

10➔ Click on the new REGIS_OrgTemplate(1) XML file twice slowly, and rename it Radar_Project_Template.

TIP: To create a completely empty XML file to start your own organizational template, right-click a folder and choose New > XML Document, and give it a name.

11➔ Click on the Radar_Project_Template file to highlight it and click the Edit button on the Description tab.

11➔ Examine the outline of sections on the left side of the editing window.

The Overview section contains a general description of the data set. The Metadata section contains information about the metadata themselves, such as who created them and whom to contact with questions about them. The Resource section contains detailed information about the data set.

TIP: When you place the cursor in a box, a pop-up at the bottom of the ArcCatalog window tells you the type of information that it should contain.

11➔ The Title, Tags, and Description will be different for each data set, so leave them blank.

11➔ For the Summary, enter This data set was developed for educational use for studying the spatial distribution of rainfall events in the United States.

11➔ For the Credits, type Based on data provided by the National Weather Service. This data set is freely distributable for noncommercial purposes with inclusion of citation and full metadata.

11➔ Set the sliders for the Appropriate Scale Range from Continent to State.

11➔ Save your edits and start editing again.

TIP: A red box indicates that the item is required by the standard.

12➔ Click on Overview: Topics & Keywords in the outline.

12➔ Check the Atmospheric Sciences box.

12➔ Set the Content Type to Downloadable Data.

12➔ Click on the + New Theme Keywords text.

12➔ Enter precipitation, NEXRAD, radar as key words.

13➔ Click on + New Place Keywords and enter United States.

13➔ Click on + New Temporal Keywords and enter 2013.

Next you will enter the Citation information, showing how it should be cited by others.

14➔ Click on Overview: Citation. Notice the messages at the top, telling you that this information is required by the metadata standard.

14➔ Enter the Citation Title as Monthly Summary of Radar Precipitation for 2013.

14➔ Click on + New Presentation Form and set the Presentation Form drop-down to Digital Map. Set the FGDC Geospatial Data Presentation Form to Raster Digital Data.

14➔ Expand the Dates entry. The Created date will be unique for each data set, so leave it blank.

14➔ The data are scheduled for publishing on 3/31/2014. Click on the Published calendar icon and click the *Date, Year* heading on top of the calendar to move quickly to March 2014. Enter 3/31/2014.

14➔ Notice that the error messages at the top have disappeared, as has red X on the Citation section.

14➔ Save your edits and start editing again.

TIP: Save your edits frequently!

15➔ Click on the Citation Contacts section.

15➔ Expand the Contact and the Contact Information sections.

This information was present in the organizational template, saving you some typing. It could be customized with your name, but generally it is wiser to keep it generic, to a position rather than a person, in case of future personnel changes. The next section provides more information about the metadata themselves.

16➔ Click on the Metadata: Details section.

16➔ Click Create to create the File Identifier.

16➔ Set the Function to Information.

16➔ Set the Date Stamp to today's date.

16➔ Set the Language to English and the country to UNITED STATES.

16➔ Set the Hierarchy Level to Dataset.

17➔ Click the Metadata: Contacts section. Again, this information was provided by the organizational template.

The previous contact information you saw referenced the data set. This one references the metadata. In this case they are the same.

18➔ Click the Metadata: Constraints entry. Expand the Down arrows. The information from the organizational template is in here and does not need updating.

18➔ In the Security Constraints section, set the Classification to Unclassified.

18➔ Click Save and start editing again.

The Resource section contains information about the data set itself. Some of it will seem familiar because you already set it in the Metadata section.

19➔ Click on the Resource: Details entry.

19➔ Set the Language to *en* for English, and the Country to the United States.

19➔ Leave the other information blank, to be filled in for each data set.

20➔ Examine the Resource: Extents entry. This part will be filled in by ArcGIS for each data set.

20➔ Examine the Resource: Points of Contact. Again, the template took care of it.

TIP: Notice that the metadata require multiple contacts to be filled in, often with the same information. The Overview: Contacts Manager and the Load a Contact button let you re-use previously entered contacts to save typing. The drop-down list seems to have a bug and not show the actual names, which makes it confusing. Use it for your own metadata with caution.

21➔ Click on the Resource: Maintenance entry.

21➔ Change the Update Frequency to Not Planned. The rest of the information would be filled out if you were planning to update the data after publication, but that is not the case for this project.

22➔ Click on Resource: Constraints and expand the General Constraints entry.

22➔ Delete the first sentence of the Use Limitation, since the free distribution policy for this project is different from the usual company policy.

22➔ Save and start editing again.

23➔ Click the Resource: Spatial Reference entry. This section will be filled in by ArcGIS.

23➔ Click the Resource: Spatial Data Representation entry. This section will be filled in by ArcGIS.

The Content, Quality, and Lineage sections will be different for each data set produced by the project, so these are left blank for now. Going on, the Distribution section explains how someone can obtain the data.

23➔ Click on the Resource: Distribution section.

23➔ Click the + New Distribution Format entry.

23➔ For the Format Name, enter ArcGIS File Geodatabase.

23➔ For the Format Version, enter Version 10.1.

The actual distributor is the manager of the company web sites, so the next section is for the company web master.

24➔ Expand the Distributor entry.

24➔ Expand the Contact entry and the Contact Information entry.

24➔ Click the + New Online Resource text. You may need to expand the Distributor and Contact Information entries again.

24➔ Enter http://www.ReGIS.MonthlyRadar.com for the Linkage.

24➔ Enter ftp for the Protocol.

24➔ For Name, enter Monthly Summary of Radar Precipitation.

24➔ For Description, enter Download site for educational data sets of monthly radar precipitation.

24➔ Set the Function to Download.

25➔ Scroll down, if necessary, and click the + New Ordering Process entry. Change the Fees to Free.

25➔ You intend to release the data at the end of March 2014. Change the Available Date to 3/31/2014.

25➔ Enter Download for Ordering Instructions.

25➔ Enter Immediate for Turnaround.

25➔ Click Save.

The project template is now complete.

Importing the template

Now you will create metadata for your first data layer for the project. The nwsradar.img image in migsdata\Metadata was downloaded and georeferenced from the National Weather Service (as you did in Chapter 11). It will serve as an example of the data used to generate the final product and is to be included in the publication data set. We will create metadata for it.

IMPORTANT TIP: It is recommended to turn off background processing before running the Import/Export tools.

26➔ Choose Geoprocessing > Geoprocessing Options from the main menu bar.

26➔ Uncheck the Enable box for background processing, if necessary. Click OK.

27➔ Click on the nwsradar.img raster in the mgisdata\Metadata folder to highlight it. Note the virtually empty metadata.

27➔ Click the Import button on the Metadata toolbar. [Import]

27➔ For the Source Metadata, click the browse button and navigate to the mgisdata\Metadata folder.

27➔ Choose the Radar_Project_Template XML file and click Add.

27➔ The metadata you've been entering are stored in the internal ArcGIS format, so change the Import Type to FROM_ARCGIS.

27➔ Click OK to start importing.

28➔ Click on nwsradar.img in the Table of Contents.

28➔ Scroll down and examine the metadata now.

The metadata for this raster, which were essentially blank a moment ago, have been updated with the information from your template. ArcCatalog also filled in some information, for example, about the spatial storage format and the coordinate system.

28➔ Click on the Preview tab in ArcCatalog.

28➔ Click the Thumbnail button.

28➔ Click the Description tab again.

The metadata have been updated to include the thumbnail you just created.

Entering metadata for an item

Now we will finish entering the metadata for the nwsradar image. However, you will need to know some information about the source data. You had one of your coworkers investigate the web site where the information was obtained and talk to the people at the National Weather Service who provide the radar data. She has assembled the information in Table 15.3 for you.

29➔ Click the Edit button and click the Overview: Item Description section.

29➔ For Tags, enter radar, NEXRAD, precipitation, storms.

29➔ For the Description, enter This data set provides an example of the radar measurements used to develop the monthly precipitation and storm summaries. Each color increment represents an increase of approximately 5 DBZ (decibels). Largest scale when displaying the data is approximately 1:4 million.

29➔ Click Save and start editing again.

This is where the value of the template is obvious. Most of this section was already filled in.

Table 15.3. Information about the radar image for inclusion in the metadata

Source information for NWS radar example	
Citation title	NWS Radar Mosaic Time and date: 1718 UTC 06/20/2013
Originator	National Weather Service 1325 East-West Highway Silver Spring, MD 20910
Resolution	0.02 degrees
Scale of source	Approximately 1:4 million
Info	National mosaic map of radar sectors for the conterminous United States. Units in DBZ from -25 to 75
URL	http://www.weather.gov/radar_tab.php
Estimated accuracy	Source resolution 0.02 degrees RMS error of transformation = 0.002 degree

30➔ Click the Overview: Citation section.

30➔ Expand Titles if necessary and replace the file name with Monthly Summary of United States Precipitation 2013 for the Title.

30➔ Expand Dates and change the Created Date to the current date. The Published date is already entered from the template.

30➔ Click Save and start editing again.

The rest of the Overview section should also remain the same, as set up in the project template. Next we turn to documenting the Content, Quality, and Lineage sections. The data set is a raster representation of a map rather than a directly measured image like a Landsat scene, so we will use the Coverage type to describe the content.

31➔ Click on the Resource: Content section.

31➔ Click the red X to delete the Image Description entry.

31➔ Click the + New Coverage Description entry.

31➔ For the Attribute Description, enter NEXRAD Radar in DBZ (decibels) and set the Content Type to Image.

We now turn to the Quality section. We only need to enter a report for quality information that we actually have, which is an estimate of the positional accuracy.

32➔ Click the Resource: Quality section.

32➔ Set the Level Scope to Dataset.

32➔ Click + New Report and set the Report Type to Relative Internal Positional Accuracy.

32➔ Set the Dimension to horizontal.

33➔ Click + New Quantitative Result.

33➔ For Value Type, enter Georeferencing RMS Error.

33➔ Enter 0.02 for the Value and select *plane angle: degree deg* for the Value Unit.

33➔ Save and return to editing.

Next we need to document the Lineage of the data, which includes the sources and processing. When entering sources, one can be satisfied with a general statement and/or source description, or one can construct a detailed citation. We will do both, although some of the fields will be blank. This section is somewhat more extended than the others.

34➔ Click the Resource: Lineage section.

34➔ In the Statement box, type Radar image was downloaded from the National Weather Service, georeferenced, and projected.

34➔ Click the + New Data Source entry.

34➔ In the Source Description box, type NWS Radar Mosaic Series, downloads available every 10 minutes. This item downloaded 17:18 UTC 07/28/2013.

34➔ For Medium Name, choose On-line.

34➔ For Scale Denominator, type 4000000.

35➔ Click + New Source Citation. Enter the Title NWS Radar Mosaic.

35➔ Click + New Presentation Form and choose Digital Map.

35➔ For FGDC Geospatial Data Presentation Form, choose Map.

35➔ Click + New Identifier and enter National Weather Service/NOAA for Code.

35➔ Expand Dates and put 7/28/2013 for both Created and Published Date.

36➔ Click + New Contact. Leave the Name blank.

36➔ Enter National Weather Service for the Organization. Set the Role to Originator.

36➔ Click + New Contact Information and choose Both for the Address Type.

36➔ Enter the NWS address from Table 15.3.

36➔ Click + New Online Resource (under the Email box) and enter the URL from Table 15.3 for the Linkage.

36➔ Enter HTTP for the Protocol, and Doppler Radar National Mosaic for the Name, and select Download for the Function.

Next we enter the two processing steps: georeferencing and projecting. Although one could enter all kinds of information about who did it and when, we will just use a general statement.

37➔ Scroll down and click the + New Process Step entry. For the Process Description enter Downloaded NWS Mosaic radar image and georeferenced it to the GCS NAD 1983 coordinate system with RMSE of 0.02 degrees using ArcGIS 10.1.

37➔ Fill the Date with the processing date, 7/28/2013.

38➔ Click +New Process Step and enter the Process Description as Projected the georeferenced raster to North America Equidistant Conic coordinate system with nearest neighbor resampling and output resolution of 2000 meters.

38➔ Fill in the process date with 7/28/2013.

38➔ Click Save and start editing again.

You are almost finished, but notice that three sections still have a red X, indicating that there are a few missing items to cover.

39➔ Click the Metadata: Details section and click Create for the File Identifier.

39➔ Click the Resource: Spatial Reference section and set the Dimension of the Reference System to horizontal.

There is one more error listed. The image has a value attribute table (VAT) containing the colormap of the image values. It is used internally by the program and does not need to have a description in the metadata.

40➔ Click the Resource: Fields section.

40➔ Click the red X next to the Details entry to delete it.

40➔ Save your edits.

40➔ Scroll down and admire your metadata.

➔ Exit ArcCatalog.

Exercises

Imagine that your class requirements include a GIS project (as it well may). Go on the Internet and find a data set that you might use for your project. Print the metadata notes form in the mgisdata\Metadata folder and fill it out for that data set.

Challenge Problem

Prepare a metadata template for your real or imagined class project.

A NAP Reference Map

Use this map as a guide to creating metadata templates and filling in the Metadata sections using the ArcGIS 10 Editor using the North American Profile of ISO 19115 style.

Key

To be filled in by developer for each layer
Provided by Organizational Template but might differ for Project Template
Provided by Organizational Template
Automatically generated by ArcCatalog

1. Overview

Item Description

Title
The name of the data set
Tags
Comma-delimited list of key words to aid in searching for the data
Summary
What is the intended use for this data set? Under what conditions is it valid?
Description
Provide a brief description of the information contained in the data set. This is also a good place to put the largest scale recommendation. Largest scale when displaying the data: 1:xxxxx.
Credits
List those who created or contributed the data. This is a good place to list modifications made to imported data, such as subsetting it to a study area from an ESRI data set. Other things may be added as appropriate.

Topics & Keywords

Topic categories
Choose from a predefined list of general topics, such as environment or transportation
Theme keywords
Search key words relating to the content or theme, such as geology or census
Place keywords
Search keywords relating to the location, such as United States, Austin, or Oregon
Temporal keywords
Keywords relating to the relevant time period of the data, such as the 2010 Census, 1990, etc.

Citation

Titles
The name used to identify the data set
Identifiers
The authoritative reference, such as the USGS or the National Weather Service
Dates
Dates when the data set was created, published, and if applicable, revised
Edition, Series
Information regarding the edition of the data, if multiple versions are available, and/or a collection or series of which the data are a part
Responsible party
The person, position title, or organization responsible for or associated with the resource. At least one of the three must be provided, with appropriate contact information, including address, phone, e-mail, hours of operation, etc.

Citation Contacts

Citation Contact(s)

The person, position title, or organization responsible for or associated with the resource. At least one of the three must be provided, with appropriate contact information, including address, phone, e-mail, hours of operation, etc.

Locales

Locale

The language in which the data set is published and the country to which it pertains

2. Metadata

Details

File Identifier

Name of the metadata file and a date stamp

Language and Character Set

Language and computer character set used to create the metadata

Hierarchy Level

The scope to which the metadata apply—are they for a data set, software, model, etc.

Metadata Standard Name

The name of the metadata standard and profile, if applicable, used to generate the data. These items can be left blank if the metadata are intended to remain within ArcGIS.

Contacts

Metadata Contacts

The person, position title, or organization responsible for the metadata. At least one of the three must be listed. Contact information should also be entered.

Maintenance

Update Frequency and Scope

Are there future updates planned for the metadata? At what intervals? When is the next scheduled update? Additional contacts may also be listed here.

Constraints

Use Limitation

Describe any potential limitations as to the use of the metadata.

Legal Constraints

Specify who may legally use the metadata. Are they freely available or must they be purchased? This is the place for an organization to enter any statement limiting liability or restricting the use of the metadata to certain entities. Includes information on copyrights, trademarks, etc.

Security Constraints

Use to indicate data for which security clearance or other restrictions are needed.

3. Resource

Status

Status

Is this in progress or a final version?

Credit

List those who created or contributed the data. This is a good place to list modifications made to imported data, such as subsetting them to a study area from an ESRI data set. Other things may be added as appropriate.

Language and Character Set

Language and computer character set used to create the metadata

Spatial Representation Type

Vector data? Grid? Table? Video? What is the intended scale resolution for vector data, or distance resolution (cell size) for raster data?

Processing Environment

What operating system was used to generate the data?

Extents

This entire section is updated automatically by ArcCatalog.

Points of Contact

Contacts

The person, position title, or organization responsible for or associated with the data set. At least one of the three must be provided, with appropriate contact information, including address, phone, e-mail, hours of operation, etc. If there is an online point of contact, then the URL linkage must be listed.

Maintenance

Update Frequency and Scope

Are there future updates planned for the data set? At what intervals? When is the next scheduled update? Additional contacts may also be listed here.

Constraints

Use Limitation

Describe any potential limitations on the use of the data set.

Legal Constraints

Specify who may legally use the data set. Is it freely available or must it be purchased? This is the place for an organization to enter any statement limiting liability or restricting the use of the metadata to certain entities. Includes information on copyrights, trademarks, etc.

Security Constraints

Use to indicate data for which security clearance or other restrictions are needed.

Spatial Reference

This entire section is updated automatically by ArcCatalog.

Spatial Data Representation

This entire section is updated automatically by ArcCatalog.

Content

Coverage, Image, or Feature Description

Add a section based on whether the data set is a coverage (gridded thematic data such as a geology raster), an image, or a feature catalog (vector data set). Then details about the content will be added, dependent on the type. For example, the gridded NEXRAD radar data are a coverage with a physical measurement content type. An image description would contain information such as the number of bands, the illumination angle, quality codes, processing levels, etc.

Quality

Level Scope

Level to which the data quality reports refer, such as the data set, a table, an attribute, a model, etc.

Reports

One or more reports may be added as applicable, including common reports on:

Logical Consistency Report: *Logical consistency is largely concerned with topology. Did you test to see if the data contain dangles, gaps, or overlaps? What tests were applied, and what were the results?*

Completeness Report: *Provide information about omissions, selection criteria, generalization, and other processes that might impact how complete a data set is. Are all spatial entities included? For example, did you get all the wells or might some be missing? Did you subset the data from their original source? Were any criteria used in deciding which features to include (public versus private roads, for example)?*

Attribute Accuracy Report: *Summarize processes used to establish the accuracy of the attribute(s), for example, known detection limits of analyses. Evaluate detail or completeness of categorical classes. Provide known information about problems with any attributes. If no accuracy data are available, enter "Unknown."*

Horizontal Accuracy Report: *If data are from a standard US federal data product, enter the national map accuracy standard of 1 in 10,000. If data are surveyed or obtained via GPS, enter the known or estimated positional accuracy of the survey or GPS unit. If data have been georeferenced, transformed, or spatially adjusted, also include the RMS error.*

Lineage

Data Source(s)

A general statement or detailed citation of the source(s) of data used to create the resource. Multiple sources and citations may be added. You may include original publication details of the data set: title, originator, publication date, edition, and so on.

Process Step(s)

Describe one or more processing actions taken on the data before reaching its final form, for example, importing STDS quads and joining elevation attributes, merging into a single data set, projecting to current coordinate system, and clipping to study area boundary. Enter software version, process date, and the name of the person who did the processing. Include impacts on accuracy, such as an RMS error associated with georeferencing.

Distribution

Distribution format

Format in which the data are distributed, for example, zip file, DVD, download, specifications, etc.

Distributor

Enter the agency or person responsible for distributing the data and their contact information. It also includes information on how to order the data, download options, URL linkages, costs, etc.

Fields

Entity and Attribute Details

A listing of data fields with specific field definitions. Most of the information is supplied by ArcGIS, but the user can add important descriptions. Describe the main characteristics, such as animal counts in number of animals, stocking densities in cow/calf units per acre, or production in bushels.

Geoprocessing History

This entire section is updated automatically by ArcCatalog.

A CSDGM Reference Map

Use this map as a guide to creating or understanding FGDC CSDGM metadata.

Key

To be filled in by developer for each layer
Provided by Organizational Template but might differ for Project Template
Provided by Organizational Template
Automatically generated by ArcCatalog

1. Identification

General

Abstract
Provide a brief description of the information contained in the data set.
Purpose
What is the intended use for this data set? Under what conditions is it valid?
Supplemental Information
This is a good place to list modifications made to imported data, such as subsetting it to a study area from an ESRI data set. This is also a good place to put the largest scale recommendation. Largest scale when displaying the data: 1:xxxxx. Other things may be added as appropriate.
Access Constraints
Who is allowed to access the data? The public? Only those who have licensed it?
Use Constraints
Generally, this contains a legal statement concerning those who may access the data, whether and under what conditions they may be redistributed, and a liability clause.
Data Set Credit
Credit the originator of the data set (USGS? Contractor? Map author?).
Native Data Set Environment
Native Data Set Format

Contact

Point of Contact Details
Provide the person or organization to be contacted with questions about the data set.

Citation

Citation Details
Provide information on the originator of the data set or publication as completely as possible, including publication references, URLs, agency address/phone, and so on.

Time Period

Currentness Reference
For what time period is this layer valid? Is it unlikely to change (topography), good only for a short time period (evapotranspiration), or somewhere in between (population)?

Status

Status
Is this in progress or a final version? How often are updates, if any, intended?

Spatial Domain

Bounding coordinates

Keywords

Keyword and Thesaurus
Provide at least two keywords for searching, including place name keywords.

2. Data Quality

General

Logical Consistency Report

Logical consistency is largely concerned with topology. Did you test to see if the data contain dangles, gaps, or overlaps? What tests were applied, and what were the results?

Completeness Report

Provide information about omissions, selection criteria, generalization, and other processes that might impact how complete a data set is. Are all spatial entities included? For example, did you get all the wells or might some be missing? Did you subset the data from their original source? Were any criteria used in deciding which features to include (public versus private roads, for example)?

Attribute Accuracy

Accuracy Report

Summarize processes used to establish the accuracy of the attribute(s), for example, known detection limits of analyses. Evaluate detail or completeness of categorical classes. Provide known information about problems with any attributes. If no accuracy data are available, enter "Unknown".

Value/Explanation

For each attribute with a known accuracy, enter the data value and any explanation. For example, for a TMDL reading, enter "TMDL" as the value and "Detection limit xx mg/l, accuracy +/1 yy mg/l" for the Explanation.

Positional Accuracy

Horizontal Accuracy Report/Value/Explanation

If data are from a standard US federal data product, enter the national map accuracy standard of 1 in 10,000. If data are surveyed or obtained via GPS, enter the known or estimated positional accuracy of the survey or GPS unit. If data have been georeferenced, transformed, or spatially adjusted, also include the RMS error.

Source Information

General

The Source Scale Denominator is the original scale of the data set, such as 24,000 for quad data. Type of source media would be paper for a digitized map, data download/CD for a DLG, GPS unit, and so on. The Source Citation Abbreviation would be DLG, DRG, and so on.

Source Citation

Enter the original publication details of the data set: title, originator, publication date, edition, and so on.

Source Time Period of Content

Enter Currentness reference as Ground Condition (at time of measurement) or Publication Date. Also enter whether data were collected all at one time, at multiple times, or in a range of times and enter dates. For example, if you sampled wells from January 08 to March 09, enter Range and the dates.

Process Step

Process Description

Describe one or more processing actions taken on the data before reaching their final form, for example, importing STDS quads and joining elevation attributes, merging into a single data set, projecting to current coordinate system, and clipping to study area boundary. Enter software version, process date, and the name of the person who did the processing. Include impacts on accuracy, such as an RMS error associated with georeferencing.

3. Data Organization

General

This entire section is updated automatically by ArcCatalog.

4. Spatial Reference

This entire section is updated automatically by ArcCatalog.

5. Entity Attribute

Fill out either the Detailed Description tab or the Overview Description tab, whichever makes the most sense for documenting the attribute values of that particular data set.

Detailed Description

Entity Type

Attribute

General

This section is mostly filled in for you, but, for each attribute important to the data set, you should enter a ***Definition****, including units. For a field titled "TMDL", you should enter "Total Maximum Daily Load in mg/l". For an Area field, you might enter "Feature area in sq km". If you wish to enter accuracies for fields, you can, but this is optional. The* ***Definition Source*** *describes who defined a particular attribute. If you are using Anderson landcover categories Level I or Level II, for example, you would enter it. If it is very generic or obvious, such as acres or population, just leave this part blank.*

You don't have to fill in a Definition for every field, but you should do it for anything that is not obvious and/or needs interpretation, units, and so on. Think about the information YOU would want to know if you had to use this data set.

Overview Description

This is the place to save time if you have many fields in a data set with the same type of data. For example, an agricultural table might have 20 fields of different animal counts and 30 fields of agricultural production for different crops in acres. In some cases, you might have a large table of values that are not obvious. If you are using zoning codes and the table does not include a text description, you will need to provide that information as a separate table in the database. Here you could reference the name of that table so that people know where to find out what the codes mean.

Data Set Overview

Describe the general purpose of the table, for example, "This data set lists animal counts and production of various crops by county."

Entity and Attribute Overview

Describe the main characteristics, such as animal counts in number of animals, stocking densities in cow/calf units per acre, or production in bushels. You can put them all in one entry or add multiple entries using the + button at the bottom, such as one for animal counts, one for production, and so on.

6. Distribution

This section describes whether and how the data could be ordered from a vendor/distributor, if applicable. It also is where you put your standard liability clause.

General

Provide general instructions on how to access the data, costs, if any, and what form the distribution takes.

Distributor

Enter the agency or person responsible for distributing the data and the contact information.

Standard Order Process

Enter the procedures to be followed to order or access the data.

Available Time Period

This is the time when the data will be available to the clients or public. Enter starting date of data availability and end date of availability, if known.

7. Metadata Reference

General

Provide metadata date, contact info of person or organization that created the metadata, and access and redistribution limits on metadata, if any.

Skills Reference

BASICS

General

Starting ArcMap or ArcCatalog

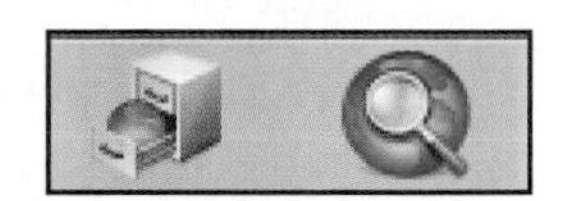

1. Look on the computer desktop or taskbar for an icon for ArcCatalog (left) or ArcMap (right). Double-click it to start the program.
2. If no icon is present on the desktop, click the Start button on the computer's menu bar. Navigate to All Programs > ArcGIS and choose the name of the program desired.

3. From ArcCatalog, launch ArcMap by clicking the ArcMap button in the menu bar.

4. From ArcMap, launch the ArcCatalog window, which sits inside ArcMap and provides access to some of the ArcCatalog functions but does not open the standalone program.

Managing windows

Both ArcMap and ArcGIS have numerous windows that can be opened when needed and closed or tucked away at other times. Windows may float above the main program window, or it may be docked in one of several docking locations. When docked, they may also be pinned, which causes them to open when the mouse is placed on them and to automatically hide when not in use. A window will generally return to the place where it was previously docked or pinned after it is closed and reopened.

Docking a window

Docking is lengthy to explain because there are so many options, but it is really relatively simple. Watch the video, or experiment until you are comfortable using the docking functions.

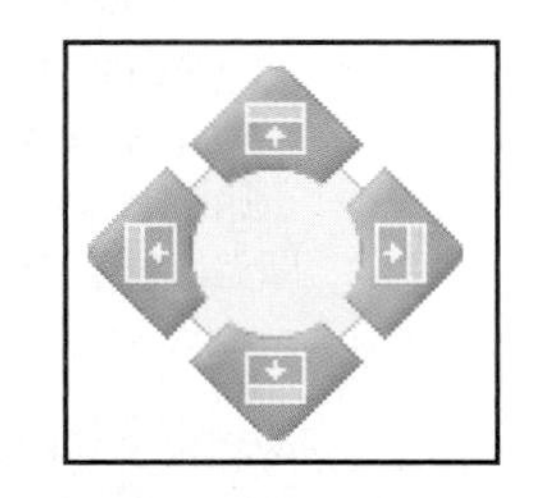

1. Click on the bar at the top of the window and drag it to one of the blue docking icons that will automatically appear. One icon will turn darker blue to show in which location the window will dock when you release the mouse.
2. If a window is already docked at the location, another set of four docking icons will appear next to the original. The second set shows where the new window will be docked relative to the existing window (left, right, below, or above) on that side of the main window. If the window is dragged to the center icon, then both old and new windows will be placed in the same spot with tabs underneath for switching between them.

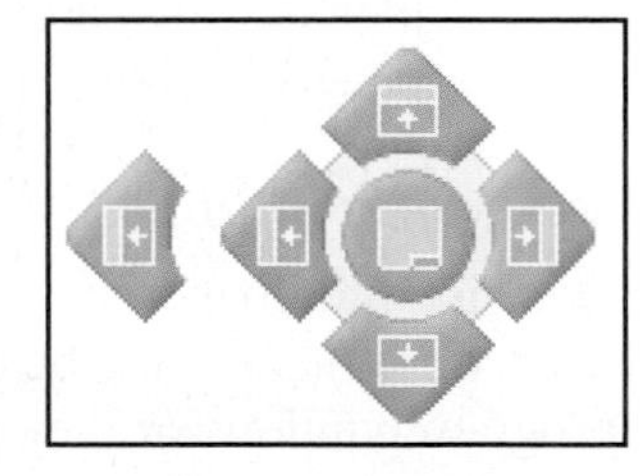

3. To undock a window, click on its top bar and drag it away from the docked location. Dock it somewhere else.
4. To make a window float again, click its top bar, drag it off any of the docking icons, and release the mouse.

Auto Hide a window

A docked window will have an Auto Hide button on its bar.

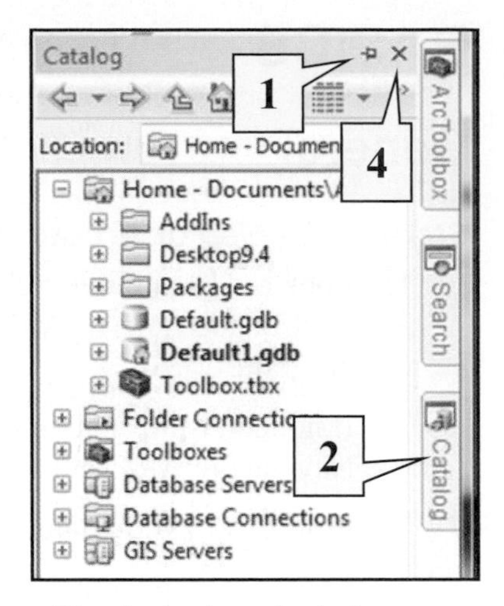

Fig. 1. A pinned window

1. Click the Auto Hide button to pin the window to the docking location (Fig. 1). It will become a tab on that side of the main window.
2. To access the window, hover the cursor on top of the tab. The window should open automatically, but you can click it if it is stubborn.
3. When you move the cursor off the window, it will automatically hide in the tab. Sometimes you must actually click elsewhere before it hides.
4. Click the X on the window bar to close the window and remove the tab.

5. Click the pin again to unpin the window. It will remain open in the docked location.

Adjusting window sizes

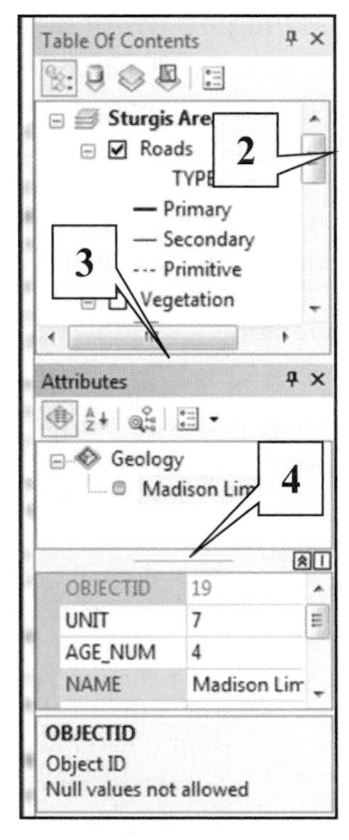

Fig. 2. Adjusting windows

1. The size of a floating window can be adjusted by clicking and dragging the corners or edges to the desired size.
2. The width of a docked toolbar can be adjusted by placing the cursor on the edge until a double-headed arrow appears (Fig. 2). Click and drag the edge to the desired width.
3. The height of a docked toolbar matches the window edge available. However, if two or more windows are docked above/below each other, then the relative width may be adjusted by placing the cursor on the edge between the windows. When the cursor changes to the double-headed arrow, click and drag the edge to the desired position.
4. Some windows have multiple panels, each of which is individually adjustable. Look for the gray line in between the panels. Click and drag it to adjust the relative size of the panels.

Managing toolbars

Dozens of toolbars are available and may be opened and closed as needed. Toolbars may be docked on any of the four edges of the program window, or they may float above the window (Fig. 3).

1. To move a floating toolbar, click and drag on the bar at the top to move it to a new location. To dock it, move it to one of the edges of the program and it will dock when it finds an open spot.

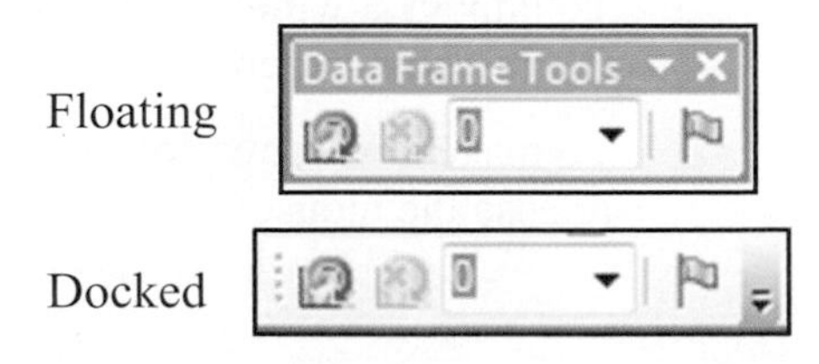

Fig. 3. Toolbars

2. To move a docked toolbar, click on its handle (the dashed gray line on its left edge) and drag it to the desired position elsewhere on the GUI.
3. To move a docked toolbar off the GUI, use the handle to drag it off the GUI. It then becomes a floating toolbar. It can be docked again by dragging it to a location on the GUI.
4. To open a toolbar that is not present, choose Customize > Toolbars from the main menu and select the desired toolbar.
5. To close a floating toolbar, click on the X box in its upper-right corner.
6. To close a docked toolbar, right-click on a toolbar or empty space and choose the toolbar to close from the list.
7. To find out the name of a tool or button, hold the cursor on top of it for a moment without clicking, and the name will pop up.

Creating/deleting folder connections

In order to access data files, you must set up a connection to the folder in which they reside. Connections may be established from ArcCatalog or ArcMap and will be available to both programs once created. Connections can be deleted when they are no longer in use.

The Connect to Folder button may be found on the Standard toolbar in ArcCatalog, on the Catalog tab in ArcMap, or on the Add Data window in ArcMap.

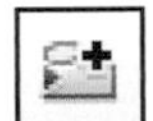

1. To connect to a folder, click on the Connect to Folder button.
2. Navigate down the directory tree and click on the folder (or drive letter) to connect to, and click OK (Fig. 4).
3. The connections are stored in ArcCatalog in an entry called Folder Connections.
4. To delete a folder connection, right-click the folder connection in ArcCatalog or in the Catalog tab and choose Disconnect Folder. This step only removes the connection; it does not delete the data from the disk.

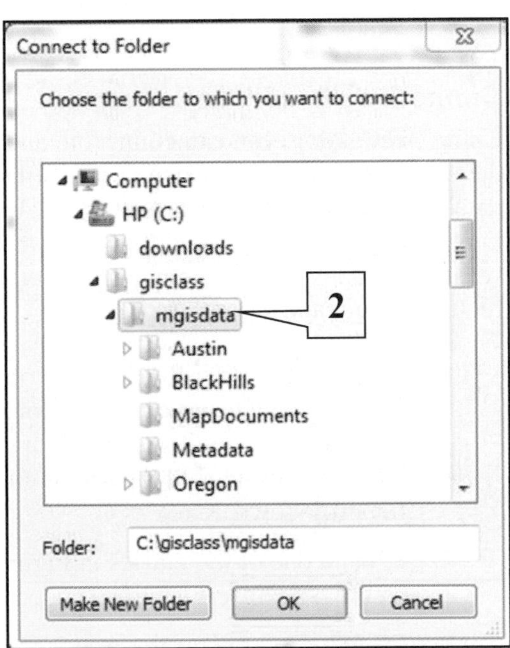

Fig. 4. Connecting to a folder

Setting options

You can control the way the programs perform various actions, display information, and other options. Each program has different options to set. Specific settings for various tasks are discussed under the appropriate headings.

ArcCatalog Options

1. Click Customize >ArcCatalog Options from the main menu bar to access the following option tabs: General, File Types, Contents, Connections, Metadata, Tables, Raster, CAD.
2. Click Geoprocessing > Geoprocessing Options from the main menu bar to access options such as background processing, overwriting of existing files, and how long results are kept.

ArcMap Options

1. Click Customize >ArcMap Options from the main menu bar to access the following option tabs: General, Data View, Layout View, Metadata, Tables, Raster, CAD, Display Cache.
2. Click Geoprocessing > Geoprocessing Options from the main menu bar to access options such as background processing, overwriting of existing files, and how long results are kept.

Connecting to an Internet service

Using ArcCatalog, you can connect to and access data from external databases or from spatial database services. A Database Server or Database Connection provides access to a tabular database such as Oracle. An ArcGIS or ArcIMS Server provides access to spatial data. You must know the type of service and the URL, such as www.geographynetwork.com. Secure services will also require a login and password.

1. In ArcCatalog, scroll to the bottom of the Catalog Tree to find the server connections (Fig. 5). In ArcMap, open the Catalog tab or click the Add Data button and find the server connections in the file drop-down.
2. Expand one of the entries in Figure 5, depending on the type of service to be added
3. Double-click one of the Add icons.
4. Type in the URL of the service in the top box.
5. If the server requires a user login and password, enter these at the bottom of the window and click OK.
6. Predefined connections are provided to ArcGIS Online and Geography Network.

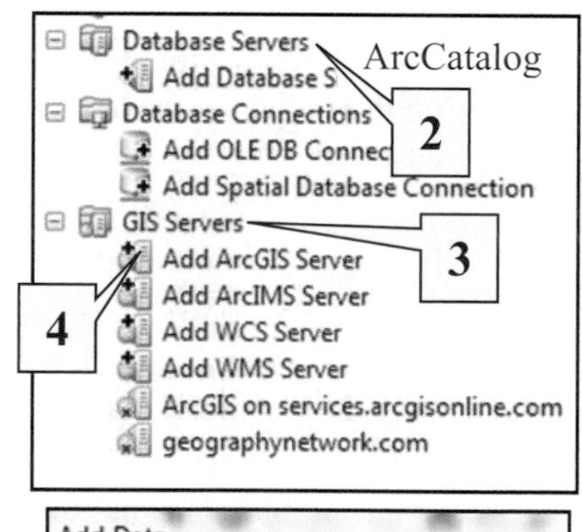

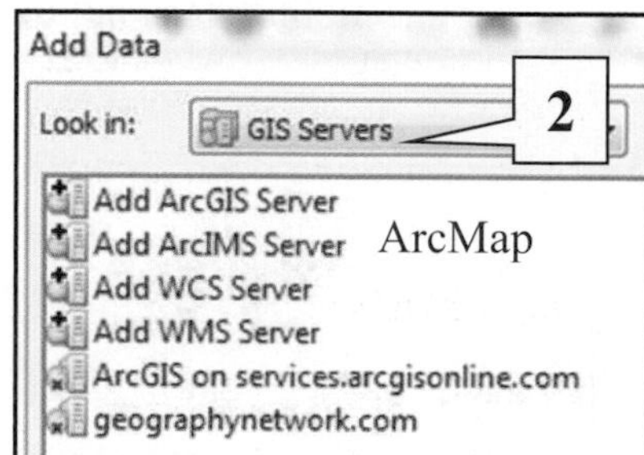

Fig. 5. Connecting to a service

Opening/saving data sets

Opening a data set

1. Click the Browse button to access the Input/Output data window (Fig. 6). A few buttons are different between ArcMap (shown) and ArcCatalog.
2. Click the drop-down to pick from a list of other folders or connections.
3. Click the Up One Level button to move up a folder.
4. Click the Go To Home Folder button to go where the map document is stored.
5. Click the Default Geodatabase button to navigate to the default geodatabase.
6. Click the Contents View Type button to change how the listed files appear: as large icons, small icons, detailed list, and so on.
7. Click the Folder Connections button to add a new folder connection.
8. Click the New Folder button to create a new folder in the current folder.
9. Click the New File Geodatabase button to create a new geodatabase in the current folder.
10. Click the New Toolbox button to create a new toolbox in the current folder.
11. Change the Show of type: box if the desired type of data is not being displayed.
12. Click the desired data set to highlight it, or type its name in the Name box, and click Add.

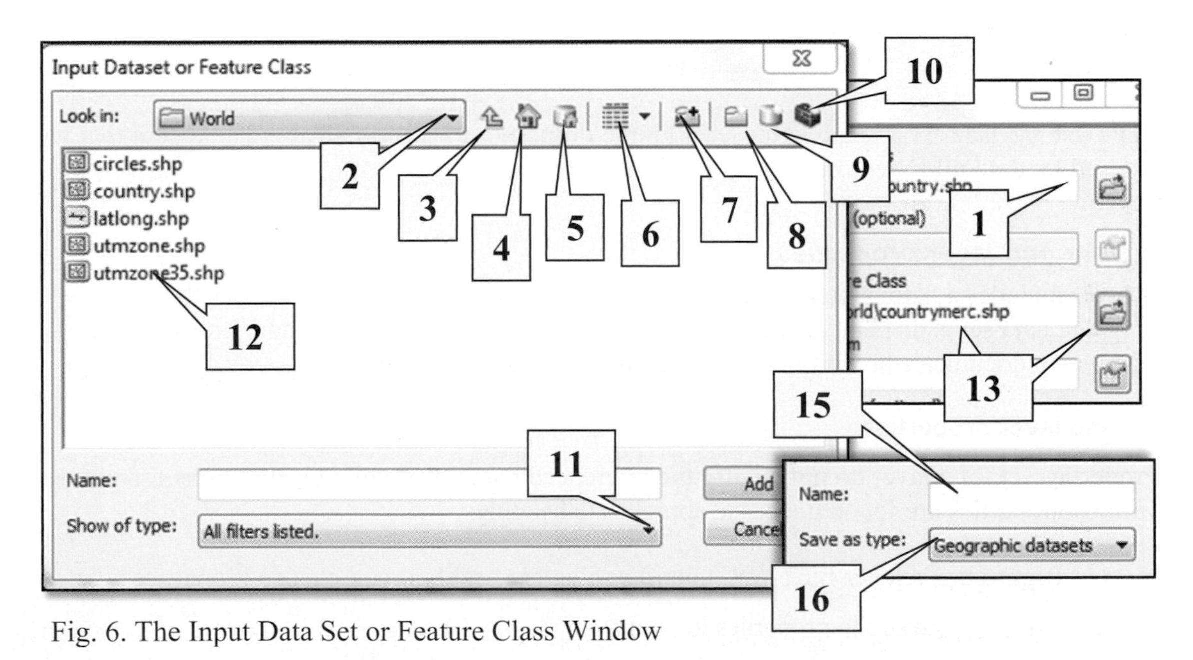

Fig. 6. The Input Data Set or Feature Class Window

Saving a data set

13. Type or edit the pathname of the data set to be saved, or click the Browse button.
14. Use buttons 2–10, if necessary, to navigate to or create the space for the file to be saved.
15. Type the name for the data set in the Name box. Extensions usually aren't needed.
16. Change the Save as type: box to change the type of data set saved, if necessary.

Using the Zoom/Pan tools

The Zoom/Pan tools are available in Preview mode in ArcCatalog on the Geography toolbar, and all are available in ArcMap on the Tools toolbar.

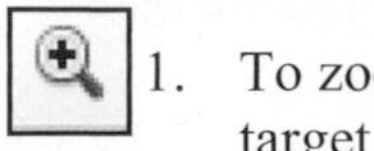

1. To zoom in to an area, click the Zoom In tool. Position the cursor to the upper left of the target area and then click and drag a box to encompass the area desired. Release the mouse button to finish.

2. To zoom out, click the Zoom Out tool and click once at the center zoom point. Draw a large box to zoom out slightly or a small box to zoom out a great deal.

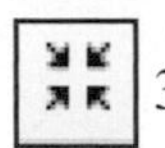

3. To zoom in by a fixed amount while remaining centered within the current extent, click the Fixed Zoom In tool (ArcMap only).

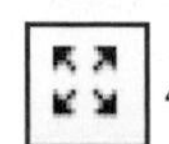

4. To zoom out by a fixed amount while remaining centered within the current extent, click the Fixed Zoom Out tool (ArcMap only).

5. To move the map around in the window, click the Pan tool, click on the map, and drag it to the desired area.

6. Click the Previous Extent button to return to the extent just before the current one. This one is useful if you make a mistake when panning or zooming.

7. Click the Next Extent button to return to the extent after the current one (this button is dimmed unless the Previous Extent button has been clicked).

8. Click the Full Extent button to view the full area of all the layers, or to an extent specified by the user.

TIP: Use the Data Frame tab in the Data Frame Properties to specify an alternate extent that will be used by the Full Extent button.

Layers and layer properties

A layer includes a pathname reference to a spatial data set plus various properties that control how that data set is displayed and used. In ArcMap, layers are held in memory and saved as part of the map document. In ArcCatalog, layers are stored in layer files.

Setting layer properties

Properties set for a layer do not modify the referenced data set. Detailed instructions for setting various properties are found under the appropriate headings.

1. Right-click a layer file (ArcCatalog) or a layer (ArcMap) and choose Properties.
2. Click the tab of the properties to be managed (Fig. 7).
3. Set the properties and click Apply to set the properties and leave the window open, or OK to apply the settings and close the window.

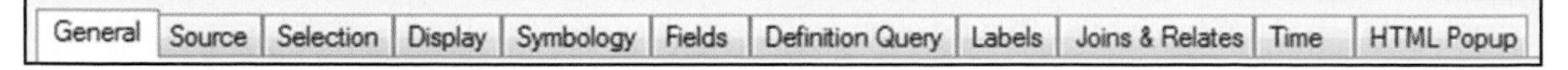

Fig. 7. Tabs for layer properties

Creating a layer file in ArcCatalog

1. Right-click the reference feature class and choose Create Layer.
2. Specify the name and location of the file to be saved. The .lyr extension will be added if necessary. Click Save.
3. Set the properties of the layer file.

Creating a layer file in ArcMap

1. Set the symbology, labeling, and other layer properties of the data or group layer.
2. Right-click the data layer name in the Table of Contents and choose Save As Layer File from the context menu.

3. Specify the name and location of the file to be stored.

Group layers and layer properties

Creating a group layer file in ArcCatalog

1. In the Catalog Tree, click the folder or geodatabase containing at least two of the feature classes to be part of the group layer.
2. Click the Contents tab to show the data in the folder. Hold down the Ctrl key and click two or more feature classes that will be included in the group layer to highlight them all.
3. Right-click one of the highlighted feature classes and choose Create Layer.
4. Specify the name and location of the file to be saved. The .lyr extension will be added.

Creating a group layer in ArcMap

1. Set the symbols, labels, and other display properties of all the layers to be included in the group. Set the Table of Contents to Display by Draw Order.
2. Click the first group member in the Table of Contents to highlight it. Use Ctrl-click to also highlight the other layers in the group (Fig. 8).
3. Right-click one of the group members and choose Group from the context menu.
4. Click on the group name to highlight it and then click it again to type in a name for it.
5. Right-click the group name and choose Save As Layer File from the context menu.
6. Enter the location and file name of the group layer and click Save.

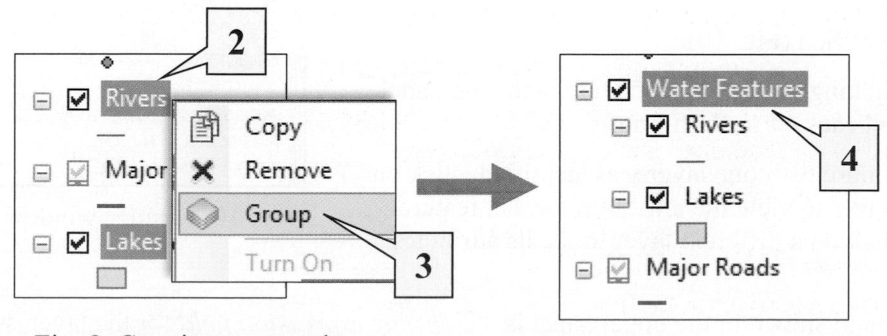

Fig. 8. Creating a group layer

Managing group layer properties

1. Right-click the new group layer (ArcMap) or layer file (ArcCatalog) to open its properties, and click the Group tab (Fig. 9).
2. Click to highlight one of the feature classes and choose Properties to open and set its layer properties.
3. Click Add to add a new layer to the group.
4. Highlight the layer and choose Remove to remove it from the group layer (the file will remain on the disk).
5. Use the Up/Down arrows to change the draw order of the layers. Drawing always starts with the bottom layer.
6. Use the General and Display tabs to set properties for the entire group, such as a scale range or transparency.

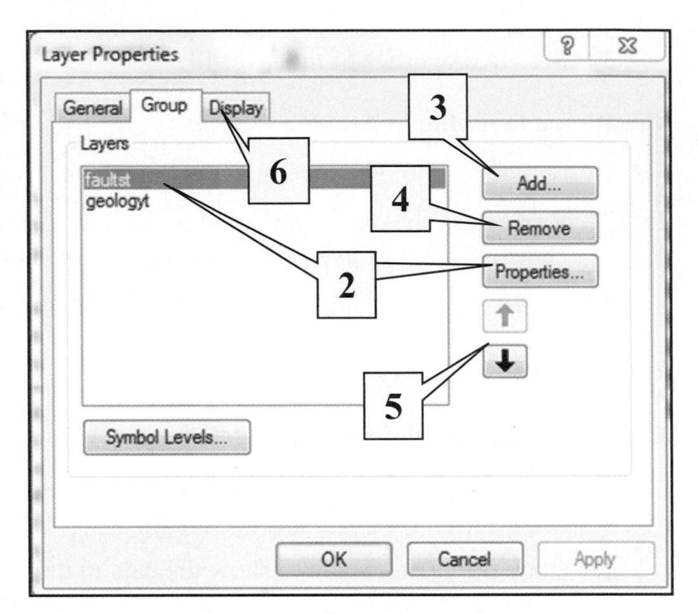

Fig. 9. Managing group layer properties

Identifying features

Identify is used to quickly view attribute information for one or several features. The windows are slightly different in ArcMap and ArcCatalog but they function much the same way. You must be using the Preview tab to use Identify in ArcCatalog.

1. Click on the Identify tool on the Geography toolbar (ArcCatalog) or the Tools toolbar (ArcMap). The Identify window will open.
2. Click on the map feature or raster cell to identify.
3. If more than one layer is present in the map, you can set which layer or layers will be identified (Fig. 10).
4. Clicking an identified feature in the top panel will cause it to flash briefly.
5. If more than one layer was identified, click on a layer to view the attributes for that feature. Click on a different layer to see its attributes.

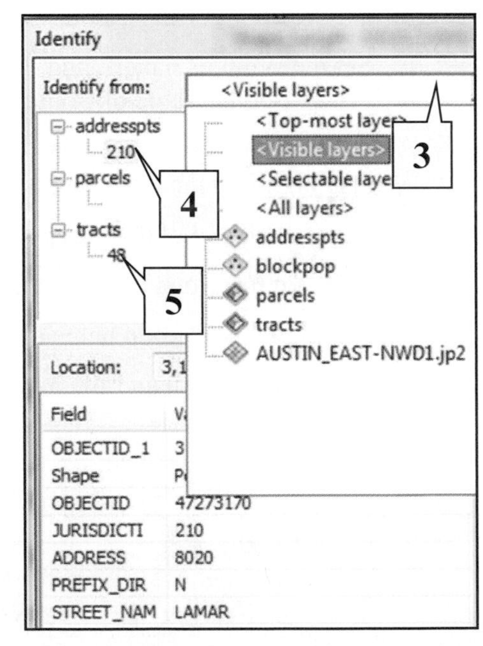

Fig. 10. Identify window (ArcMap)

TIP: The field shown in the upper panel is the *display expression field* for the layer, which can be changed in the layer properties on the Display tab.

Finding features

1. In ArcMap, click the Find tool on the Tools toolbar or in the Table Options menu. In ArcCatalog, choose Find from the Table Options menu when previewing a table.
2. Type the search text in the first box (Fig. 11a).
3. Choose the layer or set of layers to be searched.
4. Check the box for an exact match or for features that are similar to or contain the string.
5. Choose to search all fields, or set the field to search.
6. Click Find to begin the search. Records that match will be listed below.
7. Right-click one of the found records to display a context menu with options for flashing, zooming to, identifying, bookmarking, or selecting the found feature (Fig. 11b).

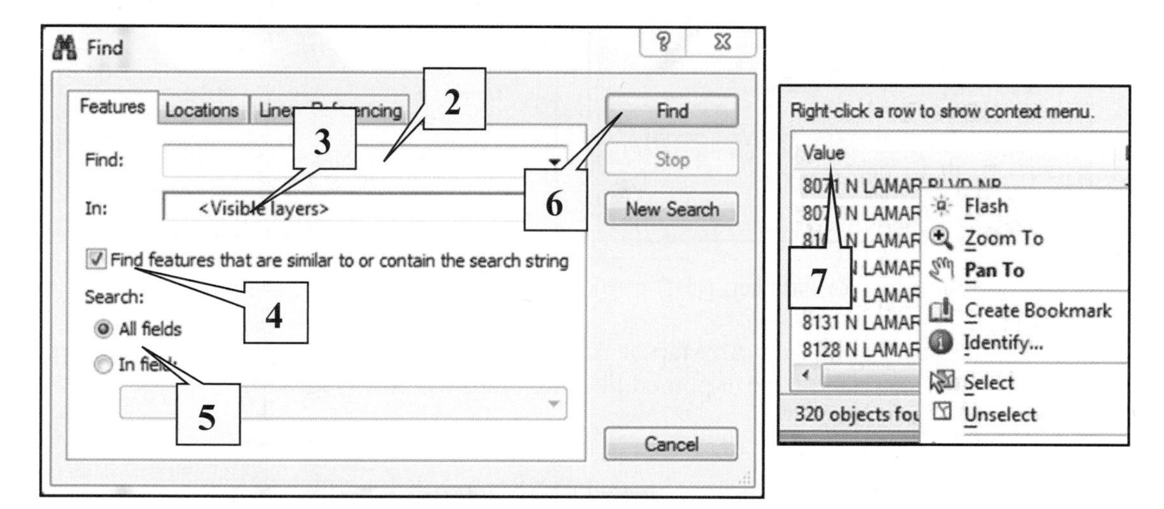

Fig. 11a. Finding features

Fig. 11b. The Find list

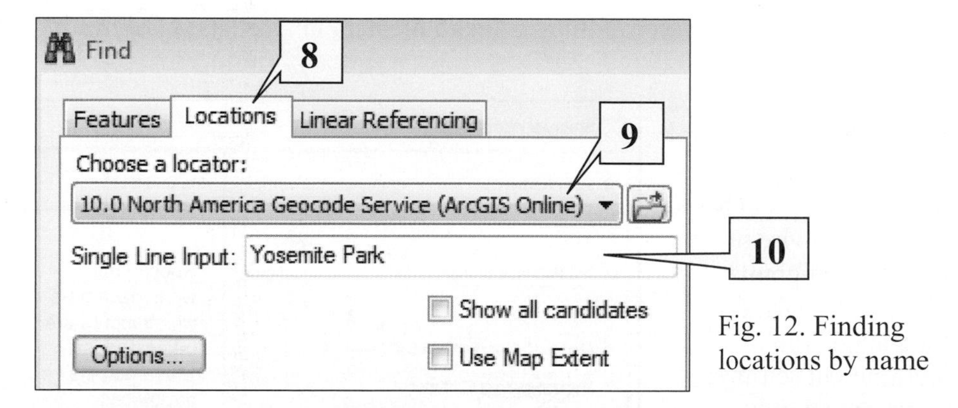

Fig. 12. Finding locations by name

8. Click the Locations tab to find a location based on a place name (Fig. 12).
9. Select one of the locator services provided by ArcGIS Online. You must have an Internet connection.
10. Type the name of the location and click Find.

Using the Search window

The same search window can be used to look for map documents, data sets, and tools. It can also be set to search the local drives only, or assigned to search a server or other Internet service.

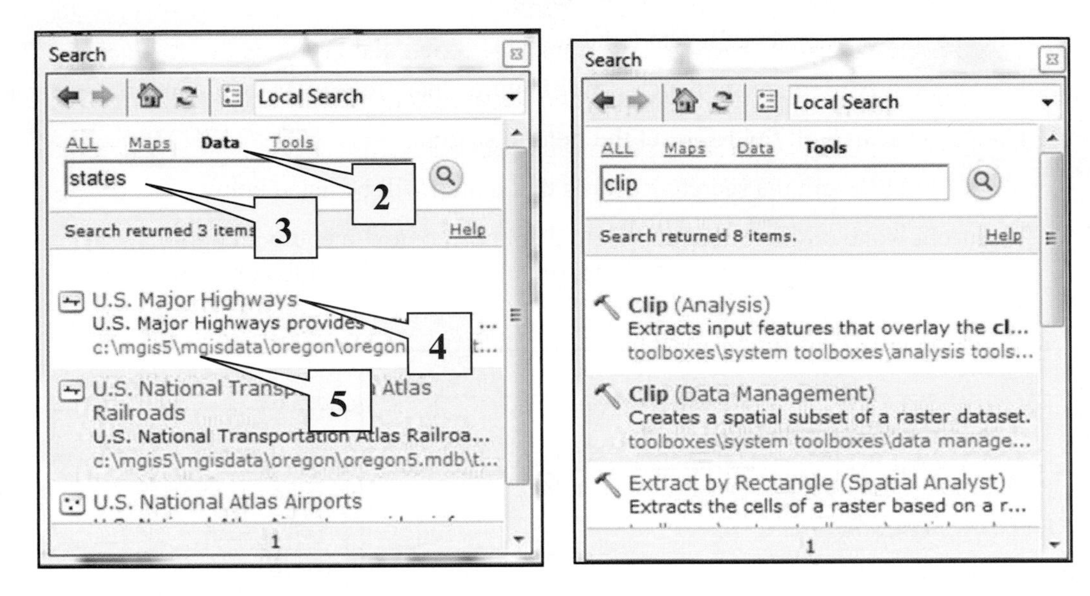

Fig. 13. Searching for data sets (left) or tools (right)

1. Click the Search button in ArcMap or ArcCatalog from the Standard toolbar. Once open, the window can be docked or pinned like any other window (Fig. 13).
2. Click the type of information to search for: All, Maps, Data, or Tools.
3. Type the search string indicating the name or key words for which to search.
4. Hover over blue text to find out more about the item or to view its item description.
5. Click on the green underlined text to follow a link to the item in ArcCatalog or the Catalog tab.

Environment settings

Environment settings control the operation of tools and functions in both ArcMap and ArcCatalog. Environment settings may be set at the program level, where they affect all tools, or they may be set for a single run of a particular tool. Environment settings are organized into functional groups, some of which are shown in Figure 14.

Use ArcCatalog to specify environments that will be used as defaults by all tools and map documents. Environments changed within a map document in ArcMap will be saved with that map document.

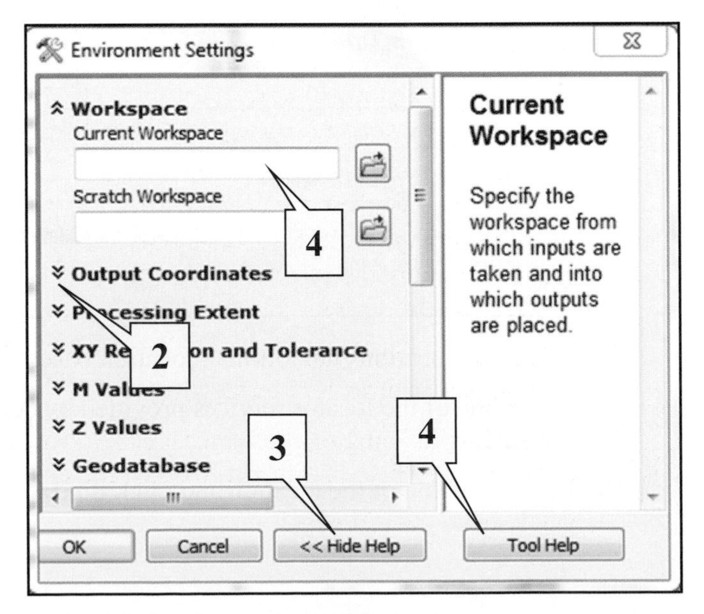

Fig. 14. Environment settings

Environments set for a tool will temporarily override the Catalog and map document settings but will cease as soon as the tool is finished.

1. In ArcCatalog or ArcMap, choose Geoprocessing > Environments from the main menu bar. In a tool, click the Environments button.
2. Use the double-arrow switch to open/close the functional group you want to set.
3. Click Show/Hide Help to get information about settings.
4. Click in a setting box to find information about that setting, and Tool Help for more detailed information about that setting.

Set the desired settings; leave the others blank or set to their defaults. Click OK.

Starting and using ArcToolbox

ArcToolbox is a dockable window that sits inside ArcCatalog or ArcMap.

1. Click on the ArcToolbox icon in either ArcCatalog or ArcMap to open the window.
2. After being opened, it may be docked or pinned like any other window.

Using a tool

3. Double-click the tool in ArcToolbox to open it (Fig. 15).
4. If using a tool for the first time, click the Show Help button to read about it.
5. With the Help showing, click on a parameter box to get more information about it.
6. Click the Tool Help button to get full help information about the tool.
7. Green dots indicate a required parameter.
8. Input/output features may be set by clicking the Browse button and locating the spatial data set, or by making a choice from the drop-down list (ArcMap only). If the drop-down list is used, only the selected features will be used in the tool.
9. Click the Environments button to change the Environment settings.

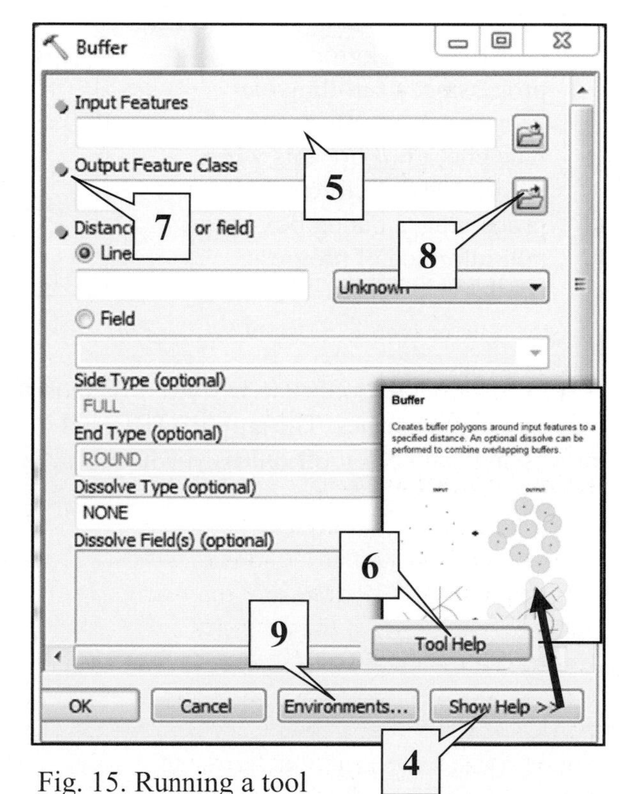

Fig. 15. Running a tool

10. If a yellow warning or a red error icon appears after entering a parameter, place the cursor over it to see the message about the error or potential error.
11. Click OK to start the tool.

TIP: Tools may be run in background (default) or foreground (while you wait). Use the Geoprocessing Options in ArcMap or ArcCatalog to enable/disable background processing.

12. If running in background, a blue scrolling status bar will appear at the bottom of the program (Fig. 16), and a notification will appear to indicate that the tool has finished.
13. In foreground processing, a status window will appear with messages indicating progress. Close the window when it is finished, or check the box to close the window automatically on successful completion. If errors occur, the box will remain open so you can read the messages.

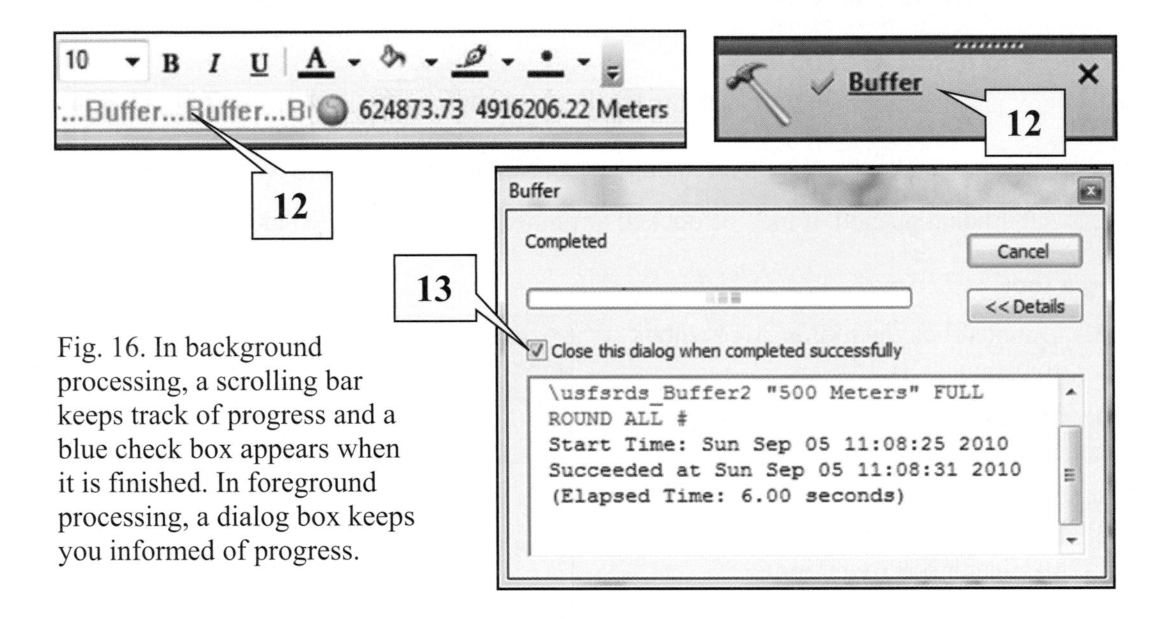

Fig. 16. In background processing, a scrolling bar keeps track of progress and a blue check box appears when it is finished. In foreground processing, a dialog box keeps you informed of progress.

TIP: Background processing is slower for short tasks than foreground processing, and it is more likely to cause crashes. The author STRONGLY recommends that you turn off background processing and keep it off unless you have a need for it.

ArcMap Basics

Opening an ArcMap document on start-up

Start ArcMap using the methods described in ***Starting ArcMap or ArcCatalog***.

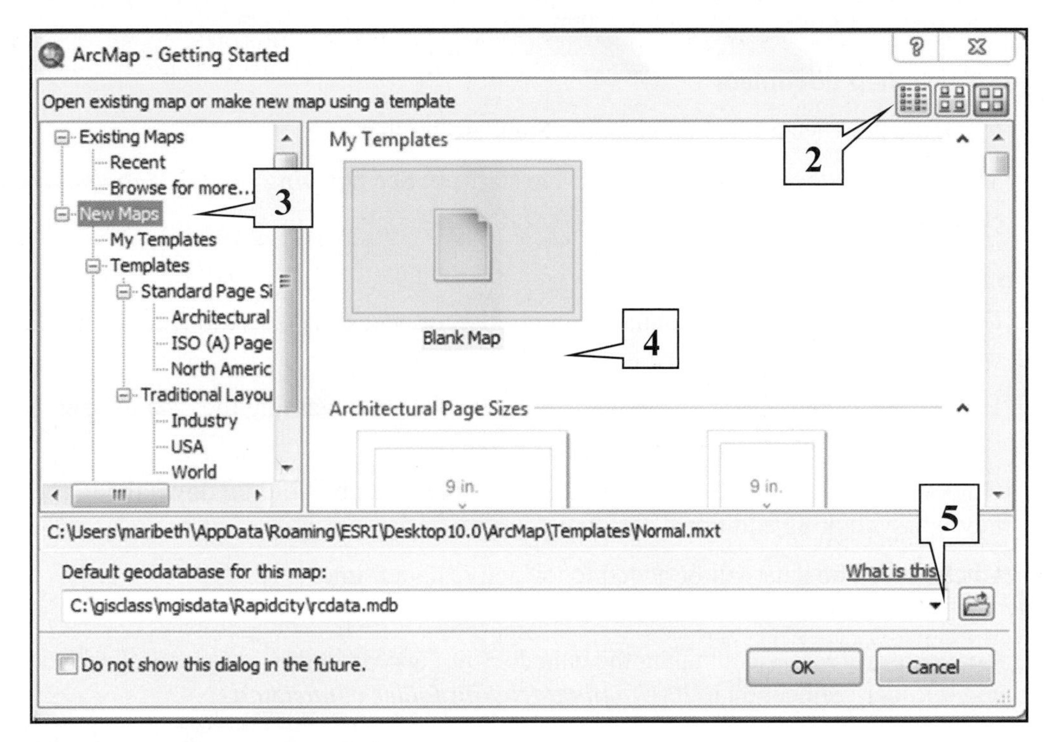

Fig. 17. ArcMap start-up dialog

1. As ArcMap starts, it displays a dialog screen (Fig. 17). Choose to start with a blank map, open a map template, or open an existing map.
2. Click to change the appearance of the templates as a list, small icons, or large icons.

To open a new map

3. Click the New Maps text, or on one of the template headings below it.
4. Click on the Blank Map icon under My Templates, or scroll down and choose one of the predefined template sizes or layouts.
5. Click the Browse button to assign a default geodatabase for the map document if you want it to have a home location to look for and save files. The drop-down shows a list of recently used geodatabases.
6. Click OK.

Switching to a different map document

To open an existing map document

1. Click the Open button on the Standard toolbar.
2. In the Open window, navigate to the right folder and double-click the desired map document, or click it and choose Open.

To open a new map document

1. Click the New Map File button on the Standard toolbar.
2. The same window appears as at ArcMap start-up. See ***Opening an ArcMap document on start-up***.

Adding data

1. Click the Add Data button on the main toolbar, or choose File > Add Data from the main menu bar.
2. Use the Add Data dialog box to navigate to the folder containing the data to open (see ***Opening/Saving data sets***).
3. Click on the data item to select it. Select multiple items by holding down the Ctrl or Shift key while clicking additional data sets.
4. Click Add. The data will be added to the active data frame.

TIP: If the disk or directory containing the data does not appear in the list, you probably need to add a folder connection to it (***Creating/deleting folder connections***).

Removing data

Removing data removes them from the map document but does not delete them from the disk.

1. Right-click the data item in the Table of Contents and choose Remove.

Saving a map document

1. Click on the Save button, or choose File > Save from the menu bar.
2. If the document has not previously been saved, a dialog box appears for the user to select a location and a name for the map document. Navigate to a directory in which to save it and then enter a name for the document.
3. Choose File > Save As from the menu bar to save the map document under a new name, leaving the original version unchanged. Specify the location and name as for Step 2.

TIP: The names of map documents, spatial data sets, and folders should not contain spaces or special characters such as #, @, &, *. Use the underscore character _ to create spaces in names, if necessary. Although ArcCatalog and ArcMap allow spaces in names, spaces can create problems for certain advanced functions. It is wise to avoid them entirely.

The Table of Contents window

The Table of Contents controls many of the display and other functions available in the program. It contains the layers and data frames that make up the map document. It offers four different views of the data, which are controlled by a set of icons on the top of the window. Each view has different settings and different options in the context menus. A few of the more common tasks are described below for each view.

List By Drawing Order

List By Drawing Order lists the layers according to the order in which they are drawn, from bottom to top (Fig. 18). It is the default view.

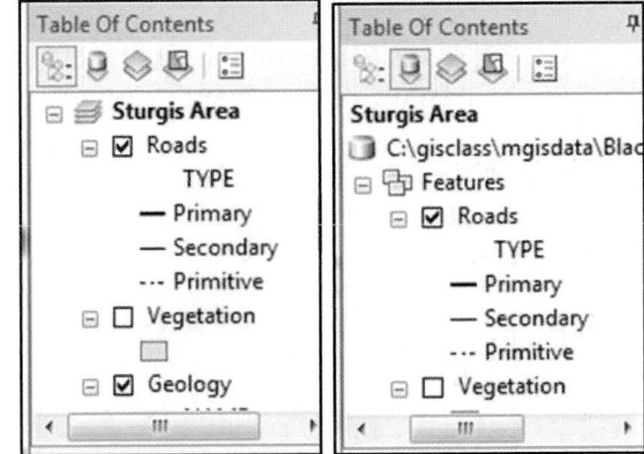

Fig. 18. Table of Contents views

1. To turn a layer on/off, click the check box.
2. Right-click an item name to open a context menu. The menu choices depend on the type of item clicked.
3. Double-click the item name to open its properties, or right-click it and choose Properties from the context menu.
4. Click twice slowly on a layer or legend item to rename it. Click Enter when done.
5. Hide the legend for a data layer by clicking the minus sign in the box next to the layer. Click the plus sign to expand it again.
6. Click on a layer symbol to change the symbol using the Symbol Selector.
7. Right-click on a layer symbol to change the color of the symbol.
8. Right-click a layer and choose Copy to make a copy of it. Then right-click one of the data frames and choose Paste Layer(s) to paste the copy into a data frame.
9. Right-click a layer and choose Remove to remove it from the map.
10. To change the draw order, click and hold on the layer name and drag it higher or lower in the list. Release the mouse button to finish.

TIP: If the check box is dimmed, then the current map scale is outside the display range set for the layer. Try zooming to another scale or reset the scale range in the layer properties.

TIP: If the layers don't seem to go anywhere when you drag them, check to make sure that the List By Draw Order view is selected at the top of the Table of Contents.

List By Source

List By Source organizes layers and tables according to the folder or geodatabase that contains them (Fig. 18). The full pathname of the folder is followed by the data items from that folder. This is the only view in which standalone tables can be accessed. Tasks 1–9 listed for the List By Drawing Order view also work in List By Source view.

List By Visibility

List By Visibility groups layers according to whether they are visible or not (Fig. 19). It also allows you to view groups of features that have the same symbol.

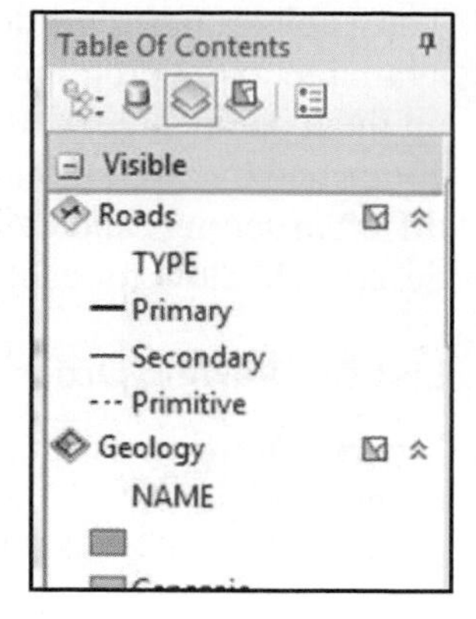

Fig. 19. List By Visibility

1. Click the icon to make a layer visible or not visible.

2. Click the Selection toggle to make a layer selectable or not.

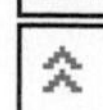
3. Click the double-arrow to expand or collapse the categories/classes of a layer.
4. Click a category or class and the corresponding features will be flashed on the screen.
5. Right-click a layer to open a context menu.

List By Selection

List By Selection provides tools for working with queries (Fig. 20).

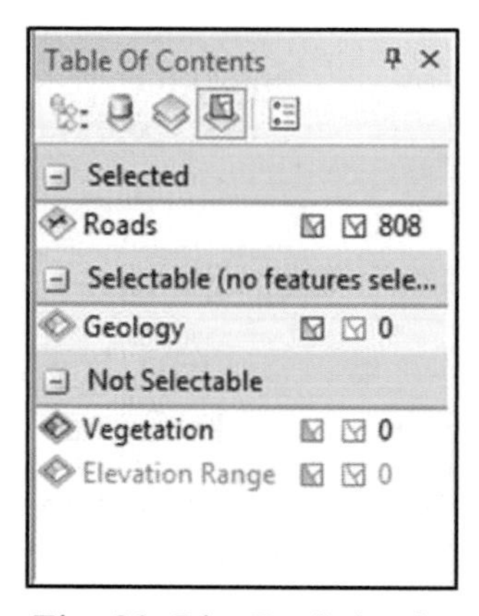

Fig. 20. List By Selection

1. Click the Visibility icon to make a layer visible or not.

2. Click the toggle switch to make a layer selectable or not.

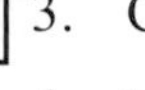
3. Click the clear button to clear the selected features.
4. Right-click a layer to open a context menu.

Setting a scale range

The visible scale range controls at which scales a layer will be drawn.

1. Right-click the layer name in the Table of Contents and choose Properties.
2. Click the General tab.
3. Fill the button to *Don't show layer when zoomed* (Fig. 21).
4. Select one of the scales using a drop-down, or type the scale denominator in the box. Set a minimum scale, maximum scale, or both.
5. To clear a scale range, fill the button to *Show layer at all scales*.

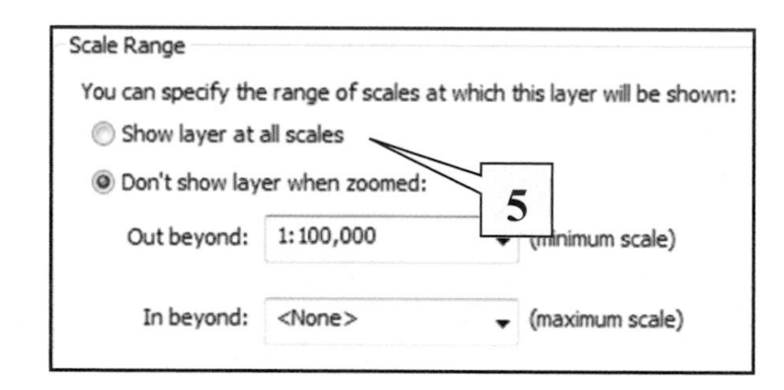

Fig. 21. Setting the scale range

TIP: You can also set or clear the scale ranges by right-clicking the layer in the Table of Contents and choosing Visible Scale Range.

Using bookmarks

A bookmark lets you return to a predefined map extent many times.

Fig. 22. Zooming to a bookmark

1. To set a **bookmark**, zoom to the desired area and select Bookmarks > Create from the main menu. Type in a name for the bookmark.
2. To zoom to an existing bookmark, click Bookmarks and choose the desired bookmark (Fig. 22).
3. To delete an existing bookmark, choose Bookmarks > Manage. Select the bookmark name by clicking on it, and then click the Remove button.
4. To zoom to the reference scale, right-click the frame name and choose Reference Scale > Zoom to Reference Scale from the menu.

Measuring features

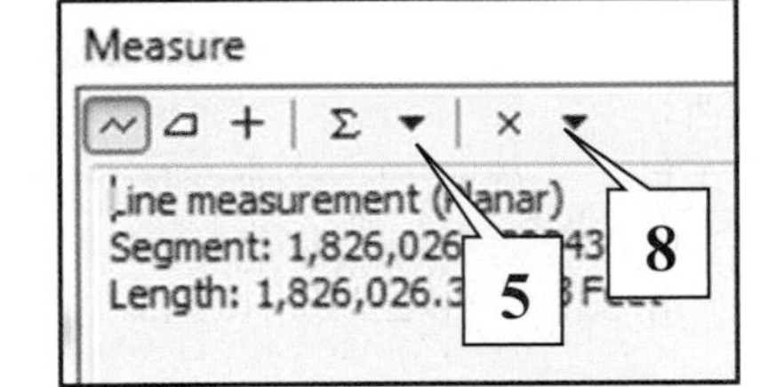

Fig. 23. The Measure tool

1. Click the Measure tool (Fig. 23). By default it measures distance.
2. Click on the map to start the line. Click along the path to be measured. Double-click to end the line. Both total length and the length of the last segment are shown.
3. Click the Measure Area button to measure a polygon area. Click to create vertices for the polygon and double-click when finished.
4. Click the Measure Feature button, and click on a feature to get its measurement.
5. Use the Choose Units button to set the units for length and area.
6. Use the Show Total button to keep a running count of all features measured.
7. Use the Clear/Reset Results button to clear running totals and reset the measurements.
8. Use the Measurement Type button to change the type of distance measure from Planar (the default).

TIP: Snapping will normally be on and will snap to features when measuring. Use Snapping > Snapping Options on the Editor toolbar to turn it off.

Setting Map Tips

1. Right-click the layer name in the Table of Contents and choose Properties from the context menu.

Display Expression
Field: NAME
Show MapTips using the display expression

2. Click the Display tab and check the box to *Show Map Tips using the display expression*.
3. Change the Field box to the desired attribute field to show in the Map Tips.
4. Uncheck the box to turn Map Tips off.

Setting data frame properties

1. Right-click the data frame name and choose Properties from the context menu.
2. Click the tab containing the properties to set (Fig. 24).
3. Enter the appropriate information and settings.

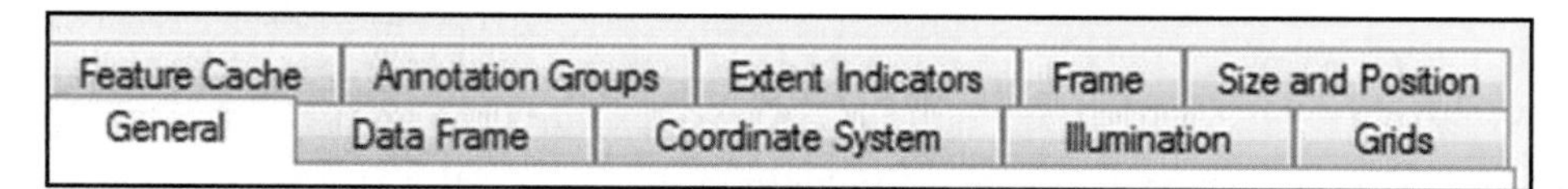

Fig. 24. Data frame properties tabs

ArcCatalog Basics

The Catalog Tree

The Catalog Tree is a dockable window in ArcCatalog that lists folders and other items in the catalog, such as database connections and toolboxes (Fig. 25). Many types of items are visible in the Catalog Tree, and the options and actions of the Catalog depend on the type of item.

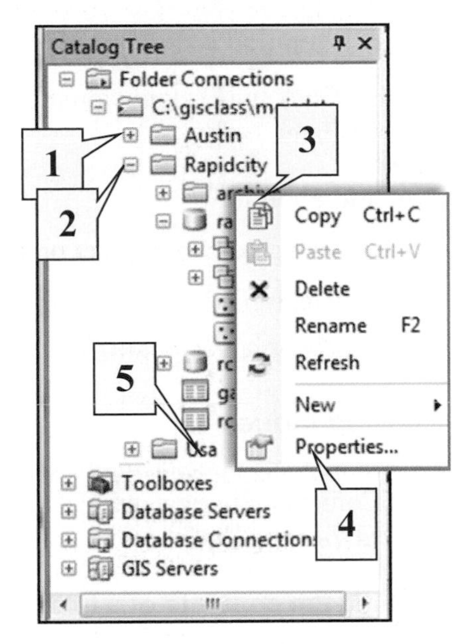

Fig. 25. The Catalog Tree

1. Click the + box to expand the contents of an item, such as a folder or geodatabase.
2. Click the – box to collapse the contents of an item.
3. Right-click an item to open a context menu with various actions. The list of actions will vary depending on the type of item.
4. Right-click an item and choose Properties to view or set its properties. Double-clicking an item is a shortcut to opening its properties.
5. To rename an item, click twice slowly on the item name. It will be highlighted and allow you to type or edit the name. Click Enter when done.

TIP: You cannot select multiple items in the Catalog Tree, such as for copying or deleting. Use the Contents tab for this purpose.

TIP: You must establish folder connections before you can see hard drives with GIS data.

Managing ArcCatalog items

Most of these options can be performed in ArcCatalog or on the Catalog tab in ArcMap. ArcToolbox also contains tools for accomplishing these tasks.

Creating new items

1. Right-click on the folder or geodatabase to contain the new item.
2. Choose New > and the type of item to create (Fig. 26). The options will depend on what item was right-clicked.
3. A window or wizard will appear. Follow the instructions for creating the item. Different items require different parameters.

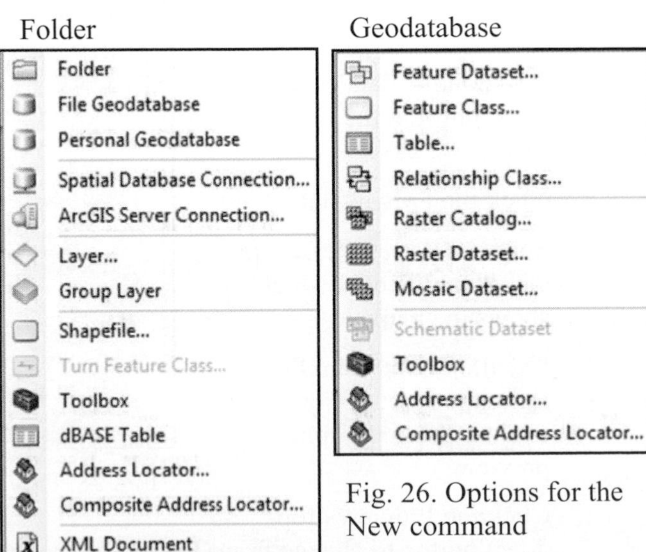

Fig. 26. Options for the New command

Copy and paste items

1. Right-click the object to be copied and choose Copy.
2. Right-click the folder or geodatabase to place the object in and choose Paste.

Renaming files

1. Click twice slowly on the item name to highlight it. Type in the new name and click Enter.

Deleting an item

1. Navigate to the folder or geodatabase containing the item to be deleted.
2. Right-click the item and choose Delete.
3. Say Yes, when prompted, to confirm deleting the item.

Viewing modes

ArcCatalog has three different viewing modes for exploring data sets: Contents, Preview, and Description. Click the tab to use the desired viewing mode. Each mode is described in detail below.

Viewing the contents of a folder

1. Click on a folder or geodatabase in the Catalog Tree.
2. Click on the Contents tab in the content window.
3. Choose one of the display options from the toolbar: Large icons, List, Details, or Thumbnails (Fig. 27).

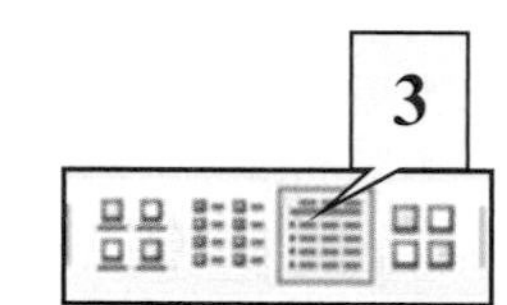

Fig. 27. Display options

Previewing data sets

Preview mode allows you to examine data to see what they look like. You can zoom/pan around them and even perform some actions on tables such as sorting. Only feature classes, rasters, layer files, and tables can be previewed. Items with multiple items inside, such as folders or geodatabases, cannot be previewed.

1. Click on the data set to highlight it in the Catalog Tree (Fig. 28).
2. Click the Preview tab.
3. Spatial data sets default to Geography mode, which shows the map.
4. Use the Zoom, Pan, Full Extent, or Identify button to explore the geography preview.
5. Click the Preview drop-down button to change from Geography mode to Table mode.

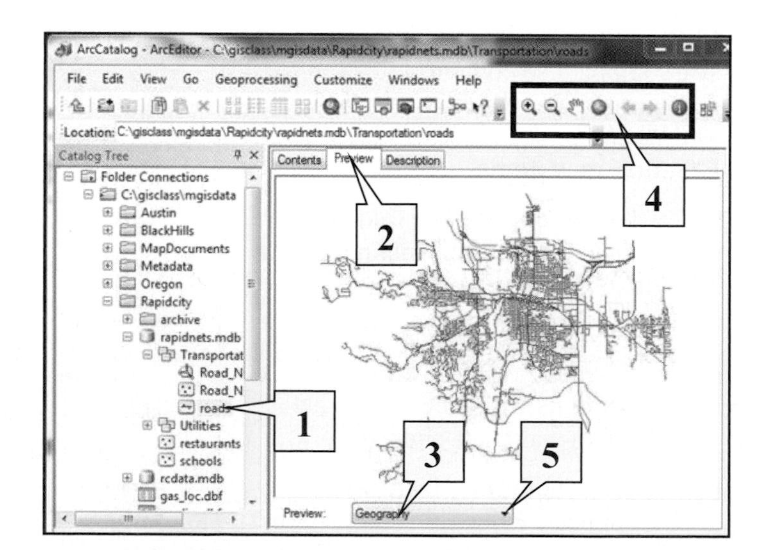

Fig. 28. Previewing a layer

6. In Table mode, additional options are provided by a Table Options menu (Fig. 29). See the Tables section for more information about these options.
7. In Table mode, navigation buttons can be used to go to the beginning or end of a table, or to step through the rows.
8. In Table mode, additional options are available by right-clicking a field name. The options are a subset of those available when working with tables in ArcMap. See the Tables section for more information about these options.

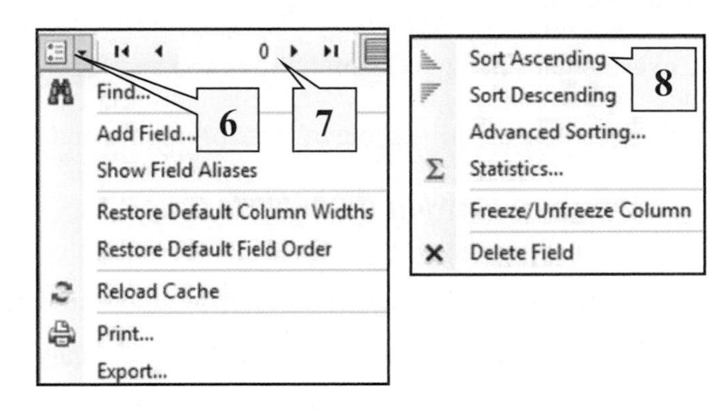

Fig. 29. Options available when previewing tables

Viewing the item description

The item description is a brief overview to give the user basic information about an item (Fig. 30). A description may be created for many kinds of items, including data sets, tools, scripts, and more. Users can view and edit descriptions in ArcCatalog.

By default, only a brief description is shown and available for editing. However, the Description tab is also the portal for developing full-fledged metadata. See the Metadata section for more information.

1. Click on the item in the Catalog Tree for which you wish to view the description.
2. Click the Description tab.
3. Click the Edit button to edit an item's description, or Print to print it.
4. The Description toolbar has other buttons that allow you to validate, export, and import descriptions. See the Metadata section and/or Chapter 15 for more information on how to use these tools.

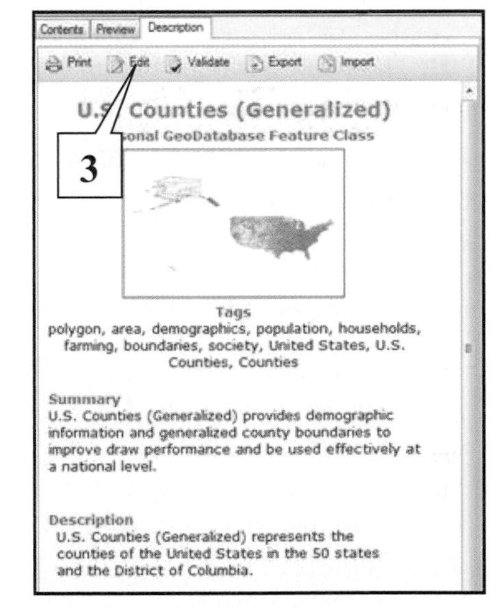

Fig. 30. Item description

Viewing and setting data set properties

Data sets, including geodatabases, feature classes, rasters, and tables, have properties that can be modified in ArcCatalog. The types of properties will vary with the type of data set.

1. To access the data set properties, right-click on the layer/table name in the Catalog Tree and choose Properties. Double-clicking the layer/table also opens its properties.
2. Click the tab containing the properties to change.

Geodatabase tabs include General and Domains.

Shapefile tabs include General, XY Coordinate System, Fields, and Indexes.

Geodatabase feature class tabs include General, Fields, XY Coordinate System, Domain/Resolution/Tolerance, Indexes, Subtypes, Relationships, Representations, and Editor Tracking.

Coverage tabs include General, Projection and Extent.

Other sections describe some of these properties and how to set them.

Creating and viewing thumbnails

1. Click the layer in the Catalog Tree to highlight it and then click the Preview tab.
2. If desired, use the Zoom and Pan buttons to modify the appearance of the layer.

3. Click the Thumbnail button in the Geography toolbar.

4. To view all the thumbnails in a folder, click on the folder, make sure the Contents tab is clicked, and choose the Thumbnail button from the Standard toolbar.
5. To view the thumbnail for a single layer, click on the layer in the Catalog Tree to highlight it and make sure the Contents tab is clicked.

TIP: Thumbnails will be placed into an item's description or metadata.

Dragging and dropping files to ArcMap

After searching through data in ArcCatalog, or on the Catalog tab, you can drag data into ArcMap for use.

1. Position the ArcCatalog and ArcMap windows so that both are visible.
2. Click the file in ArcCatalog that you want to open in ArcMap. Hold down the mouse button and drag it into the ArcMap window (you must be in the map window or the Table of Contents—not on a menu bar). Release the mouse button to drop the file into ArcMap.
3. You can also drag files from the Catalog tab to the map window in ArcMap.

TIP: If ArcMap is not visible on the screen, drag the file onto the ArcMap icon on the computer's Start Menu bar. Hold it there until ArcMap opens and then drag it into ArcMap.

DATA MANAGEMENT

Files and Geodatabases

Creating a geodatabase

1. In ArcCatalog, or in the Catalog tab, navigate to the folder in which the geodatabase will be created.
2. Right-click the folder and choose New > Personal Geodatabase or New > File Geodatabase from the menu.
3. The geodatabase appears in the display window. Type in the name of the new geodatabase and press Enter.

Creating a shapefile

1. In ArcCatalog, or in the Catalog tab, right-click the folder in which you wish to create the shapefile and choose New > Shapefile.
2. Type the name of the new shapefile.
3. Select the feature type from the drop-down button (Fig. 31).
4. Click the Edit button to specify the spatial reference.
5. Select a spatial reference from one of the folders.

TIP: For more information and details about the spatial reference, see ***Specifying a spatial reference***.

6. Click OK to create the shapefile.

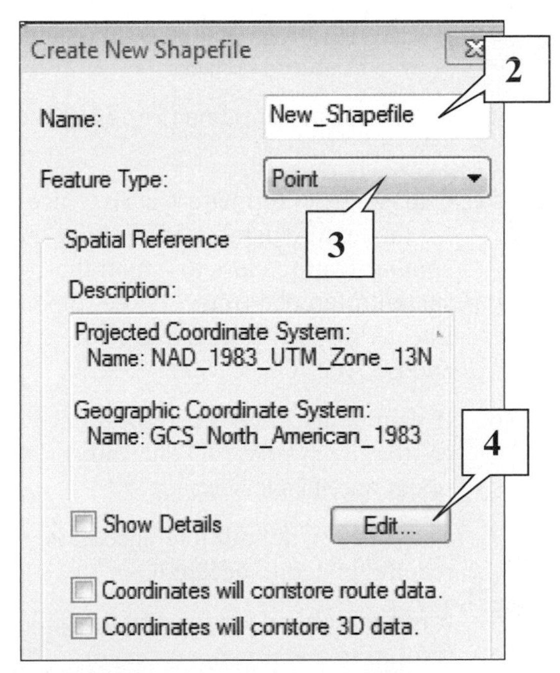

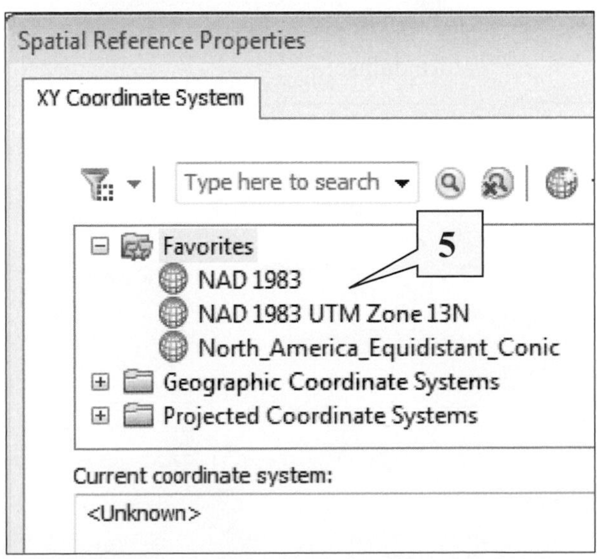

Fig. 31. Creating a new shapefile

Creating a feature dataset

1. Determine the desired coordinate system and resolution for the feature dataset since it can't be changed once defined. ArcGIS will assign default values for the resolution and domain based on the coordinate system. If you have another feature class or feature dataset with the same properties, you can import it from that one.
2. In ArcCatalog, navigate to the geodatabase to contain the feature dataset.
3. Right-click the geodatabase and choose New > Feature Dataset.
4. Enter the dataset name (Fig. 32). Click Next.
5. Choose the coordinate system to use for the feature dataset. If you have another feature class to import the spatial reference from, choose Import. Locate the feature class, select it, and choose Add. Click Next.
6. If your dataset needs a vertical coordinate system (3D data only), select one. Click Next.
7. Examine the default tolerance. It is recommended to accept it.
8. It is recommended to accept the default resolution and domain extent.
9. Click Finish to complete creating the feature dataset.

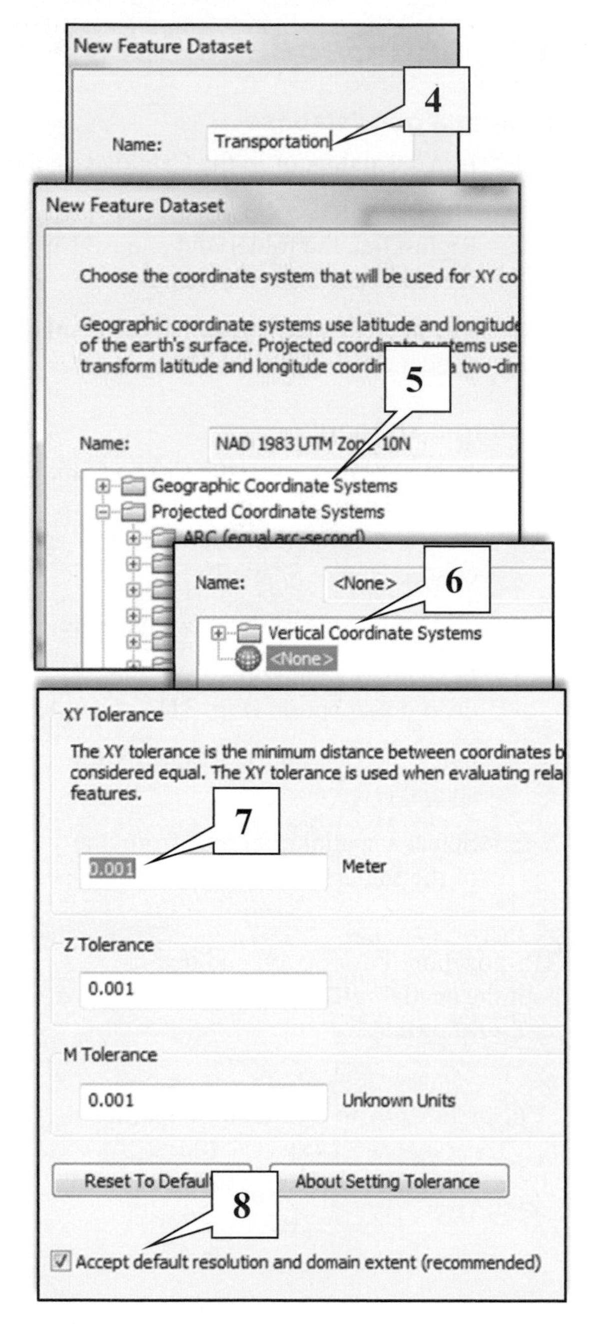

Fig. 32. Creating a new feature dataset

Creating a feature class

1. In ArcCatalog, or the Catalog tab, right-click the geodatabase or feature dataset that will contain the feature class and choose New > Feature Class.
2. Enter the name of the feature class (Fig. 33).
3. Choose the Type of feature. Click Next.

If you are creating the feature class inside a feature dataset, the spatial reference will already be defined. If it is a standalone feature class, you must define the spatial reference.

4. Specify the coordinate system for the feature class. Click Next.
5. Examine the XY Tolerance. The default is fine for most applications unless very high precision is needed.
6. Accepting the default resolution is usually recommended. If you uncheck the box, you'll get another panel to enter the resolution. Click Next.
7. To add new attribute fields, click in the first available box in the Field Name list and type in the field name.
8. Use the drop-down to choose one of the data types for the field.
9. Change the field properties as necessary.
10. To add fields from another feature class or shapefile, click the Import button. Locate the feature class and select it. All fields in the layer will be added to the new one.
11. Click Finish to create the feature class.

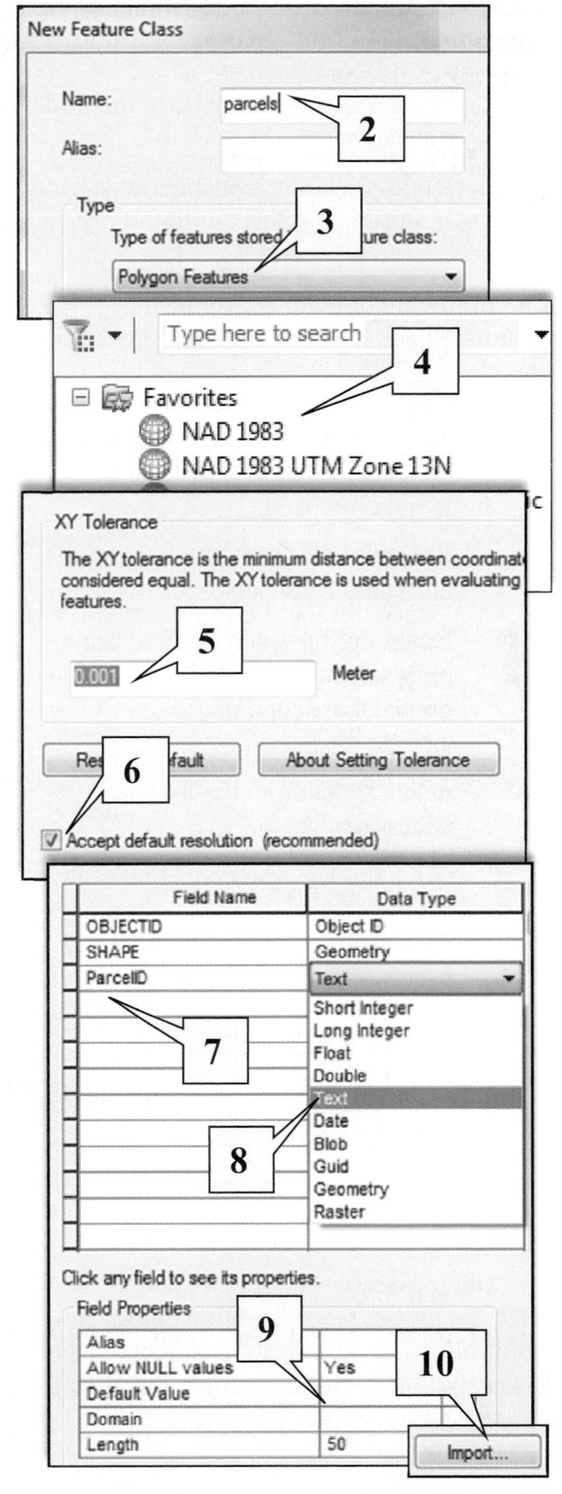

Fig. 33. Creating a new feature class

Importing feature classes

The following directions apply to importing feature classes as standalone classes or as parts of a feature dataset.

1. In ArcCatalog, navigate to the folder containing the geodatabase.
2. Right-click the geodatabase or feature dataset, choose Import, and select an import option based on the type of data to bring in (Fig. 34).

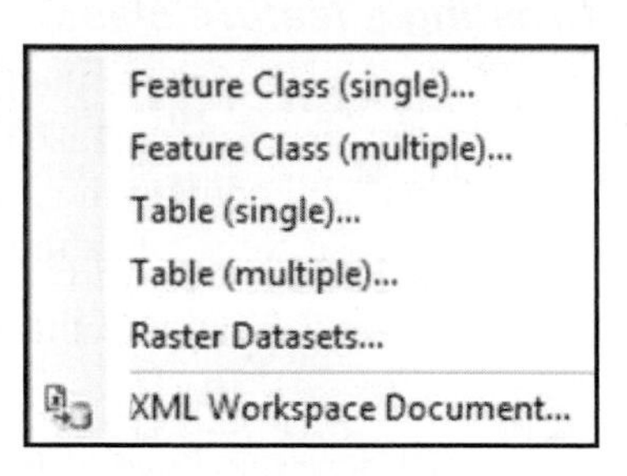

Fig. 34. Options for importing items

The following directions assume that you have chosen the Feature Class (single) tool. The other choices will be similar.

3. Click the Browse button and navigate to the directory containing the data to import (Fig. 35). Select it and click Add.
4. Enter the name of the new feature class in the geodatabase.
5. Enter an SQL expression to bring only selected records into the geodatabase (optional).
6. Examine the fields list. Add or delete fields from the list, if necessary.
7. Click the Environments button to set the geoprocessing environment defaults, if desired. For example, use the coordinate system setting to project the data to a new coordinate system as they are imported.
8. Click OK to start importing the feature class.

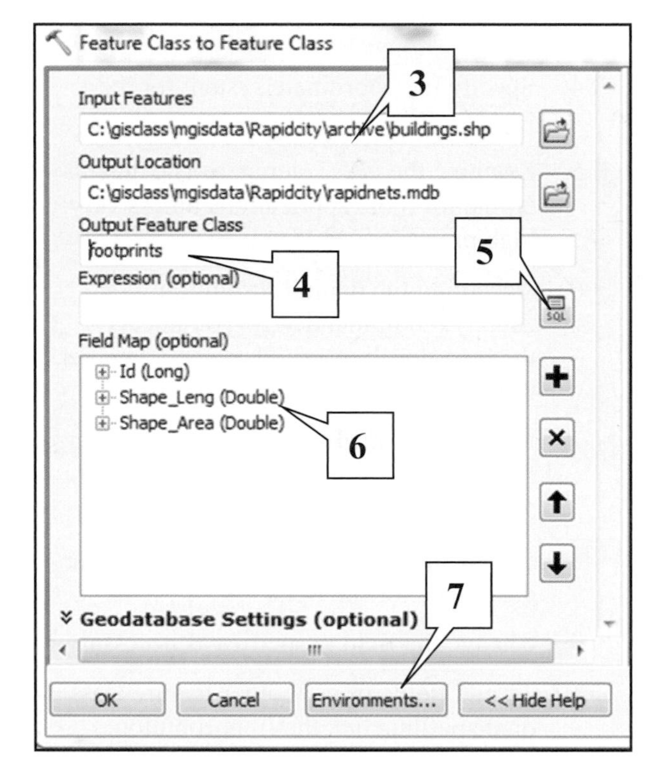

Fig. 35. Importing a shapefile into a geodatabase

TIP: To import several feature classes at the same time, choose Import > Feature Class (multiple). The options to rename the feature class, perform selections, and edit field names are not accessible with the multiple feature class tool, however.

Exporting features

Exporting a layer takes the selected records or features and creates a new data set on the disk with them.

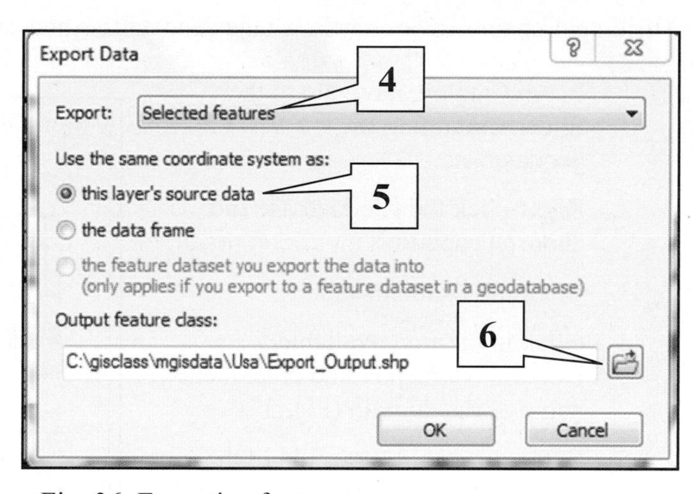

Fig. 36. Exporting features

1. In ArcMap, perform a query, if desired, to select the desired records.
2. Switch to View by Drawing Order or View by Source mode in the Table of Contents, if necessary.
3. Right-click the layer and choose Data > Export Data.
4. Choose what to export (Fig. 36). You may export all features, the selected features, or the features in the current view extent.
5. Choose which coordinate system to use.
6. Click the Browse button to specify a name and location for the output feature class. Click OK.
7. Change the *Save as type* to a shapefile or geodatabase feature class, as desired. Click Save and OK.

Creating attribute domains

Attribute domains are properties of the geodatabase and are set at that level.

1. In ArcCatalog, navigate to the folder containing the geodatabase.
2. Right-click the geodatabase and choose Properties.
3. Click the Domains tab.
4. Click in the first available Domain Name box and enter the name of the domain (Fig. 37).
5. Enter a longer description of the domain for documentation purposes.
6. Set the attribute type (Short Integer, Long Integer, Text, etc.). The domain can only be applied to attributes of the same type.
7. Click in the Domain Type box and choose a Range or a Coded Value domain.
8. For a Range domain, enter the minimum and maximum range values.

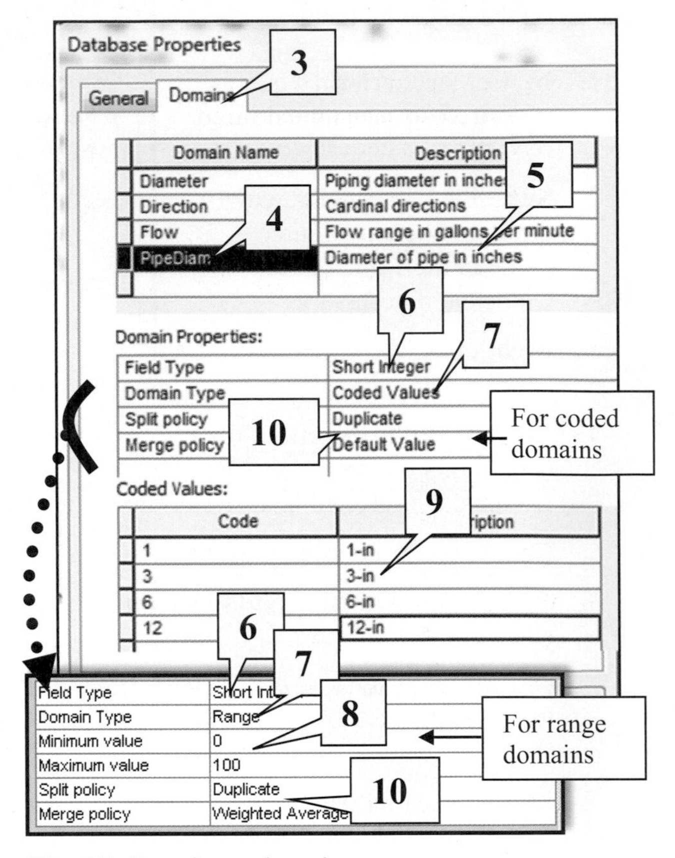

Fig. 37. Creating a domain

9. For a Coded Value domain, enter each value code and description in the lowest set of boxes. The descriptions should not be too long (< 15 characters is probably best).
10. Click in the Split and Merge policy boxes and choose the type of policy desired.
11. Click Apply to create the domain. It will be checked for errors and added if OK.
12. To enter additional domains, repeat Steps 3–10 or click OK if finished.
13. To delete a domain, click on the gray area to the left of its name to select it and press the Delete key.

TIP: Domain policies and codes may be edited after the domain is created. However, the field type and domain type, once set, cannot be altered. To change them, you must delete the domain and create a new one with the desired properties.

TIP: Always click Apply between one domain and the next. That way if you make a mistake, you know which domain to delete and reenter.

Assigning default values or domains

Default values will automatically be entered into attribute fields when a new feature is created during editing. Domains must already exist in the geodatabase before they can be applied to a feature class.

1. In ArcCatalog, right-click the feature class containing the attribute and choose Properties.
2. Click the Fields tab (Fig. 38).
3. Select the field by clicking on the gray area to the left of its name. Its properties appear underneath.
4. Click the box for Null values and choose Yes to allow null values or No to disallow them.
5. Click in the Default Value box and type in the default value.
6. To assign a domain to an attribute, click in the Domain box and select a domain from the list.
7. For best results, click Apply after completing the entries for one field, before doing another.
8. If desired, enter default values for additional items by repeating Steps 3–7.
9. Click OK to effect the changes and close the window.

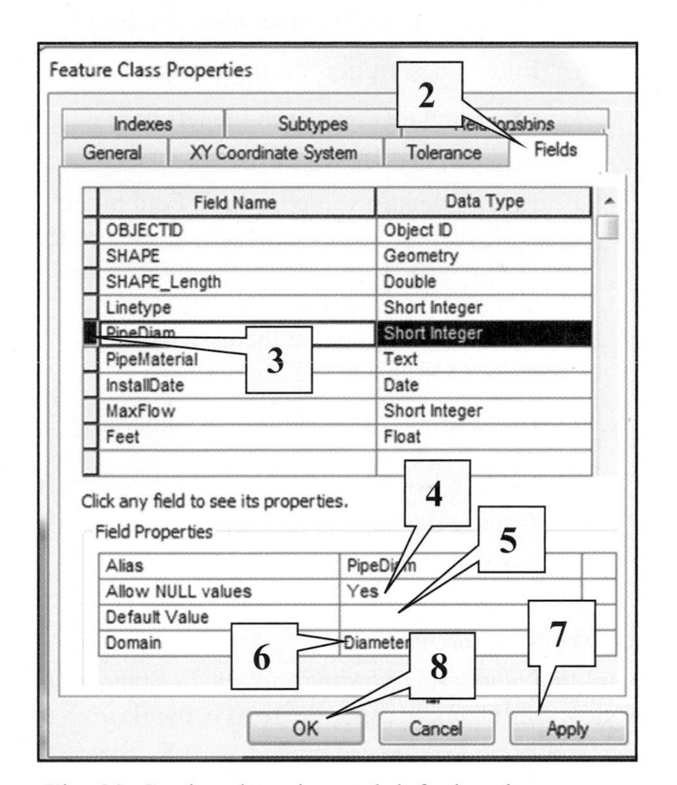

Fig. 38. Setting domains and default values for attributes

TIP: The domain type must match the field type. A long integer domain cannot be applied to a short integer field, for example. Only valid domains will be listed for a field.

Creating subtypes

Subtypes allow you to specify different defaults and behaviors for groups of features in a single feature class. Before starting, make sure that you have planned the subtype categories, created the other fields contained in the feature class table, and entered any domains to be used by the fields. Also plan the default values to be defined for each subtype.

1. In ArcCatalog, open the properties of the feature class to be given the subtypes.
2. Click the Fields tab. Create a short integer field that will be used to define the subtype categories. See ***Adding or deleting fields***. Do not set any defaults here. Click Apply to ensure that the field is created before going on to the next step.

3. Click the Subtypes tab (Fig. 39).
4. Set the Subtype Field to the short integer field created in Step 2.
5. Click in the first empty Code box and enter the value 1 (or any other short integer).
6. Enter a description for the first subtype.
7. Click in the gray area to the left to highlight the subtype row.
8. Enter a default value for each field in the table that will use one. If a field uses coded domains, be sure to enter the code value actually stored in the field, not the description value that the user sees (e.g., COP instead of copper).
9. Click Apply to finish the first subtype.
10. Repeat Steps 5–9 to add more subtypes.
11. Choose the default subtype.
12. Click OK when finished.

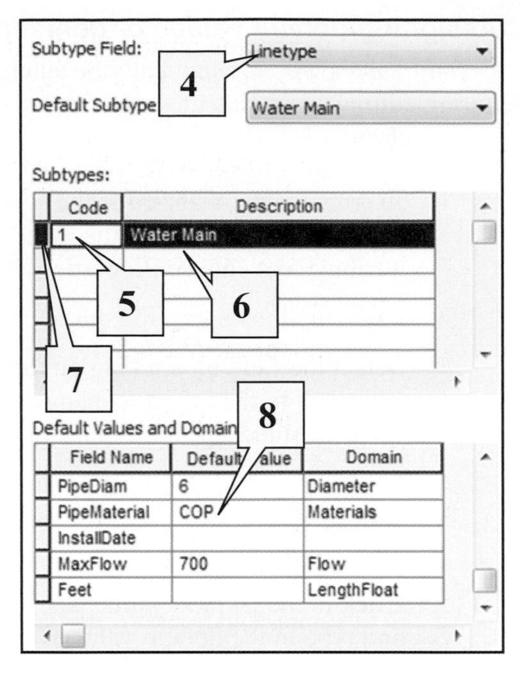

Fig. 39. Entering the first subtype

Creating a network

Building a network requires an ArcGIS Standard or Advanced license. If you want to try it, follow this example using the road network from the usdata geodatabase.

1. The features forming the network must be feature classes in a feature dataset in a geodatabase.
2. Right-click the Transportation feature dataset in the mgisdata\Usa\usdata.mdb geodatabase and choose New Geometric Network. A wizard will appear. Read the information and click Next.
3. Enter a name for the network (Fig. 40).
4. Enter No to indicate that the features should not be snapped. This data set already has the ends of roads snapped to each other. See the tip about snapping below. Click Next.
5. Check the box to include majroads in the network. Do NOT include interstates. Click Next.
6. Choose No so that all network features are enabled. Click Next.
7. Leave the role for majroads set to Simple Edge. Leave Sources & Sinks set to <None>. Click Next.

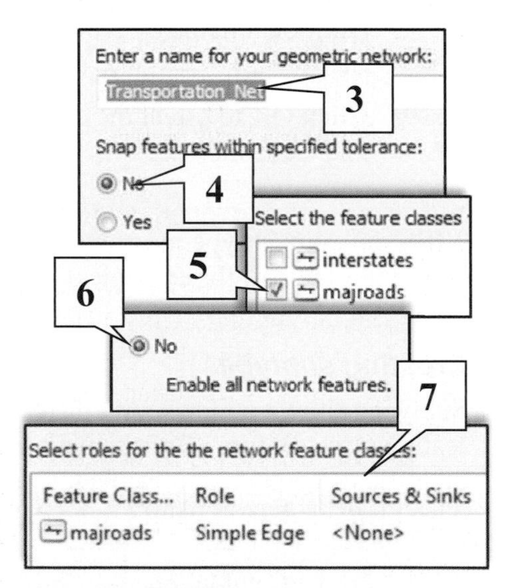

Fig. 40. Building a network

TIP: To set weights, you must know the data type of the field used for the weight. Examine the fields in ArcCatalog before creating the network to obtain this information.

We will assign two distance weights based on both miles and kilometers (Fig. 41). They are single-precision Float fields named MILES and KM.

8. Click New to add the first weight (Fig. 41).
9. Enter **Distance-MI** for the weight name and choose the Single Type. Leave the Bitgate Size alone. Click OK.
10. The Distance-MI weight is entered in the weights list. Click on it to highlight it.
11. In the panel underneath, click in the box under Field Name and choose MILES as the attribute field to associate with the weight.
12. Click New to add the second weight. Name it **Distance-KM** and set the type to Single. Click OK.

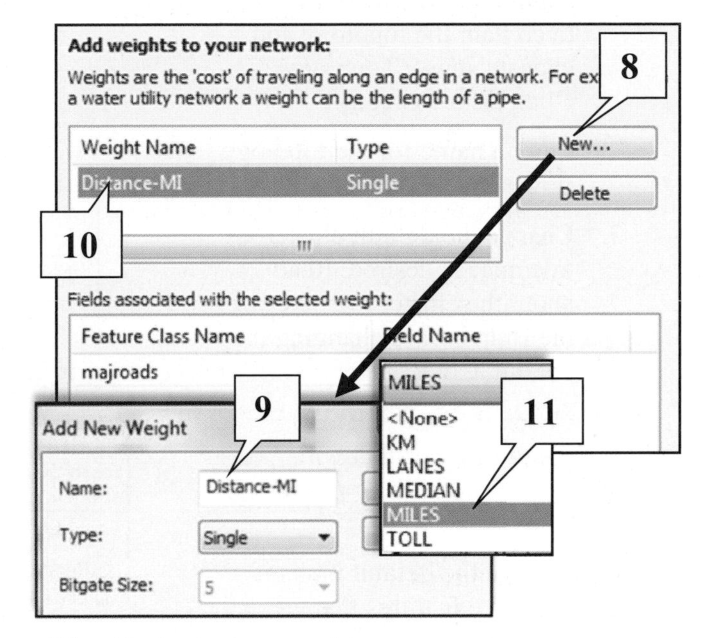

Fig. 41. Assigning weights

13. Highlight the Distance-KM weight in the upper box. Select KM in the lower field name box for the attribute field. Click Next.
14. Review the summary and choose Finish.

TIP: For the network to operate correctly, all edges must meet each other at junctions with no gaps or dangles. This adjustment may have already been done using planar topology, in which case you don't need to snap when creating the network. If not, you may be able to automatically fix the dangles by indicating a snap tolerance in map units. It must be large enough to fix the errors but not so large that features are incorrectly snapped together. Setting a good tolerance can be tricky, which is why snapping is often done interactively using topology. Choosing Yes may make changes to your data, so be careful and have a backup copy just in case.

TIP: If you make a mistake or need to change the network, the easiest way is to delete it and start over. Right-click the Transportation_Net entry in the feature dataset and choose Delete. It will remove the logical network but leave the majroads feature class as is.

Creating a planar topology

A planar topology uses rules to define the desired spatial relationships within and between feature classes. It can test features for adherence to the rules, and it provides tools for finding and fixing errors. A planar topology can only be created for a feature dataset, and only feature classes in the feature dataset can participate.

1. Right-click the feature dataset to contain the topology and choose New > Topology. Click Next to begin.
2. Enter a name for the topology (Fig. 42).
3. Change the default cluster tolerance if desired. Read about the cluster tolerance or the Help before changing the default. Click Next.
4. Check the boxes for the feature classes that will participate in the topology. Click Next.
5. Change the default ranks if you have features that are more accurate than others that should not be moved during clustering. More accurate features can be ranked higher and will not move. Click Next.
6. Click Add Rule. For a single feature class rule, choose the feature class and the rule. For a two-layer rule, choose the first feature class, the rule, and the second feature class. Click OK to add the rule. Continue until all rules are added.

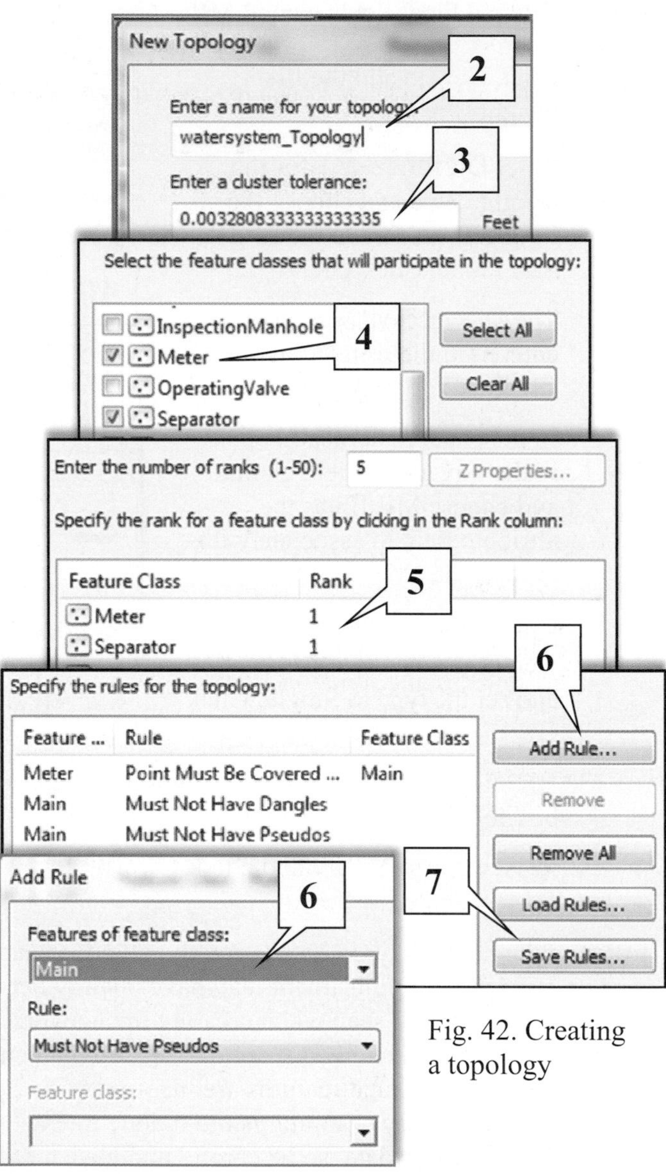

Fig. 42. Creating a topology

7. Rule sets can be saved and used again for another topology. Click Next.
8. Read the summary and click Finish to create the topology.
9. Choose Yes if you want to validate the topology immediately.

TIP: Validation includes clustering, which can move features and degrade feature accuracy depending on the cluster tolerance. Always have a backup before you validate.

Tables

Opening a table

In ArcCatalog

1. In ArcCatalog, tables may be viewed using the Preview tab. Some, but not all, of the functions described below are available in Preview mode.

In ArcMap

1. Right-click on the spatial data set name in the Table of Contents and choose Open Attribute Table from the context menu.
2. To open a standalone table, first make sure that the table is visible by clicking on the List By Source tab at the top of the Table of Contents.
3. Right-click on the table name and choose Open from the context menu.

The Table window

The Table window shows the contents of tables and provides buttons and commands for working with them.

1. The Table Options menu provides access to many commands and functions that affect tables (Fig. 43).
2. Buttons at the top of the table access the most commonly used commands, including, from left to right, Related Tables, Select By Attributes, Switch Selection, Clear Selection, Zoom To Selection, and Delete Selection.
3. Right-click a field heading to open a context menu with actions that use or affect the field (see Fig. 44).

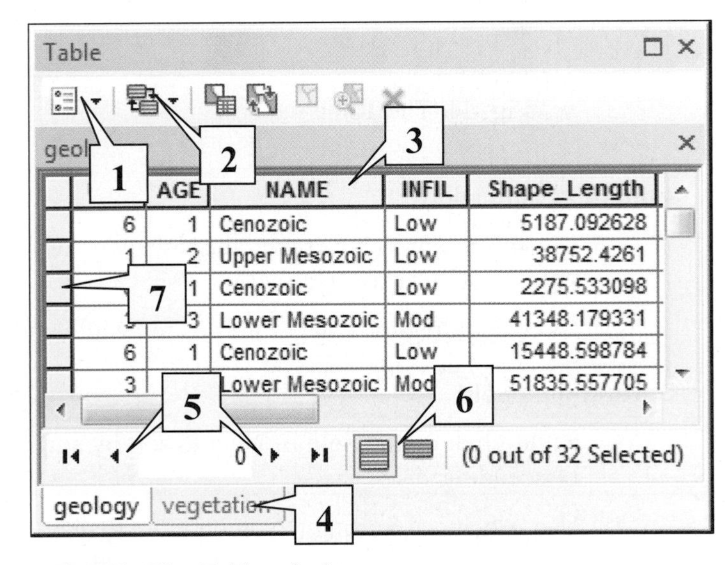

Fig. 43. The Table window

4. Normally the window shows one table. If you open more, they will appear as tabs in the lower-left corner. Click on a tab to view that table.
5. The navigation buttons can take you to the beginning, the previous record, the next record, and the end of the table.
6. These buttons toggle between showing all records and showing only selected records.
7. Click the gray area to select a record from the table.

Changing a table's appearance

You can change certain ways in which the table displays data. Note that none of these operations modifies the actual data on the disk, only the way in which the data are presented.

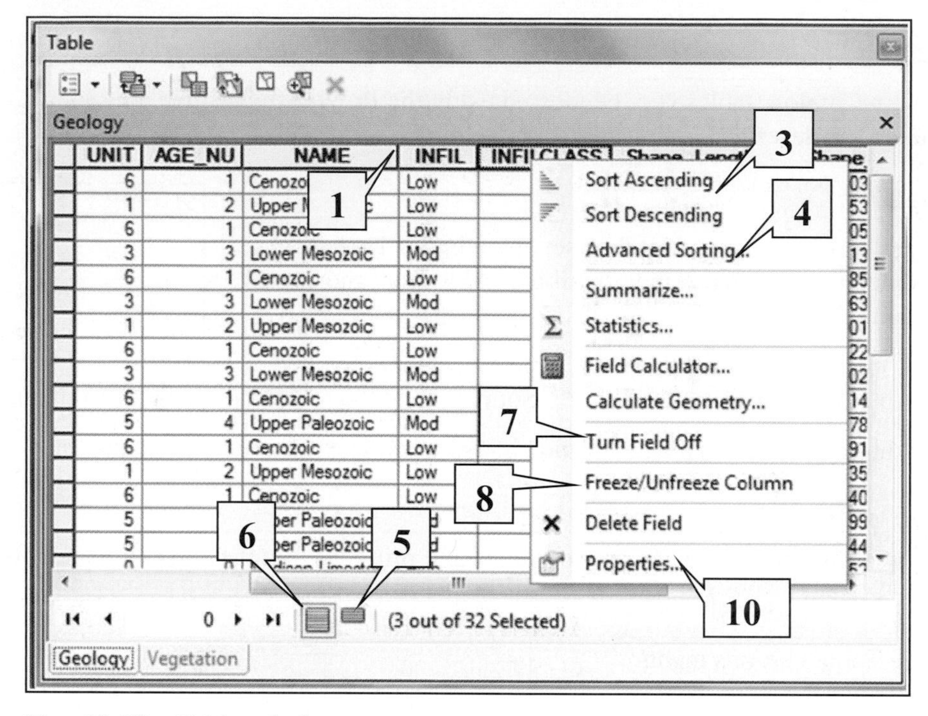

Fig. 44. The Table window

Adjusting field width

1. Use the cursor to hover over the right edge of the field to resize until it changes to a bar with a double arrow (Fig. 44).
2. Click and drag the edge to the desired width and then release the mouse button.

Sorting on a field

3. Right-click on the field heading to sort by and choose either Sort Ascending or Sort Descending.
4. Use Advanced Sorting to sort on more than one field at a time.

Displaying selected records

A table can show all records with the selected ones highlighted, or it can show only the selected records.

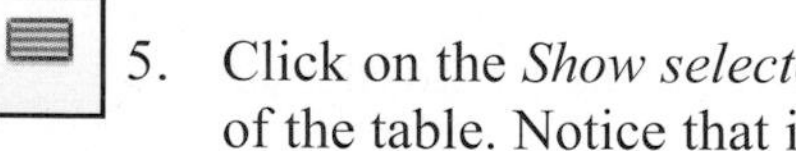

5. Click on the *Show selected records* button at the bottom of the table. Notice that it also reports how many records are currently selected.

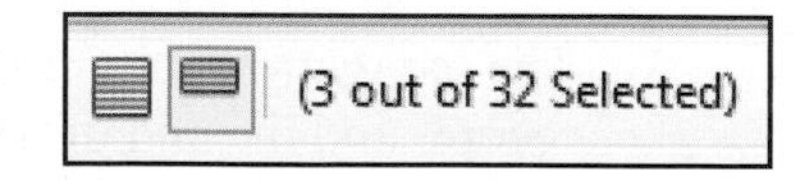

6. Click on the *Show all records* button to show all records again.

Turning one field off

7. Right-click the field heading and choose Turn Field Off. The field will no longer be shown in the table.

TIP: To turn a field back on, you must use the Fields tab in the layer properties.

To freeze/unfreeze a field

Freezing a field anchors it to the left side of the Table window so that it remains there when the table scrolls to the right.

8. Right-click the field heading and choose Freeze/Unfreeze Column.
9. To unfreeze a frozen column, right-click it and choose Freeze/Unfreeze Column.

TIP: An unfrozen column will not go back to its original place in the table automatically. Use the Fields tab of the layer properties to reestablish the default field order.

Changing field properties

10. Right-click the field heading and choose Properties (Fig. 45).
11. Change the alias of the field, if desired.
12. Change the display characteristics.
13. Change the formatting properties of numeric fields.

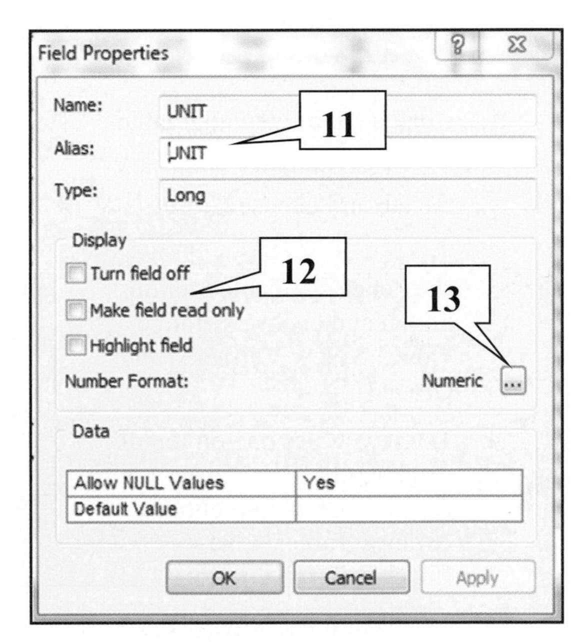

Fig. 45. Field properties

Changing table colors and fonts

The user can change how a table looks by changing its font and the colors that are used to highlight selected records.

1. With the table open, click the Table Options menu in the upper-left corner of the table and choose Appearance.
2. Click the color squares to choose a new color from the drop-down palette for the selection and highlight colors.
3. Choose a font, font size, and color.

Formatting field values

You can change the alignment, decimal places, and other formatting characteristics of numeric fields.

1. Right-click a field heading and choose Properties.
2. Click on the ellipses next to the Numeric text in the Field Properties window.
3. Choose the numeric category, such as currency or percentage (Fig. 46).
4. Specify the number of decimals or significant digits, the alignment, and other options.
5. Click OK.

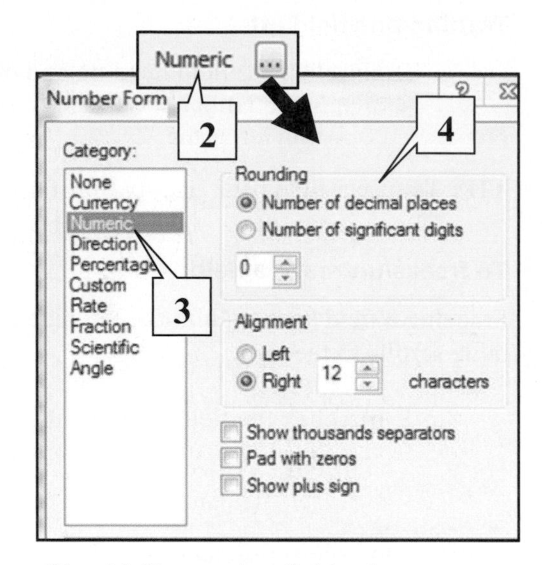

Fig. 46. Formatting field values

Managing multiple tables

The Table window shows one table at a time by default. If more tables are opened, they appear as tabs in the bottom of the window. You can also create tab groups to view multiple tables in the window. A tab group can itself contain multiple tables.

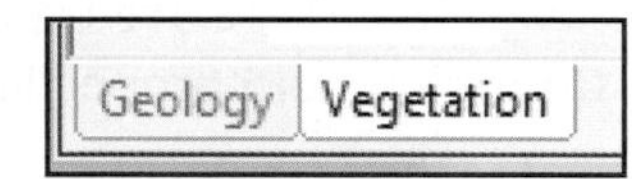

1. Click a tab to view that table.
2. To close one table, right-click the tab and choose Close.
3. To view two tables side-by-side, click the Table Options button and choose Arrange Tables > New Vertical Tab Group (Fig. 47).
4. To view tables one on top of another, click the Table Options button and choose Arrange Tables > New Horizontal Tab Group.
5. To adjust the relative size of the tables within the window, click and drag on the bar in between them.

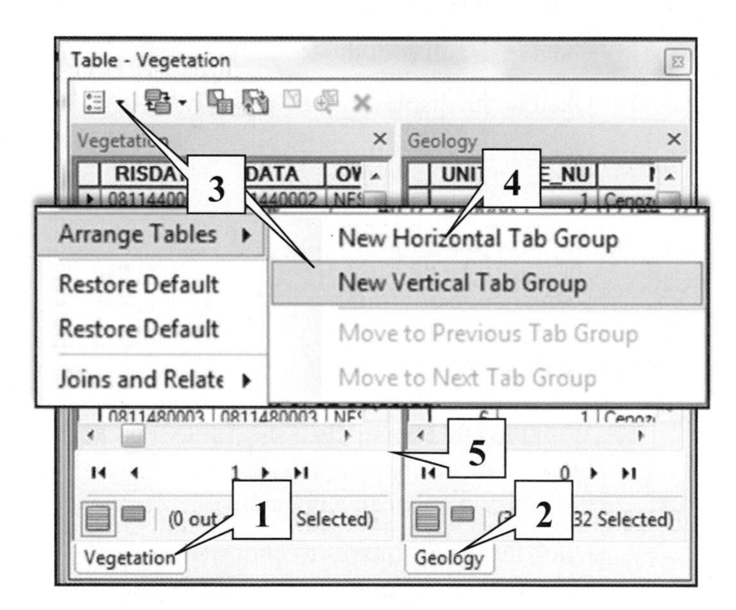

Fig. 47. Using a vertical tab group

When more than one tab group is present, one must be the active one, which is designated by a darker blue title bar. Actions taken using the Table Options menu or buttons will affect the table visible in the active tab group. Newly added tables will be placed in the active tab group.

6. Click the blue title bar of the table in the tab group to make it the active tab group.
7. To move a table from one tab group to another, click on its tab to make it visible and to make its tab group active. Choose Table Options > Move to Previous Tab Group (or Move to Next Tab Group).

The Fields tab

The Layer Properties window has a Fields tab that allows the user to modify the appearance, visible fields, field order, and other properties. Like all layer properties, they affect the appearance of the layer in the map document and do not affect the original file.

1. If the table is an attribute table, right-click on the data set name and choose Properties from the context menu. If it is a standalone table, right-click on the table name and choose Properties.
2. Click the Fields tab (Fig. 48).
3. To create an alias, click the field name in the left panel to highlight it, and type the alias in the panel on the right.
4. To hide a field, uncheck the box next to its name. Check the box to show it again.
5. To show all fields in the table, click the *Turn all fields on* button.
6. To turn off all fields, click the *Turn all fields off* button.
7. To change the order of a single field, click to highlight it and use the Move Up or Move Down button to change its position. Use the drop-down on the Move Up or Move Down button to move a field to the top or bottom of the table.

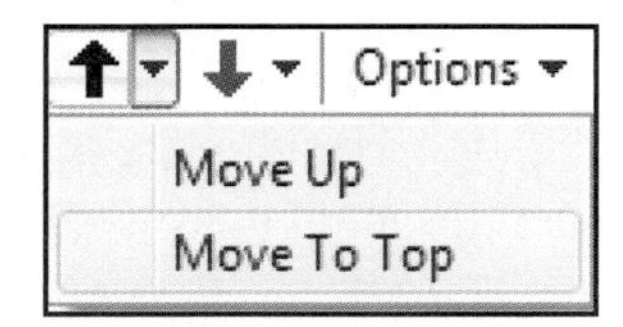

8. Choose Options > Sort Ascending or Sort Descending to sort the fields alphabetically by field name.
9. Choose Options > Reset Field Order to put the fields back in the original order as stored in the file (a good way to put an unfrozen field back in position).
10. Select whether field names or field aliases will be shown in the table and in menu items that list field names (such as Select By Attributes).

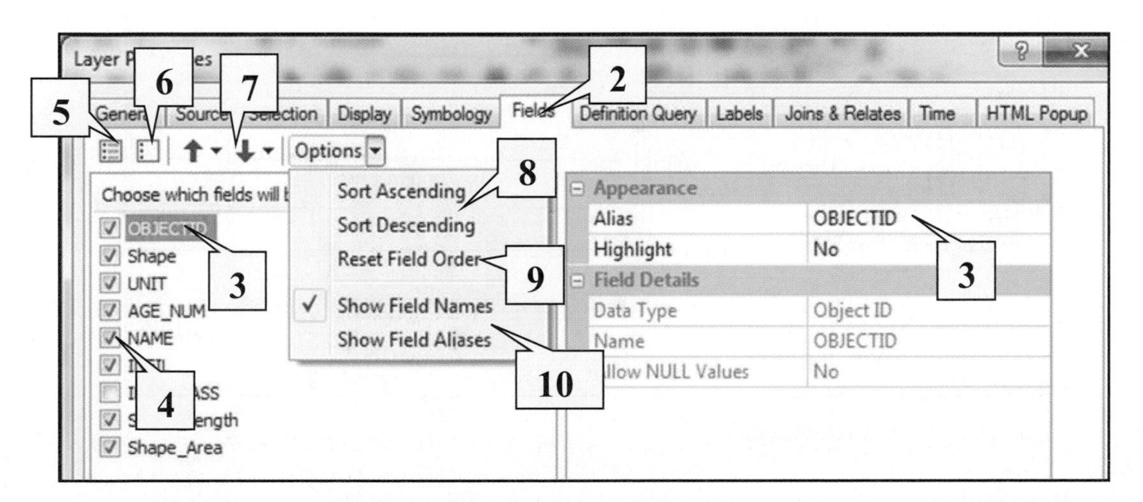

Fig. 48. The Fields tab of the Layer Properties window

Exporting a table

Exporting a table means saving it under a new name. There are two ways to export a table.

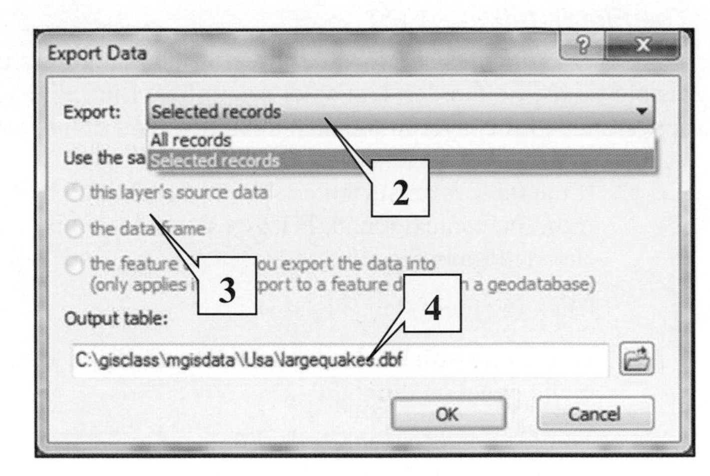

Fig. 49. Exporting records from a table

1. If the table is open, click the Options button and choose Export. If it is a standalone table, you can right-click the table name in the Table of Contents and choose Data > Export.
2. To export the entire table, choose *All records* from the drop-down box. To export a previously selected subset, change the drop-down box to say Selected records (Fig. 49).
3. The choices regarding the coordinate system are dimmed because you are exporting a table instead of a spatial data set.
4. Choose the directory and name of the output file by typing it in or clicking the Browse button. Click OK.

Getting statistics for a field

Statistics are useful for learning more about the data in fields. Statistics may be calculated for a selected set of records or for the entire table (Fig. 50).

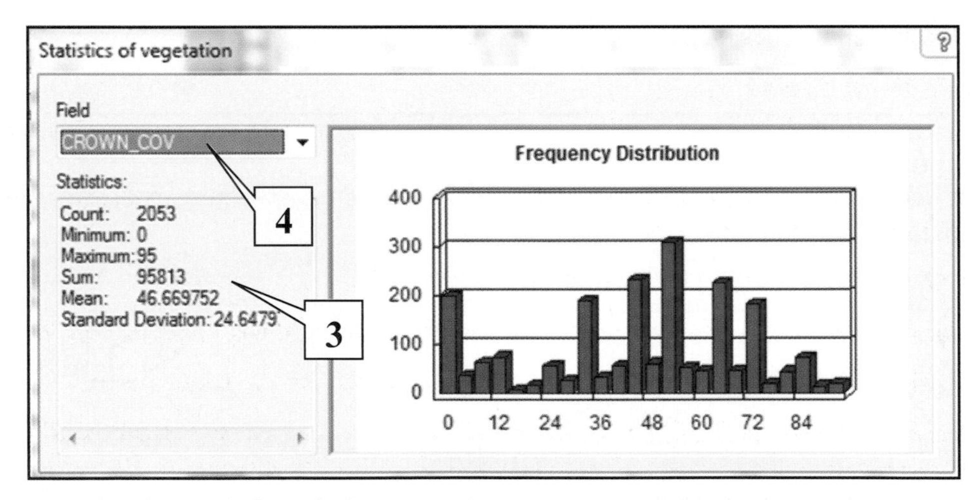

Fig. 50. The Statistics window

1. If desired, use a query to select a subset of records requiring statistical analysis.
2. With the table open, right-click on a numeric field on which to get statistics and choose Statistics from the context menu.
3. Examine the statistics and frequency diagram in the Statistics dialog box (Fig. 50). If the table has records selected, then statistics will be calculated only for the selected records.
4. Look at statistics for a different field by choosing it from the Field drop-down box.

Summarizing on a field

Summarize generates statistics for groups of features in a given field, such as finding the average magnitude of earthquakes occurring in each state.

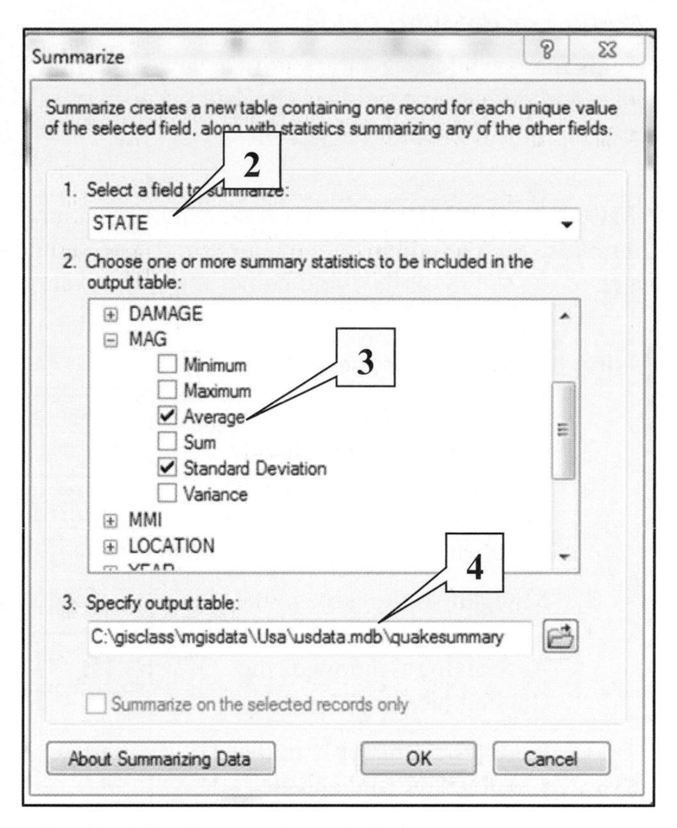

Fig. 51. The Summarize window

1. With the table open, right-click on a categorical field that will be used to group the records and select Summarize from the context menu.
2. In the Summarize dialog box, verify that the chosen field appears in the first box (Fig. 51).
3. In the second box, choose one or more statistics to calculate. Click on the plus sign to expand the list of statistics for a field, and then check the box next to the desired statistic(s).
4. Specify the name of the output table. Click on the Browse button to change the directory where the file will be saved. Click OK.
5. Click Yes to add the table to the map.

Editing fields in a table

Fields can be edited only during an editing session. For more information, consult Chapter 12.

1. If not already in an editing session, click the Editor Toolbar button on the Standard toolbar to open the Editor toolbar.
2. Choose Editor > Start Editing from the Editor toolbar.
3. Open the table, if it is not already open, by right-clicking it and choosing Open.
4. Click in the field space of the record to edit and start typing in the changes.
5. Choose Editor > Save Edits from the Editor toolbar to save during an edit session. When finished typing the changes, choose Editor > Stop Editing and answer Yes when asked whether to save your changes.

Adding or deleting fields

Fields may be added to tables in two ways. The first way uses ArcCatalog or the Catalog tab to enter and define new fields in the data set properties. This way is most efficient when you are entering many fields at once. You can also add a field in the Table window in ArcMap.

IMPORTANT TIP: Note that field names must have 13 or fewer characters; may include letters, numbers, and the underscore character (_); and should not contain spaces or special characters such as @, #, !, $, or %. Field names must also start with a letter, not a number.

Using the data set properties

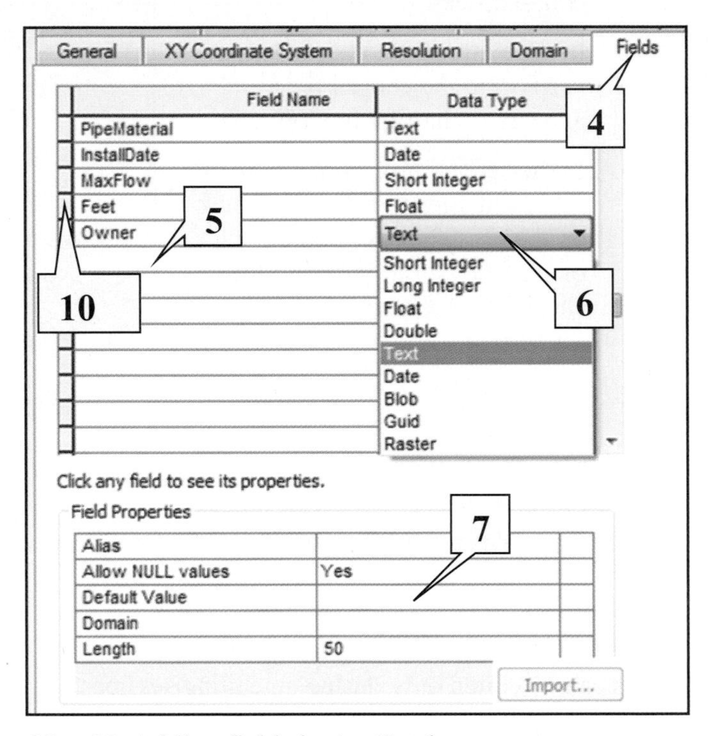

Fig. 52. Adding fields in ArcCatalog

1. If the table is currently open in ArcMap, use the Catalog tab. Or use ArcCatalog after making sure that the table is not open in ArcMap.
2. Navigate to the table in the Catalog Tree in the ArcCatalog window or the Catalog tab.
3. Right-click the table in the Catalog Tree and choose Properties from the context menu.
4. Click the Fields tab (Fig. 52).
5. Scroll down the list of fields, if necessary, to the first empty space at the bottom. Click on the empty space and type in the name of the new field.
6. Click on the Data Type box to the right of the new field, and a drop-down list of field types will appear. Choose the appropriate type.
7. View the field properties in the lower part of the dialog box and change any that need it, such as the length of a text field, a default value, or a domain.
8. Repeat Steps 5–8 until all the fields have been added.
9. Click OK or Apply to enter the changes in the Properties dialog box.
10. To delete a field, follow Steps 1–4 to bring up the table's Fields tab. Select the field by clicking on the gray tab next to its name, and press the Delete key. Warning: You cannot undo this step.

Managing fields in ArcMap

1. With the table open, click the Table Options button and choose Add Field from the menu.
2. Type in the name of the field (Fig. 53).
3. Choose the type of field from the drop-down box.
4. View the field properties in the lower part of the dialog box and change any that need it, such as the length of a text field, a default value, or a domain.
5. Click OK. The field will be added to the end of the table.
6. To delete a field, right-click on the field name in the Table window and choose Delete Field from the context menu. Because this action cannot be undone, you will be prompted to confirm the action. Click Yes to delete the field.

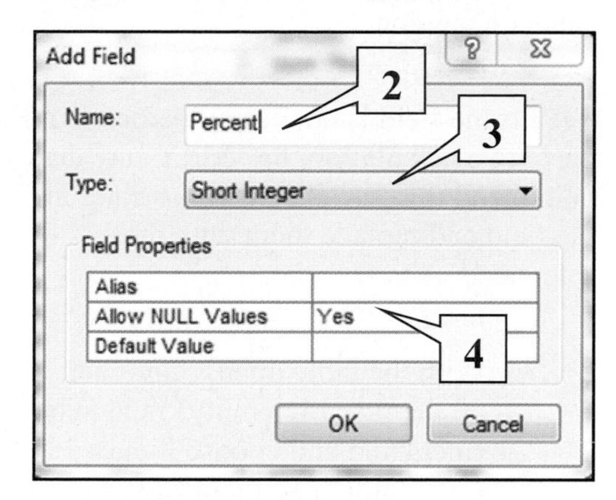

Fig. 53. Adding a field in ArcMap

TIP: You cannot add or delete a field during an edit session; you must stop editing first.

Creating a new table

New standalone tables are best created in ArcCatalog or the Catalog tab. Attribute tables are generated automatically when the program creates the associated spatial data set.

Creating a dBase table

1. In the Catalog Tree, right-click on the folder that is to contain the new file and select New > dBase Table.
2. Type in the name of the new table. Make sure the table name does not contain spaces or special characters such as @, #, %, ^, or &. Underscore characters are permitted.
3. To add fields to the table, see the instructions under ***Adding or deleting fields***.

Creating a table in a geodatabase

1. Right-click the geodatabase in the Catalog Tree and choose New > Table. A wizard will appear.
2. Enter the name of the table in the box and click Next.
3. Add fields to the table following the instructions under ***Adding or deleting fields***. Click Finish when done.

Calculating fields

Use the Field Calculator in ArcMap to enter information into many records at once, either the entire table or a selected set. The information entered can be as a single number, a text value, or a mathematical expression. Calculations may be done outside of an edit session, but then they cannot be undone.

TIP: If the field is not empty before you perform a calculation, be VERY sure that you intend to erase the existing values in the field and replace them with the calculated values.

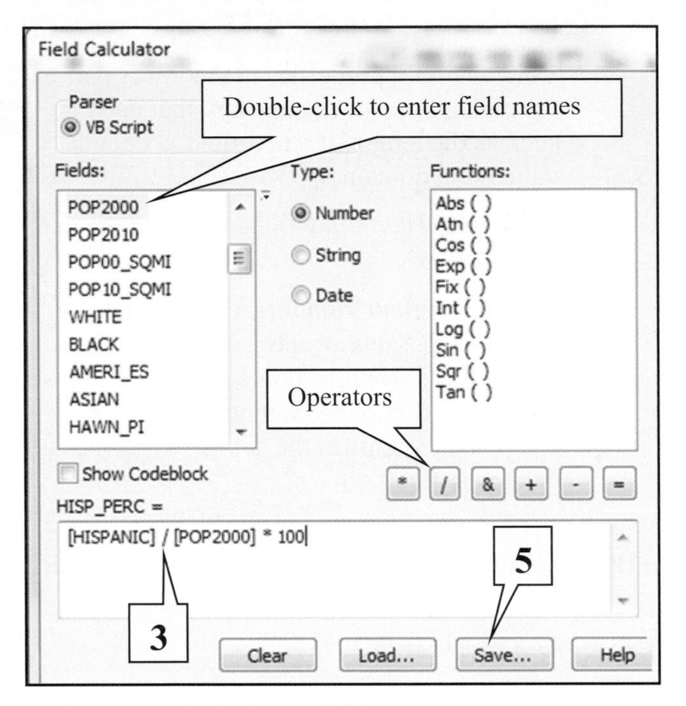

Fig. 54. Calculating a field

1. With the table open, right-click on the (usually empty) field to be calculated and choose Field Calculator. The Field Calculator will appear (Fig. 54).
2. Enter the expression in the box. Double-click a field name to enter it in the box. Click on operators to include those. Double-click functions to enter them. Numbers may be entered from the keyboard.
3. For example, the expression in Figure 54 was created by double-clicking [HISPANIC] in the Fields box, clicking the / operator, double-clicking [POP2000], clicking the * operator, and typing 100.
4. Click OK.
5. Complex expressions that are used frequently may be saved as *.cal files and loaded later. Press Save, navigate to the desired directory, and enter a file name. To reload the expression, click Load, navigate to the file location, and select the desired expression file.

Creating joins and relates

Joins and relates are used to link information from two different tables, based on a field that is common to both. The fields may be numeric or string fields, and they may have different names.

In a join, information from the source table is appended to the destination table. The destination table will then have its original fields plus the fields from the source table.

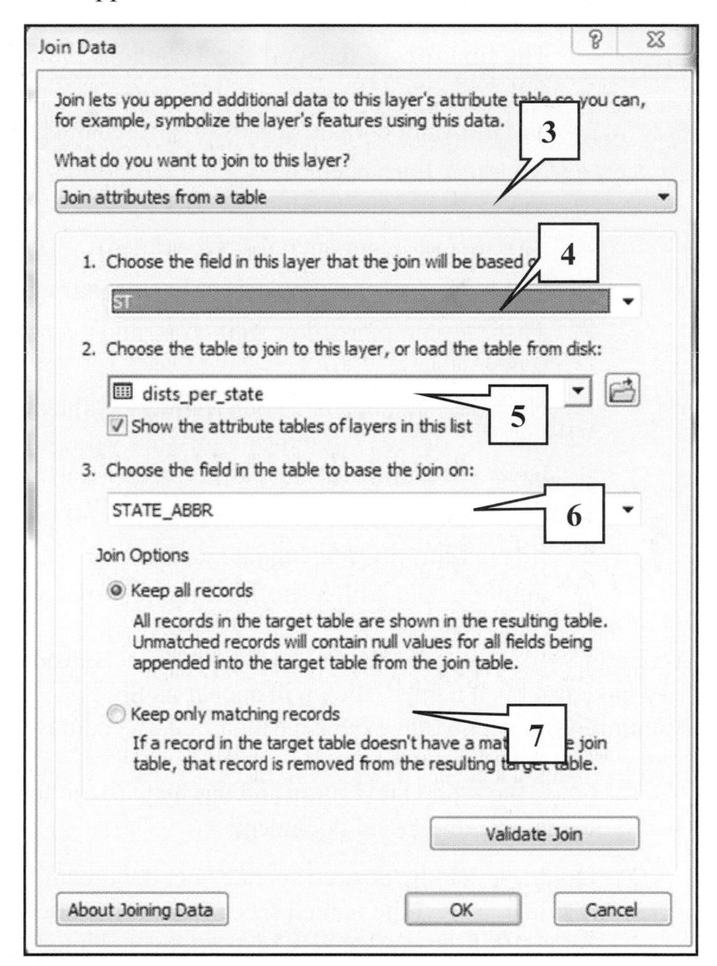

Fig. 55. The Join Data window

1. In the Table of Contents, right-click on the destination layer or table, OR click the Table Options button in the Table window.
2. Choose Joins and Relates > Joins from the menu. The Join Data dialog box appears (Fig. 55).
3. Make sure that Join Attributes from a Table is the selection action in the first drop-down box.
4. Enter the field in the destination table that will be used as the key to join the records.
5. Choose the source table from the list of tables in the map document, or use the Browse button to locate a file on disk.
6. Choose the field in the source table that will be used as the key.
7. Choose to keep unmatched records (in which case the appended fields will contain <Null> values) or drop the unmatched records from the table.
8. Click OK.
9. To remove a join, right-click the destination table or layer in the Table of Contents or the Table Options menu and choose Joins and Relates > Remove Join(s). Then select a specific join to remove, or remove all joins on that table.

TIP: Exporting a joined table will put both the destination and the source table information in the new table.

Opening Excel Data

Preparing the data

Using Excel data is not difficult, but it requires careful preparation. First, make sure that the spreadsheet meets the following requirements.

- The first row of the worksheet contains the field headings.
- Field headings must start with a letter, must have no more than 13 characters, and must not contain spaces or shift characters such as %, $, #, and so on.
- No blank lines occur between the headings and the data or within the data rows.
- The spreadsheet contains no formulas.
- The bottom of the spreadsheet has no extraneous data, such as column totals.
- There are no merged or split cells, and every column contains consistent data types.
- Numeric columns have only numeric values in them, without text characters like "x" or "n/a" to indicate missing values. If present, missing values should have a numeric NoData marker value such as –99.
- Text fields contain no commas, unless all the text is enclosed in quotes.
- It is helpful if each column has been specifically formatted as text data or as numeric data with a specified number of decimal places.

An Excel workbook may have multiple worksheets inside. You must open each one separately. If they have not been named, they will appear as Sheet1$, Sheet2$, and Sheet3$, with Sheet1$ containing the data. If they have been named, the names will appear.

1. Click the Add Data button and navigate to the folder containing the Excel document.
2. Double-click the desired spreadsheet document to look inside it. Select the named sheet you want, or select Sheet1$ if the worksheets have not been named (Fig. 56). Click Add.
3. Use the spreadsheet as if it were any other table (except it is read-only).
4. To convert it to dBase format or save it in a geodatabase, right-click the spreadsheet table and choose Export. See the Skill Reference entry in this chapter on Exporting.

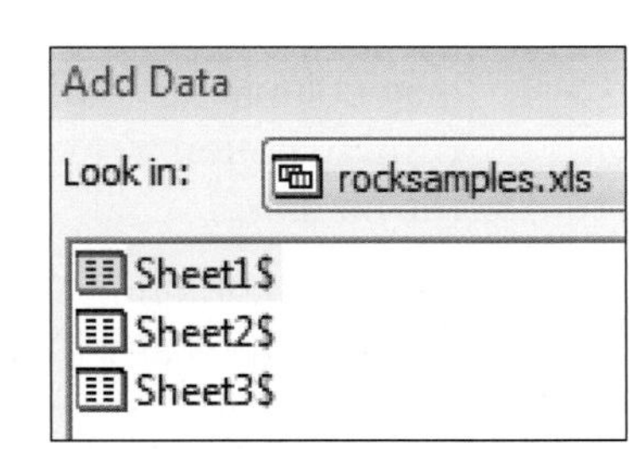

Fig. 56. An Excel workbook with multiple worksheets

TIP: You must CLOSE the spreadsheet in Excel before you can open it in ArcMap.

TIP: If you are unable to open a file in ArcMap, try saving it in an earlier Excel format.

You can also save a spreadsheet in .csv or .dbf format and open it from ArcMap that way.

Displaying x-y data from a table

Displaying *x-y* data from a table creates an event theme, which is a temporary display rather than a permanent file. To save the event layer permanently as a shapefile or feature class in a geodatabase, you need to export it. See ***Exporting features***.

1. Make sure the table has fields for *x* and *y* coordinates and that you know the coordinate system and datum of the coordinates.
2. Right-click the table in the Table of Contents and choose Display XY Data.
3. Specify which fields contain the *x-y* coordinates (Fig. 57). Recall that longitude is the *x* coordinate and latitude is the *y* coordinate.
4. Click the Edit button to specify the coordinate system of the points in the table. See ***Specifying a spatial reference***.
5. Click OK to create the points.

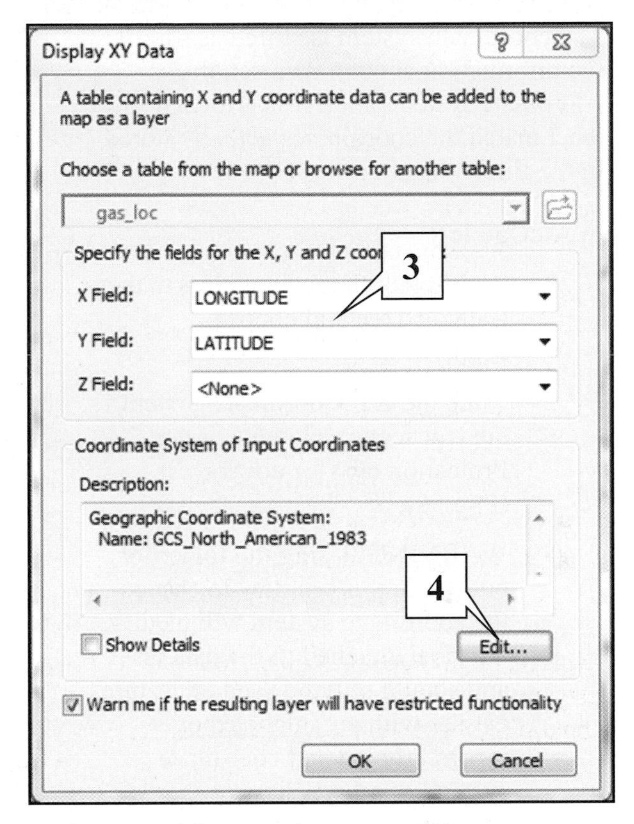

Fig. 57. Adding *x-y* data to a map

TIP: Be sure to set the coordinate system to match the *x-y* values stored in the table, not the map currently in the data frame. For example, if the table contains latitudes and longitudes, the coordinate system must be set to a GCS even though the map you are viewing may be in UTM or State Plane.

Coordinate Systems

Examining the coordinate system

Every data layer has *x-y* coordinates that must be in some type of coordinate system. The coordinate system is stored as part of the information about a data set, the coordinate system label. It is critical that the label match the coordinates actually stored in the file.

In ArcCatalog

1. Right-click the feature class in the Catalog Tree and choose Properties.
2. Click the XY Coordinate System tab (for feature classes) or the Projection tab (for coverages) (Fig. 58).
3. WARNING: Using the folders or buttons in this window to change the coordinate system will modify the label attached to the data set and should only be used on feature classes with an unlabeled or incorrectly labeled coordinate system. See ***Specifying a spatial reference***.

XY Coordinate System
2
Type here to search
Favorites
NAD 1983
3
NAD 1983 UTM Zone 13N
North_America_Equidistant_Conic
Geographic Coordinate Systems
Projected Coordinate Systems
Layers
Current coordinate system:
North_America_Equidistant_Conic
WKID: 102010 Authority: ESRI

Projection: Equidistant_Conic
False_Easting: 0.0
False_Northing: 0.0
Central_Meridian: -96.0
Standard_Parallel_1: 20.0
Standard_Parallel_2: 60.0
Latitude_Of_Origin: 40.0
Linear Unit: Meter (1.0)

Fig. 58. The XY Coordinate System tab

In ArcMap

1. Right-click the layer name and choose Properties.
2. Click the Source tab.
3. Read the coordinate system information (Fig. 59).

The top panel shows the extent, or range of *x-y* coordinates stored in the file. The lower panel shows the source data set for the layer and the coordinate system label.

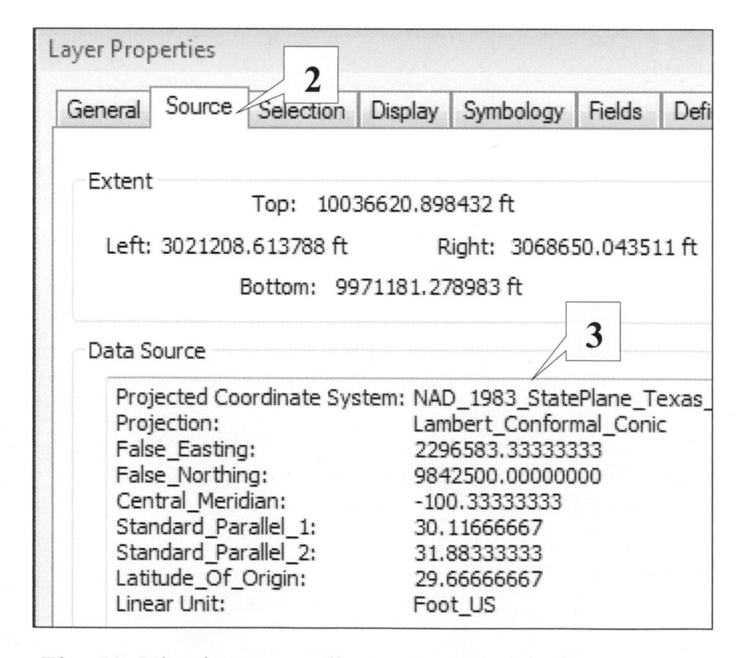

Fig. 59. Viewing a coordinate system in ArcMap

Specifying a spatial reference

The spatial reference is the complete description of a data set's coordinate system (CS). The same sets of windows are used in several places when a coordinate system needs to be specified. Coordinate system definitions are stored as text files with a .prj extension.

1. Examine the current spatial reference name and details.

The Details panel shows the parameters defined for the selected CS (Fig. 60).

2. Browse through the folders to choose the desired CS.The files are organized into geographic and projected coordinate systems (Fig. 61).
3. In ArcMap, the Layers folder lists the CS's of the current layers.
4. Once a CS is seleced, you may click the Favorites button to add it to the Favorites folder. Right-click one in the Favorites folder to remove it.
5. Type a search term and click the Search button to search for a CS.
6. Click Clear Search to clear the search term.
7. Click this drop-down to set or clear a spatial filter, which looks for a CS appropriate to the current geographic extent.
8. Click Add Coordinate System > Import to copy a CS from another data set.
9. Click Add Coordinate System > New to define a custom CS. See ***Defining a custom coordinate system***.
10. Click Add Coordinate System > Clear to set the CS to Unknown.

TIP: Always choose the plain version of a datum unless you have reason to do otherwise, such as NAD 1983 instead of NAD 1983 (HARN).

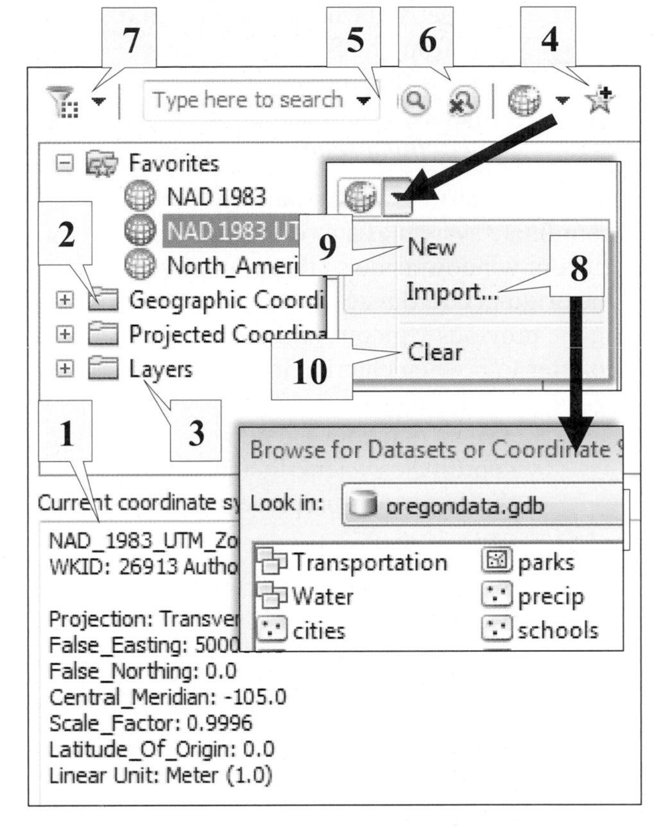

Fig. 60. Specifying a spatial reference

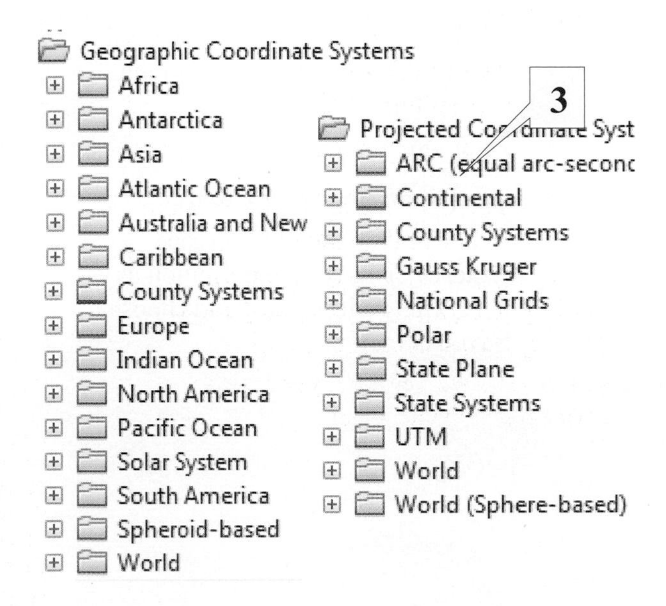

Fig. 61. Organization of coordinate system folders

Defining a custom coordinate system

ArcGIS comes with many predefined coordinate system files, but sometimes it is necessary to define a custom projected coordinate system. (Defining a custom geographic coordinate system is rare and not recommended for beginners.)

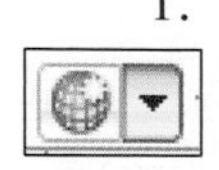

1. Choose Add Coordinate System > New from the Spatial Reference window (Fig. 60) and select Projected Coordinate System. See ***Specifying a spatial reference***.

If you are starting with a standard coordinate system as a model, shift the custom window aside so that you can see the original coordinate system parameters in the previous window, and you'll be able to refer to it when entering the parameters.

2. Type a name for the new coordinate system. Include both the datum and projection in the name (Fig. 62).
3. Select the type of projection to be used.
4. Set the parameters such as the central meridan or false easting. See Chapter 11 for a discussion of these parameters.
5. Choose the linear unit to be used to store the x-y coordinates.
6. Click Change to choose a different geographic coordinate system than the default.
7. Click OK.
8. The CS is added to the Custom folder in the Spatial Reference window. To save the definition as a .prj file, right-click it and choose Save As.

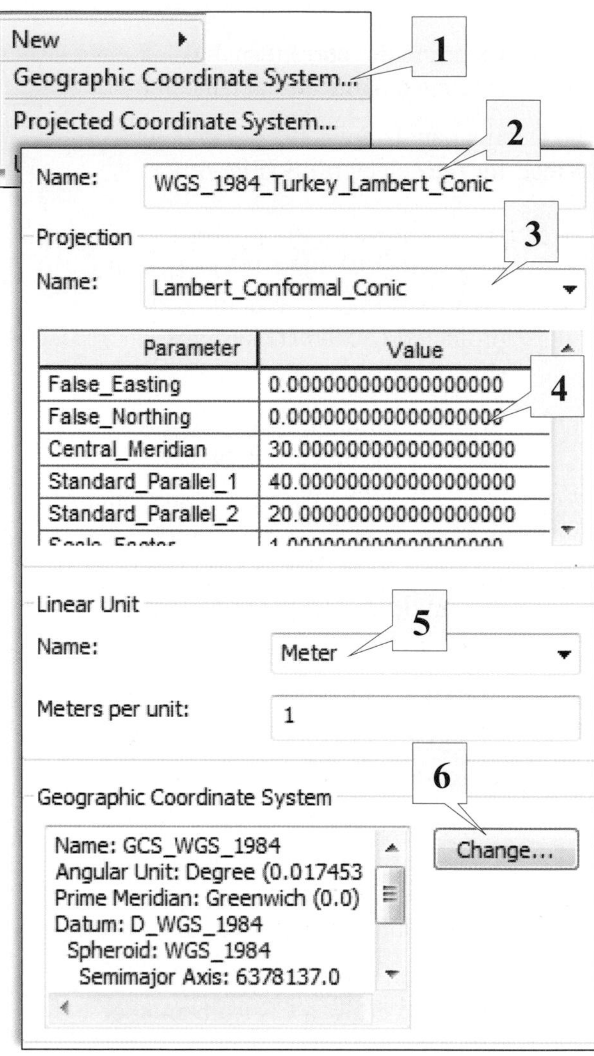

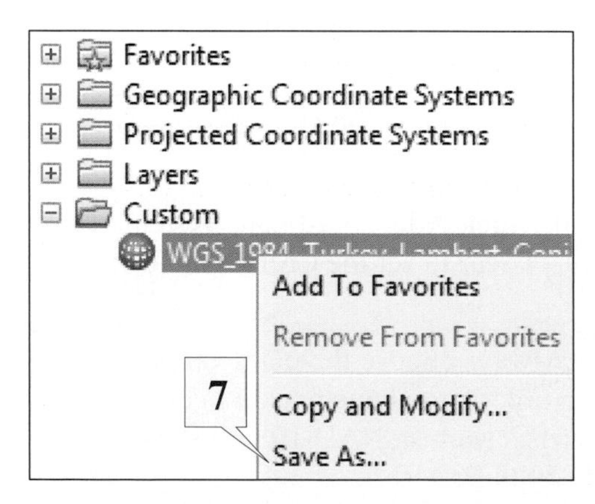

Fig. 62. Defining a custom coordinate system

Setting the coordinate system of a data frame

Use this feature to display data sets with different coordinate systems in the coordinate system of your choosing. The data sets are projected on the fly to whatever coordinate system you specify. The original data are not affected.

1. In ArcMap, right-click the data frame name and choose Properties from the context menu.
2. Click the Coordinate System tab (Fig. 63).
3. Use the folder tree provided to select a coordinate system.
4. You may use the Layers folder to select a coordinate system from the layers already in the data frame.
5. If you use the chosen coordinate system frequently, you can choose Add to Favorites to add it to the Favorites folder for quicker access next time. Highlight it in the Favorites folder and click Remove From Favorites to get it out of the folder.
6. Type a search term and click the Search button to search for a coordinate system. Click the Clear Search button to cancel the search.
7. Click the Spatial filter button to look for a CS based on the current extent; click again to turn the filter off.
8. If more than one GCS is being used in the data frame, you may click the Transformations button to select the method used to transform one datum to another.

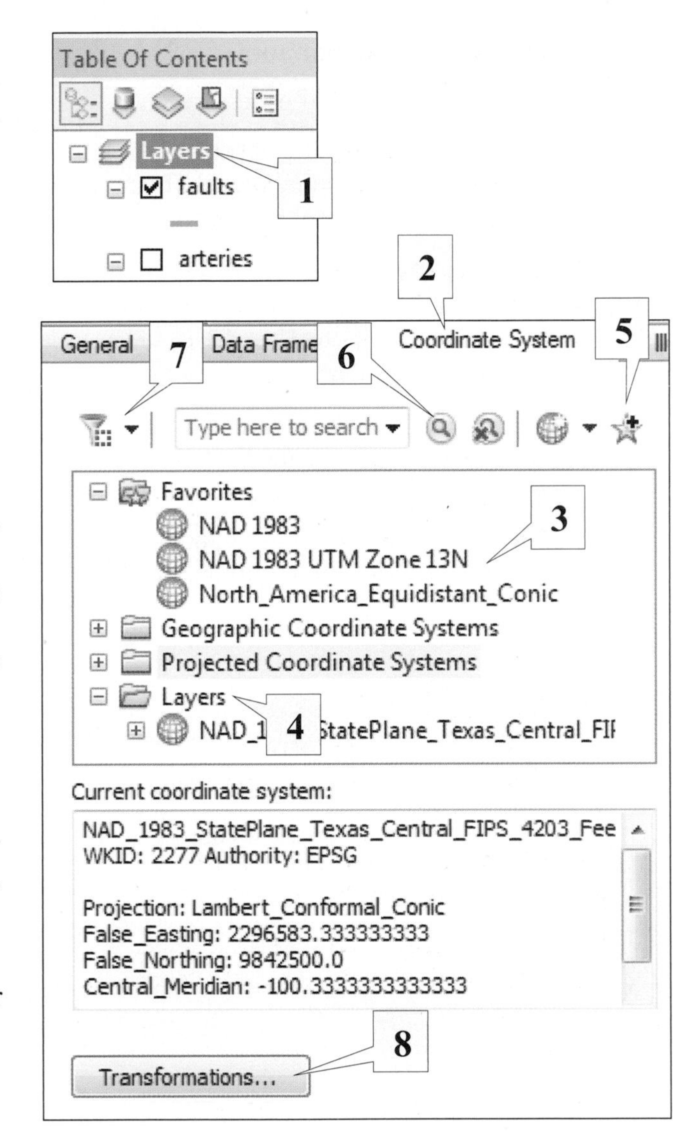

Fig. 63. Setting the data frame CS

9. For more information on using this window, see ***Specifying a spatial reference***.

Defining coordinate systems of data sets

Defining a coordinate system for a data set allows you to display it with other data. It is important to set the coordinate system to match the actual *x-y* coordinates contained in the file. This tool should be used only if the data set coordinate system is undefined or incorrectly defined.

Using the Define Projection tool

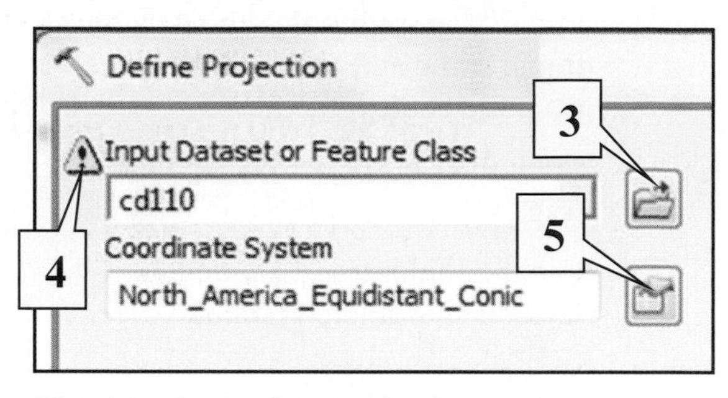

Fig. 64. The Define Projection tool

1. Open ArcToolbox, if necessary.
2. Choose the Define Projection tool from the section Data Management Tools > Projections and Transformations.
3. Click the drop-down button (ArcMap only) or the Browse button and select the data set to define (Fig. 64).
4. A warning symbol will appear if the layer already has a spatial reference. You should only change the spatial reference of a data set if the coordinate system label is currently Unknown or incorrect.

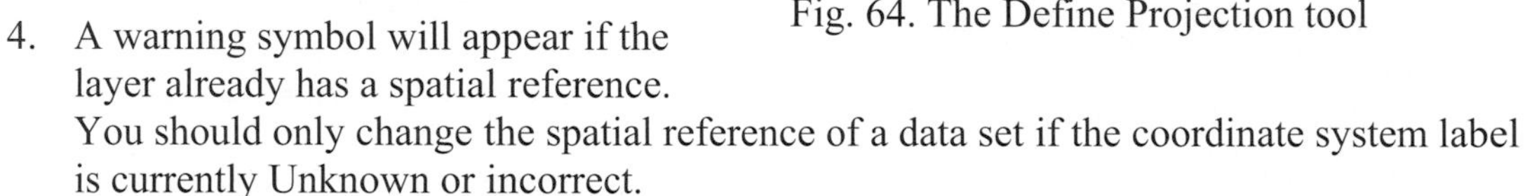

5. Click the icon to open the Spatial Reference Properties window. Select, import, or create a coordinate system definition. See ***Specifying a spatial reference***.
6. Click OK.

Using ArcCatalog

1. Right-click the data set in the Catalog Tree and choose Properties from the menu.
2. Click the XY Coordinate System tab.
3. Select, import, or create a coordinate system definition. See ***Specifying a spatial reference***.

TIP: If a layer is open in ArcMap, then use ArcToolbox or the Catalog tab to define the coordinate system. A file lock will prevent ArcCatalog from changing a data set that is open in ArcMap.

Projecting feature classes

Projecting data changes both the coordinate system definition AND the *x-y* coordinates of the features, creating a new file in the process and leaving the original unchanged.

1. In ArcToolbox choose the Project tool from the Data Management Tools > Projections and Transformations > Feature section (Fig. 65).
2. Click the drop-down (ArcMap only) or the Browse button to select the data set to be projected.
3. Click the second Browse button to specify a folder, a geodatabase, or a feature dataset to contain the output and give the output shapefile or feature class a name. Click Save to enter the new name. Click Next.
4. Click the button to open the Spatial Reference Properties window to select a coordinate system to which to project the data. See ***Specifying the spatial reference***.
5. A green dot will appear by the Geographic Transformation box if one is required to convert from one datum to another. Select one or more transformations from the drop-down box.
6. Click OK to project the data.

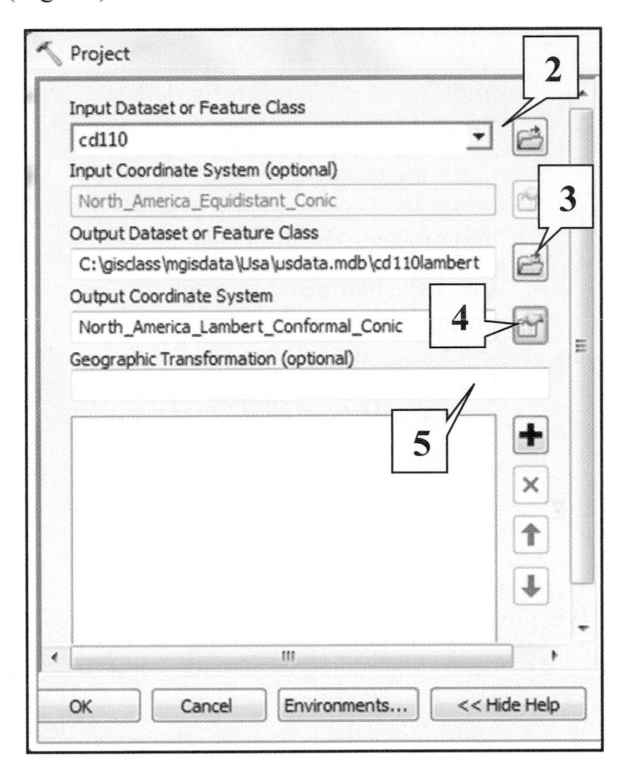

Fig. 65. The Project tool

TIP: Projecting coverages requires an ArcGIS Advanced license. If you only have an ArcGIS Basic license, a red X button will appear by the Input Dataset box.

Projecting rasters

1. If converting from a GCS to a projected coordinate system in meters or feet, calculate the target output cell size in meters and convert to feet, if necessary. Round the result to an even number. You can also usually take the cell size suggested by the tool and round it.

Cell size in degrees × 111.3 km/degree × 1000 = output cell size in meters

2. Open the ArcToolbox > Data Management Tools > Projections and Transformations > Rasters > Project Raster tool (Fig. 66).
3. Click the drop-down (ArcMap only) or use the Browse button to select the input raster to be projected.
4. Enter the name and location for the output raster.
5. Select the output coordinate system. See ***Specifying a spatial reference***.
6. If a green button appears by the Geographic Transformation box, a datum transformation is required. Choose one from the drop-down list.
7. Set the resampling technique to NEAREST if the raster contains discrete or categorical data. Set it to BILINEAR if the raster contains continuous data. See the discussion in Chapter 11 for more information.
8. Set the output cell size as a value, or use the Browse button to match it to an existing data set.
9. Select a registration point if you wish for the cell grid to be snapped to a specific location.
10. Click OK.

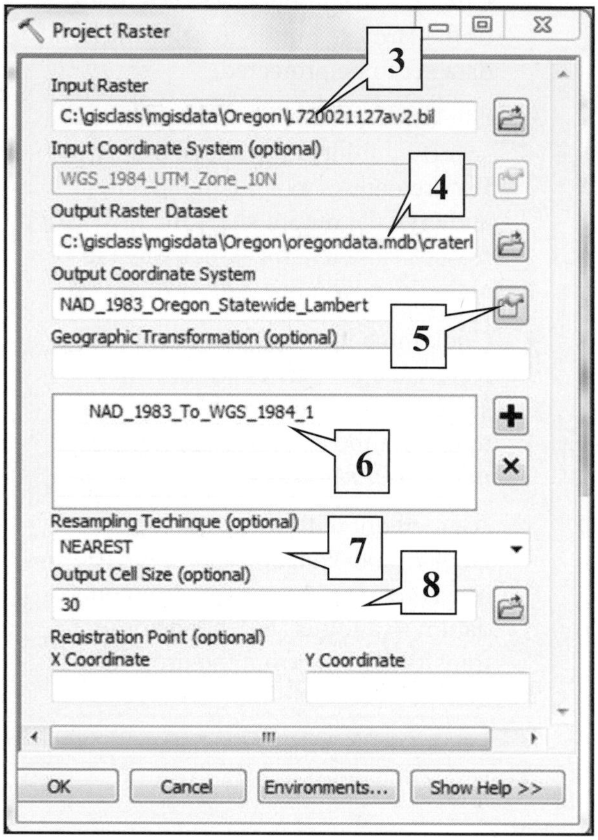

Fig. 66. The Project Raster tool

Georeferencing rasters

Georeferencing takes a raster with an undefined and unknown coordinate system and assigns a coordinate system to it by means of matching points with a reference data set.

1. Determine the likely projection of the map in the raster. It need not be exact, but the closer it is to the correct shape the more accurate the georeferencing will be.
2. Find a reference data set that has a known coordinate system, with features that can be identified and matched to locations on the raster.
3. Load the reference data set into ArcMap and set the data frame to the coordinate system chosen in Step 1.
4. Zoom and pan until the reference data set occupies approximately the same position and region in the ArcMap window as the raster map does in its image.

5. Add the image to the map document. It will usually not be visible yet.

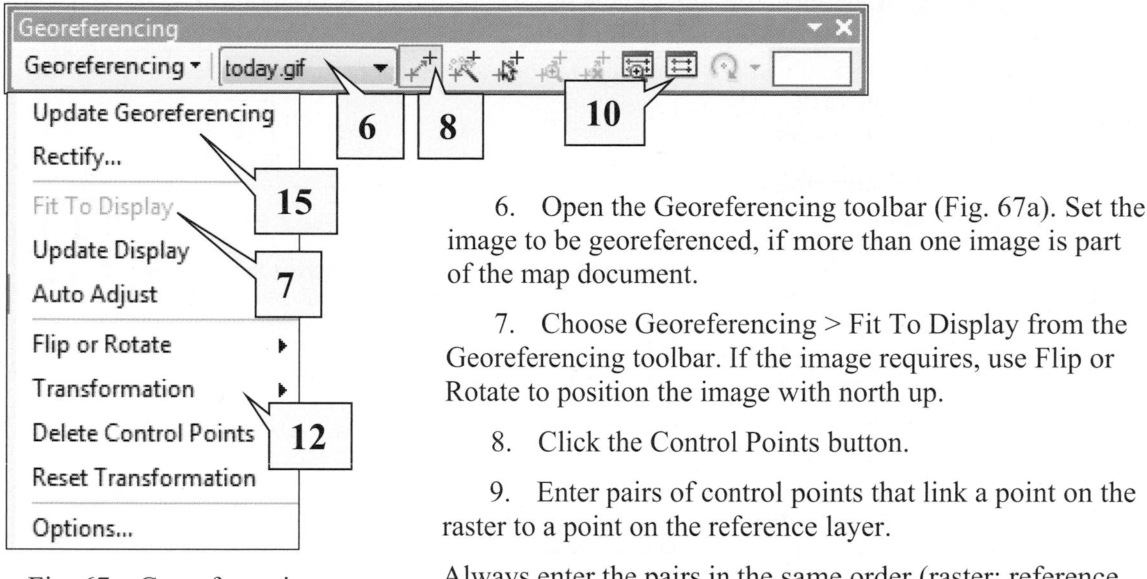

Fig. 67a. Georeferencing toolbar

6. Open the Georeferencing toolbar (Fig. 67a). Set the image to be georeferenced, if more than one image is part of the map document.

7. Choose Georeferencing > Fit To Display from the Georeferencing toolbar. If the image requires, use Flip or Rotate to position the image with north up.

8. Click the Control Points button.

9. Enter pairs of control points that link a point on the raster to a point on the reference layer.

Always enter the pairs in the same order (raster: reference or vice versa). The display will update as points are entered. Zoom and pan between pairs to get the points as accurate as possible. Place points near the four corners of the image at minimum, and more if needed.

10. When no further improvement is seen as points are entered, open the Link Table to examine the control points and residual errors (Fig. 67b).
11. A point with a high residual error relative to the others can be highlighted and deleted.
12. Change the Transformation method if a higher-order transformation is needed.
13. Save the control points as a text file, if desired, for future reference.
14. Record the total RMS error for inclusion in the data set metadata.
15. Click Update Georeferencing. The image transformation will be saved as a world file with the image. Use Rectify to save a new copy of the raster. See ***Rectifying rasters***.

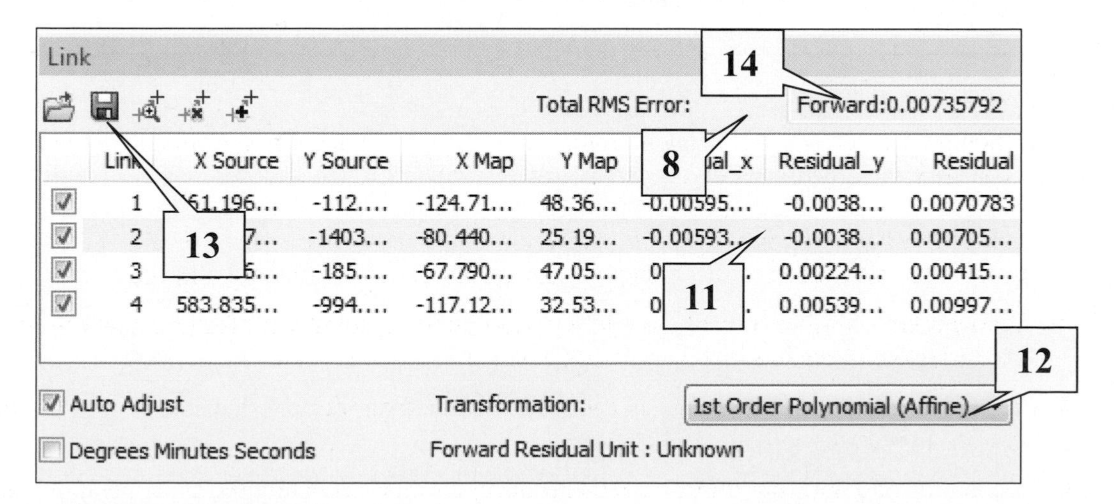

Fig. 67b. Georeferencing a raster

Rectifying rasters

Rectifying saves a georeferenced raster as a new raster file with the georeferenced coordinate system.

1. Georeference the raster. See ***Georeferencing rasters***.
2. Choose Georeferencing > Rectify from the Georeferencing toolbar.
3. Change the cell size, if desired (Fig. 68).

The suggested size is given in the raster coordinate system units (degrees in this case) and is an exact transformation of the cell size of the original raster to the new coordinate system. You can use it as is, or round it slightly if you wish.

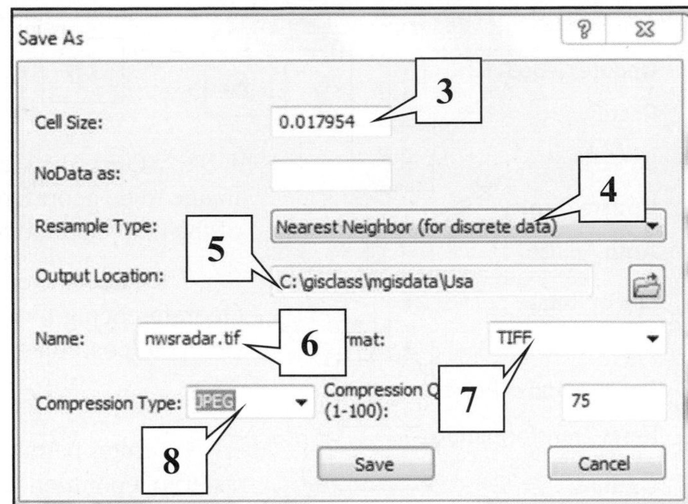

Fig. 68. Rectifying a raster

4. Set the resampling technique to NEAREST if the raster contains discrete or categorical data, or if it is an image. Set it to BILINEAR if the raster contains continuous data. See the discussion in Chapter 11 for more information.
5. Specify the folder or geodatabase in which to place the output raster.
6. Give the output raster a name. Leave the extension blank.
7. Set the output format if the raster is being saved in a folder.
8. Adjust the compression type and quality if desired.
9. Click Save.

TIP: Rectifying often changes the colors shown in an image raster. To reset it, use ArcToolbox > Data Management Tools > Raster > Raster Properties > Add Colormap to copy the colormap from the original image to the new image.

Metadata

Setting the metadata style

The default metadata style is the simple Item Description. To view or edit metadata based on a metadata standard, the style must be changed.

1. In ArcCatalog, choose Customize >ArcCatalog Options from the main menu bar.
2. Click the Metadata tab (Fig. 69).
3. Choose the desired metadata style from the drop-down list.
4. Usually, you will leave the box checked to *Automatically update when metadata is viewed.*
5. Click OK.

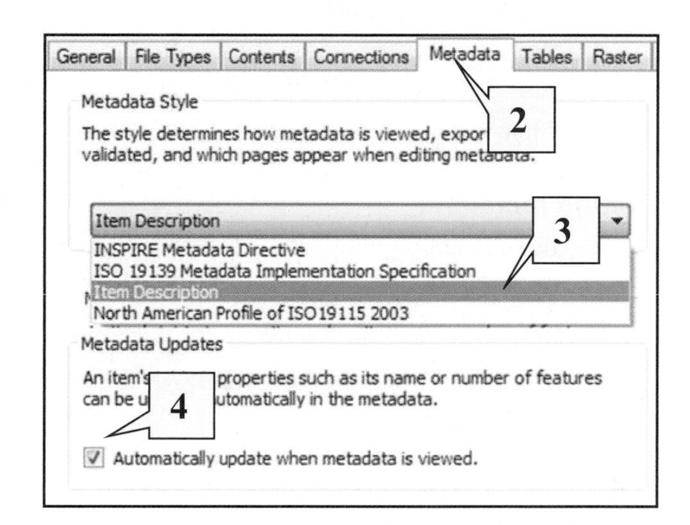

Fig. 69. Setting the metadata style

Viewing metadata

1. In ArcCatalog, navigate to the data item in the Catalog Tree and click on it to highlight it.
2. Click the Description tab.
3. If a standard-based metadata style is being used, click the text to expand the additional sections of metadata, including the ArcGIS Metadata or FGDC Metadata.

Creating a metadata template

1. Determine the fields that will be common to all the data sets for which the template will be used and decide what will be put into them.
2. In ArcCatalog, navigate to the desired folder in the Catalog Tree.
3. Right-click it and choose New > XML Document.

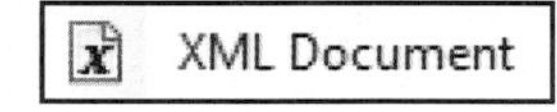

4. Click to highlight the XML file in the Catalog Tree and click the Description tab. Click Edit to begin editing the template. See ***Editing metadata***.

Importing or exporting metadata

TIP: Because of a problem in the initial release, you must turn off background processing to import/export metadata. Use Geoprocessing > Geoprocessing Options from the main menu bar.

1. In ArcCatalog, navigate to the data set to receive the metadata. Highlight it in the Catalog Tree.
2. Click the Description tab, if necessary.
3. Click Import from the menu bar in the Description panel.

4. Click the Browse button and select the data set or XML file from which the metadata are to be imported (Fig. 70).
5. Set the Import Type. If you are importing from an XML file or a data set produced using ArcCatalog 10, choose FROM_ARCGIS.
6. The target metadata should already be set.
7. Click OK.

Fig. 70. Importing metadata

Editing metadata

1. In ArcCatalog, navigate to the data set to be documented and click to highlight it in the Catalog Tree.
2. Click the Description tab.
3. Click the Edit button to open the Metadata Editor (Fig. 71).
4. Click one of the sections in the outline on the left panel. A green check means the section is complete; a red x means it is missing one or more items.
5. Fill in the desired metadata for each field. Fields with red outlines are required for the metadata according to the standard specified in the metadata style.

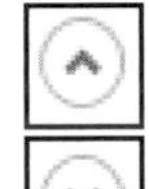

6. Use the arrow buttons to expand or collapse items within a section.

7. Use the + button to add an item container such as a report.

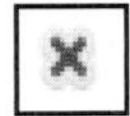

8. Use the X button to delete an item container.
9. Click Exit to exit without saving changes.
10. Click Save to exit and save changes.

TIP: Click Save and start editing again before switching to a new section.

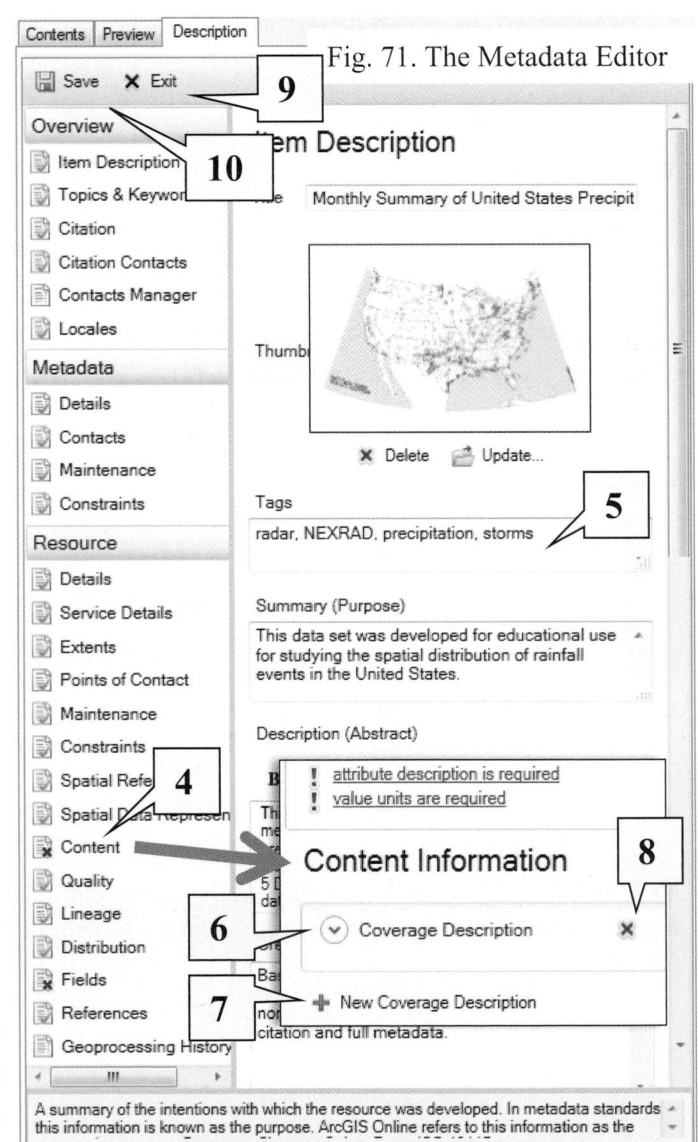

Fig. 71. The Metadata Editor

Editing

Begin an editing session

1. Open ArcMap and add the data layers you wish to edit. For best results, add only as many layers as needed.

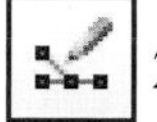

2. If the Editor toolbar is not already displayed, click the Editor Toolbar button on the Standard toolbar.
3. Choose Editor > Start Editing from the Editor toolbar.
4. If data from more than one folder or geodatabase appear in the map document, you will be prompted to choose which one to open for editing (Fig. 72). Choose a layer from the upper panel or a workspace from the lower panel.

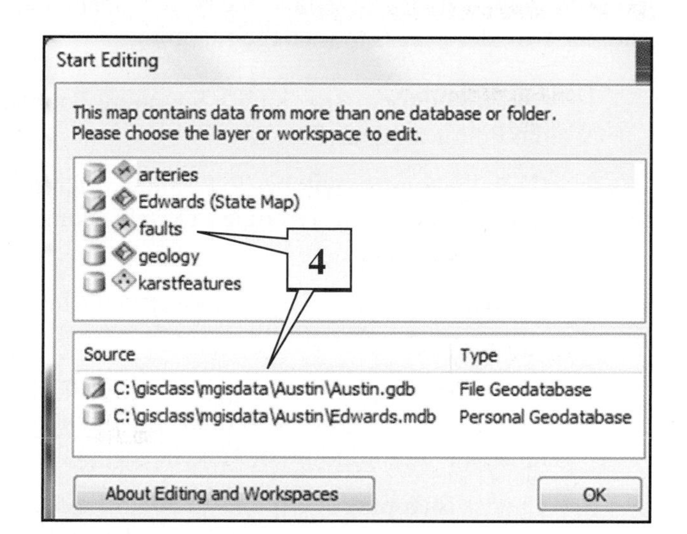

Fig. 72. Choosing an edit workspace

5. If more than one coordinate system is represented in the Edit folder or geodatabase, a warning will be given. Generally, it is best to stick to one coordinate system; however, there are some situations where it is appropriate to have more than one. Just be careful.

Saving edits

1. Choose Editor > Save Edits from the Editor toolbar.
2. Continue editing.

Stopping an edit session

1. Choose Editor > Stop Editing from the Editor toolbar.
2. When prompted, indicate whether to save any current changes or discard them.

Switching to another folder or geodatabase

1. Choose Editor > Stop Editing from the Editor toolbar.
2. Choose whether to save the changes from the current session.
3. Make sure that the layers from the new folder or geodatabase to be edited appear in the Table of Contents.
4. Choose Editor > Start Editing from the Editor toolbar.
5. Select the new folder or geodatabase to edit. Click OK.

Controlling snapping

Snapping is turned on by default. When the cursor gets close to an existing feature, a Snap Tip pops up to indicate the layer and the type of snapping that will occur on the click. The Snapping toolbar provides access to manage snapping.

1. Choose Editor > Snapping > Snapping Toolbar from the Editor toolbar (Fig. 73).
2. Click the toggle switches to turn the different types of snapping on or off: (from left to right) point, end, vertex, and edge.
3. Choose Snapping > Use Snapping to uncheck it and turn snapping off entirely.
4. Choose one of the less frequently used snapping methods by selecting it from the Snapping menu: intersection, midpoint, or tangent snapping.
5. Choose Snapping > Options to control more aspects of snapping. You can also access this menu using Editor > Snapping > Options from the Editor toolbar.
6. Increase or decrease the snapping Tolerance from the default of 10 pixels, if desired (Fig. 74).
7. Check/uncheck the Show Tips box to control whether the Snap Tips are displayed.
8. Check/uncheck the boxes to control which parts of the Snap Tips are shown.
9. Click the Text Symbol box to specify a different font for the Snap Tips.

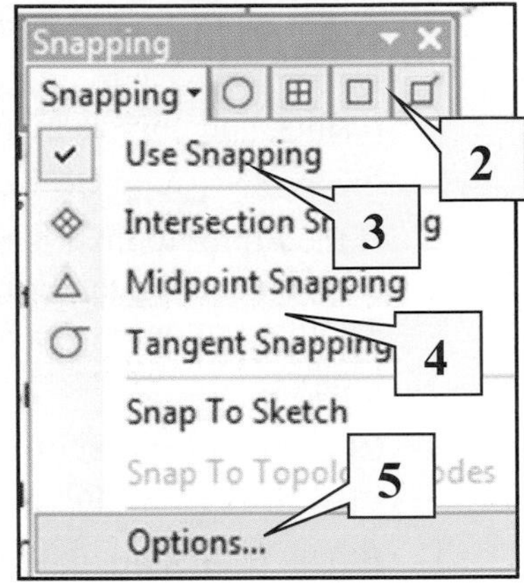

Fig. 73. The Snapping toolbar

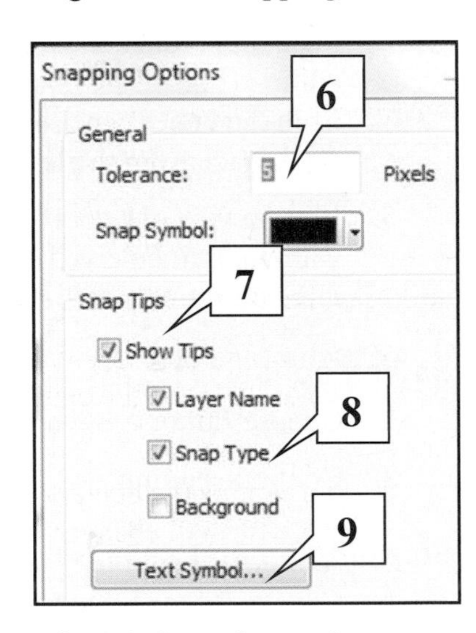

Fig. 74. Snapping options

TIP: Keep the Snapping Environment window open to facilitate making changes to the snapping settings as you edit.

Selecting things to edit

Several tools are used for selecting the features or other items to be changed during editing. Features may also be selected using the Select By Attributes and Select By Location.

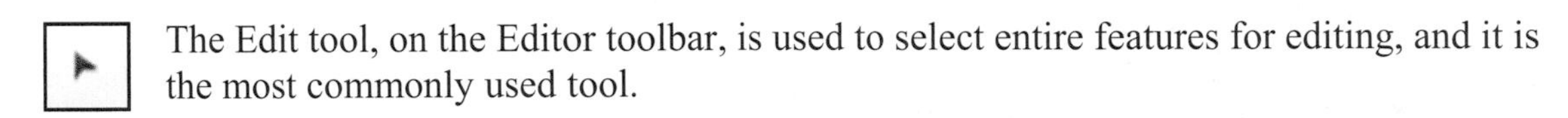

The Edit tool, on the Editor toolbar, is used to select entire features for editing, and it is the most commonly used tool.

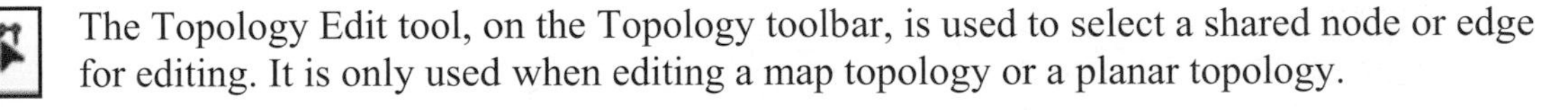

The Topology Edit tool, on the Topology toolbar, is used to select a shared node or edge for editing. It is only used when editing a map topology or a planar topology.

The Fix Topology Error tool, on the Topology toolbar, is used to select topology errors and choose from a set of actions for fixing them. It can only be used when editing a planar topology.

The Edit Annotation tool, on the Annotation toolbar, is used to select annotation.

Using an editing tool

The editing tools are sensitive to many of the same techniques used for interactive selection, including setting the selectable layers, changing the selection method, and changing the selection options, as described in Chapter 5.

1. Click the editing tool on the appropriate toolbar.
2. Click the feature to be selected or click and hold down the mouse button to drag a box around the features to be selected (not advised for use with the Topology Edit tool).
3. To select multiple features, hold down the Shift key and click on each feature in turn.
4. Click off of any feature to clear the selection, or click the Clear Selection button on the Tools toolbar.

Moving features

1. Select the feature(s) to be moved by using the Edit tool on the Editor toolbar.
2. Place the cursor over the objects until crosshairs appear.
3. Click and drag the feature to a new location.

Rotating features

1. Select the feature(s) to be rotated by using the Edit tool on the Editor toolbar or by using another selection method (Fig. 75).

2. Click the Rotate tool on the Editor toolbar.
3. Click and drag anywhere on the screen to rotate the features about the center of rotation (marked with an X).
4. To change the center of rotation, place the cursor over the X. It will change to a crosshairs symbol. Click and drag the center to a new location.
5. Click and drag anywhere outside the features to rotate them about the new center.

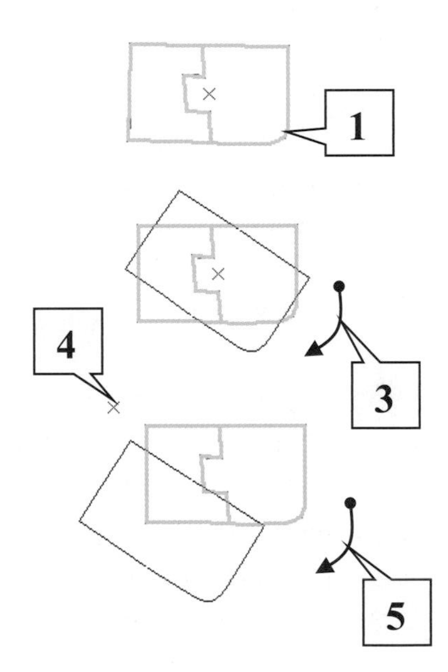

Fig. 75. Rotating features

Deleting features

1. Select the feature(s) to be deleted by using the Edit tool on the Editor toolbar or by using another selection method.

2. Make sure the Edit tool on the Editor toolbar is clicked.
3. Press the Delete key to delete the features.

Creating features with templates

Choose Editor > Editing Windows > Create Features to open the window containing the templates (Fig. 76). Templates usually appear for each layer in the Table of Contents. If a layer is displayed with a unique values map, a template will be created for each category.

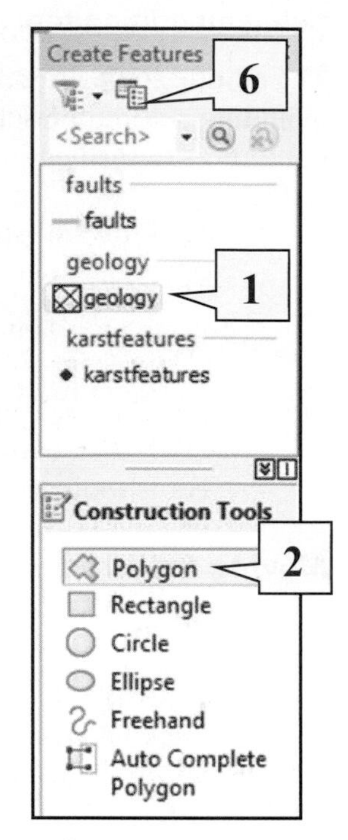

Fig. 76. The Create Features window

Using a template

1. Click the desired template for the new feature to be created.
2. Click a different construction tool than the default one, if desired.
3. Begin adding points or vertices of the feature to be added.
4. If creating lines or polygons, a provisional sketch will be created. Keep adding vertices until the sketch is complete. Double-click it, or right-click and choose Finish Sketch.
5. To create features in a different layer, change to a different template and select the construction tool.

Creating a template

6. Click the Organize Templates button at the top of the Create Features window (Fig. 76).
7. Click the New Template button (Fig. 77).
8. Check the box for the feature class for which to create templates.
9. Click Finish and Close.

Editing template properties

1. Right-click a template in the Create Features window and choose Properties.
2. Change the name, description, or tags.
3. Select the default construction tool for the template.
4. Use the buttons to change the order of the fields, either in layer order or sorted alphabetically. The View button controls whether field names or aliases are shown.
5. Enter values for any attribute fields to be placed in the table when the feature is created.

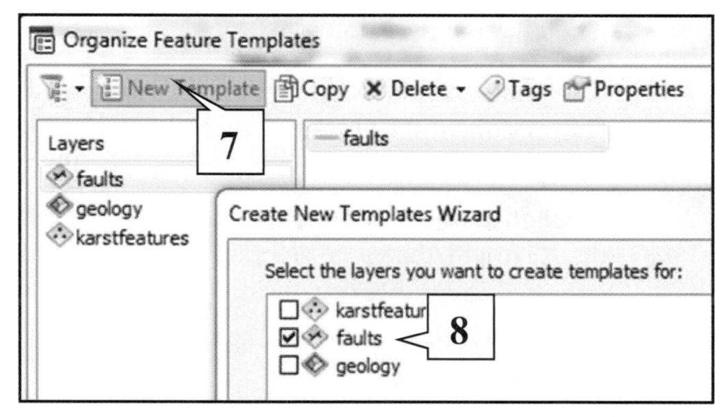

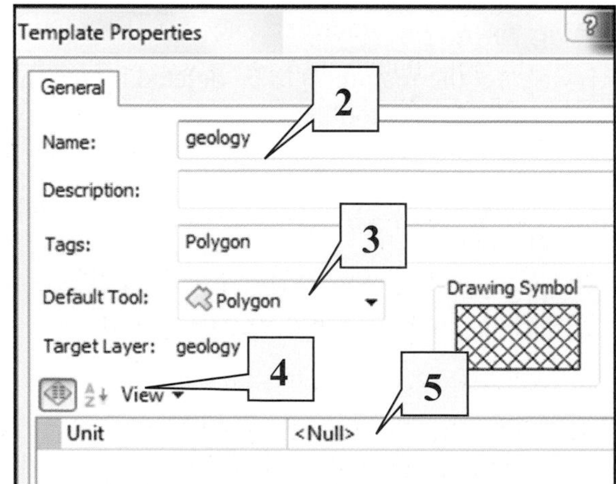

Fig. 77. Creating and modifying templates

Using the sketch context menus

Two context menus can be accessed during sketching. The Vertex menu is accessed by right-clicking on a vertex or segment of the sketch. The Sketch menu is accessed by right-clicking off the sketch.

1. To delete or edit the vertex of a sketch, right-click on a vertex to bring up the Vertex menu (Fig. 78).
2. To add a vertex to a sketch, right-click on a segment of the sketch between two vertices.
3. To bring up the Sketch menu, right-click the screen at a location off the sketch.
4. To delete the current sketch and start again, right-click to raise one of the menus and choose Delete Sketch.
5. When the sketch is complete, right-click to raise one of the menus and choose Finish Sketch, or simply double-click on the last vertex.

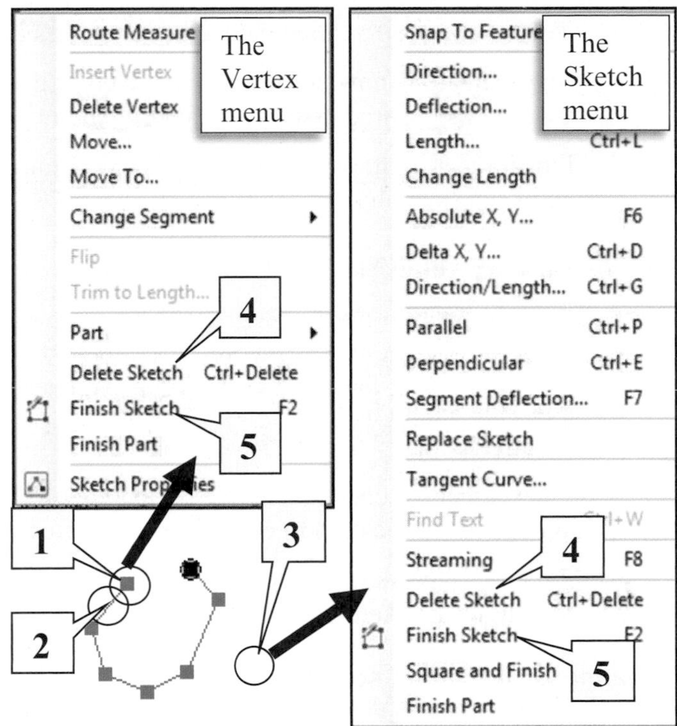

Fig. 78. The Vertex and Sketch menus

Using Sketch menu tools

The following functions can be accessed during a sketch by right-clicking anywhere off the sketch to open the Sketch menu and select the function.

Entering absolute *x-y* locations

1. Right-click off the sketch to open the Sketch menu.
2. Choose Absolute X,Y from the menu.
3. Enter the *x-y* coordinates and press Enter (Fig. 79). The coordinates must be in the same coordinate system as the data frame.
4. Finish the sketch, or switch to another sketching tool and keep sketching.

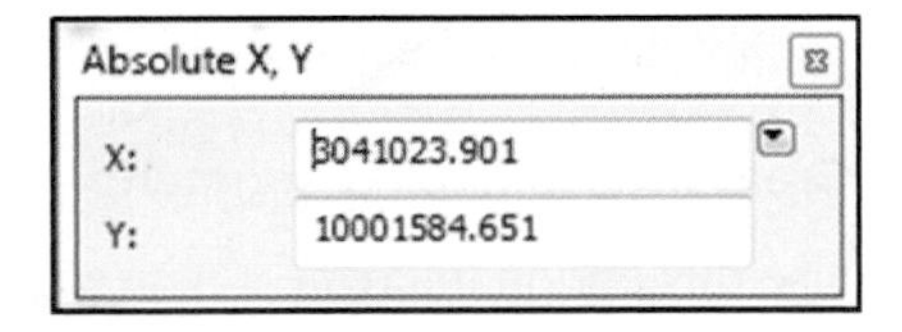

Fig. 79. Absolute X,Y

Using offsets from a previous location

1. Enter at least one vertex of the sketch.
2. Right-click off the sketch to open the Sketch menu.
3. Choose Delta X,Y from the menu.
4. Enter the distance change in the *x* and *y* directions in map units (Fig. 80). Press Enter.

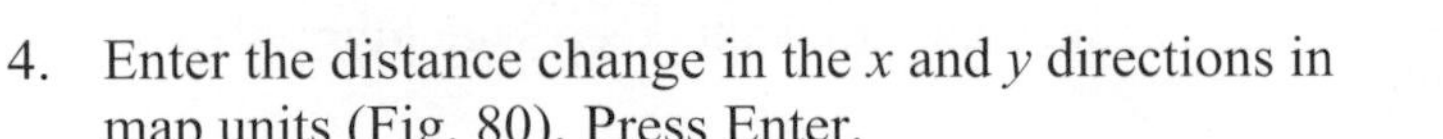

5. Finish the sketch, or switch to another sketching tool and keep sketching.

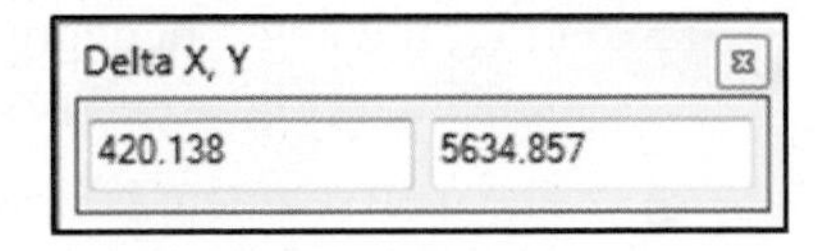

Fig. 80. Delta X,Y

Creating a segment in a specified direction

1. Enter at least one vertex of the sketch.
2. Open the Sketch menu and choose Direction (Fig. 81).
3. Type in the desired angle in degrees and press Enter.
4. The segment is now constrained at the desired angle. Click at the desired distance to enter the point.
5. Finish the sketch, or switch to another sketching tool and keep sketching.

Fig. 81. Constraining the segment direction

Creating a segment of specific length

1. Enter at least one vertex of the sketch.
2. Open the Sketch menu and choose Length.
3. Type the length in map units in the box and press Enter (Fig. 82).
4. The segment is constrained to a set length. Click at the desired location to create the next vertex.
5. Finish the sketch, or switch to another sketching tool and keep sketching.

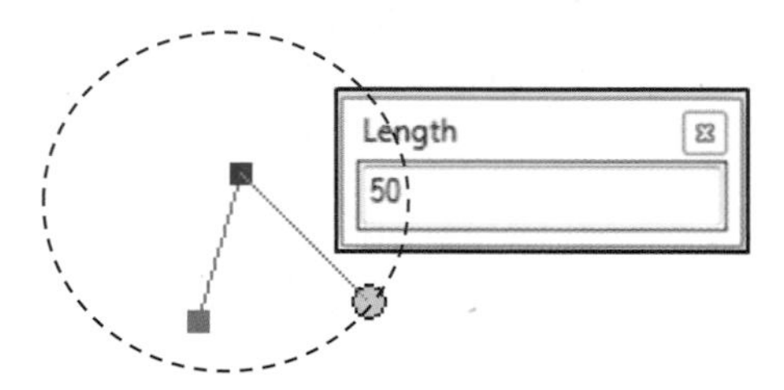

Fig. 82. Constraining the sketch segment to a set length

Creating a segment of set angle and length

1. Enter at least one vertex of the sketch.
2. Open the Sketch menu and choose Direction/Length.
3. Type in the desired angle and length and press Enter (Fig. 83).
4. Finish the sketch, or switch to another sketching tool and keep sketching.

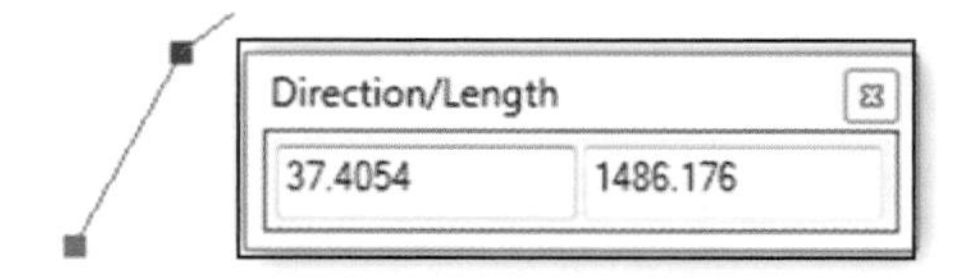

Fig. 83. Constraining the direction and the length of a segment

Creating a deflected segment

Use this option to create a new segment at a specified angle from the last segment in the sketch.

1. Enter at least one segment of the sketch.
2. Open the Sketch menu and choose Deflection.
3. Enter the angle in degrees that the new segment will be deflected (Fig. 84). Press Enter.
4. The new segment is now constrained along a line. Click at the desired distance to enter the new vertex.
5. Finish the sketch, or switch to another sketching tool and keep sketching.

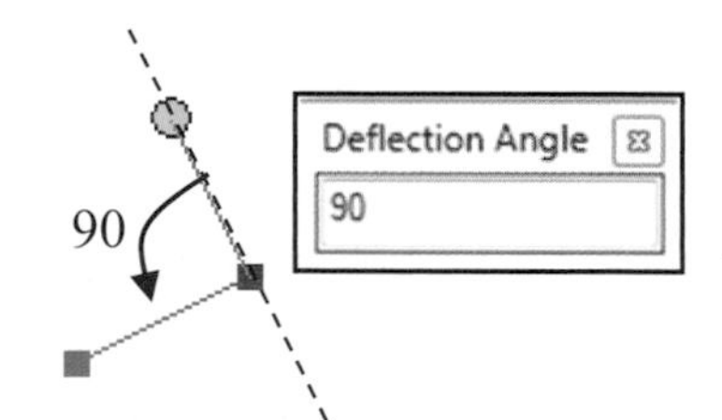

Fig. 84. Deflecting a vertex from the previous segment

Deflecting a segment from a feature

Use this option to create a new segment at a specified angle to an existing segment.

1. Enter at least one segment (two vertices) of the sketch.
2. Right-click *on an existing feature segment* (the side of this building) to open the context menu and then choose Segment Deflection.
3. Type in the angle of deflection from the existing feature and press Enter (Fig. 85).
4. The new segment is now constrained at the specified angle (45 degrees) to the existing segment (the south side of the building). Click at the desired length to create the new vertex.
5. Finish the sketch, or switch to another sketching tool and keep sketching.

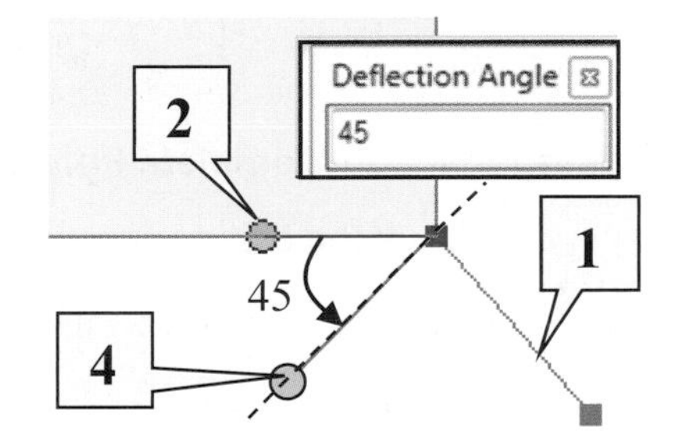

Fig. 85. Deflecting at an angle to an existing segment

Creating parallel or perpendicular segments

1. Enter at least one vertex of the sketch (Fig. 86).
2. Right-click *on an existing feature segment* (the side of this building) to open the context menu and then choose Parallel or Perpendicular.
3. The new segment will be constrained in the chosen direction. Click to enter the new vertex at the desired location.
4. Finish the sketch, or switch to another sketching tool and keep sketching.

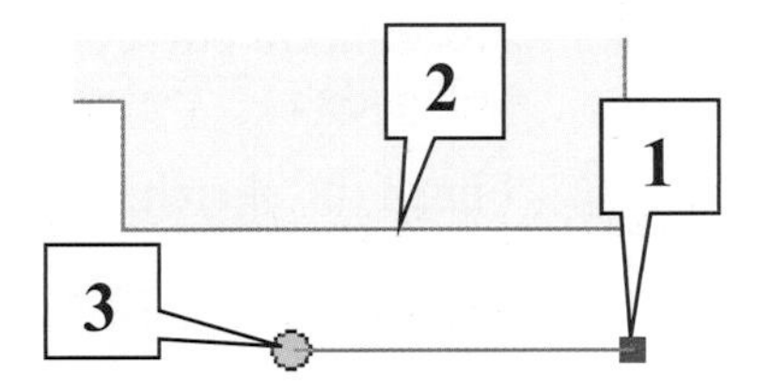

Fig. 86. Creating a segment parallel to an existing feature

Creating multipart features

Multipart features include more than one contiguous feature. This technique is used to create a single feature from multiple polygon areas such as Hawaii, from multiple segments forming a single street, or from multiple points forming a single feature.

1. Sketch the first part of the feature using any of the sketching tools and context menus (Fig. 87).
2. Right-click on or off the sketch and choose Finish Part.
3. Add as many parts as desired, ending each with a Finish Part.
4. On the last part, right-click on or off the final sketch and choose Finish Sketch.

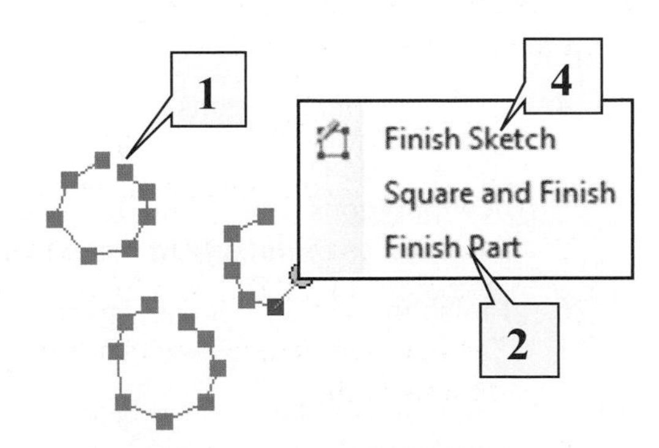

Fig. 87. Creating multipart features

Creating adjacent polygons

It is very important that adjacent polygons have identical shared boundaries. You can create adjacent polygons with coincident boundaries in two different ways by using the AutoComplete Polygons tool or by creating a large polygon boundary and cutting it up into sections.

Using the AutoComplete Polygons tool

1. Click the editing template for the layer in the Create Features window.
2. Choose the Polygon construction tool, or one of the other construction tools (Fig. 88).
3. Create the first polygon.
4. Choose the Auto Complete Polygon tool from the Construction tools list.
5. Sketch only the undefined boundary of the next polygon. Be sure to start and end inside the adjacent polygon or inside two adjacent polygons.
6. Finish the sketch to create the new polygon.

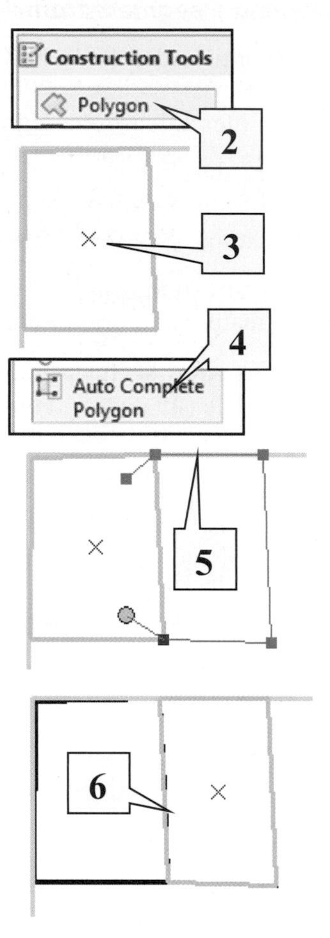

Fig. 88. Creating adjacent polygons with the Auto Complete task

Cutting polygon features

1. Click the editing template for the layer in the Create Features window.
2. Choose one of the polygon construction tools.
3. Create the polygon around the outer boundary (Fig. 89).

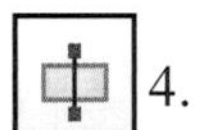

4. Click the Cut Polygons tool on the Editor toolbar.
5. Sketch a line that will form the boundary between the new polygons. Important: The line must start and end **snapped to** or **outside of** the existing polygons.
6. Finish the sketch. The new polygons will be created.

TIP: A polygon must be selected before it can be cut with the Cut Polygons tool.

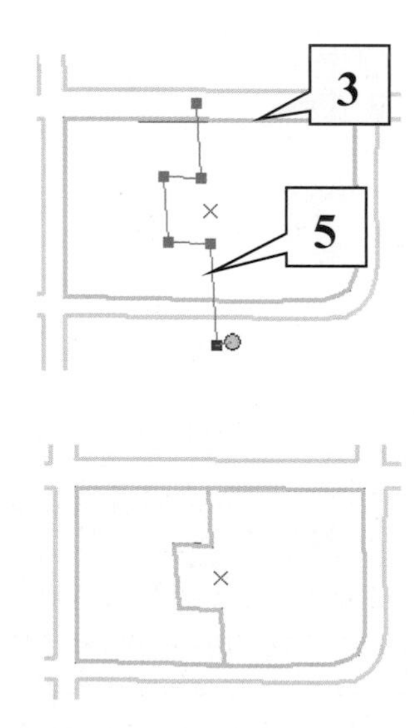

Fig. 89. Cutting a polygon

Using the Attributes window

The Attributes window is an interface for entering and editing attributes fields of features. It is convenient to dock this window during editing—under the Table of Contents is a good place.

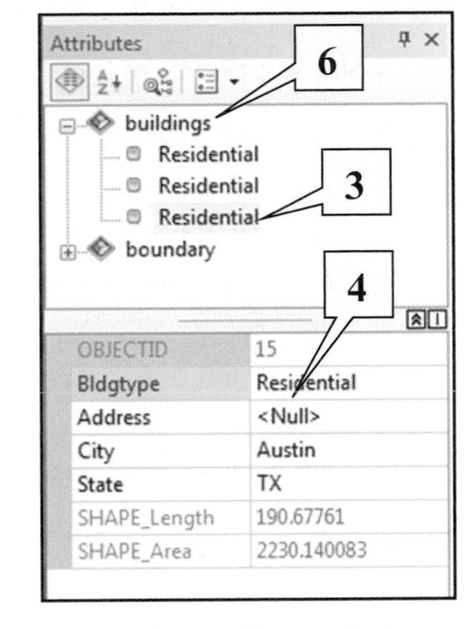

Fig. 90. The Attributes window

1. Select the feature(s) to edit.

2. Click the Attributes button on the Editor toolbar. It may come up as a tab in the Create Features window.

To edit a single feature

3. Click the record to edit in the top panel (Fig. 90). Its attributes will be shown in the lower panel.
4. Click in the attribute field and type the new value. Press Enter or tab, or move to a different field.
5. Keep editing fields until you are satisfied.

To edit all features

6. Click the layer name at the top of the list. The fields on the right will go blank.
7. Click the field to edit and type the new value. The new value will replace the existing values for each record in the selected set.

Managing records

8. Clicking an entry in the list causes the feature to flash briefly on the map so that you can locate it.
9. Right-click one entry in the list to bring up a context menu with many useful functions (Fig. 91).

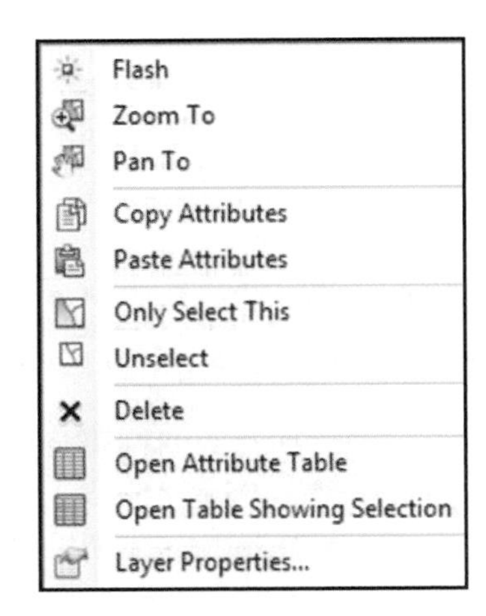

Fig. 91. The context menu

Changing the display field

10. Right-click one of the records in the Attributes window and choose Layer Properties.
11. Click the Display tab.
12. Choose the Display Expression Field from the drop-down box. Click OK.
13. The change will be applied to the Attributes window when you make the next selection.

Using the sketching tools

By default, a straight segment connects adjacent vertices entered by the user. The sketching tools provide additional options for entering a vertex or designating the shape of the connecting segment. Each sketching tool can be used at the beginning, middle, or end of a sketch. You can use multiple tools in succession during a single sketch. The sketching tools are selected from the Editor menu. The Straight Segment and Endpoint Arc Segment tools have their own icons on the Editor toolbar. The other tools are accessed from a drop-down on the Editor toolbar.

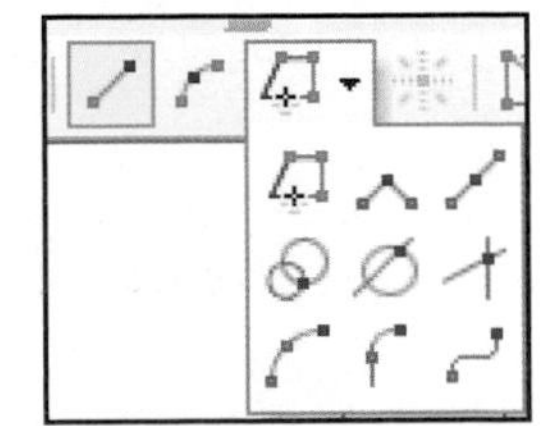

Straight Segment tool

1. Select an editing template.
2. Click on the Straight Segment tool if necessary (it is the default tool).
3. Enter vertices by clicking on the screen.

Endpoint Arc Segment tool

With this tool the user specifies the beginning, end, and radius of an arc.

1. Click on the Endpoint Arc tool. The current vertex will be the start of the arc.
2. Click on the endpoint of the arc (Fig. 92).
3. Move the mouse to visually set the desired radius of the arc. Click to enter the vertex.
4. To specify an exact radius, press the R key, type in the radius, and press Enter.

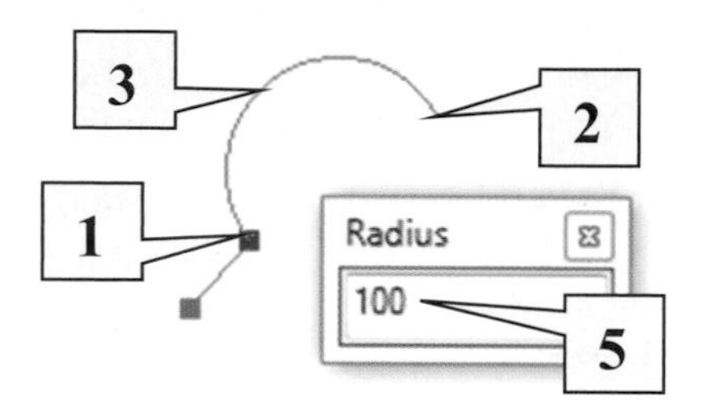

Fig. 92. The Endpoint Arc Segment tool

Tracing tool

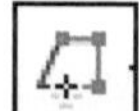

1. Select the feature to be traced (Fig. 93).
2. Select the Trace tool.
3. Press the letter O key on the keyboard to specify the offset distance in map units. Type the distance into the box. Change the corner style if desired. Press Enter.
4. Click on the feature to begin tracing and move the mouse along the feature to create the trace. (If the trace appears on the wrong side, press O again and make the distance negative.)
5. Click again to finish the trace and create the sketch.
6. Finish the sketch, or switch to another sketching tool and keep sketching. The offset will remain at the same value until you change it.

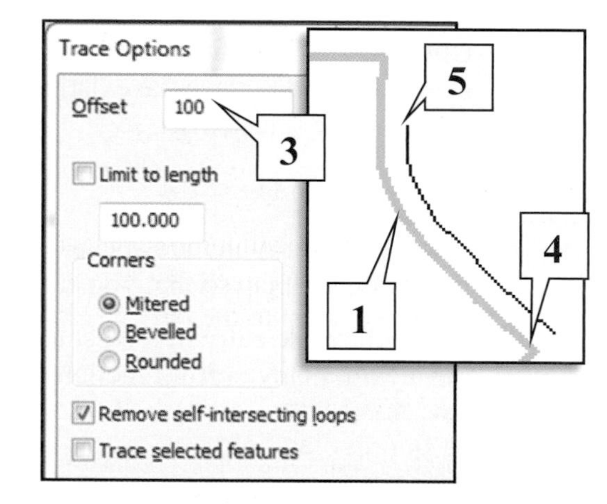

Fig. 93. Tracing a feature

TIP: Features have ending locations according to how they were digitized. A trace stops when it reaches the end of the feature. To continue tracing, click at the stopping point to end the current trace and then click again at the same point to begin a new trace.

Right Angle tool

This tool is useful for creating buildings and other features with square corners.

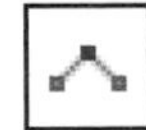

1. Select the Right Angle tool.
2. Carefully enter the first segment along the edge of the feature (Fig. 94). All corners will be perpendicular to this segment.
3. Enter the vertices of the other corners. Each one will be perpendicular to the previous segment.
4. Instead of adding the final vertex with a click, right-click OFF the sketch and choose Square and Finish from the Sketch menu.

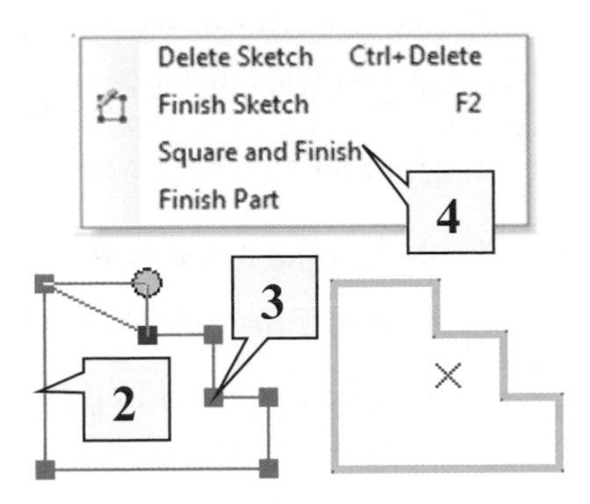

Fig. 94. Creating features with square corners

Midpoint tool

The Midpoint tool is useful for entering a vertex at the halfway point of a specified distance. Imagine that you wish to split an existing parcel exactly in half.

1. Turn on vertex snapping to ensure that the exact midpoint is found.

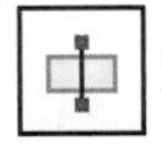

2. Select the parcel polygon to split and click the Cut Polygons tool on the Editor toolbar.

3. Choose the Midpoint tool.
4. Click on the location representing the start of the measuring line to be halved (Fig. 95). Be sure to snap to the parcel vertex.

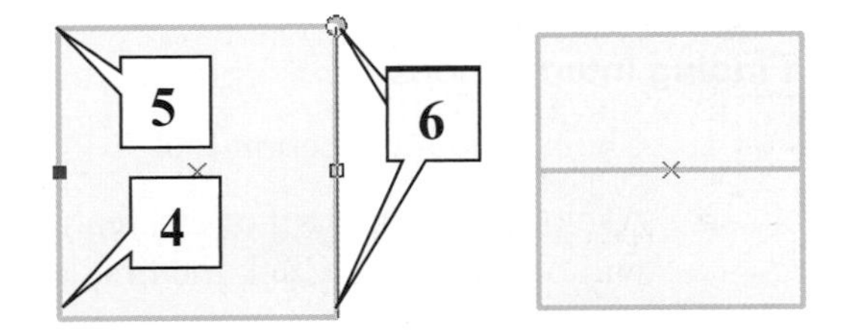

Fig. 95. Using the Midpoint tool to split a parcel in half

5. Click on the other end of the measuring line. A vertex is entered at the midpoint of the measuring line.
6. Click to start and end another measuring line on the other side of the parcel, thereby entering a second vertex.
7. Right-click and choose Finish Sketch.

Distance-Distance tool

1. Choose the Distance-Distance tool.
2. Click at the center of the first distance circle and move the mouse out to expand the circle (Fig. 96).
3. Press the D key to bring up the Distance window, and type in the exact distance in map units. Press Enter.
4. Click at the center of the second circle, and type D to enter a distance for it. Press Enter to close the window.
5. Move the cursor over one of the two intersection points (a blue circle will appear). Click to create the point or vertex.
6. Finish the sketch, or switch to another sketching tool and keep sketching.

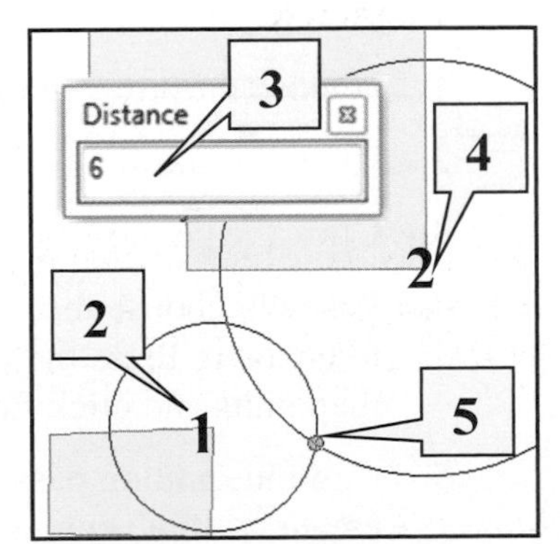

Fig. 96. Finding the intersection point of two distance circles

Direction-Distance tool

This tool creates a new vertex using a bearing direction from a known point plus a distance from another point. Imagine digitizing a tree location that has a known bearing and distance from a building corner.

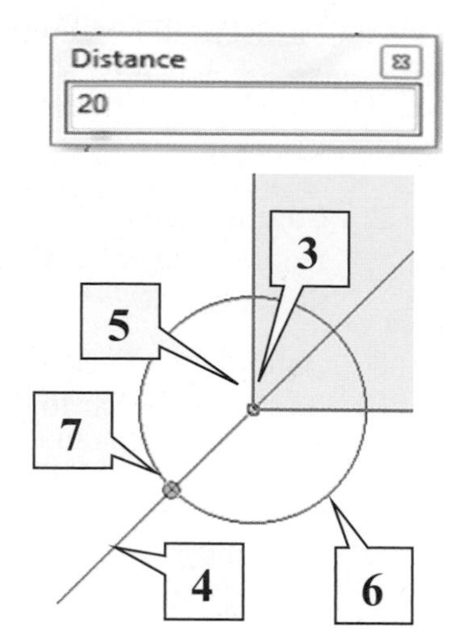

Fig. 97. Using the Direction-Distance tool

1. Select the feature template.

2. Choose the Direction-Distance tool (Fig. 97).
3. Click on the building corner to establish the rotation point of the direction segment. A temporary line appears.
4. Rotate the line to the desired direction and click, or press the D key to type a specific direction angle. Press Enter.
5. Click on the corner again to define the center of the circle.
6. Move the mouse to the desired radius and click, or press R to type a specific radius and press Enter.
7. Click on one of the two possible intersections to create the vertex.

Finding intersections

1. Choose the Intersection tool.
2. As you move the tool over a feature edge, its extension will be indicated. Click the first line on the south side (1 in Fig. 98).
3. Click the second line on the west side. Now the point of intersection will be visible where the extensions of the two lines cross.
4. Click near the intersection to enter the point or vertex.
5. Finish the sketch, or switch to another sketching tool.

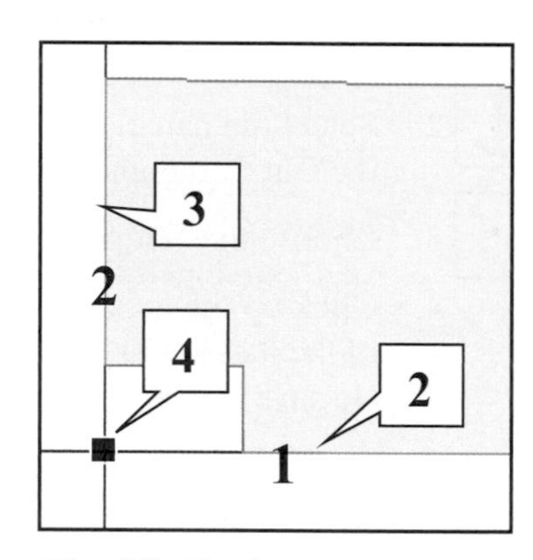

Fig. 98. Placing a vertex at the intersection of two lines

Arc Segment tool

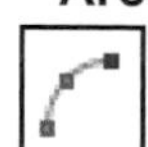

1. Choose the Arc Segment tool.
2. Click the starting point of the curve (Fig. 99).
3. Click the point through which the curve will pass.
4. Visually choose the desired curve and click the endpoint to create it. Press the R key to enter a specific radius in map units and click Enter.
5. Continue adding pass-through points and endpoints to create multiple curves, if desired.
6. Finish the sketch, or switch to another sketching tool and keep sketching.

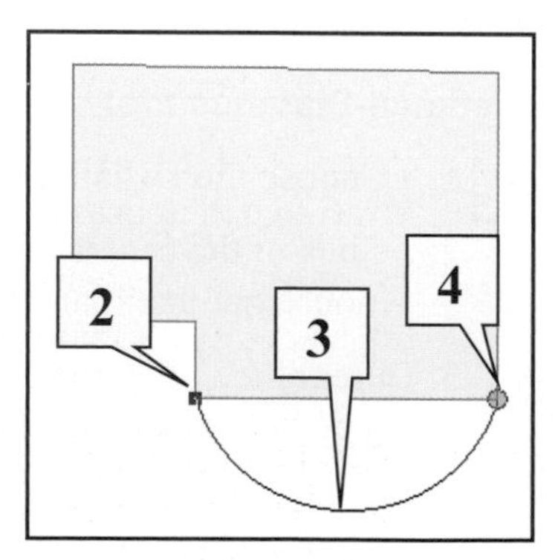

Fig. 99. Sketching a parametric curve

Tangent Curve Segment tool

This tool creates an arc that is tangential to the previous sketch segment and is useful for creating smooth curves for roads. At least one sketch segment must be present first.

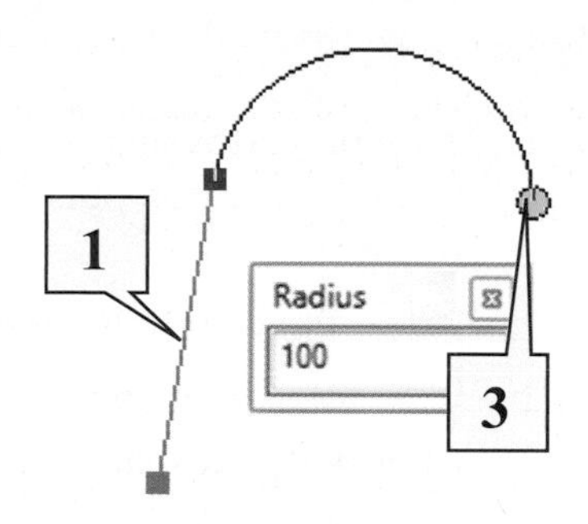

Fig. 100. Creating a tangential curve

1. Use Sketch or another sketching tool to create one or more sketch segments (Fig. 100).

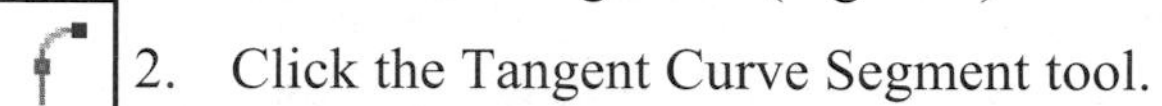

2. Click the Tangent Curve Segment tool.
3. Move the mouse to the desired endpoint of the tangential curve and click to create the curve. Click R to enter a radius in the box and press Enter.
4. Finish the sketch, or switch to another sketching tool and keep sketching.

Bezier Curve Segment tool

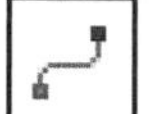

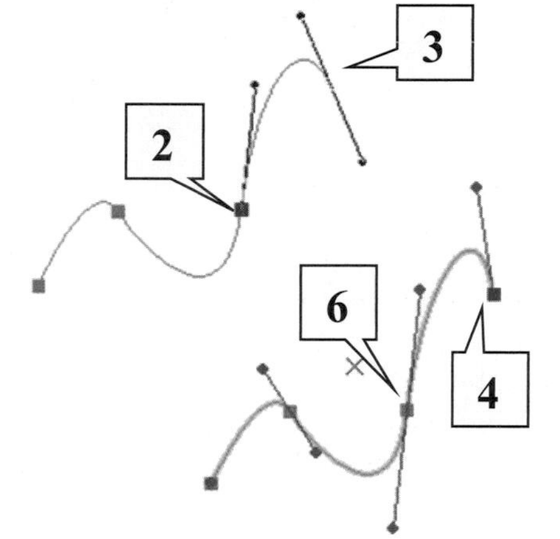

Fig. 101. Creating Bezier curves

1. Choose the Bezier Curve Segment tool.
2. Add a vertex where the curve will begin (Fig. 101). The first guide will appear. Enter a second vertex to establish the length and angle of the guide.
3. Add another vertex for the second guide. Move the cursor to set the distance and angle of the curve. Make the guide longer for more curve; rotate the line to change the curve shape.
4. Click to add a vertex at the curve's end.

5. Move the cursor to modify the handle and complete the curve's shape.

6. To change the shape later, select the feature with the Edit tool. Click the Edit Vertices tool on the Editor toolbar. Move the guides from the center, or rotate them from the ends, to change the curve shape.

Modifying a feature

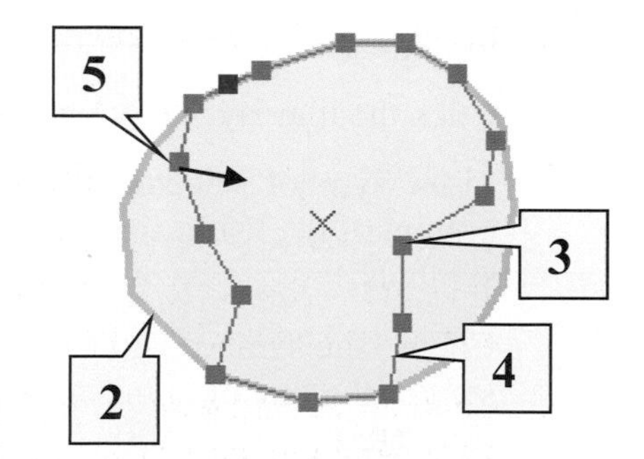

Fig. 102. Modifying a feature

1. Select a feature to modify with the Edit tool.

2. Click the Edit Vertices tool. The sketch of the feature will appear (Fig. 102).
3. To delete a vertex, place the cursor on the vertex, right-click, and choose Delete Vertex from the Sketch menu.
4. To add a vertex, place the cursor over the spot to add the vertex, right-click, and choose Add Vertex.
5. To move a vertex, place the cursor on top of the vertex to be moved until a crosshair symbol appears. Click and drag the vertex to a new location.
6. When done modifying the vertices, right-click ON the sketch and choose Finish Sketch.

Reshaping a feature

Reshaping works well for reentering part of a feature without having to redo the entire thing.

1. Turn on vertex and/or edge snapping to ensure that the start/stop points are snapped to the feature.
2. Select the feature to reshaped using the Edit tool (Fig. 103).

3. Choose the Reshape Feature tool from the Editor toolbar.
4. Click ON the feature at the start of the section to be reshaped, being sure to snap to an edge or a vertex.
5. Enter vertices defining the new shape. Use the other sketching tools and context menus as needed.
6. Double-click ON the feature at the end of the section being reshaped, or right-click and choose Finish Sketch.

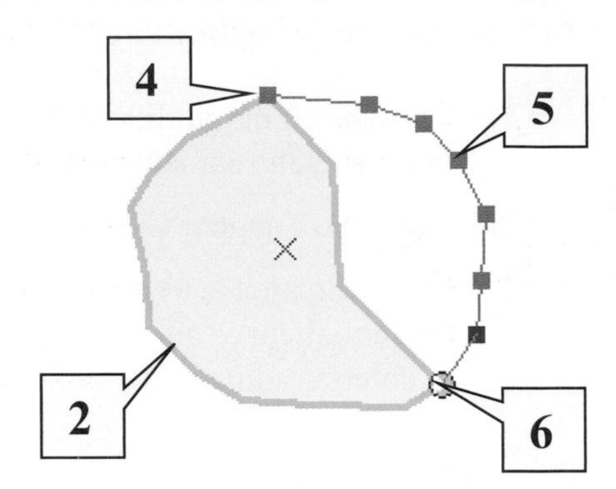

Fig. 103. Reshaping part of a polygon boundary

TIP: If the start points and endpoints of the new sketch do not fall exactly on the feature, the reshaping will not be performed.

Flipping a line

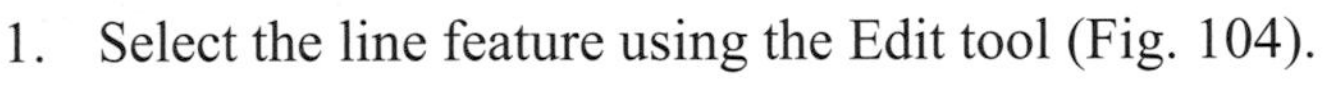

1. Select the line feature using the Edit tool (Fig. 104).

2. Click the Edit Vertices tool on the Editor toolbar. The sketch will appear.

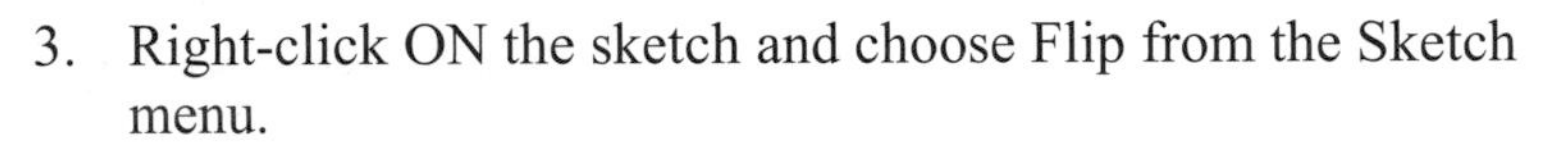

3. Right-click ON the sketch and choose Flip from the Sketch menu.
4. When done, either right-click the sketch to finish it or clear the selection by clicking on the screen.

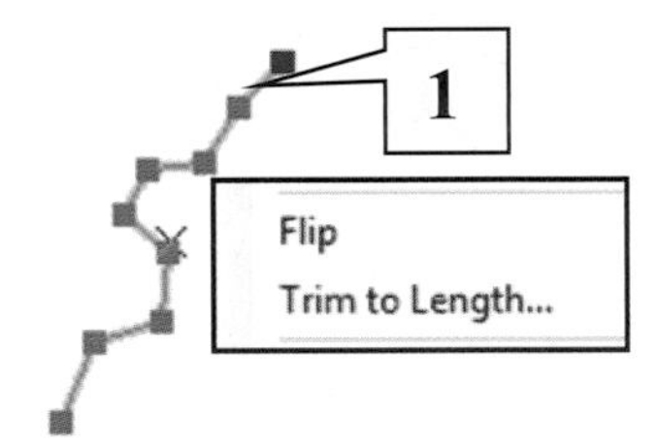

Fig. 104. Flipping a line

Displaying line directions

1. Click the line layer's symbol in the Table of Contents.
2. In the Symbol Selector, scroll to the bottom of the list to find the arrow symbols.

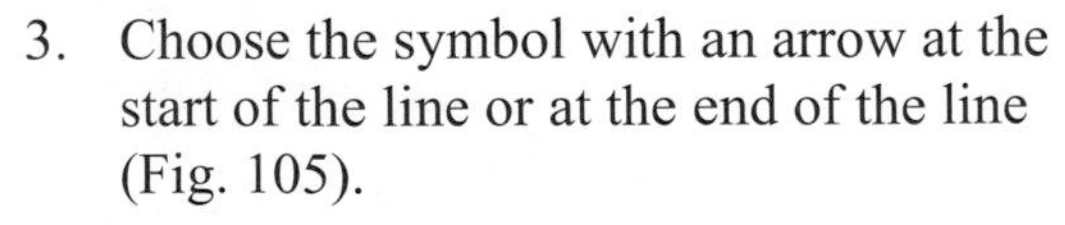

3. Choose the symbol with an arrow at the start of the line or at the end of the line (Fig. 105).
4. Change the color or width as desired. Click OK.

Fig. 105. Arrows in the Symbol Selector for showing line directions

Splitting a line

With the Split tool

Use the Split tool to split a single line in two pieces at the location you click.

1. Turn on vertex snapping if you wish to ensure that the line is split at an existing vertex.
2. Select the line to be split with the Edit tool.
3.  Click the Split tool on the Editor toolbar.
4. Click the location on the line where the split should occur.

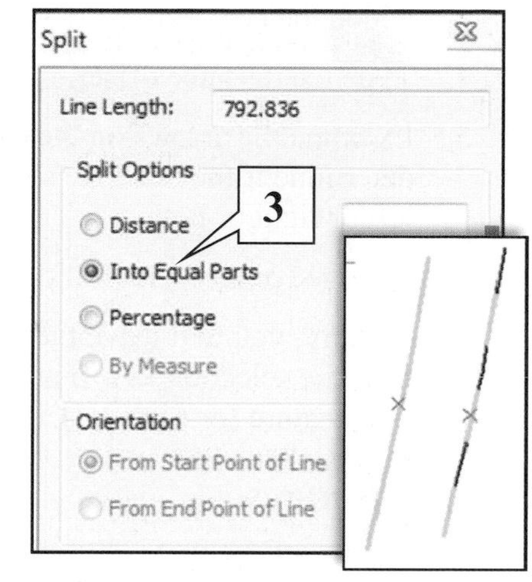

Fig. 106. Splitting a line

Into multiple lines

This task divides a line at equal intervals defined by the user or evenly spaced along the line.

1. Select the line to divide with the Edit tool (Fig. 106).
2. Choose Editor > Split from the Editor toolbar.
3. Choose the split option based on distances, equal parts, or percentages, and enter the values.
4. The new features will replace the original feature.

Creating evenly spaced points

1. Use the Edit tool to select a line along which the points will be placed (Fig. 107).
2. Choose Editor > Construct Points from the Editor toolbar. Arrows will appear on the line to show its direction.
3. Choose the template to specify the layer to which the points will be added.
4. Select the construction option as a number of evenly spaced points or by distance. Enter the values.
5. Click to place additional points at start and end, if desired.
6. Click OK.

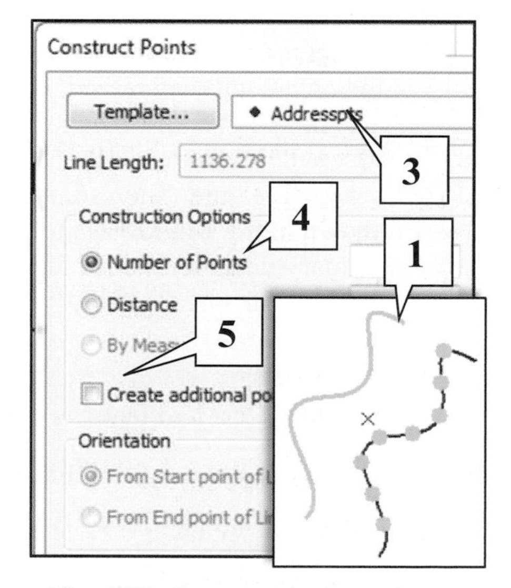

Fig. 107. Constructing evenly spaced points

Moving a feature an exact distance

1. Select the feature to be moved (Fig. 108).
2. Choose Editor > Move from the Editor toolbar.
3. Type the change in *x*- and *y*-coordinates as an offset from the current location and press Enter.

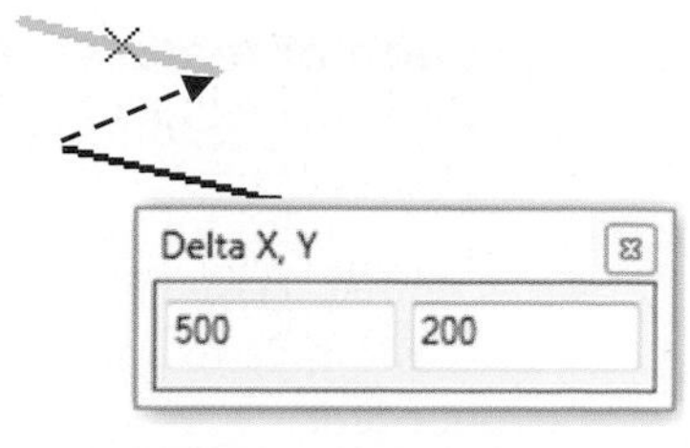

Fig. 108. Moving a line an exact offset

Using Copy Parallel

1. Select the feature to copy (Fig. 109).
2. Choose Editor > Copy Parallel from the Editor toolbar.
3. Choose the template specifying the layer into which the copy will be placed.
4. Type the distance offset.
5. Examine the arrows on the original line and determine if the new one should be placed to the left or right or both.
6. Change corner style if desired.
7. Usually, you will leave the boxes checked to treat the selection as a single line and to create a new feature for each selected feature. Click OK.

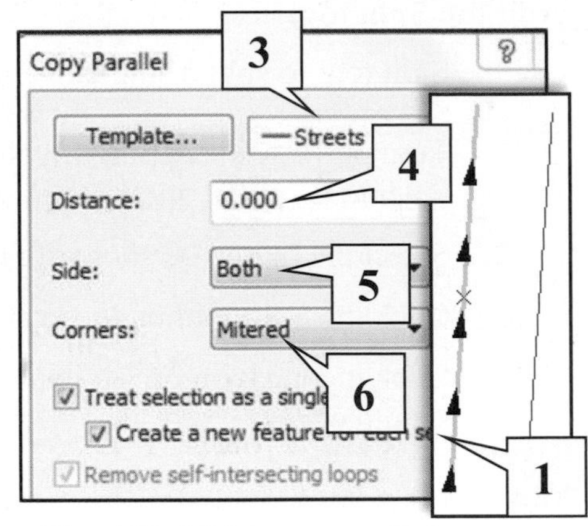

Fig. 109. Making a parallel copy

Merging features

1. Select at least two features to be merged (Fig. 110).
2. Choose Editor > Merge from the Editor toolbar.
3. Select the feature into which the others will be merged. The merged feature will be given the attributes of the selected feature. The value shown in the window is the primary display field for the layer. Click OK.

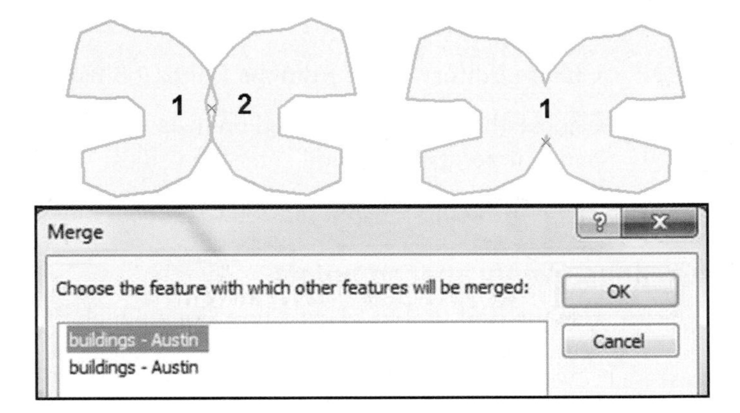

Fig. 110. Merging features

Union of features

1. Select at least two features to union (Fig. 111).
2. Choose Editor > Union from the Editor toolbar.
3. Select the template specifying the layer to which the new feature will be added.
4. The union feature remains selected and on top of the original figures, creating an overlap.

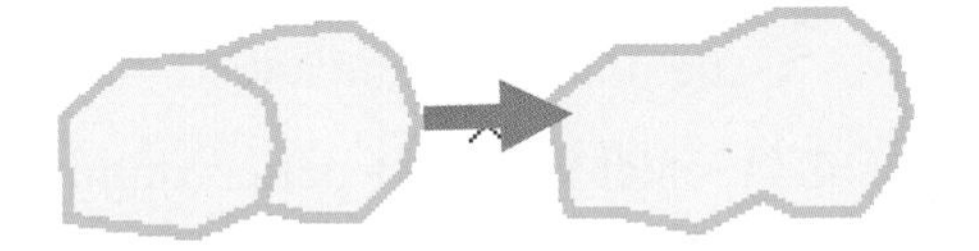

Fig. 111.Union of features

Intersection of features

1. Select the features to intersect (Fig. 112).
2. Choose Editor > Intersect from the Editor toolbar. (Use Customize > Toolbars > Customize to add it to the Editor menu, if needed.)
3. The new feature remains selected on top of the original polygons, creating an overlap.

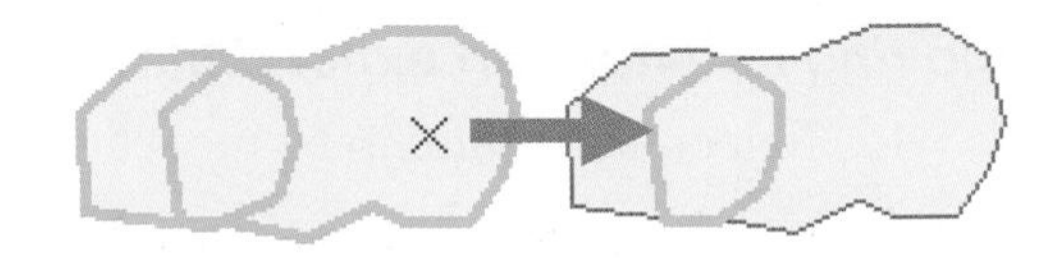

Fig. 112. Intersecting two polygons

Clipping features

Overlaps of one polygon on another are considered topological errors. Clip can be used to remove the overlap.

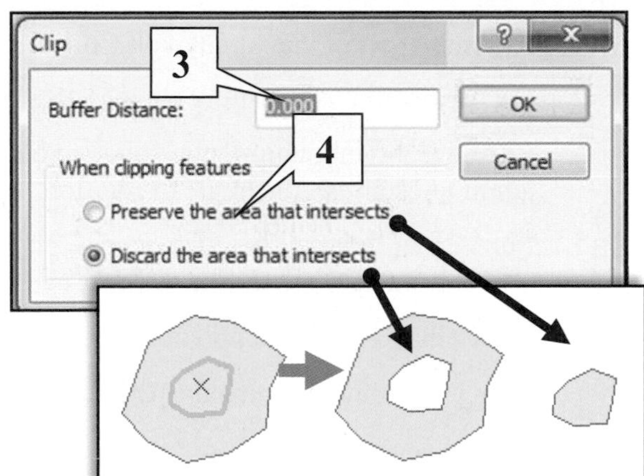

Fig. 113. Clipping with a polygon

1. Select a polygon that lies at least partially on top of another polygon (Fig. 113).
2. Choose Editor > Clip from the Editor toolbar.
3. If desired, apply a buffer to the clipping polygon by entering the value in map units.
4. Choose whether to preserve the overlapping area or to discard the overlapping area.
5. Click OK.
6. The clipping polygon will remain on top of the original feature. You may move or delete it.

Buffering features

1. Select the feature(s) to be buffered. It may be a point, a line, or a polygon (Fig. 114).
2. Choose Editor > Buffer from the Editor toolbar.
3. Select the template to indicate in which layer the buffer will be placed.
4. Type the buffer distance in map units and press Enter.
5. The new buffer will remain selected on top of the original feature, creating an overlap.

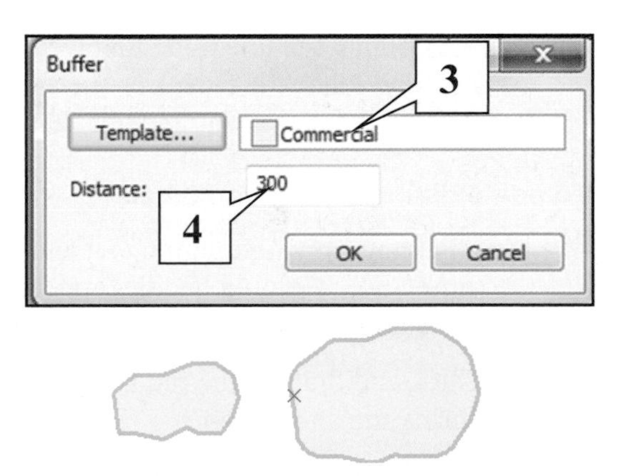

Fig. 114. Buffering a polygon

Editing annotation

Annotation stored as features in a geodatabase must be edited in an edit session.

Preparing to edit annotation

1. Make sure that the annotation layer is turned on in the Table of Contents.
2. Choose Editor > Start Editing from the Editor toolbar.
3. Click the annotation template and select a construction tool.

Adding new annotation labels

1. Click the desired construction tool in the Create Features window (Fig. 115).
2. Enter the text in the Annotation Construction window.
3. For Horizontal annotation, click on the map once to place it.
4. For Straight annotation, click once to enter the anchor point and again to establish the angle of rotation.
5. For Leader annotation, click once to place the far end of the pointer and again to define the near end and place the label.
6. For Follow Feature annotation, click on the feature to follow.
7. For Curved annotation, enter the start, mid, and endpoint of a curve.

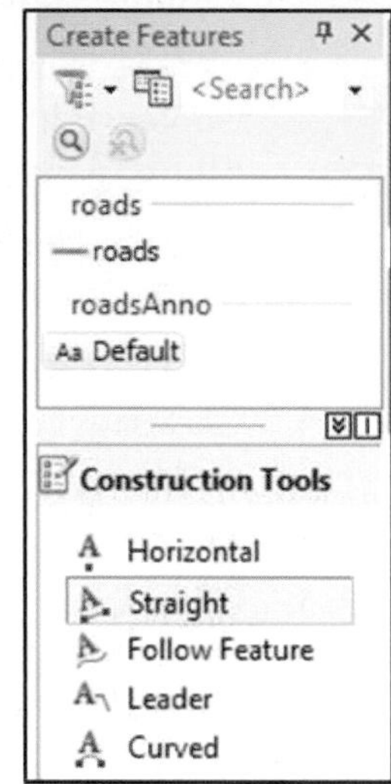

Fig. 115. Annotation construction tools

Make existing labels follow features

1. Click the Edit Annotation tool and select the annotation.
2. Right-click the annotation and choose Follow > Follow Feature Options and choose Straight or Curved depending on the desired result.
3. Right-click the line to be followed and choose Follow This Feature. Place the cursor on the annotation and move it to the desired location along the line feature.

To edit existing annotation labels

1. Click the Edit Annotation tool and select the annotation. Colored handles appear on the annotation, as shown in Figure 116.
2. Click and drag on the red handle to enlarge or shrink the annotation.
3. Click on one of the two blue handles to rotate the annotation.
4. Click on the black handle to move the annotation to a different location.

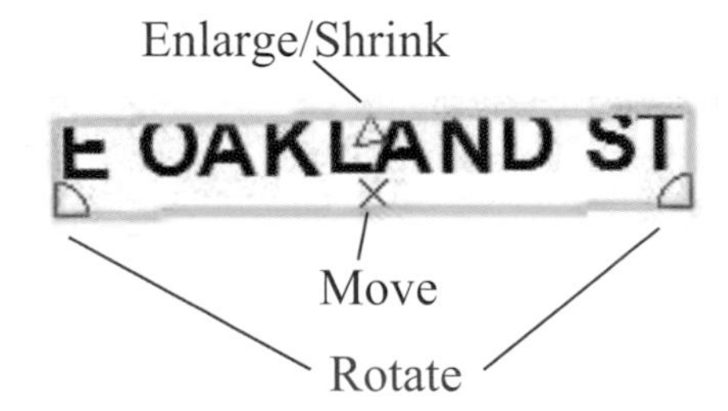

Fig. 116. Action handles on selected annotation

Modifying annotation properties

1. Select one or more pieces of annotation using the Edit Annotation tool.
2. Right-click the feature and choose Attributes from the Context menu, or click the Attributes button on the Editor toolbar.
3. Change one or more of the attributes for the annotation and click Apply.

Editing with map topology

Map topology is used to edit shared features in shapefiles or when a planar topology is not implemented. It uses the Topology toolbar.

Creating map topology

1. Choose Editor > More Editing Tools > Topology from the Editor toolbar. The Topology toolbar will appear.
2. Click on the Select Topology icon on the Topology toolbar.
3. Choose a Map Topology and check the boxes for the layers to participate (Fig. 117).
4. Click the Options arrow if you wish to set a different cluster tolerance from the default.

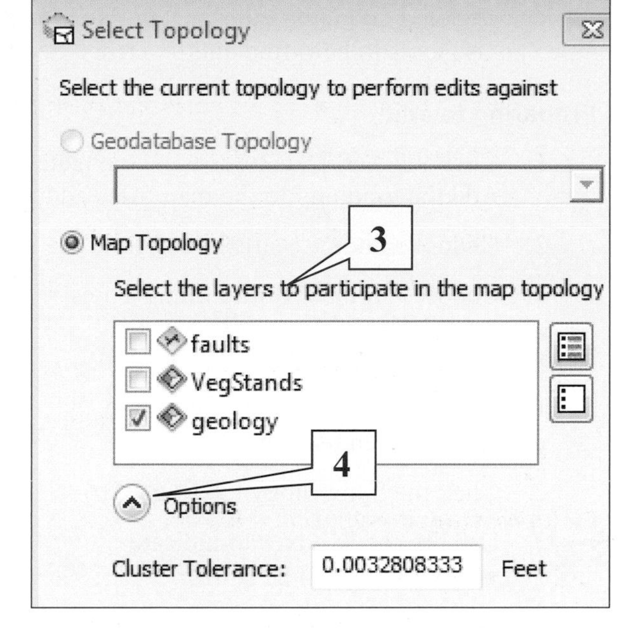

Fig. 117. Creating map topology

Reshaping a common boundary

1. Turn on vertex or edge snapping, and click on the Topology Edit tool.
2. Click the shared boundary to edit. It will be selected in a purple color to show that it is shared (Fig. 118).

3. Choose Reshape Edge tool from the Topology toolbar.
4. Click on the shared boundary at one end of the section to be reshaped.
5. Enter vertices defining the new edge.
6. End the sketch on the other end of the section and double-click to finish it.
7. The boundary will change to the shape of the sketch for both polygons.

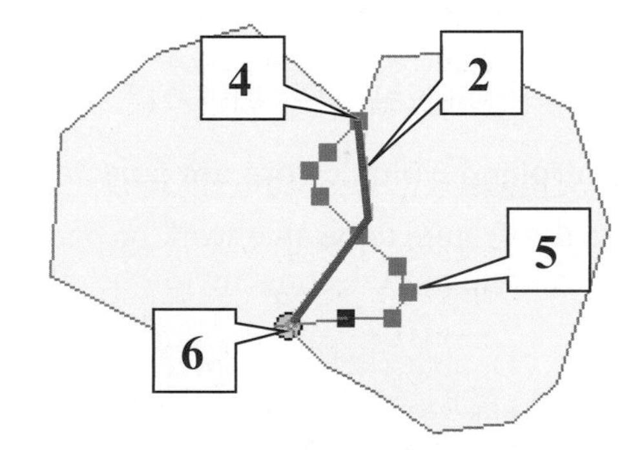

Fig. 118. Reshaping a shared boundary

Moving shared nodes

1. Click the Topology Edit tool.
2. Click on a node at the intersection of the lines to select it (Fig. 119).
3. Click and drag the node to its new location.
4. As you move the node, the movement of the shared features will be shown.
5. When you release the mouse button, the node and its attached lines will go to their new places.

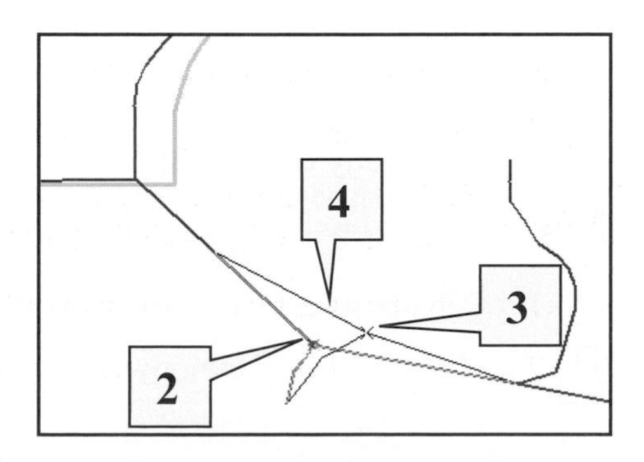

Fig. 119. Moving shared lines

Editing with planar topology

A planar topology contains rules that define how features may be spatially related to each other. You can use tools to help find and fix errors. See ***Creating a planar topology***.

Preparing to edit

1. Click the Add Data button and navigate to the feature dataset containing the topology. Add the topology to the map. Also add the participating feature classes to be edited.
2. Choose Editor > Start Editing from the Editor toolbar.
3. Choose Editor > More Editing Tools > Topology to open the Topology toolbar.

Displaying topology errors

1. Double-click the topology feature class to open its properties.
2. Click the Symbology tab (Fig. 120).
3. Use the check boxes to indicate which errors will be shown.
4. Click on an error type to highlight it and see its symbols.
5. Choose to show with a single symbol or symbolized by error type.
6. Repeat for each error type as desired. Click OK.

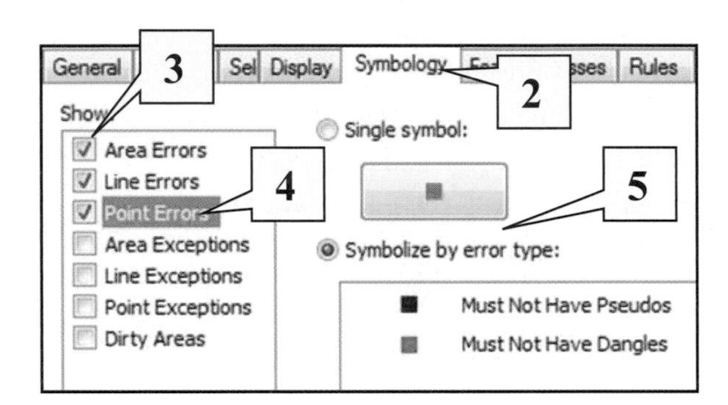

Fig. 120. Symbolizing topology errors

Controlling which errors are selectable

Use this feature to be able work on one type of error without selecting any of the others.

1. Double-click the topology feature class to open its properties.
2. Click the Selection tab (Fig. 121).
3. Check the boxes for selecting errors and exceptions as desired.
4. Check the boxes to indicate the types of errors you want to be able to select.
5. Click OK.

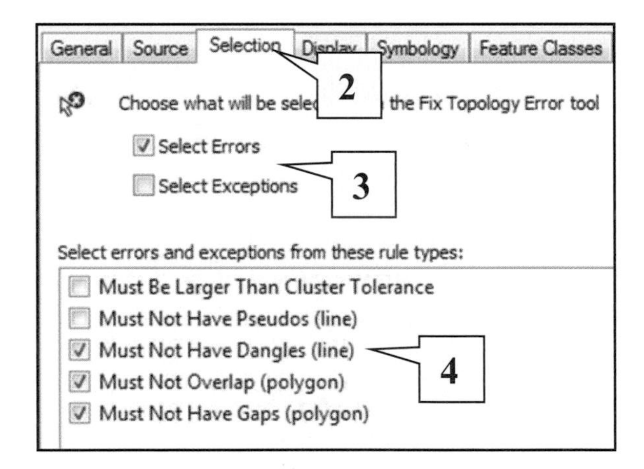

Fig. 121. Choosing selectable errors

The Fix Topology Error tool

This tool is used to manually select errors on the screen and apply suggested fixes.

1. Pan or zoom to a set of errors on the screen, if necessary.

2. Click the Fix Topology Error tool on the Topology toolbar.

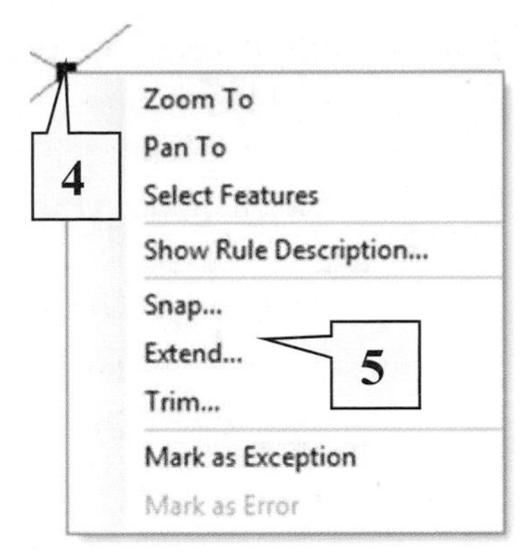

Fig. 122. Selecting a fix

3. Click on an error to select it. If you have several errors of the same type, you can hold down the Shift key to select additional errors, and/or draw a box around the errors to select them. The selected errors will turn black.
4. Right-click a selected error to open a context menu (Fig. 122).
5. Choose an action to fix the error, Snap, Extend, or Trim. Enter the tolerance to use for the item selected. Or mark the error as an exception.
6. All of the selected errors will have the same fix applied.

The Error Inspector

The Error Inspector (Fig. 123) is a dockable window that allows you to search for specific errors or groups of errors and apply suggested fixes.

1. Click the Error Inspector button on the Topology toolbar. Dock the window in a convenient place, such as the bottom of the ArcMap window. Pin it if desired.
2. Set the Show drop-down to the desired error type or leave it to show all errors.
3. Check the boxes to search for errors or exceptions or both.
4. Check the box to search only in the visible extent or not.
5. Click Search Now to search for the errors.
6. Click an error to select it, or right-click an error to get a context menu.
7. Zoom to an error for a better look.
8. Select one of the fixes for the error, or mark it as an exception.
9. Continue searching for and fixing errors as needed.

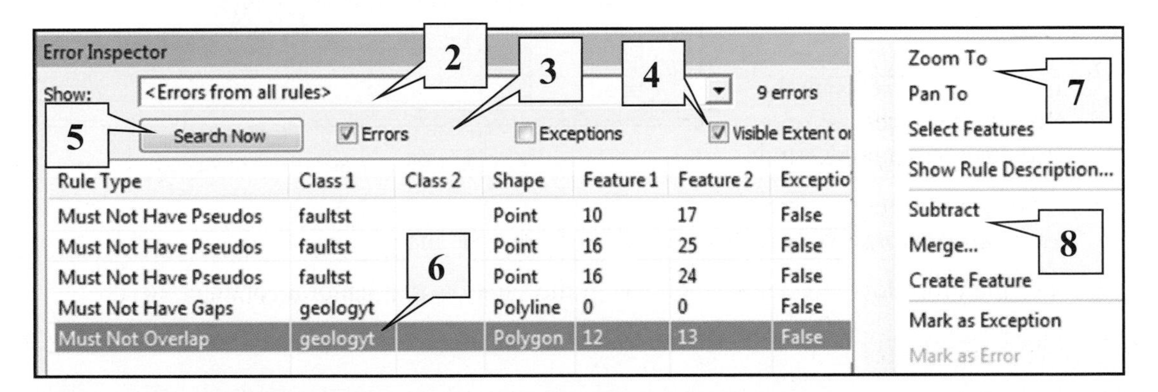

Fig. 123. The Error Inspector and the Overlap context menu

TIP: You can select multiple errors in the window by clicking one, scrolling down to the last one, and holding the Shift key while you click it. Right-click one of the selected errors for the menu.

MAPPING

Maps and Symbols

Setting symbols for a layer

You can set symbols a number of ways in ArcMap. Here are a few of the most common ways—all of these assume that all features in the layer are being drawn with the same symbol.

Changing the color of the current symbol

1. Right-click on the layer symbol in the Table of Contents and choose a color.

Changing the symbol

1. Click on the layer symbol in the Table of Contents to open the Symbol Selector window (Fig. 124).
2. Choose a symbol from the scroll box.
3. Modify the symbol's color, size, thickness, outline, or other attributes by setting the options provided. Make additional changes using the Properties button.
4. Click Edit Symbol for detailed editing of symbols, which can be saved.
5. To load additional symbols in the scroll box, click the Style References button and choose from the list of Styles.
6. Enter a key word to search for symbols and click the magnifying glass.
7. Use the drop-down to change the view of the symbols to a list or a compact view.
8. Click OK when finished.

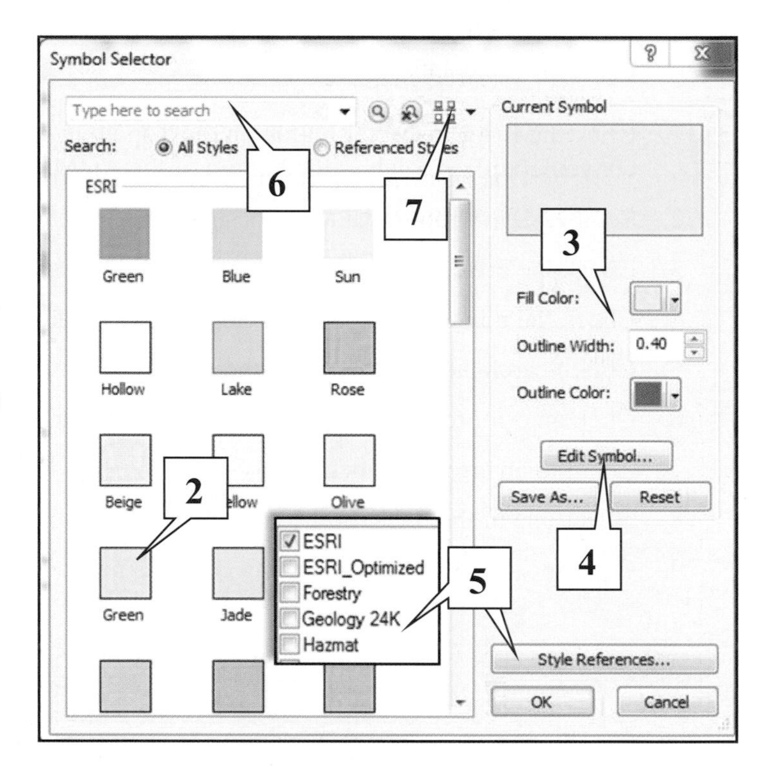

Fig. 124. The Symbol Selector

TIP: You can create your own symbols and groups of symbols, called styles, and work with them in the Style Manager. Read about it in the Help.

Creating new symbols

Symbols are created from one or more layers of symbol objects, as shown in Figure 125. Users can create new layers, put predefined symbols in them, and modify their colors and other properties to create new symbols. Symbols can also be created from imported bitmap images.

1. Click the Edit Symbol button in the Symbol Selector to access the Symbol Property Editor (Fig. 125).
2. Add new layers using the + button in the Layer area of the window. Delete layers using the X. Change the order of layers with the arrow buttons.
3. Select the type of symbol character to put in the layer. Markers, arrows, and 3D symbols are some of the types available.
4. Select the font and the subset. Many different fonts are available, giving many choices for symbols and text.
5. Select the desired character.
6. Modify the size, color, thickness, and other properties of the character.
7. Switch to the Mask tab to create a mask or halo around the symbol. This tab is useful for creating halos around text to make it easier to read.
8. Click OK when finished creating the symbol.

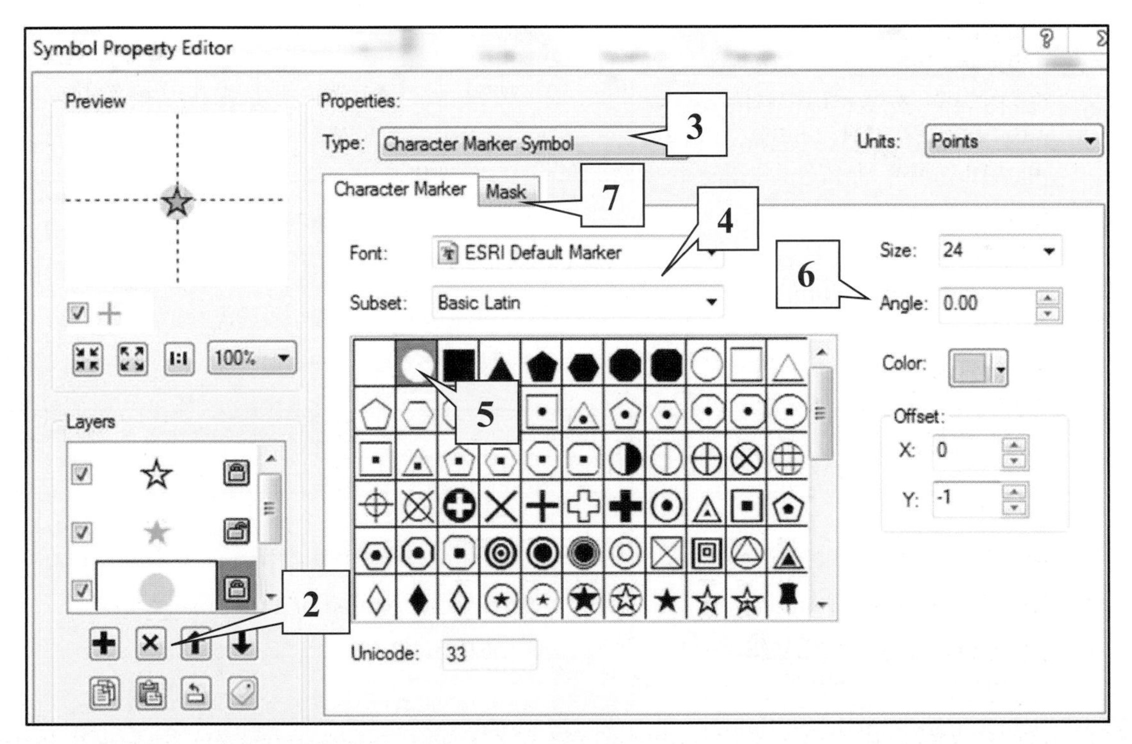

Fig. 125. Changing symbol properties

TIP: You can save a symbol as part of a style, and manage them using the Style Manager. See Help for details.

Creating maps based on attributes

Several different types of maps are available, but they all start the same way.

1. Right-click the layer name in the Table of Contents and choose Properties. Click the Symbology tab.
2. In the Show box, click on a heading to expand it and select the type of map desired.
3. Set the other map settings for each map type as described below.

Show:
Features
Categories
Unique values
Unique values, many
Match to symbols in a
Quantities
Charts
Multiple Attributes

Unique values map

1. Choose Categories: Unique values for the map type (Fig. 126).
2. Select the Value field on which the map will be based. It should contain categorical or ordinal data.
3. Choose a Color Scheme.
4. Click Add All Values to add the categories and symbols.
5. Double-click a symbol in the list to change its properties using the Symbol Selector.
6. Highlight a row by clicking, if desired, and choose Remove to place it in the <all other values> group.
7. Use the check box to choose to display the other values or not.
8. Click OK.

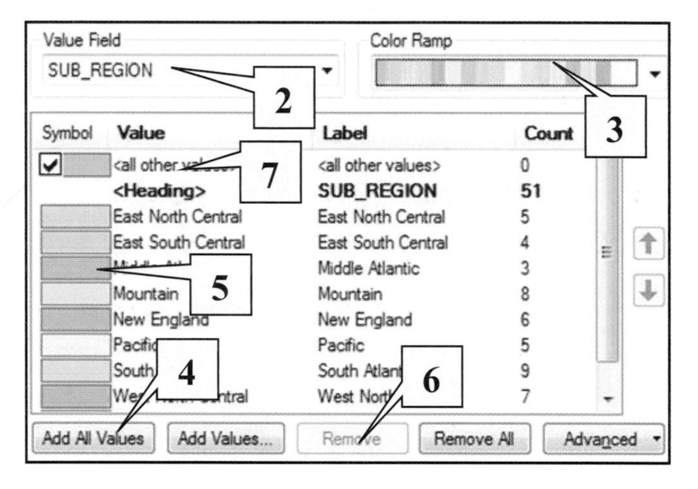

Fig. 126. Making a unique values map

Graduated color maps

Graduated color maps are for polygons.

1. Choose Quantities: Graduated colors for the map type.
2. Set the Value field on which the map will be based (Fig. 127). It must contain numbers.
3. Select a normalization field if one is desired.
4. Choose a Color Ramp using the drop-down box.
5. Change the number of classes or classification scheme, if desired. See ***Classifying data***.
6. Click OK.

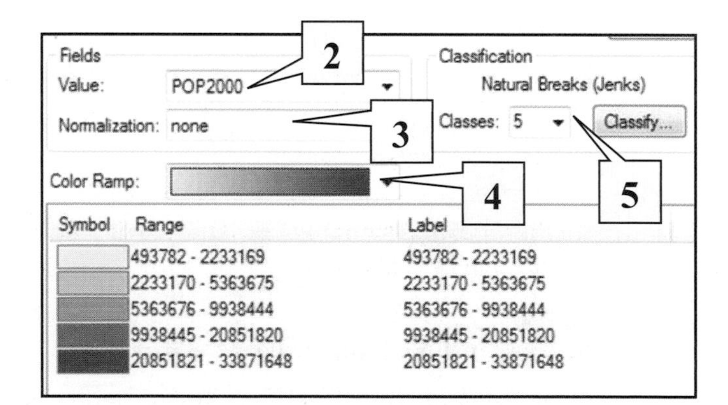

Fig. 127. Making a graduated color map

Graduated symbols map

Graduated symbols maps are for point or line data.

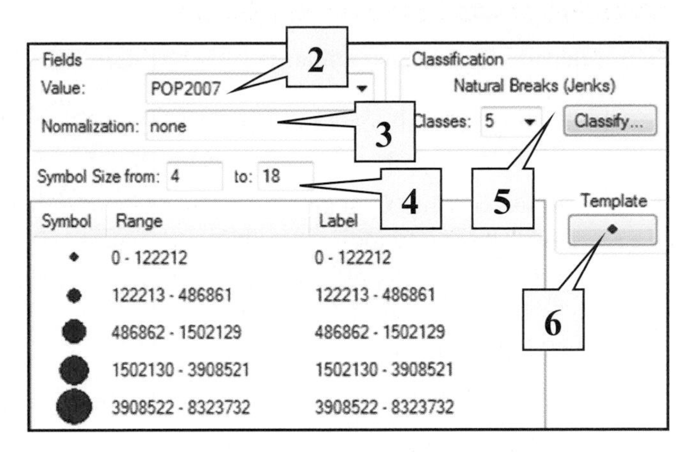

Fig. 128. Making a graduated symbols map

1. Choose Quantities: Graduated symbol for the map type.
2. Set the Value field on which the map will be based (Fig. 128). It must contain numeric data.
3. Select a normalization field if one is desired.
4. Set the sizes of the smallest and largest symbols.
5. Change the number of classes or classification scheme, if desired. See ***Classifying data***.
6. Click the Template button to open the Symbol Selector and change the shape or color of the symbol being used.
7. Click OK.

Proportional symbols map

A proportional symbols map is called an unclassed map.

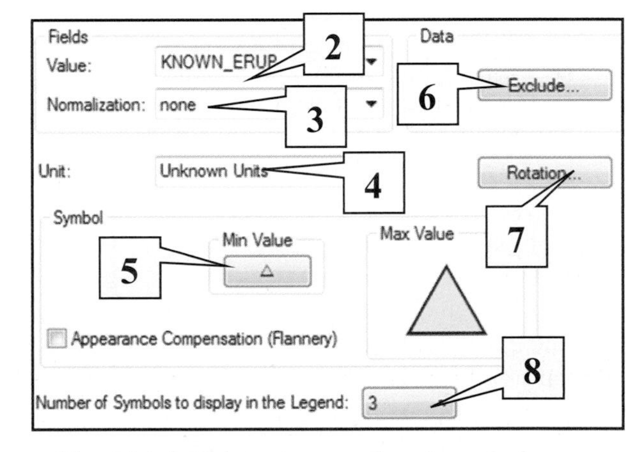

Fig. 129. Making a proportional symbols map

1. Choose Quantities: Proportional symbols for the map type.
2. Set the Value field on which the map will be based (Fig. 129). It must contain numeric data.
3. Select a normalization field if one is desired.
4. Set the units of the values, if known.
5. Click the Min Value button to define the shape and color of the symbol in the Symbol Selector. Set the size to be the smallest desired symbol. The larger symbols are based on a proportion of the data values to the smallest data value.
6. Click the Exclude button to define an expression excluding certain values from the map, for example, No Data values such as -99.
7. Click Rotation to rotate the symbol according to an angle specified by a field.
8. Select the number of symbols to display in the legend.
9. Click OK.

Dot density map

A dot density map is also unclassed.

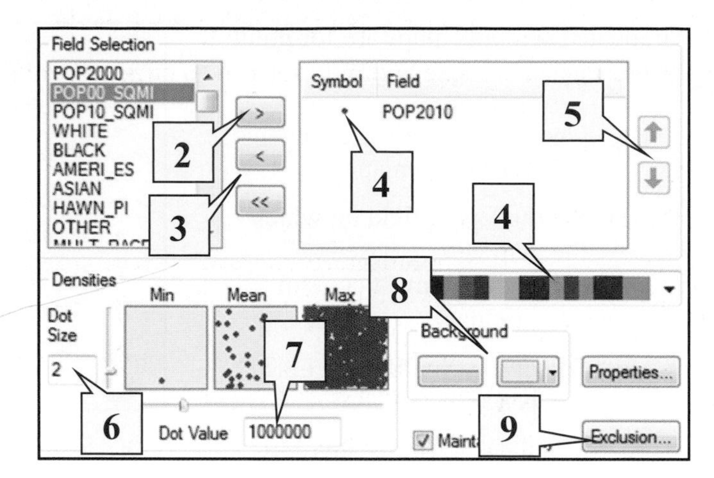

Fig. 130. Making a dot density map

1. Choose Quantities: Dot density for the map type.
2. Select a value field and click the > button to move it to the active list (Fig. 130). More than one field may be chosen.
3. Use the < button to remove a single field from the list or << to remove them all.
4. Right-click the symbol to change its color, or select a Color Scheme to assign colors based on a ramp or symbol set.
5. If more than one field is used, select one and use the arrows to change its position.
6. Set the dot symbol size.
7. Set the equivalence value for the dot. In this case, one dot represents 1 million people.
8. Use these buttons to set the color and outline symbol for the background.
9. Click the Exclusion button to define an expression excluding certain values from the map, for example, No Data values such as -99.
10. Click OK.

Chart map

A chart map is useful for comparing multiple fields. Pie, bar, or stacked chart maps can be made. The directions assume a pie chart, but the others are similar.

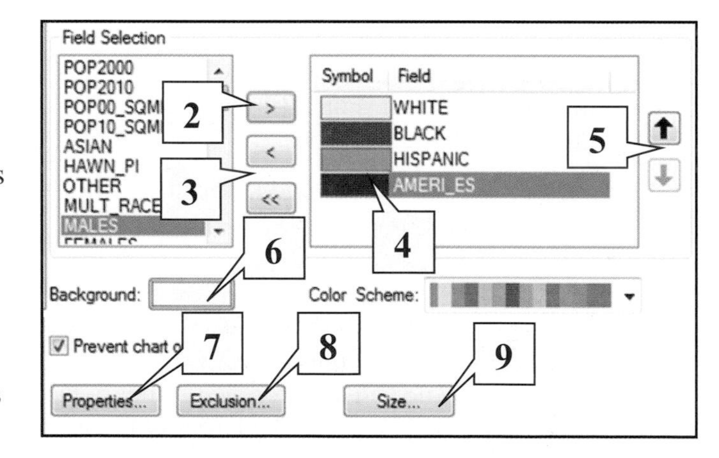

Fig. 131. Making a chart map

1. Choose Chart: Pie for the map type (Fig. 131).
2. Select two or more value fields and click the > button to move them to the list to be used.
3. Use the < button to remove a single field from the list or << to remove them all.
4. Right-click the symbol to change its color, or select a Color Scheme to assign colors based on a ramp or symbol set.
5. If more than one field is used, select one and use the arrows to change its position.
6. Set the background color.
7. Click Properties to set other chart properties such as 2D or 3D symbols.

8. Click the Exclusion button to define an expression excluding certain values from the map, for example, No Data values such as -99.
9. Click the Size button to set a fixed chart size, or set the chart size to vary based on an attribute field.
10. Click OK.

Classifying data

For quantities data, both feature and raster, ArcMap applies a default classification strategy. Use the following procedure to modify the classification.

1. On a map symbology tab, a classification box will appear if the map type uses a classification.

2. Use the drop-down box to change the number of classes.
3. For other changes, click the Classify button to open the Classification dialog box (Fig. 132). Examine the classification statistics.
4. To exclude certain records, such as those containing zeros or NoData values, click the Exclusion button and enter an expression defining the records to be excluded. (Bug in 10.1: exclusion doesn't work the first time; reopen the window and repeat.)
5. Change the classification method and number of classes or class values.
6. The Sampling button is used to reduce the number of features used to calculate statistics, which is useful when data sets are very large. The default sample is 10,000 records.
7. To manually set class breaks, either click and drag on the blue lines in the graph or type the break values directly into the box on the right.
8. Click OK when done setting the classes.

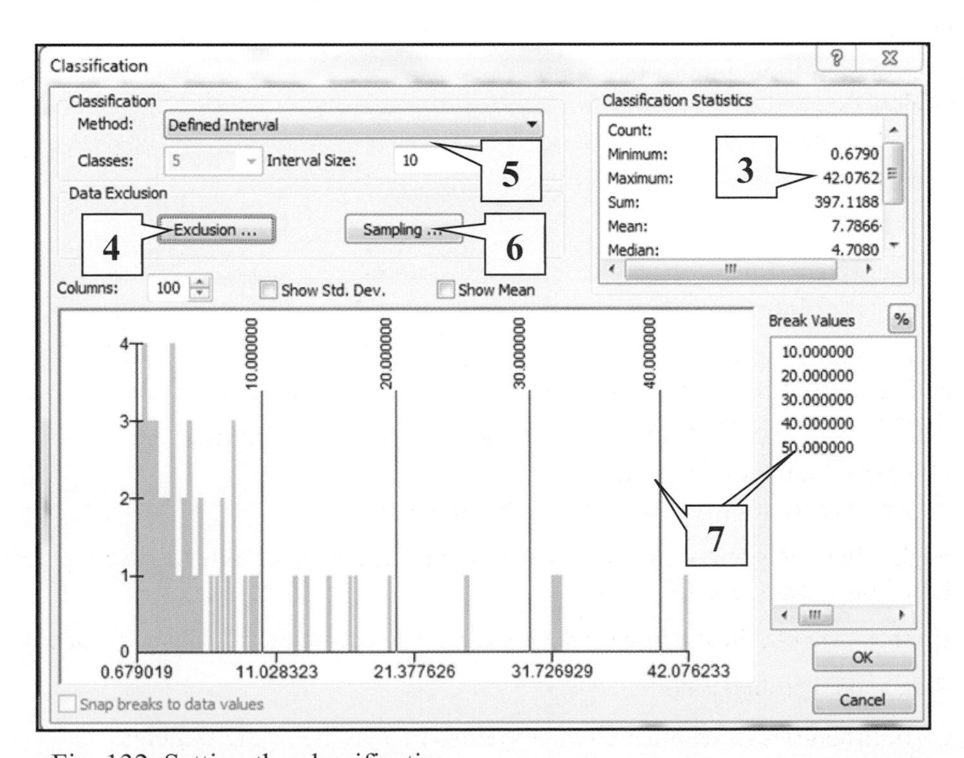

Fig. 132. Setting the classification

Modifying the appearance of the legend

Right-clicking the legend area provides a context menu with several options for modifying the legend (Fig. 133).

1. Use Flip Symbols to reverse the order of the symbols only.
2. Select a symbol by clicking on it (or multiple symbols using Ctrl-click) and choose Properties for Selected Symbols to change their appearance with the Symbol Selector. Use Properties for All Symbols to launch the Symbol Selector and modify the appearance of all of the symbols at once, such as changing them all from red to green.
3. Use Reverse Sorting to sort the classes and symbols in the reverse direction.
4. Select one or more classes and delete them with Remove Class(es).
5. Select two or more classes and merge them with Combine Classes.
6. Click Format Labels to apply numeric formatting options to the labels, such as changing the number of decimal places or including thousands separators.
7. Click Edit Description to enter or change a descriptive sentence or paragraph that can be used in the legend.
8. Click on any Range (except the lowest value of the first range) or Label in the legend to type in new values.
9. To create a new color ramp, set the first and last symbols to the desired end colors, and then choose the Ramp Colors option.

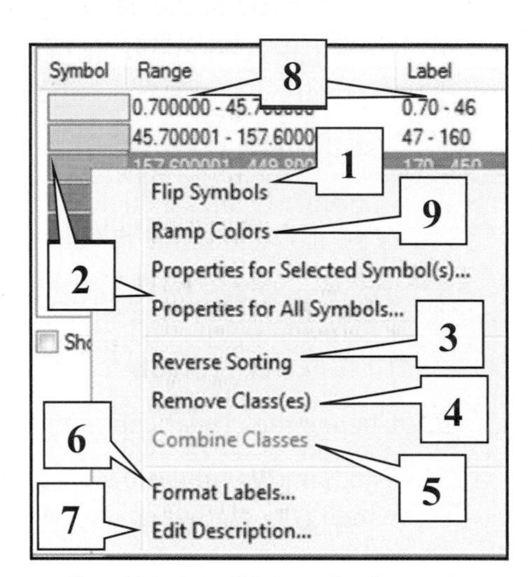

Fig. 133. Modifying the legend

Labels and Annotation

Using graphic text

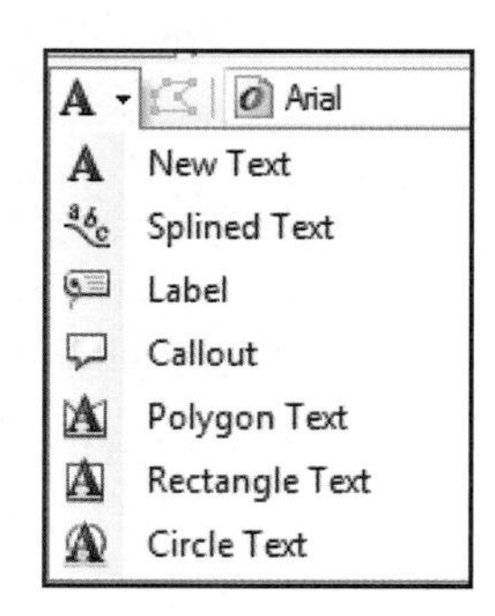

Graphic text is placed on a map document interactively and becomes a graphic on the map that is unique to the map document. Most graphic text uses strings typed by the user.

1. Make sure that you are in Layout view.
2. Look for one of the labeling tools on the Draw toolbar. Click on the black arrow for a drop-down menu to choose one of the tools.

Adding text to the map

1. In Layout view, click the New Text tool.
2. Click the desired location on the map and enter the text. Press Enter when finished.
3. Newly entered text is always selected, as shown by the dashed blue box around it. At this point you can change its font, size, or color using the menus on the Draw toolbar, or you can use the cursor to click and drag it to a new location.

TIP: To delete text, click the black arrow on the Draw toolbar, select the text, and press the Delete key. To delete all graphic elements, choose Edit > Select All Elements from the main menu bar and press the Delete key. Be careful! This option also deletes annotation.

Labeling a feature with an attribute

The Label tool, unlike the other graphic text tools, must be used in Data view. It uses the display expression in the layer properties for the text string.

1. Open the layer properties and click the Display tab. Set the Display Expression Field to the desired attribute to appear in the label. Click OK.
2. Make sure that you are in Data view.

3. Choose the Label tool from the labeling drop-down button. The Label Tool Options window will appear (Fig. 134).
4. Choose whether to *Automatically find best placement* or to *Place label at position clicked.*
5. Also choose whether to use the Label properties already set on the Labels tab or to choose a symbol from the window.
6. Click on the feature to be labeled. ArcMap will label the topmost layer clicked if more than one layer is present.

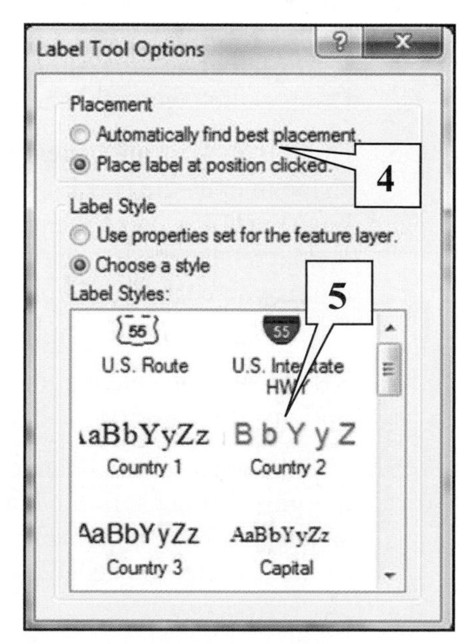

Fig. 134. Label Tool Options

Splining text along a line

Splining text makes it follow along a linear feature such as a road or stream.

1. Set the font size and style options desired using the Draw toolbar.
2. Choose the Splined Text tool from the labeling drop-down button.
3. Click vertices to define the line along which the text will appear. Double-click to end the line.
4. Type the text in the box and press Enter.
5. Use the Draw toolbar to modify the position or font characteristics as needed.

Adding a callout label

A callout places text in a box with a pointer to the feature of interest.

1. Choose the Callout tool from the labeling drop-down button.
2. Click on the feature to be labeled and drag the cursor to define the direction and length of the callout pointer.
3. Enter the text into the box and press Enter.
4. Click and drag on the text box to change its location, if desired. Click and drag on the blue dot to change the location of the pointer. Use the tools on the Draw toolbar to modify the callout's font, style, size, and so on.

Creating wrapped text boxes

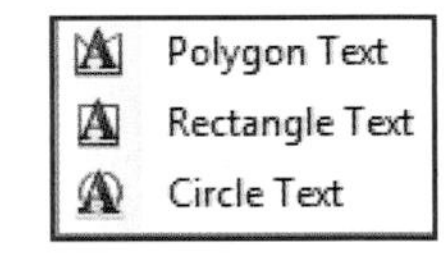

1. Choose one of the wrapped text tools: the Polygon Text, the Rectangle Text, or the Circle Text.
2. For the circle or rectangle text tool, click and drag to draw the circle/box, releasing the mouse when it reaches the desired size and shape. For the polygon tool, click on each vertex to define the desired shape. Double-click when finished.
3. Double-click inside the empty shape to open the text Properties box (Fig. 135). Type the text to be displayed. Do not use the Enter key unless you wish to enforce a new line within the text.
4. Set the symbol, spacing, or other options, if necessary.
5. Use the other tabs to change the margins, columns, frame border, size, position, or area background of the text box as needed.
6. Click OK to place the text.
7. To modify the text later, double-click it to open its properties box.

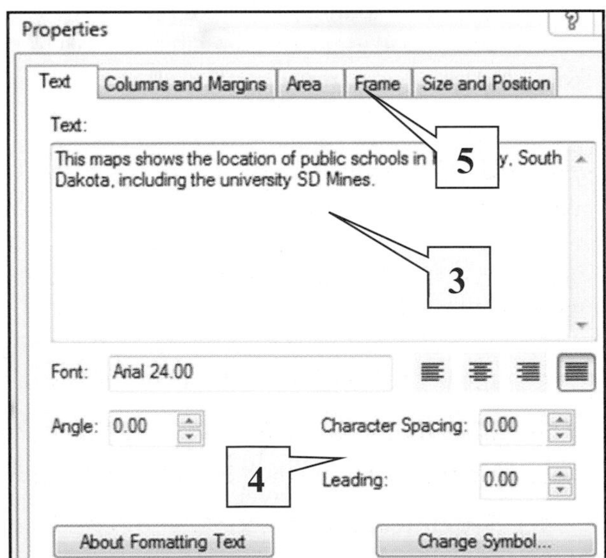

Fig. 135. Wrapped text properties window

Using dynamic labels

Dynamic labels are used to quickly label all the features in a layer.

1. Open the layer Properties and click the Labels tab.
2. Check the *Label features in this layer* box (Fig. 136).
3. Make sure the method is set to *Label all the features the same way*.
4. Choose the Label Field. Click the Expression button to enter a VBA script.
5. Edit the font settings, or select a predefined text symbol by clicking the Symbol button and choosing a predefined symbol style.
6. For detailed control of label placement, click the Placement Properties button.
7. Set the scale range, if desired, by using the label's scale range or by typing in new values. If the map scale is outside the specified range, the labels will not be drawn.
8. Select a label style, if desired. A label style includes BOTH a text symbol and predefined label placement options.
9. Click OK to place the labels.

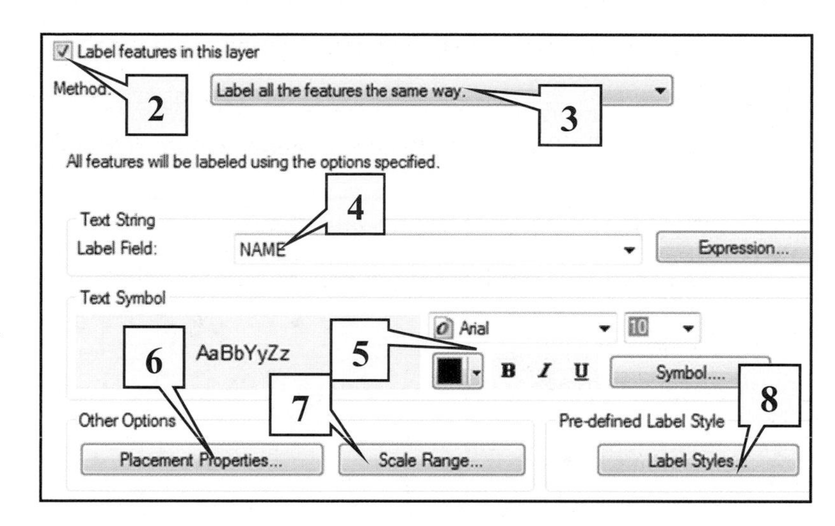

Fig. 136. The Label Properties window settings

TIP: Turn labels on and off for a layer by right-clicking the layer name in the Table of Contents and choosing the Label Features option. If the menu choice is checked, the labels are on, and choosing it will turn them off. If it is unchecked, choosing it will turn them on.

Creating label classes

Label classes are used to assign different label styles to groups within a layer, such as giving towns small labels and large cities large labels. We will designate towns as having fewer than 100,000 people. The instructions below use this example, but they can be customized for as many classes as you wish.

1. Open the layer Properties and click the Labels tab.
2. Check the box to *Label features in this layer* (Fig. 137).
3. Make sure the method is set to *Define classes of features and label each class differently*.
4. The current class is the Default class. Uncheck the box to *Label features in this class*.
5. Click Add to add a new class. Name it `Towns`. Click OK.

When the class name is showing in the Class box, then all the settings in the window will be applied to that class. We continue now by defining the Towns class and setting the label properties.

6. Click the SQL Query button.
7. Enter an expression that defines the Towns class. In this case we use:
   [POP2010] <100000. See ***Entering an SQL query***.
8. Make sure that the box is checked to *Label features in this class*.
9. Select the font to be used for this class.
10. Set the placement properties or scale range for this class, if desired.

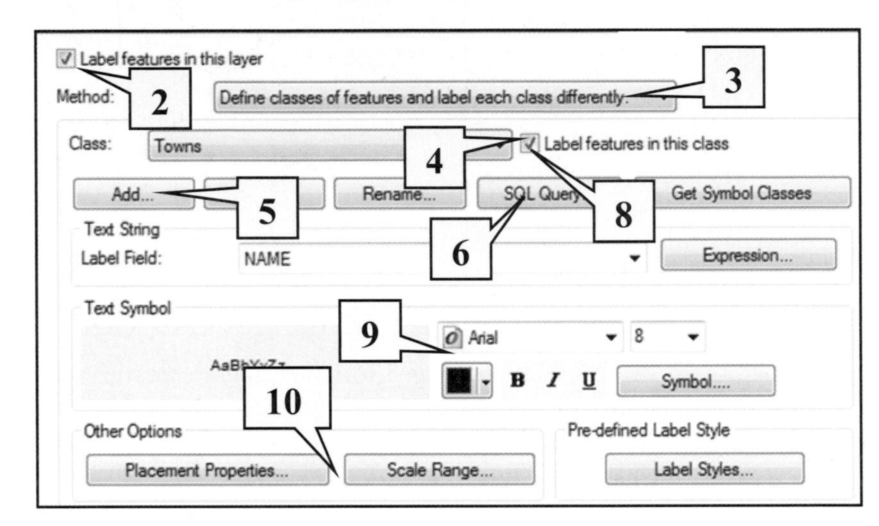

Fig. 137. Creating a label class

You are finished defining and setting properties for the Towns class. The next step is to add the Cities class.

11. Repeat Steps 5–10 for the Cities class. This time the expression will read:
    [POP2010] >= 30000.
12. Continue until all classes have been added.
13. Click OK.

TIP: If you see double labels on your features, you probably forgot to uncheck the box for the Default class. If you don't see any labels, or see only some classes, check your SQL queries.

Using the Label Manager

The Label Manager makes it easier to work with labels from many layers at one time. You can choose layers to modify with a single click rather than having to open and close label properties for each one.

1. Choose Customize > Toolbars from the main menu bar and select the Labeling toolbar.

2. Click the Label Manager button on the Labeling toolbar.
3. Use the check boxes to turn labels for layers on and off (Fig. 138). Notice that both default labels and label classes are shown in this view.

4. Click one of the label classes to highlight it. While it is highlighted, the settings in the window refer to that class.
5. Make changes to the settings. When you click Apply, the changes are applied to the highlighted class.
6. Use the Options button to modify the display of the label class tree.
7. Click OK when finished setting the properties of the labels.

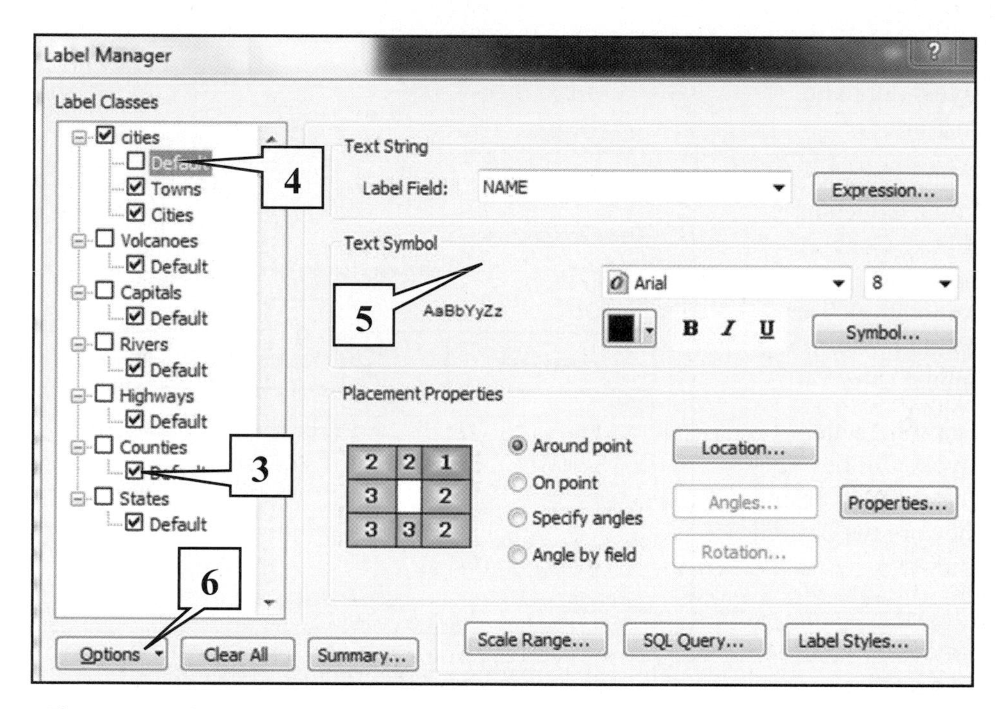

Fig. 138. Using the Label Manager

Using the Labeling toolbar

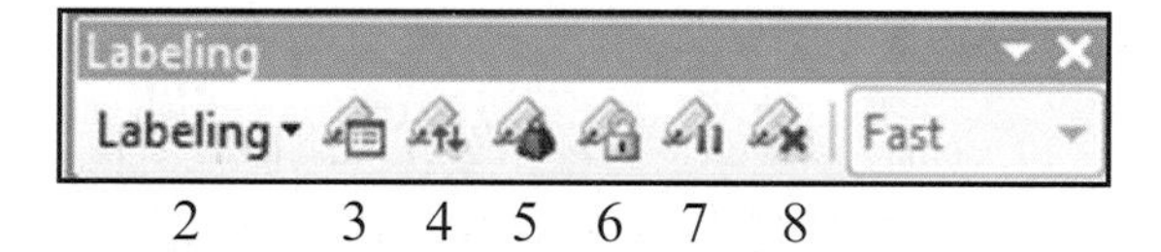

Fig. 139. Labeling toolbar buttons

1. Choose Customize > Toolbars > Labeling from the main menu bar to open the toolbar.
2. Choose Labeling > Options to set several options such as the color of unplaced labels (Fig. 139).
3. Choose the Label Manager button to open the Label Manager.
4. Use the Label Priority Ranking button to assign priorities for labels. Labels with a higher rank will have priority and will be drawn if they conflict with labels of a lower rank.
5. Use the Label Weights button to control which labels will be placed when there are overlaps between features and labels.
6. Use the Lock Labels toggle button to lock/unlock the labels against changes.
7. Use the Pause Labeling button to suspend drawing of labels if you accidently set properties that produced too many labels.
8. Use the View Unplaced Labels button to show the unplaced labels in red so that you can see which ones are missing.

Creating annotation

These directions show how to create annotation stored as text graphics in a map. Consult the Help files for information on creating annotation as a geodatabase feature class.

1. Use the Layer Properties to create dynamic labels for all the desired layers. Take care in setting the properties, weights, and so on because these will control the labels that appear.
2. Annotation will be created for all layers in the data frame with dynamic labels turned on. Turn off any dynamic labels for layers not to be converted.
3. Right-click the data frame name and choose Convert Labels to Annotation. The dialog box will appear (Fig. 140).
4. Examine which layers will be converted to ensure they are the desired ones.
5. Choose to create annotation in the map or in a geodatabase.

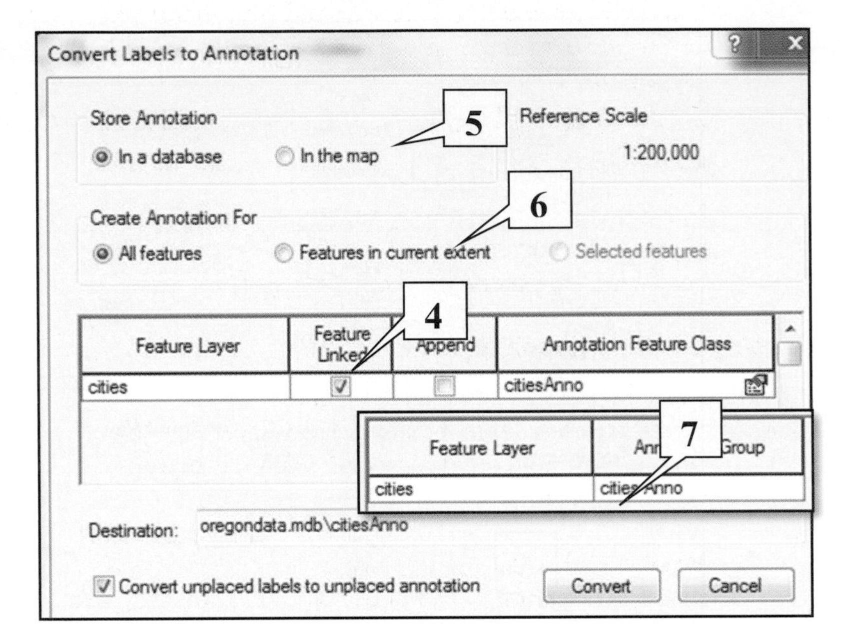

Fig. 140. Creating annotation in a geodatabase

6. Choose whether to create annotation for all features, features in the current extent, or selected features.
7. If creating geodatabase annotation, check whether to create feature-linked annotation or not. This requires an ArcGIS Standard license. Note the name of the feature class being created. If map annotation is being created, the annotation will be placed in a group.
8. Check the destination for the geodatabase annotation and change it if necessary.
9. Check the box to Convert unplaced labels if you want to place overlapping labels interactively.
10. Click Convert to create the annotation.
11. Place map overflow annotation immediately (see below). Geodatabase overflow annotation may be placed at any time.

Placing overflow map annotation

Map annotation from overlapping dynamic labels will not be placed on the map. Instead they will be placed in an overflow window and must be placed interactively.

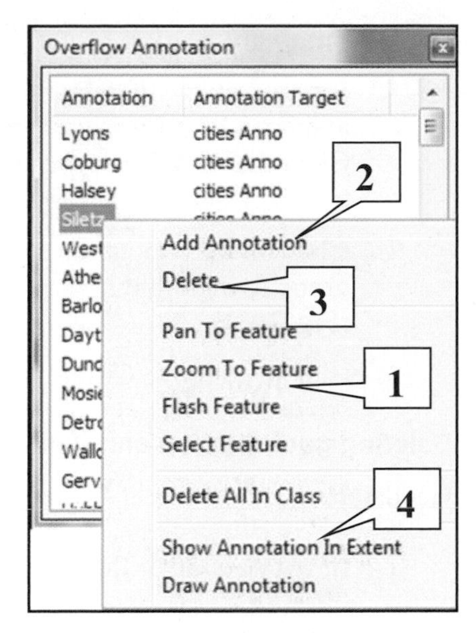

Fig. 141. Placing overflow annotation

1. Right-click a label in the overflow list and choose a method to locate it, if necessary (Fig. 141).
2. Right-click it again and choose Add Annotation to add the label to the map. Click and drag it to adjust its location, if necessary.
3. If you decide not to place the label, delete it from the list using Delete.
4. Use Show Annotation In Extent to list only the labels in the current view. Place all of these before zooming to another location—it saves time.

Placing overflow geodatabase annotation

Overflow annotation in a geodatabase is placed in the annotation feature class table along with the placed labels, but with an attribute field indicating that it is not placed. It can be placed at any time.

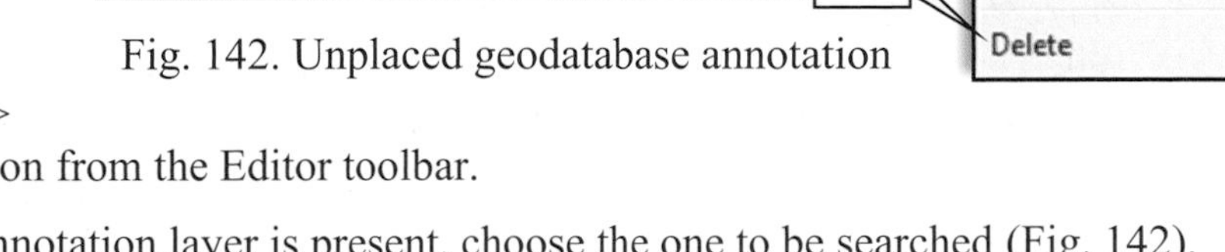

Fig. 142. Unplaced geodatabase annotation

1. Make sure that the annotation feature class is in the Table of Contents and is turned on.
2. Start an edit session. See ***Begin an editing session***.
3. Choose Editor > Editing Windows > Unplaced Annotation from the Editor toolbar.
4. If more than one annotation layer is present, choose the one to be searched (Fig. 142), and whether to search in the visible extent or the entire layer.
5. Click Search Now to find the unplaced labels.
6. Clicking a row will cause that annotation to flash on the screen so it can be located. Right-click an unplaced label to open the context menu.
7. Right-click the annotation and choose pan or zoom, if necessary.
8. Right-click the annotation and choose Place Annotation to put it on the map. Adjust its location or properties if needed.
9. Right-click the annotation and choose Delete if you decide not to place it.
10. Close the Unplaced Annotation window when done. You can return and do more later if any are left. Remember to save your edits, and stop editing.

Deleting annotation

Deleting map annotation

1. Right-click the data frame and open its properties. Click the Annotation Groups tab.
2. Click on the annotation group to highlight it (Fig. 143).
3. Choose Remove Group.

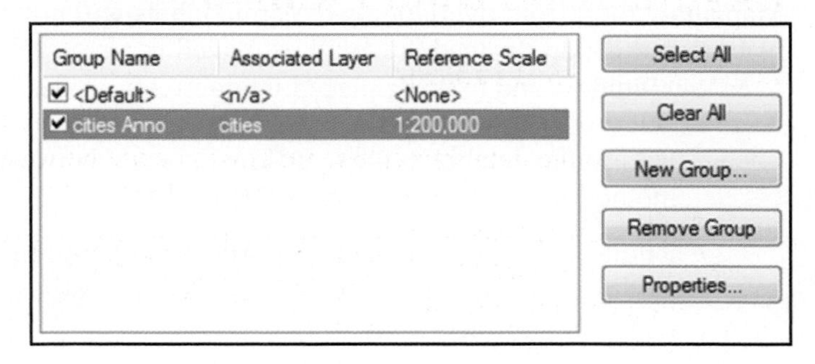

Fig. 143. Managing annotation groups

Deleting geodatabase annotation

Geodatabase annotation is deleted like any other feature class.

1. In ArcCatalog or the Catalog tab, navigate to the geodatabase containing the annotation.
2. Right-click the annotation feature class and choose Delete.

Layouts and Data Frames

Using the Layout toolbar

The Layout toolbar provides tools for zooming around the layout page (Fig. 144). It has no effect on the zoom in the data frame. The functions of the buttons, from left to right, are as follows:

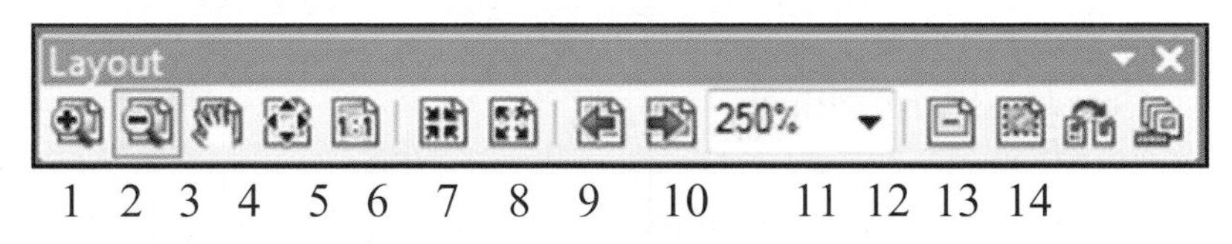

Fig. 144. The Layout toolbar

1. The Layout Zoom In tool is used to enlarge a portion of the layout. Click the tool and then click and drag a box around the desired area.
2. The Layout Zoom Out tool zooms out a specified distance from the layout, centered on the clicked point.
3. The Layout Pan tool moves the layout within the window. Click and drag on the layout to move it.
4. The Zoom Whole Page tool zooms so that the entire layout page can be seen.
5. The Zoom to 100% tool shows the layout at the same scale as it will be printed.
6. The Fixed Zoom In tool zooms in a specified amount, centered on the current display, when you click anywhere on the layout.
7. The Fixed Zoom Out tool zooms out a specified amount, centered on the current display, when you click anywhere on the layout.
8. The Go Back to Extent tool returns to the previous extent.
9. The Go Forward to Extent goes to the next extent. This button is only available if you have clicked the Go Back to Extent tool at least once.
10. The Zoom Control box sets a particular percent enlargement for the layout.
11. The Toggle Draft Mode allows you to display each element as a simple labeled box. This feature can make setting up the layout easier without waiting for each element or data frame to redraw each time a change is made.
12. The Focus Data Frame mode switches to Data View but lets you keep editing any text placed on the layout. It's a hybrid between Data View and Layout View.
13. The Change Layout button launches the Template window so you can add or change the layout of the map using a predefined template.
14. The Data Driven Pages button opens the Data Driven Pages toolbar, which is used for working with map tiles used to produce map books. See Help for more information.

Setting up the map page

1. Choose File > Page and Print Setup from the main menu bar.
2. Check the printer and use Properties to change it, if necessary (Fig. 145).
3. Set the printer paper size, source, or orientation options, if necessary.
4. Set the map page width, height, and orientation.
5. To use the settings from the current printer for the map page size, check the box. Otherwise, set the page size. See the tip below.
6. If you plan to change the map size and want the map elements to be resized along with the page, check the box.
7. If desired, check the box to show the printer margins in the layout.
8. Examine the page preview to make sure that your map (the colored picture) will fit on the paper (the white area). Adjust the settings, if necessary.

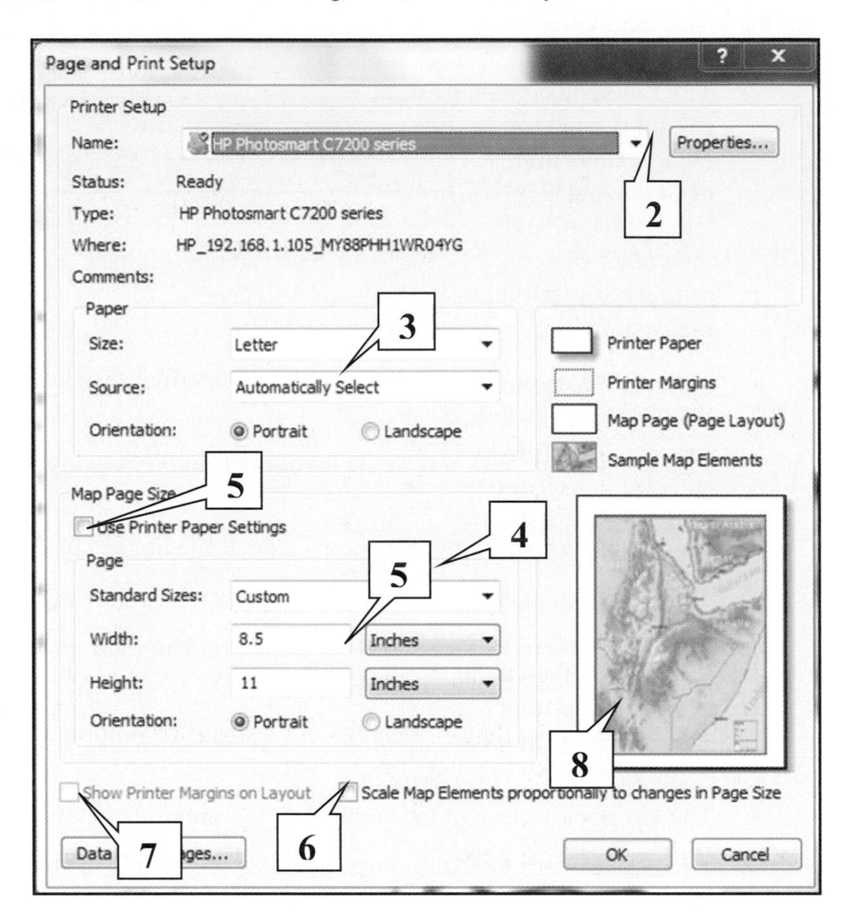

Fig. 145. The Page and Print Setup window

TIP: Using the Printer Paper Settings may cause a map document to issue a warning when it is opened on a system where the original printer is not available or to make unwanted changes to the layout when the printer is changed. The author does not recommend using this check box.

Setting the scale or extent

The Data Frame properties tab provides three options for setting the map scale of the frame: automatic, fixed scale, or fixed extent. Automatic scaling is the default. Fixed scale or fixed extent will deactivate the Zoom/Pan tools for the frame.

1. Right-click the data frame name to open its properties. Click the Data Frame tab. Choose the desired scaling method.

Using Automatic

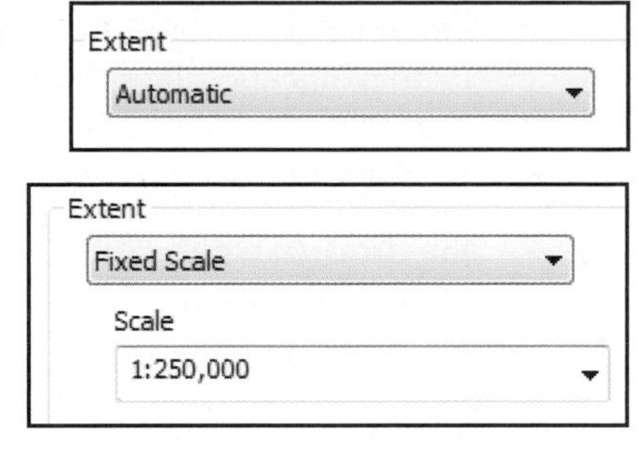

2. If you set the extent to Automatic, no other settings are needed. The scale will be determined using the zoom/pan tools in the data frame.

Using Fixed Scale

3. If you set the scale to Fixed Scale, specify the desired scale for the map. The zoom tools will be disabled, but the Pan tool will still work.
4. Check the map borders, as parts of the data frame may be cropped if this method is used.

Using Fixed Extent

5. First, use the Zoom/Pan tools to set the Data view to the desired extent. Alternatively, determine the *x-y* coordinates of the desired extent.

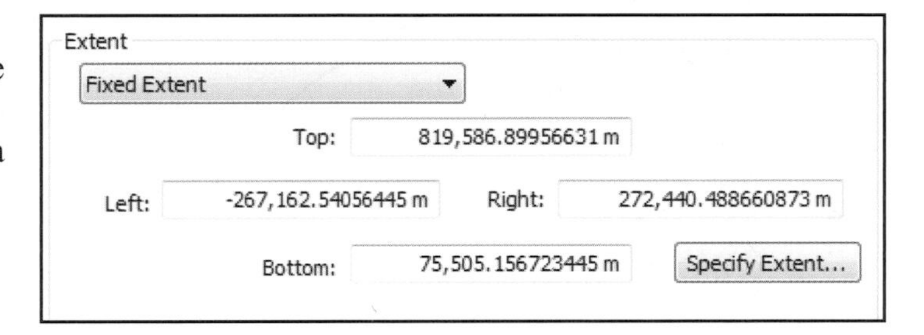

6. In the Data Frame tab, fill the Fixed Extent button. The boxes will be filled with the current *x-y* extent of the data frame. Change the *x-y* coordinates, if desired.
7. Click the Specify Extent button for more options.
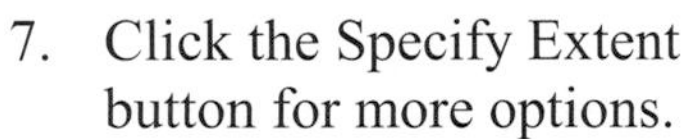
   You can set the extent to match the current visible extent, set it to match a particular data layer, set it to match an existing graphic such as a rectangle, or specify the *x-y* extent values in longitude-latitude degrees instead of map units.
8. Click OK.

Setting the reference scale

1. Use the Pan/Zoom, the Scale readout, or Bookmark tools to zoom to the desired scale.
2. Right-click the data frame name in the Table of Contents and choose Reference Scale > Set Reference Scale.

TIP: To set an exact scale value such as 1:24,000, right-click the data frame name, open the data frame Properties, click the General tab, and type a specific reference scale in the appropriate box.

3. To zoom to the reference scale, right-click the data frame name and choose Reference Scale > Zoom to Reference Scale.
4. To remove the reference scale, right-click the data frame name in the Table of Contents and choose Reference Scale > Clear Reference Scale.

TIP: Annotation, once created, retains its original reference scale set when it was created, even if the reference scale of the data frame is changed later.

Clipping to a layer

1. Open the Data Frame Properties and click the Data Frame tab.
2. Under Clip Options, change the drop-down to Clip to shape (Fig. 146).
3. Click the Specify Shape button.
4. Choose Outline of Features and select the layer defining the outer boundary.
5. Choose to clip to all features or only the visible ones.
6. Alternatively, enter coordinates for a custom rectangle.

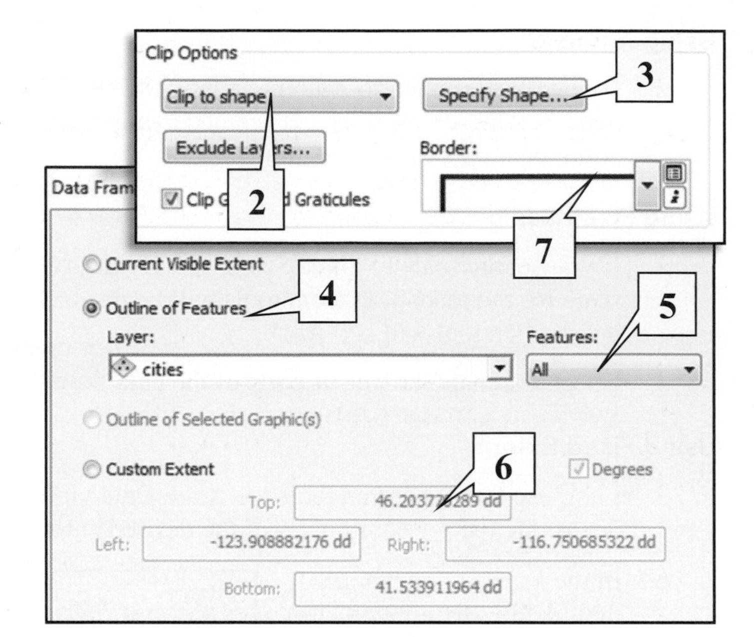

Fig. 146. Clipping layers in a data frame

7. Select a border to outline the clip feature, if desired.

Composing the data frames

The map layout has at least one data frame that can be dragged and resized on the page.

1. Make sure that the Layout view button at the bottom left of the display window is clicked.

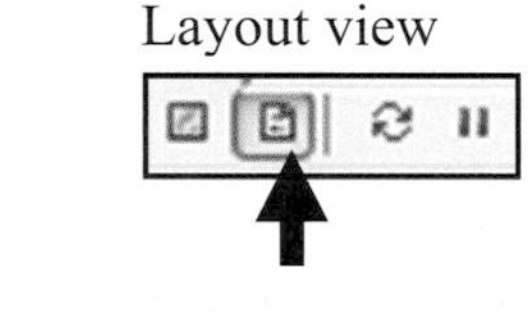

2. Choose the Select Elements tool from the Draw toolbar or the Standard toolbar.

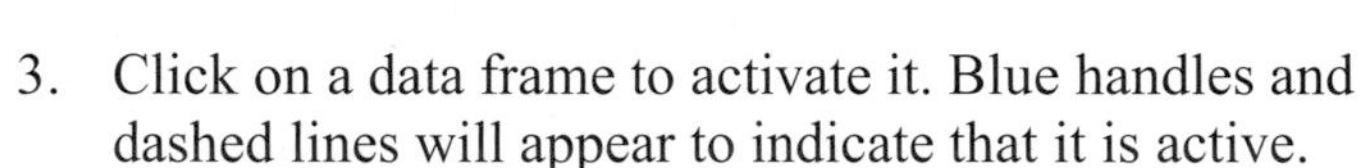

3. Click on a data frame to activate it. Blue handles and dashed lines will appear to indicate that it is active.
4. Click and drag on the data frame to move it to a new location (Fig. 147).
5. Click and drag a side handle to increase or decrease the size in one direction.
6. Click and drag on a corner handle to increase or decrease the size in two directions.

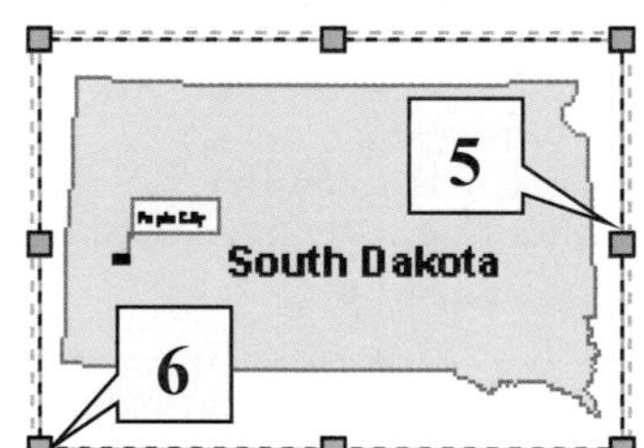

Fig. 147. Click and drag the active frame to move it or change its size.

Adding a north arrow

1. Choose Insert > North Arrow from the main menu and choose the desired symbol.
2. Click the Properties button to modify the symbol. Click OK to place it.
3. If necessary, click and drag it to the desired location, or resize it.

TIP: In some map projections, north is not straight up. Set the north arrow marker angle manually to point true north, if necessary.

Adding a title

1. On the main menu bar, choose File > Map Document Properties.
2. In the Title box, enter the title of the map. Click OK.
3. Choose Insert > Title from the main menu bar. The title appears at the top of the map, in a dashed blue line to indicate that it is selected.
4. The text remains selected. If necessary, change its size or font using the Draw toolbar or click and drag it to a new location.
5. To modify the text after it has been created, double-click it with the Select Elements tool, or right-click it and choose Properties. You can change the text, set the font and size, or specify a particular position for the text.

Adding text

1. Choose Insert > Text from the main menu bar. Type the desired text in the box on the map and press Enter.
2. The text remains selected. If necessary, change its size or font using the Draw toolbar or click and drag it to a new location.
3. To modify the text after it has been created, double-click it with the Select Elements tool, or right-click it and choose Properties. You can change the text, set the font and size, or specify a particular position for the text.

Adding graphics to layouts

The Draw toolbar (Fig. 148) provides functions for creating and modifying objects on a layout. These objects may also be created within a data frame itself, in which case they will be scaled if the map changes size. If they are in the layout, they will be unaffected by scale changes.

The Draw toolbar contains common functions found in other programs. A small black triangle on a button indicates that the button contains a menu.

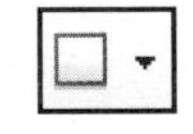

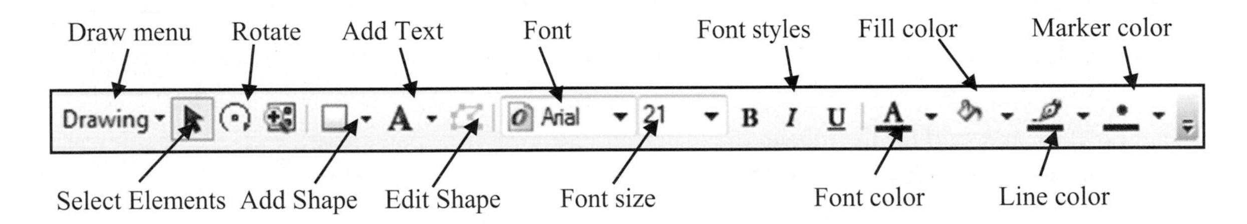

Fig. 148. The Draw toolbar

Adding a legend

1. Use the Select Elements tool to click on the data frame containing the layers to appear in the legend. The legend is always created from, and placed in, the active frame.
2. Choose Insert > Legend from the main menu bar.
3. Choose which layers will be included in the legend. To add a layer, click it in the box on the left and click the > button. To remove a layer from the legend, click it in the box on the right and click the < button. Choose the number of columns in the legend. Click Next.
4. Modify the legend title text and formatting to desired settings. Click Next.

5. Specify a legend border, background, and drop shadow, if necessary. Click Next.
6. Click a layer to modify its symbol size and patch style. Click Next.
7. Modify the spacing, if necessary (usually not necessary). Click Finish.
8. The legend appears in the map, selected. Click and drag the legend to the desired location and resize it, if desired. Resizing will change the size of the text and boxes.

TIP: To edit the individual elements of a legend, right-click the legend and choose Convert to Graphics. To work with each element, choose Draw > Ungroup from the Draw toolbar.

Modifying a legend

Right-click the legend and choose Properties. The window has four tabs: Legend, Items, Frame, and Size/Position.

The General tab

1. Change the title of the legend (Fig. 149) or its symbol.
2. Choose which layers will appear in the legend by selecting one and clicking the arrows to move them back and forth.
3. Select a layer and use the arrows to its placement in the order of items in the legend.
4. Change the default Map Connection options if desired.

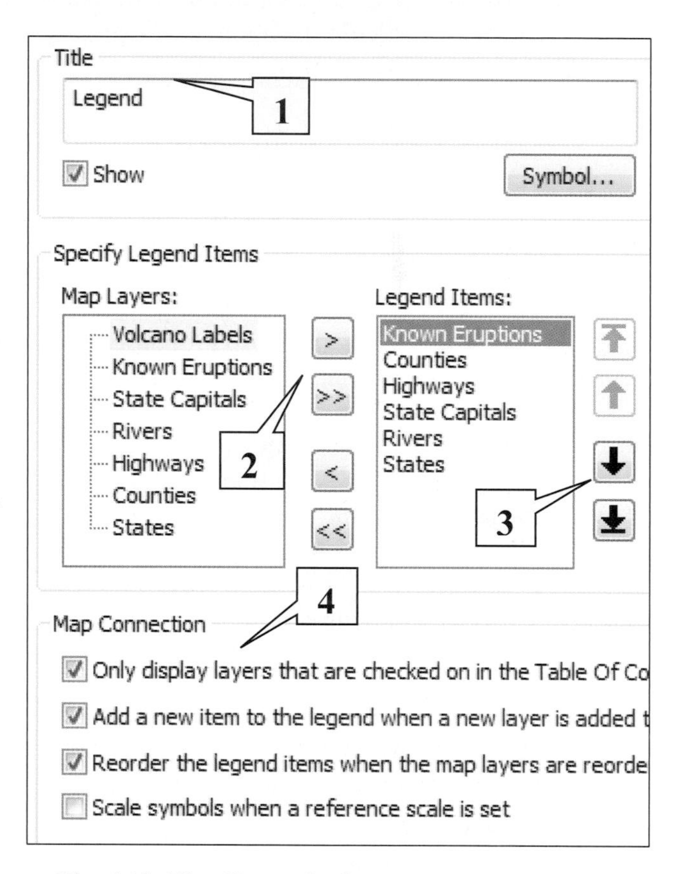

Fig. 149. The General tab

The Size and Position tab

The Size and Position tab specifies an exact position and size for the legend in page units (inches by default) (Fig. 150).

1. Set the position in page units, relative to the lower-right corner of the page.
2. The anchor point indicates which part of the legend sits at the XY distance. To place the lower-right corner of the legend at 3 inches from the left and 3 inches from the bottom, enter 3,3 and click the lower-left anchor point. To center the legend at 3,3, enter 3,3 and click the center anchor point.

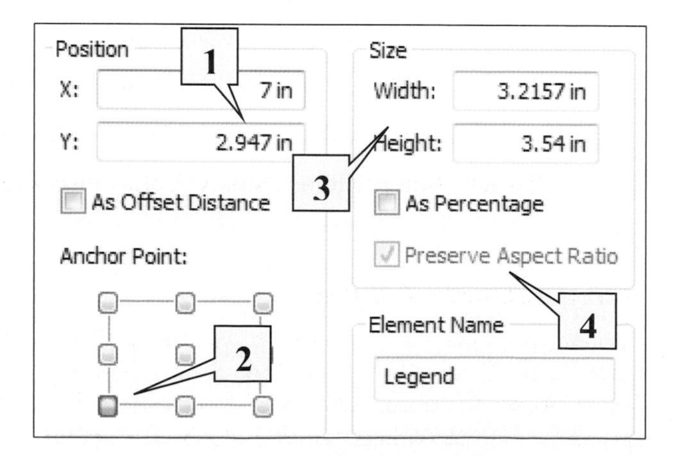

Fig. 150. The Size and Position tab

3. Set a specific width and height of the legend, if desired, either in page units or as a percentage of the page size.
4. If Preserve Aspect Ratio is checked, the shape of the legend will remain constant if it is resized.

The Items tab

This tab gives you individual control of every layer in the legend.

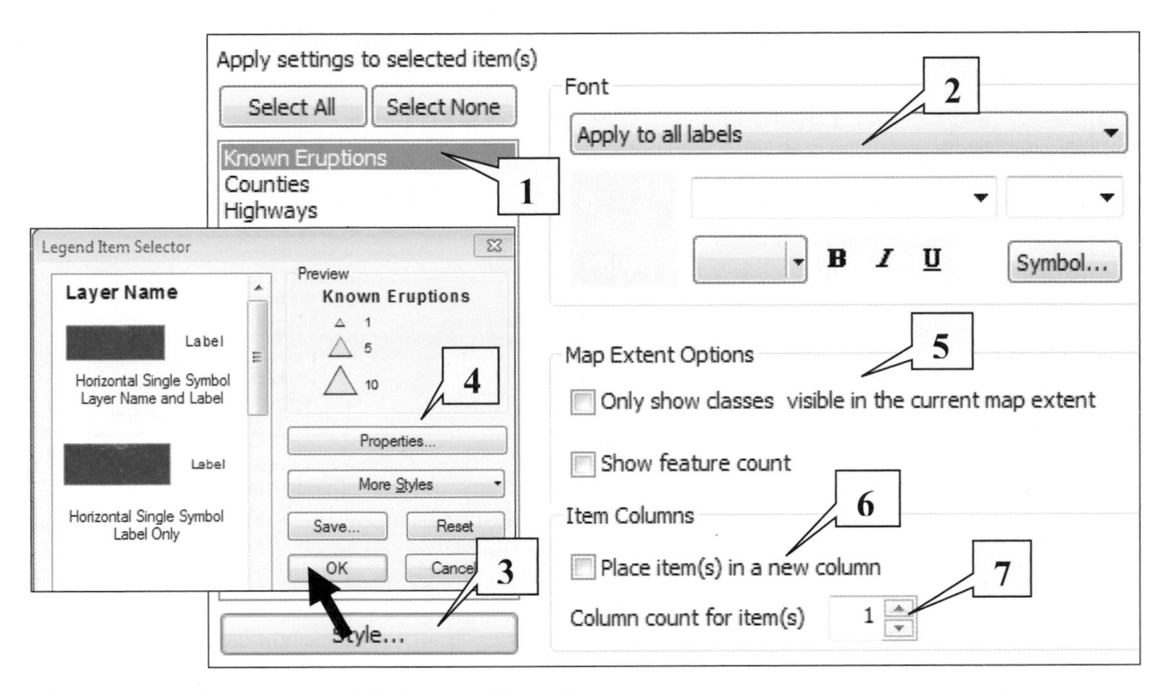

Fig. 151. The Items tab and the Legend Item Selector

1. Select the layer with the properties you wish to change (Fig. 151).
2. To change the label style, choose whether it applies to the entire legend or a single layer then set the desired font, style, size, symbol, and so on.
3. To change the headings, names, and labels included for the item, click the Style button and select one of the styles from the Legend Item Selector window.
4. Click the Properties button on the Style Selector window to edit every detail of the legend style, including the patch symbol. See ***Managing legend styles.***
5. Check the boxes for any of the desire Map Extent options.
6. To place the selected layer in a new column, check the box.
7. To set the number of columns for one or a set of items, set the column count.

Managing legend styles

The Legend Item Selector allows the user to choose from a set of legend styles, which vary in what elements are shown and how. Four different elements can be included in the legend: the Layer Name, Heading, Label, and Description. Figure 152a shows how these elements might appear in a legend that contains all four elements.

The values given to the four elements are controlled in several places. Layer names and Headings are edited in the Table of Contents (Fig. 152b). Labels can be edited on the layer's Symbology tab, and Descriptions are created by right-clicking the class in the Symbology tab and entering the desired information (Fig. 152c).

The Legend Item Selector (Fig. 152d), accessed from the Items tab, can be used to select a different style with different elements present. The Properties button on it opens the Legend Item window, which allows the user to adjust every detail of the legend, if desired (Fig. 152e).

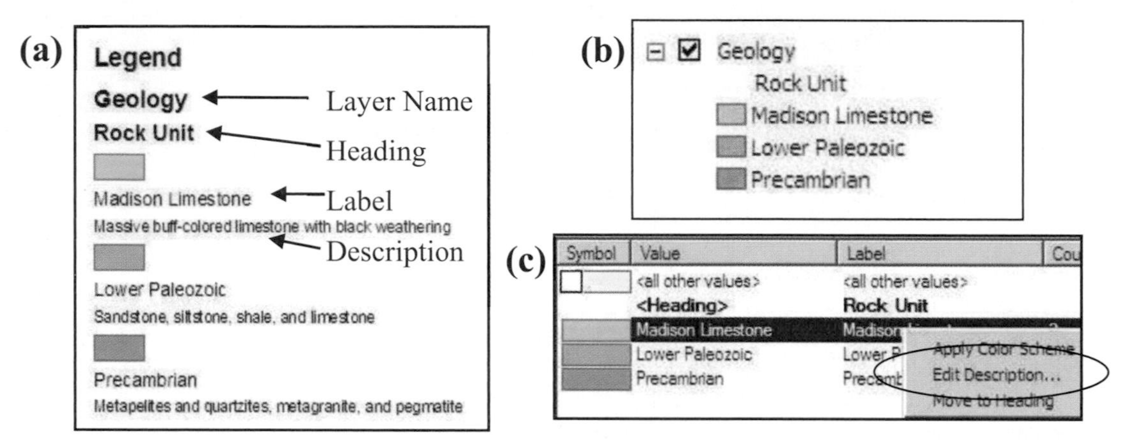

Fig. 152. Editing legends. (a) Legend items and their terms. (b) Edit layer names and headings in the Table of Contents. (c) Edit labels and descriptions on the Symbology tab. (d) The Legend Item Selector. (e) Edit every detail of the legend style.

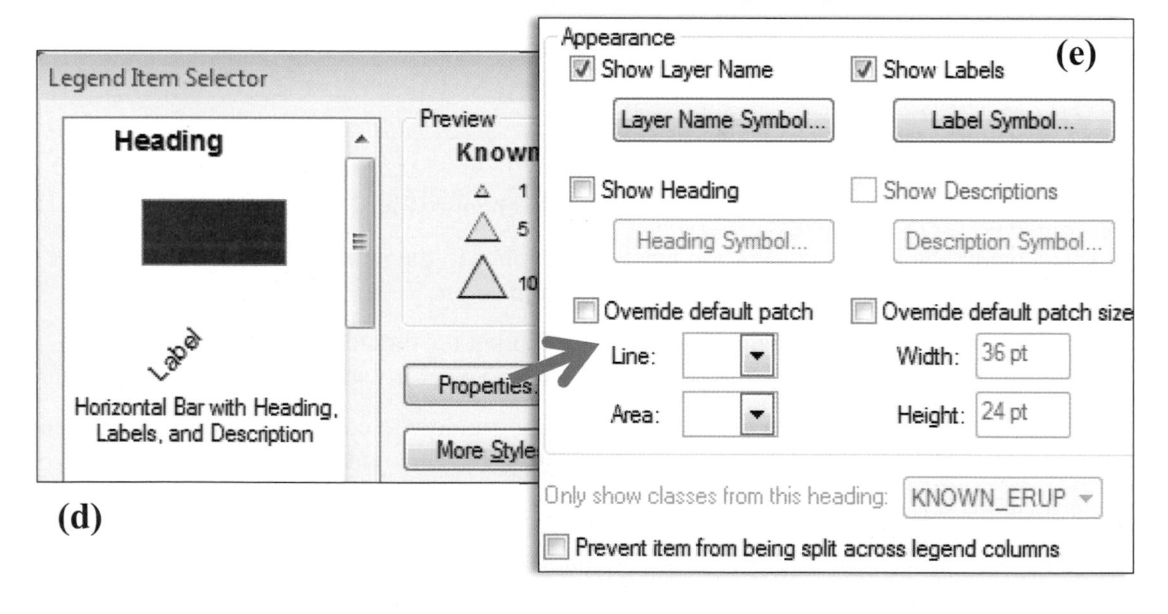

The Layout tab

This tab provides general settings for the legend layout and spacing (Fig. 153).

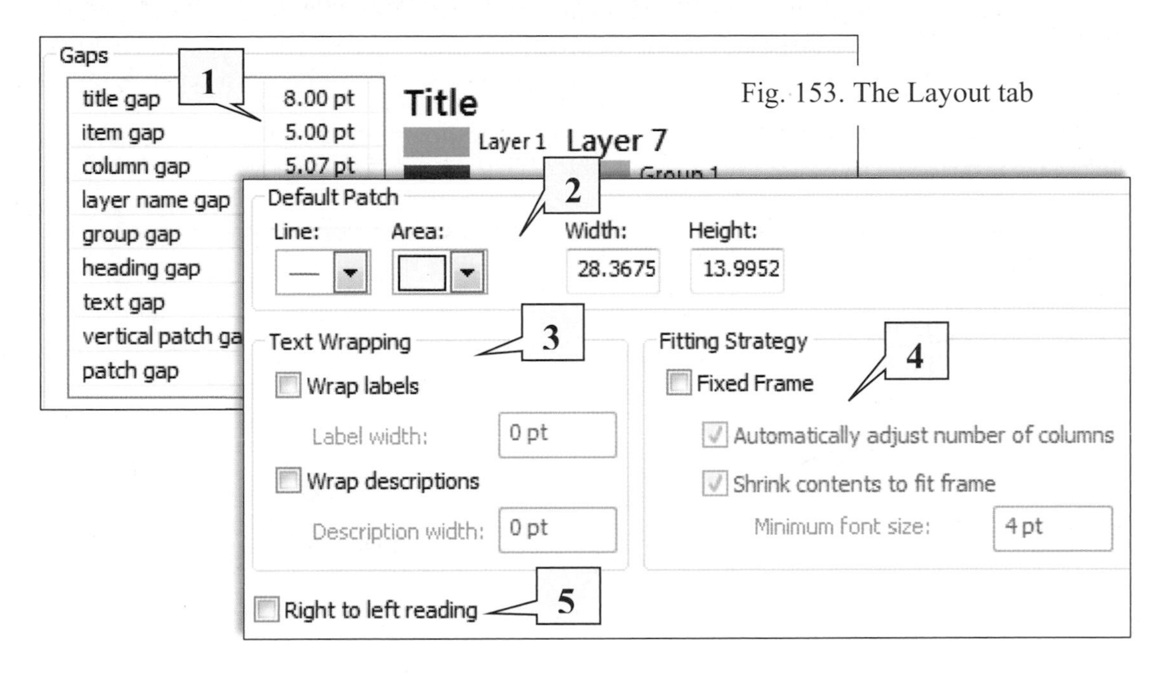

Fig. 153. The Layout tab

1. Change the spacing between the different parts of the legend.
2. Set the default patch style and size for the entire legend.
3. Set the desired text wrapping options for labels and descriptions.
4. Set the desired fitting strategy for the legend.
5. Choose Right to left reading for the legend, if desired.

The Frame tab

The Frame tab can set a border around the legend (Fig. 154).

1. Choose a border style and change its color, if desired.
2. To modify the existing border, click the Border Selector button or the Edit Border button.
3. Set the gap distance between the map and border.
4. Round the corners, if desired.
5. Set the background shade, if desired.
6. Set the Drop Shadow, if desired.

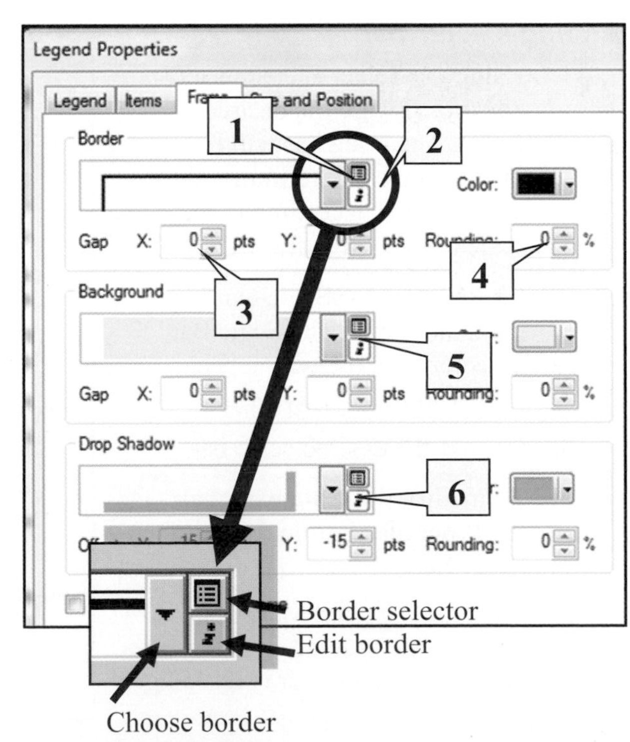

Fig. 154. The Frame tab

Adding neatlines, backgrounds, and shadows

Neatlines are lines that enclose one or more map elements. Many objects in layouts, including data frames, legends, and scale bars, have a tab in their properties to set up borders, backgrounds, and drop shadows. To access these tabs, open the element's properties by either double-clicking it or right-clicking it, and choose Properties. To create a neatline or shaded box as a separate object, do the following:

1. Choose Insert > Neatline from the main menu bar. The Neatline window appears (Fig. 155).
2. Choose the desired Placement option.
3. Set the border, background, and drop shadow styles.
4. To further modify the available styles, click the Border Selector or Edit Border buttons as shown.
5. Set the gap between the neatline and the elements and enter rounding, if desired. The higher the percentage, the more rounding occurs.
6. The Advanced button allows customized editing of the symbols, gaps, and rounding for each border, background, and shadow.

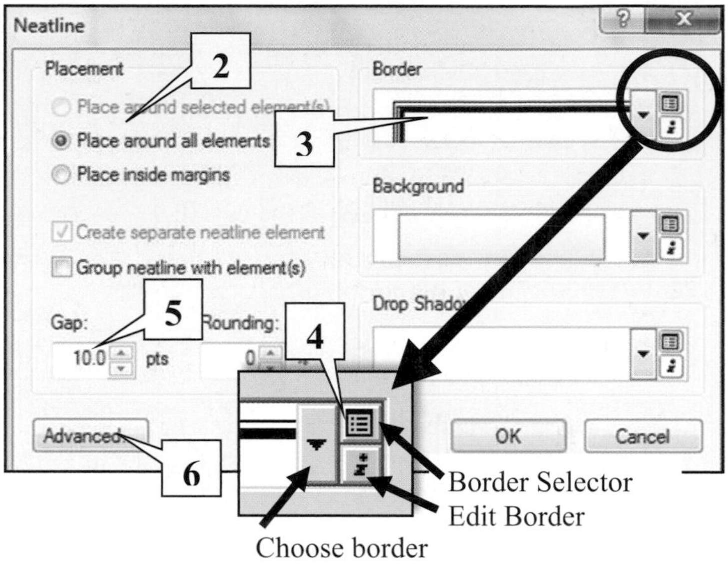

Border Selector
Edit Border
Choose border

Fig. 155. Setting up neatlines and boxes

Adding a scale bar

1. Using the Select Elements tool, click to activate the desired frame. The scale bar will be placed in the active frame and will be sized according to the scale of the active frame.
2. Choose Insert > Scale bar from the main menu. The Scale Bar Selector window will appear (Fig. 156).
3. Choose the desired scale bar.
4. To modify the scale bar, click the Properties button to open the Scale Bar.
5. The Numbers and Marks tab controls the spacing of numbers and marks on the bar (Fig. 157).

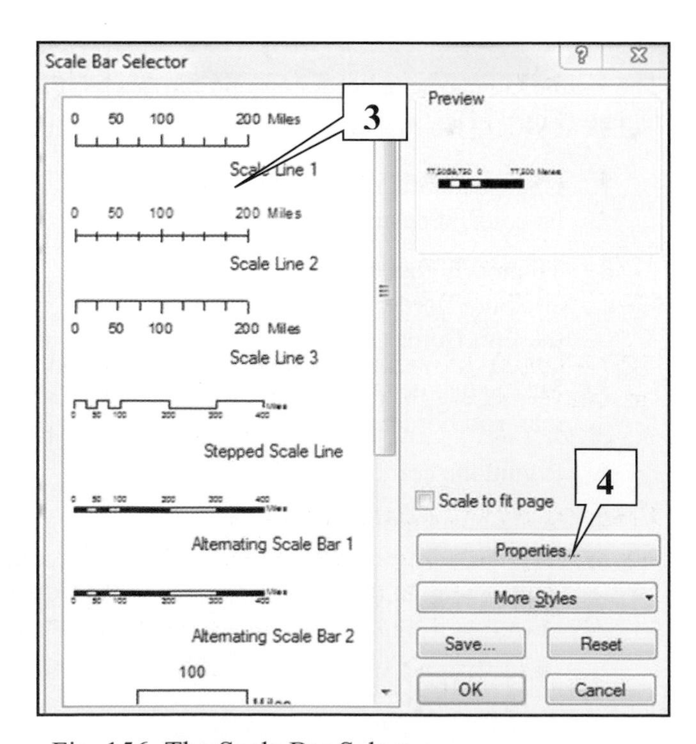

Fig. 156. The Scale Bar Selector

These values can usually be left as defaults.

6. The Format tab controls the font of the scale text and the style of the scale bar. The defaults are usually fine.
7. The Scale and Units tab controls the length and divisions of the scale bar.
8. Set the units for the scale bar to miles, kilometers, or some other unit. The default of meters is usually not desirable.
9. Choose the *When resizing…* option. One or more of the input boxes in the window may be dimmed, depending on which option is chosen.
10. Set the division value, the number of divisions, and the number of subdivisions, as applicable.
11. Check the box to place the subdivisions before the zero point rather than in the first division.
12. The chosen units will be labeled on the scale bar (e.g., Miles). Choose the label position, change the label text and font symbol, and set the gap between the scale bar and the label, if desired.

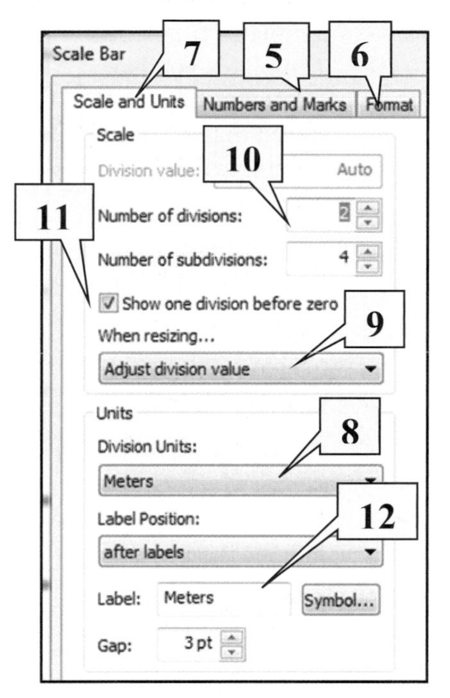

Fig. 157. Scale bar properties

TIP: To change the properties of a scale bar after it has been created, double-click the scale bar using the Select Elements tool, or right-click it and choose Properties.

Adding pictures

1. Choose Insert > Picture from the main menu bar.
2. Navigate to the folder containing the picture and click on it to select it.
3. Click Open to add the picture.
4. Resize and/or move the picture to the desired location using the Select Elements tool.

TIP: Information on allowed image formats can be found in the online help index by typing the entry "rasters" and choosing the subheading "formats, supported."

TIP: A bug in the initial release makes it difficult to resize the picture when it appears substantially larger than the layout page. A workaround is to set the layout zoom percentage to a small value so that the entire picture can be seen and resized.

Creating a map from a template

A map template is a set of data frames, titles, styles, and other map elements that are already formatted and ready to receive the data in the data frame(s). Use a map template to quickly create a map in a standard format. You can save any map as a template to create a similar map again.

1. Click the Change Layout button in the Layout toolbar.
2. Click one of the tabs to see a choice of templates (Fig. 158). Templates you have created will be stored in the My Templates tab.
3. Use the arrow buttons to scroll the tabs to see them all.
4. Click on a template to see a preview of it.
5. Use the Browse button to navigate to another directory containing more templates saved elsewhere.
6. Click the Thumbnail button to see icons of all the templates; click the List button to go back to the original list view.
7. Click the desired template and choose Next or Finish.

Fig. 158. Choosing a map template

TIP: If the template has more than one data frame, it will prompt the user to assign the data frames in the map document to the data frames in the template.

Assigning multiple frames

8. If the data frames need assigning, click each of the frames in the list on the left (Fig. 159).
9. Use the Move Up and Move Down buttons to put them in the same order as the numbered frames in the new layout.
10. Click Finish.
11. Finally, change any titles or other map elements in the template that need to be customized.

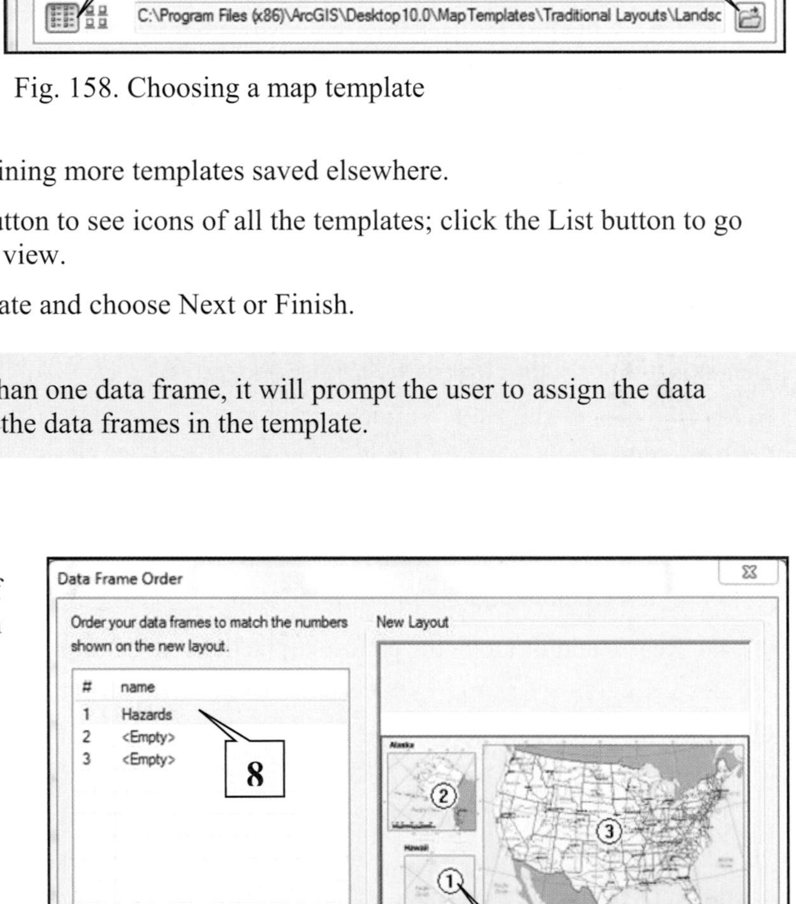

Fig. 159. Assigning data frames to the template frames

Printing a map

1. To preview a map to see how it will look on paper, choose File > Print Preview from the main menu bar.
2. To print, Choose File > Print from the main menu. The Print window appears (Fig. 160).
3. To change the printer or its properties, click the Setup button.
4. Set the number of copies to print.
5. Choose the desired tiling options if the map is larger than the printer paper.
6. Preview the layout placement on the page to ensure that the map (color) fits the paper (white). Click OK.

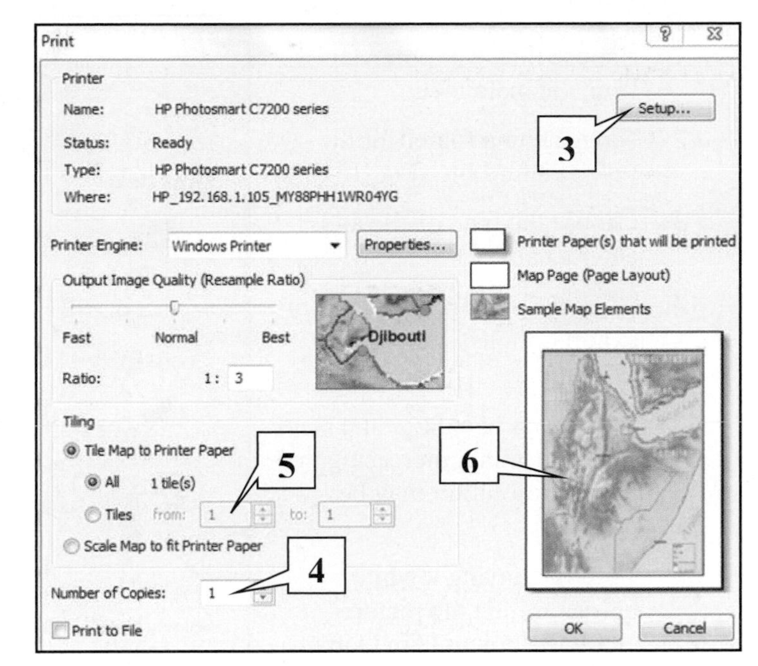

Fig. 160. The Print window

Exporting a map as a picture file

You can export a map as an image file to put it on a web page or inside another document or as a PDF to share it with others.

1. Choose File > Export Map from the main menu.
2. Navigate to a folder for the saved picture file (Fig. 161).
3. Choose the type of file and enter a name for it.
4. Click the gray arrow for more export options, such as the resolution.
5. If you plan to enlarge the map before printing, increasing the default resolution may be necessary.
6. To avoid having a white border around the picture, check the box to Clip Output to Graphics Extent.
7. Click Save.

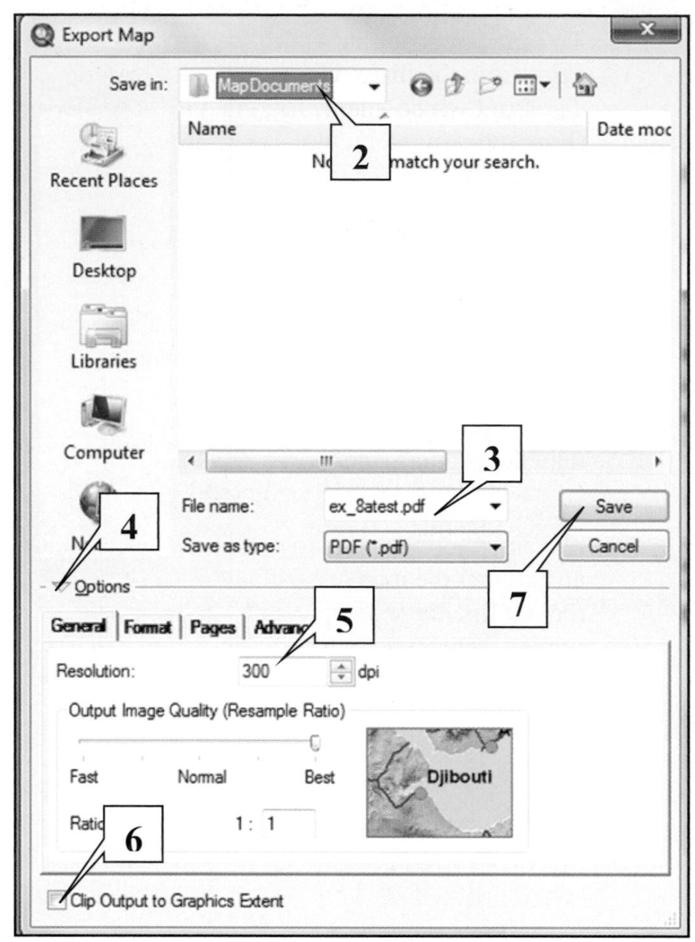

Fig. 161. Exporting a map to a PDF

Creating a simple graph

This example shows the steps to create a horizontal bar graph. Other graph types will have slightly different options. Have fun experimenting.

1. Open the table from which the graph will be created. Choose Table Options > Create Graph from the main menu bar. The Graph Wizard will appear (Fig. 162).
2. Choose the graph type.
3. Set the Value field to be graphed.
4. To sort the bars, make the Y field the same as the Value field and choose Ascending or Descending.
5. Choose the X or Y label field.
6. Uncheck the Add to legend box if you don't need a legend.
7. Set the bar color. Use Match with Layer to make it the same as the map. Choose Custom to make all the bars one color of your choosing. Choose Palette to make every bar a different color.
8. Click Next.
9. Give the graph a title and footer.
10. Select a title and position for the legend if you have one.
11. Click the Left tab and provide a title for that axis.
12. Click the Bottom tab and enter a title.
13. Click Finish.
14. To place a graph on the layout, open the graph. Right-click the blue bar at the top of the graph and choose Add to Layout.
15. To edit the graph properties, right-click on the top blue bar of the graph window and choose Properties or Advanced Properties.

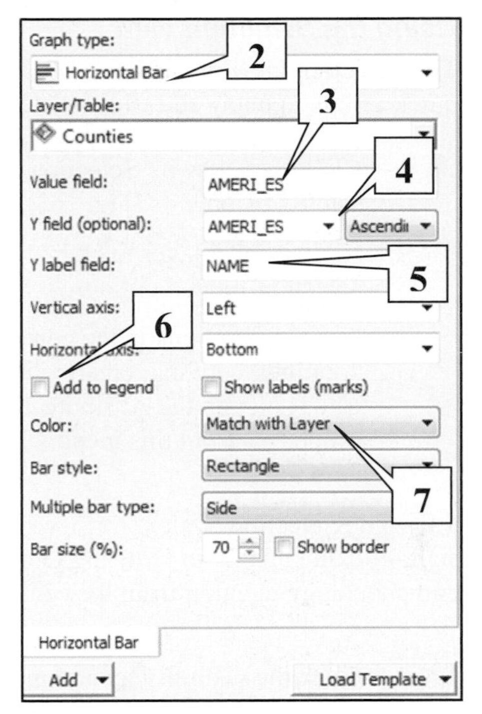

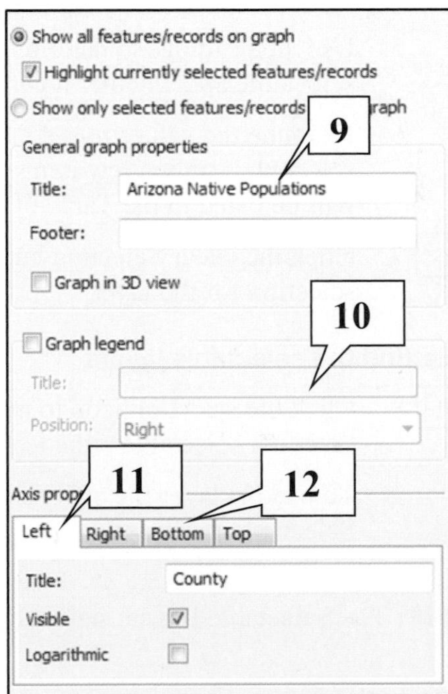

Fig. 162. The Graph Wizard

ANALYSIS

Queries

Using the Selection view

The Selection view in the Table of Contents provides quick access to many selection-related operations.

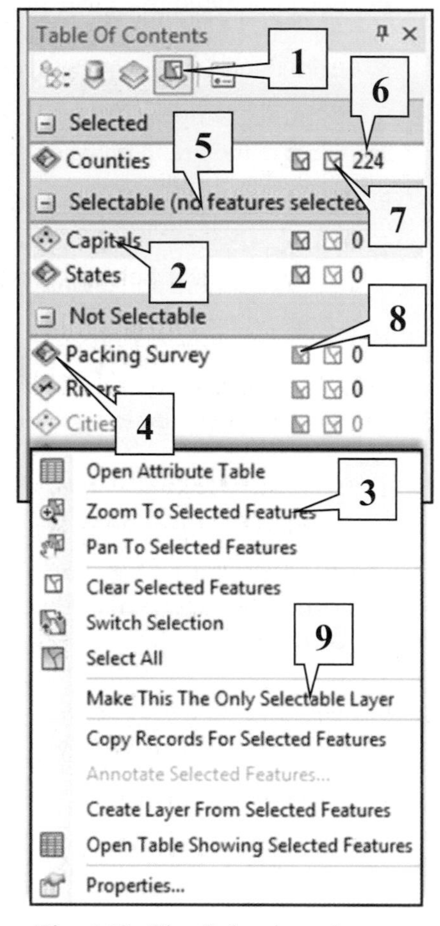

Fig. 163. The Selection view and context menu

1. Click the View by Selection icon at the top of the Table of Contents (Fig. 163).
2. Right-click a layer to open a context menu with selection-related options.
3. Use the functions in the context menu for zooming, copying, creating layers, and so on. You can also open the Attribute table and the layer properties from this menu.

TIP: The Selection context menu can also be accessed by right-clicking a layer in Drawing Order or Source view and choosing Selection from the context menu.

4. Click the visibility icon to make the layer visible in the map or not.
5. Use the headings to determine if a layer is selectable and whether it currently has a selection.
6. Examine the values to see many features are selected. If only a few items are selected, they will be listed in the Table of Contents.

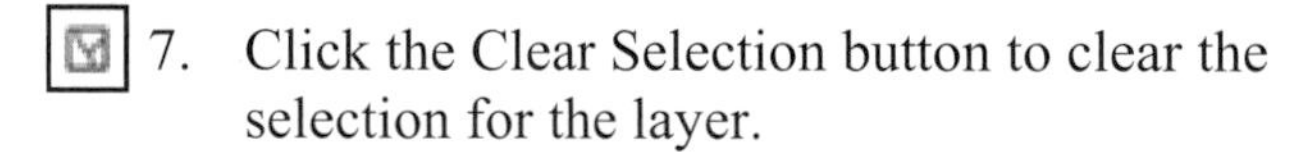

7. Click the Clear Selection button to clear the selection for the layer.

Setting the selectable layers

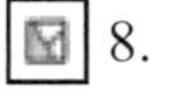

8. Click the selection icon to make the layer selectable. Click the icon again to make the layer not selectable—the icon will turn gray.
9. To make a layer the only selectable layer, right-click it and choose Make This The Only Selectable Layer.

TIP: The Selectable Layers setting does not apply to selections based on attributes or location.

Clearing selections

Selections may be cleared in a number of ways. Some clear only one layer or table; others clear selections for all layers and tables, so be careful about which you use.

Clearing the selection for a layer or table

1. Click the Clear selection button in the View by Selection view in the Table of Contents (Callout 7 in Fig. 163).
2. Right-click a layer in Selection view to open the context menu and choose Clear Selected Features. Features will be cleared for that layer only.
3. Right-click a layer in any other view and choose Selection > Clear Selected Features.
4. If a table is open, click the Table Options menu and choose Clear Selection.

Clearing all selections

5. Choose Selection > Clear Selected Features from the main menu bar.

6. Click the Clear Selection button on the Tools toolbar.

Changing the selection method

The selection method controls what happens to a previous selection when a new one is specified for the same layer.

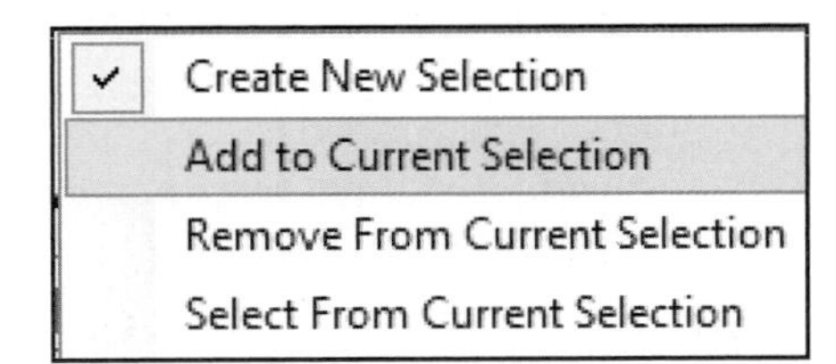

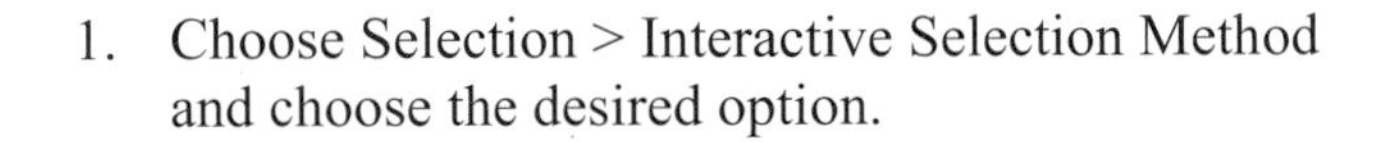

1. Choose Selection > Interactive Selection Method and choose the desired option.

TIP: The query windows also have a drop-down bar used to change the selection method for a query made in the window.

Selecting features interactively

Features may be selected visually by designating them on the screen using one of the interactive selection tools (Fig. 164). By default, any feature that touches the specified shape will be selected. To change the default, see ***Changing the selection options***.

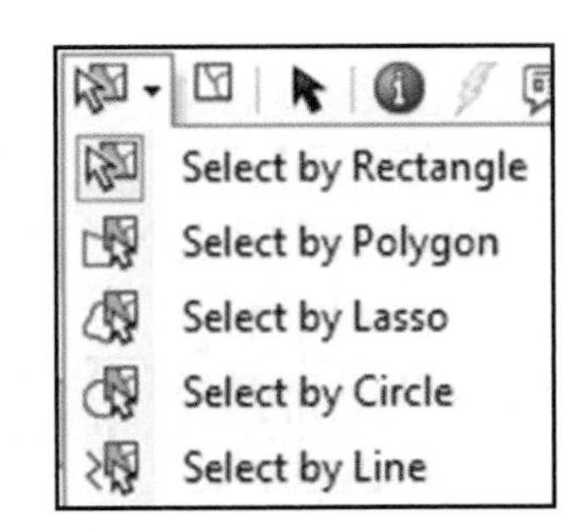

Fig. 164. The interactive selection tools

1. Set the Selectable Layers, if necessary. See ***Using the Selection Window***.
2. If necessary, change the selection method by choosing Selection > Interactive Selection Method and picking one of the methods. See ***Changing the selection method***.

3. Click the **Select by Rectangle** tool on the Tools toolbar.
4. Click on a feature in the map to select it. It will be highlighted in the selection color.
5. To add a feature to the selection, hold down the Shift key and click the feature. To remove a feature from the selection, hold down the Shift key and click the feature.
6. To add a group of features using a rectangle, click on one corner of the rectangle, drag the mouse to the desired size, and release the mouse button. All features touching the rectangle will be selected.

TIP: Hold down the Shift key while drawing the rectangle to add a group of features to the current selection.

Using the other selection tools

Click the desired tool and use the instructions below to specify the shape. By default, any feature touching the shape will be selected.

1. To **Select by Polygon**, click to enter each vertex and double-click to finish.
2. To **Select by Lasso**, place the cursor at the starting location, click and hold down the mouse button while drawing the shape, and release the mouse button to finish.
3. To **Select by Circle**, click in the center of the circle and drag out to the desired radius. You can also click R on the keyboard while the mouse button is down to specify a radius in map units. Release the mouse, type the radius, and click Enter.

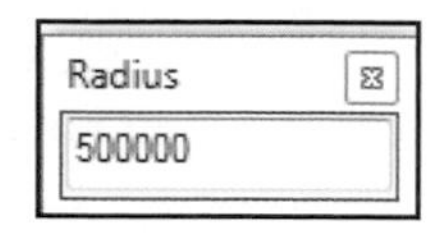

4. To **Select by Line**, click to enter each vertex and double-click the last vertex to finish.

Creating a layer from selected features

1. Perform interactive, attribute, or location queries to select the desired records.
2. Right-click the layer name and choose Selection > Create Layer from Selected Features from the context menu.
3. The new layer appears in the Table of Contents.
4. Click on the new layer name twice to give it a new name.

TIP: If the Table of Contents is in View by Selection mode, you can create the layer, but you must switch to another view or open the layer properties to name the new layer.

Creating a definition query

A definition query is a layer property that temporarily confines the features of the layer to a subset of the stored features.

1. Open the properties for the layer, such as by right-clicking it and choosing Properties.
2. Click the Definition Query tab (Fig. 165).
3. Click the Query Builder button and enter an SQL expression to define the subset. See ***Entering an SQL query***.
4. The query will be shown on the tab. Click OK.

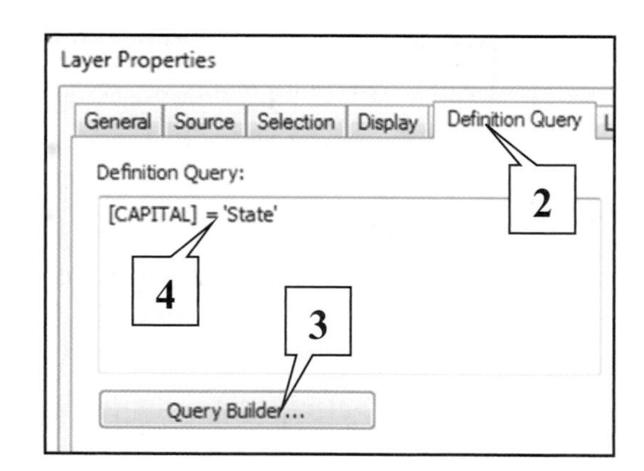

Fig. 165. Creating a definition query

TIP: Always Verify the expression in the SQL Query window to ensure that it contains no errors. An expression error would cause the layer to disappear and/or generate drawing errors.

Entering an SQL query

An SQL query is an expression to select features based on an attribute in the table. A similar window appears whenever an SQL expression is needed. The query is entered in the expression box in the lower part of the window.

1. Choose the selection method (Fig. 166). For a discussion of the different methods, see Chapter 5.
2. Enter a field name from the list into the expression box by double-clicking on the field name.
3. Enter an operator by clicking it once.
4. For categorical data, click Get Unique Values and double-click the desired value from the list.
5. Alternatively, enter a value by typing it into the box. If it is a text string, put single quotes around it.
6. Multiple queries use criteria from more than one field. The field name must be repeated for each query.
7. Click Verify to ensure that the query is correctly formulated, if desired.

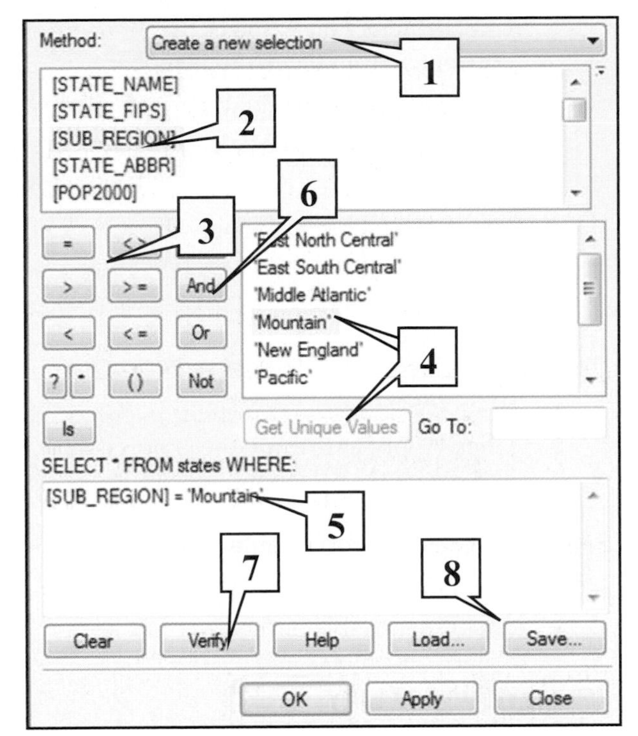

Fig. 166. Entering a query

8. Use the Save button to save a complex query or one that is frequently used. It will be saved with an .exp extension. Load it again later using the Load button.
9. Click Apply or OK to execute the query.

TIP: Many users find that typing in or editing the expression by hand tends to generate errors. Use the boxes and the buttons provided to enter expressions for quickest results.

TIP: When you are using multiple conditions, the field name must appear in every condition. To find rents between 500 and 1000 dollars, for example, enter [RENT] > 500 AND [RENT] < 1000. You cannot enter [RENT] > 500 AND < 1000.

Using Select By Attributes

You can access the Select By Attributes menu in two ways.

Selecting features from a layer

1. Choose Selection > Select By Attributes from the main menu bar.
2. Choose the layer containing the features that are to be selected (Fig. 167).
3. Check the box to show only the selectable layers in the list, if desired.
4. Choose the selection method.
5. Enter the expression. See ***Entering an SQL query***.
6. Click Apply to execute the expression and create the selection.

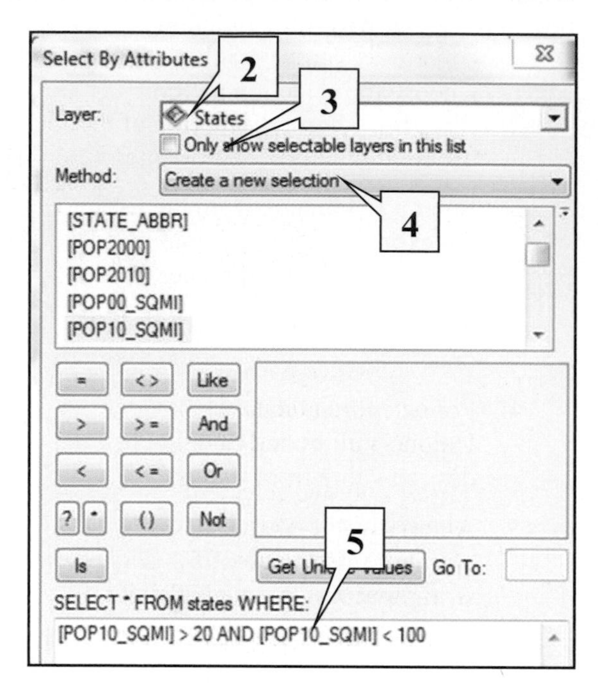

Fig. 167. The Select By Attributes window

Selecting records from a table

1. Open the table, such as by right-clicking the table or the layer in the Table of Contents and choose Open Attribute Table.
2. Choose Table Options > Select By Attributes. The window will look similar to Figure 166, but there will be no drop-down to choose the layer.
3. Choose the selection method.
4. Enter the expression. See ***Entering an SQL query***.
5. Click Apply to execute the expression and create the selection.

Using Select By Location

Select By Location compares the features to be selected (from the target layer) with features from another layer (the source layer) using a spatial condition operator such as *intersects* or *contains*.

1. Choose Selection > Select By Location from the main menu bar.
2. Choose the selection method (Fig. 168).
3. Choose the target layer(s) from which to select features. Be careful—layers from previous queries remain checked until you uncheck them.
4. Choose the source layer against which the target features will be tested.
5. To use only the selected features of the source layer, check the Use selected features box.
6. Choose the spatial condition for the test.
7. If using the condition *within distance of*, or increasing the distance over which the other criteria are applied, enter a buffer amount and distance units, and check the Apply a buffer box, if necessary.
8. Click Apply to make the selection.

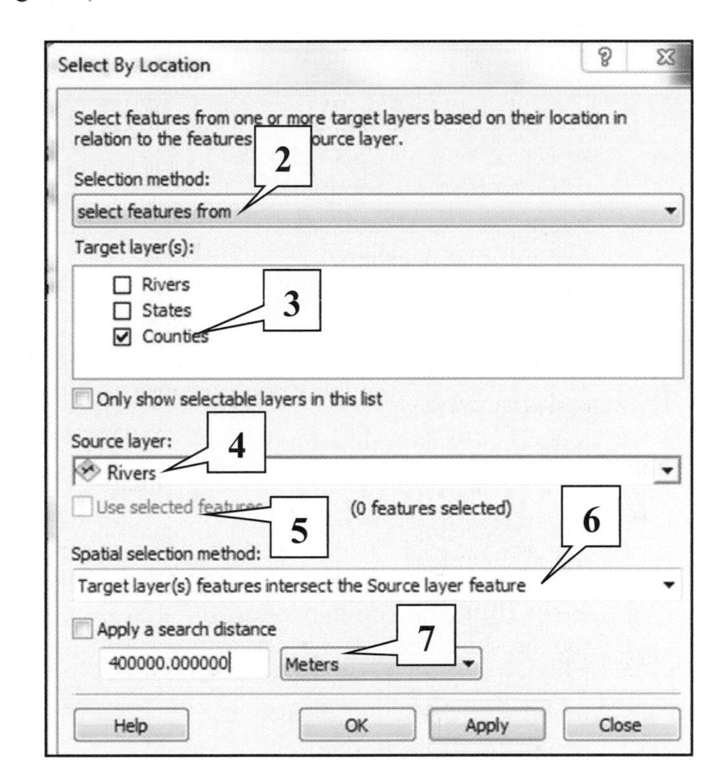

Fig. 168. The Select By Location window

TIP: To use the same layer as the selection layer and the condition layer (e.g., to select all restaurants within two miles of the restaurant you own), you must first use an attribute or an interactive query to select the feature of interest—your restaurant. Then do the Select By Location and specify the same layer for Steps 3 and 4.

Changing the selection options

The selection options control several aspects of selections and queries.

1. Choose Selection >Selection Options from the main menu bar.
2. The first box (Fig. 169) controls how shapes select features when one is using one of the interactive selection tools. By default, any feature partially or completely in the shape is selected. You may also choose to select features completely within the shape or to select features that the shape is completely within.
3. The second box sets how close you need to click to a feature before it is selected. If you often have trouble selecting points, increase this distance.
4. This box changes the selection highlight color.
5. In the fifth box, you may alter the threshold that determines how many records will be modified by a Switch Selection command. If the number exceeds the threshold, a warning prompt will be issued. This option prevents you from mistakenly doing a switch that will take a very long time to complete.
6. Finally, you can choose whether to save layers with the current selections (the default) or to clear the selections of all layers when saving the map document.

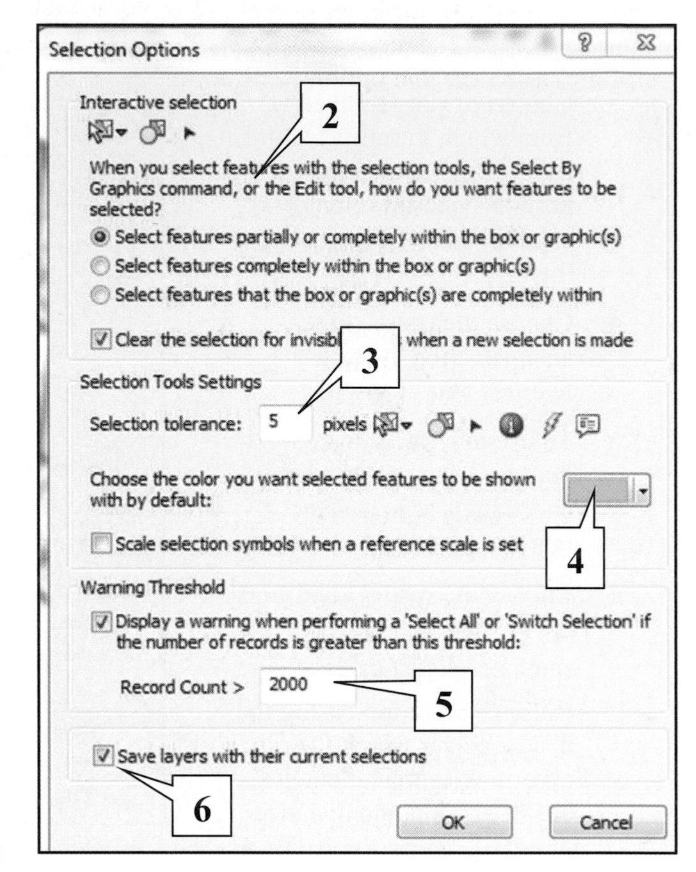

Fig. 169. Setting the selection options

Overlay and Spatial Joins

Performing a spatial join

In a spatial join, the attributes of features in the source layer are appended to the features in the destination table. The output layer always contains the same features as the destination table. When choosing the join options in Step 4, keep in mind the cardinality of the join to pick the most appropriate one.

1. Right-click the destination layer in the Table of Contents and choose Joins and Relates > Join.
2. Choose to join to another layer based on spatial location (Fig. 170).
3. Choose the source layer from the drop-down box, or click the Browse button to choose one from the disk.
4. Choose one of the two join options and specify any summary statistics desired, if applicable.
5. Specify the shapefile or feature class file to contain the new output layer.
6. For more information on joins, click the About Joining Data button.
7. Click OK to execute the join.

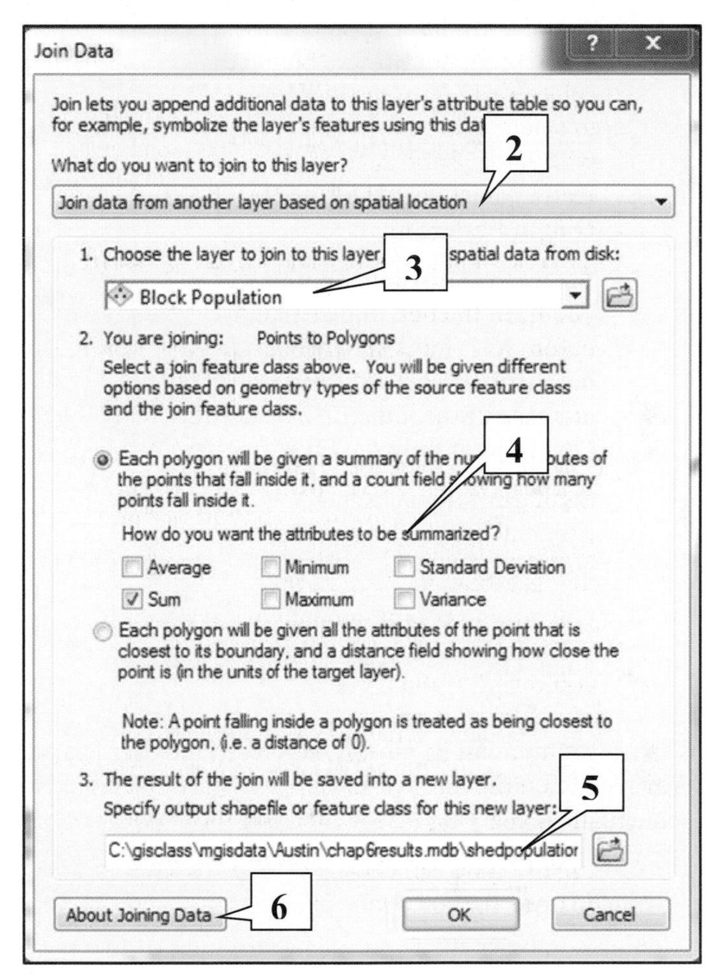

Fig. 170. Join Data window for spatial joins

Performing an intersection

Intersect combines the features and attributes of two input layers, preserving the areas that are common to both.

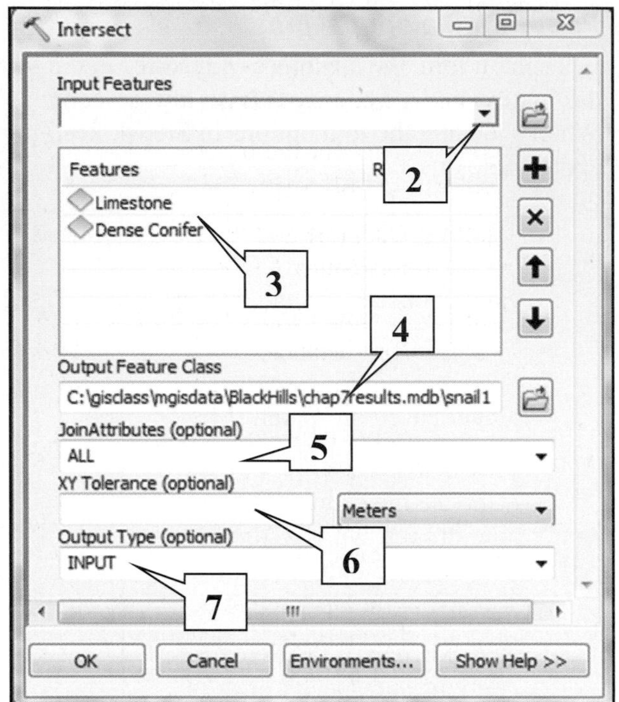

Fig. 171. The Intersect tool

1. Choose ArcToolbox > Analysis Tools > Overlay > Intersect (Fig. 171).
2. Enter the Input Features by clicking a choice from the drop-down list or by clicking the Browse button to select a file from the disk.
3. Enter a second Input Features layer.
4. Enter the name and location of the Output Feature Class.
5. Set the JoinAttributes option to NO-FID, the best choice in most cases. ALL joins all attributes from both tables, NO-FID joins all attributes except the FIDs, and FID-ONLY joins only the FIDs.
6. Set a carefully chosen XY Tolerance for help in preventing slivers.
7. Change the default output type, if desired.
8. Click OK.

TIP: Two input layers must be entered if the user has an ArcGIS Basic license. With an ArcGIS Advanced license, three or more layers may be entered and the intersections will be performed sequentially.

Performing a union

A union combines the features and attributes of two polygon layers, preserving all areas from both. The Union window is nearly identical to the Intersect window shown in Figure 171.

1. Choose ArcToolbox> Analysis Tools > Overlay > Union.
2. Enter the Input Features by clicking a choice from the drop-down list or by clicking the Browse button to select a file from the disk. Both inputs must contain polygons.
3. Enter a second Input Features layer.
4. Enter the name and location of the Output Feature Class.
5. Set the JoinAttributes option to NO-FID, the best choice in most cases.
6. Set a carefully chosen XY tolerance for help in preventing slivers.
7. Check the box if gaps are to be allowed between polygons. Click OK.

Creating buffers

Buffers create polygons delineating the area within a specified distance of a set of features. Creating buffers can be a time-intensive process.

1. Choose ArcToolbox > Analysis Tools > Proximity > Buffer.
2. Select a layer for the Input Features to buffer (Fig. 172).
3. Specify the name and location for the Output Feature Class.
4. Specify the units and value for a fixed buffer distance.
5. OR choose a field from the input layer attribute table containing the buffer distances.
6. Change the Side and End type settings, if desired.
7. Set the Dissolve Type; use ALL to dissolve boundaries between overlapping buffers or NONE to leave them as is. The ALL option is usually best.
8. Choose one or more Dissolve Field(s). Buffers sharing the same value in the field will be dissolved.
9. Click OK.

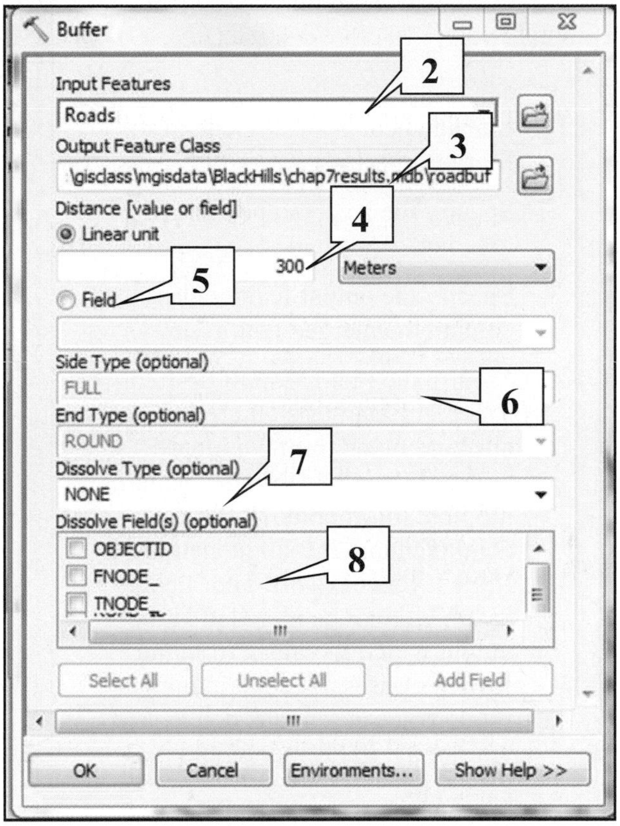

Fig. 172. The Buffer tool

Clipping a layer

Clipping removes features outside a boundary specified by a second layer.

1. Choose ArcToolbox > Analysis Tools > Extract > Clip (Fig. 173).
2. Specify the Input Features layer to be clipped.
3. Specify the Clip Features layer representing the clip boundary.
4. Specify the name and location of the Output Features.
5. Set a carefully chosen XY tolerance to help avoid slivers, if desired.
6. Click OK.

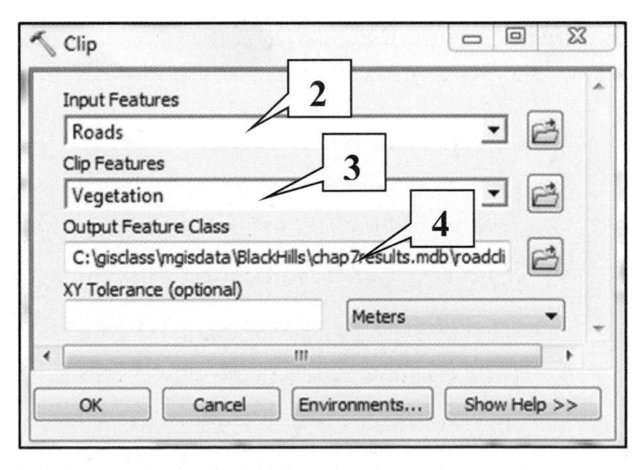

Fig. 173. Clipping a layer

Merging or appending layers

Merge combines two or more feature classes to create a new feature class. Append combines two or more feature classes and places them into an existing feature class. The data type of the inputs must match (all points, for example).

1. Choose ArcToolbox > Data Management Tools > General > Merge (Fig. 174).
2. Specify the layers to be merged together.
3. Specify the output feature class to contain the merged features.
4. Modify the Field Map, if desired. The Field Map lists all unique attribute fields from the input layers. When the attributes match, the field information will be combined into the single new field. If the layers have different fields, each field will be included in the output with null values from the feature classes not containing that field. You can rename, delete, and reorder the fields in the field map.
5. Click OK.

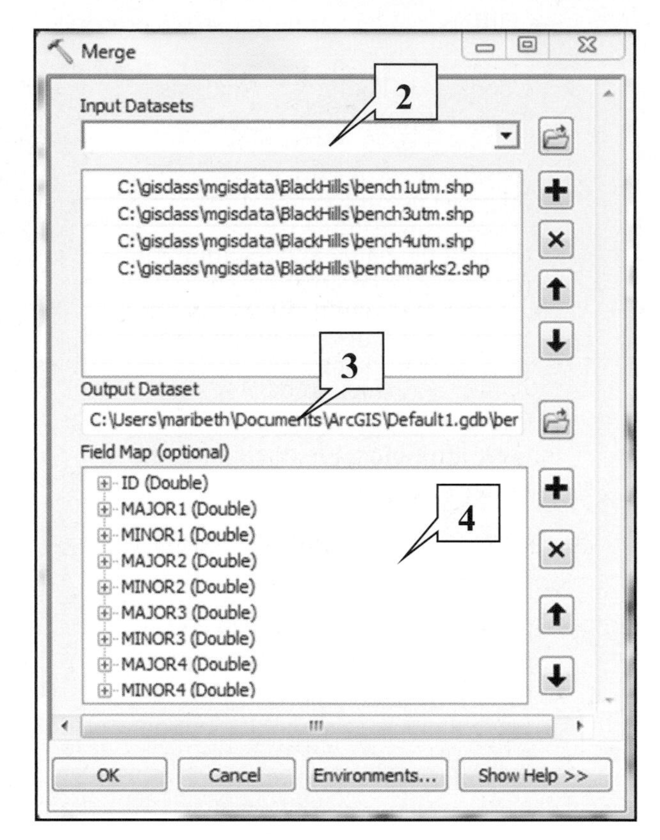

Fig. 174. Merging layers

TIP: The Append tool is nearly identical to the Merge tool window, except that there is no field map panel because the fields are defined by the existing output feature class.

Dissolving

Dissolving merges features within a feature class when they share the same value of a specified attribute.

1. Choose ArcToolbox > Data Management Tools > Generalization > Dissolve.
2. Specify the Input Features layer to be dissolved (Fig. 175).
3. Specify the name and location of the Output Features to be created.
4. Choose one or more attribute field(s) on which to base the dissolve. These fields will appear in the output layer; all other fields will be dropped.
5. Select a field to generate statistics from, if desired.
6. Ignore the warning, and click in the Statistic Type box to specify the type of statistic to use.
7. Select as many field-statistic combinations as desired.
8. Scroll down and uncheck the *Create multipart features* box if you wish to ensure that disjoint polygons remain separate features in the output.

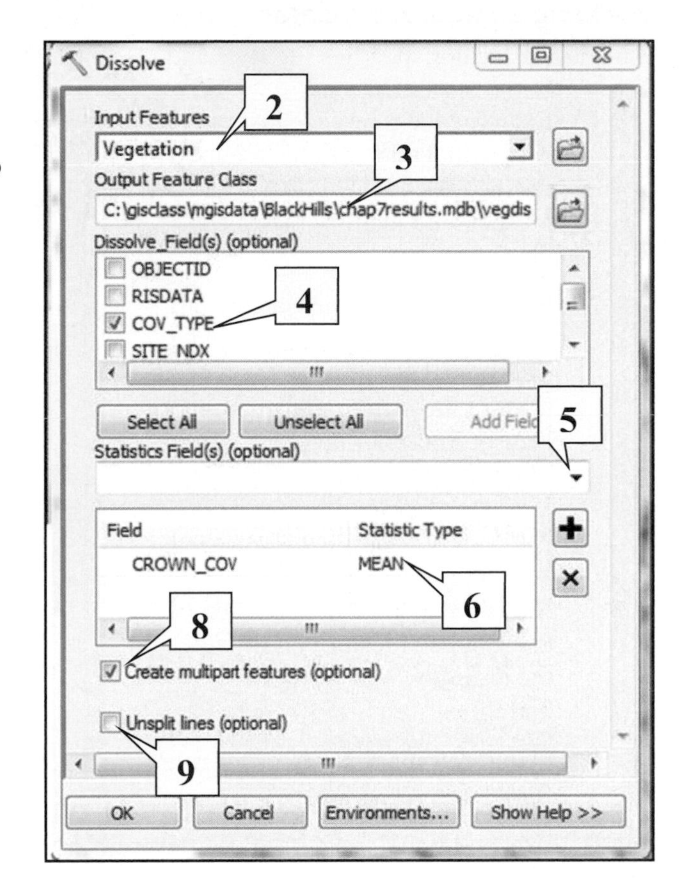

Fig. 175. Dissolving a layer

9. Check the box for *Unsplit lines* if you want lines to be dissolved only when they share an end vertex. Otherwise, disjoint lines are dissolved into a single feature.
10. Click OK to begin dissolving the layer.

Geocoding

Creating an address locator

An address locator specifies the style and reference layer used for geocoding.

1. Examine the reference layer to ensure that all the fields required for the geocoding style are present, and identify the names of the fields.
2. In ArcCatalog or the Catalog tab, right-click the name of the folder or geodatabase in which you want to store the locator and choose New > Address Locator.
3. Click the Browse button to choose the address locator style (Fig. 176).
4. Specify the name and location of the reference feature class using the drop-down or the Browse button.
5. Specify Primary Table as the role for the reference layer.
6. Add additional table(s), if desired, and specify their role(s).
7. Set the required fields (asterisk) and the optional fields to the appropriate fields in the attribute file of the reference layer.
8. Specify the name and location where the address locator will be saved.
9. Click OK to create the locator.

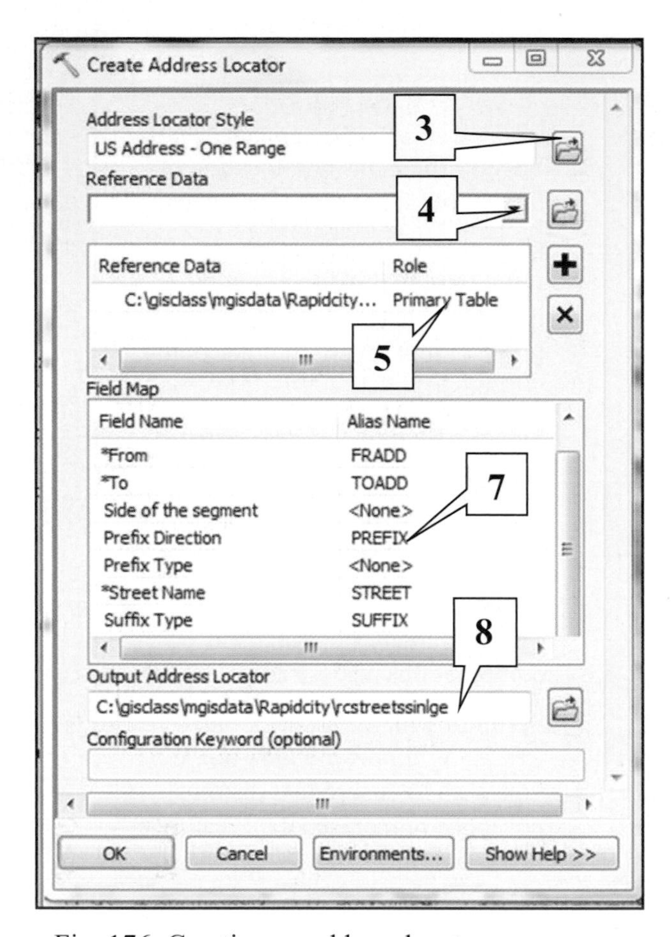

Fig. 176. Creating an address locator

TIP: Although theoretically optional, the City, State, and Zip fields must be used in the address locator for it to work. This is a bug in the initial release of ArcGIS 10.

Geocoding addresses

1. Add the address table and the reference layer (optional) to ArcMap.
2. Click the List by Source icon at the top of the Table of Contents.
3. Right-click the address table in the Table of Contents and choose Geocode Addresses.
4. If the desired locator does not appear in the menu, click Add (Fig. 177).

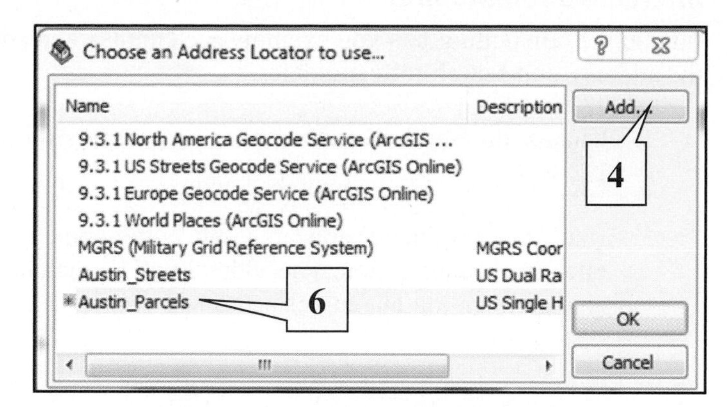

Fig. 177. Choosing an address locator

5. Navigate to the folder where the address locator is stored. Select the locator and click Add to close the menu and add the locator to the list.
6. Click the desired locator to highlight it and click OK.
7. Specify the table containing the addresses (Fig. 178).
8. Specify the fields containing the address information to be matched.
9. Specify the name and the location of the output file.
10. Click the Geocoding Options button to change the spelling sensitivity, minimum candidate score, or minimum match score.
11. Click OK to begin matching.
12. A progress bar will appear. When it is complete, click Close to finish matching or click Rematch to try matching more addresses.

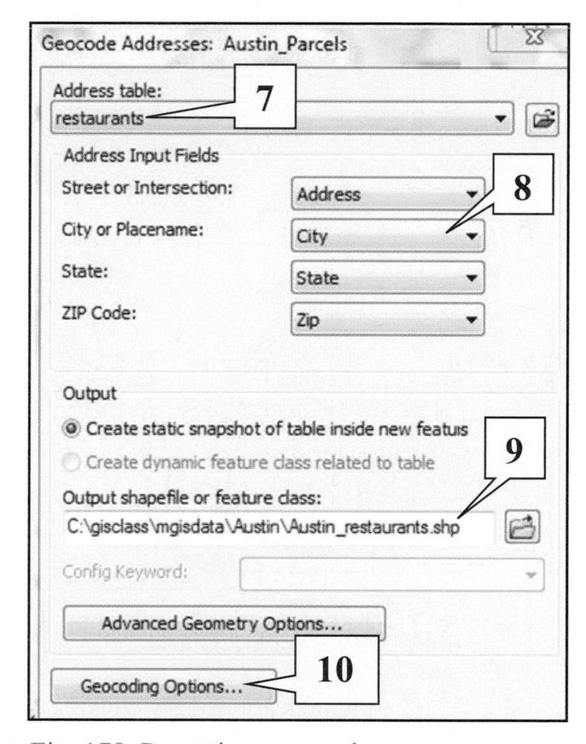

Fig. 178. Preparing to match

Rematching a geocoding result

You can reopen a completed matching session to rematch more addresses.

1. It is best to terminate any editing before rematching addresses. Choose Editing > Stop Editing from the Editor menu and save any changes, if necessary.
2. Right-click the geocoded layer in the Table of Contents and choose Data > Review/Rematch Addresses.
3. Rematch the addresses. See ***Interactive rematching***.

Interactive rematching

Interactive rematching lets you examine each unmatched address, edit errors or inconsistencies in the address, and search for candidates.

1. Change the Show Results box to Unmatched Addresses (or another group) (Fig. 179).
2. Click on one of the unmatched addresses to highlight it.
3. If a list of candidates appears in the bottom box, click the best match to choose it, and click the Match button. The address will be matched, and the U in the upper panel will change to an M. The match score will also be shown.

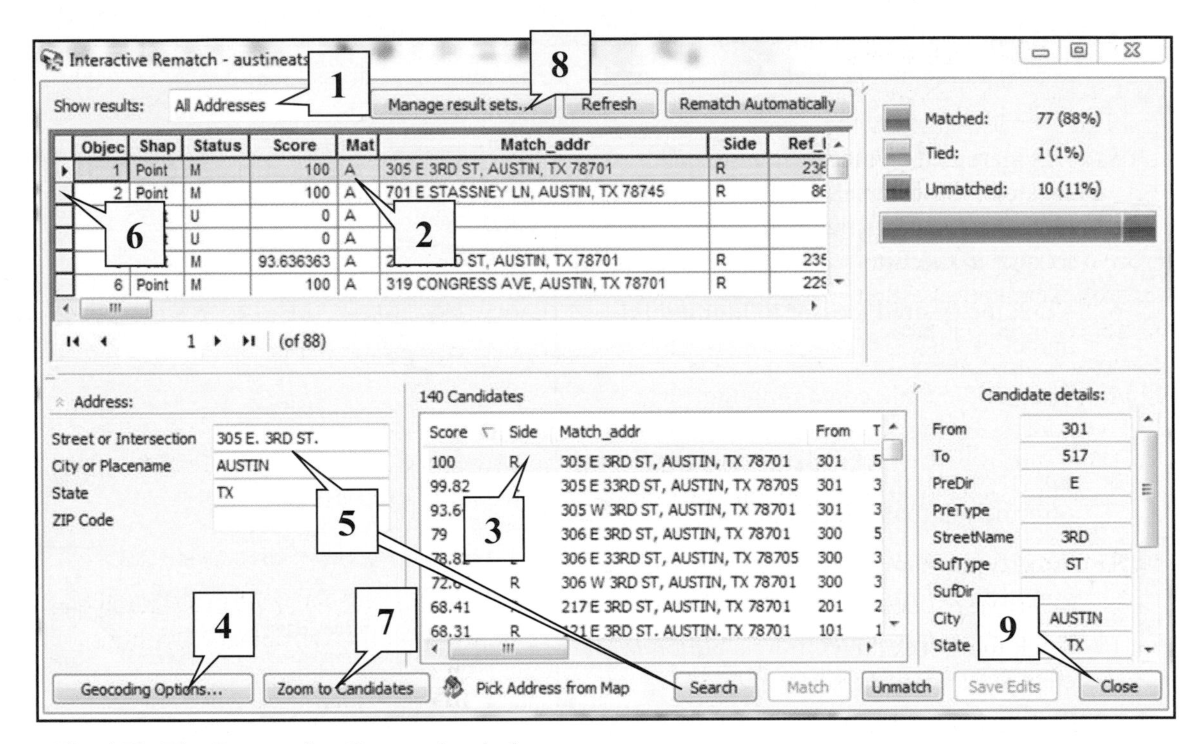

Fig. 179. The Interactive Rematch window

4. If no candidates appear, try clicking the Geocoding Options button to modify the spelling sensitivity or the minimum candidate score.
5. If still no candidates appear, examine the Address section to see if the address is being interpreted correctly. Edit the address to try alternatives, and then click Search.
6. If still no candidates appear, the address is probably unmatchable. It may be incorrect, or the reference layer may be incomplete. Click another address and begin again.
7. If you are trying to decide among several candidates, click the Zoom to Candidates button to view them on the map. Click Original Extent to go back to the previous extent.
8. You can use the *Manage result sets* button to save groups of records to work on later.
9. When you are finished rematching, click Close.

Networks

Examining network properties

1. In ArcCatalog, right-click a network in the catalog list and choose Properties.

The General tab reports which feature classes participate in the network (Fig. 180).

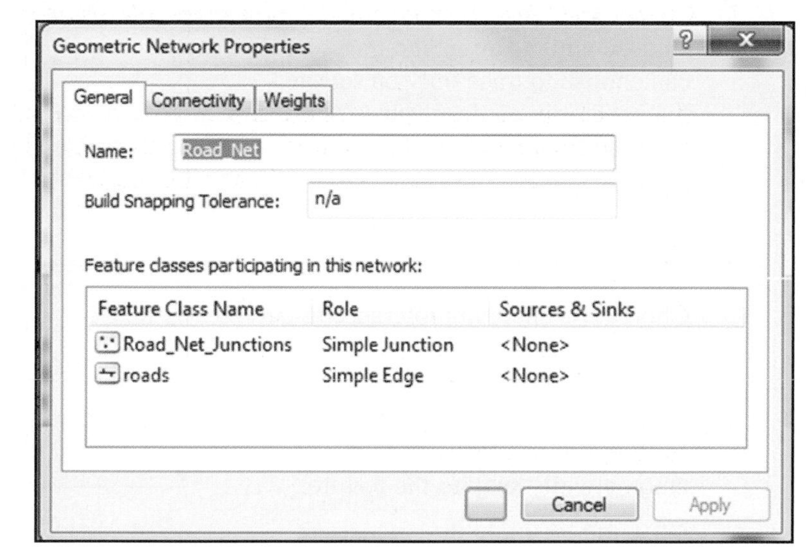

Fig. 180. General properties of a network

The Connectivity tab shows any rules that have been established regarding how the network elements can connect (e.g., a three-inch pipe can connect to a six-inch pipe only through a coupler junction feature). Connectivity rules are not required but are provided to help maintain network integrity.

The Weights tab lists weights that have been set up for the network. Weights are created at the time the network is built, but they may be used during an analysis, if desired.

Performing a trace

1. In ArcMap, click the Add Data button and choose a *Name*_Net layer, such as Road_Net, in the geodatabase containing the network. The network and its participating feature classes will be added to the map.
2. From the main menu, choose Customize > Toolbars > Utility Network Analyst.
3. Select the network to be analyzed, if more than one is available (Fig. 181).
4. Choose the Trace Task to be performed.
5. Click the Flag/Barrier tool and add desired flags and barriers at edges or junctions. For reliable results, do not mix edge and junction markers; use either but not both.
6. Click the Solve button to perform the trace.

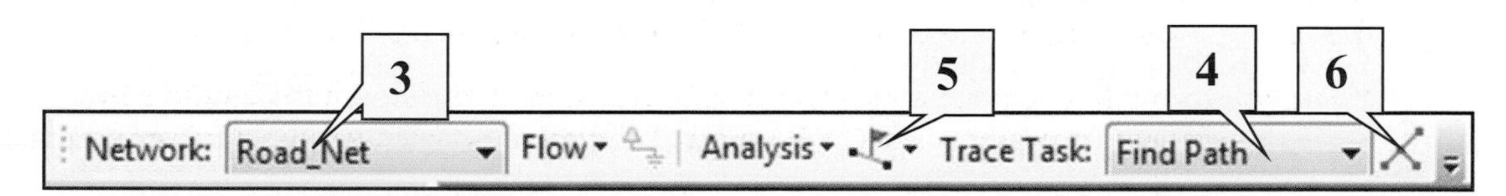

Fig. 181. Performing a simple tracing task

Setting general tracing options

1. Choose Analysis > Options from the Utility Network Analyst toolbar.
2. Click the General tab (Fig. 182).
3. Choose which features to trace. You can choose to trace only on selected features, only on unselected features, or on all features.
4. Choose whether to include edges with indeterminate or uninitialized flow when tracing.
5. Choose the snapping tolerance to be applied when adding flags and barriers. If the tool clicks a location within this distance from an edge or a junction feature, the flag or barrier will automatically snap to the feature.
6. Click OK to apply the options.

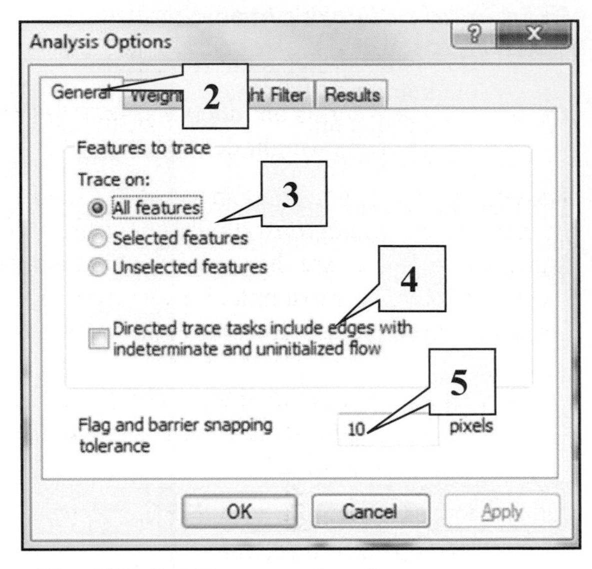

Fig. 182. Setting general options

Changing the results format

1. Choose Analysis > Options from the Utility Network Analyst toolbar.
2. Click the Results tab (Fig. 183).
3. Choose whether the trace result will take the form of a drawing or a selected set of features.
4. If you choose to return a drawing, change the color of the trace, if desired.
5. Choose whether the result will include all features included in the trace or only the features that are stopping the trace.
6. Choose whether the trace result will include edges, junctions, or both. By default, both are returned.
7. Click OK to apply the changes.

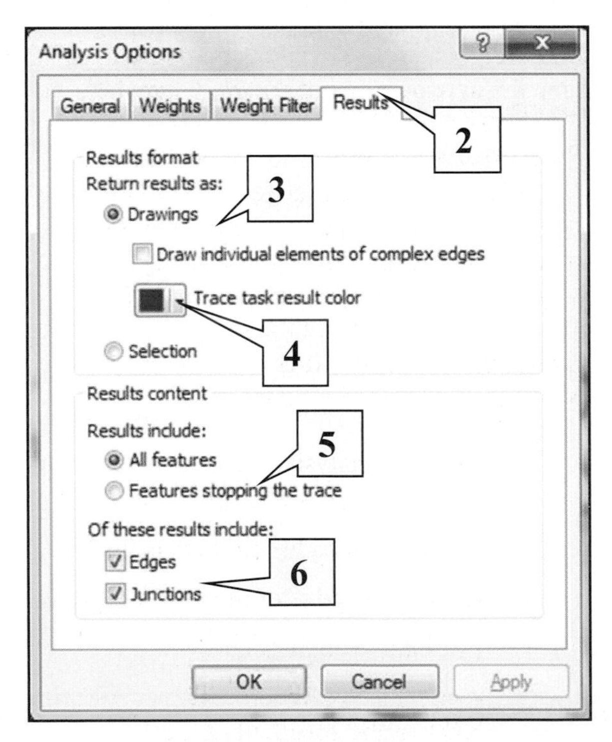

Fig. 183. Changing the results format

Using weights when tracing

1. Choose Analysis > Options from the Utility Network Analyst toolbar.
2. Click the Weights tab (Fig. 184).
3. To use weights on junctions, specify the name of the weight desired.

For edge weights, you must specify a weight in both the *along* and *against* directions of the edge. Usually, they are the same weight, but they can be different (for example, if setting a weight for uphill travel versus downhill travel).

4. Specify the name of the weight to be applied in the same direction that the line was digitized.
5. Specify a weight to be applied against the digitized direction.
6. Click OK to apply the changes.

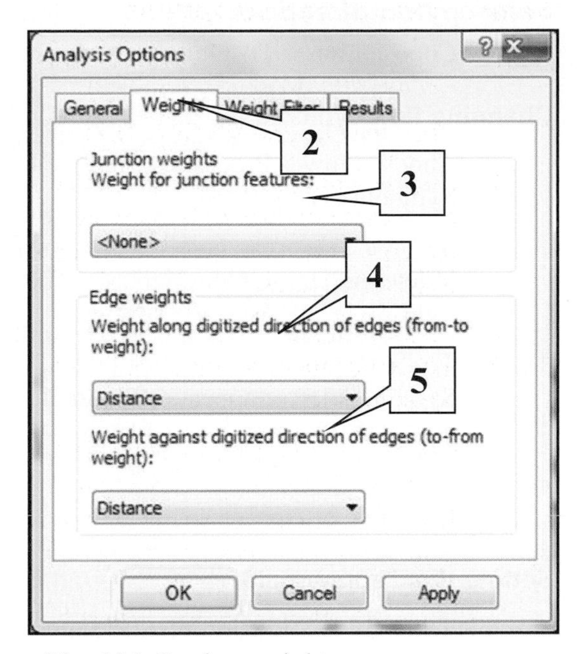

Fig. 184. Setting weights

TIP: Weights are set up at the time the network is created. If no weights are listed in the drop-down boxes, then the network contains no weights, and none can be used.

Using weight filters

Filtering weights means to apply weights that fall within a specified range. For example, a junction weight called StopTime might specify an average waiting time at intersections. You might want to sum the total time spent at traffic lights and ignore the shorter weight at stop signs. A weight filter could screen out waits of less than 15 seconds.

1. Choose Analysis > Options from the Utility Network Analyst toolbar.
2. Click the Weight Filter tab (Fig. 185).
3. Choose the edge or junction weights.
4. Enter the weight range as min – max.
5. Click the Not button to use the weights NOT in the specified range.
6. Click Verify to check that the syntax is correct.
7. Click OK or Apply.

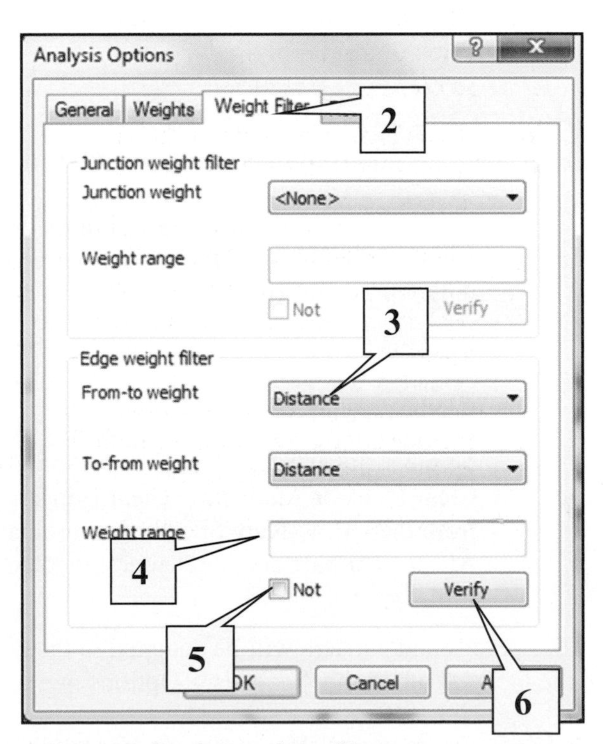

Fig. 185. Setting weight filters

Viewing flow directions

1. Choose Flow > Display Arrows from the Utility Network Analyst toolbar.

Changing flow symbols

2. Choose Flow > Properties from the Utility Network Analyst toolbar and click the Arrow Symbol tab (Fig. 186).
3. Select the flow category to see the symbol used for that category (Determinate, Indeterminate, or Uninitialized). The symbol for that category will be displayed to the right.
4. To change the symbol, click on it. The Symbol Selector appears. Choose the new symbol, set its size and color properties, and click OK.
5. Click OK or Apply.

Setting the display scale

Use this feature to avoid clutter in the map by making symbols appear only at certain scales.

6. Click the Scale tab (Fig. 186).
7. Choose to *Show arrows at all scales*, or to *Don't show arrows when zoomed* in or out beyond a specified scale.
8. Enter the scale(s).
9. Click OK or Apply.

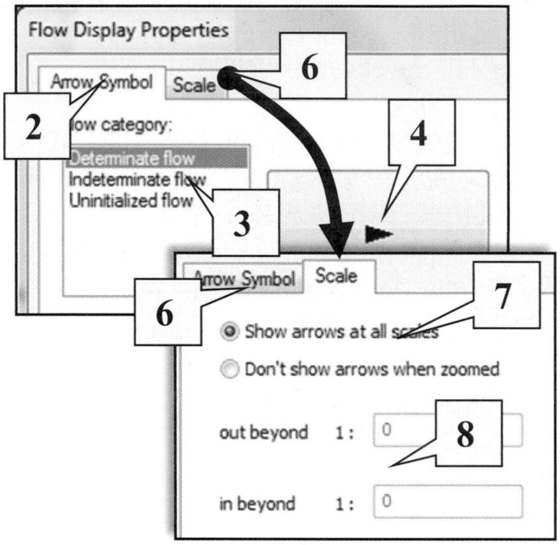

Fig. 186. Setting the flow symbols and display scale

Adding and clearing flags and barriers

Flags mark locations of interest in a network, and barriers temporarily stop flow.

1. Click the Flag tool drop-down button on the Utility Network Analyst toolbar (Fig. 187) and select the type of flag or barrier to add (edge or junction).
2. Click on a junction to add a junction flag or barrier. Click along an edge to add an edge flag or barrier.
3. To clear flags or barriers, choose Analysis > Clear Flags or Analysis > Clear Barriers from the Utility Network Analyst toolbar. All flags or barriers are cleared together.

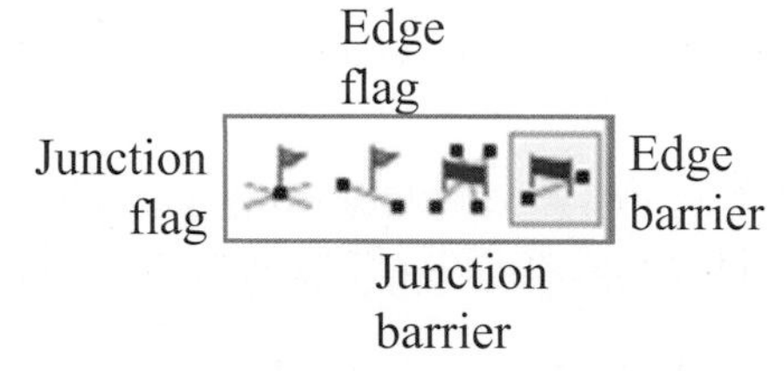

Fig. 187. Flags and barriers

TIP: Flags and barriers will be snapped to the closest feature within the snap tolerance set from the General tab in the Analysis > Options menu.

TIP: Some tracing tools will not work if the flags or the barriers are of different types. For an analysis, always use edge flags with edge barriers and junction flags with junction barriers.

Finding a path

By default, the trace will return the path that traverses the fewest number of edges. This path will not necessarily be the shortest path.

1. Choose the network to trace on.
2. Set the trace task to Find Path.
3. Choose the Junction Flag or Edge Flag tool.

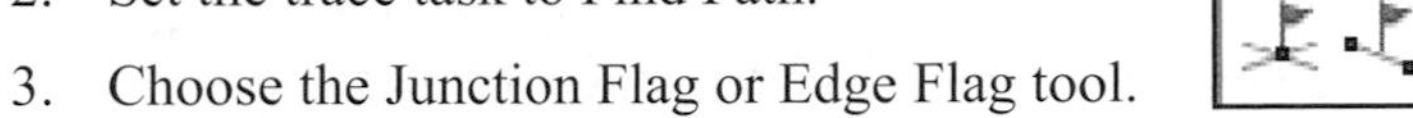

4. Click on the starting location to enter a flag.
5. Click on the end location to enter another flag.

6. Click the Solve button.

Total cost: 54

TIP: To see the cost (number of edges in the path), look in the lower-left corner of the ArcMap window before the mouse moves off the Solve button. The cost will vanish once the mouse is moved. Click Solve again to bring it back.

TIP: Enter more than two flags to visit a series of locations. The stops will be visited in the order in which they are entered.

Finding the shortest path

To find the shortest path, the network must have a Distance weight based on the length of the edges. (It may be called something other than Distance.)

1. Choose the network on which to trace.
2. Set the trace task to Find Path.
3. Choose the Junction Flag tool or the Edge Flag tool. Enter the starting, end, and any intermediate flag locations.
4. Choose Analysis > Options from the Utility Network Analyst toolbar and click the Weights tab.
5. Set both the *along* and the *against* edge weights to Distance (or whatever the weight name is). Click OK.
6. Click the Solve button.

Total cost: 18097

TIP: The distance in map units along the path will be displayed in the lower-left corner of the ArcMap window. It will vanish when the mouse is moved. Click Solve again to make it reappear.

Finding connections and loops

These trace tools can be used to find features connected to a certain location, those features disconnected from a certain location, and those features that form loops.

1. Choose the network on which to trace.
2. Set the trace task to Find Connected, Find Disconnected, or Find Loops.
3. Click the Edge or Junction Flag tool and enter the location from which to start tracing. More than one location flag may be entered, in which case tracing will commence from each location separately.
4. Click the Solve button.

Tracing upstream/downstream

These tracers work only on utility networks.

1. Choose the network on which to trace.
2. Set the trace task to Trace Upstream or Trace Downstream.
3. Click the Edge or Junction Flag tool and enter a flag at the location to begin the trace. If more than one flag is entered, tracing will begin at each location separately.
4. Click the Solve button.

Tracing with accumulation

1. Choose the network to trace on.
2. Set the trace task to Find Upstream Accumulation.
3. Choose Analysis > Options and click the Weights tab to specify the weights to be accumulated during tracing.
4. Click the Edge or Junction Flag tool and enter a flag at the location to begin the trace. If more than one flag is entered, tracing will begin at each location separately.
5. Click the Solve button.
6. Read the accumulation value in the weights' units in the lower-right corner of the ArcMap window.

Total cost: 18097

TIP: The accumulation value will vanish when the mouse is moved off the Solve button. To bring it back, click Solve again.

Rasters

Turning on Spatial Analyst

Even if Spatial Analyst is installed and registered, it must still be turned on before use.

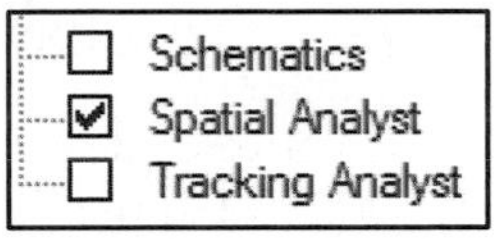

1. Open ArcMap, if necessary.
2. Choose Customize > Extensions from the main menu bar.
3. Check the box to turn on Spatial Analyst. Click Close.

☐ Schematics
☑ Spatial Analyst
☐ Tracking Analyst

Environment settings for rasters

Several of the Environment settings pertain specifically to rasters and should be checked before beginning analysis.

1. Choose Geoprocessing > Environments from the main menu bar (Fig. 188).

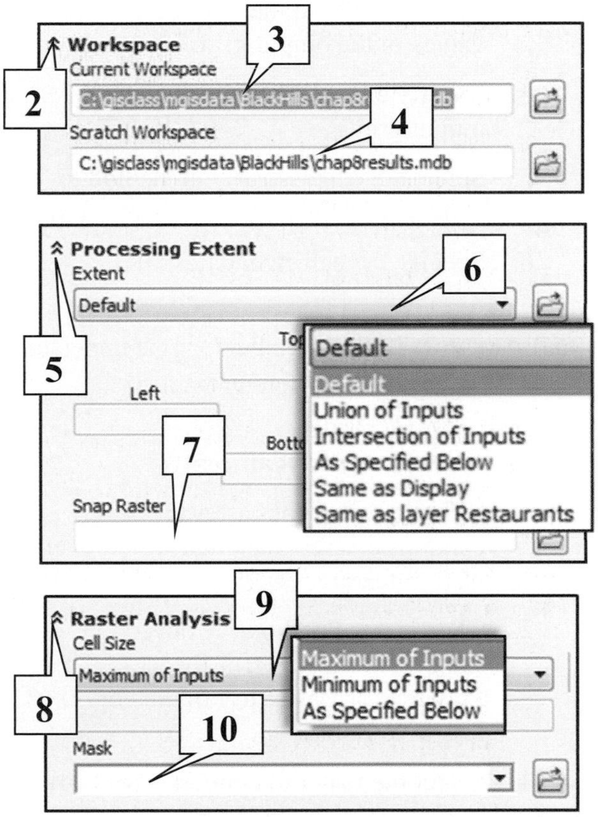

Setting the workspace

2. Expand the Workspace settings.
3. Set the Current Workspace to the folder or geodatabase that contains your data sets.
4. Set the Scratch Workspace to the same location, or another temporary folder.

Setting the extent

5. Expand the Processing Extent settings.
6. Choose one of the extent settings to control the region of the output.
7. To make the cells coincide exactly with an existing raster, select that raster in the Snap Raster box.

Setting the cell size and mask

8. Expand the Raster Analysis settings.
9. Click the drop-down box to set the cell size method: the maximum or minimum of input grids, or to a specific value.
10. Use the drop-down or Browse button to select a mask raster or polygon feature class.
11. Click OK when done setting the Environment settings.

Workspace
Current Workspace
Scratch Workspace
C:\gisclass\mgisdata\BlackHills\chap8results.mdb
Processing Extent
Extent
Default
Left
Snap Raster
Default
Union of Inputs
Intersection of Inputs
As Specified Below
Same as Display
Same as layer Restaurants
Raster Analysis
Cell Size
Maximum of Inputs
Minimum of Inputs
As Specified Below
Mask

Fig. 188. Important environment settings for raster analysis

TIP: When set from the main menu in ArcMap, these environment settings will be saved with the map document, and they must be reset if a new one is opened.

Converting between grids and features

Converting features to a raster

1. Choose ArcToolbox > Conversion Tools > To Raster > Polygon to Raster (or Polyline to Raster or Point to Raster).
2. Choose the feature class to be converted (Fig. 189). If converting from a layer in ArcMap, only the selected features will be used if a selection is present.
3. Choose the field that will become the values in the output grid.
4. Specify a location and name for the output grid.
5. Select the cell assignment method to control how the attribute value is assigned to the raster.
6. Specify the cell size. Click OK.

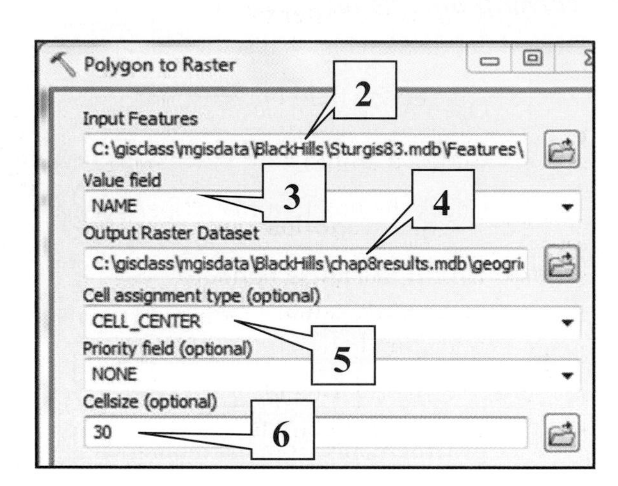

Fig. 189. Converting polygons to a raster

TIP: Use only letters, numbers, or the underscore character to name rasters. Pathnames to the raster directory should contain no spaces.

Converting a raster to features

1. Make sure that the raster to convert contains discrete data that will convert well to features.
2. Choose ArcToolbox > Conversion Tools > From Raster > Raster to Polygon (or Raster to Polyline or Raster to Point).
3. Select the raster to convert (Fig. 190).
4. Choose the field of the raster (usually the Value field) to become the defining attribute for the polygons.
5. Specify the location and name of the output feature class.
6. Check the box to *Simplify polygons*. You will usually want this option, which produces smoother boundaries that don't follow the cell edges.

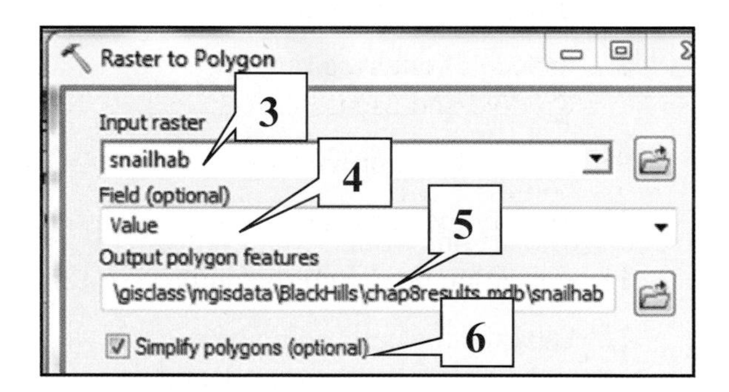

Fig. 190. Converting a raster to polygons

Reclassifying a raster

1. Choose ArcToolbox > Spatial Analyst > Reclass > Reclassify.
2. Choose the raster to reclassify (Fig. 191).
3. Choose the field in the raster to reclassify (usually Value).

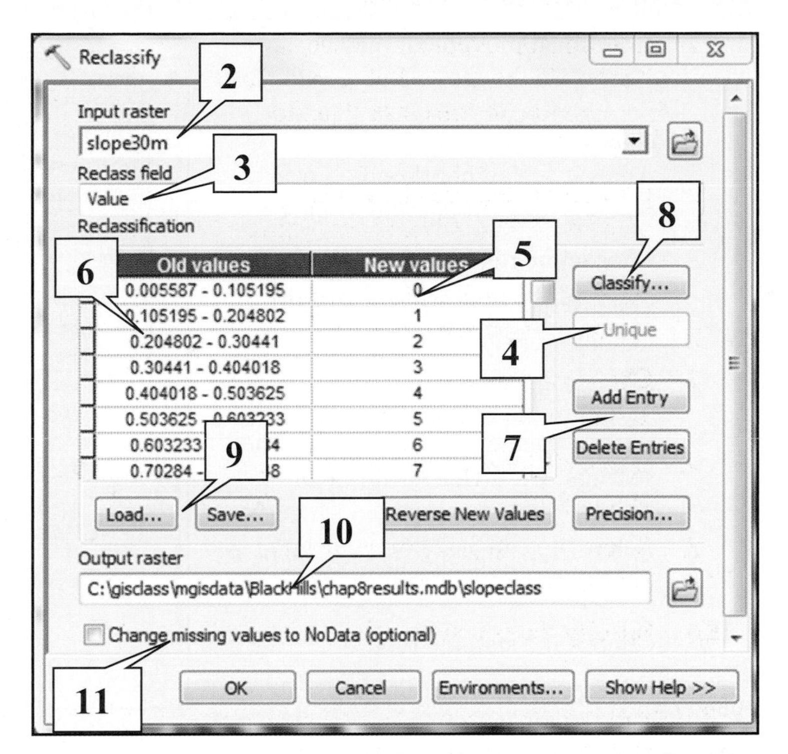

Fig. 191. Reclassifying a raster

If the grid has categories

4. Click the Unique button to see each value individually.
5. Edit the right-hand column to specify the output values.

If the grid has many values

6. Edit the ranges in the Old values column, and then enter the New values to replace them.
7. Use the Add Entry or the Delete Entries button to add or remove lines from the list of ranges.
8. Instead of entering ranges by hand, click the Classify button to open the Classify Values window and use it to set the ranges. See ***Classifying data***.
9. Save or Load previously created classifications, if desired.
10. Specify the name and location of the output grid.
11. Check the box to convert missing values to NoData, if desired.
12. Click OK.

Using neighborhood statistics

Neighborhood Statistics utilizes a roving window centered on a target cell. It calculates a statistic for all cells in the window and assigns the value to the target cell in the output grid. Either a block movement or focal movement may be specified based on the tool chosen.

1. Choose ArcToolbox > Spatial Analyst > Neighborhood > Focal Statistics (or Block Statistics).
2. Choose the raster to use (Fig. 192).
3. Specify the name and the location of the output raster.
4. Choose the neighborhood shape.
5. Specify the neighborhood size. The settings will differ for different shapes.
6. Specify whether the size is in cells or map units.
7. Choose the statistic type.
8. Click OK.

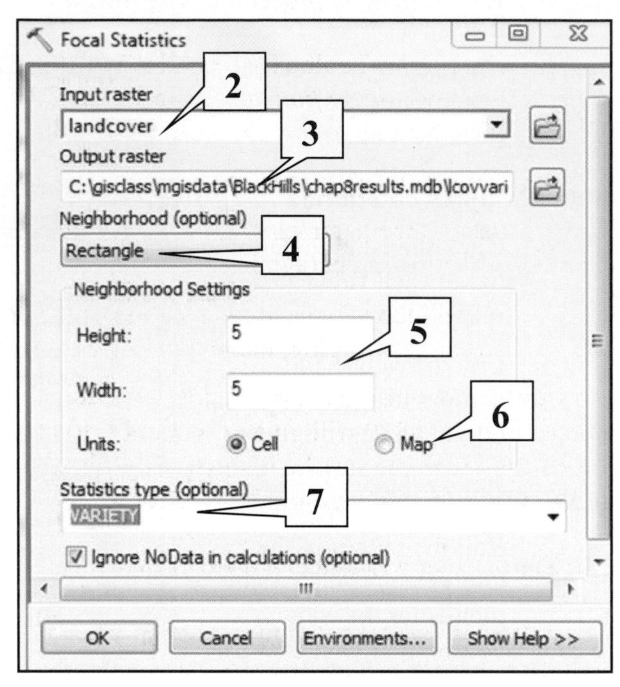

Fig. 192. Calculating neighborhood statistics

Calculating zonal statistics

Zonal Statistics calculates statistical measures for cells within zones. Two input grids are required, one defining the zones and another containing the values to use in calculating the statistics. The output is a table containing a record for each zone and a field for each statistic.

1. Choose ArcToolbox > Spatial Analyst > Zonal > Zonal Statistics.
2. Select the raster or the feature layer that defines the zones (Fig. 193).
3. Select the field from the zone layer that defines the zones.
4. Select the raster containing the values for which to calculate statistics.
5. Specify the name and the location of the output raster.
6. Select the statistic to calculate.
7. Check the box to ignore NoData cells when calculating (usually the best option).
8. Click OK.

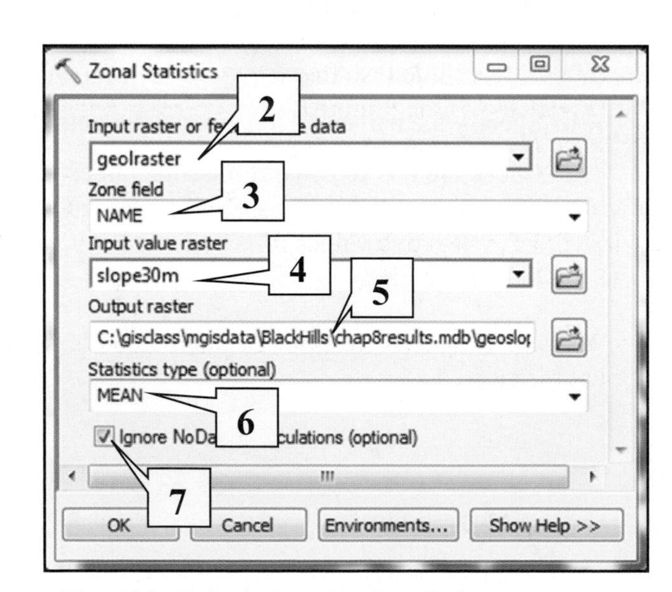

Fig. 193. Calculating zonal statistics

Using surface functions

The surface functions contour, slope, aspect, and hillshade all produce output grids based on an input surface and a few arguments. The general procedure is as follows:

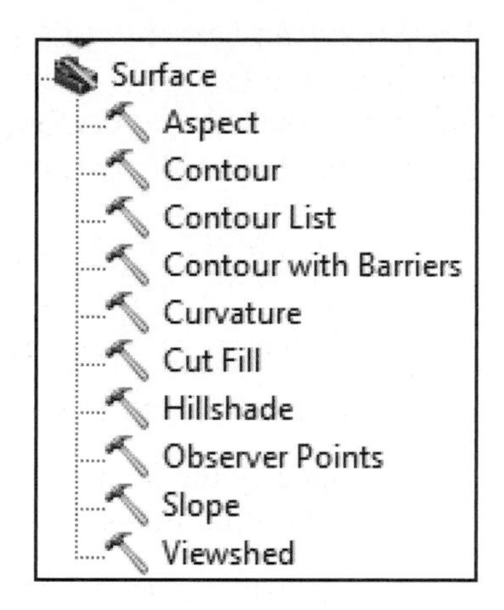

1. Choose ArcToolbox > Spatial Analyst > Surface > *function*, where *function* is the name of the desired operation.
2. Choose the surface to use as input.
3. Specify the parameters for each function as described in the following paragraphs.
4. Specify the name and the location of the output raster. (If you are contouring, the output will be a feature class.)

To **Contour**, specify the base contour value and the contour interval. For example, a base of 10 and an interval of 20 will give the contours 10, 30, 50, and so on. You can also specify a Z factor to multiply with the surface before contouring, for example, to create foot contours on a meter surface.

For **Slope**, enter a Z factor, if desired, and specify degrees or percent for the output units.

For **Aspect**, there are no additional parameters. The resulting grid contains values between 0 and 360 degrees representing the aspect, with 0 = 360 = North. A value of –1 indicates a flat area with no aspect.

For **Hillshade**, change the default values for the azimuth and altitude of the illumination source, if desired. The azimuth is given in degrees between 0 and 360 with 0 = 360 = North. The altitude is given in degrees between 0 (horizontal) and 90 (vertical). A Z factor > 1 may be specified, which will exaggerate the topographic relief. A Z factor < 1 will subdue the topographic relief. You may also check the box to model shadows. If you choose this option, cells in the shadow of another will be set to 0 so that later you can extract these to create a map of shadows. The default option calculates the illumination of all cells, including shadowed ones. The visual difference between the two options is negligible.

For **Cut Fill**, specify the before and after surfaces. Adjust the Z factor if necessary. The output grid will have negative values for areas that have been removed and positive values where material has been added.

For **Viewshed**, specify the point or line feature class containing the observer locations. Specify a Z factor if desired.

Calculating Euclidean distance

The Euclidean Distance tool finds the straight-line distance from each cell to a set of features.

1. Choose ArcToolbox > Spatial Analyst > Distance > Euclidean Distance.
2. Select the input raster of feature source data containing the features from which to calculate distances (Fig. 194).
3. Specify the name and location for the output raster.
4. Specify a maximum distance, if desired.
5. Specify the cell size.
6. Optionally specify a name and location for the output direction raster, indicating direction to the closest feature.
7. Click OK.

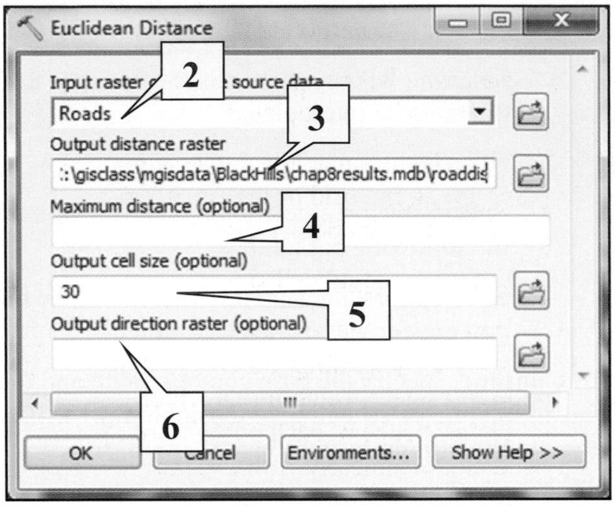

Fig. 194. Calculating Euclidean distance

Creating a density raster

1. Choose ArcToolbox > Spatial Analyst > Density > Point Density.
2. Specify the point layer (Fig. 195).
3. Choose an attribute of the point layer, such as population, on which the density will be based. To count the number of points, choose <None>.
4. Specify the name and the location of the output raster.
5. Enter the output cell size.
6. Enter the neighborhood shape and size settings. The settings will depend on the shape.
7. Enter the search radius in map units (maximum radius to include data points).
8. Click OK.

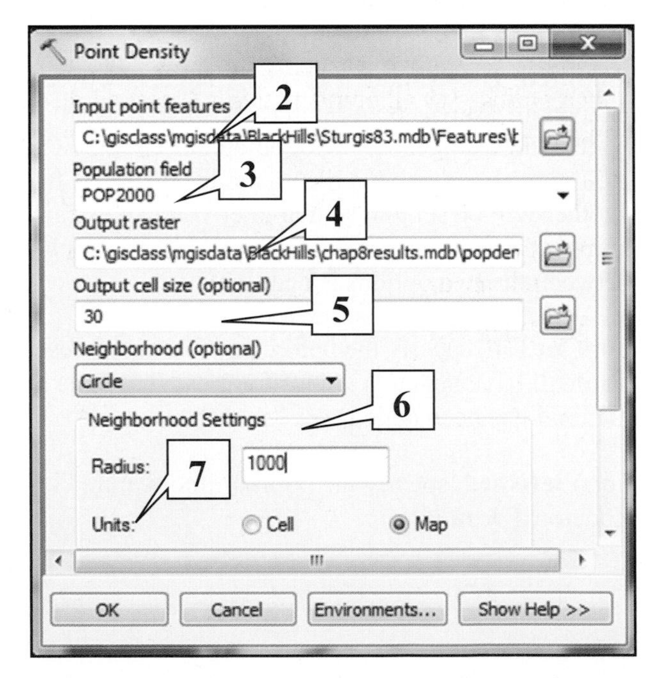

Fig. 195. Creating a density raster from points

TIP: The Kernel Density tool spreads the values around the points using a specified function. The Line Density tool calculates line densities, such as road density in a forest.

Interpolating between points

1. Choose ArcToolbox > Spatial Analyst > Interpolation > IDW.
2. Choose the point layer containing the data to interpolate (Fig. 196).
3. Enter the field containing the values to be interpolated.
4. Specify the name and the location of the output grid or leave it as is to create a temporary grid.
5. Enter the output cell size.
6. Enter the power exponent for the distance weighting. Entering 1 will give a linear weighting in which the influence of each point decreases linearly with distance. Entering 2 causes the influence to decrease as a square of the distance, and so on.
7. Enter the type of search radius.

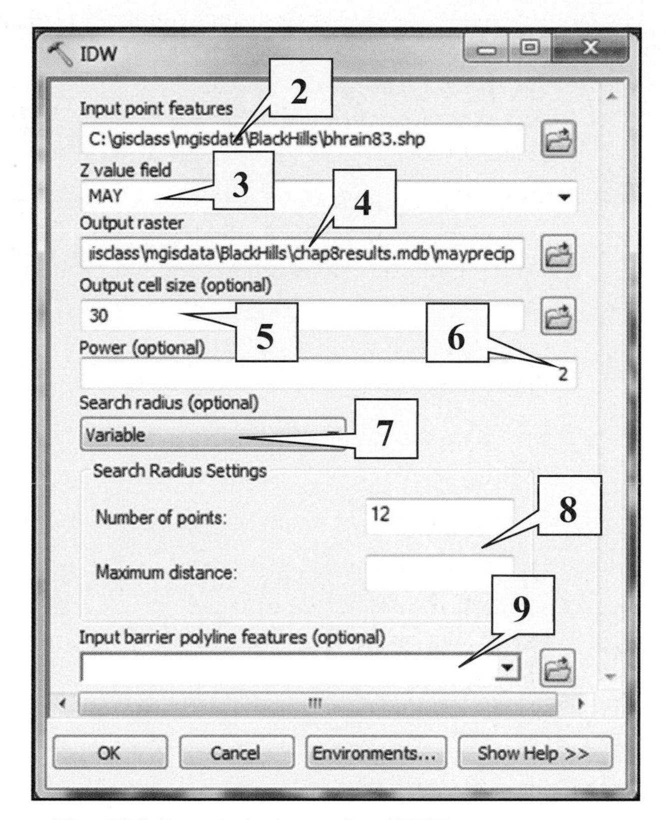

Fig. 196. Interpolation using IDW

A variable radius searches for a specified number of closest points, with an optional requirement to stop searching outside a specified maximum distance. A fixed search radius uses all points within a specified distance, and it has an optional requirement to limit the points used to a specified maximum number.

8. Enter the values for the search distance and the number of points.
9. Optionally specify a line feature class containing barriers across which interpolation will not occur.
10. Click OK.

Glossary

absolute pathname a file pathname that starts at the drive letter

active frame the data frame that is visible and responds to changes by the user

address locator a file required for geocoding, which contains the necessary style information and reference data

address standardization the conversion of a street address into specific components, such as house number, name, and type

adjacency a spatial condition that quantifies whether one feature touches another

affine transformation a process to translate, rotate, or skew an image to fit a new coordinate system

alias an alternate name displayed for a field in a table that does not have to follow length and character restrictions

analysis cell size the default cell size defined for output rasters during a raster analysis session

analysis extent the default extent defined for output rasters during a raster analysis session

analysis mask a layer applied during a raster analysis session to apply NoData values outside an area of interest

annotation labels created from map features and stored separately for detailed editing

append a function that combines features from two different feature classes

ArcToolbox a set of functions and commands for processing spatial data

ASCII an international code used to store simple text letters, numbers, and symbols; synonymous with "text file"

aspatial data data entries that are not tied to a location on the earth's surface

aspect the direction of steepest slope at a location on a surface

attribute accuracy a measure of the type or frequency of errors occurring in the attributes of a data set

attribute field a column in a table containing information about spatial features

attribute query an operation to extract specific records of information from a file based on values in the records

attribute table a table containing rows of information for features

attributes information about map features stored in columns of a table

automatic scaling fitting the contents of a data frame within a specified rectangle

azimuth a measurement of compass direction in 360 degrees with 0 = north

azimuthal projection a map projection onto a flat plane that is tangent or secant to the earth's surface

band a single array of values stored in a raster, which may store one array or many

barrier an object that prevents flow or travel in a network

bilinear a resampling method that interpolates a new cell value based on the four closest values

binary numeric data stored in base 2 as a series of ones and zeros; a computer's native data storage format

block function an analysis function that moves over a raster in adjacent, non-overlapping neighborhoods

bookmarks links to a particular map area and scale for quick access

Boolean an adjective describing data, operators, values, or files that contain or manipulate only true or false values

Boolean operators operators (AND, OR, NOT, XOR) that evaluate a pair of true or false propositions and return a true or false result

Boolean overlay evaluating rasters with Boolean operators to evaluate areas where combinations of specific conditions exist

Boolean raster a raster with cells that contain a 1 to indicate true or 0 to indicate false regarding a specified condition

buffer the delineated area within a specified distance of a feature

byte a unit of data storage in base 2 containing eight zeros or ones and holding a number from 0 to 255

capture using the Alt-PrntScrn keys to place a copy of the active window on the clipboard for pasting into another program

cardinality the relationship between records in two tables: one-to-one, one-to-many, many-to-one, or many-to-many

categorical data data that place objects into unranked groups; examples are land use and geology data

cell a square data element in a raster corresponding to one value representing conditions on the ground

cell size the dimensions of one square of map information in a raster data set

central meridian the central longitude of a map projection for which the *x* coordinate equals zero

chart a graph created in ArcMap from table data, such as bar charts or pie charts

chart map a map showing several different attributes in chart form, with one chart for each feature

choropleth map a map in which each feature, such as a state, is colored according to the values in a data field

class breaks break points used to classify values in data field groups

classification assigning features to two or more groups based on numeric values in an attribute field

classified a raster display method that divides values into two or more groups based on their numeric values

clip to remove features and portions of features that lie outside of the features of another layer

cluster tolerance a defined distance used in topological editing, causing vertices to be made coincident if they are closer together than the tolerance distance

coded domain a rule that permits only certain specific values to be assigned to an attribute, such as land use codes

colormap a set of defined colors matched to specific image pixel values, which determines how the image will appear

completeness a measure of how well a data set has captured the entire area or all representatives of a feature class

complex edge a network entity composed of multiple linear features that behaves as a single linear feature in the network

conflict detection determining which labels will overlap each other

conic projection a map projection derived by projecting latitude-longitude values on a paper cone covering a sphere

connectivity a property of linear features when they are connected to each other via junctions

containment the property of one feature including another in whole or in part

context menus computer menus that appear when certain objects on the screen are clicked with the right mouse button

continuous data that take on a variety of values and that change rapidly across a data set, such as elevation

contour a line indicating a constant value of a quantity on a surface, such as an elevation contour at 2000 ft

coordinate pair a single pair of *x* and *y* values indicating position in a planar coordinate system

coordinate space the range of *x* and *y* values onto which maps are plotted

coordinate system (1) a specified range of *x-y* values onto which a map is plotted; (2) the definition of a coordinate space used by a map layer, including the ellipsoid, datum, and projection

coverage the spatial data format created for, and used by, ArcInfo

CSDGM (Content Standard for Digital Geospatial Metadata) a metadata standard developed by the FGDC and commonly used in the United States

cubic convolution a resampling method that interpolates a new cell value from the 16 closest input cells

cut/fill a function that determines the difference between a before and after topographic surface

cylindrical projection a map projection derived by projecting spherical latitude-longitude values onto a cylinder wrapped about a sphere

dangle a line feature that fails to connect to another line feature, leaving a gap

data frame a container holding layers that are viewed and analyzed together; a map view

Data view the data frame mode optimized for the display and analysis of map data

datum a combination of an earth ellipsoid and a reference point to reduce mapping discrepancies

dBase a database program whose file format has been adopted for the shapefile data model and tables in ArcGIS

default an automatic input value assumed by a program when no value has been entered by the user

Define Projection tool a tool to guide the user through the task of assigning a coordinate system to a spatial data layer

defined interval a classification method in which the user specifies a size range for all the classes

definition query an operation to set a map layer to display only the features whose attributes meet specific criteria

degrees the measurement units used in the spherical coordinate system; a circle has 360 degrees

DEM see *digital elevation model*

destination table the table that receives data from another table during a join operation

digital elevation model (DEM) a raster array of values representing elevations at the earth's surface

digital raster graphic (DRG) a scanned image of a USGS topographic map

digitize to convert shapes on a paper map to a digital map layer by entering vertices

discrete data that take on a relatively small number of distinct values

discrete color a display option for rasters in which every different value is assigned a random color

Display tab the tab in the Table of Contents that shows layers in the order in which they are drawn from bottom to top

display units the units in which ArcMap reports the current *x-y* location of the cursor on the map

dissolve to combine features together when they share the same value for an attribute

distance join a join that combines the information from two feature tables based on the features that lie closest to each other

division units the units in which a map scale bar is measured and drawn, such as miles or kilometers

division value the length of one section of a scale bar as given in division units, such as 100 km

divisions the number of sections given to a scale bar

domain a rule that determines the values that may be entered into an attribute

dot density map a map representing attribute values by a proportional number of randomly placed dots

double-precision a numeric value stored using 16 bytes of information

DRG see *digital raster graphic*

Dual Ranges a style of geocoding that uses two ranges of addresses for each block, one on either side of the street

dynamic labels labels determined from an attribute and placed on a map automatically each time features are drawn and redrawn

edge (1) a line feature that participates in a geometric network; (2) a shared boundary between two features being edited using topology

edge snapping ensuring that new features are automatically connected to the edges of existing lines or polygons

ellipsoid a spheroidal volume with unequal axes, used to approximate the shape of the earth in map projections

end snapping ensuring that new features are automatically connected to the ends of existing line features

enterprise GIS a long-term GIS project developed by a large organization and involving many people over a long period of time

Environment settings program-level or tool-level settings that impact how tools run or set characteristics of the output, such as cell size or coordinate system

equal interval a classification method in which the user specifies a number of classes that have equal size ranges

erase an overlay function that removes features lying inside the external boundary of another polygon feature class

Euclidean distance the straight-line distance between two points

event layer a map layer of points created from a series of coordinate pairs in a table

export to create a new data file from all or a subset of features in an existing one

expression a statement containing field names, values, and/or functions used to extract records in a query or calculate values in a table

extent the range of *x-y* coordinates displayed in a map or stored in a data layer

extent rectangle the range of *x-y* coordinates occupied by the features in a data layer

false easting an arbitrary *x*-coordinate translation applied to a map projection, usually to ensure that all values are positive

false northing an arbitrary *y*-coordinate translation applied to a map projection, usually to ensure that all values are positive

feature a spatial object composed of one or more *x-y* coordinate pairs and having one or more attributes in a single record of an associated table

feature class a set of similar objects with the same attributes stored together in a spatial data file

feature dataset a set of feature classes in a geodatabase that share a common coordinate system and can participate in networks and topology

feature service an Internet map layer in which the actual features are available for display, query, and sometimes editing

feature template a set of properties and attributes that stores all the information needed to edit a layer

feature weight the priority assigned to a layer when determining which features may be drawn on top of others

FGDC (Federal Geographic Data Committee) an organization that promotes and establishes standards for exchange of data

FID (feature ID) a unique number assigned to every feature in a spatial data file and used for identification and tracking

field a single column of information in a data table

field definition parameters specified when creating an attribute field, including type of data, field length, precision, and scale

field length the maximum number of characters that can be stored in a text attribute field

filter a moving window applied to a raster data layer that calculates new values for the center pixel based on some function of the values in the window

fixed extent a constraint applied to a data frame to prevent changes in scale or extent in layout mode

fixed scale a constraint applied to a data frame to prevent the scale of the map from being changed in layout mode

flag a marker showing a point of interest in a network, such as a start point or endpoint for a path or a stop along a route

flat file database a database that stores data in simple files

flipping lines swapping the start and end nodes of a line feature so that it goes in the opposite direction

focal function raster analysis functions that determine new values for a target cell based on values in a moving neighborhood around the target

folder connection (1) a link in ArcCatalog and ArcMap that points to a folder with GIS data and serves as a shortcut to frequently used folders; (2) a link to a DBMS allowing data to be transferred

GCS see *geographic coordinate system*

geocoding the matching of a location stored in a table to a spatial point feature based on a reference spatial data layer, most often applied to converting addresses to locations

geocoding style a predetermined method for matching addresses that specifies required fields needed for matching

geodatabases a data model developed for ArcGIS 8 that employs recent database technology for storage and implements rules and topology

geographic coordinate system (GCS) a spherical coordinate system of degrees of latitude and longitude that is used to locate features on the earth's surface

geoid the shape of the earth as defined by mean sea level and affected by topographic and gravitational factors

geometric accuracy the accuracy with which the shape and position of features are represented

geometric interval a classification method that bases the class intervals on a geometric series in which each class is multiplied by a constant coefficient to produce the next higher class

geometric network a system of linear edges and point junctions used to model the flow of commodities, such as traffic or utilities

geoprocessing analysis of spatial data layers, such as dissolving, intersecting, and merging

geoprocessing service an analysis tool provided by a GIS server, in which the processing is done on the server rather than on the client's device

georeferenced a spatial data layer that is tied to a specific location on the earth's surface for display with other data

georelational a data model that links features to attributes in a separate table using a unique feature ID code

graduated color map a map that divides numeric data from a polygon feature class into classes based on value and displays the classes with different colors

graduated symbol map a map that divides numeric data from a line or point feature class into classes based on value and displays the classes with different size or thickness of symbols

graphic text text placed on a map layout that is not associated with a feature attribute; can be manipulated by the Drawing tools

graticule grid marks of latitude and longitude placed on a map boundary

grid (1) specific raster format native to GRID and Spatial Analyst; (2) generic term for a raster data set

ground control points a set of points that match easily identifiable locations on two different data layers to enable georeferencing of one layer to another

hierarchical database a database that stores information in tables with permanent links between them

hillshade a raster that displays the brightness variation of a surface as if it were illuminated by a light source at a specified azimuth and zenith angle

histogram a graph showing the number of pixels contained for each data value in a raster

image a raster data layer, usually referring to a raster that displays brightness values, as in a photograph

INFO an early database system upon which the Arc/Info software data model was based

inside join combining the information from two feature tables based on one feature that lies inside another

interactive labels simple graphic labels on a map that are placed by the user one at a time

interactive selection extracting one or more features by manually picking them from the screen with selection tools

interpolation to calculate values at locations between known measurements; to populate a raster with values extrapolated from a known set of point values

intersect an operation to overlay two spatial data layers and find the areas common to both while discarding areas unique to either

intersection the property of two features touching each other in whole or in part

interval data values that follow a regular scale but have no natural zero point, such as degrees Celsius or pH

ISO (International Organization of Standards) an international body that approves standard industry practices, including metadata standards

item description brief style of metadata offering basic information about a data set

Jenks method a way to classify numeric data into ranges defined by naturally occurring gaps in the data histogram

join the temporary combination of data from two tables based on a common attribute field or location

junction a point connecting two linear edges of a geometric network

kernel a moving window of values applied to a raster to calculate new values at a target in the middle of the window; see also *filter*

key an attribute field that is used to extract or match records in a table

label weights a priority rating assigned to values to determine which ones will be placed in case of an overlap

latitude a spherical unit measuring angular distance north or south from the equator

latitude of origin the reference latitude of a map projection where the *y* value is zero

layer a reference to a feature class and its associated properties

layer file a file that stores a pointer to spatial data along with information on how to display it

layer package a file sent from a GIS server with the symbol properties and necessary data included for use

layout the specification for a map page, including the map frames, legend title, scale bar, and so on; stored in a map document

Layout view a mode of ArcMap that is used to design and create a printed map and that allows manipulation of map layers, titles, scale bars, north arrows, and more

legend a map element that displays the names and symbols used to portray layers on a map

line a spatial feature composed of a string of *x-y* coordinate vertices and used to represent linear features such as streets

lineage the original source and processing steps behind the production of a digital geospatial data set

logical consistency a measure of how well data features represent real-world features, in particular with respect to topology

logical expression a statement composed of field names, operators, and values that specifies criteria used to select records or values from a layer or table

logical network information stored in separate tables that keeps track of relationships between elements of a network

logical operators functions, such as >, >, or =, that compare values and generate a true or false result

longitude a spherical unit measuring angular distance east or west from the Prime Meridian

loops connected circular paths in a network in which flow is indeterminate

mantissa the part of an exponential number that stores the significant digits

manual class breaks a classification scheme in which the user sets each class break to the desired value

map algebra a system that permits calculations and operations on entire raster arrays, such as adding two rasters together

map elements objects placed on a map layout, such as titles, legends, scale bars, north arrows, images, and charts

map extent the range of *x-y* values of the area being displayed in a map

map overlay to combine two spatial data layers, either for display or to evaluate the relationships between them

map package a file sent from a GIS server with the map layout and necessary data included for use

map scale the ratio of feature size on a map to its size on the ground

map service an Internet map that is available for viewing but that cannot be manipulated, symbolized, or saved by the user

map topology temporary spatial relationships developed between features during editing in ArcMap to facilitate editing of features with common nodes or boundaries

map units the units of the coordinate system in which a map is stored or displayed

mask a raster layer applied during analysis to nullify unwanted cells, such as those outside a study area boundary

measurement grid a data frame grid that provides measured *x-y* coordinate system values around the perimeter of the frame

merge (1) to combine two or more map features into one feature; (2) to combine two or more data layers into a single layer

merge policy a rule that states how the attributes of a feature should be handled if it is merged with another feature

metadata information stored about data to document its source, history, management, uses, and more

metadata standard a set of requirements laying out the types of information and organization to be used for creating metadata

metadata template a partially completed metadata document containing information repeated for many data sets and used to aid in metadata creation

minimum candidate score the lowest score at which an address may be considered to be a candidate to match an address during geocoding

minimum match score the lowest score at which a candidate will be automatically matched to an address

model a sequence of steps or calculations used to convert raw data into useful information; a scheme used to understand and predict processes in the real world based on the manipulation of data

Model Builder a graphical interface used to combine existing tools to create new tools and scripts

modify an editing technique used to edit individual vertices to change the shape of a feature

multipart feature a single feature composed of nontouching units, such as a single state feature composed of the seven Hawaiian islands

NAD see *North American Datum*

natural breaks a data classification scheme that divides values based on natural groupings or gaps between values

nearest neighbor a resampling method for discrete data that grabs the closest value to a cell center

neatline a line used to enclose one or more map elements in a rectangle

neighborhood functions calculation of a value for a target cell or feature based on surrounding values from a defined region

network or network topology an association of linear edges and connecting points, used to model the flow of a commodity, such as traffic or utilities

network weights values associated with network elements that indicate the "cost" to traverse the element

NoData a special value used to designate that a data value is absent or unknown

node the beginning and endpoint of a line feature

nominal data values that name or identify an object, such as a street name

normalized data to divide the values of an attribute field by the total of the field or by the values in another field

North American Datum (NAD) a combination of a spheroid and reference point that is used to minimize map distortion in North America

numeric data values stored as numbers rather than as names or categories

object-oriented a programming and database approach that treats software and model elements as objects with defined properties and relationships

oblique projection a map projection in which locations on a sphere are projected to a cylinder or cone of paper at an arbitrary angle

OID (ObjectID) unique number identifying a row in a table or a feature in a geodatabase feature class

One Range a style of geocoding addresses that uses a single range of addresses for each street block

ordinal data data values that indicate a rank or ordering system

origin the (0,0) point of a coordinate system

orthographic projection a map projection in which locations on a sphere are projected onto a planar surface

orthophoto an aerial photograph that has been geometrically corrected to match a map base

overlap a spatial condition that quantifies whether one feature covers all or part of another feature

overshoot a type of dangle in which a line crosses too far over a line that it is supposed to meet exactly

pan to move the display window to another part of the map without changing the map scale

parameter (1) specific value associated with map projections that define how it appears; (2) a variable that serves as an input to a model

parametric arc a line feature composed of a smooth curve derived from a given radius for each segment

pathname a list of the folders that must be traversed to locate a particular file, such as c:\mgisdata\usa\states.shp

pixel a square data element in a raster corresponding to one value representing conditions on the ground

planar topology an association of feature classes in a feature dataset, established by rules regarding the spatial relationships between features, such as not overlapping each other

point a one-dimensional feature defined by a single *x-y* coordinate pair

point snapping ensuring that new vertices are automatically connected to existing point features

polygon a closed, two-dimensional area feature defined by three or more *x-y* coordinate pairs

positional accuracy a measure of the likelihood that features on a map are actually in the locations specified

precision the number of digits allotted to store a numeric value

Prime Meridian the line of zero longitude on the earth, passing through Greenwich, England

process step data assembly and geoprocessing function performed in generating a GIS data set

profile a formal customization of an international standard developed and maintained by a community of users

Project tool a tool that converts a feature class from one coordinate system to another

projection a mathematical transformation that converts spherical units of latitude and longitude to a planar *x-y* coordinate system

project-oriented GIS a GIS project with fairly limited objectives and a finite life span

proportional symbol map a map that displays attribute values with marker or line symbols that are proportional in size relative to the value of the feature

proximity the property of one feature being close to another feature

pseudonode a type of topology error in which only two lines meet at a coincident node instead of three or more

pyramids a set of rasters with different resolutions that is calculated from a raster and used to speed displays at smaller scales

quantile a classification method that divides the data into the specified number of quantiles, so that each class has the same number of features

quantities data numeric attribute data

query an operation to extract records from a database according to a specified set of criteria

range domain a rule that stipulates the largest and smallest values that can be stored in a particular attribute

raster a data set composed of an array of numeric values, each of which represents a condition in a square element of ground

raster model a data model that uses rasters or numeric arrays to represent real-world features

ratio data data having a regular scale of measurement and a natural zero point, such as precipitation or population

reclassify to replace sets or ranges of values in a raster with different sets or ranges of values

record a row in a table containing one object

rectify to rotate, resize, or warp an image to match a map base using a selected set of ground control points

reference grid a data frame grid that provides lettered and numbered squares on a map for use with an index of the features

reference latitude the latitude of origin; see also *latitude of origin*

reference layer a layer containing special attributes needed for geocoding for which a geocoding index has been built

reference scale the scale at which text or symbols appear at their assigned size

relate a temporary association between two tables based on a common field whereby fields may be selected based on whether they match selected records in the other table

relational database a database that stores information in tables and constructs temporary relationships between them

relative pathname path to a file that starts in the current folder

resampling to change the resolution of a raster using a defined strategy to convert values from one cell grid to another

reshape an editing technique used to re-enter part of an existing line or polygon

resolution (1) the ground area represented by one cell value in a raster; (2) the default storage precision of a vector data set

RGB composite an image displayed by assigning one band of brightness information to each red, green, and blue color gun in a display monitor

RMS error (RMSE) the root mean squared differences between a set of original and translated points

rubbersheeting a process to deform a raster to fit it to a new coordinate system

Rule of Joining each record in the destination table must have one and only one matching record in the source table

scale (1) the ratio of the size of features in a map to their size on the ground; (2) the number of decimal places allotted to an attribute field for storing numbers

scale range the range of scales for which a data layer will be displayed, set by the user to avoid clutter, or the display of layers at inappropriate scales

schema the layout plan of a geodatabase, including the tables, the fields, and the relationships

script a program including ArcGIS commands and functions

secant projection a map projection in which spherical coordinates are projected onto a surface that intersects the sphere along two great circles

select to extract one or more features or records from a layer or table in preparation for another operation; to perform a query

Select By Attributes to choose a subset of features or table objects based on the values in one or more attribute fields

Select By Location to choose a subset of features based on their spatial relationship to other features

selected set the set of features that has been extracted prior to another operation; the result of a query

selection method a setting that controls what happens to previously and newly selected objects when a new selection is made

Selection tab the tab in the Table of Contents that shows the selection state and number of selected features for each layer

shapefile the spatial data model developed for, and used by, ArcView 3 and later versions

shared features features that are linked to or that share a boundary with other features

short integer an attribute field definition that uses up to five bytes to store a binary integer and that can store values up to about 62,000

simple edges linear elements in a network that always end at junctions and behave as separate entities

simple join a join that combines two layers according to common locations when a one-to-one spatial relationship exists between the layers

single symbol a map type in which every feature in a layer is displayed with the same symbol

single-precision a numeric value stored using eight bytes of information

sink a location on a network that pulls or consumes the commodity flowing through the network

sketch a provisional figure created during editing; when finished, it becomes a feature in a data layer

Sketch menu a context menu accessed by right-clicking *off* a sketch during editing

slice to divide the values in a raster into a specified number of even classes

sliver small polygon created during map overlay due to slight boundary differences in the inputs; usually considered an error

slope the drop in elevation for a specified horizontal distance, expressed as an angle or a percentage

snap tolerance a distance set during editing such that new vertices close to an existing feature are snapped to it

snapping ensuring that features within a specified distance are automatically adjusted to meet at exactly the same location to avoid gaps between features

solver program that analyzes flows or paths through a network

source (1) a location that produces or initiates the flow of a commodity through a network; (2) a spatial data file that provides the features for a map layer; (3) the original information used to develop a spatial data set

source table the table that provides the information to be appended to another table in a join

spaghetti model a model that stores spatial features as a series of *x-y* coordinates and does not store topological relationships between features

Spatial Analyst a program extension to ArcMap that is used to analyze raster data

spatial data information that is tied to a specific location on the earth's surface

spatial join a function that combines the attributes of features in two layers based on containment or distance

spatial operators a set of functions that evaluate spatial relationships, including intersection, containment, and proximity, between sets of features

spatial query an operation to extract records from a data layer based on location relative to another data layer

spatial reference the complete description of how the spatial data are stored for a feature class that includes the coordinate system, the X/Y Domain, and the precision

SPCS see *State Plane Coordinate System*

spelling sensitivity a setting used during geocoding to control the degree to which two names must resemble each other to be considered a good match

spheroid ellipsoid; see also *ellipsoid*

split policy a rule that states how attributes of a feature should be handled if it is split

SQL see *Structured Query Language*

standalone table a table of information not linked to spatial data features

standard deviation a classification scheme in which the class breaks are based on the standard deviation values of the data being mapped

standard parallel parameter of a map projection indicating the latitude(s) at which the projection surface lies tangent or secant to the sphere

State Plane Coordinate System (SPCS) a group of projections defined for different regions of the United States and designed to minimize map distortions

stereographic projection a map projection in which locations on a sphere are projected onto a planar surface

stream digitizing method of entering vertices automatically during editing, instead of clicking to create each one

street type the part of an address that indicates the type of street, such as St, Rd, Ave, and so on

stretch to spread the values of an image to cover the entire range of symbols available, often supplemented by ignoring the tails of the distribution

stretched a display method that spreads the data values over the entire range of symbols available

Structured Query Language (SQL) an established syntax for creating logical expressions to extract records from a database according to specified criteria

style a collection of map symbols and colors that are stored together and used together

Style Manager a window used to create symbols and manage them within collections of symbols called styles

subdivisions the number of units into which a single division of a scale bar is split, usually appearing on the left end

subtypes different categories created for a feature class, with each category having its own symbol and default values

suffix direction a direction appended to the end of a street name to indicate a part of the city, such as Main St *North*

Summarize a function that groups records in a table according to one categorical field and calculates statistics for each group

summarized join a join that combines the records of two attribute tables with a one-to-many cardinality by assigning each output record a single value derived statistically from the many input values

surface analysis functions designed for application to rasters representing a three-dimensional surface, such as elevation

symbol a shade, line, or marker with specified shape and color used to display map features

table data stored as an array of rows and columns, with each row representing an object or a feature and each column representing an attribute or a property of that object

Table of Contents the ArcMap window that lists the data frames and layers in the map document

tangent projection a map projection in which spherical coordinates are projected upon a surface that lies tangent to the sphere along one line of latitude

templates a map design that can be saved and applied to many different map documents

temporal accuracy a determination of the time period for which a data set is considered valid

thematic accuracy the degree to which attribute values represent the true properties in the real world

thematic mapping displaying the features of a spatial data layer based on values in its attribute table

thematic raster a raster that contains categorical or nominal data values, such as land use codes or soil types

themes terminology in ArcView 3 used to indicate a data layer containing similar features, such as roads or states

thumbnail a small snapshot showing the appearance of a data layer and displayed in ArcCatalog to aid the user in finding data

TIN (Triangulated Irregular Network) a data model for storing surfaces as triangular facets with varying orientations

topological model a data model that stores spatial relationships between features in addition to their *x-y* coordinates

topology the spatial relationships between features, such as which are connected or adjacent to each other

trace to follow an existing linear feature, as when analyzing network paths or creating a new feature that follows an existing one

trace solvers programs that analyze network flow by tracing paths in the network

transformation conversion of one geographic coordinate system (GCS) and datum to another

transverse projection a map projection in which spherical coordinates are converted to locations on a cylinder or cone tangent to the sphere along a line of longitude

undershoot a type of dangle in which a line falls short of meeting another line

union an operation to create a new feature or set of features by combining all the areas from two input layers

unique values map a map in which each attribute value is assigned its own symbol

Universal Transverse Mercator (UTM) a family of map projections defined for 60 zones around the world and based on a transverse cylindrical projection

unprojected data a spatial data layer stored in a geographic coordinate system with units of degrees of latitude and longitude

UTM see *Universal Transverse Mercator*

validation a step performed on a planar topology to test the features against the defined topology rules and identify errors

vector a spatial data storage method in which features are represented by one or more pairs of *x-y* coordinate values forming points, lines, or polygons

Venn diagram a diagram used in set theory to portray sets and subsets

vertex (pl. **vertices**) a point at which the segments of a line or polygon feature change direction

Vertex menu a short context menu accessed by clicking *on* the sketch during editing

vertex snapping ensuring that new features are automatically connected to the vertices of existing line features

viewshed the area on a three-dimensional surface that is visible from a specified point or set of points

web map an interactive map based solely on Internet data services and accessed from many platforms, including desktop software, web pages, and mobile devices

world file a text file containing the georeferencing information for a raster

XML (eXtensible Markup Language) a tagged text presentation format used to store metadata

XY tolerance a setting used for geoprocessing that defines the minimum allowable distance between vertices

X/Y Domain the maximum allowable range of values stored by a feature class

zenith angle an angular measure of the distance of an object above the horizon

zonal statistics raster analysis functions that calculate statistics from one raster for zones defined by a different raster

zone the combined areas of a raster that share the same attribute value, such as all areas with the commercial land use code

Selected Answers

Ch. 1 Tutorial

1. Wallowa
2. 36 rows
3. Wheeler county
4. NAME is the field
5. 2 coverages, 1 table, 1 raster, 1 layer file, and 7 shapefiles
6. NAD 1983 Oregon Statewide Lambert, meter
7. NOAA provides the data; USGS hosts it.

Ch. 1 Exercises

1. Method: Use the Catalog tab to examine the geodatabase.
2. GCS_WGS_1984; NAD83 Oregon Statewide Lambert
3. Method: Open Item Description tab on Catalog tab.
4. Superior, 32,213 mi^2
5. Method: Preview the table and use Find.
6. Min. 66, max. 22,658, avg. 4419.
7. Method: Open Help and search for *feature class;* choose feature class basics.
8. 1609 rows, 1964 columns, 30 meter cell, 8 bands, WGS 1984 UTM Zone 10N
9. Method: Open layer properties, click the Source tab.
10. See map.

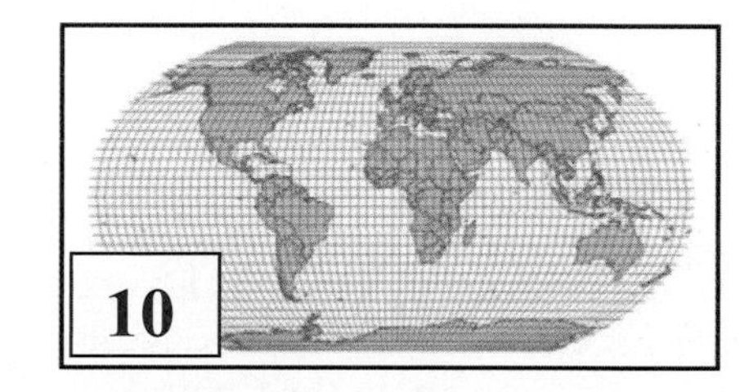

Ch 1. Challenge

Answers will vary.

Ch. 2 Tutorial

1. NAD 1983 Oregon Statewide Lambert
2. ELEVATION is interval; TYPE is categorical.
3. I for Interstates, U for US highways, S for state highways
4. Ratio data; a graduated color map
5. Interval data; stretched or classified
6. Ordinal data, unique values with monochromatic ramp

Ch. 2 Exercises

1. Volcanic hazards frame

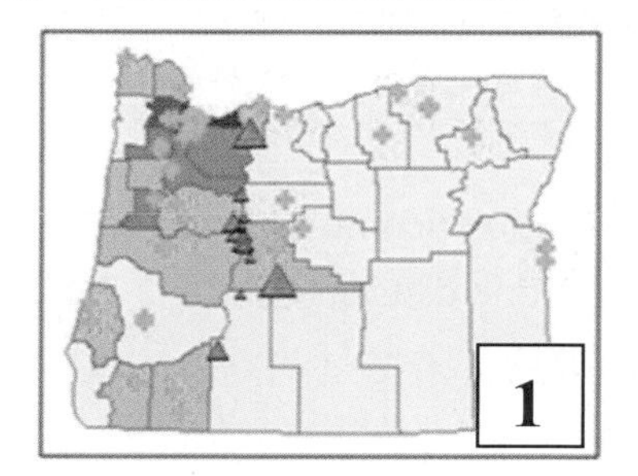

2. Farms frame

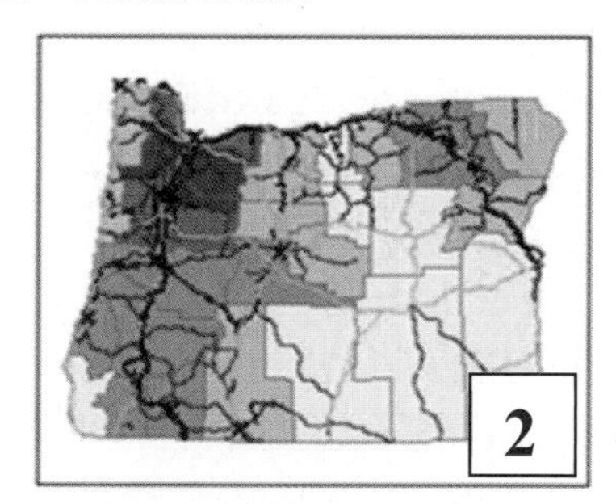

3. Housing frame

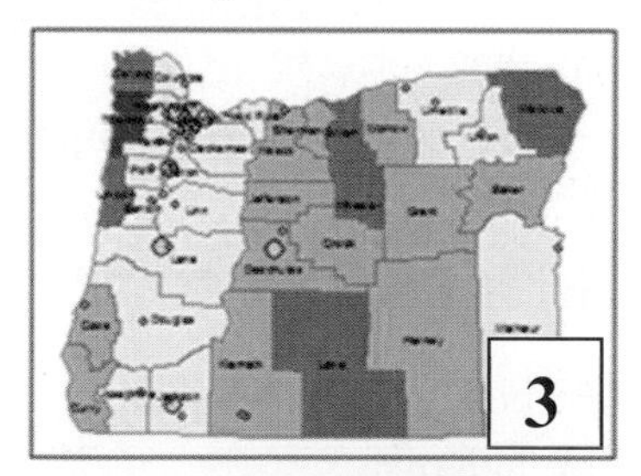

4. Physiography frame

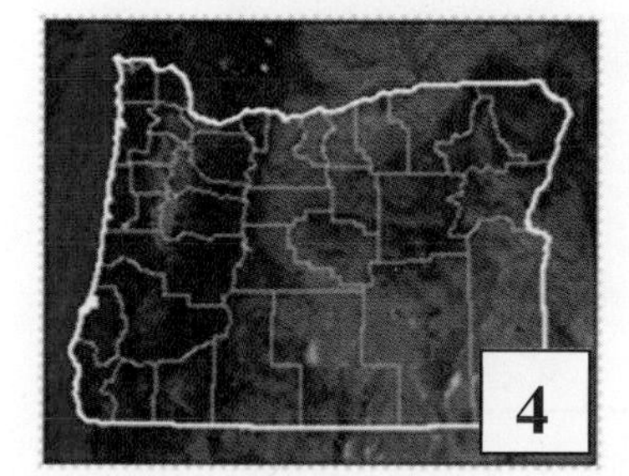

Ch. 2 Challenge
Answers will vary.

Ch. 3 Tutorial
1. Mercator and Web Mercator
2. Annotation has a reference scale.
3. Left to right: subdivision, division, division units

Ch. 3 Exercises

Answers will vary.

Ch. 3 Challenge
Answers will vary.

Ch. 4 Tutorial
1. about 38 million people
2. 256 Democratic districts
3. DISTRICTID
4. Democrats
5. Standalone table
6. Nine states have one district.
7. Republicans control nearly twice the area.
8. One-to-one relationship
9. One-to-many relationship
10. Districts is source, states is destination.
11. 22 New England reps
12. STATION_NAME and STATION

Ch. 4 Exercises
1. Method: Use Select By Attributes and Statistics and then sort.
2. New York highest number; Mississippi (or DC) highest percentage.
3. Method: Summarize by Sub_Region and sort results.
4. Use the FIPS code.
5. 839 counties. See map.

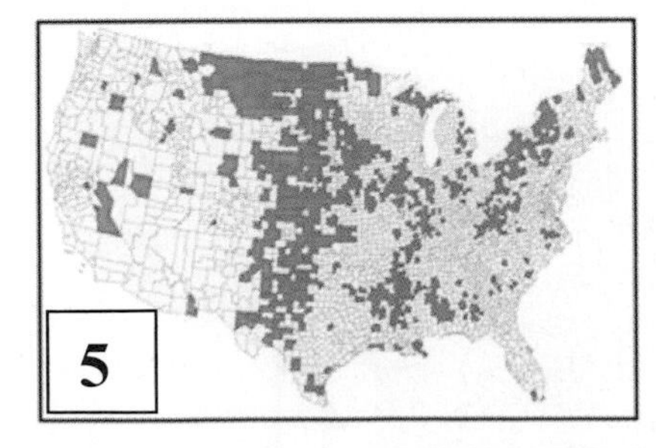

6. Texas 76, Kansas 68, Nebraska 63
7. Method: Relate US States to 111th Congress and select states. Activate related records and create layer. Query layer for Democrats.
8. 11,980,000 in state capitals; smallest 8100, largest 1,502,000, average 240,000.
9. Method: Add new field and calculate percentage. See map.

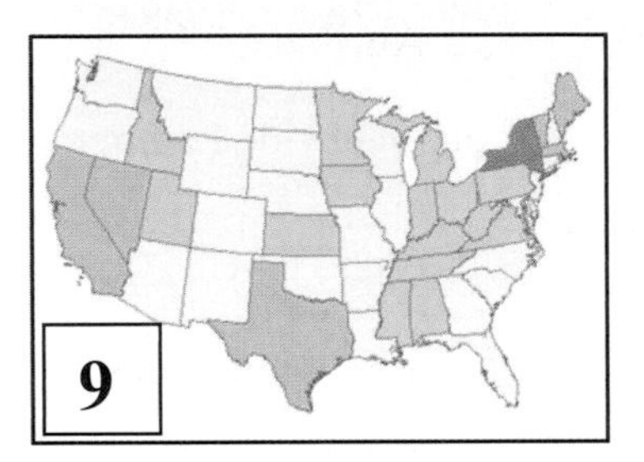

10. W N Central has 618. See map.

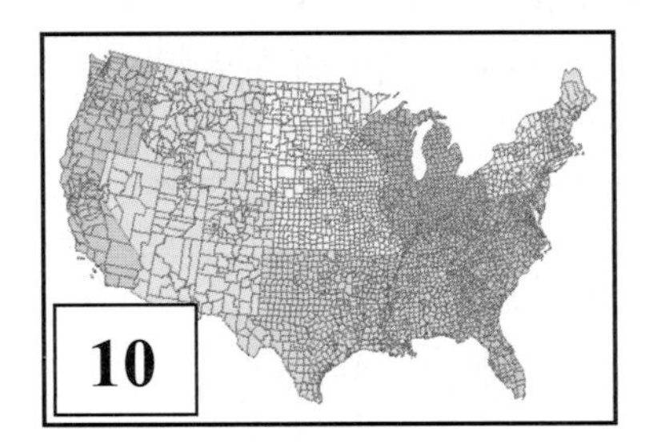

Ch. 4 Challenge

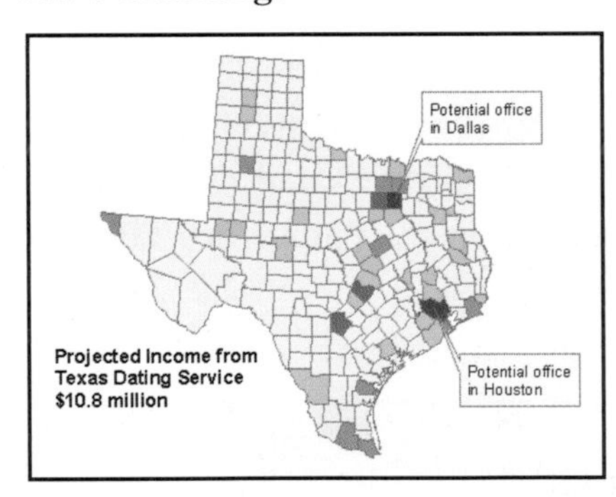

Ch. 5 Tutorial
1. 15 counties, about 309,000 people
2. 9 cities
3. 290 counties, mostly in the Central Plains states
4. 177 counties
5. 32 cities
6. 890 counties
7. Hughes County contains Pierre, SD.

Ch. 5 Tutorial, con't.

8. 3321 cities
9. 93% of cities
10. Brazos, Canadian, Pecos, Red, and Rio Grande
11. The Flathead and the Salt
12. 19 target counties, 864,000 people

Ch. 5 Exercises

1. Method: Select By Attributes and sort.
2. 822 counties out of 3141, or 26%
3. Method: Use Select By Attributes and do queries in steps.
4. 900 out of 3141 counties, or 29%; 93,914,738 out of 311,212,863 people, or 30%
5. Method: Use Select By Location first, then Select By Attributes.
6. 142 cities, about 6 million people
7. Method: Select Crater Lake and then use Select By Location twice, once for volcanoes and once for interstates.
8. 99 cities. See map.

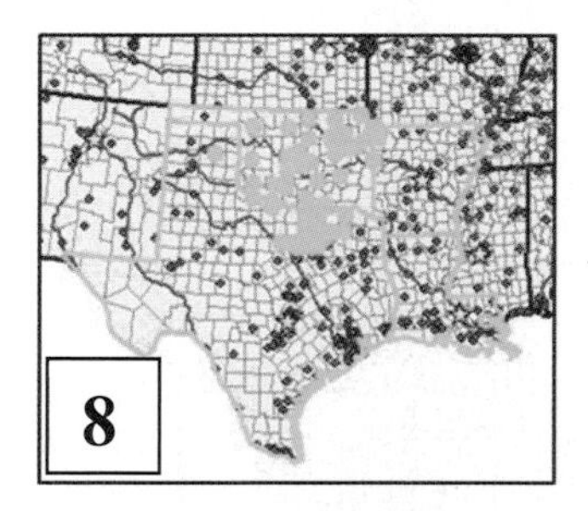

9. Method: Query for areas within 50 miles and then switch the selection.
10. Pahrump, Nevada

Ch. 6 Tutorial

1. 591 out of 1360 wells
2. About 880 feet
3. FACILITY_N
4. NAD 1983 Texas State Plane Central, in feet
5. 16,477 streets
6. Min. 907, max. 138,275, avg. 27,009
7. Little Walnut, Williamson, Walnut
8. Barton, Country Club East, Slaughter

Ch. 6 Exercises

1. Method: Summarized join of wells to destination watersheds.
2. 13 wells; six have depths < 100 feet
3. Method: Summarized join of wells to destination zone overlays.
4. Min. 19 feet, max. 15,068 feet, avg. 1543 feet
5. Method: Export elementary schools; summarized join of schools to pools.
6. Pools includes wading pools.
7. Method: Export post offices; summarized distance join of blockpop to destination offices.
8. Jordan Valley, about 334 km
9. Method: Summarized join of cities to destination airports.
10. Willamette NF and Deschutes NF have greatest number (12).Mt Hood NF has the greatest area (290 mi^2).

Ch. 6 Challenge

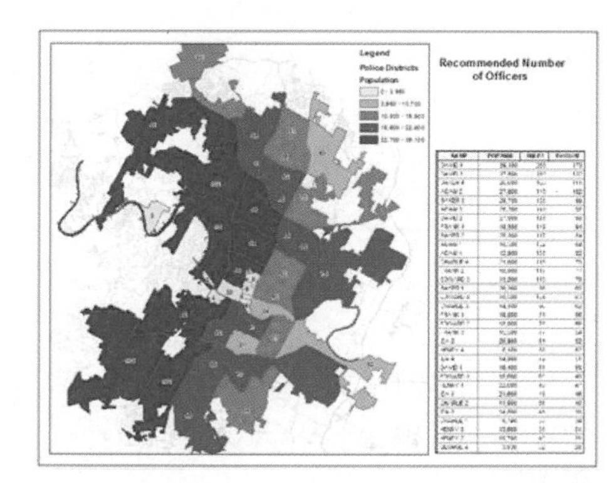

Ch. 7 Tutorial

1. ALL, NO-FID, ONLY-FID.ALL or NO-FID is best.
2. 4.9 km^2
3. 32.2 km^2
4. About 15%

Ch. 7 Exercises

1. Method: Intersect roads and geology; use statistics.
2. See map.

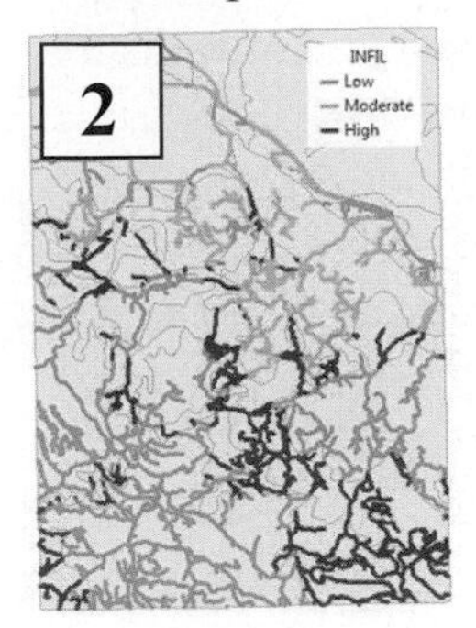

3. See map.

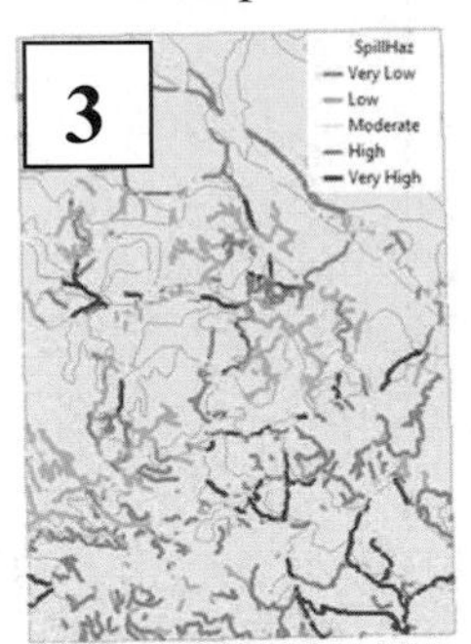

4. The answer is the table.

sumshedrange

SHEDNAME	Area
	29,813,100
Whitewood	9,055,570
False Bottom	8,385,018
Bear Butte	1,695,465

5. See map.
6. See map.

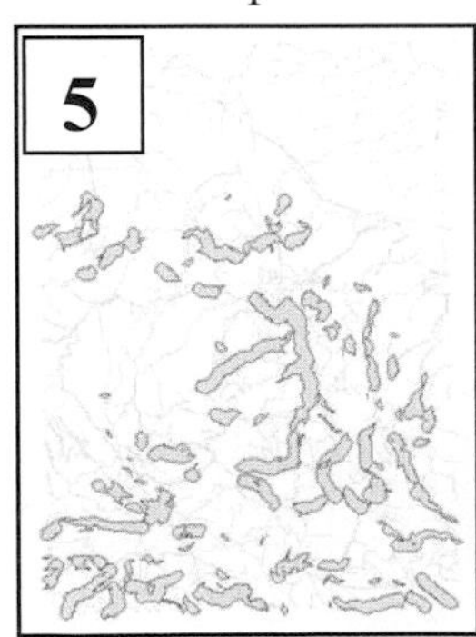

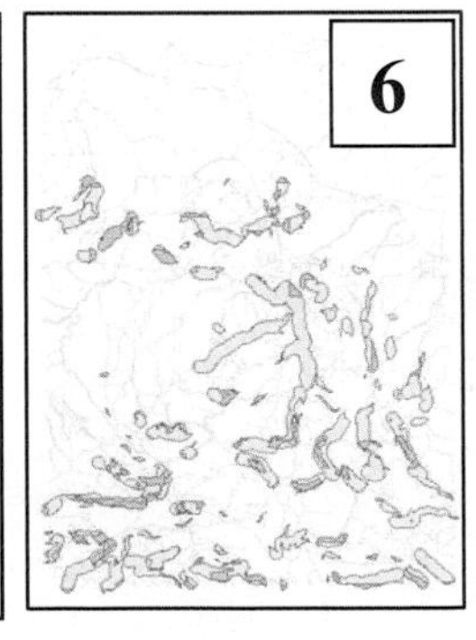

7. See map.

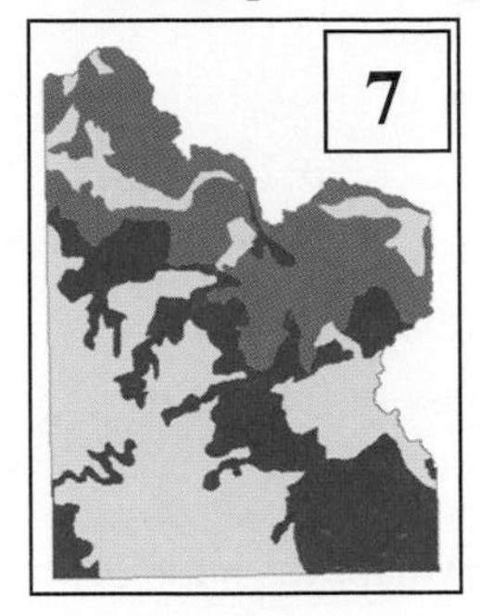

8. See map.

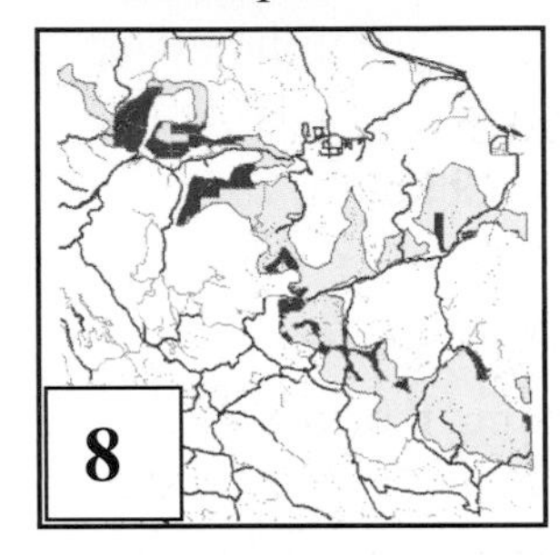

9. See map.

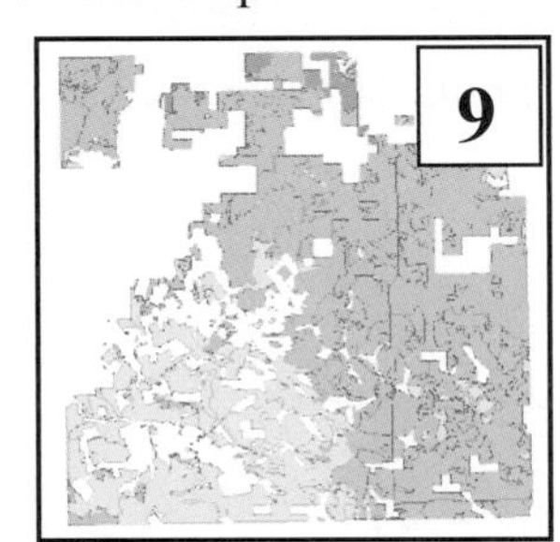

10. See map. Off-limits area is 34.3 km^2.

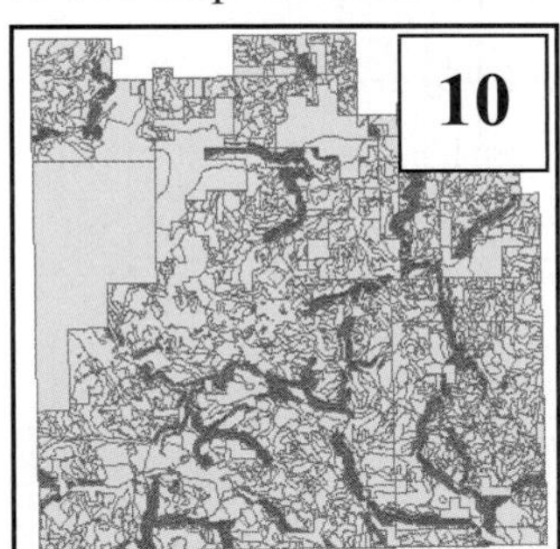

Ch. 7 Challenge

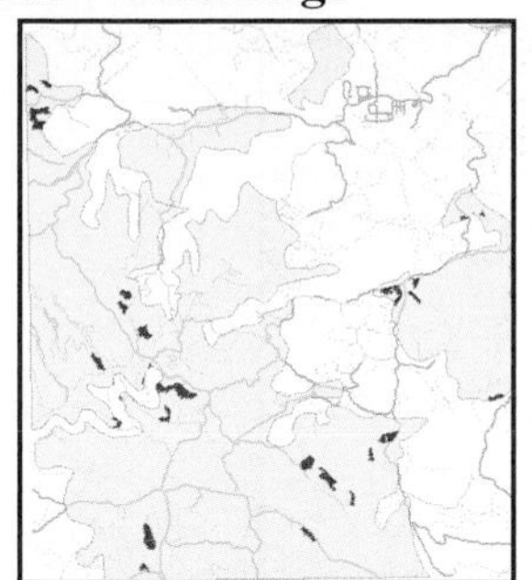

Ch. 8 Tutorial

1. About 13%
2. 354,549 cells × 30 m × 30 m/1 million m^2/km^2, or about 319 km^2
3. About 51°
4. North is red; south is light blue.
5. Stores 0-3, indicating the number of points from which the cell can be seen
6. Floating-point raster, because the precipitation values have decimals
7. Pleasant Valley, about 165 million m^3

Ch. 8 Exercises

1. See map.
2. See map.

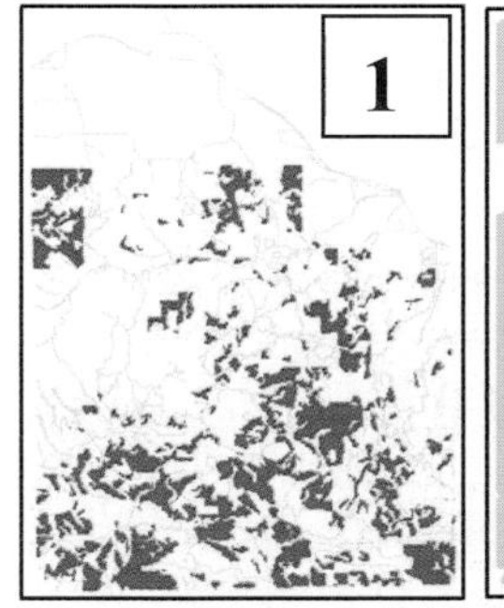

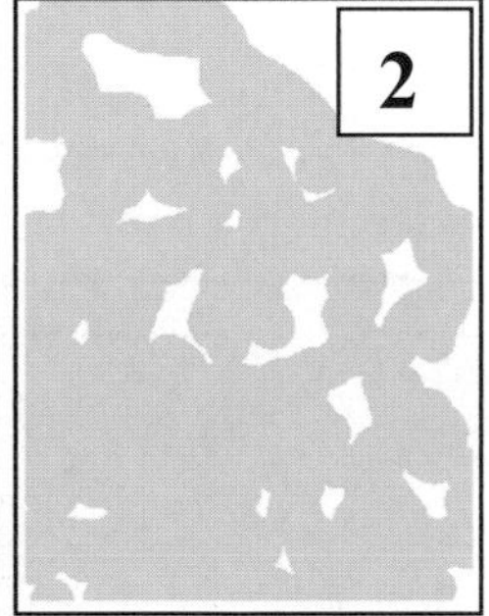

3. See map.
4. Area is 40.4 km^2. See map.

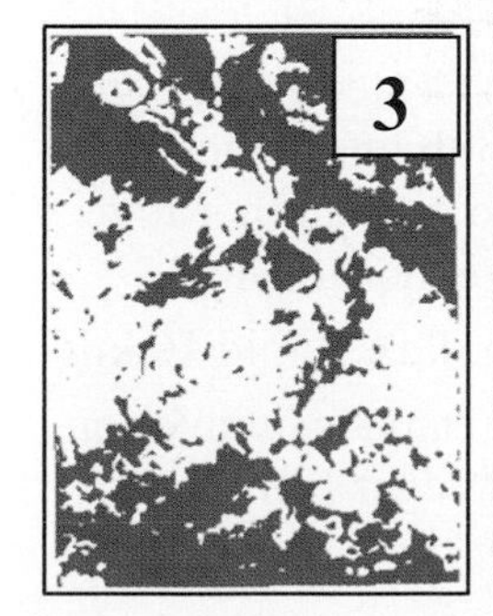

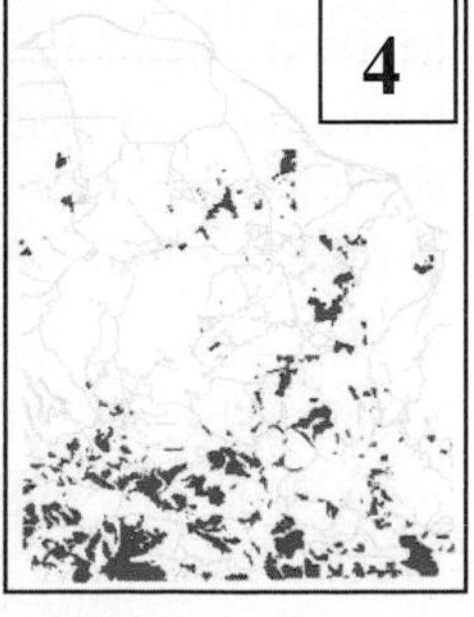

5. 12 sites. See map.

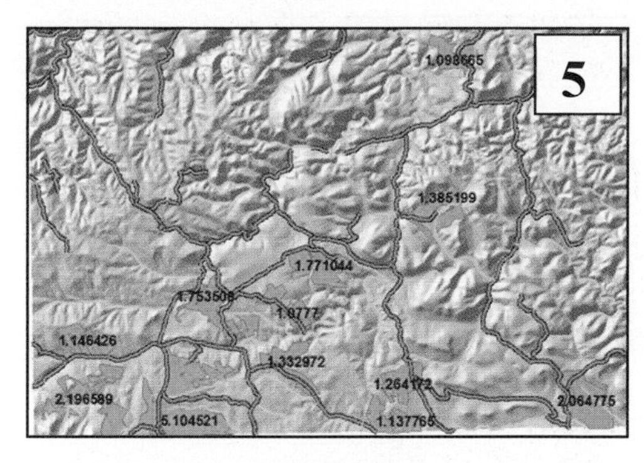

6. Highest is the Madison Limestone; lowest is the Upper Mesozoic.

NAME	MEAN
Cenozoic	5.2
Upper Mesozoic	3
Lower Mesozoic	7.6
Upper Paleozoic	12.5
Madison Limestone	15.4
Precambrian	11.7
Lower Paleozoic	11.9

7. See map.
8. See map.

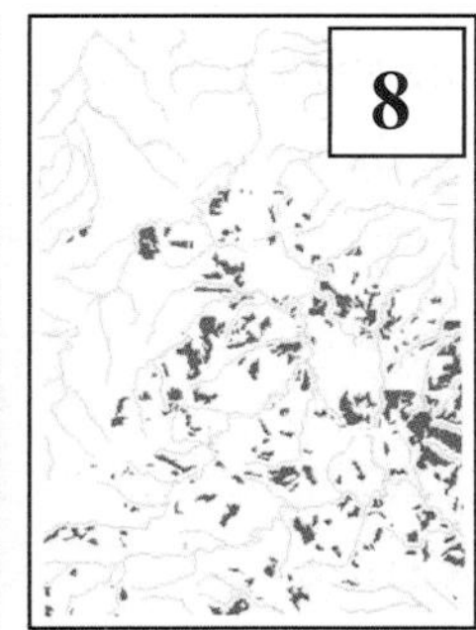

9. Highest is Anchor Hill at 1723 meters; lowest is Sly Hill at 1183 meters.
10. See map.

Ch. 8 Challenge

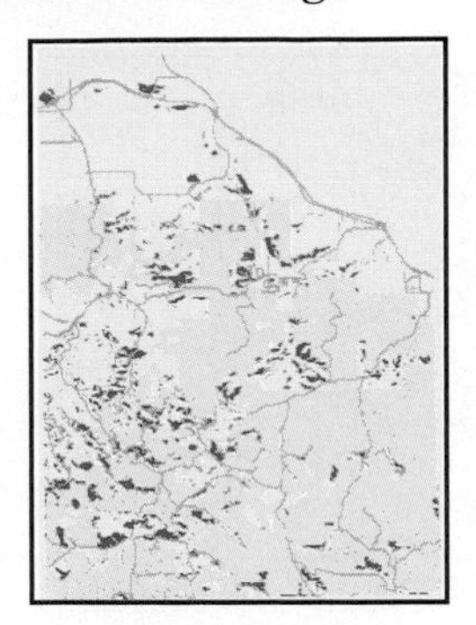

Ch. 9 Tutorial

1. The Transportation network contains a distance weight.
2. Endcaps, Galleries, Tvalves, and Water_Net_Junctions are simple junctions; Waterlines are simple edges.
3. Distance and Pressure Drop
4. About 20 edges
5. About 4000 meters
6. About 4900 meters
7. ~10 differently named streets
8. Black circle
9. 17 edges
10. About 3600 meters
11. About 53 kilometers

Ch. 9 Exercises

1. See map.

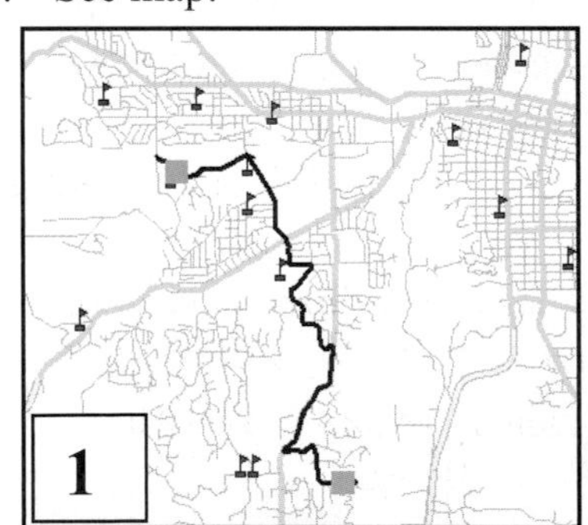

2. About 7400 meters
3. See map.

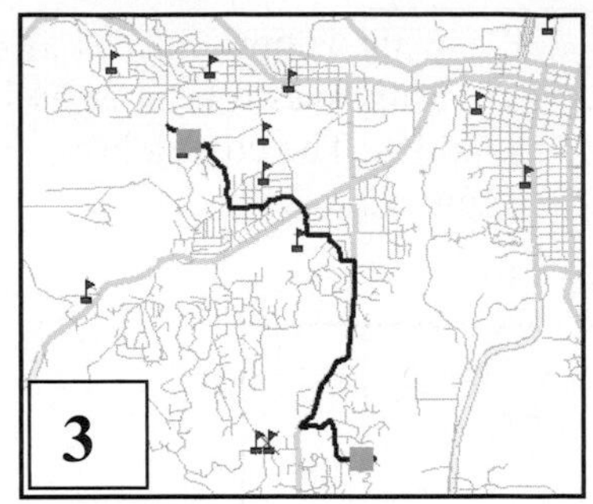

4. See map.

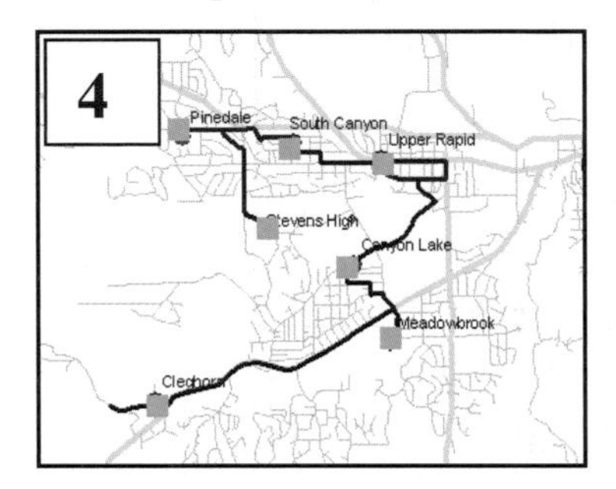

5. See map.

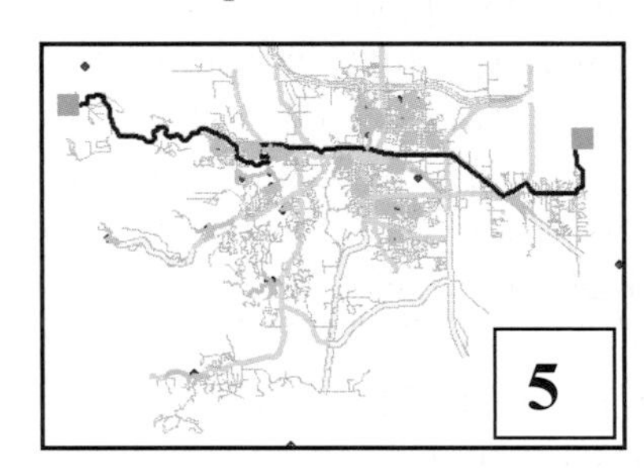

6. See map.

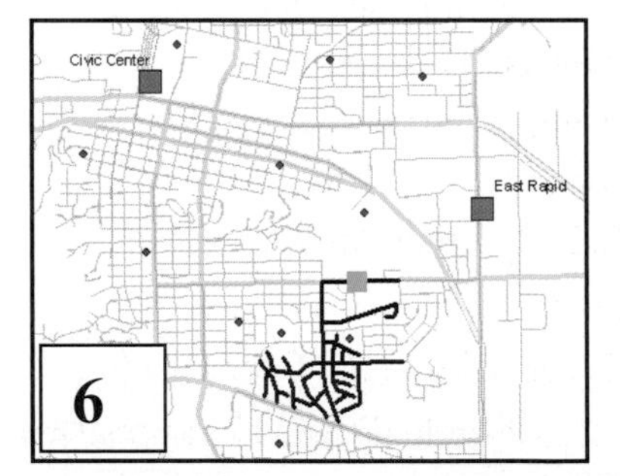

7. Method: Flag Civic Center gallery and barrier at start of north line. Trace with selection as result to get length. Move barrier to south to get north length.
8. Civic Center 17 km; East Rapid 10 km
9. Method: Find Upstream Accumulation from end and count edges (same as junctions).

Ch. 9 Exercises, con't.

10. Joe had set the analysis options to return only junctions; Mary was returning both edges and junctions. The cost in both cases is the same, 25.

Ch. 9 Challenge

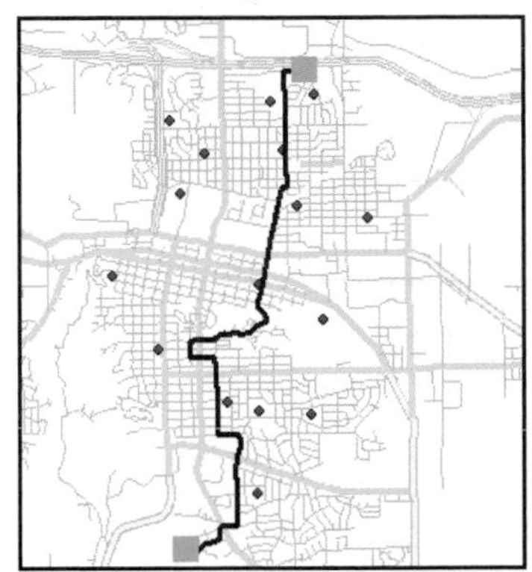

Ch. 10 Tutorial

1. Prefix direction = PREFIX_DIR, Prefix type = PRE_TYPE, Street numbers = LEFT_FRADD, LEFT_TOADD, RIGHT_FRAD, and RIGHT_TOAD, Street name = STREET_NAM, Street type = STREET_TYP, Suffix direction = SUFFIX_DIR, City = NAME, State = State, ZIP code = ZIPCODE.
2. US Address – Dual Ranges
3. ADDRESS field
4. Yes, there are intersections. The connector is the & character.
5. The streets feature class in the Austin geodatabase
6. 77 matched, 1 tied, 10 unmatched
7. City-State style
8. NAME for cities and ST for states
9. About 209 matched/tied, 17 unmatched
10. Geocoding can resolve ambiguities when two cities have the same name in two states. Interactive matching can pair cities with slightly different names.

Ch. 10 Exercises

1. Answers will vary.
2. Answers will vary.
3. See map.

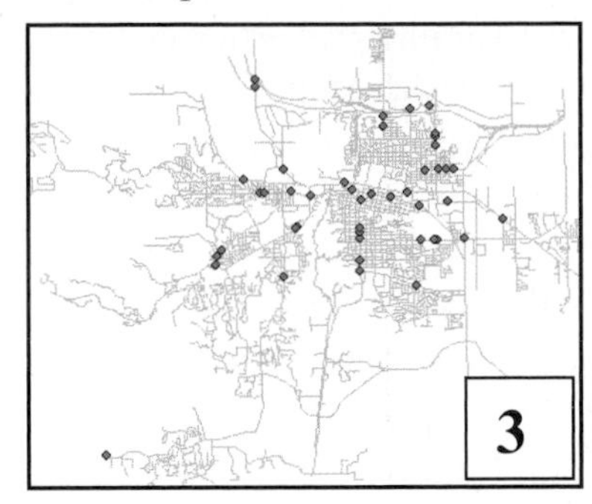

4. See map. You should have about 536 matched, 31 tied, and 20 unmatched.

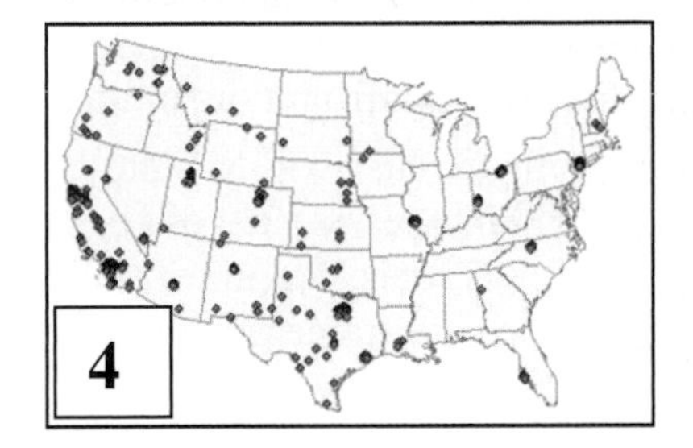

Ch. 10 Challenge

Answers will vary.

Ch. 11 Tutorial

1. GCS_WGS_1984
2. About –80° and 25°
3. Both units are decimal degrees.
4. About –5550 and 1740 miles
5. Mercator units are meters.
6. South America mostly negative; Asia mostly positive
7. Central meridian is –96° and the latitude of origin is 40°.
8. Yes, because the latitude of origin lies between the standard parallels
9. Secant conic, because it has two standard parallels
10. The new data have a different GCS than the data frame GCS.
11. It only chooses NAD27 to NAD83 conversion automatically.
12. UTM Zone 11 is ideal.
13. UTM Zone 13; it defaulted to the first data set added.
14. Geographic coordinate system (GCS)

Ch. 11 Exercises

1. Method: Examine properties in ArcCatalog or ArcMap.
2. Lambert Conformal Conic, central meridian –120.5; standard parallels 43° and 45.5°; no, it used 41.75° for the latitude of origin.
3. Method: Examine the data frame properties.
4. Humboldt County, CA: UTM Zone 10. Grafton County, NH: State Plane. Nevada: UTM Zone 11. New Jersey, UTM Zone 18 or State Plane. England: British National Grid. Antarctica: South Pole Azimuthal Equidistant or similar.
5. Method: Compare US areas to State Plane and UTM zones and choose best. For international areas, examine options available in the data frame predefined projection choices.
6. Modify Colorado State Plane Central to give parallels of 38.2° and 39.8°.
7. Method: Use Define projection to set the appropriate coordinate system.
8. Answers will vary.

Ch. 11 Challenge

Answers will vary.

Ch. 12 Tutorial

1. NAD 1983 State Plane Texas Central
2. Angle –115 degrees
3. Map units, in this case, feet
4. Use inside spatial join, but it would create additional fields, and it would not resolve one-to-many issues.

Ch. 12 Exercises

Answers will vary but should show the digitized features for the requested areas.

Ch. 12 Challenge

Answers will vary.

Ch. 13 Tutorial

1. The cluster tolerance must be smaller than the stream tolerance.

Ch. 13 Exercises

Answers will vary.

Ch. 13 Challenge

Answers will vary.

Ch. 14 Tutorial

1. NAD 1927 UTM Zone 13N
2. To save space or to be consistent with previous data or policies
3. A Float Double domain also

Ch. 14 Exercises

Answers will vary.

Ch. 14 Challenge

Answers will vary.

Ch. 15 Tutorial

1. Publication date is 6/30/2010.
2. Originated by National World Atlas and USGS; published by ESRI
3. Two steps in 2004 and 2006
4. The National Atlas
5. Source scale 1:2 million
6. Contact ESRI to purchase the software.
7. Produced by ArcGIS Content Team; follows NAP of ISO 19115 2003
8. Depth unit is kilometers.

Ch. 15 Exercises

Answers will vary.

Ch. 15 Challenge

Answers will vary.

Index

A

G

H

I

J

K

L

M

N

O

P

Q

R

Length Conversions

Multiply by the factor indicated to convert the column units to the units in the row.

	Inches	Centimeters	Feet	Yards	Meters	Miles	Kilometers	Nautical miles
Inches	1	0.3937	12	36	39.37	69,840	39,370	60,964
Centimeters	2.540	1	30.48	91.44	100	27,496	1.00×10^{5}	1.548×10^{5}
Feet	0.0833	0.0328	1	3	3.281	5,280	3,281	5,080
Yards	0.0278	0.0109	0.333	1	1.094	1,940	1,094	1,693
Meters	0.0254	0.0100	0.3048	0.914	1	1,610	1,000	1,852
Miles	1.432×10^{-5}	3.637×10^{-5}	1.718×10^{-4}	5.155×10^{-4}	6.21×10^{-4}	1	0.6215	1.151
Kilometers	2.539×10^{-5}	1.00×10^{-5}	3.048×10^{-4}	9.144×10^{-4}	0.001	1.609	1	1.852
Nautical miles	1.640×10^{-5}	6.458×10^{-6}	1.968×10^{-4}	5.905×10^{-4}	5.40×10^{-4}	0.869	0.5400	1

Area Conversions

Multiply by the factor indicated to convert the column units to the units in the row.

	Inches	Centimeters	Feet	Yards	Meters	Miles	Kilometers	Hectares	Acres
Inches	1	0.1550	144	1,296	1,550	4.878×10^{9}	1.550×10^{9}	1.550×10^{7}	6.272×10^{6}
Centimeters	6.452	1	929	8,361	10,000	7.560×10^{8}	1.00×10^{10}	1.000×10^{8}	4.046×10^{7}
Feet	0.0069	1.076×10^{-3}	1	9	10.76	3.387×10^{7}	1.076×10^{7}	1.076×10^{5}	43,560
Yards	7.716×10^{-4}	1.196×10^{-4}	0.1111	1	1.196	3.764×10^{6}	1.196×10^{6}	11,960	4,840
Meters	6.442×10^{-4}	1.00×10^{-4}	0.0929	0.8361	1	2.591×10^{6}	1,000,000	10,000	4,047
Miles	2.050×10^{-10}	1.323×10^{-9}	2.952×10^{-8}	2.657×10^{-7}	3.859×10^{-7}	1	0.3863	3.859×10^{-3}	1.563×10^{-3}
Kilometers	6.449×10^{-10}	1.00×10^{-10}	9.290×10^{-8}	8.361×10^{-7}	1.00×10^{-6}	2.589	1	0.0100	4.047×10^{-3}
Hectares	6.452×10^{-8}	1.00×10^{-8}	9.290×10^{-6}	8.361×10^{-5}	0.0001	259	100	1	0.4047
Acres	1.594×10^{-7}	2.471×10^{-8}	2.296×10^{-5}	2.066×10^{-4}	2.471×10^{-4}	640	247	2.471	1